Elementary Linear Algebra

8e Metric Version

Ron Larson
The Pennsylvania State University
The Behrend College

CENGAGE
Learning·

Australia · Brazil · Mexico · Singapore · United Kingdom · United States

Elementary Linear Algebra, **Eighth Edition, Metric Version**

Ron Larson

Metric Version Prepared by Larson Texts, Inc.

International Product Director, Global Editions:
    Timothy L. Anderson
International Services Specialist: Collette Allen
International Markets Coordinator: Tori Sitcawich
Product Assistant: Teresa Versaggi
Director, Production: Samantha Ross Miller
Content Project Manager: Rebecca Donahue
Production Service/Compositor: Larson Texts, Inc.
Senior Art Director: Vernon Boes
Cover Designer: Denise Davidson
Cover Image: qui jun peng/Shutterstock.com
Intellectual Property Analyst: Brittani Morgan
Manager, Global IP Integration: Eleanor Rummer
Manufacturing Planner: Doug Bertke

ISBN: 978-1-337-55621-7

Cengage Learning International Offices

Asia
www.cengageasia.com
tel: (65) 6410 1200

Australia/New Zealand
www.cengage.com.au
tel: (61) 3 9685 4111

Brazil
www.cengage.com.br
tel: (55) 11 3665 9900

India
www.cengage.co.in
tel: (91) 11 4364 1111

Latin America
www.cengage.com/mx
tel: (52) 55 1500 6000

**UK/Europe/Middle East/
Africa**
www.cengage.co.uk
tel: (44) 0 1264 332 424

**Represented in Canada by
Nelson Education, Ltd.**
tel: (416) 752 9100 / (800) 668 0671
www.nelson.com

Cengage Learning is a leading provider of customized learning solutions with office locations around the globe, including Singapore, the United Kingdom, Australia, Mexico, Brazil, and Japan. Locate your local office at **www.cengage.com/global**.

For product infomration: **www.cengage.com/international**
Visit your local office: **www.cengage.com/global**
Visit our corporate office: **www.cengage.com**

QR is a registered trademark of Denso Wave Incorporated

Printed in The United States of America
Print Number: 03   Print Year: 2021

Contents

*Available online at **CengageBrain.com**.

Preface

Welcome to the International Metric Version of *Elementary Linear Algebra*, Eighth Edition. For this metric version, the units of measurement used in most of the examples and exercises have been changed from U.S. Customary units to metric units. I did not convert problems that are specific to U.S. Customary units, such as dimensions of a baseball field or U.S. postal rates. As with all editions, I have been able to incorporate many useful comments from you, our user. And while much has changed in this revision, you will still find what you expect—a pedagogically sound, mathematically precise, and comprehensive textbook. Additionally, I am pleased and excited to offer you something brand new— a companion website at **LarsonLinearAlgebra.com.** My goal for every edition of this textbook is to provide students with the tools that they need to master linear algebra. I hope you find that the changes in this edition, together with **LarsonLinearAlgebra.com,** will help accomplish just that.

New To This Edition

NEW *LarsonLinearAlgebra.com*

This companion website offers multiple tools and resources to supplement your learning. Access to these features is *free*. Watch videos explaining concepts from the book, explore examples, download data sets and much more.

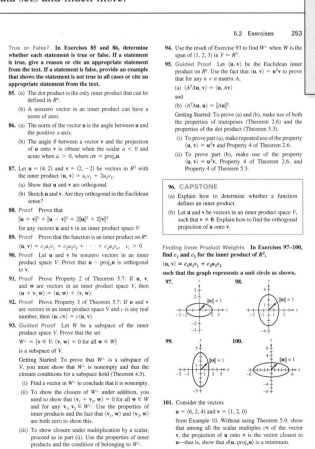

5.2 Exercises 253

True or False? In Exercises 85 and 86, determine whether each statement is true or false. If a statement is true, give a reason or cite an appropriate statement from the text. If a statement is false, provide an example that shows the statement is not true in all cases or cite an appropriate statement from the text.

85. (a) The dot product is the only inner product that can be defined in R^n.
   (b) A nonzero vector in an inner product can have a norm of zero.

86. (a) The norm of the vector $\mathbf{u}$ is the angle between $\mathbf{u}$ and the positive x-axis.
   (b) The angle θ between a vector $\mathbf{v}$ and the projection of $\mathbf{u}$ onto $\mathbf{v}$ is obtuse when the scalar $a < 0$ and acute when $a > 0$, where $a\mathbf{v} = \text{proj}_\mathbf{v}\mathbf{u}$.

87. Let $\mathbf{u} = (4, 2)$ and $\mathbf{v} = (2, -2)$ be vectors in R^2 with the inner product $\langle \mathbf{u}, \mathbf{v} \rangle = u_1v_1 + 2u_2v_2$.
   (a) Show that $\mathbf{u}$ and $\mathbf{v}$ are orthogonal.
   (b) Sketch $\mathbf{u}$ and $\mathbf{v}$. Are they orthogonal in the Euclidean sense?

88. Proof Prove that
   $$\|\mathbf{u} + \mathbf{v}\|^2 + \|\mathbf{u} - \mathbf{v}\|^2 = 2\|\mathbf{u}\|^2 + 2\|\mathbf{v}\|^2$$
   for any vectors $\mathbf{u}$ and $\mathbf{v}$ in an inner product space V.

89. Proof Prove that the function is an inner product on R^n.
   $$\langle \mathbf{u}, \mathbf{v} \rangle = c_1u_1v_1 + c_2u_2v_2 + \cdots + c_nu_nv_n, \quad c_i > 0$$

90. Proof Let $\mathbf{u}$ and $\mathbf{v}$ be nonzero vectors in an inner product space V. Prove that $\mathbf{u} - \text{proj}_\mathbf{v}\mathbf{u}$ is orthogonal to $\mathbf{v}$.

91. Proof Prove Property 2 of Theorem 5.7: If $\mathbf{u}$, $\mathbf{v}$, and $\mathbf{w}$ are vectors in an inner product space V, then $\langle \mathbf{u} + \mathbf{v}, \mathbf{w} \rangle = \langle \mathbf{u}, \mathbf{w} \rangle + \langle \mathbf{v}, \mathbf{w} \rangle$.

92. Proof Prove Property 3 of Theorem 5.7: If $\mathbf{u}$ and $\mathbf{v}$ are vectors in an inner product space V and c is any real number, then $\langle \mathbf{u}, c\mathbf{v} \rangle = c\langle \mathbf{u}, \mathbf{v} \rangle$.

93. Guided Proof Let W be a subspace of the inner product space V. Prove that the set
   $$W^\perp = \{\mathbf{v} \in V: \langle \mathbf{v}, \mathbf{w} \rangle = 0 \text{ for all } \mathbf{w} \in W\}$$
   is a subspace of V.
   Getting Started: To prove that $W^\perp$ is a subspace of V, you must show that $W^\perp$ is nonempty and that the closure conditions for a subspace hold (Theorem 4.5).
   (i) Find a vector in $W^\perp$ to conclude that it is nonempty.
   (ii) To show the closure of $W^\perp$ under addition, you need to show that $\langle \mathbf{v}_1 + \mathbf{v}_2, \mathbf{w} \rangle = 0$ for all $\mathbf{w} \in W$ and for any $\mathbf{v}_1, \mathbf{v}_2 \in W^\perp$. Use the properties of inner products and the fact that $\langle \mathbf{v}_1, \mathbf{w} \rangle$ and $\langle \mathbf{v}_2, \mathbf{w} \rangle$ are both zero to show this.
   (iii) To show closure under multiplication by a scalar, proceed as in part (ii). Use the properties of inner products and the condition of belonging to $W^\perp$.

94. Use the result of Exercise 93 to find $W^\perp$ when W is the span of $(1, 2, 3)$ in $V = R^3$.

95. Guided Proof Let $\langle \mathbf{u}, \mathbf{v} \rangle$ be the Euclidean inner product on R^n. Use the fact that $\langle \mathbf{u}, \mathbf{v} \rangle = \mathbf{u}^T\mathbf{v}$ to prove that for any $n \times n$ matrix A,
   (a) $\langle A^TA\mathbf{u}, \mathbf{v} \rangle = \langle \mathbf{u}, A\mathbf{v} \rangle$
   and
   (b) $\langle A^TA\mathbf{u}, \mathbf{u} \rangle = \|A\mathbf{u}\|^2$.
   Getting Started: To prove (a) and (b), make use of both the properties of transposes (Theorem 2.6) and the properties of the dot product (Theorem 5.3).
   (i) To prove part (a), make repeated use of the property $\langle \mathbf{u}, \mathbf{v} \rangle = \mathbf{u}^T\mathbf{v}$ and Property 4 of Theorem 2.6.
   (ii) To prove part (b), make use of the property $\langle \mathbf{u}, \mathbf{v} \rangle = \mathbf{u}^T\mathbf{v}$, Property 4 of Theorem 2.6, and Property 4 of Theorem 5.3.

96. **CAPSTONE**
   (a) Explain how to determine whether a function defines an inner product.
   (b) Let $\mathbf{u}$ and $\mathbf{v}$ be vectors in an inner product space V, such that $\mathbf{v} \neq \mathbf{0}$. Explain how to find the orthogonal projection of $\mathbf{u}$ onto $\mathbf{v}$.

Finding Inner Product Weights In Exercises 97–100, find c_1 and c_2 for the inner product of R^2,
$$\langle \mathbf{u}, \mathbf{v} \rangle = c_1u_1v_1 + c_2u_2v_2$$
such that the graph represents a unit circle as shown.

97. $\|\mathbf{u}\| = 1$

98. $\|\mathbf{u}\| = 1$

99. $\|\mathbf{u}\| = 1$

100. $\|\mathbf{u}\| = 1$

101. Consider the vectors
   $$\mathbf{u} = (6, 2, 4) \text{ and } \mathbf{v} = (1, 2, 0)$$
   from Example 10. Without using Theorem 5.9, show that among all the scalar multiples $c\mathbf{v}$ of the vector $\mathbf{v}$, the projection of $\mathbf{u}$ onto $\mathbf{v}$ is the vector closest to $\mathbf{u}$—that is, show that $d(\mathbf{u}, \text{proj}_\mathbf{v}\mathbf{u})$ is a minimum.

REVISED *Exercise Sets*

The exercise sets have been carefully and extensively examined to ensure they are rigorous, relevant, and cover all the topics necessary to understand the fundamentals of linear algebra. The exercises are ordered and titled so you can see the connections between examples and exercises. Many new skill-building, challenging, and application exercises have been added. As in earlier editions, the following pedagogically-proven types of exercises are included.

- **True or False Exercises**
- **Proofs**
- **Guided Proofs**
- **Writing Exercises**
- **Technology Exercises** (indicated throughout the text with ⌲)

Exercises utilizing **electronic data sets** are indicated by ⊡ and found at **CengageBrain.com.**

Table of Contents Changes

Based on market research and feedback from users, Section 2.5 in the previous edition (Applications of Matrix Operations) has been expanded from one section to two sections to include content on Markov chains. So now, Chapter 2 has *two* application sections: Section 2.5 (Markov Chains) and Section 2.6 (More Applications of Matrix Operations). In addition, Section 7.4 (Applications of Eigenvalues and Eigenvectors) has been expanded to include content on constrained optimization.

Trusted Features

CalcChat®

For the past several years, an independent website—CalcChat.com—has provided free solutions to all odd-numbered problems in the text. Thousands of students have visited the site for practice and help with their homework from live tutors. You can also use your smartphone's QR Code® reader to scan the icon at the beginning of each exercise set to access the solutions.

QR Code is a registered trademark of Denso Wave Incorporated

2 Matrices

- 2.1 Operations with Matrices
- 2.2 Properties of Matrix Operations
- 2.3 The Inverse of a Matrix
- 2.4 Elementary Matrices
- 2.5 Markov Chains
- 2.6 More Applications of Matrix Operations

Data Encryption (p. 94)

Computational Fluid Dynamics (p. 79)

Beam Deflection (p. 64)

Information Retrieval (p. 58)

Flight Crew Scheduling (p. 47)

Clockwise from top left, Cousin_Avi/Shutterstock.com; Gonchansk/Shutterstock.com; Gunnar Pippel/Shutterstock.com; Andresr/Shutterstock.com; nostal6ie/Shutterstock.com

39

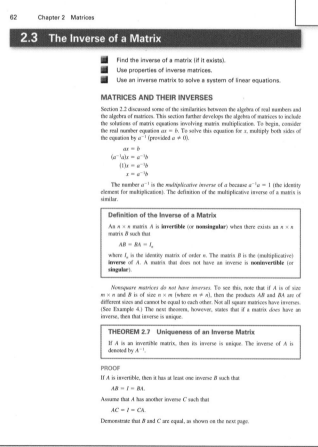

Chapter Openers

Each *Chapter Opener* highlights five real-life applications of linear algebra found throughout the chapter. Many of the applications reference the *Linear Algebra Applied* feature (discussed on the next page). You can find a full list of the applications in the *Index of Applications* on the inside front cover.

Section Objectives

A bulleted list of learning objectives, located at the beginning of each section, provides you the opportunity to preview what will be presented in the upcoming section.

Theorems, Definitions, and Properties

Presented in clear and mathematically precise language, all theorems, definitions, and properties are highlighted for emphasis and easy reference.

Proofs in Outline Form

In addition to proofs in the exercises, some proofs are presented in outline form. This omits the need for burdensome calculations.

Discovery

Using the *Discovery* feature helps you develop an intuitive understanding of mathematical concepts and relationships.

Technology Notes

Technology notes show how you can use graphing utilities and software programs appropriately in the problem-solving process. Many of the *Technology* notes reference the **Technology Guide** at **CengageBrain.com.**

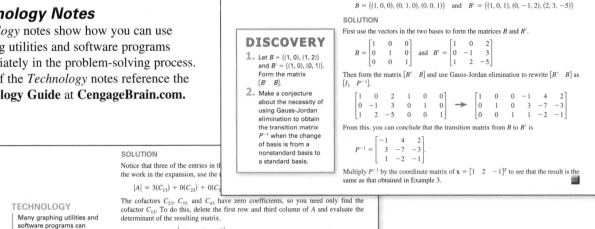

EXAMPLE 4   Finding a Transition Matrix

See LarsonLinearAlgebra.com for an interactive version of this type of example.

Find the transition matrix from B to B' for the bases for R^3 below.

$$B = \{(1, 0, 0), (0, 1, 0), (0, 0, 1)\} \quad \text{and} \quad B' = \{(1, 0, 1), (0, -1, 2), (2, 3, -5)\}$$

SOLUTION

First use the vectors in the two bases to form the matrices B and B'.

$$B = \begin{bmatrix} 1 & 0 & 0 \\ 0 & 1 & 0 \\ 0 & 0 & 1 \end{bmatrix} \quad \text{and} \quad B' = \begin{bmatrix} 1 & 0 & 2 \\ 0 & -1 & 3 \\ 1 & 2 & -5 \end{bmatrix}$$

Then form the matrix $[B' \quad B]$ and use Gauss-Jordan elimination to rewrite $[B' \quad B]$ as $[I_3 \quad P^{-1}]$.

$$\begin{bmatrix} 1 & 0 & 2 & 1 & 0 & 0 \\ 0 & -1 & 3 & 0 & 1 & 0 \\ 1 & 2 & -5 & 0 & 0 & 1 \end{bmatrix} \longrightarrow \begin{bmatrix} 1 & 0 & 0 & -1 & 4 & 2 \\ 0 & 1 & 0 & 3 & -7 & -3 \\ 0 & 0 & 1 & 1 & -2 & -1 \end{bmatrix}$$

From this, you can conclude that the transition matrix from B to B' is

$$P^{-1} = \begin{bmatrix} -1 & 4 & 2 \\ 3 & -7 & -3 \\ 1 & -2 & -1 \end{bmatrix}.$$

Multiply P^{-1} by the coordinate matrix of $\mathbf{x} = \begin{bmatrix} 1 & 2 & -1 \end{bmatrix}^T$ to see that the result is the same as that obtained in Example 3.

DISCOVERY

1. Let $B = \{(1, 0), (1, 2)\}$ and $B' = \{(1, 0), (0, 1)\}$. Form the matrix $[B' \quad B]$.

2. Make a conjecture about the necessity of using Gauss-Jordan elimination to obtain the transition matrix P^{-1} when the change of basis is from a nonstandard basis to a standard basis.

SOLUTION

Notice that three of the entries in the [   ] the work in the expansion, use the [   ]

$$|A| = 3(C_{13}) + 0(C_{23}) + 0(C_{3}$$

The cofactors C_{23}, C_{33}, and C_{43} have zero coefficients, so you need only find the cofactor C_{13}. To do this, delete the first row and third column of A and evaluate the determinant of the resulting matrix.

TECHNOLOGY

Many graphing utilities and software programs can find the determinant of a square matrix. If you use a graphing utility, then you may see something similar to the screen below for Example 4. The **Technology Guide** at *CengageBrain.com* can help you use technology to find a determinant.

```
A
   [[1  -2  3   0 ]
   [-1  1   0   2 ]
   [0   2   0   3 ]
   [3   4   0  -2]]
det A
              39
```

$$C_{13} = (-1)^{1+3} \begin{vmatrix} -1 & 1 & 2 \\ 0 & 2 & 3 \\ 3 & 4 & -2 \end{vmatrix} \qquad \text{Delete 1st row and 3rd column.}$$

$$= \begin{vmatrix} -1 & 1 & 2 \\ 0 & 2 & 3 \\ 3 & 4 & -2 \end{vmatrix} \qquad \text{Simplify.}$$

Expanding by cofactors in the second row yields

$$C_{13} = (0)(-1)^{2+1} \begin{vmatrix} 1 & 2 \\ 4 & -2 \end{vmatrix} + (2)(-1)^{2+2} \begin{vmatrix} -1 & 2 \\ 3 & -2 \end{vmatrix} +$$

$$= 0 + 2(1)(-4) + 3(-1)(-7)$$

$$= 13.$$

You obtain

$$|A| = 3(13)$$

$$= 39.$$

LINEAR ALGEBRA APPLIED

Time-frequency analysis of irregular physiological signals, such as beat-to-beat cardiac rhythm variations (also known as heart rate variability or HRV), can be difficult. This is because the structure of a signal can include multiple periodic, nonperiodic, and pseudo-periodic components. Researchers have proposed and validated a simplified HRV analysis method called orthonormal-basis partitioning and time-frequency representation (OPTR). This method can detect both abrupt and slow changes in the HRV signal's structure, divide a nonstationary HRV signal into segments that are "less nonstationary," and determine patterns in the HRV. The researchers found that although it had poor time resolution with signals that changed gradually, the OPTR method accurately represented multicomponent and abrupt changes in both real-life and simulated HRV signals. (Source: *Orthonormal-Basis Partitioning and Time-Frequency Representation of Cardiac Rhythm Dynamics, Aysin, Benhur, et al, IEEE Transactions on Biomedical Engineering, 52, no. 5*)

Sebastian Kaulitzki/Shutterstock.com

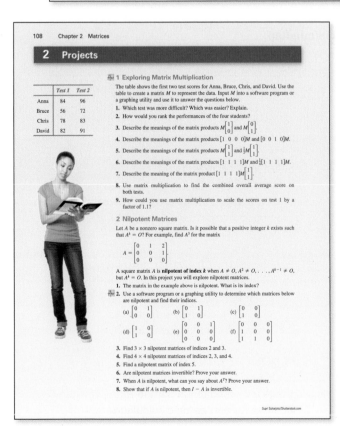

108   Chapter 2   Matrices

2   Projects

	Test 1	Test 2
Anna	84	96
Bruce	56	72
Chris	78	83
David	82	91

1 Exploring Matrix Multiplication

The table shows the first two test scores for Anna, Bruce, Chris, and David. Use the table to create a matrix M to represent the data. Input M into a software program or a graphing utility and use it to answer the questions below.

1. Which test was more difficult? Which was easier? Explain.
2. How would you rank the performances of the four students?
3. Describe the meanings of the matrix products $M \begin{bmatrix} 1 \\ 0 \end{bmatrix}$ and $M \begin{bmatrix} 0 \\ 1 \end{bmatrix}$.
4. Describe the meanings of the matrix products $\begin{bmatrix} 1 & 0 & 0 & 0 \end{bmatrix} M$ and $\begin{bmatrix} 0 & 0 & 1 & 0 \end{bmatrix} M$.
5. Describe the meanings of the matrix products $M \begin{bmatrix} 1 \\ 1 \end{bmatrix}$ and $\frac{1}{2} M \begin{bmatrix} 1 \\ 1 \end{bmatrix}$.
6. Describe the meanings of the matrix products $\begin{bmatrix} 1 & 1 & 1 & 1 \end{bmatrix} M$ and $\frac{1}{4} \begin{bmatrix} 1 & 1 & 1 & 1 \end{bmatrix} M$.
7. Describe the meaning of the matrix product $\begin{bmatrix} 1 & 1 & 1 & 1 \end{bmatrix} M \begin{bmatrix} 1 \\ 1 \end{bmatrix}$.
8. Use matrix multiplication to find the combined overall average score on both tests.
9. How could you use matrix multiplication to scale the scores on test 1 by a factor of 1.1?

2 Nilpotent Matrices

Let A be a nonzero square matrix. Is it possible that a positive integer k exists such that $A^k = O$? For example, find A^3 for the matrix

$$A = \begin{bmatrix} 0 & 1 & 2 \\ 0 & 0 & 1 \\ 0 & 0 & 0 \end{bmatrix}.$$

A square matrix A is **nilpotent of index k** when $A \neq O$, $A^2 \neq O, \ldots, A^{k-1} \neq O$, but $A^k = O$. In this project you will explore nilpotent matrices.

1. The matrix in the example above is nilpotent. What is its index?
2. Use a software program or a graphing utility to determine which matrices below are nilpotent and find their indices.

(a) $\begin{bmatrix} 0 & 1 \\ 0 & 0 \end{bmatrix}$ (b) $\begin{bmatrix} 0 & 1 \\ 1 & 0 \end{bmatrix}$ (c) $\begin{bmatrix} 0 & 0 \\ 1 & 0 \end{bmatrix}$

(d) $\begin{bmatrix} 1 & 0 \\ 1 & 0 \end{bmatrix}$ (e) $\begin{bmatrix} 0 & 0 & 1 \\ 0 & 0 & 0 \\ 0 & 0 & 0 \end{bmatrix}$ (f) $\begin{bmatrix} 0 & 0 & 0 \\ 1 & 0 & 0 \\ 1 & 1 & 0 \end{bmatrix}$

3. Find 3×3 nilpotent matrices of indices 2 and 3.
4. Find 4×4 nilpotent matrices of indices 2, 3, and 4.
5. Find a nilpotent matrix of index 5.
6. Are nilpotent matrices invertible? Prove your answer.
7. When A is nilpotent, what can you say about A^T? Prove your answer.
8. Show that if A is nilpotent, then $I - A$ is invertible.

Supri Suharjoto/Shutterstock.com

Linear Algebra Applied

The *Linear Algebra Applied* feature describes a real-life application of concepts discussed in a section. These applications include biology and life sciences, business and economics, engineering and technology, physical sciences, and statistics and probability.

Capstone Exercises

The *Capstone* is a conceptual problem that synthesizes key topics to check students' understanding of the section concepts. I recommend it.

Chapter Projects

Two per chapter, these offer the opportunity for group activities or more extensive homework assignments, and are focused on theoretical concepts or applications. Many encourage the use of technology.

Instructor Resources

Media

Instructor's Solutions Manual

The *Instructor's Solutions Manual* provides worked-out solutions for all even-numbered exercises in the text.

Cengage Learning Testing Powered by Cognero (ISBN: 978-1-305-65806-6)

is a flexible, online system that allows you to author, edit, and manage test bank content, create multiple test versions in an instant, and deliver tests from your LMS, your classroom, or wherever you want. This is available online at **cengage.com/login.**

Turn the Light On with MindTap for Larson's *Elementary Linear Algebra*

Through personalized paths of dynamic assignments and applications, MindTap is a digital learning solution and representation of your course that turns cookie cutter into cutting edge, apathy into engagement, and memorizers into higher-level thinkers.

> **The Right Content:** With MindTap's carefully curated material, you get the precise content and groundbreaking tools you need for every course you teach.
>
> **Personalization:** Customize every element of your course—from rearranging the Learning Path to inserting videos and activities.
>
> **Improved Workflow:** Save time when planning lessons with all of the trusted, most current content you need in one place in MindTap.
>
> **Tracking Students' Progress in Real Time:** Promote positive outcomes by tracking students in real time and tailoring your course as needed based on the analytics.

Learn more at **cengage.com/mindtap.**

Student Resources

Print

Student Solutions Manual
ISBN-13: 978-1-305-87658-3
The *Student Solutions Manual* provides complete worked-out solutions to all odd-numbered exercises in the text. Also included are the solutions to all Cumulative Test problems.

Media

MindTap for Larson's *Elementary Linear Algebra*
MindTap is a digital representation of your course that provides you with the tools you need to better manage your limited time, stay organized and be successful. You can complete assignments whenever and wherever you are ready to learn with course material specially customized for you by your instructor and streamlined in one proven, easy-to-use interface. With an array of study tools, you'll get a true understanding of course concepts, achieve better grades and set the groundwork for your future courses.

Learn more at **cengage.com/mindtap.**

CengageBrain.com
To access additional course materials and companion resources, please visit **CengageBrain.com.** At the **CengageBrain.com** home page, search for the ISBN of your title (from the back cover of your book) using the search box at the top of the page. This will take you to the product page where free companion resources can be found.

Acknowledgements

I would like to thank the many people who have helped me during various stages of writing this new edition. In particular, I appreciate the feedback from the dozens of instructors who took part in a detailed survey about how they teach linear algebra. I also appreciate the efforts of the following colleagues who have provided valuable suggestions throughout the life of this text:

Michael Brown, *San Diego Mesa College*

Nasser Dastrange, *Buena Vista University*

Mike Daven, *Mount Saint Mary College*

David Hemmer, *University of Buffalo, SUNY*

Wai Lau, *Seattle Pacific University*

Jorge Sarmiento, *County College of Morris.*

I would like to thank Bruce H. Edwards, University of Florida, and David C. Falvo, The Pennsylvania State University, The Behrend College, for their contributions to previous editions of *Elementary Linear Algebra*.

On a personal level, I am grateful to my spouse, Deanna Gilbert Larson, for her love, patience, and support. Also, a special thanks goes to R. Scott O'Neil.

Ron Larson, Ph.D.
Professor of Mathematics
Penn State University
www.RonLarson.com

1 Systems of Linear Equations

Electrical Network Analysis (p. 30)

Traffic Flow (p. 28)

Global Positioning System (p. 16)

Airspeed of a Plane (p. 11)

Balancing Chemical Equations (p. 4)

1.1   Introduction to Systems of Linear Equations

- Recognize a linear equation in n variables.
- Find a parametric representation of a solution set.
- Determine whether a system of linear equations is consistent or inconsistent.
- Use back-substitution and Gaussian elimination to solve a system of linear equations.

LINEAR EQUATIONS IN n VARIABLES

The study of linear algebra demands familiarity with algebra, analytic geometry, and trigonometry. Occasionally, you will find examples and exercises requiring a knowledge of calculus, and these are marked in the text.

Early in your study of linear algebra, you will discover that many of the solution methods involve multiple arithmetic steps, so it is essential that you check your work. Use software or a calculator to check your work and perform routine computations.

Although you will be familiar with some material in this chapter, you should carefully study the methods presented. This will cultivate and clarify your intuition for the more abstract material that follows.

Recall from analytic geometry that the equation of a line in two-dimensional space has the form

$$a_1 x + a_2 y = b, \quad a_1, a_2, \text{ and } b \text{ are constants.}$$

This is a **linear equation in two variables** x and y. Similarly, the equation of a plane in three-dimensional space has the form

$$a_1 x + a_2 y + a_3 z = b, \quad a_1, a_2, a_3, \text{ and } b \text{ are constants.}$$

This is a **linear equation in three variables** x, y, and z. A linear equation in n variables is defined below.

Definition of a Linear Equation in n Variables

A **linear equation in n variables** $x_1, x_2, x_3, \ldots, x_n$ has the form

$$a_1 x_1 + a_2 x_2 + a_3 x_3 + \cdots + a_n x_n = b.$$

The **coefficients** $a_1, a_2, a_3, \ldots, a_n$ are real numbers, and the **constant term** b is a real number. The number a_1 is the **leading coefficient,** and x_1 is the **leading variable.**

Linear equations have no products or roots of variables and no variables involved in trigonometric, exponential, or logarithmic functions. Variables appear only to the first power.

EXAMPLE 1   **Linear and Nonlinear Equations**

Each equation is linear.

a. $3x + 2y = 7$ **b.** $\frac{1}{2}x + y - \pi z = \sqrt{2}$ **c.** $(\sin \pi)x_1 - 4x_2 = e^2$

Each equation is not linear.

a. $xy + z = 2$ **b.** $e^x - 2y = 4$ **c.** $\sin x_1 + 2x_2 - 3x_3 = 0$

SOLUTIONS AND SOLUTION SETS

A **solution** of a linear equation in n variables is a sequence of n real numbers $s_1, s_2, s_3, \ldots, s_n$ that satisfy the equation when you substitute the values

$$x_1 = s_1, \quad x_2 = s_2, \quad x_3 = s_3, \quad \ldots, \quad x_n = s_n$$

into the equation. For example, $x_1 = 2$ and $x_2 = 1$ satisfy the equation $x_1 + 2x_2 = 4$. Some other solutions are $x_1 = -4$ and $x_2 = 4$, $x_1 = 0$ and $x_2 = 2$, and $x_1 = -2$ and $x_2 = 3$.

The set of *all* solutions of a linear equation is its **solution set,** and when you have found this set, you have **solved** the equation. To describe the entire solution set of a linear equation, use a **parametric representation,** as illustrated in Examples 2 and 3.

EXAMPLE 2   Parametric Representation of a Solution Set

Solve the linear equation $x_1 + 2x_2 = 4$.

SOLUTION

To find the solution set of an equation involving two variables, solve for one of the variables in terms of the other variable. Solving for x_1 in terms of x_2, you obtain

$$x_1 = 4 - 2x_2.$$

In this form, the variable x_2 is **free,** which means that it can take on any real value. The variable x_1 is not free because its value depends on the value assigned to x_2. To represent the infinitely many solutions of this equation, it is convenient to introduce a third variable t called a **parameter.** By letting $x_2 = t$, you can represent the solution set as

$$x_1 = 4 - 2t, \quad x_2 = t, \quad t \text{ is any real number.}$$

To obtain particular solutions, assign values to the parameter t. For instance, $t = 1$ yields the solution $x_1 = 2$ and $x_2 = 1$, and $t = 4$ yields the solution $x_1 = -4$ and $x_2 = 4$.

To parametrically represent the solution set of the linear equation in Example 2 another way, you could have chosen x_1 to be the free variable. The parametric representation of the solution set would then have taken the form

$$x_1 = s, \quad x_2 = 2 - \tfrac{1}{2}s, \quad s \text{ is any real number.}$$

For convenience, when an equation has more than one free variable, choose the variables that occur last in the equation to be the free variables.

EXAMPLE 3   Parametric Representation of a Solution Set

Solve the linear equation $3x + 2y - z = 3$.

SOLUTION

Choosing y and z to be the free variables, solve for x to obtain

$$3x = 3 - 2y + z$$
$$x = 1 - \tfrac{2}{3}y + \tfrac{1}{3}z.$$

Letting $y = s$ and $z = t$, you obtain the parametric representation

$$x = 1 - \tfrac{2}{3}s + \tfrac{1}{3}t, \quad y = s, \quad z = t$$

where s and t are any real numbers. Two particular solutions are

$$x = 1, y = 0, z = 0 \quad \text{and} \quad x = 1, y = 1, z = 2.$$

SYSTEMS OF LINEAR EQUATIONS

A **system of m linear equations in n variables** is a set of m equations, each of which is linear in the same n variables:

$$
\begin{aligned}
a_{11}x_1 + a_{12}x_2 + a_{13}x_3 + \cdots + a_{1n}x_n &= b_1 \\
a_{21}x_1 + a_{22}x_2 + a_{23}x_3 + \cdots + a_{2n}x_n &= b_2 \\
a_{31}x_1 + a_{32}x_2 + a_{33}x_3 + \cdots + a_{3n}x_n &= b_3 \\
&\ \ \vdots \\
a_{m1}x_1 + a_{m2}x_2 + a_{m3}x_3 + \cdots + a_{mn}x_n &= b_m.
\end{aligned}
$$

REMARK

The double-subscript notation indicates a_{ij} is the coefficient of x_j in the ith equation.

A system of linear equations is also called a **linear system.** A **solution** of a linear system is a sequence of numbers $s_1, s_2, s_3, \ldots, s_n$ that is a solution of each equation in the system. For example, the system

$$
\begin{aligned}
3x_1 + 2x_2 &= 3 \\
-x_1 + x_2 &= 4
\end{aligned}
$$

has $x_1 = -1$ and $x_2 = 3$ as a solution because $x_1 = -1$ and $x_2 = 3$ satisfy *both* equations. On the other hand, $x_1 = 1$ and $x_2 = 0$ is not a solution of the system because these values satisfy only the first equation in the system.

DISCOVERY

1. Graph the two lines

$$
\begin{aligned}
3x - y &= 1 \\
2x - y &= 0
\end{aligned}
$$

in the xy-plane. Where do they intersect? How many solutions does this system of linear equations have?

2. Repeat this analysis for the pairs of lines

$$
\begin{aligned}
3x - y &= 1 \\
3x - y &= 0
\end{aligned}
\quad \text{and} \quad
\begin{aligned}
3x - y &= 1 \\
6x - 2y &= 2.
\end{aligned}
$$

3. What basic types of solution sets are possible for a system of two linear equations in two variables?

See LarsonLinearAlgebra.com for an interactive version of this type of exercise.

LINEAR ALGEBRA APPLIED

In a chemical reaction, atoms reorganize in one or more substances. For example, when methane gas (CH_4) combines with oxygen (O_2) and burns, carbon dioxide (CO_2) and water (H_2O) form. Chemists represent this process by a chemical equation of the form

$$(x_1)CH_4 + (x_2)O_2 \rightarrow (x_3)CO_2 + (x_4)H_2O.$$

A chemical reaction can neither create nor destroy atoms. So, all of the atoms represented on the left side of the arrow must also be on the right side of the arrow. This is called *balancing* the chemical equation. In the above example, chemists can use a system of linear equations to find values of x_1, x_2, x_3, and x_4 that will balance the chemical equation.

It is possible for a system of linear equations to have exactly one solution, infinitely many solutions, or no solution. A system of linear equations is **consistent** when it has at least one solution and **inconsistent** when it has no solution.

EXAMPLE 4 **Systems of Two Equations in Two Variables**

Solve and graph each system of linear equations.

a. $x + y = 3$
$x - y = -1$

b. $x + y = 3$
$2x + 2y = 6$

c. $x + y = 3$
$x + y = 1$

SOLUTION

a. This system has exactly one solution, $x = 1$ and $y = 2$. One way to obtain the solution is to add the two equations to give $2x = 2$, which implies $x = 1$ and so $y = 2$. The graph of this system is two *intersecting* lines, as shown in Figure 1.1(a).

b. This system has infinitely many solutions because the second equation is the result of multiplying both sides of the first equation by 2. A parametric representation of the solution set is

$$x = 3 - t, \quad y = t, \quad t \text{ is any real number.}$$

The graph of this system is two *coincident* lines, as shown in Figure 1.1(b).

c. This system has no solution because the sum of two numbers cannot be 3 and 1 simultaneously. The graph of this system is two *parallel* lines, as shown in Figure 1.1(c).

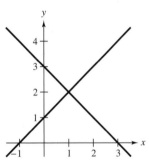 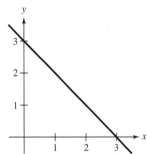 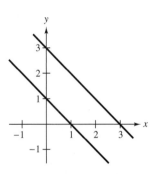

a. Two intersecting lines:
$x + y = 3$
$x - y = -1$

b. Two coincident lines:
$x + y = 3$
$2x + 2y = 6$

c. Two parallel lines:
$x + y = 3$
$x + y = 1$

Figure 1.1

Example 4 illustrates the three basic types of solution sets that are possible for a system of linear equations. This result is stated here without proof. (The proof is provided later in Theorem 2.5.)

Number of Solutions of a System of Linear Equations

For a system of linear equations, precisely one of the statements below is true.

1. The system has exactly one solution (consistent system).
2. The system has infinitely many solutions (consistent system).
3. The system has no solution (inconsistent system).

SOLVING A SYSTEM OF LINEAR EQUATIONS

Which system is easier to solve algebraically?

$$\begin{aligned} x - 2y + 3z &= 9 \\ -x + 3y \phantom{{}+ 3z} &= -4 \\ 2x - 5y + 5z &= 17 \end{aligned} \qquad \begin{aligned} x - 2y + 3z &= 9 \\ y + 3z &= 5 \\ z &= 2 \end{aligned}$$

The system on the right is clearly easier to solve. This system is in **row-echelon form**, which means that it has a "stair-step" pattern with leading coefficients of 1. To solve such a system, use **back-substitution.**

EXAMPLE 5   **Using Back-Substitution in Row-Echelon Form**

Use back-substitution to solve the system.

$$\begin{aligned} x - 2y &= 5 \qquad &\text{Equation 1} \\ y &= -2 \qquad &\text{Equation 2} \end{aligned}$$

SOLUTION

From Equation 2, you know that $y = -2$. By substituting this value of y into Equation 1, you obtain

$$\begin{aligned} x - 2(-2) &= 5 \qquad &\text{Substitute } -2 \text{ for } y. \\ x &= 1. \qquad &\text{Solve for } x. \end{aligned}$$

The system has exactly one solution: $x = 1$ and $y = -2$.

The term *back-substitution* implies that you work *backwards*. For instance, in Example 5, the second equation gives you the value of y. Then you substitute that value into the first equation to solve for x. Example 6 further demonstrates this procedure.

EXAMPLE 6   **Using Back-Substitution in Row-Echelon Form**

Solve the system.

$$\begin{aligned} x - 2y + 3z &= 9 \qquad &\text{Equation 1} \\ y + 3z &= 5 \qquad &\text{Equation 2} \\ z &= 2 \qquad &\text{Equation 3} \end{aligned}$$

SOLUTION

From Equation 3, you know the value of z. To solve for y, substitute $z = 2$ into Equation 2 to obtain

$$\begin{aligned} y + 3(2) &= 5 \qquad &\text{Substitute 2 for } z. \\ y &= -1. \qquad &\text{Solve for } y. \end{aligned}$$

Then, substitute $y = -1$ and $z = 2$ in Equation 1 to obtain

$$\begin{aligned} x - 2(-1) + 3(2) &= 9 \qquad &\text{Substitute } -1 \text{ for } y \text{ and 2 for } z. \\ x &= 1. \qquad &\text{Solve for } x. \end{aligned}$$

The solution is $x = 1$, $y = -1$, and $z = 2$.

Two systems of linear equations are **equivalent** when they have the same solution set. To solve a system that is not in row-echelon form, first rewrite it as an *equivalent* system that is in row-echelon form using the operations listed on the next page.

Operations That Produce Equivalent Systems

Each of these operations on a system of linear equations produces an *equivalent* system.

1. Interchange two equations.
2. Multiply an equation by a nonzero constant.
3. Add a multiple of an equation to another equation.

Rewriting a system of linear equations in row-echelon form usually involves a *chain* of equivalent systems, using one of the three basic operations to obtain each system. This process is called **Gaussian elimination,** after the German mathematician Carl Friedrich Gauss (1777–1855).

Carl Friedrich Gauss (1777–1855)
German mathematician Carl Friedrich Gauss is recognized, with Newton and Archimedes, as one of the three greatest mathematicians in history. Gauss used a form of what is now known as Gaussian elimination in his research. Although this method was named in his honor, the Chinese used an almost identical method some 2000 years prior to Gauss.

EXAMPLE 7   Using Elimination to Rewrite a System in Row-Echelon Form

See LarsonLinearAlgebra.com for an interactive version of this type of example.

Solve the system.

$$\begin{aligned} x - 2y + 3z &= 9 \\ -x + 3y &= -4 \\ 2x - 5y + 5z &= 17 \end{aligned}$$

SOLUTION

Although there are several ways to begin, you want to use a systematic procedure that can be applied to larger systems. Work from the upper left corner of the system, saving the x at the upper left and eliminating the other x-terms from the first column.

$$\begin{aligned} x - 2y + 3z &= 9 \\ y + 3z &= 5 \\ 2x - 5y + 5z &= 17 \end{aligned}$$ ← Adding the first equation to the second equation produces a new second equation.

$$\begin{aligned} x - 2y + 3z &= 9 \\ y + 3z &= 5 \\ -y - z &= -1 \end{aligned}$$ ← Adding -2 times the first equation to the third equation produces a new third equation.

Now that you have eliminated all but the first x from the first column, work on the second column.

$$\begin{aligned} x - 2y + 3z &= 9 \\ y + 3z &= 5 \\ 2z &= 4 \end{aligned}$$ ← Adding the second equation to the third equation produces a new third equation.

$$\begin{aligned} x - 2y + 3z &= 9 \\ y + 3z &= 5 \\ z &= 2 \end{aligned}$$ ← Multiplying the third equation by $\frac{1}{2}$ produces a new third equation.

This is the same system you solved in Example 6, and, as in that example, the solution is

$$x = 1, \quad y = -1, \quad z = 2.$$

Each of the three equations in Example 7 represents a plane in a three-dimensional coordinate system. The unique solution of the system is the point $(x, y, z) = (1, -1, 2)$, so the three planes intersect at this point, as shown in Figure 1.2.

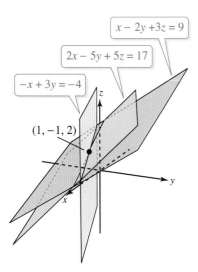

$x - 2y + 3z = 9$
$2x - 5y + 5z = 17$
$-x + 3y = -4$
$(1, -1, 2)$

Figure 1.2

Many steps are often required to solve a system of linear equations, so it is very easy to make arithmetic errors. You should develop the habit of *checking your solution by substituting it into each equation in the original system.* For instance, in Example 7, check the solution $x = 1$, $y = -1$, and $z = 2$ as shown below.

Equation 1: $(1) - 2(-1) + 3(2) = \quad 9$ Substitute the solution
Equation 2: $-(1) + 3(-1) \qquad\qquad = -4$ into each equation of the
Equation 3: $2(1) - 5(-1) + 5(2) = \quad 17$ original system.

The next example involves an inconsistent system—one that has no solution. The key to recognizing an inconsistent system is that at some stage of the Gaussian elimination process, you obtain a false statement such as $0 = -2$.

EXAMPLE 8   An Inconsistent System

Solve the system.

$$\begin{aligned} x_1 - 3x_2 + x_3 &= 1 \\ 2x_1 - x_2 - 2x_3 &= 2 \\ x_1 + 2x_2 - 3x_3 &= -1 \end{aligned}$$

SOLUTION

$$\begin{aligned} x_1 - 3x_2 + x_3 &= 1 \\ 5x_2 - 4x_3 &= 0 \\ x_1 + 2x_2 - 3x_3 &= -1 \end{aligned}$$ Adding -2 times the first equation to the second equation produces a new second equation.

$$\begin{aligned} x_1 - 3x_2 + x_3 &= 1 \\ 5x_2 - 4x_3 &= 0 \\ 5x_2 - 4x_3 &= -2 \end{aligned}$$ Adding -1 times the first equation to the third equation produces a new third equation.

(Another way of describing this operation is to say that you *subtracted* the first equation from the third equation to produce a new third equation.)

$$\begin{aligned} x_1 - 3x_2 + x_3 &= 1 \\ 5x_2 - 4x_3 &= 0 \\ 0 &= -2 \end{aligned}$$ Subtracting the second equation from the third equation produces a new third equation.

The statement $0 = -2$ is false, so this system has no solution. Moreover, this system is equivalent to the original system, so the original system also has no solution.

As in Example 7, the three equations in Example 8 represent planes in a three-dimensional coordinate system. In this example, however, the system is inconsistent. So, the planes do not have a point in common, as shown at the right.

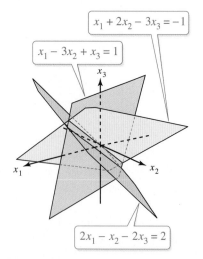

This section ends with an example of a system of linear equations that has infinitely many solutions. You can represent the solution set for such a system in parametric form, as you did in Examples 2 and 3.

EXAMPLE 9   A System with Infinitely Many Solutions

Solve the system.

$$\begin{aligned} x_2 - x_3 &= 0 \\ x_1 \quad\quad - 3x_3 &= -1 \\ -x_1 + 3x_2 \quad\quad &= 1 \end{aligned}$$

SOLUTION

Begin by rewriting the system in row-echelon form, as shown below.

$$\begin{aligned} x_1 \quad\quad - 3x_3 &= -1 \\ x_2 - x_3 &= 0 \\ -x_1 + 3x_2 \quad\quad &= 1 \end{aligned}$$
← ← Interchange the first two equations.

$$\begin{aligned} x_1 \quad\quad - 3x_3 &= -1 \\ x_2 - x_3 &= 0 \\ 3x_2 - 3x_3 &= 0 \end{aligned}$$
Adding the first equation to the third equation produces a new ← third equation.

$$\begin{aligned} x_1 \quad\quad - 3x_3 &= -1 \\ x_2 - x_3 &= 0 \\ 0 &= 0 \end{aligned}$$
Adding −3 times the second equation to the third equation ← eliminates the third equation.

The third equation is unnecessary, so omit it to obtain the system shown below.

$$\begin{aligned} x_1 \quad\quad - 3x_3 &= -1 \\ x_2 - x_3 &= 0 \end{aligned}$$

To represent the solutions, choose x_3 to be the free variable and represent it by the parameter t. Because $x_2 = x_3$ and $x_1 = 3x_3 - 1$, you can describe the solution set as

$$x_1 = 3t - 1, \quad x_2 = t, \quad x_3 = t, \quad t \text{ is any real number.}$$

DISCOVERY

1. Graph the two lines represented by the system of equations.

$$\begin{aligned} x - 2y &= 1 \\ -2x + 3y &= -3 \end{aligned}$$

2. Use Gaussian elimination to solve this system as shown below.

$$\begin{aligned} x - 2y &= 1 \\ -1y &= -1 \end{aligned}$$

$$\begin{aligned} x - 2y &= 1 \\ y &= 1 \end{aligned}$$

$$\begin{aligned} x &= 3 \\ y &= 1 \end{aligned}$$

Graph the system of equations you obtain at each step of this process. What do you observe about the lines?

See LarsonLinearAlgebra.com for an interactive version of this type of exercise.

REMARK

You are asked to repeat this graphical analysis for other systems in Exercises 91 and 92.

1.1   Exercises

See CalcChat.com for worked-out solutions to odd-numbered exercises.

Linear Equations   In Exercises 1–6, determine whether the equation is linear in the variables x and y.

1. $2x - 3y = 4$

2. $3x - 4xy = 0$

3. $\dfrac{3}{y} + \dfrac{2}{x} - 1 = 0$

4. $x^2 + y^2 = 4$

5. $2 \sin x - y = 14$

6. $(\cos 3)x + y = -16$

Parametric Representation   In Exercises 7–10, find a parametric representation of the solution set of the linear equation.

7. $2x - 4y = 0$

8. $3x - \frac{1}{2}y = 9$

9. $x + y + z = 1$

10. $12x_1 + 24x_2 - 36x_3 = 12$

Graphical Analysis   In Exercises 11–24, graph the system of linear equations. Solve the system and interpret your answer.

11. $2x + y = 4$
$x - y = 2$

12. $x + 3y = 2$
$-x + 2y = 3$

13. $-x + y = 1$
$3x - 3y = 4$

14. $\frac{1}{2}x - \frac{1}{3}y = 1$
$-2x + \frac{4}{3}y = -4$

15. $3x - 5y = 7$
$2x + y = 9$

16. $-x + 3y = 17$
$4x + 3y = 7$

17. $2x - y = 5$
$5x - y = 11$

18. $x - 5y = 21$
$6x + 5y = 21$

19. $\dfrac{x+3}{4} + \dfrac{y-1}{3} = 1$
$2x - y = 12$

20. $\dfrac{x-1}{2} + \dfrac{y+2}{3} = 4$
$x - 2y = 5$

21. $0.05x - 0.03y = 0.07$
$0.07x + 0.02y = 0.16$

22. $0.2x - 0.5y = -27.8$
$0.3x - 0.4y = 68.7$

23. $\dfrac{x}{4} + \dfrac{y}{6} = 1$
$x - y = 3$

24. $\dfrac{2x}{3} + \dfrac{y}{6} = \dfrac{2}{3}$
$4x + y = 4$

Back-Substitution   In Exercises 25–30, use back-substitution to solve the system.

25. $x_1 - x_2 = 2$
$x_2 = 3$

26. $2x_1 - 4x_2 = 6$
$3x_2 = 9$

27. $-x + y - z = 0$
$2y + z = 3$
$\frac{1}{2}z = 0$

28. $x - y = 5$
$3y + z = 11$
$4z = 8$

29. $5x_1 + 2x_2 + x_3 = 0$
$2x_1 + x_2 = 0$

30. $x_1 + x_2 + x_3 = 0$
$x_2 = 0$

Graphical Analysis   In Exercises 31–36, complete parts (a)–(e) for the system of equations.

(a) Use a graphing utility to graph the system.

(b) Use the graph to determine whether the system is consistent or inconsistent.

(c) If the system is consistent, approximate the solution.

(d) Solve the system algebraically.

(e) Compare the solution in part (d) with the approximation in part (c). What can you conclude?

31. $-3x - y = 3$
$6x + 2y = 1$

32. $4x - 5y = 3$
$-8x + 10y = 14$

33. $2x - 8y = 3$
$\frac{1}{2}x + y = 0$

34. $9x - 4y = 5$
$\frac{1}{2}x + \frac{1}{3}y = 0$

35. $4x - 8y = 9$
$0.8x - 1.6y = 1.8$

36. $-14.7x + 2.1y = 1.05$
$44.1x - 6.3y = -3.15$

System of Linear Equations   In Exercises 37–56, solve the system of linear equations.

37. $x_1 - x_2 = 0$
$3x_1 - 2x_2 = -1$

38. $3x + 2y = 2$
$6x + 4y = 14$

39. $3u + v = 240$
$u + 3v = 240$

40. $x_1 - 2x_2 = 0$
$6x_1 + 2x_2 = 0$

41. $9x - 3y = -1$
$\frac{1}{5}x + \frac{2}{5}y = -\frac{1}{3}$

42. $\frac{2}{3}x_1 + \frac{1}{6}x_2 = 0$
$4x_1 + \phantom{\frac{1}{6}}x_2 = 0$

43. $\dfrac{x-2}{4} + \dfrac{y-1}{3} = 2$
$x - 3y = 20$

44. $\dfrac{x_1+4}{3} + \dfrac{x_2+1}{2} = 1$
$3x_1 - x_2 = -2$

45. $0.02x_1 - 0.05x_2 = -0.19$
$0.03x_1 + 0.04x_2 = 0.52$

46. $0.05x_1 - 0.03x_2 = 0.21$
$0.07x_1 + 0.02x_2 = 0.17$

47. $x - y - z = 0$
$x + 2y - z = 6$
$2x - z = 5$

48. $x + y + z = 2$
$-x + 3y + 2z = 8$
$4x + y = 4$

49. $3x_1 - 2x_2 + 4x_3 = 1$
$x_1 + x_2 - 2x_3 = 3$
$2x_1 - 3x_2 + 6x_3 = 8$

The symbol ⚡ indicates an exercise in which you are instructed to use a graphing utility or software program.

50. $5x_1 - 3x_2 + 2x_3 = 3$
$2x_1 + 4x_2 - x_3 = 7$
$x_1 - 11x_2 + 4x_3 = 3$

51. $2x_1 + x_2 - 3x_3 = 4$
$4x_1 + 2x_3 = 10$
$-2x_1 + 3x_2 - 13x_3 = -8$

52. $x_1 + 4x_3 = 13$
$4x_1 - 2x_2 + x_3 = 7$
$2x_1 - 2x_2 - 7x_3 = -19$

53. $x - 3y + 2z = 18$
$5x - 15y + 10z = 18$

54. $x_1 - 2x_2 + 5x_3 = 2$
$3x_1 + 2x_2 - x_3 = -2$

55. $x + y + z + w = 6$
$2x + 3y - w = 0$
$-3x + 4y + z + 2w = 4$
$x + 2y - z + w = 0$

56. $-x_1 + 2x_4 = 1$
$4x_2 - x_3 - x_4 = 2$
$x_2 - x_4 = 0$
$3x_1 - 2x_2 + 3x_3 = 4$

System of Linear Equations In Exercises 57–62, use a software program or a graphing utility to solve the system of linear equations.

57. $123.5x + 61.3y - 32.4z = -262.74$
$54.7x - 45.6y + 98.2z = 197.4$
$42.4x - 89.3y + 12.9z = 33.66$

58. $120.2x + 62.4y - 36.5z = 258.64$
$56.8x - 42.8y + 27.3z = -71.44$
$88.1x + 72.5y - 28.5z = 225.88$

59. $x_1 + 0.5x_2 + 0.33x_3 + 0.25x_4 = 1.1$
$0.5x_1 + 0.33x_2 + 0.25x_3 + 0.21x_4 = 1.2$
$0.33x_1 + 0.25x_2 + 0.2x_3 + 0.17x_4 = 1.3$
$0.25x_1 + 0.2x_2 + 0.17x_3 + 0.14x_4 = 1.4$

60. $0.1x - 2.5y + 1.2z - 0.75w = 108$
$2.4x + 1.5y - 1.8z + 0.25w = -81$
$0.4x - 3.2y + 1.6z - 1.4w = 148.8$
$1.6x + 1.2y - 3.2z + 0.6w = -143.2$

61. $\frac{1}{2}x_1 - \frac{3}{7}x_2 + \frac{2}{9}x_3 = \frac{349}{630}$
$\frac{2}{3}x_1 + \frac{4}{9}x_2 - \frac{2}{5}x_3 = -\frac{19}{45}$
$\frac{4}{5}x_1 - \frac{1}{8}x_2 + \frac{4}{3}x_3 = \frac{139}{150}$

62. $\frac{1}{8}x - \frac{1}{7}y + \frac{1}{6}z - \frac{1}{5}w = 1$
$\frac{1}{7}x + \frac{1}{6}y - \frac{1}{5}z + \frac{1}{4}w = 1$
$\frac{1}{6}x - \frac{1}{5}y + \frac{1}{4}z - \frac{1}{3}w = 1$
$\frac{1}{5}x + \frac{1}{4}y - \frac{1}{3}z + \frac{1}{2}w = 1$

Number of Solutions In Exercises 63–66, state why the system of equations must have at least one solution. Then solve the system and determine whether it has exactly one solution or infinitely many solutions.

63. $4x + 3y + 17z = 0$
$5x + 4y + 22z = 0$
$4x + 2y + 19z = 0$

64. $2x + 3y = 0$
$4x + 3y - z = 0$
$8x + 3y + 3z = 0$

65. $5x + 5y - z = 0$
$10x + 5y + 2z = 0$
$5x + 15y - 9z = 0$

66. $16x + 3y + z = 0$
$16x + 2y - z = 0$

67. Nutrition One 240-milliliter glass of apple juice and one 240-milliliter glass of orange juice contain a total of 227 milligrams of vitamin C. Two 240-milliliter glasses of apple juice and three 240-milliliter glasses of orange juice contain a total of 578 milligrams of vitamin C. How much vitamin C is in a 240-milliliter glass of each type of juice?

68. Airplane Speed Two planes start from Los Angeles International Airport and fly in opposite directions. The second plane starts $\frac{1}{2}$ hour after the first plane, but its speed is 80 kilometers per hour faster. Two hours after the first plane departs, the planes are 3200 kilometers apart. Find the airspeed of each plane.

True or False? In Exercises 69 and 70, determine whether each statement is true or false. If a statement is true, give a reason or cite an appropriate statement from the text. If a statement is false, provide an example that shows the statement is not true in all cases or cite an appropriate statement from the text.

69. (a) A system of one linear equation in two variables is always consistent.

(b) A system of two linear equations in three variables is always consistent.

(c) If a linear system is consistent, then it has infinitely many solutions.

70. (a) A linear system can have exactly two solutions.

(b) Two systems of linear equations are equivalent when they have the same solution set.

(c) A system of three linear equations in two variables is always inconsistent.

71. Find a system of two equations in two variables, x_1 and x_2, that has the solution set given by the parametric representation $x_1 = t$ and $x_2 = 3t - 4$, where t is any real number. Then show that the solutions to the system can also be written as

$$x_1 = \frac{4}{3} + \frac{t}{3} \quad \text{and} \quad x_2 = t.$$

The symbol ⬛ indicates that electronic data sets for these exercises are available at *LarsonLinearAlgebra.com*. The data sets are compatible with MATLAB, *Mathematica*, *Maple*, TI-83 Plus, TI-84 Plus, TI-89, and Voyage 200.

72. Find a system of two equations in three variables, x_1, x_2, and x_3, that has the solution set given by the parametric representation

$$x_1 = t, \quad x_2 = s, \quad \text{and} \quad x_3 = 3 + s - t$$

where s and t are any real numbers. Then show that the solutions to the system can also be written as

$$x_1 = 3 + s - t, \quad x_2 = s, \quad \text{and} \quad x_3 = t.$$

Substitution In Exercises 73–76, solve the system of equations by first letting $A = 1/x$, $B = 1/y$, and $C = 1/z$.

73. $\dfrac{12}{x} - \dfrac{12}{y} = 7$

$\dfrac{3}{x} + \dfrac{4}{y} = 0$

74. $\dfrac{3}{x} + \dfrac{2}{y} = -1$

$\dfrac{2}{x} - \dfrac{3}{y} = -\dfrac{17}{6}$

75. $\dfrac{2}{x} + \dfrac{1}{y} - \dfrac{3}{z} = 4$

$\dfrac{4}{x} \qquad + \dfrac{2}{z} = 10$

$-\dfrac{2}{x} + \dfrac{3}{y} - \dfrac{13}{z} = -8$

76. $\dfrac{2}{x} + \dfrac{1}{y} - \dfrac{2}{z} = 5$

$\dfrac{3}{x} - \dfrac{4}{y} \qquad = -1$

$\dfrac{2}{x} + \dfrac{1}{y} + \dfrac{3}{z} = 0$

Trigonometric Coefficients In Exercises 77 and 78, solve the system of linear equations for x and y.

77. $(\cos\theta)x + (\sin\theta)y = 1$
$(-\sin\theta)x + (\cos\theta)y = 0$

78. $(\cos\theta)x + (\sin\theta)y = 1$
$(-\sin\theta)x + (\cos\theta)y = 1$

Coefficient Design In Exercises 79–84, determine the value(s) of k such that the system of linear equations has the indicated number of solutions.

79. No solution

$x + ky = 2$
$kx + y = 4$

80. Exactly one solution

$x + ky = 0$
$kx + y = 0$

81. Exactly one solution

$kx + 2ky + 3kz = 4k$
$x + y + z = 0$
$2x - y + z = 1$

82. No solution

$x + 2y + kz = 6$
$3x + 6y + 8z = 4$

83. Infinitely many solutions

$4x + ky = 6$
$kx + y = -3$

84. Infinitely many solutions

$kx + y = 16$
$3x - 4y = -64$

85. Determine the values of k such that the system of linear equations does not have a unique solution.

$x + y + kz = 3$
$x + ky + z = 2$
$kx + y + z = 1$

86. CAPSTONE Find values of a, b, and c such that the system of linear equations has (a) exactly one solution, (b) infinitely many solutions, and (c) no solution. Explain.

$x + 5y + z = 0$
$x + 6y - z = 0$
$2x + ay + bz = c$

87. Writing Consider the system of linear equations in x and y.

$a_1x + b_1y = c_1$
$a_2x + b_2y = c_2$
$a_3x + b_3y = c_3$

Describe the graphs of these three equations in the xy-plane when the system has (a) exactly one solution, (b) infinitely many solutions, and (c) no solution.

88. Writing Explain why the system of linear equations in Exercise 87 must be consistent when the constant terms c_1, c_2, and c_3 are all zero.

89. Show that if $ax^2 + bx + c = 0$ for all x, then $a = b = c = 0$.

90. Consider the system of linear equations in x and y.

$ax + by = e$
$cx + dy = f$

Under what conditions will the system have exactly one solution?

Discovery In Exercises 91 and 92, sketch the lines represented by the system of equations. Then use Gaussian elimination to solve the system. At each step of the elimination process, sketch the corresponding lines. What do you observe about the lines?

91. $x - 4y = -3$
$5x - 6y = 13$

92. $2x - 3y = 7$
$-4x + 6y = -14$

Writing In Exercises 93 and 94, the graphs of the two equations appear to be parallel. Solve the system of equations algebraically. Explain why the graphs are misleading.

93. $100y - x = 200$
$99y - x = -198$

94. $21x - 20y = 0$
$13x - 12y = 120$

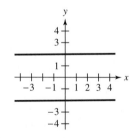

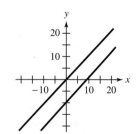

1.2 Gaussian Elimination and Gauss-Jordan Elimination

■ Determine the size of a matrix and write an augmented or coefficient matrix from a system of linear equations.

■ Use matrices and Gaussian elimination with back-substitution to solve a system of linear equations.

■ Use matrices and Gauss-Jordan elimination to solve a system of linear equations.

■ Solve a homogeneous system of linear equations.

MATRICES

Section 1.1 introduced Gaussian elimination as a procedure for solving a system of linear equations. In this section, you will study this procedure more thoroughly, beginning with some definitions. The first is the definition of a **matrix.**

REMARK

The plural of matrix is *matrices*. When each entry of a matrix is a *real* number, the matrix is a **real matrix**. Unless stated otherwise, assume all matrices in this text are real matrices.

Definition of a Matrix

If m and n are positive integers, then an $m \times n$ (read "m by n") matrix is a rectangular array

$$
\begin{array}{ccccc}
 & \text{Column 1} & \text{Column 2} & \text{Column 3} & \cdots & \text{Column } n \\
\text{Row 1} & a_{11} & a_{12} & a_{13} & \cdots & a_{1n} \\
\text{Row 2} & a_{21} & a_{22} & a_{23} & \cdots & a_{2n} \\
\text{Row 3} & a_{31} & a_{32} & a_{33} & \cdots & a_{3n} \\
\vdots & \vdots & \vdots & \vdots & & \vdots \\
\text{Row } m & a_{m1} & a_{m2} & a_{m3} & \cdots & a_{mn}
\end{array}
$$

in which each **entry,** a_{ij}, of the matrix is a number. An $m \times n$ matrix has m rows and n columns. Matrices are usually denoted by capital letters.

The entry a_{ij} is located in the ith row and the jth column. The index i is called the **row subscript** because it identifies the row in which the entry lies, and the index j is called the **column subscript** because it identifies the column in which the entry lies.

A matrix with m rows and n columns is of **size** $m \times n$. When $m = n$, the matrix is **square** of **order** n and the entries $a_{11}, a_{22}, a_{33}, \ldots, a_{nn}$ are the **main diagonal** entries.

> **EXAMPLE 1**   Sizes of Matrices

Each matrix has the indicated size.

a. $[2]$ Size: 1×1 **b.** $\begin{bmatrix} 0 & 0 \\ 0 & 0 \end{bmatrix}$ Size: 2×2 **c.** $\begin{bmatrix} e & 2 & -7 \\ \pi & \sqrt{2} & 4 \end{bmatrix}$ Size: 2×3

REMARK

Begin by aligning the variables in the equations vertically. Use 0 to show coefficients of zero in the matrix. Note the fourth column of constant terms in the augmented matrix.

One common use of matrices is to represent systems of linear equations. The matrix derived from the coefficients and constant terms of a system of linear equations is the **augmented matrix** of the system. The matrix containing only the coefficients of the system is the **coefficient matrix** of the system. Here is an example.

System	Augmented Matrix	Cofficient Matrix

$$
\begin{array}{rl}
x - 4y + 3z = 5 \\
-x + 3y - z = -3 \\
2x - 4z = 6
\end{array}
\qquad
\begin{bmatrix} 1 & -4 & 3 & 5 \\ -1 & 3 & -1 & -3 \\ 2 & 0 & -4 & 6 \end{bmatrix}
\qquad
\begin{bmatrix} 1 & -4 & 3 \\ -1 & 3 & -1 \\ 2 & 0 & -4 \end{bmatrix}
$$

ELEMENTARY ROW OPERATIONS

In the previous section, you studied three operations that produce equivalent systems of linear equations.

1. Interchange two equations.

2. Multiply an equation by a nonzero constant.

3. Add a multiple of an equation to another equation.

In matrix terminology, these three operations correspond to **elementary row operations.** An elementary row operation on an augmented matrix produces a new augmented matrix corresponding to a new (but equivalent) system of linear equations. Two matrices are **row-equivalent** when one can be obtained from the other by a finite sequence of elementary row operations.

Elementary Row Operations

1. Interchange two rows.
2. Multiply a row by a nonzero constant.
3. Add a multiple of a row to another row.

Although elementary row operations are relatively simple to perform, they can involve a lot of arithmetic, so it is easy to make a mistake. Noting the elementary row operations performed in each step can make checking your work easier.

Solving some systems involves many steps, so it is helpful to use a shorthand method of notation to keep track of each elementary row operation you perform. The next example introduces this notation.

TECHNOLOGY

Many graphing utilities and software programs can perform elementary row operations on matrices. If you use a graphing utility, you may see something similar to the screen below for Example 2(c). The **Technology Guide** at *CengageBrain.com* can help you use technology to perform elementary row operations.

```
A
       [[1 2   -4  3 ]
        [0 3   -2 -1 ]
        [2 1    5 -2 ]]

mRAdd(-2,A,1,3)
       [[1 2   -4  3 ]
        [0 3   -2 -1 ]
        [0 -3  13 -8 ]]
```

EXAMPLE 2   Elementary Row Operations

a. Interchange the first and second rows.

Original Matrix   New Row-Equivalent Matrix   Notation

$$\begin{bmatrix} 0 & 1 & 3 & 4 \\ -1 & 2 & 0 & 3 \\ 2 & -3 & 4 & 1 \end{bmatrix} \qquad \begin{bmatrix} -1 & 2 & 0 & 3 \\ 0 & 1 & 3 & 4 \\ 2 & -3 & 4 & 1 \end{bmatrix} \qquad R_1 \leftrightarrow R_2$$

b. Multiply the first row by $\frac{1}{2}$ to produce a new first row.

Original Matrix   New Row-Equivalent Matrix   Notation

$$\begin{bmatrix} 2 & -4 & 6 & -2 \\ 1 & 3 & -3 & 0 \\ 5 & -2 & 1 & 2 \end{bmatrix} \qquad \begin{bmatrix} 1 & -2 & 3 & -1 \\ 1 & 3 & -3 & 0 \\ 5 & -2 & 1 & 2 \end{bmatrix} \qquad \left(\tfrac{1}{2}\right)R_1 \to R_1$$

c. Add -2 times the first row to the third row to produce a new third row.

Original Matrix   New Row-Equivalent Matrix   Notation

$$\begin{bmatrix} 1 & 2 & -4 & 3 \\ 0 & 3 & -2 & -1 \\ 2 & 1 & 5 & -2 \end{bmatrix} \qquad \begin{bmatrix} 1 & 2 & -4 & 3 \\ 0 & 3 & -2 & -1 \\ 0 & -3 & 13 & -8 \end{bmatrix} \qquad R_3 + (-2)R_1 \to R_3$$

Notice that adding -2 times row 1 to row 3 does not change row 1.

In Example 7 in Section 1.1, you used Gaussian elimination with back-substitution to solve a system of linear equations. The next example demonstrates the matrix version of Gaussian elimination. The two methods are essentially the same. The basic difference is that with matrices you do not need to keep writing the variables.

EXAMPLE 3 **Using Elementary Row Operations to Solve a System**

Linear System

$$x - 2y + 3z = 9$$
$$-x + 3y = -4$$
$$2x - 5y + 5z = 17$$

Associated Augmented Matrix

$$\begin{bmatrix} 1 & -2 & 3 & 9 \\ -1 & 3 & 0 & -4 \\ 2 & -5 & 5 & 17 \end{bmatrix}$$

Add the first equation to the second equation.

$$x - 2y + 3z = 9$$
$$y + 3z = 5$$
$$2x - 5y + 5z = 17$$

Add the first row to the second row to produce a new second row.

$$\begin{bmatrix} 1 & -2 & 3 & 9 \\ 0 & 1 & 3 & 5 \\ 2 & -5 & 5 & 17 \end{bmatrix} R_2 + R_1 \to R_2$$

Add −2 times the first equation to the third equation.

$$x - 2y + 3z = 9$$
$$y + 3z = 5$$
$$-y - z = -1$$

Add −2 times the first row to the third row to produce a new third row.

$$\begin{bmatrix} 1 & -2 & 3 & 9 \\ 0 & 1 & 3 & 5 \\ 0 & -1 & -1 & -1 \end{bmatrix} R_3 + (-2)R_1 \to R_3$$

Add the second equation to the third equation.

$$x - 2y + 3z = 9$$
$$y + 3z = 5$$
$$2z = 4$$

Add the second row to the third row to produce a new third row.

$$\begin{bmatrix} 1 & -2 & 3 & 9 \\ 0 & 1 & 3 & 5 \\ 0 & 0 & 2 & 4 \end{bmatrix} R_3 + R_2 \to R_3$$

Multiply the third equation by $\frac{1}{2}$.

$$x - 2y + 3z = 9$$
$$y + 3z = 5$$
$$z = 2$$

Multiply the third row by $\frac{1}{2}$ to produce a new third row.

$$\begin{bmatrix} 1 & -2 & 3 & 9 \\ 0 & 1 & 3 & 5 \\ 0 & 0 & 1 & 2 \end{bmatrix} \left(\tfrac{1}{2}\right)R_3 \to R_3$$

REMARK

The term *echelon* refers to the stair-step pattern formed by the nonzero elements of the matrix.

Use back-substitution to find the solution, as in Example 6 in Section 1.1. The solution is $x = 1$, $y = -1$, and $z = 2$.

The last matrix in Example 3 is in **row-echelon** form. To be in this form, a matrix must have the properties listed below.

Row-Echelon Form and Reduced Row-Echelon Form

A matrix in **row-echelon form** has the properties below.

1. Any rows consisting entirely of zeros occur at the bottom of the matrix.
2. For each row that does not consist entirely of zeros, the first nonzero entry is 1 (called a **leading 1**).
3. For two successive (nonzero) rows, the leading 1 in the higher row is farther to the left than the leading 1 in the lower row.

A matrix in row-echelon form is in **reduced row-echelon form** when every column that has a leading 1 has zeros in every position above and below its leading 1.

EXAMPLE 4   Row-Echelon Form

Determine whether each matrix is in row-echelon form. If it is, determine whether the matrix is also in reduced row-echelon form.

a. $\begin{bmatrix} 1 & 2 & -1 & 4 \\ 0 & 1 & 0 & 3 \\ 0 & 0 & 1 & -2 \end{bmatrix}$ b. $\begin{bmatrix} 1 & 2 & -1 & 2 \\ 0 & 0 & 0 & 0 \\ 0 & 1 & 2 & -4 \end{bmatrix}$

c. $\begin{bmatrix} 1 & -5 & 2 & -1 & 3 \\ 0 & 0 & 1 & 3 & -2 \\ 0 & 0 & 0 & 1 & 4 \\ 0 & 0 & 0 & 0 & 1 \end{bmatrix}$ d. $\begin{bmatrix} 1 & 0 & 0 & -1 \\ 0 & 1 & 0 & 2 \\ 0 & 0 & 1 & 3 \\ 0 & 0 & 0 & 0 \end{bmatrix}$

e. $\begin{bmatrix} 1 & 2 & -3 & 4 \\ 0 & 2 & 1 & -1 \\ 0 & 0 & 1 & -3 \end{bmatrix}$ f. $\begin{bmatrix} 0 & 1 & 0 & 5 \\ 0 & 0 & 1 & 3 \\ 0 & 0 & 0 & 0 \end{bmatrix}$

SOLUTION

The matrices in (a), (c), (d), and (f) are in row-echelon form. The matrices in (d) and (f) are in *reduced* row-echelon form because every column that has a leading 1 has zeros in every position above and below its leading 1. The matrix in (b) is not in row-echelon form because the row of all zeros does not occur at the bottom of the matrix. The matrix in (e) is not in row-echelon form because the first nonzero entry in Row 2 is not 1.

Every matrix is row-equivalent to a matrix in row-echelon form. For instance, in Example 4(e), multiplying the second row in the matrix by $\frac{1}{2}$ changes the matrix to row-echelon form.

The procedure for using Gaussian elimination with back-substitution is summarized below.

Gaussian Elimination with Back-Substitution

1. Write the augmented matrix of the system of linear equations.
2. Use elementary row operations to rewrite the matrix in row-echelon form.
3. Write the system of linear equations corresponding to the matrix in row-echelon form, and use back-substitution to find the solution.

Gaussian elimination with back-substitution works well for solving systems of linear equations by hand or with a computer. For this algorithm, the order in which you perform the elementary row operations is important. Operate from *left to right by columns*, using elementary row operations to obtain zeros in all entries directly below the leading 1's.

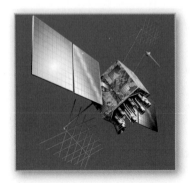

LINEAR ALGEBRA APPLIED

The Global Positioning System (GPS) is a network of 24 satellites originally developed by the U.S. military as a navigational tool. Today, GPS technology is used in a wide variety of civilian applications, such as package delivery, farming, mining, surveying, construction, banking, weather forecasting, and disaster relief. A GPS receiver works by using satellite readings to calculate its location. In three dimensions, the receiver uses signals from at least four satellites to "trilaterate" its position. In a simplified mathematical model, a system of three linear equations in four unknowns (three dimensions and time) is used to determine the coordinates of the receiver as functions of time.

EXAMPLE 5   **Gaussian Elimination with Back-Substitution**

Solve the system.

$$\begin{aligned} x_2 + x_3 - 2x_4 &= -3 \\ x_1 + 2x_2 - x_3 &= 2 \\ 2x_1 + 4x_2 + x_3 - 3x_4 &= -2 \\ x_1 - 4x_2 - 7x_3 - x_4 &= -19 \end{aligned}$$

SOLUTION

The augmented matrix for this system is

$$\begin{bmatrix} 0 & 1 & 1 & -2 & -3 \\ 1 & 2 & -1 & 0 & 2 \\ 2 & 4 & 1 & -3 & -2 \\ 1 & -4 & -7 & -1 & -19 \end{bmatrix}.$$

Obtain a leading 1 in the upper left corner and zeros elsewhere in the first column.

$$\begin{bmatrix} 1 & 2 & -1 & 0 & 2 \\ 0 & 1 & 1 & -2 & -3 \\ 2 & 4 & 1 & -3 & -2 \\ 1 & -4 & -7 & -1 & -19 \end{bmatrix}$$

Interchange the first two rows. $R_1 \leftrightarrow R_2$

$$\begin{bmatrix} 1 & 2 & -1 & 0 & 2 \\ 0 & 1 & 1 & -2 & -3 \\ 0 & 0 & 3 & -3 & -6 \\ 1 & -4 & -7 & -1 & -19 \end{bmatrix}$$

Adding -2 times the first row to the third row produces a new third row. $R_3 + (-2)R_1 \rightarrow R_3$

$$\begin{bmatrix} 1 & 2 & -1 & 0 & 2 \\ 0 & 1 & 1 & -2 & -3 \\ 0 & 0 & 3 & -3 & -6 \\ 0 & -6 & -6 & -1 & -21 \end{bmatrix}$$

Adding -1 times the first row to the fourth row produces a new fourth row. $R_4 + (-1)R_1 \rightarrow R_4$

Now that the first column is in the desired form, change the second column as shown below.

$$\begin{bmatrix} 1 & 2 & -1 & 0 & 2 \\ 0 & 1 & 1 & -2 & -3 \\ 0 & 0 & 3 & -3 & -6 \\ 0 & 0 & 0 & -13 & -39 \end{bmatrix}$$

Adding 6 times the second row to the fourth row produces a new fourth row. $R_4 + (6)R_2 \rightarrow R_4$

To write the third and fourth columns in proper form, multiply the third row by $\frac{1}{3}$ and the fourth row by $-\frac{1}{13}$.

$$\begin{bmatrix} 1 & 2 & -1 & 0 & 2 \\ 0 & 1 & 1 & -2 & -3 \\ 0 & 0 & 1 & -1 & -2 \\ 0 & 0 & 0 & 1 & 3 \end{bmatrix}$$

Multiplying the third row by $\frac{1}{3}$ and the fourth row by $-\frac{1}{13}$ produces new third and fourth rows. $\left(\frac{1}{3}\right)R_3 \rightarrow R_3$ $\left(-\frac{1}{13}\right)R_4 \rightarrow R_4$

The matrix is now in row-echelon form, and the corresponding system is shown below.

$$\begin{aligned} x_1 + 2x_2 - x_3 &= 2 \\ x_2 + x_3 - 2x_4 &= -3 \\ x_3 - x_4 &= -2 \\ x_4 &= 3 \end{aligned}$$

Use back-substitution to find that the solution is $x_1 = -1$, $x_2 = 2$, $x_3 = 1$, and $x_4 = 3$.

When solving a system of linear equations, remember that it is possible for the system to have no solution. If, in the elimination process, you obtain a row of all zeros except for the last entry, then it is unnecessary to continue the process. Simply conclude that the system has no solution, or is *inconsistent*.

EXAMPLE 6   **A System with No Solution**

Solve the system.

$$\begin{aligned} x_1 - x_2 + 2x_3 &= 4 \\ x_1 + x_3 &= 6 \\ 2x_1 - 3x_2 + 5x_3 &= 4 \\ 3x_1 + 2x_2 - x_3 &= 1 \end{aligned}$$

SOLUTION

The augmented matrix for this system is

$$\begin{bmatrix} 1 & -1 & 2 & 4 \\ 1 & 0 & 1 & 6 \\ 2 & -3 & 5 & 4 \\ 3 & 2 & -1 & 1 \end{bmatrix}.$$

Apply Gaussian elimination to the augmented matrix.

$$\begin{bmatrix} 1 & -1 & 2 & 4 \\ 0 & 1 & -1 & 2 \\ 2 & -3 & 5 & 4 \\ 3 & 2 & -1 & 1 \end{bmatrix} \quad R_2 + (-1)R_1 \to R_2$$

$$\begin{bmatrix} 1 & -1 & 2 & 4 \\ 0 & 1 & -1 & 2 \\ 0 & -1 & 1 & -4 \\ 3 & 2 & -1 & 1 \end{bmatrix} \quad R_3 + (-2)R_1 \to R_3$$

$$\begin{bmatrix} 1 & -1 & 2 & 4 \\ 0 & 1 & -1 & 2 \\ 0 & -1 & 1 & -4 \\ 0 & 5 & -7 & -11 \end{bmatrix} \quad R_4 + (-3)R_1 \to R_4$$

$$\begin{bmatrix} 1 & -1 & 2 & 4 \\ 0 & 1 & -1 & 2 \\ 0 & 0 & 0 & -2 \\ 0 & 5 & -7 & -11 \end{bmatrix} \quad R_3 + R_2 \to R_3$$

Note that the third row of this matrix consists entirely of zeros except for the last entry. This means that the original system of linear equations is *inconsistent*. To see why this is true, convert back to a system of linear equations.

$$\begin{aligned} x_1 - x_2 + 2x_3 &= 4 \\ x_2 - x_3 &= 2 \\ 0 &= -2 \\ 5x_2 - 7x_3 &= -11 \end{aligned}$$

The third equation is not possible, so the system has no solution.

GAUSS-JORDAN ELIMINATION

With Gaussian elimination, you apply elementary row operations to a matrix to obtain a (row-equivalent) row-echelon form. A second method of elimination, called **Gauss-Jordan elimination** after Carl Friedrich Gauss and Wilhelm Jordan (1842–1899), continues the reduction process until a *reduced* row-echelon form is obtained. Example 7 demonstrates this procedure.

EXAMPLE 7   Gauss-Jordan Elimination

See LarsonLinearAlgebra.com for an interactive version of this type of example.

Use Gauss-Jordan elimination to solve the system.

$$
\begin{aligned}
x - 2y + 3z &= 9 \\
-x + 3y &= -4 \\
2x - 5y + 5z &= 17
\end{aligned}
$$

SOLUTION

In Example 3, you used Gaussian elimination to obtain the row-echelon form

$$
\begin{bmatrix}
1 & -2 & 3 & 9 \\
0 & 1 & 3 & 5 \\
0 & 0 & 1 & 2
\end{bmatrix}.
$$

Now, apply elementary row operations until you obtain zeros above each of the leading 1's, as shown below.

$$
\begin{bmatrix}
1 & 0 & 9 & 19 \\
0 & 1 & 3 & 5 \\
0 & 0 & 1 & 2
\end{bmatrix}
\qquad R_1 + (2)R_2 \to R_1
$$

$$
\begin{bmatrix}
1 & 0 & 9 & 19 \\
0 & 1 & 0 & -1 \\
0 & 0 & 1 & 2
\end{bmatrix}
\qquad R_2 + (-3)R_3 \to R_2
$$

$$
\begin{bmatrix}
1 & 0 & 0 & 1 \\
0 & 1 & 0 & -1 \\
0 & 0 & 1 & 2
\end{bmatrix}
\qquad R_1 + (-9)R_3 \to R_1
$$

The matrix is now in reduced row-echelon form. Converting back to a system of linear equations, you have

$$
\begin{aligned}
x &= 1 \\
y &= -1 \\
z &= 2.
\end{aligned}
$$

The elimination procedures described in this section can sometimes result in fractional coefficients. For example, in the elimination procedure for the system

$$
\begin{aligned}
2x - 5y + 5z &= 14 \\
3x - 2y + 3z &= 9 \\
-3x + 4y &= -18
\end{aligned}
$$

you may be inclined to first multiply Row 1 by $\frac{1}{2}$ to produce a leading 1, which will result in working with fractional coefficients. Sometimes, judiciously choosing which elementary row operations you apply, and the order in which you apply them, enables you to avoid fractions.

REMARK

No matter which elementary row operations or order you use, the reduced row-echelon form of a matrix is the same.

DISCOVERY

1. Without performing any row operations, explain why the system of linear equations below is consistent.

$$
\begin{aligned}
2x_1 + 3x_2 + 5x_3 &= 0 \\
-5x_1 + 6x_2 - 17x_3 &= 0 \\
7x_1 - 4x_2 + 3x_3 &= 0
\end{aligned}
$$

2. The system below has more variables than equations. Why does it have an infinite number of solutions?

$$
\begin{aligned}
2x_1 + 3x_2 + 5x_3 + 2x_4 &= 0 \\
-5x_1 + 6x_2 - 17x_3 - 3x_4 &= 0 \\
7x_1 - 4x_2 + 3x_3 + 13x_4 &= 0
\end{aligned}
$$

The next example demonstrates how Gauss-Jordan elimination can be used to solve a system with infinitely many solutions.

EXAMPLE 8   **A System with Infinitely Many Solutions**

Solve the system of linear equations.

$$
\begin{aligned}
2x_1 + 4x_2 - 2x_3 &= 0 \\
3x_1 + 5x_2 &= 1
\end{aligned}
$$

SOLUTION

The augmented matrix for this system is

$$
\begin{bmatrix} 2 & 4 & -2 & 0 \\ 3 & 5 & 0 & 1 \end{bmatrix}.
$$

Using a graphing utility, a software program, or Gauss-Jordan elimination, verify that the reduced row-echelon form of the matrix is

$$
\begin{bmatrix} 1 & 0 & 5 & 2 \\ 0 & 1 & -3 & -1 \end{bmatrix}.
$$

The corresponding system of equations is

$$
\begin{aligned}
x_1 + 5x_3 &= 2 \\
x_2 - 3x_3 &= -1.
\end{aligned}
$$

Now, using the parameter t to represent x_3, you have

$$
x_1 = 2 - 5t, \quad x_2 = -1 + 3t, \quad x_3 = t, \quad t \text{ is any real number.}
$$

Note in Example 8 that the arbitrary parameter t represents the *nonleading* variable x_3. The variables x_1 and x_2 are written as functions of t.

You have looked at two elimination methods for solving a system of linear equations. Which is better? To some degree the answer depends on personal preference. In real-life applications of linear algebra, systems of linear equations are usually solved by computer. Most software uses a form of Gaussian elimination, with special emphasis on ways to reduce rounding errors and minimize storage of data. The examples and exercises in this text focus on the underlying concepts, so you should know both elimination methods.

HOMOGENEOUS SYSTEMS OF LINEAR EQUATIONS

Systems of linear equations in which each of the constant terms is zero are called **homogeneous.** A homogeneous system of m equations in n variables has the form

$$
\begin{aligned}
a_{11}x_1 + a_{12}x_2 + a_{13}x_3 + \cdots + a_{1n}x_n &= 0 \\
a_{21}x_1 + a_{22}x_2 + a_{23}x_3 + \cdots + a_{2n}x_n &= 0 \\
&\vdots \\
a_{m1}x_1 + a_{m2}x_2 + a_{m3}x_3 + \cdots + a_{mn}x_n &= 0.
\end{aligned}
$$

A homogeneous system must have at least one solution. Specifically, if all variables in a homogeneous system have the value zero, then each of the equations is satisfied. Such a solution is **trivial** (or obvious).

REMARK

A homogeneous system of three equations in the three variables x_1, x_2, and x_3 has the trivial solution $x_1 = 0$, $x_2 = 0$, and $x_3 = 0$.

EXAMPLE 9   Solving a Homogeneous System of Linear Equations

Solve the system of linear equations.

$$
\begin{aligned}
x_1 - x_2 + 3x_3 &= 0 \\
2x_1 + x_2 + 3x_3 &= 0
\end{aligned}
$$

SOLUTION

Applying Gauss-Jordan elimination to the augmented matrix

$$
\begin{bmatrix} 1 & -1 & 3 & 0 \\ 2 & 1 & 3 & 0 \end{bmatrix}
$$

yields the matrices shown below.

$$
\begin{bmatrix} 1 & -1 & 3 & 0 \\ 0 & 3 & -3 & 0 \end{bmatrix} \qquad R_2 + (-2)R_1 \rightarrow R_2
$$

$$
\begin{bmatrix} 1 & -1 & 3 & 0 \\ 0 & 1 & -1 & 0 \end{bmatrix} \qquad \left(\tfrac{1}{3}\right)R_2 \rightarrow R_2
$$

$$
\begin{bmatrix} 1 & 0 & 2 & 0 \\ 0 & 1 & -1 & 0 \end{bmatrix} \qquad R_1 + R_2 \rightarrow R_1
$$

The system of equations corresponding to this matrix is

$$
\begin{aligned}
x_1 \quad + 2x_3 &= 0 \\
x_2 - x_3 &= 0.
\end{aligned}
$$

Using the parameter $t = x_3$, the solution set is $x_1 = -2t$, $x_2 = t$, and $x_3 = t$, where t is any real number. This system has infinitely many solutions, one of which is the trivial solution ($t = 0$).

As illustrated in Example 9, a homogeneous system with fewer equations than variables has infinitely many solutions.

THEOREM 1.1   The Number of Solutions of a Homogeneous System

Every homogeneous system of linear equations is consistent. Moreover, if the system has fewer equations than variables, then it must have infinitely many solutions.

To prove Theorem 1.1, use the procedure in Example 9, but for a general matrix.

1.2 Exercises

See CalcChat.com for worked-out solutions to odd-numbered exercises.

Matrix Size In Exercises 1–6, determine the size of the matrix.

1. $\begin{bmatrix} 1 & 2 & -4 \\ 3 & -4 & 6 \\ 0 & 1 & 2 \end{bmatrix}$

2. $\begin{bmatrix} -2 \\ -1 \\ 1 \\ 2 \end{bmatrix}$

3. $\begin{bmatrix} 2 & -1 & -1 & 1 \\ -6 & 2 & 0 & 1 \end{bmatrix}$

4. $\begin{bmatrix} -1 \end{bmatrix}$

5. $\begin{bmatrix} 8 & 6 & 4 & 1 & 3 \\ 2 & 1 & -7 & 4 & 1 \\ 1 & 1 & -1 & 2 & 1 \\ 1 & -1 & 2 & 0 & 0 \end{bmatrix}$

6. $\begin{bmatrix} 1 & 2 & 3 & 4 & -10 \end{bmatrix}$

Elementary Row Operations In Exercises 7–10, identify the elementary row operation(s) being performed to obtain the new row-equivalent matrix.

Original Matrix

7. $\begin{bmatrix} -2 & 5 & 1 \\ 3 & -1 & -8 \end{bmatrix}$

New Row-Equivalent Matrix

$\begin{bmatrix} 13 & 0 & -39 \\ 3 & -1 & -8 \end{bmatrix}$

Original Matrix

8. $\begin{bmatrix} 3 & -1 & -4 \\ -4 & 3 & 7 \end{bmatrix}$

New Row-Equivalent Matrix

$\begin{bmatrix} 3 & -1 & -4 \\ 5 & 0 & -5 \end{bmatrix}$

Original Matrix

9. $\begin{bmatrix} 0 & -1 & -7 & 7 \\ -1 & 5 & -8 & 7 \\ 3 & -2 & 1 & 2 \end{bmatrix}$

New Row-Equivalent Matrix

$\begin{bmatrix} -1 & 5 & -8 & 7 \\ 0 & -1 & -7 & 7 \\ 0 & 13 & -23 & 23 \end{bmatrix}$

Original Matrix

10. $\begin{bmatrix} -1 & -2 & 3 & -2 \\ 2 & -5 & 1 & -7 \\ 5 & 4 & -7 & 6 \end{bmatrix}$

New Row-Equivalent Matrix

$\begin{bmatrix} -1 & -2 & 3 & -2 \\ 0 & -9 & 7 & -11 \\ 0 & -6 & 8 & -4 \end{bmatrix}$

Augmented Matrix In Exercises 11–18, find the solution set of the system of linear equations represented by the augmented matrix.

11. $\begin{bmatrix} 1 & 0 & 0 \\ 0 & 1 & 2 \end{bmatrix}$

12. $\begin{bmatrix} 1 & 0 & 2 \\ 0 & 1 & 3 \end{bmatrix}$

13. $\begin{bmatrix} 1 & -1 & 0 & 3 \\ 0 & 1 & -2 & 1 \\ 0 & 0 & 1 & -1 \end{bmatrix}$

14. $\begin{bmatrix} 1 & 2 & 1 & 0 \\ 0 & 0 & 1 & -1 \\ 0 & 0 & 0 & 0 \end{bmatrix}$

15. $\begin{bmatrix} 2 & 1 & -1 & 3 \\ 1 & -1 & 1 & 0 \\ 0 & 1 & 2 & 1 \end{bmatrix}$

16. $\begin{bmatrix} 3 & -1 & 1 & 5 \\ 1 & 2 & 1 & 0 \\ 1 & 0 & 1 & 2 \end{bmatrix}$

17. $\begin{bmatrix} 1 & 2 & 0 & 1 & 4 \\ 0 & 1 & 2 & 1 & 3 \\ 0 & 0 & 1 & 2 & 1 \\ 0 & 0 & 0 & 1 & 4 \end{bmatrix}$

18. $\begin{bmatrix} 1 & 2 & 0 & 1 & 3 \\ 0 & 1 & 3 & 0 & 1 \\ 0 & 0 & 1 & 2 & 0 \\ 0 & 0 & 0 & 0 & 2 \end{bmatrix}$

Row-Echelon Form In Exercises 19–24, determine whether the matrix is in row-echelon form. If it is, determine whether it is also in reduced row-echelon form.

19. $\begin{bmatrix} 1 & 0 & 0 & 0 \\ 0 & 1 & 1 & 2 \\ 0 & 0 & 0 & 0 \end{bmatrix}$

20. $\begin{bmatrix} 0 & 1 & 0 & 0 \\ 1 & 0 & 2 & 1 \end{bmatrix}$

21. $\begin{bmatrix} -2 & 0 & 1 & 5 \\ 0 & -1 & 2 & 1 \\ 0 & 0 & 0 & 2 \end{bmatrix}$

22. $\begin{bmatrix} 1 & 0 & 2 & 1 \\ 0 & 1 & 3 & 4 \\ 0 & 0 & 1 & 0 \end{bmatrix}$

23. $\begin{bmatrix} 0 & 0 & 1 & 0 & 0 \\ 0 & 0 & 0 & 1 & 0 \\ 0 & 0 & 0 & 2 & 0 \end{bmatrix}$

24. $\begin{bmatrix} 1 & 0 & 0 & 0 \\ 0 & 0 & 0 & 1 \\ 0 & 0 & 0 & 0 \end{bmatrix}$

System of Linear Equations In Exercises 25–38, solve the system using either Gaussian elimination with back-substitution or Gauss-Jordan elimination.

25. $x + 3y = 11$
    $3x + y = 9$

26. $2x + 6y = 16$
    $-2x - 6y = -16$

27. $-x + 2y = 1.5$
    $2x - 4y = 3$

28. $2x - y = -0.1$
    $3x + 2y = 1.6$

29. $-3x + 5y = -22$
    $3x + 4y = 4$
    $4x - 8y = 32$

30. $x + 2y = 0$
    $x + y = 6$
    $3x - 2y = 8$

31.
$$\begin{aligned} x_1 \quad\quad - 3x_3 &= -2 \\ 3x_1 + x_2 - 2x_3 &= 5 \\ 2x_1 + 2x_2 + x_3 &= 4 \end{aligned}$$

32.
$$\begin{aligned} 3x_1 - 2x_2 + 3x_3 &= 22 \\ 3x_2 - x_3 &= 24 \\ 6x_1 - 7x_2 \quad\quad &= -22 \end{aligned}$$

33.
$$\begin{aligned} 2x_1 + \quad\quad 3x_3 &= 3 \\ 4x_1 - 3x_2 + 7x_3 &= 5 \\ 8x_1 - 9x_2 + 15x_3 &= 10 \end{aligned}$$

34.
$$\begin{aligned} x_1 + x_2 - 5x_3 &= 3 \\ x_1 \quad\quad - 2x_3 &= 1 \\ 2x_1 - x_2 - x_3 &= 0 \end{aligned}$$

35.
$$\begin{aligned} 4x + 12y - 7z - 20w &= 22 \\ 3x + 9y - 5z - 28w &= 30 \end{aligned}$$

36.
$$\begin{aligned} x + 2y + z &= 8 \\ -3x - 6y - 3z &= -21 \end{aligned}$$

37.
$$\begin{aligned} 3x + 3y + 12z &= 6 \\ x + y + 4z &= 2 \\ 2x + 5y + 20z &= 10 \\ -x + 2y + 8z &= 4 \end{aligned}$$

38.
$$\begin{aligned} 2x + y - z + 2w &= -6 \\ 3x + 4y \quad\quad + w &= 1 \\ x + 5y + 2z + 6w &= -3 \\ 5x + 2y - z - w &= 3 \end{aligned}$$

System of Linear Equations In Exercises 39–42, use a software program or a graphing utility to solve the system of linear equations.

39.
$$\begin{aligned} x_1 - 2x_2 + 5x_3 - 3x_4 &= 23.6 \\ x_1 + 4x_2 - 7x_3 - 2x_4 &= 45.7 \\ 3x_1 - 5x_2 + 7x_3 + 4x_4 &= 29.9 \end{aligned}$$

40.
$$\begin{aligned} x_1 + x_2 - 2x_3 + 3x_4 + 2x_5 &= 9 \\ 3x_1 + 3x_2 - x_3 + x_4 + x_5 &= 5 \\ 2x_1 + 2x_2 - x_3 + x_4 - 2x_5 &= 1 \\ 4x_1 + 4x_2 + x_3 \quad\quad - 3x_5 &= 4 \\ 8x_1 + 5x_2 - 2x_3 - x_4 + 2x_5 &= 3 \end{aligned}$$

41.
$$\begin{aligned} x_1 - x_2 + 2x_3 + 2x_4 + 6x_5 &= 6 \\ 3x_1 - 2x_2 + 4x_3 + 4x_4 + 12x_5 &= 14 \\ x_2 - x_3 - x_4 - 3x_5 &= -3 \\ 2x_1 - 2x_2 + 4x_3 + 5x_4 + 15x_5 &= 10 \\ 2x_1 - 2x_2 + 4x_3 + 4x_4 + 13x_5 &= 13 \end{aligned}$$

42.
$$\begin{aligned} x_1 + 2x_2 - 2x_3 + 2x_4 - x_5 + 3x_6 &= 0 \\ 2x_1 - x_2 + 3x_3 + x_4 - 3x_5 + 2x_6 &= 17 \\ x_1 + 3x_2 - 2x_3 + x_4 - 2x_5 - 3x_6 &= -5 \\ 3x_1 - 2x_2 + x_3 - x_4 + 3x_5 - 2x_6 &= -1 \\ -x_1 - 2x_2 + x_3 + 2x_4 - 2x_5 + 3x_6 &= 10 \\ x_1 - 3x_2 + x_3 + 3x_4 - 2x_5 + x_6 &= 11 \end{aligned}$$

Homogeneous System In Exercises 43–46, solve the homogeneous linear system corresponding to the given coefficient matrix.

43. $\begin{bmatrix} 1 & 0 & 0 \\ 0 & 1 & 1 \\ 0 & 0 & 0 \end{bmatrix}$
44. $\begin{bmatrix} 1 & 0 & 0 & 0 \\ 0 & 1 & 1 & 0 \end{bmatrix}$

45. $\begin{bmatrix} 1 & 0 & 0 & 1 \\ 0 & 0 & 1 & 0 \\ 0 & 0 & 0 & 0 \end{bmatrix}$
46. $\begin{bmatrix} 0 & 0 & 0 \\ 0 & 0 & 0 \\ 0 & 0 & 0 \end{bmatrix}$

47. Finance A small software corporation borrowed $500,000 to expand its software line. The corporation borrowed some of the money at 3%, some at 4%, and some at 5%. Use a system of equations to determine how much was borrowed at each rate when the annual interest was $20,500 and the amount borrowed at 4% was $2\frac{1}{2}$ times the amount borrowed at 3%. Solve the system using matrices.

48. Tips A food server examines the amount of money earned in tips after working an 8-hour shift. The server has a total of $95 in denominations of $1, $5, $10, and $20 bills. The total number of paper bills is 26. The number of $5 bills is 4 times the number of $10 bills, and the number of $1 bills is 1 less than twice the number of $5 bills. Write a system of linear equations to represent the situation. Then use matrices to find the number of each denomination.

Matrix Representation In Exercises 49 and 50, assume that the matrix is the *augmented* matrix of a system of linear equations, and (a) determine the number of equations and the number of variables, and (b) find the value(s) of k such that the system is consistent. Then assume that the matrix is the *coefficient* matrix of a *homogeneous* system of linear equations, and repeat parts (a) and (b).

49. $A = \begin{bmatrix} 1 & k & 2 \\ -3 & 4 & 1 \end{bmatrix}$

50. $A = \begin{bmatrix} 2 & -1 & 3 \\ -4 & 2 & k \\ 4 & -2 & 6 \end{bmatrix}$

Coefficient Design In Exercises 51 and 52, find values of a, b, and c (if possible) such that the system of linear equations has (a) a unique solution, (b) no solution, and (c) infinitely many solutions.

51.
$$\begin{aligned} x + y \quad\quad &= 2 \\ y + z &= 2 \\ x \quad\quad + z &= 2 \\ ax + by + cz &= 0 \end{aligned}$$

52.
$$\begin{aligned} x + y \quad\quad &= 0 \\ y + z &= 0 \\ x \quad\quad + z &= 0 \\ ax + by + cz &= 0 \end{aligned}$$

53. The system below has one solution: $x = 1$, $y = -1$, and $z = 2$.

$$\begin{array}{ll} 4x - 2y + 5z = 16 & \text{Equation 1} \\ x + y \quad\;\; = 0 & \text{Equation 2} \\ -x - 3y + 2z = 6 & \text{Equation 3} \end{array}$$

Solve the systems provided by (a) Equations 1 and 2, (b) Equations 1 and 3, and (c) Equations 2 and 3. (d) How many solutions does each of these systems have?

54. Assume the system below has a unique solution.

$$\begin{array}{ll} a_{11}x_1 + a_{12}x_2 + a_{13}x_3 = b_1 & \text{Equation 1} \\ a_{21}x_1 + a_{22}x_2 + a_{23}x_3 = b_2 & \text{Equation 2} \\ a_{31}x_1 + a_{32}x_2 + a_{33}x_3 = b_3 & \text{Equation 3} \end{array}$$

Does the system composed of Equations 1 and 2 have a unique solution, no solution, or infinitely many solutions?

Row Equivalence **In Exercises 55 and 56, find the reduced row-echelon matrix that is row-equivalent to the given matrix.**

55. $\begin{bmatrix} 1 & 2 \\ -1 & 2 \end{bmatrix}$ **56.** $\begin{bmatrix} 1 & 2 & 3 \\ 4 & 5 & 6 \\ 7 & 8 & 9 \end{bmatrix}$

57. Writing Describe all possible 2×2 reduced row-echelon matrices. Support your answer with examples.

58. Writing Describe all possible 3×3 reduced row-echelon matrices. Support your answer with examples.

True or False? **In Exercises 59 and 60, determine whether each statement is true or false. If a statement is true, give a reason or cite an appropriate statement from the text. If a statement is false, provide an example that shows the statement is not true in all cases or cite an appropriate statement from the text.**

59. (a) A 6×3 matrix has six rows.

(b) Every matrix is row-equivalent to a matrix in row-echelon form.

(c) If the row-echelon form of the augmented matrix of a system of linear equations contains the row $[1\ 0\ 0\ 0\ 0]$, then the original system is inconsistent.

(d) A homogeneous system of four linear equations in six variables has infinitely many solutions.

60. (a) A 4×7 matrix has four columns.

(b) Every matrix has a unique reduced row-echelon form.

(c) A homogeneous system of four linear equations in four variables is always consistent.

(d) Multiplying a row of a matrix by a constant is one of the elementary row operations.

61. Writing Is it possible for a system of linear equations with fewer equations than variables to have no solution? If so, give an example.

62. Writing Does a matrix have a unique row-echelon form? Illustrate your answer with examples.

Row Equivalence **In Exercises 63 and 64, determine conditions on a, b, c, and d such that the matrix**

$$\begin{bmatrix} a & b \\ c & d \end{bmatrix}$$

will be row-equivalent to the given matrix.

63. $\begin{bmatrix} 1 & 0 \\ 0 & 1 \end{bmatrix}$ **64.** $\begin{bmatrix} 1 & 0 \\ 0 & 0 \end{bmatrix}$

Homogeneous System **In Exercises 65 and 66, find all values of λ (the Greek letter lambda) for which the homogeneous linear system has nontrivial solutions.**

65. $\begin{array}{l} (\lambda - 2)x + \quad\quad y = 0 \\ \quad\quad x + (\lambda - 2)y = 0 \end{array}$

66. $\begin{array}{l} (2\lambda + 9)x - 5y = 0 \\ \quad\quad x - \lambda y = 0 \end{array}$

67. The augmented matrix represents a system of linear equations that has been reduced using Gauss-Jordan elimination. Write a system of equations with nonzero coefficients that the reduced matrix could represent.

$$\begin{bmatrix} 1 & 0 & 3 & -2 \\ 0 & 1 & 4 & 1 \\ 0 & 0 & 0 & 0 \end{bmatrix}$$

There are many correct answers.

68. CAPSTONE In your own words, describe the difference between a matrix in row-echelon form and a matrix in reduced row-echelon form. Include an example of each to support your explanation.

69. Writing Consider the 2×2 matrix $\begin{bmatrix} a & b \\ c & d \end{bmatrix}$. Perform the sequence of row operations.

(a) Add (-1) times the second row to the first row.

(b) Add 1 times the first row to the second row.

(c) Add (-1) times the second row to the first row.

(d) Multiply the first row by (-1).

What happened to the original matrix? Describe, in general, how to interchange two rows of a matrix using only the second and third elementary row operations.

70. Writing Describe the row-echelon form of an augmented matrix that corresponds to a linear system that (a) is inconsistent, and (b) has infinitely many solutions.

1.3 Applications of Systems of Linear Equations

■ Set up and solve a system of equations to fit a polynomial function to a set of data points.

■ Set up and solve a system of equations to represent a network.

Systems of linear equations arise in a wide variety of applications. In this section you will look at two applications, and you will see more in subsequent chapters. The first application shows how to fit a polynomial function to a set of data points in the plane. The second application focuses on networks and Kirchhoff's Laws for electricity.

POLYNOMIAL CURVE FITTING

Consider n points in the xy-plane

$$(x_1, y_1), (x_2, y_2), \ldots, (x_n, y_n)$$

that represent a collection of data, and you want to find a polynomial function of degree $n - 1$

$$p(x) = a_0 + a_1 x + a_2 x^2 + \cdots + a_{n-1} x^{n-1}$$

whose graph passes through the points. This procedure is called **polynomial curve fitting.** When all x-coordinates of the points are distinct, there is precisely one polynomial function of degree $n - 1$ (or less) that fits the n points, as shown in Figure 1.3.

To solve for the n coefficients of $p(x)$, substitute each of the n points into the polynomial function and obtain n linear equations in n variables $a_0, a_1, a_2, \ldots, a_{n-1}$.

$$a_0 + a_1 x_1 + a_2 x_1^2 + \cdots + a_{n-1} x_1^{n-1} = y_1$$
$$a_0 + a_1 x_2 + a_2 x_2^2 + \cdots + a_{n-1} x_2^{n-1} = y_2$$
$$\vdots$$
$$a_0 + a_1 x_n + a_2 x_n^2 + \cdots + a_{n-1} x_n^{n-1} = y_n$$

Example 1 demonstrates this procedure with a second-degree polynomial.

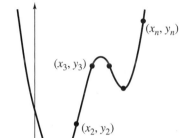

Polynomial Curve Fitting

Figure 1.3

EXAMPLE 1 Polynomial Curve Fitting

Determine the polynomial $p(x) = a_0 + a_1 x + a_2 x^2$ whose graph passes through the points $(1, 4)$, $(2, 0)$, and $(3, 12)$.

SOLUTION

Substituting $x = 1$, 2, and 3 into $p(x)$ and equating the results to the respective y-values produces the system of linear equations in the variables a_0, a_1, and a_2 shown below.

$$p(1) = a_0 + a_1(1) + a_2(1)^2 = a_0 + a_1 + a_2 = 4$$
$$p(2) = a_0 + a_1(2) + a_2(2)^2 = a_0 + 2a_1 + 4a_2 = 0$$
$$p(3) = a_0 + a_1(3) + a_2(3)^2 = a_0 + 3a_1 + 9a_2 = 12$$

The solution of this system is

$$a_0 = 24, \ a_1 = -28, \text{ and } a_2 = 8$$

so the polynomial function is

$$p(x) = 24 - 28x + 8x^2.$$

Figure 1.4 shows the graph of p.

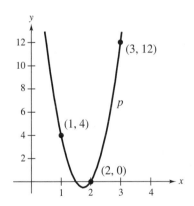

Figure 1.4

EXAMPLE 2   **Polynomial Curve Fitting**

See LarsonLinearAlgebra.com for an interactive version of this type of example.

Find a polynomial that fits the points

$$(-2, 3), (-1, 5), (0, 1), (1, 4), \text{ and } (2, 10).$$

SOLUTION

You are given five points, so choose a fourth-degree polynomial function

$$p(x) = a_0 + a_1 x + a_2 x^2 + a_3 x^3 + a_4 x^4.$$

Substituting the points into $p(x)$ produces the system of linear equations shown below.

$$
\begin{aligned}
a_0 - 2a_1 + 4a_2 - 8a_3 + 16a_4 &= 3 \\
a_0 - a_1 + a_2 - a_3 +   a_4 &= 5 \\
a_0 &= 1 \\
a_0 + a_1 + a_2 + a_3 +   a_4 &= 4 \\
a_0 + 2a_1 + 4a_2 + 8a_3 + 16a_4 &= 10
\end{aligned}
$$

The solution of these equations is

$$a_0 = 1, \quad a_1 = -\tfrac{5}{4}, \quad a_2 = \tfrac{101}{24}, \quad a_3 = \tfrac{3}{4}, \quad a_4 = -\tfrac{17}{24}$$

which means the polynomial function is

$$p(x) = 1 - \tfrac{5}{4}x + \tfrac{101}{24}x^2 + \tfrac{3}{4}x^3 - \tfrac{17}{24}x^4.$$

Figure 1.5 shows the graph of p.

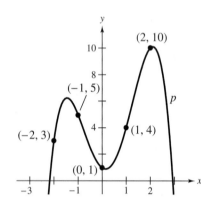

Figure 1.5

The system of linear equations in Example 2 is relatively easy to solve because the x-values are small. For a set of points with large x-values, it is usually best to *translate* the values before attempting the curve-fitting procedure. The next example demonstrates this approach.

EXAMPLE 3   **Translating Large *x*-Values Before Curve Fitting**

Find a polynomial that fits the points

$$
\begin{array}{ccccc}
(x_1, y_1) & (x_2, y_2) & (x_3, y_3) & (x_4, y_4) & (x_5, y_5) \\
(2011, 3), & (2012, 5), & (2013, 1), & (2014, 4), & (2015, 10).
\end{array}
$$

SOLUTION

The given x-values are large, so use the translation

$$z = x - 2013$$

to obtain

$$
\begin{array}{ccccc}
(z_1, y_1) & (z_2, y_2) & (z_3, y_3) & (z_4, y_4) & (z_5, y_5) \\
(-2, 3), & (-1, 5), & (0, 1), & (1, 4), & (2, 10).
\end{array}
$$

This is the same set of points as in Example 2. So, the polynomial that fits these points is

$$p(z) = 1 - \tfrac{5}{4}z + \tfrac{101}{24}z^2 + \tfrac{3}{4}z^3 - \tfrac{17}{24}z^4.$$

Letting $z = x - 2013$, you have

$$p(x) = 1 - \tfrac{5}{4}(x - 2013) + \tfrac{101}{24}(x - 2013)^2 + \tfrac{3}{4}(x - 2013)^3 - \tfrac{17}{24}(x - 2013)^4.$$

EXAMPLE 4   **An Application of Curve Fitting**

Find a polynomial that relates the periods of the three planets that are closest to the Sun to their mean distances from the Sun, as shown in the table. Then use the polynomial to calculate the period of Mars, and compare it to the value shown in the table. (The mean distances are in astronomical units, and the periods are in years.)

Planet	Mercury	Venus	Earth	Mars
Mean Distance	0.387	0.723	1.000	1.524
Period	0.241	0.615	1.000	1.881

SOLUTION

Begin by fitting a quadratic polynomial function

$$p(x) = a_0 + a_1x + a_2x^2$$

to the points

$$(0.387, 0.241), (0.723, 0.615), \text{ and } (1, 1).$$

The system of linear equations obtained by substituting these points into $p(x)$ is

$$a_0 + 0.387a_1 + (0.387)^2a_2 = 0.241$$
$$a_0 + 0.723a_1 + (0.723)^2a_2 = 0.615$$
$$a_0 + \quad a_1 + \quad a_2 = 1.$$

The approximate solution of the system is

$$a_0 \approx -0.0634, \quad a_1 \approx 0.6119, \quad a_2 \approx 0.4515$$

which means that an approximation of the polynomial function is

$$p(x) = -0.0634 + 0.6119x + 0.4515x^2.$$

Using $p(x)$ to evaluate the period of Mars produces

$$p(1.524) \approx 1.918 \text{ years.}$$

Note that the period of Mars is shown in the table as 1.881 years. The figure below provides a graphical comparison of the polynomial function to the values shown in the table.

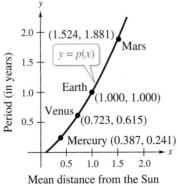

Mean distance from the Sun
(in astronomical units)

As illustrated in Example 4, a polynomial that fits some of the points in a data set is not necessarily an accurate model for other points in the data set. Generally, the farther the other points are from those used to fit the polynomial, the worse the fit. For instance, the mean distance of Jupiter from the Sun is 5.203 astronomical units. Using $p(x)$ in Example 4 to approximate the period gives 15.343 years—a poor estimate of Jupiter's actual period of 11.862 years.

The problem of curve fitting can be difficult. Types of functions other than polynomial functions may provide better fits. For instance, look again at the curve-fitting problem in Example 4. Taking the natural logarithms of the distances and periods produces the results shown in the table.

Planet	Mercury	Venus	Earth	Mars
Mean Distance, x	0.387	0.723	1.000	1.524
ln x	−0.949	−0.324	0.0	0.421
Period, y	0.241	0.615	1.000	1.881
ln y	−1.423	−0.486	0.0	0.632

Now, fitting a polynomial to the logarithms of the distances and periods produces the *linear relationship*

$$\ln y = \tfrac{3}{2} \ln x$$

which is shown graphically below.

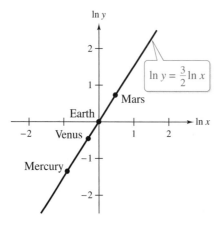

From $\ln y = \tfrac{3}{2} \ln x$, it follows that $y = x^{3/2}$, or $y^2 = x^3$. In other words, the square of the period (in years) of each planet is equal to the cube of its mean distance (in astronomical units) from the Sun. Johannes Kepler first discovered this relationship in 1619.

LINEAR ALGEBRA APPLIED

Researchers in Italy studying the acoustical noise levels from vehicular traffic at a busy three-way intersection used a system of linear equations to model the traffic flow at the intersection. To help formulate the system of equations, "operators" stationed themselves at various locations along the intersection and counted the numbers of vehicles that passed them. *(Source: Acoustical Noise Analysis in Road Intersections: A Case Study, Guarnaccia, Claudio, Recent Advances in Acoustics & Music, Proceedings of the 11th WSEAS International Conference on Acoustics & Music: Theory & Applications)*

iStockphoto.com/Nikada

NETWORK ANALYSIS

Networks composed of branches and junctions are used as models in such fields as economics, traffic analysis, and electrical engineering. In a network model, you assume that the total flow into a junction is equal to the total flow out of the junction. For example, the junction shown below has 25 units flowing into it, so there must be 25 units flowing out of it. You can represent this with the linear equation

$$x_1 + x_2 = 25.$$

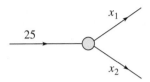

Each junction in a network gives rise to a linear equation, so you can analyze the flow through a network composed of several junctions by solving a system of linear equations. Example 5 illustrates this procedure.

EXAMPLE 5   Analysis of a Network

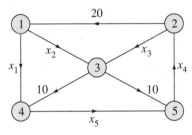

Figure 1.6

Set up a system of linear equations to represent the network shown in Figure 1.6. Then solve the system.

SOLUTION

Each of the network's five junctions gives rise to a linear equation, as shown below.

$$
\begin{aligned}
x_1 + x_2 &= 20 && \text{Junction 1} \\
x_3 - x_4 &= -20 && \text{Junction 2} \\
x_2 + x_3 &= 20 && \text{Junction 3} \\
x_1 \qquad\quad - x_5 &= -10 && \text{Junction 4} \\
- x_4 + x_5 &= -10 && \text{Junction 5}
\end{aligned}
$$

The augmented matrix for this system is

$$
\begin{bmatrix}
1 & 1 & 0 & 0 & 0 & 20 \\
0 & 0 & 1 & -1 & 0 & -20 \\
0 & 1 & 1 & 0 & 0 & 20 \\
1 & 0 & 0 & 0 & -1 & -10 \\
0 & 0 & 0 & -1 & 1 & -10
\end{bmatrix}.
$$

Gauss-Jordan elimination produces the matrix

$$
\begin{bmatrix}
1 & 0 & 0 & 0 & -1 & -10 \\
0 & 1 & 0 & 0 & 1 & 30 \\
0 & 0 & 1 & 0 & -1 & -10 \\
0 & 0 & 0 & 1 & -1 & 10 \\
0 & 0 & 0 & 0 & 0 & 0
\end{bmatrix}.
$$

From the matrix above,

$$x_1 - x_5 = -10, \quad x_2 + x_5 = 30, \quad x_3 - x_5 = -10, \quad \text{and} \quad x_4 - x_5 = 10.$$

Letting $t = x_5$, you have

$$x_1 = t - 10, \quad x_2 = -t + 30, \quad x_3 = t - 10, \quad x_4 = t + 10, \quad x_5 = t$$

where t is any real number, so this system has infinitely many solutions.

In Example 5, if you could control the amount of flow along the branch labeled x_5, then you could also control the flow represented by each of the other variables. For example, letting $t = 10$ results in the flows shown in the figure at the right. (Verify this.)

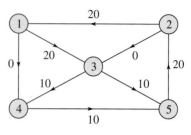

You may be able to see how the type of network analysis demonstrated in Example 5 could be used in problems dealing with the flow of traffic through the streets of a city or the flow of water through an irrigation system.

An electrical network is another type of network where analysis is commonly applied. An analysis of such a system uses two properties of electrical networks known as **Kirchhoff's Laws.**

REMARK

A closed path is a sequence of branches such that the beginning point of the first branch coincides with the ending point of the last branch.

1. All the current flowing into a junction must flow out of it.

2. The sum of the products IR (I is current and R is resistance) around a closed path is equal to the total voltage in the path.

In an electrical network, current is measured in amperes, or amps (A), resistance is measured in ohms (Ω, the Greek letter omega), and the product of current and resistance is measured in volts (V). The symbol ⊣⊢ represents a battery. The larger vertical bar denotes where the current flows out of the terminal. The symbol ⌁⌁⌁ denotes resistance. An arrow in the branch shows the direction of the current.

EXAMPLE 6   Analysis of an Electrical Network

Determine the currents I_1, I_2, and I_3 for the electrical network shown in Figure 1.7.

SOLUTION

Applying Kirchhoff's first law to either junction produces

$$I_1 + I_3 = I_2 \qquad \text{Junction 1 or Junction 2}$$

and applying Kirchhoff's second law to the two paths produces

$$R_1 I_1 + R_2 I_2 = 3I_1 + 2I_2 = 7 \qquad \text{Path 1}$$
$$R_2 I_2 + R_3 I_3 = 2I_2 + 4I_3 = 8. \qquad \text{Path 2}$$

So, you have the system of three linear equations in the variables I_1, I_2, and I_3 shown below.

$$
\begin{aligned}
I_1 - I_2 + I_3 &= 0 \\
3I_1 + 2I_2 &= 7 \\
2I_2 + 4I_3 &= 8
\end{aligned}
$$

Applying Gauss-Jordan elimination to the augmented matrix

$$
\begin{bmatrix}
1 & -1 & 1 & 0 \\
3 & 2 & 0 & 7 \\
0 & 2 & 4 & 8
\end{bmatrix}
$$

produces the reduced row-echelon form

$$
\begin{bmatrix}
1 & 0 & 0 & 1 \\
0 & 1 & 0 & 2 \\
0 & 0 & 1 & 1
\end{bmatrix}
$$

which means $I_1 = 1$ amp, $I_2 = 2$ amps, and $I_3 = 1$ amp.

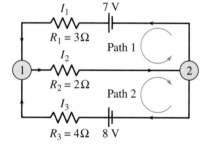

Figure 1.7

EXAMPLE 7   **Analysis of an Electrical Network**

Determine the currents I_1, I_2, I_3, I_4, I_5, and I_6 for the electrical network shown below.

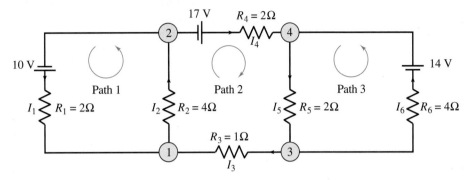

SOLUTION

Applying Kirchhoff's first law to the four junctions produces

$$I_1 + I_3 = I_2 \qquad \text{Junction 1}$$
$$I_1 + I_4 = I_2 \qquad \text{Junction 2}$$
$$I_3 + I_6 = I_5 \qquad \text{Junction 3}$$
$$I_4 + I_6 = I_5 \qquad \text{Junction 4}$$

and applying Kirchhoff's second law to the three paths produces

$$2I_1 + 4I_2 \qquad\qquad\qquad = 10 \qquad \text{Path 1}$$
$$4I_2 + I_3 + 2I_4 + 2I_5 = 17 \qquad \text{Path 2}$$
$$2I_5 + 4I_6 = 14. \qquad \text{Path 3}$$

You now have the system of seven linear equations in the variables I_1, I_2, I_3, I_4, I_5, and I_6 shown below.

$$I_1 - I_2 + I_3 \qquad\qquad\qquad = 0$$
$$I_1 - I_2 \qquad + I_4 \qquad\qquad = 0$$
$$I_3 \qquad - I_5 + I_6 = 0$$
$$I_4 - I_5 + I_6 = 0$$
$$2I_1 + 4I_2 \qquad\qquad\qquad = 10$$
$$4I_2 + I_3 + 2I_4 + 2I_5 = 17$$
$$2I_5 + 4I_6 = 14$$

The augmented matrix for this system is

$$\begin{bmatrix} 1 & -1 & 1 & 0 & 0 & 0 & 0 \\ 1 & -1 & 0 & 1 & 0 & 0 & 0 \\ 0 & 0 & 1 & 0 & -1 & 1 & 0 \\ 0 & 0 & 0 & 1 & -1 & 1 & 0 \\ 2 & 4 & 0 & 0 & 0 & 0 & 10 \\ 0 & 4 & 1 & 2 & 2 & 0 & 17 \\ 0 & 0 & 0 & 0 & 2 & 4 & 14 \end{bmatrix}.$$

Using a graphing utility, a software program, or Gauss-Jordan elimination, solve this system to obtain

$$I_1 = 1, \quad I_2 = 2, \quad I_3 = 1, \quad I_4 = 1, \quad I_5 = 3, \quad \text{and} \quad I_6 = 2.$$

So, $I_1 = 1$ amp, $I_2 = 2$ amps, $I_3 = 1$ amp, $I_4 = 1$ amp, $I_5 = 3$ amps, and $I_6 = 2$ amps.

1.3 Exercises

See CalcChat.com for worked-out solutions to odd-numbered exercises.

Polynomial Curve Fitting In Exercises 1–12, (a) determine the polynomial function whose graph passes through the points, and (b) sketch the graph of the polynomial function, showing the points.

1. $(2, 5)$, $(3, 2)$, $(4, 5)$
2. $(0, 0)$, $(2, -2)$, $(4, 0)$
3. $(2, 4)$, $(3, 6)$, $(5, 10)$
4. $(2, 4)$, $(3, 4)$, $(4, 4)$
5. $(-1, 3)$, $(0, 0)$, $(1, 1)$, $(4, 58)$
6. $(0, 42)$, $(1, 0)$, $(2, -40)$, $(3, -72)$
7. $(-2, 28)$, $(-1, 0)$, $(0, -6)$, $(1, -8)$, $(2, 0)$
8. $(-4, 18)$, $(0, 1)$, $(4, 0)$, $(6, 28)$, $(8, 135)$
9. $(2013, 5)$, $(2014, 7)$, $(2015, 12)$
10. $(2012, 150)$, $(2013, 180)$, $(2014, 240)$, $(2015, 360)$
11. $(0.072, 0.203)$, $(0.120, 0.238)$, $(0.148, 0.284)$
12. $(1, 1)$, $(1.189, 1.587)$, $(1.316, 2.080)$, $(1.414, 2.520)$

13. Use $\sin 0 = 0$, $\sin \frac{\pi}{2} = 1$, and $\sin \pi = 0$ to estimate $\sin \frac{\pi}{3}$.

14. Use $\log_2 1 = 0$, $\log_2 2 = 1$, and $\log_2 4 = 2$ to estimate $\log_2 3$.

Equation of a Circle In Exercises 15 and 16, find an equation of the circle that passes through the points.

15. $(1, 3)$, $(-2, 6)$, $(4, 2)$
16. $(-5, 1)$, $(-3, 2)$, $(-1, 1)$

17. **Population** The U.S. census lists the population of the United States as 249 million in 1990, 282 million in 2000, and 309 million in 2010. Fit a second-degree polynomial passing through these three points and use it to predict the populations in 2020 and 2030. (Source: U.S. Census Bureau)

18. **Population** The table shows the U.S. populations for the years 1970, 1980, 1990, and 2000. (Source: U.S. Census Bureau)

Year	1970	1980	1990	2000
Population (in millions)	205	227	249	282

(a) Find a cubic polynomial that fits the data and use it to estimate the population in 2010.

(b) The actual population in 2010 was 309 million. How does your estimate compare?

19. **Net Profit** The table shows the net profits (in millions of dollars) for Microsoft from 2007 through 2014. (Source: Microsoft Corp.)

Year	2007	2008	2009	2010
Net Profit	14,065	17,681	14,569	18,760

Year	2011	2012	2013	2014
Net Profit	23,150	23,171	22,453	22,074

(a) Set up a system of equations to fit the data for the years 2007, 2008, 2009, and 2010 to a cubic model.

(b) Solve the system. Does the solution produce a reasonable model for determining net profits after 2010? Explain.

20. **Sales** The table shows the sales (in billions of dollars) for Wal-Mart stores from 2006 through 2013. (Source: Wal-Mart Stores, Inc.)

Year	2006	2007	2008	2009
Sales	348.7	378.8	405.6	408.2

Year	2010	2011	2012	2013
Sales	421.8	447.0	469.2	476.2

(a) Set up a system of equations to fit the data for the years 2006, 2007, 2008, 2009, and 2010 to a quartic model.

(b) Solve the system. Does the solution produce a reasonable model for determining sales after 2010? Explain.

21. **Network Analysis** The figure shows the flow of traffic (in vehicles per hour) through a network of streets.

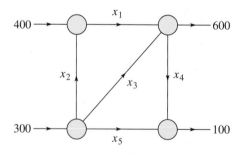

(a) Solve this system for x_i, $i = 1, 2, \ldots, 5$.

(b) Find the traffic flow when $x_3 = 0$ and $x_5 = 100$.

(c) Find the traffic flow when $x_3 = x_5 = 100$.

22. Network Analysis   The figure shows the flow of traffic (in vehicles per hour) through a network of streets.

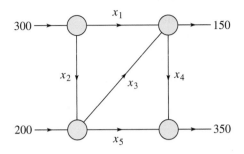

(a) Solve this system for x_i, $i = 1, 2, \ldots, 5$.

(b) Find the traffic flow when $x_2 = 200$ and $x_3 = 50$.

(c) Find the traffic flow when $x_2 = 150$ and $x_3 = 0$.

23. Network Analysis   The figure shows the flow of traffic (in vehicles per hour) through a network of streets.

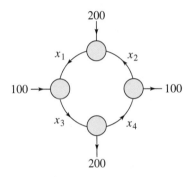

(a) Solve this system for x_i, $i = 1, 2, 3, 4$.

(b) Find the traffic flow when $x_4 = 0$.

(c) Find the traffic flow when $x_4 = 100$.

(d) Find the traffic flow when $x_1 = 2x_2$.

24. Network Analysis   Water is flowing through a network of pipes (in thousands of cubic meters per hour), as shown in the figure.

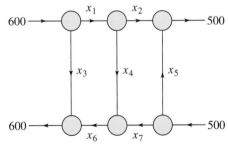

(a) Solve this system for the water flow represented by x_i, $i = 1, 2, \ldots, 7$.

(b) Find the water flow when $x_1 = x_2 = 100$.

(c) Find the water flow when $x_6 = x_7 = 0$.

(d) Find the water flow when $x_5 = 1000$ and $x_6 = 0$.

25. Network Analysis   Determine the currents I_1, I_2, and I_3 for the electrical network shown in the figure.

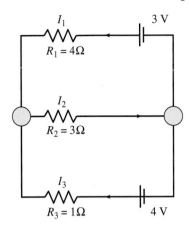

26. Network Analysis   Determine the currents I_1, I_2, I_3, I_4, I_5, and I_6 for the electrical network shown in the figure.

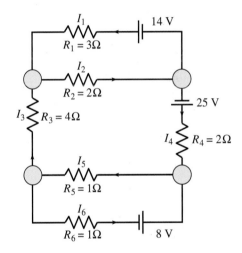

27. Network Analysis

(a) Determine the currents I_1, I_2, and I_3 for the electrical network shown in the figure.

(b) How is the result affected when A is changed to 2 volts and B is changed to 6 volts?

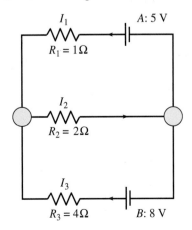

28. CAPSTONE

(a) Explain how to use systems of linear equations for polynomial curve fitting.

(b) Explain how to use systems of linear equations to perform network analysis.

Temperature In Exercises 29 and 30, the figure shows the boundary temperatures (in degrees Celsius) of an insulated thin metal plate. The steady-state temperature at an interior junction is approximately equal to the mean of the temperatures at the four surrounding junctions. Use a system of linear equations to approximate the interior temperatures T_1, T_2, T_3, and T_4.

29.

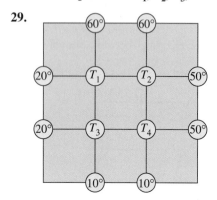

30.

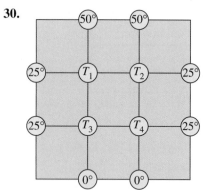

Partial Fraction Decomposition In Exercises 31–34, use a system of equations to find the partial fraction decomposition of the rational expression. Solve the system using matrices.

31. $\dfrac{4x^2}{(x+1)^2(x-1)} = \dfrac{A}{x-1} + \dfrac{B}{x+1} + \dfrac{C}{(x+1)^2}$

32. $\dfrac{3x^2 - 7x - 12}{(x+4)(x-4)^2} = \dfrac{A}{x+4} + \dfrac{B}{x-4} + \dfrac{C}{(x-4)^2}$

33. $\dfrac{3x^2 - 3x - 2}{(x+2)(x-2)^2} = \dfrac{A}{x+2} + \dfrac{B}{x-2} + \dfrac{C}{(x-2)^2}$

34. $\dfrac{20 - x^2}{(x+2)(x-2)^2} = \dfrac{A}{x+2} + \dfrac{B}{x-2} + \dfrac{C}{(x-2)^2}$

Calculus In Exercises 35 and 36, find the values of x, y, and λ that satisfy the system of equations. Such systems arise in certain problems of calculus, and λ is called the Lagrange multiplier.

35. $\begin{aligned} 2x \quad\quad + \lambda \quad\quad &= 0 \\ 2y + \lambda \quad\quad &= 0 \\ x + \ y \quad\quad - 4 &= 0 \end{aligned}$

36. $\begin{aligned} 2y + 2\lambda + \quad 2 &= 0 \\ 2x \quad\quad + \ \lambda + \quad 1 &= 0 \\ 2x + \ y \quad\quad - 100 &= 0 \end{aligned}$

37. **Calculus** The graph of a parabola passes through the points $(0, 1)$ and $\left(\frac{1}{2}, \frac{1}{2}\right)$ and has a horizontal tangent line at $\left(\frac{1}{2}, \frac{1}{2}\right)$. Find an equation for the parabola and sketch its graph.

38. **Calculus** The graph of a cubic polynomial function has horizontal tangent lines at $(1, -2)$ and $(-1, 2)$. Find an equation for the function and sketch its graph.

39. **Guided Proof** Prove that if a polynomial function $p(x) = a_0 + a_1 x + a_2 x^2$ is zero for $x = -1$, $x = 0$, and $x = 1$, then $a_0 = a_1 = a_2 = 0$.

Getting Started: Write a system of linear equations and solve the system for a_0, a_1, and a_2.

  (i) Substitute $x = -1$, 0, and 1 into $p(x)$.

  (ii) Set each result equal to 0.

  (iii) Solve the resulting system of linear equations in the variables a_0, a_1, and a_2.

40. **Proof** Generalizing the statement in Exercise 39, if a polynomial function

$$p(x) = a_0 + a_1 x + \cdots + a_{n-1} x^{n-1}$$

is zero for more than $n - 1$ x-values, then

$$a_0 = a_1 = \cdots = a_{n-1} = 0.$$

Use this result to prove that there is at most one polynomial function of degree $n - 1$ (or less) whose graph passes through n points in the plane with distinct x-coordinates.

41. (a) The graph of a function f passes through the points $(0, 1)$, $\left(2, \frac{1}{3}\right)$, and $\left(4, \frac{1}{5}\right)$. Find a quadratic function whose graph passes through these points.

(b) Find a polynomial function p of degree 2 or less that passes through the points $(0, 1)$, $(2, 3)$, and $(4, 5)$. Then sketch the graph of $y = 1/p(x)$ and compare this graph with the graph of the polynomial function found in part (a).

42. **Writing** Try to find a polynomial to fit the data shown in the table. What happens, and why?

x	1	2	3	3	4
y	1	1	2	3	4

1 Review Exercises

See CalcChat.com for worked-out solutions to odd-numbered exercises.

Linear Equations In Exercises 1–6, determine whether the equation is linear in the variables x and y.

1. $2x - y^2 = 4$
2. $2xy - 6y = 0$
3. $(\cot 5)x - y = 3$
4. $e^{-2}x + 5y = 8$
5. $\dfrac{2}{x} + 4y = 3$
6. $\dfrac{x}{2} - \dfrac{y}{4} = 0$

Parametric Representation In Exercises 7 and 8, find a parametric representation of the solution set of the linear equation.

7. $-3x + 4y - 2z = 1$
8. $3x_1 + 2x_2 - 4x_3 = 0$

System of Linear Equations In Exercises 9–20, solve the system of linear equations.

9. $x + y = 2$
    $3x - y = 0$

10. $x + y = -1$
     $3x + 2y = 0$

11. $3y = 2x$
    $y = x + 4$

12. $x = y + 3$
    $4x = y + 10$

13. $y + x = 0$
    $2x + y = 0$

14. $y = 5x$
    $y = -x$

15. $x - y = 9$
    $-x + y = 1$

16. $40x_1 + 30x_2 = 24$
    $20x_1 + 15x_2 = -14$

17. $\frac{1}{2}x - \frac{1}{3}y = 0$
    $3x + 2(y + 5) = 10$

18. $\frac{1}{3}x + \frac{4}{7}y = 3$
    $2x + 3y = 15$

19. $0.2x_1 + 0.3x_2 = 0.14$
    $0.4x_1 + 0.5x_2 = 0.20$

20. $0.2x - 0.1y = 0.07$
    $0.4x - 0.5y = -0.01$

Matrix Size In Exercises 21 and 22, determine the size of the matrix.

21. $\begin{bmatrix} 2 & 3 & -1 \\ 0 & 5 & 1 \end{bmatrix}$

22. $\begin{bmatrix} 2 & 1 \\ -4 & -1 \\ 0 & 5 \end{bmatrix}$

Augmented Matrix In Exercises 23–26, find the solution set of the system of linear equations represented by the augmented matrix.

23. $\begin{bmatrix} 1 & 2 & -5 \\ 2 & 1 & 5 \end{bmatrix}$

24. $\begin{bmatrix} -2 & 3 & 0 \\ 0 & 0 & 0 \end{bmatrix}$

25. $\begin{bmatrix} 1 & 2 & 0 & 0 \\ 0 & 0 & 1 & 0 \\ 0 & 0 & 0 & 0 \end{bmatrix}$

26. $\begin{bmatrix} 1 & 2 & 3 & 0 \\ 0 & 0 & 0 & 1 \\ 0 & 0 & 0 & 0 \end{bmatrix}$

Row-Echelon Form In Exercises 27–30, determine whether the matrix is in row-echelon form. If it is, determine whether it is also in reduced row-echelon form.

27. $\begin{bmatrix} 1 & 2 & -3 & 0 \\ 0 & 0 & 0 & 1 \\ 0 & 0 & 0 & 0 \end{bmatrix}$

28. $\begin{bmatrix} 1 & 0 & 1 & 1 \\ 0 & 1 & 2 & 1 \\ 0 & 0 & 0 & 1 \end{bmatrix}$

29. $\begin{bmatrix} -1 & 2 & 1 \\ 0 & 1 & 0 \\ 0 & 0 & 1 \end{bmatrix}$

30. $\begin{bmatrix} 0 & 1 & 0 & 0 \\ 0 & 0 & 1 & 2 \\ 0 & 0 & 0 & 0 \end{bmatrix}$

System of Linear Equations In Exercises 31–40, solve the system using either Gaussian elimination with back-substitution or Gauss-Jordan elimination.

31. $-x + y + 2z = 1$
    $2x + 3y + z = -2$
    $5x + 4y + 2z = 4$

32. $4x + 2y + z = 18$
    $4x - 2y - 2z = 28$
    $2x - 3y + 2z = -8$

33. $2x + 3y + 3z = 3$
    $6x + 6y + 12z = 13$
    $12x + 9y - z = 2$

34. $2x + y + 2z = 4$
    $2x + 2y = 5$
    $2x - y + 6z = 2$

35. $x - 2y + z = -6$
    $2x - 3y = -7$
    $-x + 3y - 3z = 11$

36. $2x + 6z = -9$
    $3x - 2y + 11z = -16$
    $3x - y + 7z = -11$

37. $x + 2y + 6z = 1$
    $2x + 5y + 15z = 4$
    $3x + y + 3z = -6$

38. $2x_1 + 5x_2 - 19x_3 = 34$
    $3x_1 + 8x_2 - 31x_3 = 54$

39. $2x_1 + x_2 + x_3 + 2x_4 = -1$
    $5x_1 - 2x_2 + x_3 - 3x_4 = 0$
    $-x_1 + 3x_2 + 2x_3 + 2x_4 = 1$
    $3x_1 + 2x_2 + 3x_3 - 5x_4 = 12$

40. $x_1 + 5x_2 + 3x_3 = 14$
    $4x_2 + 2x_3 + 5x_4 = 3$
    $3x_3 + 8x_4 + 6x_5 = 16$
    $2x_1 + 4x_2 - 2x_5 = 0$
    $2x_1 - x_3 = 0$

System of Linear Equations   **In Exercises 41–46, use a software program or a graphing utility to solve the system of linear equations.**

41.
$$x_1 + x_2 + x_3 = 15.4$$
$$x_1 - x_2 - 2x_3 = 27.9$$
$$3x_1 - 2x_2 + x_3 = 76.9$$

42.
$$1.1x_1 + 2.3x_2 + 3.4x_3 = 0$$
$$1.1x_1 - 2.2x_2 - 4.4x_3 = 0$$
$$-1.7x_1 + 3.4x_2 + 6.8x_3 = 1$$

43.
$$3x + 3y + 12z = 6$$
$$x + y + 4z = 2$$
$$2x + 5y + 20z = 10$$
$$-x + 2y + 8z = 4$$

44.
$$x + 2y + z + 3w = 0$$
$$x - y \quad + w = 0$$
$$5y - z + 2w = 0$$

45.
$$2x + 10y + 2z = 6$$
$$x + 5y + 2z = 6$$
$$x + 5y + z = 3$$
$$-3x - 15y + 3z = -9$$

46.
$$2x + y - z + 2w = -6$$
$$3x + 4y \quad + w = 1$$
$$x + 5y + 2z + 6w = -3$$
$$5x + 2y - z - w = 3$$

Homogeneous System   **In Exercises 47–50, solve the homogeneous system of linear equations.**

47.
$$x_1 - 2x_2 - 8x_3 = 0$$
$$3x_1 + 2x_2 \quad = 0$$

48.
$$2x_1 + 4x_2 - 7x_3 = 0$$
$$x_1 - 3x_2 + 9x_3 = 0$$

49.
$$-2x_1 + 7x_2 - 3x_3 = 0$$
$$4x_1 - 12x_2 + 5x_3 = 0$$
$$12x_2 + 7x_3 = 0$$

50.
$$x_1 + 3x_2 + 5x_3 = 0$$
$$x_1 + 4x_2 + \tfrac{1}{2}x_3 = 0$$

51. Determine the values of k such that the system of linear equations is inconsistent.
$$kx + y = 0$$
$$x + ky = 1$$

52. Determine the values of k such that the system of linear equations has exactly one solution.
$$x - y + 2z = 0$$
$$-x + y - z = 0$$
$$x + ky + z = 0$$

53. Find values of a and b such that the system of linear equations has (a) no solution, (b) exactly one solution, and (c) infinitely many solutions.
$$x + 2y = 3$$
$$ax + by = -9$$

54. Find (if possible) values of a, b, and c such that the system of linear equations has (a) no solution, (b) exactly one solution, and (c) infinitely many solutions.
$$2x - y + z = a$$
$$x + y + 2z = b$$
$$3y + 3z = c$$

55. **Writing**   Describe a method for showing that two matrices are row-equivalent. Are the two matrices below row-equivalent?
$$\begin{bmatrix} 1 & 1 & 2 \\ 0 & -1 & 2 \\ 3 & 1 & 2 \end{bmatrix} \text{ and } \begin{bmatrix} 1 & 2 & 3 \\ 4 & 3 & 6 \\ 5 & 5 & 10 \end{bmatrix}$$

56. **Writing**   Describe all possible 2×3 reduced row-echelon matrices. Support your answer with examples.

57. Let $n \geq 3$. Find the reduced row-echelon form of the $n \times n$ matrix.
$$\begin{bmatrix} 1 & 2 & 3 & \cdots & n \\ n+1 & n+2 & n+3 & \cdots & 2n \\ 2n+1 & 2n+2 & 2n+3 & \cdots & 3n \\ \vdots & \vdots & \vdots & & \vdots \\ n^2-n+1 & n^2-n+2 & n^2-n+3 & \cdots & n^2 \end{bmatrix}$$

58. Find all values of λ for which the homogeneous system of linear equations has nontrivial solutions.
$$(\lambda + 2)x_1 - 2x_2 + 3x_3 = 0$$
$$-2x_1 + (\lambda - 1)x_2 + 6x_3 = 0$$
$$x_1 + 2x_2 + \lambda x_3 = 0$$

True or False?   **In Exercises 59 and 60, determine whether each statement is true or false. If a statement is true, give a reason or cite an appropriate statement from the text. If a statement is false, provide an example that shows the statement is not true in all cases or cite an appropriate statement from the text.**

59. (a) There is only one way to parametrically represent the solution set of a linear equation.

(b) A consistent system of linear equations can have infinitely many solutions.

60. (a) A homogeneous system of linear equations must have at least one solution.

(b) A system of linear equations with fewer equations than variables always has at least one solution.

61. **Sports**   In Super Bowl I, on January 15, 1967, the Green Bay Packers defeated the Kansas City Chiefs by a score of 35 to 10. The total points scored came from a combination of touchdowns, extra-point kicks, and field goals, worth 6, 1, and 3 points, respectively. The numbers of touchdowns and extra-point kicks were equal. There were six times as many touchdowns as field goals. Find the numbers of touchdowns, extra-point kicks, and field goals scored. (Source: National Football League)

62. Agriculture A mixture of 24 liters of chemical A, 32 liters of chemical B, and 52 liters of chemical C is required to kill a destructive crop insect. Commercial spray X contains 1, 2, and 2 parts, respectively, of these chemicals. Commercial spray Y contains only chemical C. Commercial spray Z contains chemicals A, B, and C in equal amounts. How much of each type of commercial spray is needed to get the desired mixture?

Partial Fraction Decomposition **In Exercises 63 and 64, use a system of equations to find the partial fraction decomposition of the rational expression. Solve the system using matrices.**

63. $\dfrac{8x^2}{(x-1)^2(x+1)} = \dfrac{A}{x+1} + \dfrac{B}{x-1} + \dfrac{C}{(x-1)^2}$

64. $\dfrac{3x^2+3x-2}{(x+1)^2(x-1)} = \dfrac{A}{x+1} + \dfrac{B}{x-1} + \dfrac{C}{(x+1)^2}$

Polynomial Curve Fitting **In Exercises 65 and 66, (a) determine the polynomial function whose graph passes through the points, and (b) sketch the graph of the polynomial function, showing the points.**

65. $(2, 5), (3, 0), (4, 20)$

66. $(-1, -1), (0, 0), (1, 1), (2, 4)$

67. Sales A company has sales (measured in millions) of \$50, \$60, and \$75 during three consecutive years. Find a quadratic function that fits the data, and use it to predict the sales during the fourth year.

68. The polynomial function

$$p(x) = a_0 + a_1x + a_2x^2 + a_3x^3$$

is zero when $x = 1, 2, 3,$ and 4. What are the values of $a_0, a_1, a_2,$ and a_3?

69. Deer Population A wildlife management team studied the population of deer in one small tract of a wildlife preserve. The table shows the population and the number of years since the study began.

Year	0	4	80
Population	80	68	30

(a) Set up a system of equations to fit the data to a quadratic function.

(b) Solve the system.

(c) Use a graphing utility to fit the data to a quadratic model.

(d) Compare the quadratic function in part (b) with the model in part (c).

(e) Cite the statement from the text that verifies your results.

70. Vertical Motion An object moving vertically is at the given heights at the specified times. Find the position equation

$$s = \tfrac{1}{2}at^2 + v_0t + s_0$$

for the object.

(a) At $t = 0$ seconds, $s = 50$ meters
At $t = 1$ second, $s = 29.9$ meters
At $t = 2$ seconds, $s = 0$ meters

(b) At $t = 1$ second, $s = 40$ meters
At $t = 2$ seconds, $s = 25.3$ meters
At $t = 3$ seconds, $s = 0.8$ meter

(c) At $t = 1$ second, $s = 56$ meters
At $t = 2$ seconds, $s = 35.2$ meters
At $t = 3$ seconds, $s = 4.6$ meters

71. Network Analysis The figure shows the flow through a network.

(a) Solve the system for x_i, $i = 1, 2, \ldots, 6$.

(b) Find the flow when $x_3 = 100$, $x_5 = 50$, and $x_6 = 50$.

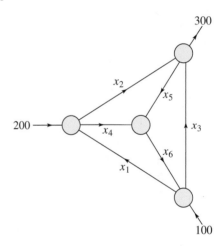

72. Network Analysis Determine the currents I_1, I_2, and I_3 for the electrical network shown in the figure.

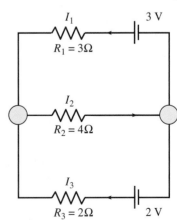

1   Projects

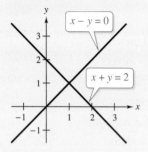

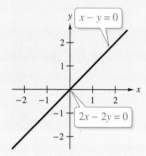

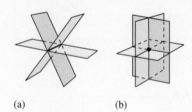

Figure 1.8

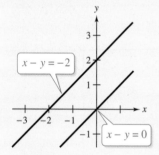

(a)   (b)

(c)

Figure 1.9

1 Graphing Linear Equations

You saw in Section 1.1 that you can represent a system of two linear equations in two variables x and y geometrically as two lines in the plane. These lines can intersect at a point, coincide, or be parallel, as shown in Figure 1.8.

1. Consider the system below, where a and b are constants.

$$x - 2y = 5$$
$$ax + by = 10$$

   (a) Find values of a and b for which the resulting system has a unique solution.

   (b) Find values of a and b for which the resulting system has infinitely many solutions.

   (c) Find values of a and b for which the resulting system has no solution.

   (d) Graph the lines for each of the systems in parts (a), (b), and (c).

2. Now consider a system of three linear equations in x, y, and z. Each equation represents a plane in the three-dimensional coordinate system.

   (a) Find an example of a system represented by three planes intersecting in a line, as shown in Figure 1.9(a).

   (b) Find an example of a system represented by three planes intersecting at a point, as shown in Figure 1.9(b).

   (c) Find an example of a system represented by three planes with no common intersection, as shown in Figure 1.9(c).

   (d) Are there other configurations of three planes in addition to those given in Figure 1.9? Explain.

2 Underdetermined and Overdetermined Systems

The system of linear equations below is **underdetermined** because there are more variables than equations.

$$2x_1 + 4x_2 - x_3 = 3$$
$$x_1 - 3x_2 + 2x_3 = -1$$

Similarly, the system below is **overdetermined** because there are more equations than variables.

$$2x_1 + x_2 = 3$$
$$3x_1 - 3x_2 = -2$$
$$5x_1 - x_2 = 1$$

Explore whether the number of variables and the number of equations have any bearing on the consistency of a system of linear equations. For Exercises 1–4, if an answer is yes, give an example. Otherwise, explain why the answer is no.

1. Can you find a consistent underdetermined linear system?

2. Can you find a consistent overdetermined linear system?

3. Can you find an inconsistent underdetermined linear system?

4. Can you find an inconsistent overdetermined linear system?

5. Explain why you would expect an overdetermined linear system to be inconsistent. Must this always be the case?

6. Explain why you would expect an underdetermined linear system to have infinitely many solutions. Must this always be the case?

2 Matrices

Data Encryption (p. 94)

Computational Fluid Dynamics (p. 79)

Beam Deflection (p. 64)

Information Retrieval (p. 58)

Flight Crew Scheduling (p. 47)

2.1   Operations with Matrices

■ Determine whether two matrices are equal.

■ Add and subtract matrices and multiply a matrix by a scalar.

■ Multiply two matrices.

■ Use matrices to solve a system of linear equations.

■ Partition a matrix and write a linear combination of column vectors.

EQUALITY OF MATRICES

In Section 1.2, you used matrices to solve systems of linear equations. This chapter introduces some fundamentals of matrix theory and further applications of matrices.

It is standard mathematical convention to represent matrices in any one of the three ways listed below.

1. An uppercase letter such as A, B, or C

2. A representative element enclosed in brackets, such as $[a_{ij}]$, $[b_{ij}]$, or $[c_{ij}]$

3. A rectangular array of numbers

$$\begin{bmatrix} a_{11} & a_{12} & \cdots & a_{1n} \\ a_{21} & a_{22} & \cdots & a_{2n} \\ \vdots & \vdots & & \vdots \\ a_{m1} & a_{m2} & \cdots & a_{mn} \end{bmatrix}$$

As mentioned in Chapter 1, the matrices in this text are primarily *real matrices*. That is, their entries are real numbers.

Two matrices are *equal* when their corresponding entries are equal.

Definition of Equality of Matrices

Two matrices $A = [a_{ij}]$ and $B = [b_{ij}]$ are **equal** when they have the same size $(m \times n)$ and $a_{ij} = b_{ij}$ for $1 \le i \le m$ and $1 \le j \le n$.

REMARK

The phrase "if and only if" means the statement is true in both directions. For example, "*p* if and only if *q*" means that *p* implies *q* and *q* implies *p*.

EXAMPLE 1   **Equality of Matrices**

Consider the four matrices

$$A = \begin{bmatrix} 1 & 2 \\ 3 & 4 \end{bmatrix}, \quad B = \begin{bmatrix} 1 \\ 3 \end{bmatrix}, \quad C = \begin{bmatrix} 1 & 3 \end{bmatrix}, \quad \text{and} \quad D = \begin{bmatrix} 1 & 2 \\ x & 4 \end{bmatrix}.$$

Matrices A and B are **not** equal because they are of different sizes. Similarly, B and C are not equal. Matrices A and D are equal if and only if $x = 3$.   ■

A matrix that has only one column, such as matrix B in Example 1, is a **column matrix** or **column vector**. Similarly, a matrix that has only one row, such as matrix C in Example 1, is a **row matrix** or **row vector**. Boldface lowercase letters often designate column matrices and row matrices. For instance, matrix A in Example 1 can be partitioned into the two column matrices $\mathbf{a}_1 = \begin{bmatrix} 1 \\ 3 \end{bmatrix}$ and $\mathbf{a}_2 = \begin{bmatrix} 2 \\ 4 \end{bmatrix}$ as shown below.

$$A = \begin{bmatrix} 1 & 2 \\ 3 & 4 \end{bmatrix} = \left[\begin{array}{c|c} 1 & 2 \\ 3 & 4 \end{array} \right] = [\mathbf{a}_1 \mid \mathbf{a}_2]$$

MATRIX ADDITION, SUBTRACTION, AND SCALAR MULTIPLICATION

To **add** two matrices (of the same size), add their corresponding entries.

Definition of Matrix Addition

If $A = [a_{ij}]$ and $B = [b_{ij}]$ are matrices of size $m \times n$, then their **sum** is the $m \times n$ matrix $A + B = [a_{ij} + b_{ij}]$.

The sum of two matrices of different sizes is undefined.

EXAMPLE 2   Addition of Matrices

a. $\begin{bmatrix} -1 & 2 \\ 0 & 1 \end{bmatrix} + \begin{bmatrix} 1 & 3 \\ -1 & 2 \end{bmatrix} = \begin{bmatrix} -1+1 & 2+3 \\ 0+(-1) & 1+2 \end{bmatrix} = \begin{bmatrix} 0 & 5 \\ -1 & 3 \end{bmatrix}$

b. $\begin{bmatrix} 0 & 1 & -2 \\ 1 & 2 & 3 \end{bmatrix} + \begin{bmatrix} 0 & 0 & 0 \\ 0 & 0 & 0 \end{bmatrix} = \begin{bmatrix} 0 & 1 & -2 \\ 1 & 2 & 3 \end{bmatrix}$

c. $\begin{bmatrix} 1 \\ -3 \\ -2 \end{bmatrix} + \begin{bmatrix} -1 \\ 3 \\ 2 \end{bmatrix} = \begin{bmatrix} 0 \\ 0 \\ 0 \end{bmatrix}$ **d.** $\begin{bmatrix} 2 & 1 & 0 \\ 4 & 0 & -1 \end{bmatrix} + \begin{bmatrix} 0 & 1 \\ -1 & 3 \end{bmatrix}$ is undefined.

REMARK

It is often convenient to rewrite the scalar multiple cA by factoring c out of every entry in the matrix. For example, factoring the scalar $\frac{1}{2}$ out of the matrix below gives

$$\begin{bmatrix} \frac{1}{2} & -\frac{3}{2} \\ \frac{5}{2} & \frac{1}{2} \end{bmatrix} = \frac{1}{2}\begin{bmatrix} 1 & -3 \\ 5 & 1 \end{bmatrix}.$$

When working with matrices, real numbers are referred to as **scalars**. To multiply a matrix A by a scalar c, multiply each entry in A by c.

Definition of Scalar Multiplication

If $A = [a_{ij}]$ is an $m \times n$ matrix and c is a scalar, then the **scalar multiple** of A by c is the $m \times n$ matrix $cA = [ca_{ij}]$.

You can use $-A$ to represent the scalar product $(-1)A$. If A and B are of the same size, then $A - B$ represents the sum of A and $(-1)B$. That is, $A - B = A + (-1)B$.

EXAMPLE 3   Scalar Multiplication and Matrix Subtraction

For the matrices A and B, find (a) $3A$, (b) $-B$, and (c) $3A - B$.

$$A = \begin{bmatrix} 1 & 2 & 4 \\ -3 & 0 & -1 \\ 2 & 1 & 2 \end{bmatrix} \quad \text{and} \quad B = \begin{bmatrix} 2 & 0 & 0 \\ 1 & -4 & 3 \\ -1 & 3 & 2 \end{bmatrix}$$

SOLUTION

a. $3A = 3\begin{bmatrix} 1 & 2 & 4 \\ -3 & 0 & -1 \\ 2 & 1 & 2 \end{bmatrix} = \begin{bmatrix} 3(1) & 3(2) & 3(4) \\ 3(-3) & 3(0) & 3(-1) \\ 3(2) & 3(1) & 3(2) \end{bmatrix} = \begin{bmatrix} 3 & 6 & 12 \\ -9 & 0 & -3 \\ 6 & 3 & 6 \end{bmatrix}$

b. $-B = (-1)\begin{bmatrix} 2 & 0 & 0 \\ 1 & -4 & 3 \\ -1 & 3 & 2 \end{bmatrix} = \begin{bmatrix} -2 & 0 & 0 \\ -1 & 4 & -3 \\ 1 & -3 & -2 \end{bmatrix}$

c. $3A - B = \begin{bmatrix} 3 & 6 & 12 \\ -9 & 0 & -3 \\ 6 & 3 & 6 \end{bmatrix} - \begin{bmatrix} 2 & 0 & 0 \\ 1 & -4 & 3 \\ -1 & 3 & 2 \end{bmatrix} = \begin{bmatrix} 1 & 6 & 12 \\ -10 & 4 & -6 \\ 7 & 0 & 4 \end{bmatrix}$

MATRIX MULTIPLICATION

Another basic matrix operation is **matrix multiplication.** To see the usefulness of this operation, consider the application below, in which matrices are helpful for organizing information.

A football stadium has three concession areas, located in the south, north, and west stands. The top-selling items are peanuts, hot dogs, and soda. Sales for one day are given in the first matrix below, and the prices (in dollars) of the three items are given in the second matrix.

Numbers of Items Sold

	Peanuts	Hot Dogs	Sodas		Selling Price	
South Stand	120	250	305		2.00	Peanuts
North Stand	207	140	419		3.00	Hot Dogs
West Stand	29	120	190		2.75	Soda

To calculate the total sales of the three top-selling items at the south stand, multiply each entry in the first row of the matrix on the left by the corresponding entry in the price column matrix on the right and add the results. The south stand sales are

$$(120)(2.00) + (250)(3.00) + (305)(2.75) = \$1828.75 \qquad \text{South stand sales}$$

Similarly, the sales for the other two stands are shown below.

$$(207)(2.00) + (140)(3.00) + (419)(2.75) = \$1986.25 \qquad \text{North stand sales}$$
$$(29)(2.00) + (120)(3.00) + (190)(2.75) = \$940.50 \qquad \text{West stand sales}$$

The preceding computations are examples of matrix multiplication. You can write the product of the 3×3 matrix indicating the number of items sold and the 3×1 matrix indicating the selling prices as shown below.

$$\begin{bmatrix} 120 & 250 & 305 \\ 207 & 140 & 419 \\ 29 & 120 & 190 \end{bmatrix} \begin{bmatrix} 2.00 \\ 3.00 \\ 2.75 \end{bmatrix} = \begin{bmatrix} 1828.75 \\ 1986.25 \\ 940.50 \end{bmatrix}$$

The product of these matrices is the 3×1 matrix giving the total sales for each of the three stands.

The definition of the product of two matrices shown below is based on the ideas just developed. Although at first glance this definition may seem unusual, you will see that it has many practical applications.

Definition of Matrix Multiplication

If $A = [a_{ij}]$ is an $m \times n$ matrix and $B = [b_{ij}]$ is an $n \times p$ matrix, then the **product** AB is an $m \times p$ matrix

$$AB = [c_{ij}]$$

where

$$c_{ij} = \sum_{k=1}^{n} a_{ik}b_{kj}$$
$$= a_{i1}b_{1j} + a_{i2}b_{2j} + a_{i3}b_{3j} + \cdots + a_{in}b_{nj}.$$

This definition means that to find the entry in the ith row and the jth column of the product AB, multiply the entries in the ith row of A by the corresponding entries in the jth column of B and then add the results. The next example illustrates this process.

| EXAMPLE 4 | **Finding the Product of Two Matrices** |

Find the product AB, where

$$A = \begin{bmatrix} -1 & 3 \\ 4 & -2 \\ 5 & 0 \end{bmatrix} \quad \text{and} \quad B = \begin{bmatrix} -3 & 2 \\ -4 & 1 \end{bmatrix}.$$

SOLUTION

First, note that the product AB is defined because A has size 3×2 and B has size 2×2. Moreover, the product AB has size 3×2, and will take the form

$$\begin{bmatrix} -1 & 3 \\ 4 & -2 \\ 5 & 0 \end{bmatrix} \begin{bmatrix} -3 & 2 \\ -4 & 1 \end{bmatrix} = \begin{bmatrix} c_{11} & c_{12} \\ c_{21} & c_{22} \\ c_{31} & c_{32} \end{bmatrix}.$$

To find c_{11} (the entry in the first row and first column of the product), multiply corresponding entries in the first row of A and the first column of B. That is,

$$c_{11} = (-1)(-3) + (3)(-4) = -9$$

$$\begin{bmatrix} -1 & 3 \\ 4 & -2 \\ 5 & 0 \end{bmatrix} \begin{bmatrix} -3 & 2 \\ -4 & 1 \end{bmatrix} = \begin{bmatrix} -9 & c_{12} \\ c_{21} & c_{22} \\ c_{31} & c_{32} \end{bmatrix}.$$

Similarly, to find c_{12}, multiply corresponding entries in the first row of A and the second column of B to obtain

$$c_{12} = (-1)(2) + (3)(1) = 1$$

$$\begin{bmatrix} -1 & 3 \\ 4 & -2 \\ 5 & 0 \end{bmatrix} \begin{bmatrix} -3 & 2 \\ -4 & 1 \end{bmatrix} = \begin{bmatrix} -9 & 1 \\ c_{21} & c_{22} \\ c_{31} & c_{32} \end{bmatrix}.$$

Continuing this pattern produces the results shown below.

$$
\begin{aligned}
c_{21} &= (4)(-3) + (-2)(-4) &=& -4 \\
c_{22} &= (4)(2) + (-2)(1) &=& 6 \\
c_{31} &= (5)(-3) + (0)(-4) &=& -15 \\
c_{32} &= (5)(2) + (0)(1) &=& 10
\end{aligned}
$$

The product is

$$AB = \begin{bmatrix} -1 & 3 \\ 4 & -2 \\ 5 & 0 \end{bmatrix} \begin{bmatrix} -3 & 2 \\ -4 & 1 \end{bmatrix} = \begin{bmatrix} -9 & 1 \\ -4 & 6 \\ -15 & 10 \end{bmatrix}.$$

Be sure you understand that for the product of two matrices to be defined, the number of columns of the first matrix must equal the number of rows of the second matrix. That is,

$$
\begin{array}{ccc}
A & B & = & AB. \\
m \times n & n \times p & & m \times p
\end{array}
$$

Equal

Size of AB

So, the product BA is not defined for matrices such as A and B in Example 4.

Arthur Cayley
(1821–1895)
British mathematician Arthur Cayley is credited with giving an abstract definition of a matrix. Cayley was a Cambridge University graduate and a lawyer by profession. He began his groundbreaking work on matrices as he studied the theory of transformations. Cayley also was instrumental in the development of determinants (discussed in Chapter 3). Cayley and two American mathematicians, Benjamin Peirce (1809–1880) and his son, Charles S. Peirce (1839–1914), are credited with developing "matrix algebra."

The general pattern for matrix multiplication is shown below. To obtain the element in the ith row and the jth column of the product AB, use the ith row of A and the jth column of B.

$$\begin{bmatrix} a_{11} & a_{12} & a_{13} & \cdots & a_{1n} \\ a_{21} & a_{22} & a_{23} & \cdots & a_{2n} \\ \vdots & \vdots & \vdots & & \vdots \\ a_{i1} & a_{i2} & a_{i3} & \cdots & a_{in} \\ \vdots & \vdots & \vdots & & \vdots \\ a_{m1} & a_{m2} & a_{m3} & \cdots & a_{mn} \end{bmatrix} \begin{bmatrix} b_{11} & b_{12} & \cdots & b_{1j} & \cdots & b_{1p} \\ b_{21} & b_{22} & \cdots & b_{2j} & \cdots & b_{2p} \\ b_{31} & b_{32} & \cdots & b_{3j} & \cdots & b_{3p} \\ \vdots & \vdots & & \vdots & & \vdots \\ b_{n1} & b_{n2} & \cdots & b_{nj} & \cdots & b_{np} \end{bmatrix} = \begin{bmatrix} c_{11} & c_{12} & \cdots & c_{1j} & \cdots & c_{1p} \\ c_{21} & c_{22} & \cdots & c_{2j} & \cdots & c_{2p} \\ \vdots & \vdots & & \vdots & & \vdots \\ c_{i1} & c_{i2} & \cdots & c_{ij} & \cdots & c_{ip} \\ \vdots & \vdots & & \vdots & & \vdots \\ c_{m1} & c_{m2} & \cdots & c_{mj} & \cdots & c_{mp} \end{bmatrix}$$

$$a_{i1}b_{1j} + a_{i2}b_{2j} + a_{i3}b_{3j} + \cdots + a_{in}b_{nj} = c_{ij}$$

DISCOVERY

Let

$$A = \begin{bmatrix} 1 & 2 \\ 3 & 4 \end{bmatrix} \quad \text{and} \quad B = \begin{bmatrix} 0 & 1 \\ 1 & 2 \end{bmatrix}.$$

Find $A + B$ and $B + A$. Is matrix addition commutative?

Find AB and BA. Is matrix multiplication commutative?

EXAMPLE 5   Matrix Multiplication

See LarsonLinearAlgebra.com for an interactive version of this type of example.

a. $\begin{bmatrix} 1 & 0 & 3 \\ 2 & -1 & -2 \end{bmatrix} \begin{bmatrix} -2 & 4 & 2 \\ 1 & 0 & 0 \\ -1 & 1 & -1 \end{bmatrix} = \begin{bmatrix} -5 & 7 & -1 \\ -3 & 6 & 6 \end{bmatrix}$

$\quad\quad 2 \times 3 \quad\quad\quad 3 \times 3 \quad\quad\quad 2 \times 3$

b. $\begin{bmatrix} 3 & 4 \\ -2 & 5 \end{bmatrix} \begin{bmatrix} 1 & 0 \\ 0 & 1 \end{bmatrix} = \begin{bmatrix} 3 & 4 \\ -2 & 5 \end{bmatrix}$

$\quad\quad 2 \times 2 \quad\quad 2 \times 2 \quad\quad 2 \times 2$

c. $\begin{bmatrix} 1 & 2 \\ 1 & 1 \end{bmatrix} \begin{bmatrix} -1 & 2 \\ 1 & -1 \end{bmatrix} = \begin{bmatrix} 1 & 0 \\ 0 & 1 \end{bmatrix}$

$\quad\quad 2 \times 2 \quad\quad 2 \times 2 \quad\quad 2 \times 2$

d. $[1 \quad -2 \quad -3] \begin{bmatrix} 2 \\ -1 \\ 1 \end{bmatrix} = [1]$

$\quad\quad 1 \times 3 \quad\quad 3 \times 1 \quad\quad 1 \times 1$

e. $\begin{bmatrix} 2 \\ -1 \\ 1 \end{bmatrix} [1 \quad -2 \quad -3] = \begin{bmatrix} 2 & -4 & -6 \\ -1 & 2 & 3 \\ 1 & -2 & -3 \end{bmatrix}$

$\quad 3 \times 1 \quad\quad 1 \times 3 \quad\quad\quad 3 \times 3$

Note the difference between the two products in parts (d) and (e) of Example 5. In general, matrix multiplication is not commutative. It is usually not true that the product AB is equal to the product BA. (See Section 2.2 for further discussion of the noncommutativity of matrix multiplication.)

SYSTEMS OF LINEAR EQUATIONS

One practical application of matrix multiplication is representing a system of linear equations. Note how the system

$$a_{11}x_1 + a_{12}x_2 + a_{13}x_3 = b_1$$
$$a_{21}x_1 + a_{22}x_2 + a_{23}x_3 = b_2$$
$$a_{31}x_1 + a_{32}x_2 + a_{33}x_3 = b_3$$

can be written as the matrix equation $A\mathbf{x} = \mathbf{b}$, where A is the coefficient matrix of the system, and $\mathbf{x}$ and $\mathbf{b}$ are column matrices.

$$\underbrace{\begin{bmatrix} a_{11} & a_{12} & a_{13} \\ a_{21} & a_{22} & a_{23} \\ a_{31} & a_{32} & a_{33} \end{bmatrix}}_{A} \underbrace{\begin{bmatrix} x_1 \\ x_2 \\ x_3 \end{bmatrix}}_{\mathbf{x}} = \underbrace{\begin{bmatrix} b_1 \\ b_2 \\ b_3 \end{bmatrix}}_{\mathbf{b}}$$

EXAMPLE 6   Solving a System of Linear Equations

Solve the matrix equation $A\mathbf{x} = \mathbf{0}$, where

$$A = \begin{bmatrix} 1 & -2 & 1 \\ 2 & 3 & -2 \end{bmatrix}, \quad \mathbf{x} = \begin{bmatrix} x_1 \\ x_2 \\ x_3 \end{bmatrix}, \quad \text{and} \quad \mathbf{0} = \begin{bmatrix} 0 \\ 0 \end{bmatrix}.$$

SOLUTION

As a system of linear equations, $A\mathbf{x} = \mathbf{0}$ is

$$x_1 - 2x_2 + x_3 = 0$$
$$2x_1 + 3x_2 - 2x_3 = 0.$$

Using Gauss-Jordan elimination on the augmented matrix of this system, you obtain

$$\begin{bmatrix} 1 & 0 & -\frac{1}{7} & 0 \\ 0 & 1 & -\frac{4}{7} & 0 \end{bmatrix}.$$

So, the system has infinitely many solutions. Here a convenient choice of a parameter is $x_3 = 7t$, and you can write the solution set as

$$x_1 = t, \quad x_2 = 4t, \quad x_3 = 7t, \quad t \text{ is any real number.}$$

In matrix terminology, you have found that the matrix equation

$$\begin{bmatrix} 1 & -2 & 1 \\ 2 & 3 & -2 \end{bmatrix} \begin{bmatrix} x_1 \\ x_2 \\ x_3 \end{bmatrix} = \begin{bmatrix} 0 \\ 0 \end{bmatrix}$$

has infinitely many solutions represented by

$$\mathbf{x} = \begin{bmatrix} x_1 \\ x_2 \\ x_3 \end{bmatrix} = \begin{bmatrix} t \\ 4t \\ 7t \end{bmatrix} = t \begin{bmatrix} 1 \\ 4 \\ 7 \end{bmatrix}, \quad t \text{ is any scalar.}$$

That is, any scalar multiple of the column matrix on the right is a solution. Here are some sample solutions:

$$\begin{bmatrix} 1 \\ 4 \\ 7 \end{bmatrix}, \quad \begin{bmatrix} 2 \\ 8 \\ 14 \end{bmatrix}, \quad \begin{bmatrix} 0 \\ 0 \\ 0 \end{bmatrix}, \quad \text{and} \quad \begin{bmatrix} -1 \\ -4 \\ -7 \end{bmatrix}.$$

TECHNOLOGY

Many graphing utilities and software programs can perform matrix addition, scalar multiplication, and matrix multiplication. When you use a graphing utility to check one of the solutions in Example 6, you may see something similar to the screen below.

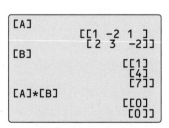

The **Technology Guide** at *CengageBrain.com* can help you use technology to perform matrix operations.

PARTITIONED MATRICES

The system $A\mathbf{x} = \mathbf{b}$ can be represented in a more convenient way by partitioning the matrices A and $\mathbf{x}$ in the manner shown below. If

$$A = \begin{bmatrix} a_{11} & a_{12} & \cdots & a_{1n} \\ a_{21} & a_{22} & \cdots & a_{2n} \\ \vdots & \vdots & & \vdots \\ a_{m1} & a_{m2} & \cdots & a_{mn} \end{bmatrix}, \quad \mathbf{x} = \begin{bmatrix} x_1 \\ x_2 \\ \vdots \\ x_n \end{bmatrix}, \quad \text{and} \quad \mathbf{b} = \begin{bmatrix} b_1 \\ b_2 \\ \vdots \\ b_m \end{bmatrix}$$

are the coefficient matrix, the column matrix of unknowns, and the right-hand side, respectively, of the $m \times n$ linear system $A\mathbf{x} = \mathbf{b}$, then

$$\begin{bmatrix} a_{11} & a_{12} & \cdots & a_{1n} \\ a_{21} & a_{22} & \cdots & a_{2n} \\ \vdots & \vdots & & \vdots \\ a_{m1} & a_{m2} & \cdots & a_{mn} \end{bmatrix}\begin{bmatrix} x_1 \\ x_2 \\ \vdots \\ x_n \end{bmatrix} = \mathbf{b}$$

$$\begin{bmatrix} a_{11}x_1 + a_{12}x_2 + \cdots + a_{1n}x_n \\ a_{21}x_1 + a_{22}x_2 + \cdots + a_{2n}x_n \\ \vdots \\ a_{m1}x_1 + a_{m2}x_2 + \cdots + a_{mn}x_n \end{bmatrix} = \mathbf{b}$$

$$x_1 \begin{bmatrix} a_{11} \\ a_{21} \\ \vdots \\ a_{m1} \end{bmatrix} + x_2 \begin{bmatrix} a_{12} \\ a_{22} \\ \vdots \\ a_{m2} \end{bmatrix} + \cdots + x_n \begin{bmatrix} a_{1n} \\ a_{2n} \\ \vdots \\ a_{mn} \end{bmatrix} = \mathbf{b}.$$

In other words,

$$A\mathbf{x} = x_1\mathbf{a}_1 + x_2\mathbf{a}_2 + \cdots + x_n\mathbf{a}_n = \mathbf{b}$$

where $\mathbf{a}_1, \mathbf{a}_2, \ldots, \mathbf{a}_n$ are the columns of the matrix A. The expression

$$x_1 \begin{bmatrix} a_{11} \\ a_{21} \\ \vdots \\ a_{m1} \end{bmatrix} + x_2 \begin{bmatrix} a_{12} \\ a_{22} \\ \vdots \\ a_{m2} \end{bmatrix} + \cdots + x_n \begin{bmatrix} a_{1n} \\ a_{2n} \\ \vdots \\ a_{mn} \end{bmatrix}$$

is called a **linear combination** of the column matrices $\mathbf{a}_1, \mathbf{a}_2, \ldots, \mathbf{a}_n$ with **coefficients** $x_1, x_2, \ldots, x_n$.

Linear Combinations of Column Vectors

The matrix product $A\mathbf{x}$ is a linear combination of the column vectors $\mathbf{a}_1$, $\mathbf{a}_2, \ldots, \mathbf{a}_n$ that form the coefficient matrix A.

$$x_1 \begin{bmatrix} a_{11} \\ a_{21} \\ \vdots \\ a_{m1} \end{bmatrix} + x_2 \begin{bmatrix} a_{12} \\ a_{22} \\ \vdots \\ a_{m2} \end{bmatrix} + \cdots + x_n \begin{bmatrix} a_{1n} \\ a_{2n} \\ \vdots \\ a_{mn} \end{bmatrix}$$

Furthermore, the system

$$A\mathbf{x} = \mathbf{b}$$

is consistent if and only if $\mathbf{b}$ can be expressed as such a linear combination, where the coefficients of the linear combination are a solution of the system.

EXAMPLE 7   **Solving a System of Linear Equations**

The linear system

$$x_1 + 2x_2 + 3x_3 = 0$$
$$4x_1 + 5x_2 + 6x_3 = 3$$
$$7x_1 + 8x_2 + 9x_3 = 6$$

can be rewritten as a matrix equation $A\mathbf{x} = \mathbf{b}$, as shown below.

$$x_1\begin{bmatrix} 1 \\ 4 \\ 7 \end{bmatrix} + x_2\begin{bmatrix} 2 \\ 5 \\ 8 \end{bmatrix} + x_3\begin{bmatrix} 3 \\ 6 \\ 9 \end{bmatrix} = \begin{bmatrix} 0 \\ 3 \\ 6 \end{bmatrix}$$

Using Gaussian elimination, you can show that this system has infinitely many solutions, one of which is $x_1 = 1$, $x_2 = 1$, $x_3 = -1$.

$$1\begin{bmatrix} 1 \\ 4 \\ 7 \end{bmatrix} + 1\begin{bmatrix} 2 \\ 5 \\ 8 \end{bmatrix} + (-1)\begin{bmatrix} 3 \\ 6 \\ 9 \end{bmatrix} = \begin{bmatrix} 0 \\ 3 \\ 6 \end{bmatrix}$$

That is, $\mathbf{b}$ can be expressed as a linear combination of the columns of A. This representation of one column vector in terms of others is a fundamental theme of linear algebra.

Just as you partition A into columns and $\mathbf{x}$ into rows, it is often useful to consider an $m \times n$ matrix partitioned into smaller matrices. For example, you can partition the matrix below as shown.

$$\begin{bmatrix} 1 & 2 & 0 & 0 \\ 3 & 4 & 0 & 0 \\ -1 & -2 & 2 & 1 \end{bmatrix} \quad \left[\begin{array}{cc|cc} 1 & 2 & 0 & 0 \\ 3 & 4 & 0 & 0 \\ \hline -1 & -2 & 2 & 1 \end{array}\right]$$

You can also partition the matrix into column matrices

$$\left[\begin{array}{c|c|c|c} 1 & 2 & 0 & 0 \\ 3 & 4 & 0 & 0 \\ -1 & -2 & 2 & 1 \end{array}\right] = \begin{bmatrix} \mathbf{c}_1 & \mathbf{c}_2 & \mathbf{c}_3 & \mathbf{c}_4 \end{bmatrix}$$

or row matrices

$$\left[\begin{array}{cccc} 1 & 2 & 0 & 0 \\ \hline 3 & 4 & 0 & 0 \\ \hline -1 & -2 & 2 & 1 \end{array}\right] = \begin{bmatrix} \mathbf{r}_1 \\ \mathbf{r}_2 \\ \mathbf{r}_3 \end{bmatrix}.$$

LINEAR ALGEBRA APPLIED

Many real-life applications of linear systems involve enormous numbers of equations and variables. For example, a flight crew scheduling problem for American Airlines required the manipulation of a matrix with 837 rows and more than 12,750,000 columns. To solve this application of *linear programming*, researchers partitioned the problem into smaller pieces and solved it on a computer. *(Source: Very Large-Scale Linear Programming. A Case Study in Combining Interior Point and Simplex Methods, Bixby, Robert E., et al., Operations Research, 40, no. 5)*

2.1 Exercises

See CalcChat.com for worked-out solutions to odd-numbered exercises.

Equality of Matrices   In Exercises 1–4, find x and y.

1. $\begin{bmatrix} x & -2 \\ 7 & y \end{bmatrix} = \begin{bmatrix} -4 & -2 \\ 7 & 22 \end{bmatrix}$

2. $\begin{bmatrix} -5 & x \\ y & 8 \end{bmatrix} = \begin{bmatrix} -5 & 13 \\ 12 & 8 \end{bmatrix}$

3. $\begin{bmatrix} 16 & 4 & 5 & 4 \\ -3 & 13 & 15 & 6 \\ 0 & 2 & 4 & 0 \end{bmatrix} = \begin{bmatrix} 16 & 4 & 2x+1 & 4 \\ -3 & 13 & 15 & 3x \\ 0 & 2 & 3y-5 & 0 \end{bmatrix}$

4. $\begin{bmatrix} x+2 & 8 & -3 \\ 1 & 2y & 2x \\ 7 & -2 & y+2 \end{bmatrix} = \begin{bmatrix} 2x+6 & 8 & -3 \\ 1 & 18 & -8 \\ 7 & -2 & 11 \end{bmatrix}$

Operations with Matrices   In Exercises 5–10, find, if possible, (a) $A + B$, (b) $A - B$, (c) $2A$, (d) $2A - B$, and (e) $B + \frac{1}{2}A$.

5. $A = \begin{bmatrix} 1 & 2 \\ 2 & 1 \end{bmatrix}, \quad B = \begin{bmatrix} -3 & -2 \\ 4 & 2 \end{bmatrix}$

6. $A = \begin{bmatrix} 6 & -1 \\ 2 & 4 \\ -3 & 5 \end{bmatrix}, \quad B = \begin{bmatrix} 1 & 4 \\ -1 & 5 \\ 1 & 10 \end{bmatrix}$

7. $A = \begin{bmatrix} 2 & 1 & 1 \\ -1 & -1 & 4 \end{bmatrix}, \quad B = \begin{bmatrix} 2 & -3 & 4 \\ -3 & 1 & -2 \end{bmatrix}$

8. $A = \begin{bmatrix} 3 & 2 & -1 \\ 2 & 4 & 5 \\ 0 & 1 & 2 \end{bmatrix}, \quad B = \begin{bmatrix} 0 & 2 & 1 \\ 5 & 4 & 2 \\ 2 & 1 & 0 \end{bmatrix}$

9. $A = \begin{bmatrix} 6 & 0 & 3 \\ -1 & -4 & 0 \end{bmatrix}, \quad B = \begin{bmatrix} 8 & -1 \\ 4 & -3 \end{bmatrix}$

10. $A = \begin{bmatrix} 3 \\ 2 \\ -1 \end{bmatrix}, \quad B = \begin{bmatrix} -4 & 6 & 2 \end{bmatrix}$

11. Find (a) c_{21} and (b) c_{13}, where $C = 2A - 3B$,
$$A = \begin{bmatrix} 5 & 4 & 4 \\ -3 & 1 & 2 \end{bmatrix}, \quad \text{and} \quad B = \begin{bmatrix} 1 & 2 & -7 \\ 0 & -5 & 1 \end{bmatrix}.$$

12. Find (a) c_{23} and (b) c_{32}, where $C = 5A + 2B$,
$$A = \begin{bmatrix} 4 & 11 & -9 \\ 0 & 3 & 2 \\ -3 & 1 & 1 \end{bmatrix}, \quad \text{and} \quad B = \begin{bmatrix} 1 & 0 & 5 \\ -4 & 6 & 11 \\ -6 & 4 & 9 \end{bmatrix}.$$

13. Solve for x, y, and z in the matrix equation
$$4\begin{bmatrix} x & y \\ z & -1 \end{bmatrix} = 2\begin{bmatrix} y & z \\ -x & 1 \end{bmatrix} + 2\begin{bmatrix} 4 & x \\ 5 & -x \end{bmatrix}.$$

14. Solve for x, y, z, and w in the matrix equation
$$\begin{bmatrix} w & x \\ y & x \end{bmatrix} = \begin{bmatrix} -4 & 3 \\ 2 & -1 \end{bmatrix} + 2\begin{bmatrix} y & w \\ z & x \end{bmatrix}.$$

Finding Products of Two Matrices   In Exercises 15–28, find, if possible, (a) AB and (b) BA.

15. $A = \begin{bmatrix} 1 & 2 \\ 4 & 2 \end{bmatrix}, \quad B = \begin{bmatrix} 2 & -1 \\ -1 & 8 \end{bmatrix}$

16. $A = \begin{bmatrix} 2 & -2 \\ -1 & 4 \end{bmatrix}, \quad B = \begin{bmatrix} 4 & 1 \\ 2 & -2 \end{bmatrix}$

17. $A = \begin{bmatrix} 2 & -1 & 3 \\ 5 & 1 & -2 \\ 2 & 2 & 3 \end{bmatrix}, \quad B = \begin{bmatrix} 0 & 1 & 2 \\ -4 & 1 & 3 \\ -4 & -1 & -2 \end{bmatrix}$

18. $A = \begin{bmatrix} 1 & -1 & 7 \\ 2 & -1 & 8 \\ 3 & 1 & -1 \end{bmatrix}, \quad B = \begin{bmatrix} 1 & 1 & 2 \\ 2 & 1 & 1 \\ 1 & -3 & 2 \end{bmatrix}$

19. $A = \begin{bmatrix} 2 & 1 \\ -3 & 4 \\ 1 & 6 \end{bmatrix}, \quad B = \begin{bmatrix} 0 & -1 & 0 \\ 4 & 0 & 2 \\ 8 & -1 & 7 \end{bmatrix}$

20. $A = \begin{bmatrix} 3 & 2 & 1 \\ -3 & 0 & 4 \\ 4 & -2 & -4 \end{bmatrix}, \quad B = \begin{bmatrix} 1 & 2 \\ 2 & -1 \\ 1 & -2 \end{bmatrix}$

21. $A = \begin{bmatrix} 3 & 2 & 1 \end{bmatrix}, \quad B = \begin{bmatrix} 2 \\ 3 \\ 0 \end{bmatrix}$

22. $A = \begin{bmatrix} -1 \\ 2 \\ -2 \\ 1 \end{bmatrix}, \quad B = \begin{bmatrix} 2 & 1 & 3 & 2 \end{bmatrix}$

23. $A = \begin{bmatrix} -1 & 3 \\ 4 & -5 \\ 0 & 2 \end{bmatrix}, \quad B = \begin{bmatrix} 1 & 2 \\ 0 & 7 \end{bmatrix}$

24. $A = \begin{bmatrix} 2 & -3 \\ 5 & 2 \end{bmatrix}, \quad B = \begin{bmatrix} 2 & 1 \\ 1 & 3 \\ 2 & -1 \end{bmatrix}$

25. $A = \begin{bmatrix} 0 & -1 & 0 \\ 4 & 0 & 2 \\ 8 & -1 & 7 \end{bmatrix}, \quad B = \begin{bmatrix} 2 \\ -3 \\ 1 \end{bmatrix}$

26. $A = \begin{bmatrix} 2 & 1 & 2 \\ 3 & -1 & -2 \\ -2 & 1 & -2 \end{bmatrix}, \quad B = \begin{bmatrix} 4 & 0 & 1 & 3 \\ -1 & 2 & -3 & -1 \\ -2 & 1 & 4 & 3 \end{bmatrix}$

27. $A = \begin{bmatrix} 6 \\ -2 \\ 1 \\ 6 \end{bmatrix}, \quad B = \begin{bmatrix} 10 & 12 \end{bmatrix}$

28. $A = \begin{bmatrix} 1 & 0 & 3 & -2 & 4 \\ 6 & 13 & 8 & -17 & 20 \end{bmatrix}, \quad B = \begin{bmatrix} 1 & 6 \\ 4 & 2 \end{bmatrix}$

Matrix Size In Exercises 29–36, let $A, B, C, D,$ and E be matrices with the sizes shown below.

A: 3×4 B: 3×4 C: 4×2 D: 4×2 E: 4×3

If defined, determine the size of the matrix. If not defined, explain why.

29. $A + B$ **30.** $C + E$

31. $\frac{1}{2}D$ **32.** $-4A$

33. AC **34.** BE

35. $E - 2A$ **36.** $2D + C$

Solving a Matrix Equation In Exercises 37 and 38, solve the matrix equation $A\mathbf{x} = \mathbf{0}$.

37. $A = \begin{bmatrix} 2 & -1 & -1 \\ 1 & -2 & 2 \end{bmatrix}$, $\mathbf{x} = \begin{bmatrix} x_1 \\ x_2 \\ x_3 \end{bmatrix}$, $\mathbf{0} = \begin{bmatrix} 0 \\ 0 \end{bmatrix}$

38. $A = \begin{bmatrix} 1 & 2 & 1 & 3 \\ 1 & -1 & 0 & 1 \\ 0 & 1 & -1 & 2 \end{bmatrix}$, $\mathbf{x} = \begin{bmatrix} x_1 \\ x_2 \\ x_3 \\ x_4 \end{bmatrix}$, $\mathbf{0} = \begin{bmatrix} 0 \\ 0 \\ 0 \end{bmatrix}$

Solving a System of Linear Equations In Exercises 39–48, write the system of linear equations in the form $A\mathbf{x} = \mathbf{b}$ and solve this matrix equation for $\mathbf{x}$.

39. $\begin{aligned} -x_1 + x_2 &= 4 \\ -2x_1 + x_2 &= 0 \end{aligned}$ **40.** $\begin{aligned} 2x_1 + 3x_2 &= 5 \\ x_1 + 4x_2 &= 10 \end{aligned}$

41. $\begin{aligned} -2x_1 - 3x_2 &= -4 \\ 6x_1 + x_2 &= -36 \end{aligned}$ **42.** $\begin{aligned} -4x_1 + 9x_2 &= -13 \\ x_1 - 3x_2 &= 12 \end{aligned}$

43. $\begin{aligned} x_1 - 2x_2 + 3x_3 &= 9 \\ -x_1 + 3x_2 - x_3 &= -6 \\ 2x_1 - 5x_2 + 5x_3 &= 17 \end{aligned}$

44. $\begin{aligned} x_1 + x_2 - 3x_3 &= -1 \\ -x_1 + 2x_2 &= 1 \\ x_1 - x_2 + x_3 &= 2 \end{aligned}$

45. $\begin{aligned} x_1 - 5x_2 + 2x_3 &= -20 \\ -3x_1 + x_2 - x_3 &= 8 \\ -2x_2 + 5x_3 &= -16 \end{aligned}$

46. $\begin{aligned} x_1 - x_2 + 4x_3 &= 17 \\ x_1 + 3x_2 &= -11 \\ -6x_2 + 5x_3 &= 40 \end{aligned}$

47. $\begin{aligned} 2x_1 - x_2 + x_4 &= 3 \\ 3x_2 - x_3 - x_4 &= -3 \\ x_1 + x_3 - 3x_4 &= -4 \\ x_1 + x_2 + 2x_3 &= 0 \end{aligned}$

48. $\begin{aligned} x_1 + x_2 &= 0 \\ x_2 + x_3 &= 0 \\ x_3 + x_4 &= 0 \\ x_4 + x_5 &= 0 \\ -x_1 + x_2 - x_3 + x_4 - x_5 &= 5 \end{aligned}$

Writing a Linear Combination In Exercises 49–52, write the column matrix b as a linear combination of the columns of A.

49. $A = \begin{bmatrix} 1 & -1 & 2 \\ 3 & -3 & 1 \end{bmatrix}$, $\mathbf{b} = \begin{bmatrix} -1 \\ 7 \end{bmatrix}$

50. $A = \begin{bmatrix} 1 & 2 & 4 \\ -1 & 0 & 2 \\ 0 & 1 & 3 \end{bmatrix}$, $\mathbf{b} = \begin{bmatrix} 1 \\ 3 \\ 2 \end{bmatrix}$

51. $A = \begin{bmatrix} 1 & 1 & -5 \\ 1 & 0 & -1 \\ 2 & -1 & -1 \end{bmatrix}$, $\mathbf{b} = \begin{bmatrix} 3 \\ 1 \\ 0 \end{bmatrix}$

52. $A = \begin{bmatrix} -3 & 5 \\ 3 & 4 \\ 4 & -8 \end{bmatrix}$, $\mathbf{b} = \begin{bmatrix} -22 \\ 4 \\ 32 \end{bmatrix}$

Solving a Matrix Equation In Exercises 53 and 54, solve for A.

53. $\begin{bmatrix} 1 & 2 \\ 3 & 5 \end{bmatrix} A = \begin{bmatrix} 1 & 0 \\ 0 & 1 \end{bmatrix}$

54. $\begin{bmatrix} 2 & -1 \\ 3 & -2 \end{bmatrix} A = \begin{bmatrix} 1 & 0 \\ 0 & 1 \end{bmatrix}$

Solving a Matrix Equation In Exercises 55 and 56, solve the matrix equation for $a, b, c,$ and d.

55. $\begin{bmatrix} 1 & 2 \\ 3 & 4 \end{bmatrix} \begin{bmatrix} a & b \\ c & d \end{bmatrix} = \begin{bmatrix} 6 & 3 \\ 19 & 2 \end{bmatrix}$

56. $\begin{bmatrix} a & b \\ c & d \end{bmatrix} \begin{bmatrix} 2 & 1 \\ 3 & 1 \end{bmatrix} = \begin{bmatrix} 3 & 17 \\ 4 & -1 \end{bmatrix}$

Diagonal Matrix In Exercises 57 and 58, find the product AA for the diagonal matrix. A square matrix

$$A = \begin{bmatrix} a_{11} & 0 & 0 & \cdots & 0 \\ 0 & a_{22} & 0 & \cdots & 0 \\ 0 & 0 & a_{33} & \cdots & 0 \\ \vdots & \vdots & \vdots & & \vdots \\ 0 & 0 & 0 & \cdots & a_{nn} \end{bmatrix}$$

is a diagonal matrix when all entries that are not on the main diagonal are zero.

57. $A = \begin{bmatrix} -1 & 0 & 0 \\ 0 & 2 & 0 \\ 0 & 0 & 3 \end{bmatrix}$ **58.** $A = \begin{bmatrix} 2 & 0 & 0 \\ 0 & -3 & 0 \\ 0 & 0 & 0 \end{bmatrix}$

Finding Products of Diagonal Matrices In Exercises 59 and 60, find the products AB and BA for the diagonal matrices.

59. $A = \begin{bmatrix} 2 & 0 \\ 0 & -3 \end{bmatrix}$, $B = \begin{bmatrix} -5 & 0 \\ 0 & 4 \end{bmatrix}$

60. $A = \begin{bmatrix} 3 & 0 & 0 \\ 0 & -5 & 0 \\ 0 & 0 & 0 \end{bmatrix}$, $B = \begin{bmatrix} -7 & 0 & 0 \\ 0 & 4 & 0 \\ 0 & 0 & 12 \end{bmatrix}$

61. Guided Proof Prove that if A and B are diagonal matrices (of the same size), then $AB = BA$.

Getting Started: To prove that the matrices AB and BA are equal, you need to show that their corresponding entries are equal.

   (i) Begin your proof by letting $A = [a_{ij}]$ and $B = [b_{ij}]$ be two diagonal $n \times n$ matrices.

   (ii) The ijth entry of the product AB is

   $$c_{ij} = \sum_{k=1}^{n} a_{ik}b_{kj}.$$

   (iii) Evaluate the entries c_{ij} for the two cases $i \neq j$ and $i = j$.

   (iv) Repeat this analysis for the product BA.

62. Writing Let A and B be 3×3 matrices, where A is diagonal.

   (a) Describe the product AB. Illustrate your answer with examples.

   (b) Describe the product BA. Illustrate your answer with examples.

   (c) How do the results in parts (a) and (b) change when the diagonal entries of A are all equal?

Trace of a Matrix In Exercises 63–66, find the trace of the matrix. The trace of an $n \times n$ matrix A is the sum of the main diagonal entries. That is, $\text{Tr}(A) = a_{11} + a_{22} + \cdots + a_{nn}$.

63. $\begin{bmatrix} 1 & 2 & 3 \\ 0 & -2 & 4 \\ 3 & 1 & 3 \end{bmatrix}$ **64.** $\begin{bmatrix} 1 & 0 & 0 \\ 0 & 1 & 0 \\ 0 & 0 & 1 \end{bmatrix}$

65. $\begin{bmatrix} 1 & 0 & 2 & 1 \\ 0 & 1 & -1 & 2 \\ 4 & 2 & 1 & 0 \\ 0 & 0 & 5 & 1 \end{bmatrix}$ **66.** $\begin{bmatrix} 1 & 4 & 3 & 2 \\ 4 & 0 & 6 & 1 \\ 3 & 6 & 2 & 1 \\ 2 & 1 & 1 & -3 \end{bmatrix}$

67. Proof Prove that each statement is true when A and B are square matrices of order n and c is a scalar.

   (a) $\text{Tr}(A + B) = \text{Tr}(A) + \text{Tr}(B)$

   (b) $\text{Tr}(cA) = c\text{Tr}(A)$

68. Proof Prove that if A and B are square matrices of order n, then $\text{Tr}(AB) = \text{Tr}(BA)$.

69. Find conditions on w, x, y, and z such that $AB = BA$ for the matrices below.

$$A = \begin{bmatrix} w & x \\ y & z \end{bmatrix} \quad \text{and} \quad B = \begin{bmatrix} 1 & 1 \\ -1 & 1 \end{bmatrix}$$

70. Verify $AB = BA$ for the matrices below.

$$A = \begin{bmatrix} \cos \alpha & -\sin \alpha \\ \sin \alpha & \cos \alpha \end{bmatrix} \quad \text{and} \quad B = \begin{bmatrix} \cos \beta & -\sin \beta \\ \sin \beta & \cos \beta \end{bmatrix}$$

71. Show that the matrix equation has no solution.

$$\begin{bmatrix} 1 & 1 \\ 1 & 1 \end{bmatrix} A = \begin{bmatrix} 1 & 0 \\ 0 & 1 \end{bmatrix}$$

72. Show that no 2×2 matrices A and B exist that satisfy the matrix equation

$$AB - BA = \begin{bmatrix} 1 & 0 \\ 0 & 1 \end{bmatrix}.$$

73. Exploration Let $i = \sqrt{-1}$ and let

$$A = \begin{bmatrix} i & 0 \\ 0 & i \end{bmatrix} \quad \text{and} \quad B = \begin{bmatrix} 0 & -i \\ i & 0 \end{bmatrix}.$$

   (a) Find A^2, A^3, and A^4. (*Note:* $A^2 = AA$, $A^3 = AAA = A^2A$, and so on.) Identify any similarities with i^2, i^3, and i^4.

   (b) Find and identify B^2.

74. Guided Proof Prove that if the product AB is a square matrix, then the product BA is defined.

Getting Started: To prove that the product BA is defined, you need to show that the number of columns of B equals the number of rows of A.

   (i) Begin your proof by noting that the number of columns of A equals the number of rows of B.

   (ii) Then assume that A has size $m \times n$ and B has size $n \times p$.

   (iii) Use the hypothesis that the product AB is a square matrix.

75. Proof Prove that if both products AB and BA are defined, then AB and BA are square matrices.

76. Let A and B be matrices such that the product AB is defined. Show that if A has two identical rows, then the corresponding two rows of AB are also identical.

77. Let A and B be $n \times n$ matrices. Show that if the ith row of A has all zero entries, then the ith row of AB will have all zero entries. Give an example using 2×2 matrices to show that the converse is not true.

78. CAPSTONE Let matrices A and B be of sizes 3×2 and 2×2, respectively. Answer each question and explain your answers.

   (a) Is it possible that $A = B$?

   (b) Is $A + B$ defined?

   (c) Is AB defined? If so, is it possible that $AB = BA$?

79. Agriculture A fruit grower raises two crops, apples and peaches. The grower ships each of these crops to three different outlets. In the matrix

$$A = \begin{bmatrix} 125 & 100 & 75 \\ 100 & 175 & 125 \end{bmatrix}$$

a_{ij} represents the number of units of crop i that the grower ships to outlet j. The matrix

$$B = [\$3.50 \quad \$6.00]$$

represents the profit per unit. Find the product BA and state what each entry of the matrix represents.

80. Manufacturing A corporation has three factories, each of which manufactures acoustic guitars and electric guitars. In the matrix

$$A = \begin{bmatrix} 70 & 50 & 25 \\ 35 & 100 & 70 \end{bmatrix}$$

a_{ij} represents the number of guitars of type i produced at factory j in one day. Find the production levels when production increases by 20%.

81. Politics In the matrix

From

R   D   I

$$P = \begin{bmatrix} 0.6 & 0.1 & 0.1 \\ 0.2 & 0.7 & 0.1 \\ 0.2 & 0.2 & 0.8 \end{bmatrix} \begin{matrix} R \\ D \\ I \end{matrix} \Big\} \text{ To}$$

each entry p_{ij} ($i \ne j$) represents the proportion of the voting population that changes from party j to party i, and p_{ii} represents the proportion that remains loyal to party i from one election to the next. Find and interpret the product of P with itself.

82. Population The matrices show the numbers of people (in thousands) who lived in each region of the United States in 2010 and 2013. The regional populations are separated into three age categories. (Source: U.S. Census Bureau)

2010

	0–17	18–64	65+
Northeast	12,306	35,240	7830
Midwest	16,095	41,830	9051
South	27,799	72,075	14,985
Mountain	5698	13,717	2710
Pacific	12,222	31,867	5901

2013

	0–17	18–64	65+
Northeast	12,026	35,471	8446
Midwest	15,772	41,985	9791
South	27,954	73,703	16,727
Mountain	5710	14,067	3104
Pacific	12,124	32,614	6636

(a) The total population in 2010 was approximately 309 million and the total population in 2013 was about 316 million. Rewrite the matrices to give the information as percents of the total population.

(b) Write a matrix that gives the changes in the percents of the population in each region and age group from 2010 to 2013.

(c) Based on the result of part (b), which age group(s) show relative growth from 2010 to 2013?

Block Multiplication In Exercises 83 and 84, perform the block multiplication of matrices A and B. If matrices A and B are each partitioned into four submatrices

$$A = \begin{bmatrix} A_{11} & A_{12} \\ A_{21} & A_{22} \end{bmatrix} \quad \text{and} \quad B = \begin{bmatrix} B_{11} & B_{12} \\ B_{21} & B_{22} \end{bmatrix}$$

then you can block multiply A and B, provided the sizes of the submatrices are such that the matrix multiplications and additions are defined.

$$AB = \begin{bmatrix} A_{11} & A_{12} \\ A_{21} & A_{22} \end{bmatrix} \begin{bmatrix} B_{11} & B_{12} \\ B_{21} & B_{22} \end{bmatrix}$$

$$= \begin{bmatrix} A_{11}B_{11} + A_{12}B_{21} & A_{11}B_{12} + A_{12}B_{22} \\ A_{21}B_{11} + A_{22}B_{21} & A_{21}B_{12} + A_{22}B_{22} \end{bmatrix}$$

83. $A = \begin{bmatrix} 1 & 2 & 0 & 0 \\ 0 & 1 & 0 & 0 \\ 0 & 0 & 2 & 1 \end{bmatrix}$, $B = \begin{bmatrix} 1 & 2 & 0 \\ -1 & 1 & 0 \\ 0 & 0 & 1 \\ 0 & 0 & 3 \end{bmatrix}$

84. $A = \begin{bmatrix} 0 & 0 & 1 & 0 \\ 0 & 0 & 0 & 1 \\ -1 & 0 & 0 & 0 \\ 0 & -1 & 0 & 0 \end{bmatrix}$, $B = \begin{bmatrix} 1 & 2 & 3 & 4 \\ 5 & 6 & 7 & 8 \\ 1 & 2 & 3 & 4 \\ 5 & 6 & 7 & 8 \end{bmatrix}$

True or False? In Exercises 85 and 86, determine whether each statement is true or false. If a statement is true, give a reason or cite an appropriate statement from the text. If a statement is false, provide an example that shows the statement is not true in all cases or cite an appropriate statement from the text.

85. (a) For the product of two matrices to be defined, the number of columns of the first matrix must equal the number of rows of the second matrix.

(b) The system $A\mathbf{x} = \mathbf{b}$ is consistent if and only if $\mathbf{b}$ can be expressed as a linear combination of the columns of A, where the coefficients of the linear combination are a solution of the system.

86. (a) If A is an $m \times n$ matrix and B is an $n \times r$ matrix, then the product AB is an $m \times r$ matrix.

(b) The matrix equation $A\mathbf{x} = \mathbf{b}$, where A is the coefficient matrix and $\mathbf{x}$ and $\mathbf{b}$ are column matrices, can be used to represent a system of linear equations.

87. The columns of matrix T show the coordinates of the vertices of a triangle. Matrix A is a transformation matrix.

$$A = \begin{bmatrix} 0 & -1 \\ 1 & 0 \end{bmatrix}, \quad T = \begin{bmatrix} 1 & 2 & 3 \\ 1 & 4 & 2 \end{bmatrix}$$

(a) Find AT and AAT. Then sketch the original triangle and the two transformed triangles. What transformation does A represent?

(b) A triangle is determined by AAT. Describe the transformation process that produces the triangle determined by AT and then the triangle determined by T.

2.2   Properties of Matrix Operations

■ Use the properties of matrix addition, scalar multiplication, and zero matrices.

■ Use the properties of matrix multiplication and the identity matrix.

■ Find the transpose of a matrix.

ALGEBRA OF MATRICES

In Section 2.1, you concentrated on the mechanics of the three basic matrix operations: matrix addition, scalar multiplication, and matrix multiplication. This section begins to develop the **algebra of matrices.** You will see that this algebra shares many (but not all) of the properties of the algebra of real numbers. Theorem 2.1 lists several properties of matrix addition and scalar multiplication.

**THEOREM 2.1   Properties of Matrix Addition
and Scalar Multiplication**

If A, B, and C are $m \times n$ matrices, and c and d are scalars, then the properties below are true.

1. $A + B = B + A$ Commutative property of addition
2. $A + (B + C) = (A + B) + C$ Associative property of addition
3. $(cd)A = c(dA)$ Associative property of multiplication
4. $1A = A$ Multiplicative identity
5. $c(A + B) = cA + cB$ Distributive property
6. $(c + d)A = cA + dA$ Distributive property

PROOF

The proofs of these six properties follow directly from the definitions of matrix addition, scalar multiplication, and the corresponding properties of real numbers. For example, to prove the commutative property of *matrix addition*, let $A = [a_{ij}]$ and $B = [b_{ij}]$. Then, using the commutative property of *addition of real numbers,* write

$$A + B = [a_{ij} + b_{ij}] = [b_{ij} + a_{ij}] = B + A.$$

Similarly, to prove Property 5, use the distributive property (for real numbers) of multiplication over addition to write

$$c(A + B) = [c(a_{ij} + b_{ij})] = [ca_{ij} + cb_{ij}] = cA + cB.$$

The proofs of the remaining four properties are left as exercises. (See Exercises 61–64.)

The preceding section defined matrix addition as the sum of *two* matrices, making it a binary operation. The associative property of matrix addition now allows you to write expressions such as $A + B + C$ as $(A + B) + C$ or as $A + (B + C)$. This same reasoning applies to sums of four or more matrices.

EXAMPLE 1   Addition of More than Two Matrices

To obtain the sum of four matrices, add corresponding entries as shown below.

$$\begin{bmatrix} 1 \\ 2 \end{bmatrix} + \begin{bmatrix} -1 \\ -1 \end{bmatrix} + \begin{bmatrix} 0 \\ 1 \end{bmatrix} + \begin{bmatrix} 2 \\ -3 \end{bmatrix} = \begin{bmatrix} 2 \\ -1 \end{bmatrix}$$

One important property of the addition of real numbers is that the number 0 is the additive identity. That is, $c + 0 = c$ for any real number c. For matrices, a similar property holds. Specifically, if A is an $m \times n$ matrix and O_{mn} is the $m \times n$ matrix consisting entirely of zeros, then $A + O_{mn} = A$. The matrix O_{mn} is a **zero matrix**, and it is the **additive identity** for the set of all $m \times n$ matrices. For example, the matrix below is the additive identity for the set of all 2×3 matrices.

$$O_{23} = \begin{bmatrix} 0 & 0 & 0 \\ 0 & 0 & 0 \end{bmatrix}$$

When the size of the matrix is understood, you may denote a zero matrix simply by O or $\mathbf{0}$.

The properties of zero matrices listed below are relatively easy to prove, and their proofs are left as an exercise. (See Exercise 65.)

REMARK

Property 2 can be described by saying that matrix $-A$ is the **additive inverse** of A.

THEOREM 2.2   Properties of Zero Matrices

If A is an $m \times n$ matrix and c is a scalar, then the properties below are true.
1. $A + O_{mn} = A$
2. $A + (-A) = O_{mn}$
3. If $cA = O_{mn}$, then $c = 0$ or $A = O_{mn}$.

The algebra of real numbers and the algebra of matrices have many similarities. For example, compare the two solutions below.

Real Numbers (Solve for x.)	$m \times n$ Matrices (Solve for X.)
$x + a = b$	$X + A = B$
$x + a + (-a) = b + (-a)$	$X + A + (-A) = B + (-A)$
$x + 0 = b - a$	$X + O = B - A$
$x = b - a$	$X = B - A$

Example 2 demonstrates the process of solving a matrix equation.

EXAMPLE 2   Solving a Matrix Equation

Solve for X in the equation $3X + A = B$, where

$$A = \begin{bmatrix} 1 & -2 \\ 0 & 3 \end{bmatrix} \quad \text{and} \quad B = \begin{bmatrix} -3 & 4 \\ 2 & 1 \end{bmatrix}.$$

SOLUTION

Begin by solving the equation for X to obtain

$$3X = B - A$$
$$X = \tfrac{1}{3}(B - A).$$

Now, using the matrices A and B, you have

$$X = \tfrac{1}{3}\left(\begin{bmatrix} -3 & 4 \\ 2 & 1 \end{bmatrix} - \begin{bmatrix} 1 & -2 \\ 0 & 3 \end{bmatrix} \right)$$
$$= \tfrac{1}{3}\begin{bmatrix} -4 & 6 \\ 2 & -2 \end{bmatrix}$$
$$= \begin{bmatrix} -\tfrac{4}{3} & 2 \\ \tfrac{2}{3} & -\tfrac{2}{3} \end{bmatrix}.$$

PROPERTIES OF MATRIX MULTIPLICATION

The next theorem extends the algebra of matrices to include some useful properties of matrix multiplication. The proof of Property 2 is below. The proofs of the remaining properties are left as an exercise. (See Exercise 66.)

REMARK

Note that no commutative property of matrix multiplication is listed in Theorem 2.3. The product AB may not be equal to the product BA, as illustrated in Example 4 on the next page.

THEOREM 2.3   Properties of Matrix Multiplication

If A, B, and C are matrices (with sizes such that the matrix products are defined), and c is a scalar, then the properties below are true.
1. $A(BC) = (AB)C$ Associative property of multiplication
2. $A(B + C) = AB + AC$ Distributive property
3. $(A + B)C = AC + BC$ Distributive property
4. $c(AB) = (cA)B = A(cB)$

PROOF

To prove Property 2, show that the corresponding entries of matrices $A(B + C)$ and $AB + AC$ are equal. Assume A has size $m \times n$, B has size $n \times p$, and C has size $n \times p$. Using the definition of matrix multiplication, the entry in the ith row and jth column of $A(B + C)$ is $a_{i1}(b_{1j} + c_{1j}) + a_{i2}(b_{2j} + c_{2j}) + \cdots + a_{in}(b_{nj} + c_{nj})$. Moreover, the entry in the ith row and jth column of $AB + AC$ is

$$(a_{i1}b_{1j} + a_{i2}b_{2j} + \cdots + a_{in}b_{nj}) + (a_{i1}c_{1j} + a_{i2}c_{2j} + \cdots + a_{in}c_{nj}).$$

By distributing and regrouping, you can see that these two ijth entries are equal. So,

$$A(B + C) = AB + AC.$$

The associative property of matrix multiplication permits you to write such matrix products as ABC without ambiguity, as demonstrated in Example 3.

EXAMPLE 3   Matrix Multiplication Is Associative

Find the matrix product ABC by grouping the factors first as $(AB)C$ and then as $A(BC)$. Show that you obtain the same result from both processes.

$$A = \begin{bmatrix} 1 & -2 \\ 2 & -1 \end{bmatrix}, \quad B = \begin{bmatrix} 1 & 0 & 2 \\ 3 & -2 & 1 \end{bmatrix}, \quad C = \begin{bmatrix} -1 & 0 \\ 3 & 1 \\ 2 & 4 \end{bmatrix}$$

SOLUTION

Grouping the factors as $(AB)C$, you have

$$(AB)C = \left(\begin{bmatrix} 1 & -2 \\ 2 & -1 \end{bmatrix} \begin{bmatrix} 1 & 0 & 2 \\ 3 & -2 & 1 \end{bmatrix} \right) \begin{bmatrix} -1 & 0 \\ 3 & 1 \\ 2 & 4 \end{bmatrix}$$

$$= \begin{bmatrix} -5 & 4 & 0 \\ -1 & 2 & 3 \end{bmatrix} \begin{bmatrix} -1 & 0 \\ 3 & 1 \\ 2 & 4 \end{bmatrix} = \begin{bmatrix} 17 & 4 \\ 13 & 14 \end{bmatrix}.$$

Grouping the factors as $A(BC)$, you obtain the same result.

$$A(BC) = \begin{bmatrix} 1 & -2 \\ 2 & -1 \end{bmatrix} \left(\begin{bmatrix} 1 & 0 & 2 \\ 3 & -2 & 1 \end{bmatrix} \begin{bmatrix} -1 & 0 \\ 3 & 1 \\ 2 & 4 \end{bmatrix} \right)$$

$$= \begin{bmatrix} 1 & -2 \\ 2 & -1 \end{bmatrix} \begin{bmatrix} 3 & 8 \\ -7 & 2 \end{bmatrix} = \begin{bmatrix} 17 & 4 \\ 13 & 14 \end{bmatrix}$$

The next example shows that even when both products AB and BA are defined, they may not be equal.

EXAMPLE 4   **Noncommutativity of Matrix Multiplication**

Show that AB and BA are not equal for the matrices

$$A = \begin{bmatrix} 1 & 3 \\ 2 & -1 \end{bmatrix} \quad \text{and} \quad B = \begin{bmatrix} 2 & -1 \\ 0 & 2 \end{bmatrix}.$$

SOLUTION

$$AB = \begin{bmatrix} 1 & 3 \\ 2 & -1 \end{bmatrix}\begin{bmatrix} 2 & -1 \\ 0 & 2 \end{bmatrix} = \begin{bmatrix} 2 & 5 \\ 4 & -4 \end{bmatrix}, \quad BA = \begin{bmatrix} 2 & -1 \\ 0 & 2 \end{bmatrix}\begin{bmatrix} 1 & 3 \\ 2 & -1 \end{bmatrix} = \begin{bmatrix} 0 & 7 \\ 4 & -2 \end{bmatrix}$$

$AB \neq BA$

Do not conclude from Example 4 that the matrix products AB and BA are *never* equal. Sometimes they are equal. For example, find AB and BA for the matrices below.

$$A = \begin{bmatrix} 1 & 2 \\ 1 & 1 \end{bmatrix} \quad \text{and} \quad B = \begin{bmatrix} -2 & 4 \\ 2 & -2 \end{bmatrix}$$

You will see that the two products are equal. The point is that although AB and BA are sometimes equal, AB and BA are usually not equal.

Another important quality of matrix algebra is that it does not have a general cancellation property for matrix multiplication. That is, when $AC = BC$, it is not necessarily true that $A = B$. Example 5 demonstrates this. (In the next section you will see that, for some special types of matrices, cancellation is valid.)

EXAMPLE 5   **An Example in Which Cancellation Is Not Valid**

Show that $AC = BC$.

$$A = \begin{bmatrix} 1 & 3 \\ 0 & 1 \end{bmatrix}, \quad B = \begin{bmatrix} 2 & 4 \\ 2 & 3 \end{bmatrix}, \quad C = \begin{bmatrix} 1 & -2 \\ -1 & 2 \end{bmatrix}$$

SOLUTION

$$AC = \begin{bmatrix} 1 & 3 \\ 0 & 1 \end{bmatrix}\begin{bmatrix} 1 & -2 \\ -1 & 2 \end{bmatrix} = \begin{bmatrix} -2 & 4 \\ -1 & 2 \end{bmatrix}, \quad BC = \begin{bmatrix} 2 & 4 \\ 2 & 3 \end{bmatrix}\begin{bmatrix} 1 & -2 \\ -1 & 2 \end{bmatrix} = \begin{bmatrix} -2 & 4 \\ -1 & 2 \end{bmatrix}$$

$AC = BC$, even though $A \neq B$.

You will now look at a special type of *square* matrix that has 1's on the main diagonal and 0's elsewhere.

$$I_n = \begin{bmatrix} 1 & 0 & \cdots & 0 \\ 0 & 1 & \cdots & 0 \\ \vdots & \vdots & & \vdots \\ 0 & 0 & \cdots & 1 \end{bmatrix}$$
$$n \times n$$

For instance, for $n = 1$, 2, and 3,

$$I_1 = [1], \quad I_2 = \begin{bmatrix} 1 & 0 \\ 0 & 1 \end{bmatrix}, \quad I_3 = \begin{bmatrix} 1 & 0 & 0 \\ 0 & 1 & 0 \\ 0 & 0 & 1 \end{bmatrix}.$$

When the order of the matrix is understood to be n, you may denote I_n simply as I.

As stated in Theorem 2.4 on the next page, the matrix I_n serves as the **identity** for matrix multiplication; it is the **identity matrix of order n.** The proof of this theorem is left as an exercise. (See Exercise 67.)

REMARK

Note that if A is a *square* matrix of order n, then $AI_n = I_nA = A$.

THEOREM 2.4   Properties of the Identity Matrix

If A is a matrix of size $m \times n$, then the properties below are true.
1. $AI_n = A$ 2. $I_mA = A$

EXAMPLE 6   **Multiplication by an Identity Matrix**

a. $\begin{bmatrix} 3 & -2 \\ 4 & 0 \\ -1 & 1 \end{bmatrix} \begin{bmatrix} 1 & 0 \\ 0 & 1 \end{bmatrix} = \begin{bmatrix} 3 & -2 \\ 4 & 0 \\ -1 & 1 \end{bmatrix}$ b. $\begin{bmatrix} 1 & 0 & 0 \\ 0 & 1 & 0 \\ 0 & 0 & 1 \end{bmatrix} \begin{bmatrix} -2 \\ 1 \\ 4 \end{bmatrix} = \begin{bmatrix} -2 \\ 1 \\ 4 \end{bmatrix}$

For repeated multiplication of *square* matrices, use the same exponential notation used with real numbers. That is, $A^1 = A$, $A^2 = AA$, and for a positive integer k, A^k is

$$A^k = \underbrace{AA \cdots A}_{k \text{ factors}}.$$

It is convenient also to define $A^0 = I_n$ (where A is a square matrix of order n). These definitions allow you to establish the properties (1) $A^jA^k = A^{j+k}$ and (2) $(A^j)^k = A^{jk}$, where j and k are nonnegative integers.

EXAMPLE 7   **Repeated Multiplication of a Square Matrix**

For the matrix $A = \begin{bmatrix} 2 & -1 \\ 3 & 0 \end{bmatrix}$,

$$A^3 = \left(\begin{bmatrix} 2 & -1 \\ 3 & 0 \end{bmatrix}\begin{bmatrix} 2 & -1 \\ 3 & 0 \end{bmatrix}\right)\begin{bmatrix} 2 & -1 \\ 3 & 0 \end{bmatrix} = \begin{bmatrix} 1 & -2 \\ 6 & -3 \end{bmatrix}\begin{bmatrix} 2 & -1 \\ 3 & 0 \end{bmatrix} = \begin{bmatrix} -4 & -1 \\ 3 & -6 \end{bmatrix}.$$

In Section 1.1, you saw that a system of linear equations has exactly one solution, infinitely many solutions, or no solution. You can use matrix algebra to prove this.

THEOREM 2.5   Number of Solutions of a Linear System

For a system of linear equations, precisely one of the statements below is true.
1. The system has exactly one solution.
2. The system has infinitely many solutions.
3. The system has no solution.

PROOF

Represent the system by the matrix equation $A\mathbf{x} = \mathbf{b}$. If the system has exactly one solution or no solution, then there is nothing to prove. So, assume that the system has at least two distinct solutions $\mathbf{x}_1$ and $\mathbf{x}_2$. If you show that this assumption implies that the system has infinitely many solutions, then the proof will be complete. When $\mathbf{x}_1$ and $\mathbf{x}_2$ are solutions, you have $A\mathbf{x}_1 = A\mathbf{x}_2 = \mathbf{b}$ and $A(\mathbf{x}_1 - \mathbf{x}_2) = O$. This implies that the (nonzero) column matrix $\mathbf{x}_h = \mathbf{x}_1 - \mathbf{x}_2$ is a solution of the homogeneous system of linear equations $A\mathbf{x} = O$. So, for any scalar c,

$$A(\mathbf{x}_1 + c\mathbf{x}_h) = A\mathbf{x}_1 + A(c\mathbf{x}_h) = \mathbf{b} + c(A\mathbf{x}_h) = \mathbf{b} + cO = \mathbf{b}.$$

Then $\mathbf{x}_1 + c\mathbf{x}_h$ is a solution of $A\mathbf{x} = \mathbf{b}$ for any scalar c. There are infinitely many possible values of c and each value produces a different solution, so the system has infinitely many solutions.

THE TRANSPOSE OF A MATRIX

The **transpose** of a matrix is formed by writing its rows as columns. For example, if A is the $m \times n$ matrix

$$A = \begin{bmatrix} a_{11} & a_{12} & a_{13} & \cdots & a_{1n} \\ a_{21} & a_{22} & a_{23} & \cdots & a_{2n} \\ a_{31} & a_{32} & a_{33} & \cdots & a_{3n} \\ \vdots & \vdots & \vdots & & \vdots \\ a_{m1} & a_{m2} & a_{m3} & \cdots & a_{mn} \end{bmatrix}$$

Size: $m \times n$

then the transpose, denoted by A^T, is the $n \times m$ matrix

$$A^T = \begin{bmatrix} a_{11} & a_{21} & a_{31} & \cdots & a_{m1} \\ a_{12} & a_{22} & a_{32} & \cdots & a_{m2} \\ a_{13} & a_{23} & a_{33} & \cdots & a_{m3} \\ \vdots & \vdots & \vdots & & \vdots \\ a_{1n} & a_{2n} & a_{3n} & \cdots & a_{mn} \end{bmatrix}.$$

Size: $n \times m$

EXAMPLE 8 Transposes of Matrices

Find the transpose of each matrix.

a. $A = \begin{bmatrix} 2 \\ 8 \end{bmatrix}$ **b.** $B = \begin{bmatrix} 1 & 2 & 3 \\ 4 & 5 & 6 \\ 7 & 8 & 9 \end{bmatrix}$ **c.** $C = \begin{bmatrix} 1 & 2 & 0 \\ 2 & 1 & 0 \\ 0 & 0 & 1 \end{bmatrix}$ **d.** $D = \begin{bmatrix} 0 & 1 \\ 2 & 4 \\ 1 & -1 \end{bmatrix}$

SOLUTION

a. $A^T = \begin{bmatrix} 2 & 8 \end{bmatrix}$

b. $B^T = \begin{bmatrix} 1 & 4 & 7 \\ 2 & 5 & 8 \\ 3 & 6 & 9 \end{bmatrix}$

c. $C^T = \begin{bmatrix} 1 & 2 & 0 \\ 2 & 1 & 0 \\ 0 & 0 & 1 \end{bmatrix}$

d. $D^T = \begin{bmatrix} 0 & 2 & 1 \\ 1 & 4 & -1 \end{bmatrix}$

THEOREM 2.6 Properties of Transposes

If A and B are matrices (with sizes such that the matrix operations are defined) and c is a scalar, then the properties below are true.

1. $(A^T)^T = A$ Transpose of a transpose
2. $(A + B)^T = A^T + B^T$ Transpose of a sum
3. $(cA)^T = c(A^T)$ Transpose of a scalar multiple
4. $(AB)^T = B^T A^T$ Transpose of a product

PROOF

The transpose operation interchanges rows and columns, so Property 1 seems to make sense. To prove Property 1, let A be an $m \times n$ matrix. Observe that A^T has size $n \times m$ and $(A^T)^T$ has size $m \times n$, the same as A. To show that $(A^T)^T = A$, you must show that the ijth entries are the same. Let a_{ij} be the ijth entry of A. Then a_{ij} is the jith entry of A^T, and the ijth entry of $(A^T)^T$. This proves Property 1. The proofs of the remaining properties are left as an exercise. (See Exercise 68.)

DISCOVERY

Let $A = \begin{bmatrix} 1 & 2 \\ 3 & 4 \end{bmatrix}$ and $B = \begin{bmatrix} 3 & 5 \\ 1 & -1 \end{bmatrix}$.

Find $(AB)^T$, $A^T B^T$, and $B^T A^T$.

Make a conjecture about the transpose of a product of two square matrices.

Select two other square matrices to check your conjecture.

REMARK

Note that the square matrix in part (c) is equal to its transpose. Such a matrix is **symmetric**. A matrix A is symmetric when $A = A^T$. From this definition it should be clear that a symmetric matrix must be square. Also, if $A = [a_{ij}]$ is a symmetric matrix, then $a_{ij} = a_{ji}$ for all $i \neq j$.

REMARK

Remember that you *reverse the order* of multiplication when forming the transpose of a product. That is, the transpose of AB is $(AB)^T = B^T A^T$ and is usually *not* equal to $A^T B^T$.

Properties 2 and 4 can be generalized to cover sums or products of any finite number of matrices. For instance, the transpose of the sum of three matrices is $(A + B + C)^T = A^T + B^T + C^T$ and the transpose of the product of three matrices is $(ABC)^T = C^T B^T A^T$.

EXAMPLE 9   **Finding the Transpose of a Product**

See LarsonLinearAlgebra.com for an interactive version of this type of example.

Show that $(AB)^T$ and $B^T A^T$ are equal.

$$A = \begin{bmatrix} 2 & 1 & -2 \\ -1 & 0 & 3 \\ 0 & -2 & 1 \end{bmatrix} \quad \text{and} \quad B = \begin{bmatrix} 3 & 1 \\ 2 & -1 \\ 3 & 0 \end{bmatrix}$$

SOLUTION

$$AB = \begin{bmatrix} 2 & 1 & -2 \\ -1 & 0 & 3 \\ 0 & -2 & 1 \end{bmatrix} \begin{bmatrix} 3 & 1 \\ 2 & -1 \\ 3 & 0 \end{bmatrix} = \begin{bmatrix} 2 & 1 \\ 6 & -1 \\ -1 & 2 \end{bmatrix}$$

$$(AB)^T = \begin{bmatrix} 2 & 6 & -1 \\ 1 & -1 & 2 \end{bmatrix}$$

$$B^T A^T = \begin{bmatrix} 3 & 2 & 3 \\ 1 & -1 & 0 \end{bmatrix} \begin{bmatrix} 2 & -1 & 0 \\ 1 & 0 & -2 \\ -2 & 3 & 1 \end{bmatrix} = \begin{bmatrix} 2 & 6 & -1 \\ 1 & -1 & 2 \end{bmatrix}$$

$$(AB)^T = B^T A^T$$

EXAMPLE 10   **The Product of a Matrix and Its Transpose**

For the matrix $A = \begin{bmatrix} 1 & 3 \\ 0 & -2 \\ -2 & -1 \end{bmatrix}$, find the product AA^T and show that it is symmetric.

REMARK

The property demonstrated in Example 10 is true in general. That is, for any matrix A, the matrix AA^T is symmetric. The matrix $A^T A$ is also symmetric. You are asked to prove these properties in Exercise 69.

SOLUTION

$$AA^T = \begin{bmatrix} 1 & 3 \\ 0 & -2 \\ -2 & -1 \end{bmatrix} \begin{bmatrix} 1 & 0 & -2 \\ 3 & -2 & -1 \end{bmatrix} = \begin{bmatrix} 10 & -6 & -5 \\ -6 & 4 & 2 \\ -5 & 2 & 5 \end{bmatrix}$$

It follows that $AA^T = (AA^T)^T$, so AA^T is symmetric.

LINEAR ALGEBRA APPLIED

Information retrieval systems such as Internet search engines make use of matrix theory and linear algebra to keep track of information. To illustrate, consider a simplified example. You could represent the occurrences of m available keywords in a database of n documents with A, an $m \times n$ matrix in which an entry is 1 when the keyword occurs in the document and 0 when it does not occur in the document. You could represent a search with the $m \times 1$ column matrix **x**, in which a 1 entry represents a keyword you are searching and 0 represents a keyword you are not searching. Then, the $n \times 1$ matrix product $A^T \mathbf{x}$ would represent the number of keywords in your search that occur in each of the n documents. For a discussion on the PageRank algorithm that is used in Google's search engine, see Section 2.5 (page 86).

2.2 Exercises

See CalcChat.com for worked-out solutions to odd-numbered exercises.

Evaluating an Expression In Exercises 1–6, evaluate the expression.

1. $\begin{bmatrix} -5 & 0 \\ 3 & -6 \end{bmatrix} + \begin{bmatrix} 7 & 1 \\ -2 & -1 \end{bmatrix} + \begin{bmatrix} -10 & -8 \\ 14 & 6 \end{bmatrix}$

2. $\begin{bmatrix} 6 & 8 \\ -1 & 0 \end{bmatrix} + \begin{bmatrix} 0 & 5 \\ -3 & -1 \end{bmatrix} + \begin{bmatrix} -11 & -7 \\ 2 & -1 \end{bmatrix}$

3. $4\left(\begin{bmatrix} -4 & 0 & 1 \\ 0 & 2 & 3 \end{bmatrix} - \begin{bmatrix} 2 & 1 & -2 \\ 3 & -6 & 0 \end{bmatrix} \right)$

4. $\frac{1}{2}([5 \quad -2 \quad 4 \quad 0] + [14 \quad 6 \quad -18 \quad 9])$

5. $-3\left(\begin{bmatrix} 0 & -3 \\ 7 & 2 \end{bmatrix} + \begin{bmatrix} -6 & 3 \\ 8 & 1 \end{bmatrix} \right) - 2\begin{bmatrix} 4 & -4 \\ 7 & -9 \end{bmatrix}$

6. $-\begin{bmatrix} 4 & 11 \\ -2 & -1 \\ 9 & 3 \end{bmatrix} + \frac{1}{6}\left(\begin{bmatrix} -5 & -1 \\ 3 & 4 \\ 0 & 13 \end{bmatrix} + \begin{bmatrix} 7 & 5 \\ -9 & -1 \\ 6 & -1 \end{bmatrix} \right)$

Operations with Matrices In Exercises 7–12, perform the operations, given $a = 3$, $b = -4$, and

$$A = \begin{bmatrix} 1 & 2 \\ 3 & 4 \end{bmatrix}, \quad B = \begin{bmatrix} 0 & 1 \\ -1 & 2 \end{bmatrix}, \quad O = \begin{bmatrix} 0 & 0 \\ 0 & 0 \end{bmatrix}.$$

7. $aA + bB$
8. $A + B$
9. $ab(B)$
10. $(a + b)B$
11. $(a - b)(A - B)$
12. $(ab)O$

13. Solve for X in the equation, given

$$A = \begin{bmatrix} -4 & 0 \\ 1 & -5 \\ -3 & 2 \end{bmatrix} \quad \text{and} \quad B = \begin{bmatrix} 1 & 2 \\ -2 & 1 \\ 4 & 4 \end{bmatrix}.$$

(a) $3X + 2A = B$ (b) $2A - 5B = 3X$
(c) $X - 3A + 2B = O$ (d) $6X - 4A - 3B = O$

14. Solve for X in the equation, given

$$A = \begin{bmatrix} -2 & -1 \\ 1 & 0 \\ 3 & -4 \end{bmatrix} \quad \text{and} \quad B = \begin{bmatrix} 0 & 3 \\ 2 & 0 \\ -4 & -1 \end{bmatrix}.$$

(a) $X = 3A - 2B$ (b) $2X = 2A - B$
(c) $2X + 3A = B$ (d) $2A + 4B = -2X$

Operations with Matrices In Exercises 15–22, perform the operations, given $c = -2$ and

$$A = \begin{bmatrix} 1 & 2 & 3 \\ 0 & 1 & -1 \end{bmatrix}, B = \begin{bmatrix} 1 & 3 \\ -1 & 2 \end{bmatrix}, C = \begin{bmatrix} 0 & 1 \\ -1 & 0 \end{bmatrix}.$$

15. $c(BA)$
16. $c(CB)$
17. $B(CA)$
18. $C(BC)$
19. $(B + C)A$
20. $B(C + O)$
21. $cB(C + C)$
22. $B(cA)$

Associativity of Matrix Multiplication In Exercises 23 and 24, find the matrix product ABC by (a) grouping the factors as $(AB)C$, and (b) grouping the factors as $A(BC)$. Show that you obtain the same result from both processes.

23. $A = \begin{bmatrix} 1 & 2 \\ 3 & 4 \end{bmatrix}$, $B = \begin{bmatrix} 0 & 1 \\ 2 & 3 \end{bmatrix}$, $C = \begin{bmatrix} 3 & 0 \\ 0 & 1 \end{bmatrix}$

24. $A = \begin{bmatrix} -4 & 2 \\ 1 & -3 \end{bmatrix}$, $B = \begin{bmatrix} 1 & -5 & 0 \\ -2 & 3 & 3 \end{bmatrix}$,

$C = \begin{bmatrix} -3 & 4 \\ 0 & 1 \\ -1 & 1 \end{bmatrix}$

Noncommutativity of Matrix Multiplication In Exercises 25 and 26, show that AB and BA are not equal for the given matrices.

25. $A = \begin{bmatrix} -2 & 1 \\ 0 & 3 \end{bmatrix}$, $B = \begin{bmatrix} 4 & 0 \\ -1 & 2 \end{bmatrix}$

26. $A = \begin{bmatrix} \frac{1}{4} & \frac{1}{2} \\ \frac{1}{2} & \frac{1}{2} \end{bmatrix}$, $B = \begin{bmatrix} \frac{1}{2} & \frac{1}{2} \\ \frac{1}{2} & \frac{1}{4} \end{bmatrix}$

Equal Matrix Products In Exercises 27 and 28, show that $AC = BC$, even though $A \neq B$.

27. $A = \begin{bmatrix} 0 & 1 \\ 0 & 1 \end{bmatrix}$, $B = \begin{bmatrix} 1 & 0 \\ 1 & 0 \end{bmatrix}$, $C = \begin{bmatrix} 2 & 3 \\ 2 & 3 \end{bmatrix}$

28. $A = \begin{bmatrix} 1 & 2 & 3 \\ 0 & 5 & 4 \\ 3 & -2 & 1 \end{bmatrix}$, $B = \begin{bmatrix} 4 & -6 & 3 \\ 5 & 4 & 4 \\ -1 & 0 & 1 \end{bmatrix}$,

$C = \begin{bmatrix} 0 & 0 & 0 \\ 0 & 0 & 0 \\ 4 & -2 & 3 \end{bmatrix}$

Zero Matrix Product In Exercises 29 and 30, show that $AB = O$, even though $A \neq O$ and $B \neq O$.

29. $A = \begin{bmatrix} 3 & 3 \\ 4 & 4 \end{bmatrix}$ and $B = \begin{bmatrix} 1 & -1 \\ -1 & 1 \end{bmatrix}$

30. $A = \begin{bmatrix} 2 & 4 \\ 2 & 4 \end{bmatrix}$ and $B = \begin{bmatrix} 1 & -2 \\ -\frac{1}{2} & 1 \end{bmatrix}$

Operations with Matrices In Exercises 31–36, perform the operations when

$$A = \begin{bmatrix} 1 & 2 \\ 0 & -1 \end{bmatrix}.$$

31. IA
32. AI
33. $A(I + A)$
34. $A + IA$
35. A^2
36. A^4

Writing In Exercises 37 and 38, explain why the formula is *not* valid for matrices. Illustrate your argument with examples.

37. $(A + B)(A - B) = A^2 - B^2$

38. $(A + B)(A + B) = A^2 + 2AB + B^2$

Finding the Transpose of a Matrix In Exercises 39 and 40, find the transpose of the matrix.

39. $D = \begin{bmatrix} 1 & -2 \\ -3 & 4 \\ 5 & -1 \end{bmatrix}$ 40. $D = \begin{bmatrix} 6 & -7 & 19 \\ -7 & 0 & 23 \\ 19 & 23 & -32 \end{bmatrix}$

Finding the Transpose of a Product of Two Matrices In Exercises 41–44, verify that $(AB)^T = B^T A^T$.

41. $A = \begin{bmatrix} -1 & 1 & -2 \\ 2 & 0 & 1 \end{bmatrix}$ and $B = \begin{bmatrix} -3 & 0 \\ 1 & 2 \\ 1 & -1 \end{bmatrix}$

42. $A = \begin{bmatrix} 1 & 2 \\ 0 & -2 \end{bmatrix}$ and $B = \begin{bmatrix} -3 & -1 \\ 2 & 1 \end{bmatrix}$

43. $A = \begin{bmatrix} 2 & 1 \\ 0 & 1 \\ -2 & 1 \end{bmatrix}$ and $B = \begin{bmatrix} 2 & 3 & 1 \\ 0 & 4 & -1 \end{bmatrix}$

44. $A = \begin{bmatrix} 2 & 1 & -1 \\ 0 & 1 & 3 \\ 4 & 0 & 2 \end{bmatrix}$ and $B = \begin{bmatrix} 1 & 0 & -1 \\ 2 & 1 & -2 \\ 0 & 1 & 3 \end{bmatrix}$

Multiplication with the Transpose of a Matrix In Exercises 45–48, find (a) $A^T A$ and (b) AA^T. Show that each of these products is symmetric.

45. $A = \begin{bmatrix} 4 & 2 & 1 \\ 0 & 2 & -1 \end{bmatrix}$ 46. $A = \begin{bmatrix} 1 & -1 \\ 3 & 4 \\ 0 & -2 \end{bmatrix}$

47. $A = \begin{bmatrix} 0 & -4 & 3 & 2 \\ 8 & 4 & 0 & 1 \\ -2 & 3 & 5 & 1 \\ 0 & 0 & -3 & 2 \end{bmatrix}$

48. $A = \begin{bmatrix} 4 & -3 & 2 & 0 \\ 2 & 0 & 11 & -1 \\ -1 & -2 & 0 & 3 \\ 14 & -2 & 12 & -9 \\ 6 & 8 & -5 & 4 \end{bmatrix}$

Finding a Power of a Matrix In Exercises 49–52, find the power of A for the matrix

$$A = \begin{bmatrix} 1 & 0 & 0 & 0 & 0 \\ 0 & -1 & 0 & 0 & 0 \\ 0 & 0 & 1 & 0 & 0 \\ 0 & 0 & 0 & -1 & 0 \\ 0 & 0 & 0 & 0 & 1 \end{bmatrix}.$$

49. A^{16} 50. A^{17}

51. A^{19} 52. A^{20}

Finding an *n*th Root of a Matrix In Exercises 53 and 54, find the *n*th root of the matrix B. An *n*th root of a matrix B is a matrix A such that $A^n = B$.

53. $B = \begin{bmatrix} 9 & 0 \\ 0 & 4 \end{bmatrix}$, $n = 2$

54. $B = \begin{bmatrix} 8 & 0 & 0 \\ 0 & -1 & 0 \\ 0 & 0 & 27 \end{bmatrix}$, $n = 3$

True or False? In Exercises 55 and 56, determine whether each statement is true or false. If a statement is true, give a reason or cite an appropriate statement from the text. If a statement is false, provide an example that shows the statement is not true in all cases or cite an appropriate statement from the text.

55. (a) Matrix addition is commutative.

(b) The transpose of the product of two matrices equals the product of their transposes; that is, $(AB)^T = A^T B^T$.

(c) For any matrix C the matrix CC^T is symmetric.

56. (a) Matrix multiplication is commutative.

(b) If the matrices A, B, and C satisfy $AB = AC$, then $B = C$.

(c) The transpose of the sum of two matrices equals the sum of their transposes.

57. Consider the matrices below.

$$X = \begin{bmatrix} 1 \\ 0 \\ 1 \end{bmatrix}, \quad Y = \begin{bmatrix} 1 \\ 1 \\ 0 \end{bmatrix}, \quad Z = \begin{bmatrix} 2 \\ -1 \\ 3 \end{bmatrix}, \quad W = \begin{bmatrix} 1 \\ 1 \\ 1 \end{bmatrix}$$

(a) Find scalars a and b such that $Z = aX + bY$.

(b) Show that there do not exist scalars a and b such that $W = aX + bY$.

(c) Show that if $aX + bY + cW = O$, then $a = 0$, $b = 0$, and $c = 0$.

(d) Find scalars a, b, and c, not all equal to zero, such that $aX + bY + cZ = O$.

58. **CAPSTONE** In the matrix equation

$$aX + A(bB) = b(AB + IB)$$

X, A, B, and I are square matrices, and a and b are nonzero scalars. Justify each step in the solution below.

$$aX + (Ab)B = b(AB + B)$$
$$aX + bAB = bAB + bB$$
$$aX + bAB + (-bAB) = bAB + bB + (-bAB)$$
$$aX = bAB + bB + (-bAB)$$
$$aX = bAB + (-bAB) + bB$$
$$aX = bB$$
$$X = \frac{b}{a}B$$

Polynomial Function In Exercises 59 and 60, find $f(A)$ using the definition below.

If $f(x) = a_0 + a_1x + a_2x^2 + \cdots + a_nx^n$ is a polynomial function, then for a square matrix A,

$$f(A) = a_0I + a_1A + a_2A^2 + \cdots + a_nA^n.$$

59. $f(x) = 2 - 5x + x^2, \quad A = \begin{bmatrix} 2 & 0 \\ 4 & 5 \end{bmatrix}$

60. $f(x) = -10 + 5x - 2x^2 + x^3, \quad A = \begin{bmatrix} 2 & 1 & -1 \\ 1 & 0 & 2 \\ -1 & 1 & 3 \end{bmatrix}$

61. Guided Proof Prove the associative property of matrix addition: $A + (B + C) = (A + B) + C$.

Getting Started: To prove that $A + (B + C)$ and $(A + B) + C$ are equal, show that their corresponding entries are equal.

(i) Begin your proof by letting A, B, and C be $m \times n$ matrices.

(ii) Observe that the ijth entry of $B + C$ is $b_{ij} + c_{ij}$.

(iii) Furthermore, the ijth entry of $A + (B + C)$ is $a_{ij} + (b_{ij} + c_{ij})$.

(iv) Determine the ijth entry of $(A + B) + C$.

62. Proof Prove the associative property of multiplication: $(cd)A + c(dA)$.

63. Proof Prove that the scalar 1 is the identity for scalar multiplication: $1A = A$.

64. Proof Prove the distributive property:

$(c + d)A = cA + dA$.

65. Proof Prove Theorem 2.2.

66. Proof Complete the proof of Theorem 2.3.

(a) Prove the associative property of multiplication:

$A(BC) = (AB)C$.

(b) Prove the distributive property:

$(A + B)C = AC + BC$.

(c) Prove the property:

$c(AB) = (cA)B = A(cB)$.

67. Proof Prove Theorem 2.4.

68. Proof Prove Properties 2, 3, and 4 of Theorem 2.6.

69. Guided Proof Prove that if A is an $m \times n$ matrix, then AA^T and A^TA are symmetric matrices.

Getting Started: To prove that AA^T is symmetric, you need to show that it is equal to its transpose, $(AA^T)^T = AA^T$.

(i) Begin your proof with the left-hand matrix expression $(AA^T)^T$.

(ii) Use the properties of the transpose operation to show that $(AA^T)^T$ can be simplified to equal the right-hand expression, AA^T.

(iii) Repeat this analysis for the product A^TA.

70. Proof Let A and B be two $n \times n$ symmetric matrices.

(a) Give an example to show that the product AB is not necessarily symmetric.

(b) Prove that the product AB is symmetric if and only if $AB = BA$.

Symmetric and Skew-Symmetric Matrices In Exercises 71–74, determine whether the matrix is symmetric, skew-symmetric, or neither. A square matrix is skew-symmetric when $A^T = -A$.

71. $A = \begin{bmatrix} 0 & 2 \\ -2 & 0 \end{bmatrix}$ **72.** $A = \begin{bmatrix} 2 & 1 \\ 1 & 3 \end{bmatrix}$

73. $A = \begin{bmatrix} 0 & 2 & 1 \\ 2 & 0 & 3 \\ 1 & 3 & 0 \end{bmatrix}$ **74.** $A = \begin{bmatrix} 0 & 2 & -1 \\ -2 & 0 & -3 \\ 1 & 3 & 0 \end{bmatrix}$

75. Proof Prove that the main diagonal of a skew-symmetric matrix consists entirely of zeros.

76. Proof Prove that if A and B are $n \times n$ skew-symmetric matrices, then $A + B$ is skew-symmetric.

77. Proof Let A be a square matrix of order n.

(a) Show that $\frac{1}{2}(A + A^T)$ is symmetric.

(b) Show that $\frac{1}{2}(A - A^T)$ is skew-symmetric.

(c) Prove that A can be written as the sum of a symmetric matrix B and a skew-symmetric matrix C, $A = B + C$.

(d) Write the matrix below as the sum of a symmetric matrix and a skew-symmetric matrix.

$$A = \begin{bmatrix} 2 & 5 & 3 \\ -3 & 6 & 0 \\ 4 & 1 & 1 \end{bmatrix}$$

78. Proof Prove that if A is an $n \times n$ matrix, then $A - A^T$ is skew-symmetric.

79. Consider matrices of the form

$$A = \begin{bmatrix} 0 & a_{12} & a_{13} & \cdots & a_{1n} \\ 0 & 0 & a_{23} & \cdots & a_{2n} \\ \vdots & \vdots & \vdots & & \vdots \\ 0 & 0 & 0 & \cdots & a_{n-1,n} \\ 0 & 0 & 0 & \cdots & 0 \end{bmatrix}.$$

(a) Write a 2×2 matrix and a 3×3 matrix in the form of A.

(b) Use a graphing utility to raise each of the matrices to higher powers. Describe the result.

(c) Use the result of part (b) to make a conjecture about powers of A when A is a 4×4 matrix. Use a graphing utility to test your conjecture.

(d) Use the results of parts (b) and (c) to make a conjecture about powers of A when A is an $n \times n$ matrix.

2.3   The Inverse of a Matrix

■   Find the inverse of a matrix (if it exists).

■   Use properties of inverse matrices.

■   Use an inverse matrix to solve a system of linear equations.

MATRICES AND THEIR INVERSES

Section 2.2 discussed some of the similarities between the algebra of real numbers and the algebra of matrices. This section further develops the algebra of matrices to include the solutions of matrix equations involving matrix multiplication. To begin, consider the real number equation $ax = b$. To solve this equation for x, multiply both sides of the equation by a^{-1} (provided $a \neq 0$).

$$ax = b$$
$$(a^{-1}a)x = a^{-1}b$$
$$(1)x = a^{-1}b$$
$$x = a^{-1}b$$

The number a^{-1} is the *multiplicative inverse* of a because $a^{-1}a = 1$ (the identity element for multiplication). The definition of the multiplicative inverse of a matrix is similar.

Definition of the Inverse of a Matrix

An $n \times n$ matrix A is **invertible** (or **nonsingular**) when there exists an $n \times n$ matrix B such that

$$AB = BA = I_n$$

where I_n is the identity matrix of order n. The matrix B is the (multiplicative) **inverse** of A. A matrix that does not have an inverse is **noninvertible** (or **singular**).

Nonsquare matrices do not have inverses. To see this, note that if A is of size $m \times n$ and B is of size $n \times m$ (where $m \neq n$), then the products AB and BA are of different sizes and cannot be equal to each other. Not all square matrices have inverses. (See Example 4.) The next theorem, however, states that if a matrix *does* have an inverse, then that inverse is unique.

THEOREM 2.7   Uniqueness of an Inverse Matrix

If A is an invertible matrix, then its inverse is unique. The inverse of A is denoted by A^{-1}.

PROOF

If A is invertible, then it has at least one inverse B such that

$$AB = I = BA.$$

Assume that A has another inverse C such that

$$AC = I = CA.$$

Demonstrate that B and C are equal, as shown on the next page.

$$AB = I$$
$$C(AB) = CI$$
$$(CA)B = C$$
$$IB = C$$
$$B = C$$

Consequently $B = C$, and it follows that the inverse of a matrix is unique.

The inverse A^{-1} of an invertible matrix A is unique, so you can call it *the* inverse of A and write $AA^{-1} = A^{-1}A = I$.

EXAMPLE 1　The Inverse of a Matrix

Show that B is the inverse of A, where

$$A = \begin{bmatrix} -1 & 2 \\ -1 & 1 \end{bmatrix} \quad \text{and} \quad B = \begin{bmatrix} 1 & -2 \\ 1 & -1 \end{bmatrix}.$$

SOLUTION

Using the definition of an inverse matrix, show that B is the inverse of A by showing that $AB = I = BA$.

$$AB = \begin{bmatrix} -1 & 2 \\ -1 & 1 \end{bmatrix}\begin{bmatrix} 1 & -2 \\ 1 & -1 \end{bmatrix} = \begin{bmatrix} -1+2 & 2-2 \\ -1+1 & 2-1 \end{bmatrix} = \begin{bmatrix} 1 & 0 \\ 0 & 1 \end{bmatrix}$$

$$BA = \begin{bmatrix} 1 & -2 \\ 1 & -1 \end{bmatrix}\begin{bmatrix} -1 & 2 \\ -1 & 1 \end{bmatrix} = \begin{bmatrix} -1+2 & 2-2 \\ -1+1 & 2-1 \end{bmatrix} = \begin{bmatrix} 1 & 0 \\ 0 & 1 \end{bmatrix}$$

REMARK

Recall that it is not always true that $AB = BA$, even when both products are defined. If A and B are both square matrices and $AB = I_n$, however, then it can be shown that $BA = I_n$. (The proof of this is omitted.) So, in Example 1, you need only check that $AB = I_2$.

The next example shows how to use a system of equations to find the inverse of a matrix.

EXAMPLE 2　Finding the Inverse of a Matrix

Find the inverse of the matrix

$$A = \begin{bmatrix} 1 & 4 \\ -1 & -3 \end{bmatrix}.$$

SOLUTION

To find the inverse of A, solve the matrix equation $AX = I$ for X.

$$\begin{bmatrix} 1 & 4 \\ -1 & -3 \end{bmatrix}\begin{bmatrix} x_{11} & x_{12} \\ x_{21} & x_{22} \end{bmatrix} = \begin{bmatrix} 1 & 0 \\ 0 & 1 \end{bmatrix}$$

$$\begin{bmatrix} x_{11}+4x_{21} & x_{12}+4x_{22} \\ -x_{11}-3x_{21} & -x_{12}-3x_{22} \end{bmatrix} = \begin{bmatrix} 1 & 0 \\ 0 & 1 \end{bmatrix}$$

Equating corresponding entries, you obtain two systems of linear equations.

$$x_{11}+4x_{21} = 1 \qquad x_{12}+4x_{22} = 0$$
$$-x_{11}-3x_{21} = 0 \qquad -x_{12}-3x_{22} = 1$$

Solving the first system, you find that $x_{11} = -3$ and $x_{21} = 1$. Similarly, solving the second system, you find that $x_{12} = -4$ and $x_{22} = 1$. So, the inverse of A is

$$X = A^{-1} = \begin{bmatrix} -3 & -4 \\ 1 & 1 \end{bmatrix}.$$

Use matrix multiplication to check this result.

Generalizing the method used to solve Example 2 provides a convenient method for finding an inverse. Note that the two systems of linear equations

$$x_{11} + 4x_{21} = 1 \qquad x_{12} + 4x_{22} = 0$$
$$-x_{11} - 3x_{21} = 0 \qquad -x_{12} - 3x_{22} = 1$$

have the *same coefficient matrix*. Rather than solve the two systems represented by

$$\begin{bmatrix} 1 & 4 & 1 \\ -1 & -3 & 0 \end{bmatrix} \quad \text{and} \quad \begin{bmatrix} 1 & 4 & 0 \\ -1 & -3 & 1 \end{bmatrix}$$

separately, solve them simultaneously by **adjoining** the identity matrix to the coeficient matrix to obtain

$$\begin{bmatrix} 1 & 4 & 1 & 0 \\ -1 & -3 & 0 & 1 \end{bmatrix}.$$

By applying Gauss-Jordan elimination to this matrix, solve *both* systems with a single elimination process, as shown below.

$$\begin{bmatrix} 1 & 4 & 1 & 0 \\ 0 & 1 & 1 & 1 \end{bmatrix} \qquad R_2 + R_1 \rightarrow R_2$$

$$\begin{bmatrix} 1 & 0 & -3 & -4 \\ 0 & 1 & 1 & 1 \end{bmatrix} \qquad R_1 + (-4)R_2 \rightarrow R_1$$

Applying Gauss-Jordan elimination to the "doubly augmented" matrix $\begin{bmatrix} A & I \end{bmatrix}$, you obtain the matrix $\begin{bmatrix} I & A^{-1} \end{bmatrix}$.

$$\underbrace{\begin{bmatrix} 1 & 4 \\ -1 & -3 \end{bmatrix}}_{A} \quad \underbrace{\begin{bmatrix} 1 & 0 \\ 0 & 1 \end{bmatrix}}_{I} \quad \longrightarrow \quad \underbrace{\begin{bmatrix} 1 & 0 \\ 0 & 1 \end{bmatrix}}_{I} \quad \underbrace{\begin{bmatrix} -3 & -4 \\ 1 & 1 \end{bmatrix}}_{A^{-1}}$$

This procedure (or algorithm) works for an arbitrary $n \times n$ matrix. If A cannot be row reduced to I_n, then A is noninvertible (or singular). This procedure will be formally justified in the next section, after introducing the concept of an elementary matrix. For now, a summary of the algorithm is shown below.

Finding the Inverse of a Matrix by Gauss-Jordan Elimination

Let A be a square matrix of order n.

1. Write the $n \times 2n$ matrix that consists of A on the left and the $n \times n$ identity matrix I on the right to obtain $\begin{bmatrix} A & I \end{bmatrix}$. This process is called **adjoining** matrix I to matrix A.
2. If possible, row reduce A to I using elementary row operations on the entire matrix $\begin{bmatrix} A & I \end{bmatrix}$. The result will be the matrix $\begin{bmatrix} I & A^{-1} \end{bmatrix}$. If this is not possible, then A is noninvertible (or singular).
3. Check your work by multiplying to see that $AA^{-1} = I = A^{-1}A$.

LINEAR ALGEBRA APPLIED

Recall Hooke's law, which states that for relatively small deformations of an elastic object, the amount of deflection is directly proportional to the force causing the deformation. In a simply supported elastic beam subjected to multiple forces, deflection **d** is related to force **w** by the matrix equation

$$\mathbf{d} = F\mathbf{w}$$

where F is a *flexibility matrix* whose entries depend on the material of the beam. The inverse of the flexibility matrix, F^{-1}, is the *stiffness matrix*. In Exercises 61 and 62, you are asked to find the stiffness matrix F^{-1} and the force matrix **w** for a given set of flexibility and deflection matrices.

EXAMPLE 3   **Finding the Inverse of a Matrix**

See LarsonLinearAlgebra.com for an interactive version of this type of example.

Find the inverse of the matrix.

$$A = \begin{bmatrix} 1 & -1 & 0 \\ 1 & 0 & -1 \\ -6 & 2 & 3 \end{bmatrix}$$

SOLUTION

Begin by adjoining the identity matrix to A to form the matrix

$$[A \quad I] = \begin{bmatrix} 1 & -1 & 0 & 1 & 0 & 0 \\ 1 & 0 & -1 & 0 & 1 & 0 \\ -6 & 2 & 3 & 0 & 0 & 1 \end{bmatrix}.$$

Use elementary row operations to obtain the form

$$[I \quad A^{-1}]$$

as shown below.

$$\begin{bmatrix} 1 & -1 & 0 & 1 & 0 & 0 \\ 0 & 1 & -1 & -1 & 1 & 0 \\ -6 & 2 & 3 & 0 & 0 & 1 \end{bmatrix} \qquad R_2 + (-1)R_1 \rightarrow R_2$$

$$\begin{bmatrix} 1 & -1 & 0 & 1 & 0 & 0 \\ 0 & 1 & -1 & -1 & 1 & 0 \\ 0 & -4 & 3 & 6 & 0 & 1 \end{bmatrix} \qquad R_3 + (6)R_1 \rightarrow R_3$$

$$\begin{bmatrix} 1 & -1 & 0 & 1 & 0 & 0 \\ 0 & 1 & -1 & -1 & 1 & 0 \\ 0 & 0 & -1 & 2 & 4 & 1 \end{bmatrix} \qquad R_3 + (4)R_2 \rightarrow R_3$$

$$\begin{bmatrix} 1 & -1 & 0 & 1 & 0 & 0 \\ 0 & 1 & -1 & -1 & 1 & 0 \\ 0 & 0 & 1 & -2 & -4 & -1 \end{bmatrix} \qquad (-1)R_3 \rightarrow R_3$$

$$\begin{bmatrix} 1 & -1 & 0 & 1 & 0 & 0 \\ 0 & 1 & 0 & -3 & -3 & -1 \\ 0 & 0 & 1 & -2 & -4 & -1 \end{bmatrix} \qquad R_2 + R_3 \rightarrow R_2$$

$$\begin{bmatrix} 1 & 0 & 0 & -2 & -3 & -1 \\ 0 & 1 & 0 & -3 & -3 & -1 \\ 0 & 0 & 1 & -2 & -4 & -1 \end{bmatrix} \qquad R_1 + R_2 \rightarrow R_1$$

The matrix A is invertible, and its inverse is

$$A^{-1} = \begin{bmatrix} -2 & -3 & -1 \\ -3 & -3 & -1 \\ -2 & -4 & -1 \end{bmatrix}.$$

Confirm this by showing that

$$AA^{-1} = I = A^{-1}A.$$

TECHNOLOGY

Many graphing utilities and software programs can find the inverse of a square matrix. When you use a graphing utility, you may see something similar to the screen below for Example 3. The **Technology Guide** at *CengageBrain.com* can help you use technology to find the inverse of a matrix.

```
A
      [[1  -1  0 ]
       [1   0  -1]
       [-6  2  3 ]]
A⁻¹
      [[-2 -3 -1]
       [-3 -3 -1]
       [-2 -4 -1]]
```

The process shown in Example 3 applies to any $n \times n$ matrix A and will find the inverse of A, if it exists. When A has no inverse, the process will also tell you that. The next example applies the process to a singular matrix (one that has no inverse).

EXAMPLE 4   A Singular Matrix

Show that the matrix has no inverse.

$$A = \begin{bmatrix} 1 & 2 & 0 \\ 3 & -1 & 2 \\ -2 & 3 & -2 \end{bmatrix}$$

SOLUTION

Adjoin the identity matrix to A to form

$$[A \quad I] = \begin{bmatrix} 1 & 2 & 0 & 1 & 0 & 0 \\ 3 & -1 & 2 & 0 & 1 & 0 \\ -2 & 3 & -2 & 0 & 0 & 1 \end{bmatrix}$$

and apply Gauss-Jordan elimination to obtain

$$\begin{bmatrix} 1 & 2 & 0 & 1 & 0 & 0 \\ 0 & -7 & 2 & -3 & 1 & 0 \\ 0 & 0 & 0 & -1 & 1 & 1 \end{bmatrix}.$$

Note that the "A portion" of the matrix has a row of zeros. So it is not possible to rewrite the matrix $[A \quad I]$ in the form $[I \quad A^{-1}]$. This means that A has no inverse, or is noninvertible (or singular).

Using Gauss-Jordan elimination to find the inverse of a matrix works well (even as a computer technique) for matrices of size 3×3 or greater. For 2×2 matrices, however, you can use a formula for the inverse rather than Gauss-Jordan elimination.

If A is a 2×2 matrix

$$A = \begin{bmatrix} a & b \\ c & d \end{bmatrix}$$

REMARK

The denominator $ad - bc$ is called the **determinant** of A. You will study determinants in detail in Chapter 3.

then A is invertible if and only if $ad - bc \neq 0$. Moreover, if $ad - bc \neq 0$, then the inverse is

$$A^{-1} = \frac{1}{ad - bc}\begin{bmatrix} d & -b \\ -c & a \end{bmatrix}.$$

EXAMPLE 5   Finding Inverses of 2 × 2 Matrices

If possible, find the inverse of each matrix.

a. $A = \begin{bmatrix} 3 & -1 \\ -2 & 2 \end{bmatrix}$ **b.** $B = \begin{bmatrix} 3 & -1 \\ -6 & 2 \end{bmatrix}$

SOLUTION

a. For the matrix A, apply the formula for the inverse of a 2×2 matrix to obtain $ad - bc = (3)(2) - (-1)(-2) = 4$. This quantity is not zero, so A is invertible. Form the inverse by interchanging the entries on the main diagonal, changing the signs of the other two entries, and multiplying by the scalar $\frac{1}{4}$, as shown below.

$$A^{-1} = \frac{1}{4}\begin{bmatrix} 2 & 1 \\ 2 & 3 \end{bmatrix} = \begin{bmatrix} \frac{1}{2} & \frac{1}{4} \\ \frac{1}{2} & \frac{3}{4} \end{bmatrix}$$

b. For the matrix B, you have $ad - bc = (3)(2) - (-1)(-6) = 0$, which means that B is noninvertible.

PROPERTIES OF INVERSES

Theorem 2.8 below lists important properties of inverse matrices.

THEOREM 2.8   Properties of Inverse Matrices

If A is an invertible matrix, k is a positive integer, and c is a nonzero scalar, then A^{-1}, A^k, cA, and A^T are invertible and the statements below are true.

1. $(A^{-1})^{-1} = A$ **2.** $(A^k)^{-1} = \underbrace{A^{-1}A^{-1} \cdots A^{-1}}_{k \text{ factors}} = (A^{-1})^k$

3. $(cA)^{-1} = \dfrac{1}{c}A^{-1}$ **4.** $(A^T)^{-1} = (A^{-1})^T$

PROOF

The key to the proofs of Properties 1, 3, and 4 is the fact that the inverse of a matrix is unique (Theorem 2.7). That is, if $BC = CB = I$, then C is the inverse of B.

Property 1 states that the inverse of A^{-1} is A itself. To prove this, observe that $A^{-1}A = AA^{-1} = I$, which means that A is the inverse of A^{-1}. Thus, $A = (A^{-1})^{-1}$.

Similarly, Property 3 states that $\dfrac{1}{c}A^{-1}$ is the inverse of (cA), $c \neq 0$. To prove this, use the properties of scalar multiplication given in Theorems 2.1 and 2.3.

$$(cA)\left(\frac{1}{c}A^{-1}\right) = \left(c\frac{1}{c}\right)AA^{-1} = (1)I = I$$

$$\left(\frac{1}{c}A^{-1}\right)(cA) = \left(\frac{1}{c}c\right)A^{-1}A = (1)I = I$$

So $\dfrac{1}{c}A^{-1}$ is the inverse of (cA), which implies that $(cA)^{-1} = \dfrac{1}{c}A^{-1}$. Properties 2 and 4 are left for you to prove. (See Exercises 63 and 64.)

For nonsingular matrices, the exponential notation used for repeated multiplication of *square* matrices can be extended to include exponents that are negative integers. This may be done by defining A^{-k} to be

$$A^{-k} = \underbrace{A^{-1}A^{-1} \cdots A^{-1}}_{k \text{ factors}} = (A^{-1})^k.$$

Using this convention you can show that the properties $A^jA^k = A^{j+k}$ and $(A^j)^k = A^{jk}$ are true for any integers j and k.

DISCOVERY

Let $A = \begin{bmatrix} 1 & 2 \\ 1 & 3 \end{bmatrix}$ and $B = \begin{bmatrix} 2 & -1 \\ 1 & -1 \end{bmatrix}$.

Find $(AB)^{-1}$, $A^{-1}B^{-1}$, and $B^{-1}A^{-1}$.

Make a conjecture about the inverse of a product of two nonsingular matrices. Then select two other nonsingular matrices of the same order and see whether your conjecture holds.

See LarsonLinearAlgebra.com for an interactive version of this type of exercise.

EXAMPLE 6   **The Inverse of the Square of a Matrix**

Compute A^{-2} two different ways and show that the results are equal.

$$A = \begin{bmatrix} 1 & 1 \\ 2 & 4 \end{bmatrix}$$

SOLUTION

One way to find A^{-2} is to find $(A^2)^{-1}$ by squaring the matrix A to obtain

$$A^2 = \begin{bmatrix} 3 & 5 \\ 10 & 18 \end{bmatrix}$$

and using the formula for the inverse of a 2×2 matrix to obtain

$$(A^2)^{-1} = \tfrac{1}{4}\begin{bmatrix} 18 & -5 \\ -10 & 3 \end{bmatrix} = \begin{bmatrix} \frac{9}{2} & -\frac{5}{4} \\ -\frac{5}{2} & \frac{3}{4} \end{bmatrix}.$$

Another way to find A^{-2} is to find $(A^{-1})^2$ by finding A^{-1}

$$A^{-1} = \tfrac{1}{2}\begin{bmatrix} 4 & -1 \\ -2 & 1 \end{bmatrix} = \begin{bmatrix} 2 & -\frac{1}{2} \\ -1 & \frac{1}{2} \end{bmatrix}$$

and then squaring this matrix to obtain

$$(A^{-1})^2 = \begin{bmatrix} \frac{9}{2} & -\frac{5}{4} \\ -\frac{5}{2} & \frac{3}{4} \end{bmatrix}.$$

Note that both methods produce the same result.

The next theorem gives a formula for computing the inverse of a product of two matrices.

THEOREM 2.9   The Inverse of a Product

If A and B are invertible matrices of order n, then AB is invertible and

$$(AB)^{-1} = B^{-1}A^{-1}.$$

PROOF

To show that $B^{-1}A^{-1}$ is the inverse of AB, you need only show that it conforms to the definition of an inverse matrix. That is,

$$(AB)(B^{-1}A^{-1}) = A(BB^{-1})A^{-1} = A(I)A^{-1} = (AI)A^{-1} = AA^{-1} = I.$$

In a similar way, $(B^{-1}A^{-1})(AB) = I$. So, AB is invertible and its inverse is $B^{-1}A^{-1}$.

Theorem 2.9 states that the inverse of a product of two invertible matrices is the product of their inverses taken in the *reverse* order. This can be generalized to include the product of more than two invertible matrices:

$$(A_1A_2A_3 \cdots A_n)^{-1} = A_n^{-1} \cdots A_3^{-1}A_2^{-1}A_1^{-1}.$$

(See Example 4 in Appendix.)

| EXAMPLE 7 | Finding the Inverse of a Matrix Product |

Find $(AB)^{-1}$ for the matrices

$$A = \begin{bmatrix} 1 & 3 & 3 \\ 1 & 4 & 3 \\ 1 & 3 & 4 \end{bmatrix} \quad \text{and} \quad B = \begin{bmatrix} 1 & 2 & 3 \\ 1 & 3 & 3 \\ 2 & 4 & 3 \end{bmatrix}$$

using the fact that A^{-1} and B^{-1} are

$$A^{-1} = \begin{bmatrix} 7 & -3 & -3 \\ -1 & 1 & 0 \\ -1 & 0 & 1 \end{bmatrix} \quad \text{and} \quad B^{-1} = \begin{bmatrix} 1 & -2 & 1 \\ -1 & 1 & 0 \\ \frac{2}{3} & 0 & -\frac{1}{3} \end{bmatrix}.$$

REMARK

Note that you *reverse the order* of multiplication to find the inverse of *AB*. That is, $(AB)^{-1} = B^{-1}A^{-1}$, and the inverse of *AB* is usually *not* equal to $A^{-1}B^{-1}$.

SOLUTION

Using Theorem 2.9 produces

$$(AB)^{-1} = B^{-1}A^{-1}$$

$$= \begin{bmatrix} 1 & -2 & 1 \\ -1 & 1 & 0 \\ \frac{2}{3} & 0 & -\frac{1}{3} \end{bmatrix} \begin{bmatrix} 7 & -3 & -3 \\ -1 & 1 & 0 \\ -1 & 0 & 1 \end{bmatrix}$$

$$= \begin{bmatrix} 8 & -5 & -2 \\ -8 & 4 & 3 \\ 5 & -2 & -\frac{7}{3} \end{bmatrix}.$$

One important property in the algebra of real numbers is the cancellation property. That is, if $ac = bc$ $(c \neq 0)$, then $a = b$. *Invertible* matrices have similar cancellation properties.

THEOREM 2.10   Cancellation Properties

If C is an invertible matrix, then the properties below are true.

1. If $AC = BC$, then $A = B$.   Right cancellation property
2. If $CA = CB$, then $A = B$.   Left cancellation property

PROOF

To prove Property 1, use the fact that C is invertible and write

$$AC = BC$$
$$(AC)C^{-1} = (BC)C^{-1}$$
$$A(CC^{-1}) = B(CC^{-1})$$
$$AI = BI$$
$$A = B.$$

The second property can be proved in a similar way. (See Exercise 65.)

Be sure to remember that Theorem 2.10 can be applied only when C is an *invertible* matrix. If C is not invertible, then cancellation is not usually valid. For instance, Example 5 in Section 2.2 gives an example of a matrix equation $AC = BC$ in which $A \neq B$, because C is not invertible in the example.

SYSTEMS OF EQUATIONS

For *square* systems of equations (those having the same number of equations as variables), you can use the theorem below to determine whether the system has a unique solution.

THEOREM 2.11   Systems of Equations with Unique Solutions

If A is an invertible matrix, then the system of linear equations $A\mathbf{x} = \mathbf{b}$ has a unique solution $\mathbf{x} = A^{-1}\mathbf{b}$.

PROOF

The matrix A is nonsingular, so the steps shown below are valid.

$$A\mathbf{x} = \mathbf{b}$$
$$A^{-1}A\mathbf{x} = A^{-1}\mathbf{b}$$
$$I\mathbf{x} = A^{-1}\mathbf{b}$$
$$\mathbf{x} = A^{-1}\mathbf{b}$$

This solution is unique because if $\mathbf{x}_1$ and $\mathbf{x}_2$ were two solutions, then you could apply the cancellation property to the equation $A\mathbf{x}_1 = \mathbf{b} = A\mathbf{x}_2$ to conclude that $\mathbf{x}_1 = \mathbf{x}_2$.

One use of Theorem 2.11 is in solving *several* systems that all have the same coefficient matrix A. You could find the inverse matrix once and then solve each system by computing the product $A^{-1}\mathbf{b}$.

EXAMPLE 8   Solving Systems of Equations Using an Inverse Matrix

Use an inverse matrix to solve each system.

a. $2x + 3y + z = -1$
$3x + 3y + z = 1$
$2x + 4y + z = -2$

b. $2x + 3y + z = 4$
$3x + 3y + z = 8$
$2x + 4y + z = 5$

c. $2x + 3y + z = 0$
$3x + 3y + z = 0$
$2x + 4y + z = 0$

SOLUTION

First note that the coefficient matrix for each system is $A = \begin{bmatrix} 2 & 3 & 1 \\ 3 & 3 & 1 \\ 2 & 4 & 1 \end{bmatrix}$.

Using Gauss-Jordan elimination, $A^{-1} = \begin{bmatrix} -1 & 1 & 0 \\ -1 & 0 & 1 \\ 6 & -2 & -3 \end{bmatrix}$.

a. $\mathbf{x} = A^{-1}\mathbf{b} = \begin{bmatrix} -1 & 1 & 0 \\ -1 & 0 & 1 \\ 6 & -2 & -3 \end{bmatrix}\begin{bmatrix} -1 \\ 1 \\ -2 \end{bmatrix} = \begin{bmatrix} 2 \\ -1 \\ -2 \end{bmatrix}$ The solution is $x = 2$, $y = -1$, and $z = -2$.

b. $\mathbf{x} = A^{-1}\mathbf{b} = \begin{bmatrix} -1 & 1 & 0 \\ -1 & 0 & 1 \\ 6 & -2 & -3 \end{bmatrix}\begin{bmatrix} 4 \\ 8 \\ 5 \end{bmatrix} = \begin{bmatrix} 4 \\ 1 \\ -7 \end{bmatrix}$ The solution is $x = 4$, $y = 1$, and $z = -7$.

c. $\mathbf{x} = A^{-1}\mathbf{b} = \begin{bmatrix} -1 & 1 & 0 \\ -1 & 0 & 1 \\ 6 & -2 & -3 \end{bmatrix}\begin{bmatrix} 0 \\ 0 \\ 0 \end{bmatrix} = \begin{bmatrix} 0 \\ 0 \\ 0 \end{bmatrix}$ The solution is trivial: $x = 0$, $y = 0$, and $z = 0$.

2.3 Exercises

See CalcChat.com for worked-out solutions to odd-numbered exercises.

The Inverse of a Matrix   **In Exercises 1–6, show that** B **is the inverse of** A.

1. $A = \begin{bmatrix} 2 & 1 \\ 5 & 3 \end{bmatrix}$, $B = \begin{bmatrix} 3 & -1 \\ -5 & 2 \end{bmatrix}$

2. $A = \begin{bmatrix} 1 & -1 \\ -1 & 2 \end{bmatrix}$, $B = \begin{bmatrix} 2 & 1 \\ 1 & 1 \end{bmatrix}$

3. $A = \begin{bmatrix} 1 & 2 \\ 3 & 4 \end{bmatrix}$, $B = \begin{bmatrix} -2 & 1 \\ \frac{3}{2} & -\frac{1}{2} \end{bmatrix}$

4. $A = \begin{bmatrix} 1 & -1 \\ 2 & 3 \end{bmatrix}$, $B = \begin{bmatrix} \frac{3}{5} & \frac{1}{5} \\ -\frac{2}{5} & \frac{1}{5} \end{bmatrix}$

5. $A = \begin{bmatrix} -2 & 2 & 3 \\ 1 & -1 & 0 \\ 0 & 1 & 4 \end{bmatrix}$, $B = \frac{1}{3}\begin{bmatrix} -4 & -5 & 3 \\ -4 & -8 & 3 \\ 1 & 2 & 0 \end{bmatrix}$

6. $A = \begin{bmatrix} 2 & -17 & 11 \\ -1 & 11 & -7 \\ 0 & 3 & -2 \end{bmatrix}$, $B = \begin{bmatrix} 1 & 1 & 2 \\ 2 & 4 & -3 \\ 3 & 6 & -5 \end{bmatrix}$

Finding the Inverse of a Matrix   **In Exercises 7–30, find the inverse of the matrix (if it exists).**

7. $\begin{bmatrix} 2 & 0 \\ 0 & 3 \end{bmatrix}$

8. $\begin{bmatrix} 2 & -2 \\ 2 & 2 \end{bmatrix}$

9. $\begin{bmatrix} 1 & 2 \\ 3 & 7 \end{bmatrix}$

10. $\begin{bmatrix} 1 & -2 \\ 2 & -3 \end{bmatrix}$

11. $\begin{bmatrix} -7 & 33 \\ 4 & -19 \end{bmatrix}$

12. $\begin{bmatrix} -1 & 1 \\ 3 & -3 \end{bmatrix}$

13. $\begin{bmatrix} 1 & 1 & 1 \\ 3 & 5 & 4 \\ 3 & 6 & 5 \end{bmatrix}$

14. $\begin{bmatrix} 1 & 2 & 2 \\ 3 & 7 & 9 \\ -1 & -4 & -7 \end{bmatrix}$

15. $\begin{bmatrix} 1 & 2 & -1 \\ 3 & 7 & -10 \\ 7 & 16 & -21 \end{bmatrix}$

16. $\begin{bmatrix} 10 & 5 & -7 \\ -5 & 1 & 4 \\ 3 & 2 & -2 \end{bmatrix}$

17. $\begin{bmatrix} 1 & 1 & 2 \\ 3 & 1 & 0 \\ -2 & 0 & 3 \end{bmatrix}$

18. $\begin{bmatrix} 3 & 2 & 5 \\ 2 & 2 & 4 \\ -4 & 4 & 0 \end{bmatrix}$

19. $\begin{bmatrix} 2 & 0 & 0 \\ 0 & 3 & 0 \\ 0 & 0 & 5 \end{bmatrix}$

20. $\begin{bmatrix} -\frac{5}{6} & \frac{1}{3} & \frac{11}{6} \\ 0 & \frac{2}{3} & 2 \\ 1 & -\frac{1}{2} & -\frac{5}{2} \end{bmatrix}$

21. $\begin{bmatrix} 0.6 & 0 & -0.3 \\ 0.7 & -1 & 0.2 \\ 1 & 0 & -0.9 \end{bmatrix}$

22. $\begin{bmatrix} 0.1 & 0.2 & 0.3 \\ -0.3 & 0.2 & 0.2 \\ 0.5 & 0.5 & 0.5 \end{bmatrix}$

23. $\begin{bmatrix} 1 & 0 & 0 \\ 3 & 4 & 0 \\ 2 & 5 & 5 \end{bmatrix}$

24. $\begin{bmatrix} 1 & 0 & 0 \\ 3 & 0 & 0 \\ 2 & 5 & 5 \end{bmatrix}$

25. $\begin{bmatrix} -8 & 0 & 0 & 0 \\ 0 & 1 & 0 & 0 \\ 0 & 0 & 0 & 0 \\ 0 & 0 & 0 & -5 \end{bmatrix}$

26. $\begin{bmatrix} 1 & 0 & 0 & 0 \\ 0 & 2 & 0 & 0 \\ 0 & 0 & -2 & 0 \\ 0 & 0 & 0 & 3 \end{bmatrix}$

27. $\begin{bmatrix} 1 & -2 & -1 & -2 \\ 3 & -5 & -2 & -3 \\ 2 & -5 & -2 & -5 \\ -1 & 4 & 4 & 11 \end{bmatrix}$

28. $\begin{bmatrix} 4 & 8 & -7 & 14 \\ 2 & 5 & -4 & 6 \\ 0 & 2 & 1 & -7 \\ 3 & 6 & -5 & 10 \end{bmatrix}$

29. $\begin{bmatrix} 1 & 0 & 3 & 0 \\ 0 & 2 & 0 & 4 \\ 1 & 0 & 3 & 0 \\ 0 & 2 & 0 & 4 \end{bmatrix}$

30. $\begin{bmatrix} 1 & 3 & -2 & 0 \\ 0 & 2 & 4 & 6 \\ 0 & 0 & -2 & 1 \\ 0 & 0 & 0 & 5 \end{bmatrix}$

Finding the Inverse of a 2 × 2 Matrix   **In Exercises 31–36, use the formula on page 66 to find the inverse of the 2 × 2 matrix (if it exists).**

31. $\begin{bmatrix} 2 & 3 \\ -1 & 5 \end{bmatrix}$

32. $\begin{bmatrix} 1 & -2 \\ -3 & 2 \end{bmatrix}$

33. $\begin{bmatrix} -4 & -6 \\ 2 & 3 \end{bmatrix}$

34. $\begin{bmatrix} -12 & 3 \\ 5 & -2 \end{bmatrix}$

35. $\begin{bmatrix} \frac{7}{2} & -\frac{3}{4} \\ \frac{1}{5} & \frac{4}{5} \end{bmatrix}$

36. $\begin{bmatrix} -\frac{1}{4} & \frac{9}{4} \\ \frac{5}{3} & \frac{8}{9} \end{bmatrix}$

Finding the Inverse of the Square of a Matrix   **In Exercises 37–40, compute** A^{-2} **two different ways and show that the results are equal.**

37. $A = \begin{bmatrix} 0 & -2 \\ -1 & 3 \end{bmatrix}$

38. $A = \begin{bmatrix} 2 & 7 \\ -5 & 6 \end{bmatrix}$

39. $A = \begin{bmatrix} -2 & 0 & 0 \\ 0 & 1 & 0 \\ 0 & 0 & 3 \end{bmatrix}$

40. $A = \begin{bmatrix} 6 & 0 & 4 \\ -2 & 7 & -1 \\ 3 & 1 & 2 \end{bmatrix}$

Finding the Inverses of Products and Transposes   **In Exercises 41–44, use the inverse matrices to find (a)** $(AB)^{-1}$, **(b)** $(A^T)^{-1}$, **and (c)** $(2A)^{-1}$.

41. $A^{-1} = \begin{bmatrix} 2 & 5 \\ -7 & 6 \end{bmatrix}$, $B^{-1} = \begin{bmatrix} 7 & -3 \\ 2 & 0 \end{bmatrix}$

42. $A^{-1} = \begin{bmatrix} -\frac{2}{7} & \frac{1}{7} \\ \frac{3}{7} & \frac{2}{7} \end{bmatrix}$, $B^{-1} = \begin{bmatrix} \frac{5}{11} & \frac{2}{11} \\ \frac{3}{11} & -\frac{1}{11} \end{bmatrix}$

43. $A^{-1} = \begin{bmatrix} 1 & -\frac{1}{2} & \frac{3}{4} \\ \frac{3}{2} & \frac{1}{2} & -2 \\ \frac{1}{4} & 1 & \frac{1}{2} \end{bmatrix}$, $B^{-1} = \begin{bmatrix} 2 & 4 & \frac{5}{2} \\ -\frac{3}{4} & 2 & \frac{1}{4} \\ \frac{1}{4} & \frac{1}{2} & 2 \end{bmatrix}$

44. $A^{-1} = \begin{bmatrix} 1 & -4 & 2 \\ 0 & 1 & 3 \\ 4 & 2 & 1 \end{bmatrix}$, $B^{-1} = \begin{bmatrix} 6 & 5 & -3 \\ -2 & 4 & -1 \\ 1 & 3 & 4 \end{bmatrix}$

Solving a System of Equations Using an Inverse
In Exercises 45–48, use an inverse matrix to solve each system of linear equations.

45. (a) $x + 2y = -1$
$\quad\quad x - 2y = 3$
(b) $x + 2y = 10$
$\quad\quad x - 2y = -6$

46. (a) $2x - y = -3$
$\quad\quad 2x + y = 7$
(b) $2x - y = -1$
$\quad\quad 2x + y = -3$

47. (a) $x_1 + 2x_2 + x_3 = 2$
$\quad\quad x_1 + 2x_2 - x_3 = 4$
$\quad\quad x_1 - 2x_2 + x_3 = -2$
(b) $x_1 + 2x_2 + x_3 = 1$
$\quad\quad x_1 + 2x_2 - x_3 = 3$
$\quad\quad x_1 - 2x_2 + x_3 = -3$

48. (a) $x_1 + x_2 - 2x_3 = 0$
$\quad\quad x_1 - 2x_2 + x_3 = 0$
$\quad\quad x_1 - x_2 - x_3 = -1$
(b) $x_1 + x_2 - 2x_3 = -1$
$\quad\quad x_1 - 2x_2 + x_3 = 2$
$\quad\quad x_1 - x_2 - x_3 = 0$

Solving a System of Equations Using an Inverse
In Exercises 49–52, use a software program or a graphing utility to solve the system of linear equations using an inverse matrix.

49. $x_1 + 2x_2 - x_3 + 3x_4 - x_5 = -3$
$\quad x_1 - 3x_2 + x_3 + 2x_4 - x_5 = -3$
$\quad 2x_1 + x_2 + x_3 - 3x_4 + x_5 = 6$
$\quad x_1 - x_2 + 2x_3 + x_4 - x_5 = 2$
$\quad 2x_1 + x_2 - x_3 + 2x_4 + x_5 = -3$

50. $x_1 + x_2 - x_3 + 3x_4 - x_5 = 3$
$\quad 2x_1 + x_2 + x_3 + x_4 + x_5 = 4$
$\quad x_1 + x_2 - x_3 + 2x_4 - x_5 = 3$
$\quad 2x_1 + x_2 + 4x_3 + x_4 - x_5 = -1$
$\quad 3x_1 + x_2 + x_3 - 2x_4 + x_5 = 5$

51. $2x_1 - 3x_2 + x_3 - 2x_4 + x_5 - 4x_6 = 20$
$\quad 3x_1 + x_2 - 4x_3 + x_4 - x_5 + 2x_6 = -16$
$\quad 4x_1 + x_2 - 3x_3 + 4x_4 - x_5 + 2x_6 = -12$
$\quad -5x_1 - x_2 + 4x_3 + 2x_4 - 5x_5 + 3x_6 = -2$
$\quad x_1 + x_2 - 3x_3 + 4x_4 - 3x_5 + x_6 = -15$
$\quad 3x_1 - x_2 + 2x_3 - 3x_4 + 2x_5 - 6x_6 = 25$

52. $4x_1 - 2x_2 + 4x_3 + 2x_4 - 5x_5 - x_6 = 1$
$\quad 3x_1 + 6x_2 - 5x_3 - 6x_4 + 3x_5 + 3x_6 = -11$
$\quad 2x_1 - 3x_2 + x_3 + 3x_4 - x_5 - 2x_6 = 0$
$\quad -x_1 + 4x_2 - 4x_3 - 6x_4 + 2x_5 + 4x_6 = -9$
$\quad 3x_1 - x_2 + 5x_3 + 2x_4 - 3x_5 - 5x_6 = 1$
$\quad -2x_1 + 3x_2 - 4x_3 - 6x_4 + x_5 + 2x_6 = -12$

Matrix Equal to Its Own Inverse In Exercises 53 and 54, find x such that the matrix is equal to its own inverse.

53. $A = \begin{bmatrix} 3 & x \\ -2 & -3 \end{bmatrix}$

54. $A = \begin{bmatrix} 2 & x \\ -1 & -2 \end{bmatrix}$

Singular Matrix In Exercises 55 and 56, find x such that the matrix is singular.

55. $A = \begin{bmatrix} 4 & x \\ -2 & -3 \end{bmatrix}$

56. $A = \begin{bmatrix} x & 2 \\ -3 & 4 \end{bmatrix}$

Solving a Matrix Equation In Exercises 57 and 58, find A.

57. $(2A)^{-1} = \begin{bmatrix} 1 & 2 \\ 3 & 4 \end{bmatrix}$

58. $(4A)^{-1} = \begin{bmatrix} 2 & 4 \\ -3 & 2 \end{bmatrix}$

Finding the Inverse of a Matrix In Exercises 59 and 60, show that the matrix is invertible and find its inverse.

59. $A = \begin{bmatrix} \sin\theta & \cos\theta \\ -\cos\theta & \sin\theta \end{bmatrix}$

60. $A = \begin{bmatrix} \sec\theta & \tan\theta \\ \tan\theta & \sec\theta \end{bmatrix}$

Beam Deflection In Exercises 61 and 62, forces w_1, w_2, and w_3 (in newtons) act on a simply supported elastic beam, resulting in deflections $d_1, d_2,$ and d_3 (in millimeters) in the beam (see figure).

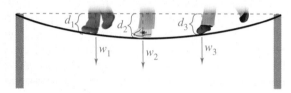

Use the matrix equation $\mathbf{d} = F\mathbf{w}$, where

$$\mathbf{d} = \begin{bmatrix} d_1 \\ d_2 \\ d_3 \end{bmatrix}, \quad \mathbf{w} = \begin{bmatrix} w_1 \\ w_2 \\ w_3 \end{bmatrix}$$

and F is the 3×3 *flexibility matrix* for the beam, to find the stiffness matrix F^{-1} and the force matrix $\mathbf{w}$. The entries of F are measured in millimeters per newton.

61. $F = \begin{bmatrix} 0.0457 & 0.0228 & 0.0171 \\ 0.0228 & 0.0343 & 0.0228 \\ 0.0171 & 0.0228 & 0.0457 \end{bmatrix}$, $\mathbf{d} = \begin{bmatrix} 14.859 \\ 16.256 \\ 21.209 \end{bmatrix}$

62. $F = \begin{bmatrix} 0.0971 & 0.0571 & 0.0457 \\ 0.0571 & 0.0685 & 0.0571 \\ 0.0457 & 0.0571 & 0.0971 \end{bmatrix}$, $\mathbf{d} = \begin{bmatrix} 0 \\ 3.810 \\ 0 \end{bmatrix}$

63. **Proof** Prove Property 2 of Theorem 2.8: If A is an invertible matrix and k is a positive integer, then

$$(A^k)^{-1} = \underbrace{A^{-1}A^{-1} \cdots A^{-1}}_{k \text{ factors}} = (A^{-1})^k$$

64. **Proof** Prove Property 4 of Theorem 2.8: If A is an invertible matrix, then $(A^T)^{-1} = (A^{-1})^T$.

65. **Proof** Prove Property 2 of Theorem 2.10: If C is an invertible matrix such that $CA = CB$, then $A = B$.

66. **Proof** Prove that if $A^2 = A$, then

$$I - 2A = (I - 2A)^{-1}.$$

67. Guided Proof Prove that the inverse of a symmetric nonsingular matrix is symmetric.

Getting Started: To prove that the inverse of A is symmetric, you need to show that $(A^{-1})^T = A^{-1}$.

(i) Let A be a symmetric, nonsingular matrix.

(ii) This means that $A^T = A$ and A^{-1} exists.

(iii) Use the properties of the transpose to show that $(A^{-1})^T$ is equal to A^{-1}.

68. Proof Prove that if A, B, and C are square matrices and $ABC = I$, then B is invertible and $B^{-1} = CA$.

69. Proof Prove that if A is invertible and $AB = O$, then $B = O$.

70. Guided Proof Prove that if $A^2 = A$, then either A is singular or $A = I$.

Getting Started: You must show that either A is singular or A equals the identity matrix.

(i) Begin your proof by observing that A is either singular or nonsingular.

(ii) If A is singular, then you are done.

(iii) If A is nonsingular, then use the inverse matrix A^{-1} and the hypothesis $A^2 = A$ to show that $A = I$.

True or False? In Exercises 71 and 72, determine whether each statement is true or false. If a statement is true, give a reason or cite an appropriate statement from the text. If a statement is false, provide an example that shows the statement is not true in all cases or cite an appropriate statement from the text.

71. (a) If the matrices A, B, and C satisfy $BA = CA$ and A is invertible, then $B = C$.

(b) The inverse of the product of two matrices is the product of their inverses; that is, $(AB)^{-1} = A^{-1}B^{-1}$.

(c) If A can be row reduced to the identity matrix, then A is nonsingular.

72. (a) The inverse of the inverse of a nonsingular matrix A, $(A^{-1})^{-1}$, is equal to A itself.

(b) The matrix $\begin{bmatrix} a & b \\ c & d \end{bmatrix}$ is invertible when $ab - dc \neq 0$.

(c) If A is a square matrix, then the system of linear equations $A\mathbf{x} = \mathbf{b}$ has a unique solution.

73. Writing Is the sum of two invertible matrices invertible? Explain why or why not. Illustrate your conclusion with appropriate examples.

74. Writing Under what conditions will the diagonal matrix

$$A = \begin{bmatrix} a_{11} & 0 & 0 & \cdots & 0 \\ 0 & a_{22} & 0 & \cdots & 0 \\ \vdots & \vdots & \vdots & & \vdots \\ 0 & 0 & 0 & \cdots & a_{nn} \end{bmatrix}$$

be invertible? Assume that A is invertible and find its inverse.

75. Use the result of Exercise 74 to find A^{-1} for each matrix.

(a) $A = \begin{bmatrix} -1 & 0 & 0 \\ 0 & 3 & 0 \\ 0 & 0 & 2 \end{bmatrix}$

(b) $A = \begin{bmatrix} \frac{1}{2} & 0 & 0 \\ 0 & \frac{1}{3} & 0 \\ 0 & 0 & \frac{1}{4} \end{bmatrix}$

76. Let $A = \begin{bmatrix} 1 & 2 \\ -2 & 1 \end{bmatrix}$.

(a) Show that $A^2 - 2A + 5I = O$, where I is the identity matrix of order 2.

(b) Show that $A^{-1} = \frac{1}{5}(2I - A)$.

(c) Show that for any square matrix satisfying $A^2 - 2A + 5I = O$, the inverse of A is $A^{-1} = \frac{1}{5}(2I - A)$.

77. Proof Let $\mathbf{u}$ be an $n \times 1$ column matrix satisfying $\mathbf{u}^T\mathbf{u} = I_1$. The $n \times n$ matrix $H = I_n - 2\mathbf{u}\mathbf{u}^T$ is called a **Householder matrix**.

(a) Prove that H is symmetric and nonsingular.

(b) Let $\mathbf{u} = \begin{bmatrix} \sqrt{2}/2 \\ \sqrt{2}/2 \\ 0 \end{bmatrix}$. Show that $\mathbf{u}^T\mathbf{u} = I_1$ and find the Householder matrix H.

78. Proof Let A and B be $n \times n$ matrices. Prove that if the matrix $I - AB$ is nonsingular, then so is $I - BA$.

79. Let A, D, and P be $n \times n$ matrices satisfying $AP = PD$. Assume that P is nonsingular and solve this equation for A. Must it be true that $A = D$?

80. Find an example of a singular 2×2 matrix satisfying $A^2 = A$.

81. Writing Explain how to determine whether the inverse of a matrix exists. If so, explain how to find the inverse.

82. CAPSTONE As mentioned on page 66, if A is a 2×2 matrix

$$A = \begin{bmatrix} a & b \\ c & d \end{bmatrix}$$

then A is invertible if and only if $ad - bc \neq 0$. Verify that the inverse of A is

$$A^{-1} = \frac{1}{ad - bc}\begin{bmatrix} d & -b \\ -c & a \end{bmatrix}.$$

83. Writing Explain in your own words how to write a system of three linear equations in three variables as a matrix equation, $A\mathbf{x} = \mathbf{b}$, as well as how to solve the system using an inverse matrix.

2.4   Elementary Matrices

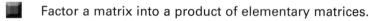

■ Factor a matrix into a product of elementary matrices.

■ Find and use an *LU*-factorization of a matrix to solve a system of linear equations.

ELEMENTARY MATRICES AND ELEMENTARY ROW OPERATIONS

Section 1.2 introduced the three elementary row operations for matrices listed below.

1. Interchange two rows.

2. Multiply a row by a nonzero constant.

3. Add a multiple of a row to another row.

In this section, you will see how to use matrix multiplication to perform these operations.

REMARK

The identity matrix I_n is elementary by this definition because it can be obtained from itself by multiplying any one of its rows by 1.

> **Definition of an Elementary Matrix**
>
> An $n \times n$ matrix is an **elementary matrix** when it can be obtained from the identity matrix I_n by a single elementary row operation.

EXAMPLE 1 | **Elementary Matrices and Nonelementary Matrices**

Which of the matrices below are elementary? For those that are, describe the corresponding elementary row operation.

a. $\begin{bmatrix} 1 & 0 & 0 \\ 0 & 3 & 0 \\ 0 & 0 & 1 \end{bmatrix}$
 b. $\begin{bmatrix} 1 & 0 & 0 \\ 0 & 1 & 0 \end{bmatrix}$

c. $\begin{bmatrix} 1 & 0 & 0 \\ 0 & 1 & 0 \\ 0 & 0 & 0 \end{bmatrix}$
 d. $\begin{bmatrix} 1 & 0 & 0 \\ 0 & 0 & 1 \\ 0 & 1 & 0 \end{bmatrix}$

e. $\begin{bmatrix} 1 & 0 \\ 2 & 1 \end{bmatrix}$
 f. $\begin{bmatrix} 1 & 0 & 0 \\ 0 & 2 & 0 \\ 0 & 0 & -1 \end{bmatrix}$

SOLUTION

a. This matrix *is* elementary. To obtain it from I_3, multiply the second row of I_3 by 3.

b. This matrix is *not* elementary because it is not square.

c. This matrix is *not* elementary because to obtain it from I_3, you must multiply the third row of I_3 by 0 (row multiplication must be by a nonzero constant).

d. This matrix *is* elementary. To obtain it from I_3, interchange the second and third rows of I_3.

e. This matrix *is* elementary. To obtain it from I_2, multiply the first row of I_2 by 2 and add the result to the second row.

f. This matrix is *not* elementary because it requires two elementary row operations to obtain from I_3.

Elementary matrices are useful because they enable you to use matrix multiplication to perform elementary row operations, as demonstrated in Example 2.

EXAMPLE 2   Elementary Matrices and Elementary Row Operations

a. In the matrix product below, E is the elementary matrix in which the first two rows of I_3 are interchanged.

$$
\overset{E}{\begin{bmatrix} 0 & 1 & 0 \\ 1 & 0 & 0 \\ 0 & 0 & 1 \end{bmatrix}} \overset{A}{\begin{bmatrix} 0 & 2 & 1 \\ 1 & -3 & 6 \\ 3 & 2 & -1 \end{bmatrix}} = \begin{bmatrix} 1 & -3 & 6 \\ 0 & 2 & 1 \\ 3 & 2 & -1 \end{bmatrix}
$$

Note that the first two rows of A are interchanged when multiplying *on the left* by E.

b. In the matrix product below, E is the elementary matrix in which the second row of I_3 is multiplied by $\frac{1}{2}$.

$$
\overset{E}{\begin{bmatrix} 1 & 0 & 0 \\ 0 & \frac{1}{2} & 0 \\ 0 & 0 & 1 \end{bmatrix}} \overset{A}{\begin{bmatrix} 1 & 0 & -4 & 1 \\ 0 & 2 & 6 & -4 \\ 0 & 1 & 3 & 1 \end{bmatrix}} = \begin{bmatrix} 1 & 0 & -4 & 1 \\ 0 & 1 & 3 & -2 \\ 0 & 1 & 3 & 1 \end{bmatrix}
$$

Note that the second row of A is multiplied by $\frac{1}{2}$ when multiplying *on the left* by E.

c. In the matrix product below, E is the elementary matrix in which 2 times the first row of I_3 is added to the second row.

$$
\overset{E}{\begin{bmatrix} 1 & 0 & 0 \\ 2 & 1 & 0 \\ 0 & 0 & 1 \end{bmatrix}} \overset{A}{\begin{bmatrix} 1 & 0 & -1 \\ -2 & -2 & 3 \\ 0 & 4 & 5 \end{bmatrix}} = \begin{bmatrix} 1 & 0 & -1 \\ 0 & -2 & 1 \\ 0 & 4 & 5 \end{bmatrix}
$$

Note that 2 times the first row of A is added to the second row when multiplying *on the left* by E.

Notice from Example 2(b) that you can use matrix multiplication to perform elementary row operations on *nonsquare* matrices. If the size of A is $n \times p$, then E must have order n.

In each of the three products in Example 2, you are able to perform elementary row operations by multiplying *on the left* by an elementary matrix. The next theorem, stated without proof, generalizes this property of elementary matrices.

REMARK

Be sure to remember in Theorem 2.12 to multiply *A* *on the left* by the elementary matrix *E*. This text does not consider right multiplication by elementary matrices, which involves column operations.

THEOREM 2.12   Representing Elementary Row Operations

Let E be the elementary matrix obtained by performing an elementary row operation on I_m. If that same elementary row operation is performed on an $m \times n$ matrix A, then the resulting matrix is the product EA.

Most applications of elementary row operations require a sequence of operations. For instance, Gaussian elimination usually requires several elementary row operations to row reduce a matrix. This translates into multiplication on the left by several elementary matrices. The order of multiplication is important; the elementary matrix immediately to the left of A corresponds to the row operation performed first. Example 3 demonstrates this process.

EXAMPLE 3 **Using Elementary Matrices**

Find a sequence of elementary matrices that can be used to write the matrix A in row-echelon form.

$$A = \begin{bmatrix} 0 & 1 & 3 & 5 \\ 1 & -3 & 0 & 2 \\ 2 & -6 & 2 & 0 \end{bmatrix}$$

SOLUTION

Matrix	Elementary Row Operation	Elementary Matrix
$\begin{bmatrix} 1 & -3 & 0 & 2 \\ 0 & 1 & 3 & 5 \\ 2 & -6 & 2 & 0 \end{bmatrix}$	$R_1 \leftrightarrow R_2$	$E_1 = \begin{bmatrix} 0 & 1 & 0 \\ 1 & 0 & 0 \\ 0 & 0 & 1 \end{bmatrix}$
$\begin{bmatrix} 1 & -3 & 0 & 2 \\ 0 & 1 & 3 & 5 \\ 0 & 0 & 2 & -4 \end{bmatrix}$	$R_3 + (-2)R_1 \rightarrow R_3$	$E_2 = \begin{bmatrix} 1 & 0 & 0 \\ 0 & 1 & 0 \\ -2 & 0 & 1 \end{bmatrix}$
$\begin{bmatrix} 1 & -3 & 0 & 2 \\ 0 & 1 & 3 & 5 \\ 0 & 0 & 1 & -2 \end{bmatrix}$	$\left(\frac{1}{2}\right)R_3 \rightarrow R_3$	$E_3 = \begin{bmatrix} 1 & 0 & 0 \\ 0 & 1 & 0 \\ 0 & 0 & \frac{1}{2} \end{bmatrix}$

The three elementary matrices E_1, E_2, and E_3 can be used to perform the same elimination.

REMARK

The procedure demonstrated in Example 3 is primarily of theoretical interest. In other words, this procedure is not a practical method for performing Gaussian elimination.

$$B = E_3 E_2 E_1 A = \begin{bmatrix} 1 & 0 & 0 \\ 0 & 1 & 0 \\ 0 & 0 & \frac{1}{2} \end{bmatrix} \begin{bmatrix} 1 & 0 & 0 \\ 0 & 1 & 0 \\ -2 & 0 & 1 \end{bmatrix} \begin{bmatrix} 0 & 1 & 0 \\ 1 & 0 & 0 \\ 0 & 0 & 1 \end{bmatrix} \begin{bmatrix} 0 & 1 & 3 & 5 \\ 1 & -3 & 0 & 2 \\ 2 & -6 & 2 & 0 \end{bmatrix}$$

$$= \begin{bmatrix} 1 & 0 & 0 \\ 0 & 1 & 0 \\ 0 & 0 & \frac{1}{2} \end{bmatrix} \begin{bmatrix} 1 & 0 & 0 \\ 0 & 1 & 0 \\ -2 & 0 & 1 \end{bmatrix} \begin{bmatrix} 1 & -3 & 0 & 2 \\ 0 & 1 & 3 & 5 \\ 2 & -6 & 2 & 0 \end{bmatrix}$$

$$= \begin{bmatrix} 1 & 0 & 0 \\ 0 & 1 & 0 \\ 0 & 0 & \frac{1}{2} \end{bmatrix} \begin{bmatrix} 1 & -3 & 0 & 2 \\ 0 & 1 & 3 & 5 \\ 0 & 0 & 2 & -4 \end{bmatrix} = \begin{bmatrix} 1 & -3 & 0 & 2 \\ 0 & 1 & 3 & 5 \\ 0 & 0 & 1 & -2 \end{bmatrix}$$

The two matrices in Example 3

$$A = \begin{bmatrix} 0 & 1 & 3 & 5 \\ 1 & -3 & 0 & 2 \\ 2 & -6 & 2 & 0 \end{bmatrix} \quad \text{and} \quad B = \begin{bmatrix} 1 & -3 & 0 & 2 \\ 0 & 1 & 3 & 5 \\ 0 & 0 & 1 & -2 \end{bmatrix}$$

are row-equivalent because you can obtain B by performing a sequence of row operations on A. That is, $B = E_3 E_2 E_1 A$.

The definition of row-equivalent matrices is restated below using elementary matrices.

Definition of Row Equivalence

Let A and B be $m \times n$ matrices. Matrix B is **row-equivalent** to A when there exists a finite number of elementary matrices E_1, E_2, . . . , E_k such that

$$B = E_k E_{k-1} \cdots E_2 E_1 A.$$

You know from Section 2.3 that not all square matrices are invertible. Every elementary matrix, however, is invertible. Moreover, the inverse of an elementary matrix is itself an elementary matrix.

THEOREM 2.13 Elementary Matrices Are Invertible

If E is an elementary matrix, then E^{-1} exists and is an elementary matrix.

The inverse of an elementary matrix E is the elementary matrix that converts E back to I_n. For instance, the inverses of the three elementary matrices in Example 3 are shown below.

Elementary Matrix   Inverse Matrix

$$E_1 = \begin{bmatrix} 0 & 1 & 0 \\ 1 & 0 & 0 \\ 0 & 0 & 1 \end{bmatrix} \begin{matrix} R_1 \leftrightarrow R_2 \\ \\ \end{matrix} \qquad E_1^{-1} = \begin{bmatrix} 0 & 1 & 0 \\ 1 & 0 & 0 \\ 0 & 0 & 1 \end{bmatrix} \begin{matrix} R_1 \leftrightarrow R_2 \\ \\ \end{matrix}$$

$$E_2 = \begin{bmatrix} 1 & 0 & 0 \\ 0 & 1 & 0 \\ -2 & 0 & 1 \end{bmatrix} \begin{matrix} \\ \\ R_3 + (-2)R_1 \rightarrow R_3 \end{matrix} \qquad E_2^{-1} = \begin{bmatrix} 1 & 0 & 0 \\ 0 & 1 & 0 \\ 2 & 0 & 1 \end{bmatrix} \begin{matrix} \\ \\ R_3 + (2)R_1 \rightarrow R_3 \end{matrix}$$

$$E_3 = \begin{bmatrix} 1 & 0 & 0 \\ 0 & 1 & 0 \\ 0 & 0 & \frac{1}{2} \end{bmatrix} \begin{matrix} \\ \\ \left(\frac{1}{2}\right)R_3 \rightarrow R_3 \end{matrix} \qquad E_3^{-1} = \begin{bmatrix} 1 & 0 & 0 \\ 0 & 1 & 0 \\ 0 & 0 & 2 \end{bmatrix} \begin{matrix} \\ \\ (2)R_3 \rightarrow R_3 \end{matrix}$$

Use matrix multiplication to check these results.

The next theorem states that every invertible matrix can be written as the product of elementary matrices.

THEOREM 2.14 A Property of Invertible Matrices

A square matrix A is invertible if and only if it can be written as the product of elementary matrices.

PROOF

The phrase "if and only if" means that there are actually two parts to the theorem. On the one hand, you have to show that *if A is invertible, then* it can be written as the product of elementary matrices. Then you have to show that *if A can be written as the product of elementary matrices, then A is invertible.*

To prove the theorem in one direction, assume A is invertible. From Theorem 2.11 you know that the system of linear equations represented by $A\mathbf{x} = O$ has only the trivial solution. But this implies that the augmented matrix $[A \quad O]$ can be rewritten in the form $[I \quad O]$ (using elementary row operations corresponding to $E_1, E_2, \ldots,$ and E_k). So, $E_k \cdots E_2 E_1 A = I$ and it follows that $A = E_1^{-1} E_2^{-1} \cdots E_k^{-1}$. A can be written as the product of elementary matrices.

To prove the theorem in the other direction, assume A is the product of elementary matrices. Every elementary matrix is invertible and the product of invertible matrices is invertible, so it follows that A is invertible. This completes the proof.

Example 4 illustrates the first part of this proof.

EXAMPLE 4	Writing a Matrix as the Product of Elementary Matrices

Find a sequence of elementary matrices whose product is the nonsingular matrix

$$A = \begin{bmatrix} -1 & -2 \\ 3 & 8 \end{bmatrix}.$$

SOLUTION

Begin by finding a sequence of elementary row operations that can be used to rewrite A in reduced row-echelon form.

Matrix	Elementary Row Operation	Elementary Matrix
$\begin{bmatrix} 1 & 2 \\ 3 & 8 \end{bmatrix}$	$(-1)R_1 \rightarrow R_1$	$E_1 = \begin{bmatrix} -1 & 0 \\ 0 & 1 \end{bmatrix}$
$\begin{bmatrix} 1 & 2 \\ 0 & 2 \end{bmatrix}$	$R_2 + (-3)R_1 \rightarrow R_2$	$E_2 = \begin{bmatrix} 1 & 0 \\ -3 & 1 \end{bmatrix}$
$\begin{bmatrix} 1 & 2 \\ 0 & 1 \end{bmatrix}$	$\left(\frac{1}{2}\right)R_2 \rightarrow R_2$	$E_3 = \begin{bmatrix} 1 & 0 \\ 0 & \frac{1}{2} \end{bmatrix}$
$\begin{bmatrix} 1 & 0 \\ 0 & 1 \end{bmatrix}$	$R_1 + (-2)R_2 \rightarrow R_1$	$E_4 = \begin{bmatrix} 1 & -2 \\ 0 & 1 \end{bmatrix}$

Now, from the matrix product $E_4E_3E_2E_1A = I$, solve for A to obtain $A = E_1^{-1}E_2^{-1}E_3^{-1}E_4^{-1}$. This implies that A is a product of elementary matrices.

$$A = \overset{E_1^{-1}}{\begin{bmatrix} -1 & 0 \\ 0 & 1 \end{bmatrix}} \overset{E_2^{-1}}{\begin{bmatrix} 1 & 0 \\ 3 & 1 \end{bmatrix}} \overset{E_3^{-1}}{\begin{bmatrix} 1 & 0 \\ 0 & 2 \end{bmatrix}} \overset{E_4^{-1}}{\begin{bmatrix} 1 & 2 \\ 0 & 1 \end{bmatrix}} = \begin{bmatrix} -1 & -2 \\ 3 & 8 \end{bmatrix}$$

In Section 2.3, you learned a process for finding the inverse of a nonsingular matrix A. There, you used Gauss-Jordan elimination to reduce the augmented matrix $[A \quad I]$ to $[I \quad A^{-1}]$. You can now use Theorem 2.14 to justify this procedure. Specifically, the proof of Theorem 2.14 allows you to write the product

$$I = E_k \cdots E_2E_1A.$$

Multiplying both sides of this equation (on the right) by A^{-1}, $A^{-1} = E_k \cdots E_2E_1I$. In other words, a sequence of elementary matrices that reduces A to the identity I also reduces the identity I to A^{-1}. Applying the corresponding sequence of elementary row operations to the matrices A and I simultaneously, you have

$$E_k \cdots E_2E_1[A \quad I] = [I \quad A^{-1}].$$

Of course, if A is singular, then no such sequence is possible.

The next theorem ties together some important relationships between $n \times n$ matrices and systems of linear equations. The essential parts of this theorem have already been proved (see Theorems 2.11 and 2.14); it is left to you to fill in the other parts of the proof.

THEOREM 2.15   Equivalent Conditions

If A is an $n \times n$ matrix, then the statements below are equivalent.

1. A is invertible.
2. $A\mathbf{x} = \mathbf{b}$ has a unique solution for every $n \times 1$ column matrix $\mathbf{b}$.
3. $A\mathbf{x} = O$ has only the trivial solution.
4. A is row-equivalent to I_n.
5. A can be written as the product of elementary matrices.

THE *LU*-FACTORIZATION

At the heart of the most efficient and modern algorithms for solving linear systems $A\mathbf{x} = \mathbf{b}$ is the *LU*-factorization, in which the square matrix A is expressed as a product, $A = LU$. In this product, the square matrix L is **lower triangular,** which means all the entries above the main diagonal are zero. The square matrix U is **upper triangular,** which means all the entries below the main diagonal are zero.

$$\begin{bmatrix} a_{11} & 0 & 0 \\ a_{21} & a_{22} & 0 \\ a_{31} & a_{32} & a_{33} \end{bmatrix} \qquad \begin{bmatrix} a_{11} & a_{12} & a_{13} \\ 0 & a_{22} & a_{23} \\ 0 & 0 & a_{33} \end{bmatrix}$$

3 × 3 lower triangular matrix   3 × 3 upper triangular matrix

---

Definition of *LU*-Factorization

If the $n \times n$ matrix A can be written as the product of a lower triangular matrix L and an upper triangular matrix U, then $A = LU$ is an ***LU*-factorization** of A.

EXAMPLE 5   *LU*-Factorizations

a. $\begin{bmatrix} 1 & 2 \\ 1 & 0 \end{bmatrix} = \begin{bmatrix} 1 & 0 \\ 1 & 1 \end{bmatrix}\begin{bmatrix} 1 & 2 \\ 0 & -2 \end{bmatrix} = LU$

is an *LU*-factorization of the matrix

$$A = \begin{bmatrix} 1 & 2 \\ 1 & 0 \end{bmatrix}$$

as the product of the lower triangular matrix

$$L = \begin{bmatrix} 1 & 0 \\ 1 & 1 \end{bmatrix}$$

and the upper triangular matrix

$$U = \begin{bmatrix} 1 & 2 \\ 0 & -2 \end{bmatrix}.$$

b. $A = \begin{bmatrix} 1 & -3 & 0 \\ 0 & 1 & 3 \\ 2 & -10 & 2 \end{bmatrix} = \begin{bmatrix} 1 & 0 & 0 \\ 0 & 1 & 0 \\ 2 & -4 & 1 \end{bmatrix}\begin{bmatrix} 1 & -3 & 0 \\ 0 & 1 & 3 \\ 0 & 0 & 14 \end{bmatrix} = LU$

is an *LU*-factorization of the matrix A.

LINEAR ALGEBRA APPLIED

Computational fluid dynamics (CFD) is the computer-based simulation of such real-life phenomena as fluid flow, heat transfer, and chemical reactions. Solving the conservation of energy, mass, and momentum equations involved in a CFD analysis can involve large systems of linear equations. So, for efficiency in computing, CFD analyses often use matrix partitioning and *LU*-factorization in their algorithms. Aerospace companies such as Boeing and Airbus have used CFD analysis in aircraft design. For instance, engineers at Boeing used CFD analysis to simulate airflow around a virtual model of their 787 aircraft to help produce a faster and more efficient design than those of earlier Boeing aircraft.

If a square matrix A row reduces to an upper triangular matrix U using only the row operation of adding a multiple of one row to another row below it, then it is relatively easy to find an LU-factorization of the matrix A. All you need to do is keep track of the individual row operations, as shown in the next example.

EXAMPLE 6   Finding an LU-Factorization of a Matrix

Find an LU-factorization of the matrix $A = \begin{bmatrix} 1 & -3 & 0 \\ 0 & 1 & 3 \\ 2 & -10 & 2 \end{bmatrix}$.

SOLUTION

Begin by row reducing A to upper triangular form while keeping track of the elementary matrices used for each row operation.

Matrix	Elementary Row Operation	Elementary Matrix
$\begin{bmatrix} 1 & -3 & 0 \\ 0 & 1 & 3 \\ 0 & -4 & 2 \end{bmatrix}$	$R_3 + (-2)R_1 \to R_3$	$E_1 = \begin{bmatrix} 1 & 0 & 0 \\ 0 & 1 & 0 \\ -2 & 0 & 1 \end{bmatrix}$
$\begin{bmatrix} 1 & -3 & 0 \\ 0 & 1 & 3 \\ 0 & 0 & 14 \end{bmatrix}$	$R_3 + (4)R_2 \to R_3$	$E_2 = \begin{bmatrix} 1 & 0 & 0 \\ 0 & 1 & 0 \\ 0 & 4 & 1 \end{bmatrix}$

The reduced matrix above is an upper triangular matrix U, and it follows that $E_2 E_1 A = U$, or $A = E_1^{-1} E_2^{-1} U$. The product of the lower triangular matrices

$$E_1^{-1} E_2^{-1} = \begin{bmatrix} 1 & 0 & 0 \\ 0 & 1 & 0 \\ 2 & 0 & 1 \end{bmatrix} \begin{bmatrix} 1 & 0 & 0 \\ 0 & 1 & 0 \\ 0 & -4 & 1 \end{bmatrix} = \begin{bmatrix} 1 & 0 & 0 \\ 0 & 1 & 0 \\ 2 & -4 & 1 \end{bmatrix}$$

is a lower triangular matrix L, so the factorization $A = LU$ is complete. Notice that this is the same LU-factorization as in Example 5(b).

If A row reduces to an upper triangular matrix U using only the row operation of adding a multiple of one row to another row below it, then A has an LU-factorization.

$$E_k \cdots E_2 E_1 A = U$$
$$A = E_1^{-1} E_2^{-1} \cdots E_k^{-1} U = LU$$

Here L is the product of the inverses of the elementary matrices used in the row reduction.

Note that the multipliers in Example 6 are -2 and 4, which are the negatives of the corresponding entries in L. This is true in general. If U can be obtained from A using only the elementary row operation of adding a multiple of one row to another row below it, then the matrix L is lower triangular (with 1's along the diagonal), and the negative of each multiplier is in the same position as that of the corresponding zero in U below the main diagonal.

Once you have obtained an LU-factorization of a matrix A, you can then solve the system of n linear equations in n variables $A\mathbf{x} = \mathbf{b}$ very efficiently in two steps.

1. Write $\mathbf{y} = U\mathbf{x}$ and solve $L\mathbf{y} = \mathbf{b}$ for $\mathbf{y}$.

2. Solve $U\mathbf{x} = \mathbf{y}$ for $\mathbf{x}$.

The column matrix $\mathbf{x}$ is the solution of the original system because $A\mathbf{x} = LU\mathbf{x} = L\mathbf{y} = \mathbf{b}$.

The second step is just back-substitution, because the matrix U is upper triangular. The first step is similar, except that it starts at the top of the matrix, because L is lower triangular. For this reason, the first step is often called **forward substitution.**

EXAMPLE 7   **Solving a Linear System Using**
_LU_-Factorization

See LarsonLinearAlgebra.com for an interactive version of this type of example.

Solve the linear system.

$$
\begin{aligned}
x_1 - \ 3x_2 \qquad\quad &= -5 \\
x_2 + 3x_3 &= -1 \\
2x_1 - 10x_2 + 2x_3 &= -20
\end{aligned}
$$

SOLUTION

You obtained an *LU*-factorization of the coefficient matrix *A* in Example 6.

$$
A = \begin{bmatrix} 1 & -3 & 0 \\ 0 & 1 & 3 \\ 2 & -10 & 2 \end{bmatrix}
$$

$$
= \begin{bmatrix} 1 & 0 & 0 \\ 0 & 1 & 0 \\ 2 & -4 & 1 \end{bmatrix}\begin{bmatrix} 1 & -3 & 0 \\ 0 & 1 & 3 \\ 0 & 0 & 14 \end{bmatrix}
$$

First, let $\mathbf{y} = U\mathbf{x}$ and solve the system $L\mathbf{y} = \mathbf{b}$ for $\mathbf{y}$.

$$
\begin{bmatrix} 1 & 0 & 0 \\ 0 & 1 & 0 \\ 2 & -4 & 1 \end{bmatrix}\begin{bmatrix} y_1 \\ y_2 \\ y_3 \end{bmatrix} = \begin{bmatrix} -5 \\ -1 \\ -20 \end{bmatrix}
$$

Solve this system using forward substitution. Starting with the first equation, you have $y_1 = -5$. The second equation gives $y_2 = -1$. Finally, from the third equation,

$$
\begin{aligned}
2y_1 - 4y_2 + y_3 &= -20 \\
y_3 &= -20 - 2y_1 + 4y_2 \\
y_3 &= -20 - 2(-5) + 4(-1) \\
y_3 &= -14.
\end{aligned}
$$

The solution of $L\mathbf{y} = \mathbf{b}$ is

$$
\mathbf{y} = \begin{bmatrix} -5 \\ -1 \\ -14 \end{bmatrix}.
$$

Now solve the system $U\mathbf{x} = \mathbf{y}$ for $\mathbf{x}$ using back-substitution.

$$
\begin{bmatrix} 1 & -3 & 0 \\ 0 & 1 & 3 \\ 0 & 0 & 14 \end{bmatrix}\begin{bmatrix} x_1 \\ x_2 \\ x_3 \end{bmatrix} = \begin{bmatrix} -5 \\ -1 \\ -14 \end{bmatrix}
$$

From the bottom equation, $x_3 = -1$. Then, the second equation gives

$$
x_2 + 3(-1) = -1
$$

or $x_2 = 2$. Finally, the first equation gives

$$
x_1 - 3(2) = -5
$$

or $x_1 = 1$. So, the solution of the original system of equations is

$$
\mathbf{x} = \begin{bmatrix} 1 \\ 2 \\ -1 \end{bmatrix}.
$$

2.4 Exercises

See CalcChat.com for worked-out solutions to odd-numbered exercises.

Elementary Matrices **In Exercises 1–8, determine whether the matrix is elementary. If it is, state the elementary row operation used to produce it.**

1. $\begin{bmatrix} 1 & 0 \\ 0 & 2 \end{bmatrix}$

2. $\begin{bmatrix} 1 & 0 & 0 \\ 0 & 0 & 1 \end{bmatrix}$

3. $\begin{bmatrix} 1 & 0 \\ 2 & 1 \end{bmatrix}$

4. $\begin{bmatrix} 0 & 1 \\ 1 & 0 \end{bmatrix}$

5. $\begin{bmatrix} 2 & 0 & 0 \\ 0 & 0 & 1 \\ 0 & 1 & 0 \end{bmatrix}$

6. $\begin{bmatrix} 1 & 0 & 0 \\ 0 & 1 & 0 \\ 2 & 0 & 1 \end{bmatrix}$

7. $\begin{bmatrix} 1 & 0 & 0 & 0 \\ 0 & 1 & 0 & 0 \\ 0 & -5 & 1 & 0 \\ 0 & 0 & 0 & 1 \end{bmatrix}$

8. $\begin{bmatrix} 1 & 0 & 0 & 0 \\ 2 & 1 & 0 & 0 \\ 0 & 0 & 1 & 0 \\ 0 & 0 & -3 & 1 \end{bmatrix}$

Finding an Elementary Matrix **In Exercises 9–12, let A, B, and C be**

$$A = \begin{bmatrix} 1 & 2 & -3 \\ 0 & 1 & 2 \\ -1 & 2 & 0 \end{bmatrix}, \quad B = \begin{bmatrix} -1 & 2 & 0 \\ 0 & 1 & 2 \\ 1 & 2 & -3 \end{bmatrix}, \text{ and}$$

$$C = \begin{bmatrix} 0 & 4 & -3 \\ 0 & 1 & 2 \\ -1 & 2 & 0 \end{bmatrix}.$$

9. Find an elementary matrix E such that $EA = B$.

10. Find an elementary matrix E such that $EA = C$.

11. Find an elementary matrix E such that $EB = A$.

12. Find an elementary matrix E such that $EC = A$.

Finding a Sequence of Elementary Matrices **In Exercises 13–18, find a sequence of elementary matrices that can be used to write the matrix in row-echelon form.**

13. $\begin{bmatrix} 0 & 1 & 7 \\ 5 & 10 & -5 \end{bmatrix}$

14. $\begin{bmatrix} 0 & 3 & -3 & 6 \\ 1 & -1 & 2 & -2 \\ 0 & 0 & 2 & 2 \end{bmatrix}$

15. $\begin{bmatrix} 1 & -2 & -1 & 0 \\ 0 & 4 & 8 & -4 \\ -6 & 12 & 8 & 1 \end{bmatrix}$

16. $\begin{bmatrix} 1 & 3 & 0 \\ 2 & 5 & -1 \\ 3 & -2 & -4 \end{bmatrix}$

17. $\begin{bmatrix} -2 & 1 & 0 \\ 3 & -4 & 0 \\ 1 & -2 & 2 \\ -1 & 2 & -2 \end{bmatrix}$

18. $\begin{bmatrix} 1 & -6 & 0 & 2 \\ 0 & -3 & 3 & 9 \\ 2 & 5 & -1 & 1 \\ 4 & 8 & -5 & 1 \end{bmatrix}$

Finding the Inverse of an Elementary Matrix **In Exercises 19–24, find the inverse of the elementary matrix.**

19. $\begin{bmatrix} 0 & 1 \\ 1 & 0 \end{bmatrix}$

20. $\begin{bmatrix} 5 & 0 \\ 0 & 1 \end{bmatrix}$

21. $\begin{bmatrix} 0 & 0 & 1 \\ 0 & 1 & 0 \\ 1 & 0 & 0 \end{bmatrix}$

22. $\begin{bmatrix} 1 & 0 & 0 \\ 0 & 1 & 0 \\ 0 & -3 & 1 \end{bmatrix}$

23. $\begin{bmatrix} k & 0 & 0 \\ 0 & 1 & 0 \\ 0 & 0 & 1 \end{bmatrix}$
$k \neq 0$

24. $\begin{bmatrix} 1 & 0 & 0 & 0 \\ 0 & 1 & k & 0 \\ 0 & 0 & 1 & 0 \\ 0 & 0 & 0 & 1 \end{bmatrix}$

Finding the Inverse of a Matrix **In Exercises 25–28, find the inverse of the matrix using elementary matrices.**

25. $\begin{bmatrix} 3 & -2 \\ 1 & 0 \end{bmatrix}$

26. $\begin{bmatrix} 2 & 0 \\ 1 & 1 \end{bmatrix}$

27. $\begin{bmatrix} 1 & 0 & -1 \\ 0 & 6 & -1 \\ 0 & 0 & 4 \end{bmatrix}$

28. $\begin{bmatrix} 1 & 0 & -2 \\ 0 & 2 & 1 \\ 0 & 0 & 1 \end{bmatrix}$

Finding a Sequence of Elementary Matrices **In Exercises 29–36, find a sequence of elementary matrices whose product is the given nonsingular matrix.**

29. $\begin{bmatrix} 1 & 2 \\ 1 & 0 \end{bmatrix}$

30. $\begin{bmatrix} 0 & 1 \\ 1 & 0 \end{bmatrix}$

31. $\begin{bmatrix} 4 & -1 \\ 3 & -1 \end{bmatrix}$

32. $\begin{bmatrix} 1 & 1 \\ 2 & 1 \end{bmatrix}$

33. $\begin{bmatrix} 1 & -2 & 0 \\ -1 & 3 & 0 \\ 0 & 0 & 1 \end{bmatrix}$

34. $\begin{bmatrix} 1 & 2 & 3 \\ 2 & 5 & 6 \\ 1 & 3 & 4 \end{bmatrix}$

35. $\begin{bmatrix} 1 & 0 & 0 & 1 \\ 0 & -1 & 3 & 0 \\ 0 & 0 & 2 & 0 \\ 0 & 0 & 1 & -1 \end{bmatrix}$

36. $\begin{bmatrix} 4 & 0 & 0 & 2 \\ 0 & 1 & 0 & 1 \\ 0 & 0 & -1 & 2 \\ 1 & 0 & 0 & -2 \end{bmatrix}$

37. Writing Is the product of two elementary matrices always elementary? Explain.

38. Writing E is the elementary matrix obtained by interchanging two rows in I_n. A is an $n \times n$ matrix.

(a) How will EA compare with A? (b) Find E^2.

39. Use elementary matrices to find the inverse of

$$A = \begin{bmatrix} 1 & 0 & 0 \\ 0 & 1 & 0 \\ a & b & c \end{bmatrix}, \quad c \neq 0.$$

40. Use elementary matrices to find the inverse of

$$A = \begin{bmatrix} 1 & a & 0 \\ 0 & 1 & 0 \\ 0 & 0 & 1 \end{bmatrix} \begin{bmatrix} 1 & 0 & 0 \\ b & 1 & 0 \\ 0 & 0 & 1 \end{bmatrix} \begin{bmatrix} 1 & 0 & 0 \\ 0 & 1 & 0 \\ 0 & 0 & c \end{bmatrix},$$

$c \neq 0.$

True or False? **In Exercises 41 and 42, determine whether each statement is true or false. If a statement is true, give a reason or cite an appropriate statement from the text. If a statement is false, provide an example that shows the statement is not true in all cases or cite an appropriate statement from the text.**

41. (a) The identity matrix is an elementary matrix.

(b) If E is an elementary matrix, then $2E$ is an elementary matrix.

(c) The inverse of an elementary matrix is an elementary matrix.

42. (a) The zero matrix is an elementary matrix.

(b) A square matrix is nonsingular when it can be written as the product of elementary matrices.

(c) $A\mathbf{x} = O$ has only the trivial solution if and only if $A\mathbf{x} = \mathbf{b}$ has a unique solution for every $n \times 1$ column matrix $\mathbf{b}$.

Finding an *LU*-Factorization of a Matrix **In Exercises 43–46, find an *LU*-factorization of the matrix.**

43. $\begin{bmatrix} 1 & 0 \\ -2 & 1 \end{bmatrix}$ **44.** $\begin{bmatrix} -2 & 1 \\ -6 & 4 \end{bmatrix}$

45. $\begin{bmatrix} 3 & 0 & 1 \\ 6 & 1 & 1 \\ -3 & 1 & 0 \end{bmatrix}$ **46.** $\begin{bmatrix} 2 & 0 & 0 \\ 0 & -3 & 1 \\ 10 & 12 & 3 \end{bmatrix}$

Solving a Linear System Using *LU*-Factorization **In Exercises 47 and 48, use an *LU*-factorization of the coefficient matrix to solve the linear system.**

47. $2x + y \quad = \quad 1$
  $y - z = \quad 2$
  $-2x + y + z = -2$

48. $2x_1 \qquad\qquad = \quad 4$
  $-2x_1 + x_2 - x_3 \quad = -4$
  $6x_1 + 2x_2 + x_3 \quad = 15$
  $\qquad\qquad -x_4 = -1$

Idempotent Matrices **In Exercises 49–52, determine whether the matrix is idempotent. A square matrix A is idempotent when $A^2 = A$.**

49. $\begin{bmatrix} 1 & 0 \\ 0 & 0 \end{bmatrix}$ **50.** $\begin{bmatrix} 0 & 1 \\ 1 & 0 \end{bmatrix}$

51. $\begin{bmatrix} 0 & 0 & 1 \\ 0 & 1 & 0 \\ 1 & 0 & 0 \end{bmatrix}$ **52.** $\begin{bmatrix} 0 & 1 & 0 \\ 1 & 0 & 0 \\ 0 & 0 & 1 \end{bmatrix}$

53. Determine a and b such that A is idempotent.

$$A = \begin{bmatrix} 1 & 0 \\ a & b \end{bmatrix}$$

54. Guided Proof Prove that A is idempotent if and only if A^T is idempotent.

Getting Started: The phrase "if and only if" means that you have to prove two statements:

1. If A is idempotent, then A^T is idempotent.

2. If A^T is idempotent, then A is idempotent.

(i) Begin your proof of the first statement by assuming that A is idempotent.

(ii) This means that $A^2 = A$.

(iii) Use the properties of the transpose to show that A^T is idempotent.

(iv) Begin your proof of the second statement by assuming that A^T is idempotent.

55. Proof Prove that if A is an $n \times n$ matrix that is idempotent and invertible, then $A = I_n$.

56. Proof Prove that if A and B are idempotent and $AB = BA$, then AB is idempotent.

57. Guided Proof Prove that if A is row-equivalent to B and B is row-equivalent to C, then A is row-equivalent to C.

Getting Started: To prove that A is row-equivalent to C, you have to find elementary matrices $E_1, E_2, \ldots, E_k$ such that $A = E_k \cdots E_2 E_1 C$.

(i) Begin by observing that A is row-equivalent to B and B is row-equivalent to C.

(ii) This means that there exist elementary matrices $F_1, F_2, \ldots, F_n$ and $G_1, G_2, \ldots, G_m$ such that $A = F_n \cdots F_2 F_1 B$ and $B = G_m \cdots G_2 G_1 C$.

(iii) Combine the matrix equations from step (ii).

58. Proof Prove that if A is row-equivalent to B, then B is row-equivalent to A.

59. Proof Let A be a nonsingular matrix. Prove that if B is row-equivalent to A, then B is also nonsingular.

60. CAPSTONE

(a) Explain how to find an elementary matrix.

(b) Explain how to use elementary matrices to find an *LU*-factorization of a matrix.

(c) Explain how to use *LU*-factorization to solve a linear system.

61. Show that the matrix below does not have an *LU*-factorization.

$$A = \begin{bmatrix} 0 & 1 \\ 1 & 0 \end{bmatrix}$$

2.5   Markov Chains

■ Use a stochastic matrix to find the nth state matrix of a Markov chain.

■ Find the steady state matrix of a Markov chain.

■ Find the steady state matrix of an absorbing Markov chain.

STOCHASTIC MATRICES AND MARKOV CHAINS

Many types of applications involve a finite set of *states* $\{S_1, S_2, \ldots, S_n\}$ of a population. For instance, residents of a city may live downtown or in the suburbs. Voters may vote Democrat, Republican, or Independent. Soft drink consumers may buy Coca-Cola, Pepsi Cola, or another brand.

The probability that a member of a population will change from the jth state to the ith state is represented by a number p_{ij}, where $0 \leq p_{ij} \leq 1$. A probability of $p_{ij} = 0$ means that the member is certain *not* to change from the jth state to the ith state, whereas a probability of $p_{ij} = 1$ means that the member is certain to change from the jth state to the ith state.

$$
\overbrace{}^{\text{From}}
$$

$$
P = \begin{bmatrix} p_{11} & p_{12} & \cdots & p_{1n} \\ p_{21} & p_{22} & \cdots & p_{2n} \\ \vdots & \vdots & & \vdots \\ p_{n1} & p_{n2} & \cdots & p_{nn} \end{bmatrix} \begin{matrix} S_1 \\ S_2 \\ \vdots \\ S_n \end{matrix} \Bigg\} \text{ To}
$$

P is called the **matrix of transition probabilities** because it gives the probabilities of each possible type of transition (or change) within the population.

At each transition, each member in a given state must either stay in that state or change to another state. For probabilities, this means that the sum of the entries in any column of P is 1. For instance, in the first column,

$$
p_{11} + p_{21} + \cdots + p_{n1} = 1.
$$

Such a matrix is called **stochastic** (the term "stochastic" means "regarding conjecture"). That is, an $n \times n$ matrix P is a **stochastic matrix** when each entry is a number between 0 and 1 inclusive, and the sum of the entries in each column of P is 1.

EXAMPLE 1

Examples of Stochastic Matrices and Nonstochastic Matrices

The matrices in parts (a), (b), and (c) are stochastic, but the matrices in parts (d), (e), and (f) are not.

a. $\begin{bmatrix} 1 & 0 & 0 \\ 0 & 1 & 0 \\ 0 & 0 & 1 \end{bmatrix}$

b. $\begin{bmatrix} \frac{1}{4} & \frac{1}{5} & \frac{1}{3} & \frac{1}{2} \\ \frac{1}{4} & \frac{13}{60} & 0 & \frac{1}{6} \\ \frac{1}{4} & \frac{1}{3} & \frac{1}{3} & \frac{1}{6} \\ \frac{1}{4} & \frac{1}{4} & \frac{1}{3} & \frac{1}{6} \end{bmatrix}$

c. $\begin{bmatrix} 0.9 & 0.8 \\ 0.1 & 0.2 \end{bmatrix}$

d. $\begin{bmatrix} \frac{1}{2} & \frac{1}{4} \\ 0 & \frac{3}{4} \end{bmatrix}$

e. $\begin{bmatrix} \frac{1}{2} & \frac{1}{4} & \frac{1}{4} \\ \frac{1}{3} & 0 & \frac{2}{3} \\ \frac{1}{4} & \frac{3}{4} & 0 \end{bmatrix}$

f. $\begin{bmatrix} 0.1 & 0.2 & 0.3 & 0.4 \\ 0.2 & 0.3 & 0.4 & 0.5 \\ 0.3 & 0.4 & 0.5 & 0.6 \\ 0.4 & 0.5 & 0.6 & 0.7 \end{bmatrix}$

Example 2 describes the use of a stochastic matrix to measure consumer preferences.

EXAMPLE 2   **A Consumer Preference Model**

Two competing companies offer satellite television service to a city with 100,000 households. Figure 2.1 shows the changes in satellite subscriptions each year. Company A now has 15,000 subscribers and Company B has 20,000 subscribers. How many subscribers will each company have in one year?

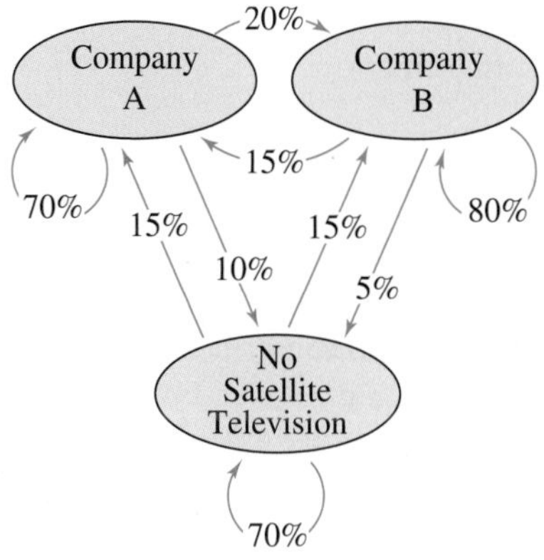

SOLUTION

The matrix of transition probabilities is

$$\overbrace{}^{\text{From}}$$

$$\begin{array}{ccc} A & B & \text{None} \end{array}$$

$$P = \begin{bmatrix} 0.70 & 0.15 & 0.15 \\ 0.20 & 0.80 & 0.15 \\ 0.10 & 0.05 & 0.70 \end{bmatrix} \begin{matrix} A \\ B \\ \text{None} \end{matrix} \bigg\} \text{To}$$

and the initial **state matrix** representing the portions of the total population in the three states is

$$X_0 = \begin{bmatrix} 0.1500 \\ 0.2000 \\ 0.6500 \end{bmatrix}. \begin{matrix} A \\ B \\ \text{None} \end{matrix}$$

To find the state matrix representing the portions of the population in the three states in one year, multiply P by X_0 to obtain

$$X_1 = PX_0 = \begin{bmatrix} 0.70 & 0.15 & 0.15 \\ 0.20 & 0.80 & 0.15 \\ 0.10 & 0.05 & 0.70 \end{bmatrix} \begin{bmatrix} 0.1500 \\ 0.2000 \\ 0.6500 \end{bmatrix} = \begin{bmatrix} 0.2325 \\ 0.2875 \\ 0.4800 \end{bmatrix}.$$

In one year, Company A will have $0.2325(100,000) = 23,250$ subscribers and Company B will have $0.2875(100,000) = 28,750$ subscribers.

A **Markov chain**, named after Russian mathematician Andrey Andreyevich Markov (1856–1922), is a sequence $\{X_n\}$ of state matrices that are related by the equation $X_{k+1} = PX_k$, where P is a stochastic matrix. For instance, consider the consumer preference model discussed in Example 2. To find the state matrix representing the portions of the population in each state in three years, repeatedly multiply the initial state matrix X_0 by the matrix of transition probabilities P.

$$X_1 = PX_0$$

$$X_2 = PX_1 = P \cdot PX_0 = P^2X_0$$

$$X_3 = PX_2 = P \cdot P^2X_0 = P^3X_0$$

In general, the nth state matrix of a Markov chain is P^nX_0, as summarized below.

---

nth State Matrix of a Markov Chain

The nth state matrix of a Markov chain for which P is the matrix of transition probabilities and X_0 is the initial state matrix is

$$X_n = P^nX_0.$$

Example 3 uses the model discussed in Example 2 to demonstrate this process.

Figure 2.1

REMARK

Always assume that the matrix P of transition probabilities in a Markov chain remains constant between states.

EXAMPLE 3   **A Consumer Preference Model**

Assuming the matrix of transition probabilities from Example 2 remains the same year after year, find the number of subscribers each satellite television company will have after (a) 3 years, (b) 5 years, (c) 10 years, and (d) 15 years.

SOLUTION

a. To find the numbers of subscribers after 3 years, first find X_3.

$$X_3 = P^3 X_0 \approx \begin{bmatrix} 0.3028 \\ 0.3904 \\ 0.3068 \end{bmatrix} \begin{matrix} \text{A} \\ \text{B} \\ \text{None} \end{matrix} \qquad \text{After 3 years}$$

After 3 years, Company A will have about $0.3028(100{,}000) = 30{,}280$ subscribers and Company B will have about $0.3904(100{,}000) = 39{,}040$ subscribers.

b. To find the numbers of subscribers after 5 years, first find X_5.

$$X_5 = P^5 X_0 \approx \begin{bmatrix} 0.3241 \\ 0.4381 \\ 0.2378 \end{bmatrix} \begin{matrix} \text{A} \\ \text{B} \\ \text{None} \end{matrix} \qquad \text{After 5 years}$$

After 5 years, Company A will have about $0.3241(100{,}000) = 32{,}410$ subscribers and Company B will have about $0.4381(100{,}000) = 43{,}810$ subscribers.

c. To find the numbers of subscribers after 10 years, first find X_{10}.

$$X_{10} = P^{10} X_0 \approx \begin{bmatrix} 0.3329 \\ 0.4715 \\ 0.1957 \end{bmatrix} \begin{matrix} \text{A} \\ \text{B} \\ \text{None} \end{matrix} \qquad \text{After 10 years}$$

After 10 years, Company A will have about $0.3329(100{,}000) = 33{,}290$ subscribers and Company B will have about $0.4715(100{,}000) = 47{,}150$ subscribers.

d. To find the numbers of subscribers after 15 years, first find X_{15}.

$$X_{15} = P^{15} X_0 \approx \begin{bmatrix} 0.3333 \\ 0.4756 \\ 0.1911 \end{bmatrix} \begin{matrix} \text{A} \\ \text{B} \\ \text{None} \end{matrix} \qquad \text{After 15 years}$$

After 15 years, Company A will have about $0.3333(100{,}000) = 33{,}330$ subscribers and Company B will have about $0.4756(100{,}000) = 47{,}560$ subscribers.

LINEAR ALGEBRA APPLIED

Google's PageRank algorithm makes use of Markov chains. For a search set that contains n web pages, define an $n \times n$ matrix A such that $a_{ij} = 1$ when page j references page i and $a_{ij} = 0$ otherwise. Adjust A to account for web pages without external references, scale each column of A so that A is stochastic, and call this matrix B. Then define

$$M = pB + \frac{1-p}{n}E$$

where p is the probability that a user follows a link on a page, $1 - p$ is the probability that the user goes to any page at random, and E is an $n \times n$ matrix whose entries are all 1. The Markov chain whose matrix of transition probabilities is M converges to a unique *steady state matrix,* which gives an estimate of page ranks. Section 10.3 discusses a method that can be used to estimate the steady state matrix.

STEADY STATE MATRIX OF A MARKOV CHAIN

In Example 3, notice that there is little difference between the numbers of subscribers after 10 years and after 15 years. If you continue the process shown in this example, then the state matrix X_n eventually reaches a **steady state.** That is, as long as the matrix P does not change, the matrix product P^nX approaches a limit $\overline{X}$. In Example 3, the limit is the *steady state matrix*

$$\overline{X} = \begin{bmatrix} \frac{1}{3} \\ \frac{10}{21} \\ \frac{4}{21} \end{bmatrix} \approx \begin{bmatrix} 0.3333 \\ 0.4762 \\ 0.1905 \end{bmatrix}. \qquad \begin{matrix} \text{A} \\ \text{B} \\ \text{None} \end{matrix} \qquad \text{Steady state matrix}$$

Check to see that $P\overline{X} = \overline{X}$, as shown below.

$$P\overline{X} = \begin{bmatrix} 0.70 & 0.15 & 0.15 \\ 0.20 & 0.80 & 0.15 \\ 0.10 & 0.05 & 0.70 \end{bmatrix} \begin{bmatrix} \frac{1}{3} \\ \frac{10}{21} \\ \frac{4}{21} \end{bmatrix} = \begin{bmatrix} \frac{1}{3} \\ \frac{10}{21} \\ \frac{4}{21} \end{bmatrix} = \overline{X}$$

In Example 5, you will verify the above result by finding the steady state matrix $\overline{X}$.

The matrix of transition probabilities P used above is an example of a *regular* stochastic matrix. A stochastic matrix P is **regular** when some power of P has only positive entries.

EXAMPLE 4   Regular Stochastic Matrices

a. The stochastic matrix

$$P = \begin{bmatrix} 0.70 & 0.15 & 0.15 \\ 0.20 & 0.80 & 0.15 \\ 0.10 & 0.05 & 0.70 \end{bmatrix}$$

is regular because P^1 has only positive entries.

b. The stochastic matrix

$$P = \begin{bmatrix} 0.50 & 1.00 \\ 0.50 & 0 \end{bmatrix}$$

is regular because

$$P^2 = \begin{bmatrix} 0.75 & 0.50 \\ 0.25 & 0.50 \end{bmatrix}$$

has only positive entries.

c. The stochastic matrix

$$P = \begin{bmatrix} \frac{1}{3} & 0 & 1 \\ \frac{1}{3} & 1 & 0 \\ \frac{1}{3} & 0 & 0 \end{bmatrix}$$

is not regular because every power of P has two zeros in its second column. (Verify this.)

REMARK

For a regular stochastic matrix P, the sequence of successive powers

$$P, P^2, P^3, \ldots$$

approaches a *stable matrix* $\overline{P}$. The entries in each column of $\overline{P}$ are equal to the corresponding entries in the steady state matrix $\overline{X}$. You are asked to show this in Exercise 55.

When P is a regular stochastic matrix, the corresponding **regular Markov chain**

$$PX_0, P^2X_0, P^3X_0, \ldots$$

approaches a unique **steady state matrix** $\overline{X}$. You are asked to prove this in Exercise 56.

EXAMPLE 5   Finding a Steady State Matrix

See LarsonLinearAlgebra.com for an interactive version of this type of example.

Find the steady state matrix $\overline{X}$ of the Markov chain whose matrix of transition probabilities is the regular matrix

$$P = \begin{bmatrix} 0.70 & 0.15 & 0.15 \\ 0.20 & 0.80 & 0.15 \\ 0.10 & 0.05 & 0.70 \end{bmatrix}.$$

SOLUTION

Note that P is the matrix of transition probabilities that you found in Example 2 and whose steady state matrix $\overline{X}$ you verified at the top of page 87. To *find* $\overline{X}$, begin by letting $\overline{X} = \begin{bmatrix} x_1 \\ x_2 \\ x_3 \end{bmatrix}$. Then use the matrix equation $P\overline{X} = \overline{X}$ to obtain

$$\begin{bmatrix} 0.70 & 0.15 & 0.15 \\ 0.20 & 0.80 & 0.15 \\ 0.10 & 0.05 & 0.70 \end{bmatrix}\begin{bmatrix} x_1 \\ x_2 \\ x_3 \end{bmatrix} = \begin{bmatrix} x_1 \\ x_2 \\ x_3 \end{bmatrix}$$

or

$$0.70x_1 + 0.15x_2 + 0.15x_3 = x_1$$
$$0.20x_1 + 0.80x_2 + 0.15x_3 = x_2$$
$$0.10x_1 + 0.05x_2 + 0.70x_3 = x_3.$$

Use these equations and the fact that $x_1 + x_2 + x_3 = 1$ to write the system of linear equations below.

$$-0.30x_1 + 0.15x_2 + 0.15x_3 = 0$$
$$0.20x_1 - 0.20x_2 + 0.15x_3 = 0$$
$$0.10x_1 + 0.05x_2 - 0.30x_3 = 0$$
$$x_1 + x_2 + x_3 = 1.$$

Use any appropriate method to verify that the solution of this system is

$$x_1 = \tfrac{1}{3}, \quad x_2 = \tfrac{10}{21}, \quad \text{and} \quad x_3 = \tfrac{4}{21}.$$

So the steady state matrix is

$$\overline{X} = \begin{bmatrix} \frac{1}{3} \\ \frac{10}{21} \\ \frac{4}{21} \end{bmatrix} \approx \begin{bmatrix} 0.3333 \\ 0.4762 \\ 0.1905 \end{bmatrix}.$$

Check that $P\overline{X} = \overline{X}$.

A summary for finding the steady state matrix $\overline{X}$ of a Markov chain is below.

REMARK

Recall from Example 2 that the state matrix consists of entries that are portions of the whole. So it should make sense that

$$x_1 + x_2 + x_3 = 1.$$

REMARK

If P is *not* regular, then the corresponding Markov chain may or may not have a unique steady state matrix.

Finding the Steady State Matrix of a Markov Chain

1. Check to see that the matrix of transition probabilities P is a regular matrix.

2. Solve the system of linear equations obtained from the matrix equation $P\overline{X} = \overline{X}$ along with the equation $x_1 + x_2 + \cdots + x_n = 1$.

3. Check the solution found in Step 2 in the matrix equation $P\overline{X} = \overline{X}$.

ABSORBING MARKOV CHAINS

The Markov chain discussed in Examples 3 and 5 is *regular*. Other types of Markov chains can be used to model real-life situations. One of these includes *absorbing* Markov chains.

Consider a Markov chain with n different states $\{S_1, S_2, \ldots, S_n\}$. The ith state S_i is an **absorbing state** when, in the matrix of transition probabilities P, $p_{ii} = 1$. That is, the entry on the main diagonal of P is 1 and all other entries in the ith column of P are 0. An **absorbing Markov chain** has the two properties listed below.

1. The Markov chain has at least one absorbing state.

2. It is possible for a member of the population to move from any nonabsorbing state to an absorbing state in a finite number of transitions.

EXAMPLE 6   **Absorbing and Nonabsorbing Markov Chains**

a. For the matrix

$$P = \begin{array}{c} \\ \end{array} \overset{\displaystyle \text{From}}{\overbrace{\begin{array}{ccc} S_1 & S_2 & S_3 \end{array}}}$$

$$P = \begin{bmatrix} 0.4 & 0 & 0 \\ 0 & 1 & 0.5 \\ 0.6 & 0 & 0.5 \end{bmatrix} \begin{array}{l} S_1 \\ S_2 \\ S_3 \end{array} \bigg\} \text{To}$$

the second state, represented by the second column, is absorbing. Moreover, the corresponding Markov chain is also absorbing because it is possible to move from S_1 to S_2 in two transitions, and it is possible to move from S_3 to S_2 in one transition. (See Figure 2.2.)

b. For the matrix

$$\overset{\displaystyle \text{From}}{\overbrace{\begin{array}{cccc} S_1 & S_2 & S_3 & S_4 \end{array}}}$$

$$P = \begin{bmatrix} 0.5 & 0 & 0 & 0 \\ 0.5 & 1 & 0 & 0 \\ 0 & 0 & 0.4 & 0.5 \\ 0 & 0 & 0.6 & 0.5 \end{bmatrix} \begin{array}{l} S_1 \\ S_2 \\ S_3 \\ S_4 \end{array} \bigg\} \text{To}$$

the second state is absorbing. However, the corresponding Markov chain is *not* absorbing because there is no way to move from state S_3 or state S_4 to state S_2. (See Figure 2.3.)

c. The matrix

$$\overset{\displaystyle \text{From}}{\overbrace{\begin{array}{cccc} S_1 & S_2 & S_3 & S_4 \end{array}}}$$

$$P = \begin{bmatrix} 0.5 & 0 & 0.2 & 0 \\ 0.2 & 1 & 0.3 & 0 \\ 0.1 & 0 & 0.4 & 0 \\ 0.2 & 0 & 0.1 & 1 \end{bmatrix} \begin{array}{l} S_1 \\ S_2 \\ S_3 \\ S_4 \end{array} \bigg\} \text{To}$$

has two absorbing states: S_2 and S_4. Moreover, the corresponding Markov chain is also absorbing because it is possible to move from either of the nonabsorbing states, S_1 or S_3, to either of the absorbing states in one step. (See Figure 2.4.)

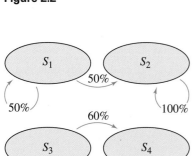

40% S_1
60%
S_2 S_3
50%
100%   50%

Figure 2.2

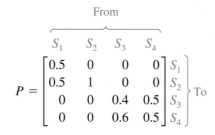

S_1 S_2
50%
50%   100%
60%
S_3 S_4
40%   50%   50%

Figure 2.3

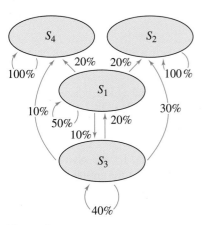

S_4 S_2
20%   20%
100%   100%
S_1
10%   30%
50%   20%
10%
S_3
40%

Figure 2.4

It is possible for some absorbing Markov chains to have a unique steady state matrix. Other absorbing Markov chains have an infinite number of steady state matrices. Example 7 demonstrates.

EXAMPLE 7 **Finding Steady State Matrices of Absorbing Markov Chains**

Find the steady state matrix $\overline{X}$ of each absorbing Markov chain with matrix of transition probabilities P.

a. $P = \begin{bmatrix} 0.4 & 0 & 0 \\ 0 & 1 & 0.5 \\ 0.6 & 0 & 0.5 \end{bmatrix}$
b. $P = \begin{bmatrix} 0.5 & 0 & 0.2 & 0 \\ 0.2 & 1 & 0.3 & 0 \\ 0.1 & 0 & 0.4 & 0 \\ 0.2 & 0 & 0.1 & 1 \end{bmatrix}$

SOLUTION

a. Use the matrix equation $P\overline{X} = \overline{X}$, or

$$\begin{bmatrix} 0.4 & 0 & 0 \\ 0 & 1 & 0.5 \\ 0.6 & 0 & 0.5 \end{bmatrix} \begin{bmatrix} x_1 \\ x_2 \\ x_3 \end{bmatrix} = \begin{bmatrix} x_1 \\ x_2 \\ x_3 \end{bmatrix}$$

along with the equation $x_1 + x_2 + x_3 = 1$ to write the system of linear equations

$$\begin{aligned} -0.6x_1 & & & = 0 \\ & & 0.5x_3 & = 0 \\ 0.6x_1 & & -0.5x_3 & = 0 \\ x_1 + x_2 & + & x_3 & = 1. \end{aligned}$$

The solution of this system is $x_1 = 0$, $x_2 = 1$, and $x_3 = 0$, so the steady state matrix is $\overline{X} = \begin{bmatrix} 0 & 1 & 0 \end{bmatrix}^T$. Note that $\overline{X}$ coincides with the second column of the matrix of transition probabilities P.

b. Use the matrix equation $P\overline{X} = \overline{X}$, or

$$\begin{bmatrix} 0.5 & 0 & 0.2 & 0 \\ 0.2 & 1 & 0.3 & 0 \\ 0.1 & 0 & 0.4 & 0 \\ 0.2 & 0 & 0.1 & 1 \end{bmatrix} \begin{bmatrix} x_1 \\ x_2 \\ x_3 \\ x_4 \end{bmatrix} = \begin{bmatrix} x_1 \\ x_2 \\ x_3 \\ x_4 \end{bmatrix}$$

along with the equation $x_1 + x_2 + x_3 + x_4 = 1$ to write the system of linear equations

$$\begin{aligned} -0.5x_1 & + 0.2x_3 & = 0 \\ 0.2x_1 & + 0.3x_3 & = 0 \\ 0.1x_1 & - 0.6x_3 & = 0 \\ 0.2x_1 & + 0.1x_3 & = 0 \\ x_1 + x_2 & + \quad x_3 + x_4 & = 1. \end{aligned}$$

REMARK

Note that the steady state matrix for an absorbing Markov chain has nonzero values only in the absorbing state(s). These states *absorb* the population.

The solution of this system is $x_1 = 0$, $x_2 = 1 - t$, $x_3 = 0$, and $x_4 = t$, where t is any real number such that $0 \le t \le 1$. So, the steady state matrix is $\overline{X} = \begin{bmatrix} 0 & 1-t & 0 & t \end{bmatrix}^T$. The Markov chain has an infinite number of steady state matrices.

In general, a regular Markov chain or an absorbing Markov chain with one absorbing state has a unique steady state matrix regardless of the initial state matrix. Further, an absorbing Markov chain with two or more absorbing states has an infinite number of steady state matrices, which depend on the initial state matrix. In Exercise 49, you are asked to show this dependence for the Markov chain whose matrix of transition probabilities is given in Example 7(b).

2.5 Exercises

Stochastic Matrices In Exercises 1–6, determine whether the matrix is stochastic.

1. $\begin{bmatrix} \frac{2}{5} & -\frac{2}{5} \\ \frac{3}{5} & \frac{7}{5} \end{bmatrix}$

2. $\begin{bmatrix} 1 + \sqrt{2} & 1 - \sqrt{2} \\ -\sqrt{2} & \sqrt{2} \end{bmatrix}$

3. $\begin{bmatrix} 0.\overline{3} & 0.1\overline{6} & 0.25 \\ 0.\overline{3} & 0.\overline{6} & 0.25 \\ 0.\overline{3} & 0.1\overline{6} & 0.5 \end{bmatrix}$

4. $\begin{bmatrix} 0.3 & 0.5 & 0.2 \\ 0.1 & 0.2 & 0.7 \\ 0.8 & 0.1 & 0.1 \end{bmatrix}$

5. $\begin{bmatrix} 1 & 0 & 0 & 0 \\ 0 & 1 & 0 & 0 \\ 0 & 0 & 1 & 0 \\ 0 & 0 & 0 & 1 \end{bmatrix}$

6. $\begin{bmatrix} \frac{1}{2} & \frac{2}{9} & \frac{1}{4} & \frac{4}{15} \\ \frac{1}{6} & \frac{1}{3} & \frac{1}{4} & \frac{4}{15} \\ \frac{1}{6} & \frac{2}{9} & \frac{1}{4} & \frac{4}{15} \\ \frac{1}{6} & \frac{2}{9} & \frac{1}{4} & \frac{1}{5} \end{bmatrix}$

7. **Airplane Allocation** An airline has 30 airplanes in Los Angeles, 12 airplanes in St. Louis, and 8 airplanes in Dallas. During an eight-hour period, 20% of the planes in Los Angeles fly to St. Louis and 10% fly to Dallas. Of the planes in St. Louis, 25% fly to Los Angeles and 50% fly to Dallas. Of the planes in Dallas, 12.5% fly to Los Angeles and 50% fly to St. Louis. How many planes are in each city after 8 hours?

8. **Chemistry** In a chemistry experiment, a test tube contains 10,000 molecules of a compound. Initially, 20% of the molecules are in a gas state, 60% are in a liquid state, and 20% are in a solid state. After introducing a catalyst, 40% of the gas molecules change to liquid, 30% of the liquid molecules change to solid, and 50% of the solid molecules change to liquid. How many molecules are in each state after introducing the catalyst?

Finding State Matrices In Exercises 9 and 10, use the matrix of transition probabilities P and initial state matrix X_0 to find the state matrices X_1, X_2, and X_3.

9. $P = \begin{bmatrix} 0.6 & 0.1 & 0.1 \\ 0.2 & 0.7 & 0.1 \\ 0.2 & 0.2 & 0.8 \end{bmatrix}$, $X_0 = \begin{bmatrix} 0.1 \\ 0.1 \\ 0.8 \end{bmatrix}$

10. $P = \begin{bmatrix} 0.6 & 0.2 & 0 \\ 0.2 & 0.7 & 0.1 \\ 0.2 & 0.1 & 0.9 \end{bmatrix}$, $X_0 = \begin{bmatrix} \frac{1}{3} \\ \frac{1}{3} \\ \frac{1}{3} \end{bmatrix}$

11. **Purchase of a Product** The market research department at a manufacturing plant determines that 20% of the people who purchase the plant's product during any month will not purchase it the next month. On the other hand, 30% of the people who do not purchase the product during any month will purchase it the next month. In a population of 1000 people, 100 people purchased the product this month. How many will purchase the product (a) next month and (b) in 2 months?

12. **Spread of a Virus** A medical researcher is studying the spread of a virus in a population of 1000 laboratory mice. During any week, there is an 80% probability that an infected mouse will overcome the virus, and during the same week there is a 10% probability that a noninfected mouse will become infected. Three hundred mice are currently infected with the virus. How many will be infected (a) next week and (b) in 3 weeks?

13. **Television Watching** A college dormitory houses 200 students. Those who watch an hour or more of television on any day always watch for less than an hour the next day. One-fourth of those who watch television for less than an hour one day will watch an hour or more the next day. Half of the students watched television for an hour or more today. How many will watch television for an hour or more (a) tomorrow, (b) in 2 days, and (c) in 30 days?

14. **Sports Activities** Students in a gym class have a choice of swimming or playing basketball each day. Thirty percent of the students who swim one day will swim the next day. Sixty percent of the students who play basketball one day will play basketball the next day. Today, 100 students swam and 150 students played basketball. How many students will swim (a) tomorrow, (b) in two days, and (c) in four days?

15. **Smokers and Nonsmokers** In a population of 10,000, there are 5000 nonsmokers, 2500 smokers of one pack or less per day, and 2500 smokers of more than one pack per day. During any month, there is a 5% probability that a nonsmoker will begin smoking a pack or less per day, and a 2% probability that a nonsmoker will begin smoking more than a pack per day. For smokers who smoke a pack or less per day, there is a 10% probability of quitting and a 10% probability of increasing to more than a pack per day. For smokers who smoke more than a pack per day, there is a 5% probability of quitting and a 10% probability of dropping to a pack or less per day. How many people will be in each group (a) in 1 month, (b) in 2 months, and (c) in 1 year?

16. **Consumer Preference**   In a population of 100,000 consumers, there are 20,000 users of Brand A, 30,000 users of Brand B, and 50,000 who use neither brand. During any month, a Brand A user has a 20% probability of switching to Brand B and a 5% probability of not using either brand. A Brand B user has a 15% probability of switching to Brand A and a 10% probability of not using either brand. A nonuser has a 10% probability of purchasing Brand A and a 15% probability of purchasing Brand B. How many people will be in each group (a) in 1 month, (b) in 2 months, and (c) in 18 months?

Regular and Steady State Matrices   **In Exercises 17–30, determine whether the stochastic matrix P is regular. Then find the steady state matrix $\overline{X}$ of the Markov chain with matrix of transition probabilities P.**

17. $P = \begin{bmatrix} 0.5 & 0.1 \\ 0.5 & 0.9 \end{bmatrix}$ 18. $P = \begin{bmatrix} 0 & 0.3 \\ 1 & 0.7 \end{bmatrix}$

19. $P = \begin{bmatrix} 1 & 0.75 \\ 0 & 0.25 \end{bmatrix}$ 20. $P = \begin{bmatrix} 0.2 & 0 \\ 0.8 & 1 \end{bmatrix}$

21. $P = \begin{bmatrix} \frac{1}{2} & \frac{1}{3} \\ \frac{1}{2} & \frac{2}{3} \end{bmatrix}$ 22. $P = \begin{bmatrix} \frac{2}{5} & \frac{7}{10} \\ \frac{3}{5} & \frac{3}{10} \end{bmatrix}$

23. $P = \begin{bmatrix} \frac{2}{5} & \frac{3}{10} & \frac{1}{2} \\ \frac{1}{5} & \frac{1}{5} & \frac{1}{10} \\ \frac{2}{5} & \frac{1}{2} & \frac{2}{5} \end{bmatrix}$ 24. $P = \begin{bmatrix} \frac{2}{9} & \frac{1}{4} & \frac{1}{3} \\ \frac{1}{3} & \frac{1}{2} & \frac{1}{3} \\ \frac{4}{9} & \frac{1}{4} & \frac{1}{3} \end{bmatrix}$

25. $P = \begin{bmatrix} 1 & 0 & 0.15 \\ 0 & 1 & 0.10 \\ 0 & 0 & 0.75 \end{bmatrix}$

26. $P = \begin{bmatrix} \frac{1}{2} & \frac{1}{5} & 1 \\ \frac{1}{3} & \frac{1}{5} & 0 \\ \frac{1}{6} & \frac{3}{5} & 0 \end{bmatrix}$

27. $P = \begin{bmatrix} 0.22 & 0.20 & 0.65 \\ 0.62 & 0.60 & 0.15 \\ 0.16 & 0.20 & 0.20 \end{bmatrix}$

28. $P = \begin{bmatrix} 0.1 & 0 & 0.3 \\ 0.7 & 1 & 0.3 \\ 0.2 & 0 & 0.4 \end{bmatrix}$

29. $P = \begin{bmatrix} \frac{1}{4} & \frac{1}{3} & \frac{1}{2} & 1 \\ \frac{1}{4} & \frac{1}{3} & \frac{1}{2} & 0 \\ \frac{1}{4} & \frac{1}{3} & 0 & 0 \\ \frac{1}{4} & 0 & 0 & 0 \end{bmatrix}$

30. $P = \begin{bmatrix} 1 & 0 & 0 & 0 \\ 0 & 0 & 1 & 0 \\ 0 & 1 & 0 & 0 \\ 0 & 0 & 0 & 1 \end{bmatrix}$

31. (a) Find the steady state matrix $\overline{X}$ using the matrix of transition probabilities P in Exercise 9.

    (b) Find the steady state matrix $\overline{X}$ using the matrix of transition probabilities P in Exercise 10.

32. Find the steady state matrix for each stochastic matrix in Exercises 1–6.

33. **Fundraising**   A nonprofit organization collects contributions from members of a community. During any year, 40% of those who make contributions will not contribute the next year. On the other hand, 10% of those who do not make contributions will contribute the next year. Find and interpret the steady state matrix for this situation.

34. **Grade Distribution**   In a college class, 70% of the students who receive an "A" on one assignment will receive an "A" on the next assignment. On the other hand, 10% of the students who do not receive an "A" on one assignment will receive an "A" on the next assignment. Find and interpret the steady state matrix for this situation.

35. **Stock Sales and Purchases**   Eight hundred fifty stockholders invest in one of three stocks. During any month, 25% of Stock A holders move their investment to Stock B and 10% to Stock C. Of Stock B holders, 10% move their investment to Stock A. Of Stock C holders, 15% move their investment to Stock A and 5% to Stock B. Find and interpret the steady state matrix for this situation.

36. **Customer Preference**   Two movie theatres that show several different movies each night compete for the same audience. Of the people who attend Theatre A one night, 10% will attend again the next night and 5% will attend Theatre B the next night. Of the people who attend Theatre B one night, 8% will attend again the next night and 6% will attend Theatre A the next night. Of the people who attend neither theatre one night, 3% will attend Theatre A the next night and 4% will attend Theatre B the next night. Find and interpret the steady state matrix for this situation.

Absorbing Markov Chains   **In Exercises 37–40, determine whether the Markov chain with matrix of transition probabilities P is absorbing. Explain.**

37. $P = \begin{bmatrix} 0.8 & 0.3 & 0 \\ 0.2 & 0.1 & 0 \\ 0 & 0.6 & 1 \end{bmatrix}$ 38. $P = \begin{bmatrix} 1 & 0 & 0 \\ 0 & 0.3 & 0.9 \\ 0 & 0.7 & 0.1 \end{bmatrix}$

39. $P = \begin{bmatrix} \frac{2}{5} & \frac{1}{5} & 0 & 0 \\ \frac{1}{5} & \frac{3}{5} & 0 & \frac{1}{2} \\ \frac{2}{5} & \frac{1}{5} & 1 & 0 \\ 0 & 0 & 0 & \frac{1}{2} \end{bmatrix}$ 40. $P = \begin{bmatrix} 0.3 & 0.7 & 0.2 & 0 \\ 0.2 & 0.1 & 0.1 & 0 \\ 0.1 & 0.1 & 0.1 & 0 \\ 0.4 & 0.1 & 0.6 & 1 \end{bmatrix}$

Finding a Steady State Matrix In Exercises 41–44, find the steady state matrix $\overline{X}$ of the absorbing Markov chain with matrix of transition probabilities P.

41. $P = \begin{bmatrix} 0.6 & 0 & 0.3 \\ 0.2 & 1 & 0.6 \\ 0.2 & 0 & 0.1 \end{bmatrix}$ **42.** $P = \begin{bmatrix} 0.1 & 0 & 0 \\ 0.2 & 1 & 0 \\ 0.7 & 0 & 1 \end{bmatrix}$

43. $P = \begin{bmatrix} 1 & 0.2 & 0.1 & 0.3 \\ 0 & 0.3 & 0.6 & 0.3 \\ 0 & 0.1 & 0.2 & 0.2 \\ 0 & 0.4 & 0.1 & 0.2 \end{bmatrix}$

44. $P = \begin{bmatrix} 0.7 & 0 & 0.2 & 0.1 \\ 0.1 & 1 & 0.5 & 0.6 \\ 0 & 0 & 0.2 & 0.2 \\ 0.2 & 0 & 0.1 & 0.1 \end{bmatrix}$

45. Epidemic Model In a population of 200,000 people, 40,000 are infected with a virus. After a person becomes infected and then recovers, the person is immune (cannot become infected again). Of the people who are infected, 5% will die each year and the others will recover. Of the people who have never been infected, 25% will become infected each year. How many people will be infected in 4 years?

46. Chess Tournament Two people are engaged in a chess tournament. Each starts with two playing chips. After each game, the loser must give the winner one chip. Player 2 is more advanced than Player 1 and has a 70% chance of winning each game. The tournament is over when one player obtains all four chips. What is the probability that Player 1 will win the tournament?

47. Explain how you can determine the steady state matrix $\overline{X}$ of an absorbing Markov chain by inspection.

---

48. CAPSTONE

(a) Explain how to find the nth state matrix of a Markov chain.

(b) Explain how to find the steady state matrix of a Markov chain.

(c) What is a regular Markov chain?

(d) What is an absorbing Markov chain?

(e) How is an absorbing Markov chain different than a regular Markov chain?

---

49. Consider the Markov chain whose matrix of transition probabilities P is given in Example 7(b). Show that the steady state matrix $\overline{X}$ depends on the initial state matrix X_0 by finding $\overline{X}$ for each X_0.

(a) $X_0 = \begin{bmatrix} 0.25 \\ 0.25 \\ 0.25 \\ 0.25 \end{bmatrix}$ (b) $X_0 = \begin{bmatrix} 0.25 \\ 0.25 \\ 0.40 \\ 0.10 \end{bmatrix}$

50. Markov Chain with Reflecting Boundaries The figure below illustrates an example of a **Markov chain with reflecting boundaries.**

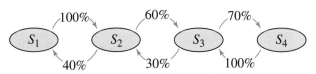

(a) Explain why it is appropriate to say that this type of Markov chain has *reflecting boundaries*.

(b) Use the figure to write the matrix of transition probabilities P for the Markov chain.

(c) Find P^{30} and P^{31}. Find several other high even powers $2n$ and odd powers $2n + 1$ of P. What do you observe?

(d) Find the steady state matrix $\overline{X}$ of the Markov chain. How are the entries in the columns of P^{2n} and P^{2n+1} related to the entries in $\overline{X}$?

Nonabsorbing Markov Chain In Exercises 51 and 52, consider the matrix P in Example 6(b).

51. Is it possible to find a steady state matrix $\overline{X}$ for the corresponding Markov chain? If so, find a steady state matrix. If not, explain why.

52. Create a new matrix P' by changing the second column of P to $\begin{bmatrix} 0.6 & 0.4 & 0 & 0 \end{bmatrix}^T$, resulting in a second state that is no longer absorbing. Determine whether each matrix X below can be a steady state matrix for the Markov chain corresponding to P'. Explain.

(a) $X = \begin{bmatrix} \frac{6}{11} \\ \frac{5}{11} \\ 0 \\ 0 \end{bmatrix}$ (b) $X = \begin{bmatrix} 0 \\ 0 \\ \frac{5}{11} \\ \frac{6}{11} \end{bmatrix}$

53. Proof Prove that the product of two 2×2 stochastic matrices is stochastic.

54. Proof Let P be a 2×2 stochastic matrix. Prove that there exists a 2×1 state matrix X with nonnegative entries such that $PX = X$.

55. In Example 5, show that for the regular stochastic matrix P, the sequence of successive powers

$$P, P^2, P^3, \ldots$$

approaches a stable matrix $\overline{P}$, where the entries in each column of $\overline{P}$ are equal to the corresponding entries in the steady state matrix $\overline{X}$. Repeat for several other regular stochastic matrices P and corresponding steady state matrices $\overline{X}$.

56. Proof Prove that when P is a regular stochastic matrix, the corresponding regular Markov chain

$$PX_0, P^2X_0, P^3X_0, \ldots$$

approaches a unique steady state matrix $\overline{X}$.

2.6   More Applications of Matrix Operations

■ Use matrix multiplication to encode and decode messages.

■ Use matrix algebra to analyze an economic system (Leontief input-output model).

■ Find the least squares regression line for a set of data.

CRYPTOGRAPHY

A **cryptogram** is a message written according to a secret code (the Greek word *kryptos* means "hidden"). One method of using matrix multiplication to **encode** and **decode** messages is introduced below.

To begin, assign a number to each letter in the alphabet (with 0 assigned to a blank space), as shown.

0 = __	9 = I	18 = R
1 = A	10 = J	19 = S
2 = B	11 = K	20 = T
3 = C	12 = L	21 = U
4 = D	13 = M	22 = V
5 = E	14 = N	23 = W
6 = F	15 = O	24 = X
7 = G	16 = P	25 = Y
8 = H	17 = Q	26 = Z

Then convert the message to numbers and partition it into **uncoded row matrices,** each having *n* entries, as demonstrated in Example 1.

EXAMPLE 1   Forming Uncoded Row Matrices

Write the uncoded row matrices of size 1×3 for the message MEET ME MONDAY.

SOLUTION

Partitioning the message (including blank spaces, but ignoring punctuation) into groups of three produces the uncoded row matrices shown below.

$$[13 \quad 5 \quad 5] \ [20 \quad 0 \quad 13] \ [5 \quad 0 \quad 13] \ [15 \quad 14 \quad 4] \ [1 \quad 25 \quad 0]$$
$$M \quad E \quad E \quad T \_ \quad M \quad E \_ \quad M \quad O \quad N \quad D \quad A \quad Y \_$$

Note the use of a blank space to fill out the last uncoded row matrix.

LINEAR ALGEBRA APPLIED

Information security is of the utmost importance when conducting business online. If a malicious party should receive confidential information such as passwords, personal identification numbers, credit card numbers, Social Security numbers, bank account details, or sensitive company information, then the effects can be damaging. To protect the confidentiality and integrity of such information, Internet security can include the use of data *encryption,* the process of encoding information so that the only way to decode it, apart from an "exhaustion attack," is to use a *key.* Data encryption technology uses algorithms based on the material presented here, but on a much more sophisticated level, to prevent malicious parties from discovering the key.

To **encode** a message, choose an $n \times n$ invertible matrix A and multiply the uncoded row matrices (on the right) by A to obtain **coded row matrices.** Example 2 demonstrates this process.

EXAMPLE 2   Encoding a Message

Use the invertible matrix

$$A = \begin{bmatrix} 1 & -2 & 2 \\ -1 & 1 & 3 \\ 1 & -1 & -4 \end{bmatrix}$$

to encode the message MEET ME MONDAY.

SOLUTION

Obtain the coded row matrices by multiplying each of the uncoded row matrices found in Example 1 by the matrix A, as shown below.

Uncoded Row Matrix	Encoding Matrix A	Coded Row Matrix

$$[13 \quad 5 \quad 5]\begin{bmatrix} 1 & -2 & 2 \\ -1 & 1 & 3 \\ 1 & -1 & -4 \end{bmatrix} = [13 \quad -26 \quad 21]$$

$$[20 \quad 0 \quad 13]\begin{bmatrix} 1 & -2 & 2 \\ -1 & 1 & 3 \\ 1 & -1 & -4 \end{bmatrix} = [33 \quad -53 \quad -12]$$

$$[5 \quad 0 \quad 13]\begin{bmatrix} 1 & -2 & 2 \\ -1 & 1 & 3 \\ 1 & -1 & -4 \end{bmatrix} = [18 \quad -23 \quad -42]$$

$$[15 \quad 14 \quad 4]\begin{bmatrix} 1 & -2 & 2 \\ -1 & 1 & 3 \\ 1 & -1 & -4 \end{bmatrix} = [5 \quad -20 \quad 56]$$

$$[1 \quad 25 \quad 0]\begin{bmatrix} 1 & -2 & 2 \\ -1 & 1 & 3 \\ 1 & -1 & -4 \end{bmatrix} = [-24 \quad 23 \quad 77]$$

The sequence of coded row matrices is

$$[13 \, -26 \quad 21][33 \, -53 \, -12][18 \, -23 \, -42][5 \, -20 \quad 56][-24 \quad 23 \quad 77].$$

Finally, removing the matrix notation produces the cryptogram

$$13 \, -26 \, 21 \, 33 \, -53 \, -12 \, 18 \, -23 \, -42 \, 5 \, -20 \, 56 \, -24 \, 23 \, 77.$$

For those who do not know the encoding matrix A, decoding the cryptogram found in Example 2 is difficult. But for an authorized receiver who knows the encoding matrix A, decoding is relatively simple. The receiver just needs to multiply the coded row matrices by A^{-1} to retrieve the uncoded row matrices. In other words, if

$$X = [x_1 \; x_2 \; \ldots \; x_n]$$

is an uncoded $1 \times n$ matrix, then $Y = XA$ is the corresponding encoded matrix. The receiver of the encoded matrix can decode Y by multiplying on the right by A^{-1} to obtain

$$YA^{-1} = (XA)A^{-1} = X.$$

Example 3 demonstrates this procedure.

EXAMPLE 3   **Decoding a Message**

Use the inverse of the matrix

$$A = \begin{bmatrix} 1 & -2 & 2 \\ -1 & 1 & 3 \\ 1 & -1 & -4 \end{bmatrix}$$

to decode the cryptogram

13 −26 21 33 −53 −12 18 −23 −42 5 −20 56 −24 23 77.

SOLUTION

Begin by using Gauss-Jordan elimination to find A^{-1}.

$$\begin{array}{cc} [A & I] \end{array}$$ $$\begin{array}{cc} [I & A^{-1}] \end{array}$$

$$\begin{bmatrix} 1 & -2 & 2 & 1 & 0 & 0 \\ -1 & 1 & 3 & 0 & 1 & 0 \\ 1 & -1 & -4 & 0 & 0 & 1 \end{bmatrix} \longrightarrow \begin{bmatrix} 1 & 0 & 0 & -1 & -10 & -8 \\ 0 & 1 & 0 & -1 & -6 & -5 \\ 0 & 0 & 1 & 0 & -1 & -1 \end{bmatrix}$$

Now, to decode the message, partition the message into groups of three to form the coded row matrices

$$\begin{bmatrix} 13 & -26 & 21 \end{bmatrix}\begin{bmatrix} 33 & -53 & -12 \end{bmatrix}\begin{bmatrix} 18 & -23 & -42 \end{bmatrix}\begin{bmatrix} 5 & -20 & 56 \end{bmatrix}\begin{bmatrix} -24 & 23 & 77 \end{bmatrix}.$$

To obtain the decoded row matrices, multiply each coded row matrix by A^{-1} (on the right).

Coded Row Matrix	Decoding Matrix A^{-1}	Decoded Row Matrix

$$\begin{bmatrix} 13 & -26 & 21 \end{bmatrix}\begin{bmatrix} -1 & -10 & -8 \\ -1 & -6 & -5 \\ 0 & -1 & -1 \end{bmatrix} = \begin{bmatrix} 13 & 5 & 5 \end{bmatrix}$$

$$\begin{bmatrix} 33 & -53 & -12 \end{bmatrix}\begin{bmatrix} -1 & -10 & -8 \\ -1 & -6 & -5 \\ 0 & -1 & -1 \end{bmatrix} = \begin{bmatrix} 20 & 0 & 13 \end{bmatrix}$$

$$\begin{bmatrix} 18 & -23 & -42 \end{bmatrix}\begin{bmatrix} -1 & -10 & -8 \\ -1 & -6 & -5 \\ 0 & -1 & -1 \end{bmatrix} = \begin{bmatrix} 5 & 0 & 13 \end{bmatrix}$$

$$\begin{bmatrix} 5 & -20 & 56 \end{bmatrix}\begin{bmatrix} -1 & -10 & -8 \\ -1 & -6 & -5 \\ 0 & -1 & -1 \end{bmatrix} = \begin{bmatrix} 15 & 14 & 4 \end{bmatrix}$$

$$\begin{bmatrix} -24 & 23 & 77 \end{bmatrix}\begin{bmatrix} -1 & -10 & -8 \\ -1 & -6 & -5 \\ 0 & -1 & -1 \end{bmatrix} = \begin{bmatrix} 1 & 25 & 0 \end{bmatrix}$$

The sequence of decoded row matrices is

$$\begin{bmatrix} 13 & 5 & 5 \end{bmatrix}\begin{bmatrix} 20 & 0 & 13 \end{bmatrix}\begin{bmatrix} 5 & 0 & 13 \end{bmatrix}\begin{bmatrix} 15 & 14 & 4 \end{bmatrix}\begin{bmatrix} 1 & 25 & 0 \end{bmatrix}$$

and the message is

13 5 5 20 0 13 5 0 13 15 14 4 1 25 0.
M   E   E   T   _   M   E   _   M   O   N   D   A   Y   _

LEONTIEF INPUT-OUTPUT MODELS

In 1936, American economist Wassily W. Leontief (1906–1999) published a model concerning the input and output of an economic system. In 1973, Leontief received a Nobel prize for his work in economics. A brief discussion of Leontief's model follows.

Consider an economic system that has n different industries $I_1, I_2, \ldots, I_n$, each having **input** needs (raw materials, utilities, etc.) and an **output** (finished product). In producing each unit of output, an industry may use the outputs of other industries, including itself. For example, an electric utility uses outputs from other industries, such as coal and water, and also uses its own electricity.

Let d_{ij} be the amount of output the jth industry needs from the ith industry to produce one unit of output per year. The matrix of these coefficients is the **input-output matrix.**

$$D = \begin{bmatrix} d_{11} & d_{12} & \cdots & d_{1n} \\ d_{21} & d_{22} & \cdots & d_{2n} \\ \vdots & \vdots & & \vdots \\ d_{n1} & d_{n2} & \cdots & d_{nn} \end{bmatrix} \begin{matrix} I_1 \\ I_2 \\ \vdots \\ I_n \end{matrix}$$

To understand how to use this matrix, consider $d_{12} = 0.4$. This means that for Industry 2 to produce one unit of its product, it must use 0.4 unit of Industry 1's product. If $d_{33} = 0.2$, then Industry 3 needs 0.2 unit of its own product to produce one unit. For this model to work, the values of d_{ij} must satisfy $0 \leq d_{ij} \leq 1$ and the sum of the entries in any column must be less than or equal to 1.

EXAMPLE 4   Forming an Input-Output Matrix

Consider a simple economic system consisting of three industries: electricity, water, and coal. Production, or output, of one unit of electricity requires 0.5 unit of itself, 0.25 unit of water, and 0.25 unit of coal. Production of one unit of water requires 0.1 unit of electricity, 0.6 unit of itself, and 0 units of coal. Production of one unit of coal requires 0.2 unit of electricity, 0.15 unit of water, and 0.5 unit of itself. Find the input-output matrix for this system.

SOLUTION

The column entries show the amounts each industry requires from the others, and from itself, to produce one unit of output.

$$\begin{bmatrix} 0.5 & 0.1 & 0.2 \\ 0.25 & 0.6 & 0.15 \\ 0.25 & 0 & 0.5 \end{bmatrix} \begin{matrix} E \\ W \\ C \end{matrix}$$

The row entries show the amounts each industry supplies to the others, and to itself, for that industry to produce one unit of output. For instance, the electricity industry supplies 0.5 unit to itself, 0.1 unit to water, and 0.2 unit to coal.

To develop the Leontief input-output model further, let the total output of the ith industry be denoted by x_i. If the economic system is **closed** (that is, the economic system sells its products only to industries within the system, as in the example above), then the total output of the ith industry is

$$x_i = d_{i1}x_1 + d_{i2}x_2 + \cdots + d_{in}x_n. \quad \text{Closed system}$$

On the other hand, if the industries within the system sell products to nonproducing groups (such as governments or charitable organizations) outside the system, then the system is **open** and the total output of the ith industry is

$$x_i = d_{i1}x_1 + d_{i2}x_2 + \cdots + d_{in}x_n + e_i \qquad \text{Open system}$$

where e_i represents the external demand for the ith industry's product. The system of n linear equations below represents the collection of total outputs for an open system.

$$x_1 = d_{11}x_1 + d_{12}x_2 + \cdots + d_{1n}x_n + e_1$$
$$x_2 = d_{21}x_1 + d_{22}x_2 + \cdots + d_{2n}x_n + e_2$$
$$\vdots$$
$$x_n = d_{n1}x_1 + d_{n2}x_2 + \cdots + d_{nn}x_n + e_n$$

The matrix form of this system is $X = DX + E$, where X is the **output matrix** and E is the **external demand matrix.**

EXAMPLE 5   Solving for the Output Matrix of an Open System

See LarsonLinearAlgebra.com for an interactive version of this type of example.

An economic system composed of three industries has the input-output matrix shown below.

User (Output)

$$D = \begin{bmatrix} 0.1 & 0.43 & 0 \\ 0.15 & 0 & 0.37 \\ 0.23 & 0.03 & 0.02 \end{bmatrix} \begin{matrix} A \\ B \\ C \end{matrix} \rbrace \text{Supplier (Input)}$$

with columns labeled A B C.

Find the output matrix X when the external demands are

$$E = \begin{bmatrix} 20{,}000 \\ 30{,}000 \\ 25{,}000 \end{bmatrix}. \begin{matrix} A \\ B \\ C \end{matrix}$$

SOLUTION

Letting I be the identity matrix, write the equation $X = DX + E$ as $IX - DX = E$, which means that $(I - D)X = E$. Using the matrix D above produces

$$I - D = \begin{bmatrix} 0.9 & -0.43 & 0 \\ -0.15 & 1 & -0.37 \\ -0.23 & -0.03 & 0.98 \end{bmatrix}.$$

Using Gauss-Jordan elimination,

$$(I - D)^{-1} \approx \begin{bmatrix} 1.25 & 0.55 & 0.21 \\ 0.30 & 1.14 & 0.43 \\ 0.30 & 0.16 & 1.08 \end{bmatrix}.$$

So, the output matrix is

$$X = (I - D)^{-1}E \approx \begin{bmatrix} 1.25 & 0.55 & 0.21 \\ 0.30 & 1.14 & 0.43 \\ 0.30 & 0.16 & 1.08 \end{bmatrix}\begin{bmatrix} 20{,}000 \\ 30{,}000 \\ 25{,}000 \end{bmatrix} = \begin{bmatrix} 46{,}750 \\ 50{,}950 \\ 37{,}800 \end{bmatrix} \begin{matrix} A \\ B \\ C \end{matrix}$$

To produce the given external demands, the outputs of the three industries must be approximately 46,750 units for industry A, 50,950 units for industry B, and 37,800 units for industry C.

REMARK

The economic systems described in Examples 4 and 5 are, of course, simple ones. In the real world, an economic system would include many industries or industrial groups. A detailed analysis using the Leontief input-output model could easily require an input-output matrix greater than 100 × 100 in size. Clearly, this type of analysis would require the aid of a computer.

LEAST SQUARES REGRESSION ANALYSIS

You will now look at a procedure used in statistics to develop linear models. The next example demonstrates a visual method for approximating a line of best fit for a set of data points.

EXAMPLE 6   **A Visual Straight-Line Approximation**

Determine a line that appears to best fit the points (1, 1), (2, 2), (3, 4), (4, 4), and (5, 6).

SOLUTION

Plot the points, as shown in Figure 2.5. It appears that a good choice would be the line whose slope is 1 and whose y-intercept is 0.5. The equation of this line is

$$y = 0.5 + x.$$

An examination of the line in Figure 2.5 reveals that you can improve the fit by rotating the line counterclockwise slightly, as shown in Figure 2.6. It seems clear that this line, whose equation is $y = 1.2x$, fits the points better than the original line.

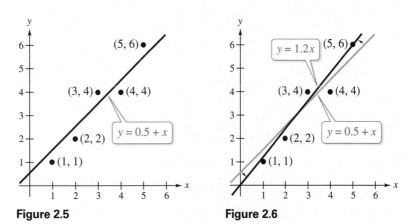

Figure 2.5   **Figure 2.6**

One way of measuring how well a function $y = f(x)$ fits a set of points

$$(x_1, y_1), (x_2, y_2), \ldots, (x_n, y_n)$$

is to compute the differences between the values from the function $f(x_i)$ and the actual values y_i. These values are shown in Figure 2.7. By squaring the differences and summing the results, you obtain a measure of error called the **sum of squared error.** The table shows the sums of squared errors for the two linear models.

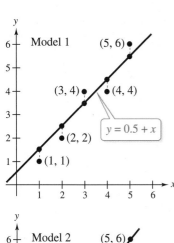

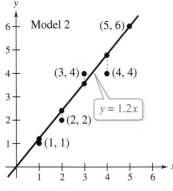

Figure 2.7

Model 1: $f(x) = 0.5 + x$				Model 2: $f(x) = 1.2x$			
x_i	y_i	$f(x_i)$	$[y_i - f(x_i)]^2$	x_i	y_i	$f(x_i)$	$[y_i - f(x_i)]^2$
1	1	1.5	$(-0.5)^2$	1	1	1.2	$(-0.2)^2$
2	2	2.5	$(-0.5)^2$	2	2	2.4	$(-0.4)^2$
3	4	3.5	$(+0.5)^2$	3	4	3.6	$(+0.4)^2$
4	4	4.5	$(-0.5)^2$	4	4	4.8	$(-0.8)^2$
5	6	5.5	$(+0.5)^2$	5	6	6.0	$(0.0)^2$
Sum			1.25	Sum			1.00

The sums of squared errors confirm that the second model fits the points better than the first model.

Of all possible linear models for a given set of points, the model that has the best fit is the one that minimizes the sum of squared error. This model is the **least squares regression line,** and the procedure for finding it is the **method of least squares.**

Definition of Least Squares Regression Line

For a set of points

$$(x_1, y_1), (x_2, y_2), \ldots, (x_n, y_n)$$

the **least squares regression line** is the linear function

$$f(x) = a_0 + a_1 x$$

that minimizes the sum of squared error

$$[y_1 - f(x_1)]^2 + [y_2 - f(x_2)]^2 + \cdots + [y_n - f(x_n)]^2.$$

To find the least squares regression line for a set of points, begin by forming the system of linear equations

$$y_1 = f(x_1) + [y_1 - f(x_1)]$$
$$y_2 = f(x_2) + [y_2 - f(x_2)]$$
$$\vdots$$
$$y_n = f(x_n) + [y_n - f(x_n)]$$

where the right-hand term

$$[y_i - f(x_i)]$$

of each equation is the error in the approximation of y_i by $f(x_i)$. Then write this error as

$$e_i = y_i - f(x_i)$$

and write the system of equations in the form

$$y_1 = (a_0 + a_1 x_1) + e_1$$
$$y_2 = (a_0 + a_1 x_2) + e_2$$
$$\vdots$$
$$y_n = (a_0 + a_1 x_n) + e_n.$$

Now, if you define $Y, X, A,$ and E as

$$Y = \begin{bmatrix} y_1 \\ y_2 \\ \vdots \\ y_n \end{bmatrix}, \quad X = \begin{bmatrix} 1 & x_1 \\ 1 & x_2 \\ \vdots & \vdots \\ 1 & x_n \end{bmatrix}, \quad A = \begin{bmatrix} a_0 \\ a_1 \end{bmatrix}, \quad E = \begin{bmatrix} e_1 \\ e_2 \\ \vdots \\ e_n \end{bmatrix}$$

then the n linear equations may be replaced by the matrix equation

$$Y = XA + E.$$

Note that the matrix X has a column of 1's (corresponding to a_0) and a column containing the x_i's. This matrix equation can be used to determine the coefficients of the least squares regression line, as shown on the next page.

REMARK

You will learn more about this
procedure in Section 5.4.

Matrix Form for Linear Regression

For the regression model $Y = XA + E$, the coefficients of the least squares
regression line are given by the matrix equation

$$A = (X^TX)^{-1}X^TY$$

and the sum of squared error is E^TE.

Example 7 demonstrates the use of this procedure to find the least squares
regression line for the set of points from Example 6.

EXAMPLE 7   **Finding the Least Squares Regression Line**

Find the least squares regression line for the points $(1, 1), (2, 2), (3, 4), (4, 4),$
and $(5, 6)$.

SOLUTION

The matrices X and Y are

$$X = \begin{bmatrix} 1 & 1 \\ 1 & 2 \\ 1 & 3 \\ 1 & 4 \\ 1 & 5 \end{bmatrix} \quad \text{and} \quad Y = \begin{bmatrix} 1 \\ 2 \\ 4 \\ 4 \\ 6 \end{bmatrix}.$$

This means that

$$X^TX = \begin{bmatrix} 1 & 1 & 1 & 1 & 1 \\ 1 & 2 & 3 & 4 & 5 \end{bmatrix} \begin{bmatrix} 1 & 1 \\ 1 & 2 \\ 1 & 3 \\ 1 & 4 \\ 1 & 5 \end{bmatrix} = \begin{bmatrix} 5 & 15 \\ 15 & 55 \end{bmatrix}$$

and

$$X^TY = \begin{bmatrix} 1 & 1 & 1 & 1 & 1 \\ 1 & 2 & 3 & 4 & 5 \end{bmatrix} \begin{bmatrix} 1 \\ 2 \\ 4 \\ 4 \\ 6 \end{bmatrix} = \begin{bmatrix} 17 \\ 63 \end{bmatrix}.$$

Now, using $(X^TX)^{-1}$ to find the coefficient matrix A, you have

$$A = (X^TX)^{-1}X^TY$$

$$= \tfrac{1}{50}\begin{bmatrix} 55 & -15 \\ -15 & 5 \end{bmatrix}\begin{bmatrix} 17 \\ 63 \end{bmatrix}$$

$$= \begin{bmatrix} -0.2 \\ 1.2 \end{bmatrix}.$$

So, the least squares regression line is

$$y = -0.2 + 1.2x$$

as shown in Figure 2.8. The sum of squared error for this line is 0.8 (verify this), which
means that this line fits the data better than either of the two experimental linear
models determined earlier.

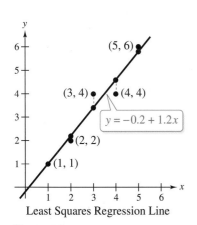

Least Squares Regression Line

Figure 2.8

2.6 Exercises

See CalcChat.com for worked-out solutions to odd-numbered exercises.

Encoding a Message In Exercises 1 and 2, write the uncoded row matrices for the message. Then encode the message using the matrix *A*.

1. *Message:*   SELL CONSOLIDATED

   Row Matrix Size: 1×3

   Encoding Matrix: $A = \begin{bmatrix} 1 & -1 & 0 \\ 1 & 0 & -1 \\ -6 & 2 & 3 \end{bmatrix}$

2. *Message:*   HELP IS COMING

   Row Matrix Size: 1×4

   Encoding Matrix: $A = \begin{bmatrix} -2 & 3 & -1 & -1 \\ -1 & 1 & 1 & 1 \\ -1 & -1 & 1 & 2 \\ 3 & 1 & -2 & -4 \end{bmatrix}$

Decoding a Message In Exercises 3–6, use A^{-1} to decode the cryptogram.

3. $A = \begin{bmatrix} 1 & 2 \\ 3 & 5 \end{bmatrix}$,

   11 21 64 112 25 50 29 53 23 46 40 75 55 92

4. $A = \begin{bmatrix} 2 & 3 \\ 3 & 4 \end{bmatrix}$,

   85 120 6 8 10 15 84 117 42 56 90 125 60 80 30 45 19 26

5. $A = \begin{bmatrix} 1 & 2 & 2 \\ 3 & 7 & 9 \\ -1 & -4 & -7 \end{bmatrix}$,

   13 19 10 −1 −33 −77 3 −2 −14 4 1 −9 −5 −25 −47 4 1 −9

6. $A = \begin{bmatrix} 3 & -4 & 2 \\ 0 & 2 & 1 \\ 4 & -5 & 3 \end{bmatrix}$,

   112 −140 83 19 −25 13 72 −76 61 95 −118 71 20 21 38 35 −23 36 42 −48 32

7. **Decoding a Message** The cryptogram below was encoded with a 2×2 matrix. The last word of the message is __RON. What is the message?

   8 21 −15 −10 −13 −13 5 10 5 25 5 19 −1 6 20 40 −18 −18 1 16

8. **Decoding a Message** The cryptogram below was encoded with a 2×2 matrix. The last word of the message is __SUE. What is the message?

   5 2 25 11 −2 −7 −15 −15 32 14 −8 −13 38 19 −19 −19 37 16

9. **Decoding a Message** Use a software program or a graphing utility to decode the cryptogram.

   $A = \begin{bmatrix} 1 & 0 & 2 \\ 2 & -1 & 1 \\ 0 & 1 & 2 \end{bmatrix}$

   38 −14 29 56 −15 62 17 3 38 18 20 76 18 −5 21 29 −7 32 32 9 77 36 −8 48 33 −5 51 41 3 79 12 1 26 58 −22 49 63 −19 69 28 8 67 31 −11 27 41 −18 28

10. **Decoding a Message** A code breaker intercepted the encoded message below.

    45 −35 38 −30 18 −18 35 −30 81 −60 42 −28 75 −55 2 −2 22 −21 15 −10

    Let the inverse of the encoding matrix be

    $A^{-1} = \begin{bmatrix} w & x \\ y & z \end{bmatrix}$.

    (a) You know that $\begin{bmatrix} 45 & -35 \end{bmatrix} A^{-1} = \begin{bmatrix} 10 & 15 \end{bmatrix}$ and $\begin{bmatrix} 38 & -30 \end{bmatrix} A^{-1} = \begin{bmatrix} 8 & 14 \end{bmatrix}$. Write and solve two systems of equations to find w, x, y, and z.

    (b) Decode the message.

11. **Industrial System** A system composed of two industries, coal and steel, has the input requirements below.

    (a) To produce \$1.00 worth of output, the coal industry requires \$0.10 of its one product and \$0.80 of steel.

    (b) To produce \$1.00 worth of output, the steel industry requires \$0.10 of its own product and \$0.20 of coal.

    Find D, the input-output matrix for this system. Then solve for the output matrix X in the equation $X = DX + E$, where E is the external demand matrix

    $E = \begin{bmatrix} 10,000 \\ 20,000 \end{bmatrix}$.

12. **Industrial System** An industrial system has two industries with the input requirements below.

    (a) To produce \$1.00 worth of output, Industry A requires \$0.30 of its own product and \$0.40 of Industry B's product.

    (b) To produce \$1.00 worth of output, Industry B requires \$0.20 of its own product and \$0.40 of Industry A's product.

    Find D, the input-output matrix for this system. Then solve for the output matrix X in the equation $X = DX + E$, where E is the external demand matrix

    $E = \begin{bmatrix} 10,000 \\ 20,000 \end{bmatrix}$.

13. Solving for the Output Matrix A small community includes a farmer, a baker, and a grocer and has the input-output matrix D and external demand matrix E below.

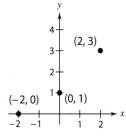

$$D = \begin{bmatrix} 0.4 & 0.5 & 0.5 \\ 0.3 & 0.0 & 0.3 \\ 0.2 & 0.2 & 0.0 \end{bmatrix} \begin{matrix} \text{Farmer} \\ \text{Baker} \\ \text{Grocer} \end{matrix} \quad \text{and} \quad E = \begin{bmatrix} 1000 \\ 1000 \\ 1000 \end{bmatrix}$$

Solve for the output matrix X in the equation $X = DX + E$.

14. Solving for the Output Matrix An industrial system with three industries has the input-output matrix D and external demand matrix E below.

$$D = \begin{bmatrix} 0.2 & 0.4 & 0.4 \\ 0.4 & 0.2 & 0.2 \\ 0.0 & 0.2 & 0.2 \end{bmatrix} \quad \text{and} \quad E = \begin{bmatrix} 5000 \\ 2000 \\ 8000 \end{bmatrix}$$

Solve for the output matrix X in the equation $X = DX + E$.

Least Squares Regression Analysis In Exercises 15–18, (a) sketch the line that appears to be the best fit for the given points, (b) find the least squares regression line, and (c) determine the sum of squared error.

15.

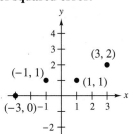

16.

17.

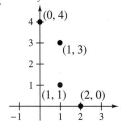

18.

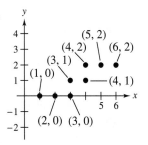

Finding the Least Squares Regression Line In Exercises 19–26, find the least squares regression line.

19. $(0, 0), (1, 1), (2, 4)$

20. $(1, 0), (3, 3), (5, 6)$

21. $(-2, 0), (-1, 1), (0, 1), (1, 2)$

22. $(-4, -1), (-2, 0), (2, 4), (4, 5)$

23. $(-5, 1), (1, 3), (2, 3), (2, 5)$

24. $(-3, 4), (-1, 2), (1, 1), (3, 0)$

25. $(-5, 10), (-1, 8), (3, 6), (7, 4), (5, 5)$

26. $(0, 6), (4, 3), (5, 0), (8, -4), (10, -5)$

27. Demand A hardware retailer wants to know the demand for a rechargeable power drill as a function of price. The ordered pairs $(25, 82)$, $(30, 75)$, $(35, 67)$, and $(40, 55)$ represent the price x (in dollars) and the corresponding monthly sales y.

(a) Find the least squares regression line for the data.

(b) Estimate the demand when the price is $32.95.

28. Wind Energy Consumption The table shows the wind energy consumptions y (in quadrillions of Btus, or British thermal units) in the United States from 2009 through 2013. Find the least squares regression line for the data. Let t represent the year, with $t = 9$ corresponding to 2009. Use the linear regression capabilities of a graphing utility to check your work. (Source: U.S. Energy Information Administration)

Year	2009	2010	2011	2012	2013
Consumption, y	0.72	0.92	1.17	1.34	1.60

29. Wildlife A wildlife management team studied the reproduction rates of deer in three tracts of a wildlife preserve. The team recorded the number of females x in each tract and the percent of females y in each tract that had offspring the following year. The table shows the results.

Number, x	100	120	140
Percent, y	75	68	55

(a) Find the least squares regression line for the data.

(b) Use a graphing utility to graph the model and the data in the same viewing window.

(c) Use the model to create a table of estimated values for y. Compare the estimated values with the actual data.

(d) Use the model to estimate the percent of females that had offspring when there were 170 females.

(e) Use the model to estimate the number of females when 40% of the females had offspring.

30. CAPSTONE

(a) Explain how to use matrix multiplication to encode and decode messages.

(b) Explain how to use a Leontief input-output model to analyze an economic system.

(c) Explain how to use matrices to find the least squares regression line for a set of data.

31. Use your school's library, the Internet, or some other reference source to derive the matrix form for linear regression given at the top of page 101.

2 Review Exercises

Operations with Matrices **In Exercises 1–6, perform the matrix operations.**

1. $\begin{bmatrix} 2 & 1 & 0 \\ 0 & 5 & -4 \end{bmatrix} - 3\begin{bmatrix} 5 & 3 & -6 \\ 0 & -2 & 5 \end{bmatrix}$

2. $-2\begin{bmatrix} 1 & 2 \\ 5 & -4 \\ 6 & 0 \end{bmatrix} + 8\begin{bmatrix} 7 & 1 \\ 1 & 2 \\ 1 & 4 \end{bmatrix}$

3. $\begin{bmatrix} 1 & 2 \\ 5 & -4 \\ 6 & 0 \end{bmatrix}\begin{bmatrix} 6 & -2 & 8 \\ 4 & 0 & 0 \end{bmatrix}$

4. $\begin{bmatrix} 1 & 5 \\ 2 & -4 \end{bmatrix}\begin{bmatrix} 6 & -2 & 8 \\ 4 & 0 & 0 \end{bmatrix}$

5. $\begin{bmatrix} 1 & 3 & 2 \\ 0 & 2 & -4 \\ 0 & 0 & 3 \end{bmatrix}\begin{bmatrix} 4 & -3 & 2 \\ 0 & 3 & -1 \\ 0 & 0 & 2 \end{bmatrix}$

6. $\begin{bmatrix} 2 & 1 \\ 6 & 0 \end{bmatrix}\begin{bmatrix} 4 & 2 \\ -3 & 1 \end{bmatrix} + \begin{bmatrix} -2 & 4 \\ 0 & 4 \end{bmatrix}$

Solving a System of Linear Equations **In Exercises 7–10, write the system of linear equations in the form $A\mathbf{x} = \mathbf{b}$. Then use Gaussian elimination to solve this matrix equation for x.**

7. $\begin{aligned} 2x_1 + x_2 &= -8 \\ x_1 + 4x_2 &= -4 \end{aligned}$

8. $\begin{aligned} 2x_1 - x_2 &= 5 \\ 3x_1 + 2x_2 &= -4 \end{aligned}$

9. $\begin{aligned} -3x_1 - x_2 + x_3 &= 0 \\ 2x_1 + 4x_2 - 5x_3 &= -3 \\ x_1 - 2x_2 + 3x_3 &= 1 \end{aligned}$

10. $\begin{aligned} 2x_1 + 3x_2 + x_3 &= 10 \\ 2x_1 - 3x_2 - 3x_3 &= 22 \\ 4x_1 - 2x_2 + 3x_3 &= -2 \end{aligned}$

Finding and Multiplying with a Transpose **In Exercises 11–14, find $A^T, A^TA,$ and AA^T.**

11. $A = \begin{bmatrix} 1 & 2 & -3 \\ 0 & 1 & 2 \end{bmatrix}$

12. $A = \begin{bmatrix} 3 & -1 \\ 2 & 0 \end{bmatrix}$

13. $A = \begin{bmatrix} 1 \\ 3 \\ -1 \end{bmatrix}$

14. $A = \begin{bmatrix} 1 & -2 & -3 \end{bmatrix}$

Finding the Inverse of a Matrix **In Exercises 15–18, find the inverse of the matrix (if it exists).**

15. $\begin{bmatrix} 3 & -1 \\ 2 & -1 \end{bmatrix}$

16. $\begin{bmatrix} 4 & -1 \\ -8 & 2 \end{bmatrix}$

17. $\begin{bmatrix} 2 & 3 & 1 \\ 2 & -3 & -3 \\ 4 & 0 & 3 \end{bmatrix}$

18. $\begin{bmatrix} 1 & 1 & 1 \\ 0 & 1 & 1 \\ 0 & 0 & 1 \end{bmatrix}$

Using the Inverse of a Matrix **In Exercises 19–26, use an inverse matrix to solve each system of linear equations or matrix equation.**

19. $\begin{aligned} 5x_1 + 4x_2 &= 2 \\ -x_1 + x_2 &= -22 \end{aligned}$

20. $\begin{aligned} 3x_1 + 2x_2 &= 1 \\ x_1 + 4x_2 &= -3 \end{aligned}$

21. $\begin{aligned} -x_1 + x_2 + 2x_3 &= 1 \\ 2x_1 + 3x_2 + x_3 &= -2 \\ 5x_1 + 4x_2 + 2x_3 &= 4 \end{aligned}$

22. $\begin{aligned} x_1 + x_2 + 2x_3 &= 0 \\ x_1 - x_2 + x_3 &= -1 \\ 2x_1 + x_2 + x_3 &= 2 \end{aligned}$

23. $\begin{bmatrix} 5 & 4 \\ -1 & 1 \end{bmatrix}\begin{bmatrix} x \\ y \end{bmatrix} = \begin{bmatrix} -15 \\ -6 \end{bmatrix}$

24. $\begin{bmatrix} 2 & -1 \\ 3 & 4 \end{bmatrix}\begin{bmatrix} x \\ y \end{bmatrix} = \begin{bmatrix} 5 \\ -2 \end{bmatrix}$

25. $\begin{bmatrix} 0 & 1 & -2 \\ -1 & 3 & 1 \\ 2 & -2 & 4 \end{bmatrix}\begin{bmatrix} x_1 \\ x_2 \\ x_3 \end{bmatrix} = \begin{bmatrix} -1 \\ 0 \\ 2 \end{bmatrix}$

26. $\begin{bmatrix} 0 & 1 & 2 \\ 3 & 2 & 1 \\ 4 & -3 & -4 \end{bmatrix}\begin{bmatrix} x \\ y \\ z \end{bmatrix} = \begin{bmatrix} 0 \\ -1 \\ -7 \end{bmatrix}$

Solving a Matrix Equation **In Exercises 27 and 28, find A.**

27. $(3A)^{-1} = \begin{bmatrix} 4 & -1 \\ 2 & 3 \end{bmatrix}$

28. $(2A)^{-1} = \begin{bmatrix} 2 & 4 \\ 0 & 1 \end{bmatrix}$

Nonsingular Matrix **In Exercises 29 and 30, find x such that the matrix A is nonsingular.**

29. $A = \begin{bmatrix} 3 & 1 \\ x & -1 \end{bmatrix}$

30. $A = \begin{bmatrix} 2 & x \\ 1 & 4 \end{bmatrix}$

Finding the Inverse of an Elementary Matrix **In Exercises 31 and 32, find the inverse of the elementary matrix.**

31. $\begin{bmatrix} 1 & 0 & 4 \\ 0 & 1 & 0 \\ 0 & 0 & 1 \end{bmatrix}$

32. $\begin{bmatrix} 1 & 0 & 0 \\ 0 & 6 & 0 \\ 0 & 0 & 1 \end{bmatrix}$

Finding a Sequence of Elementary Matrices **In Exercises 33–36, find a sequence of elementary matrices whose product is the given nonsingular matrix.**

33. $\begin{bmatrix} 2 & 3 \\ 0 & 1 \end{bmatrix}$

34. $\begin{bmatrix} -3 & 13 \\ 1 & -4 \end{bmatrix}$

35. $\begin{bmatrix} 1 & 0 & 1 \\ 0 & 1 & -2 \\ 0 & 0 & 4 \end{bmatrix}$

36. $\begin{bmatrix} 3 & 0 & 6 \\ 0 & 2 & 0 \\ 1 & 0 & 3 \end{bmatrix}$

37. Find two 2×2 matrices A such that $A^2 = I$.

38. Find two 2×2 matrices A such that $A^2 = O$.

39. Find three 2×2 idempotent matrices. (Recall that a square matrix A is *idempotent* when $A^2 = A$.)

40. Find 2×2 matrices A and B such that $AB = O$ but $BA \neq O$.

41. Consider the matrices below.

$$X = \begin{bmatrix} 1 \\ 2 \\ 0 \\ 1 \end{bmatrix}, \quad Y = \begin{bmatrix} -1 \\ 0 \\ 3 \\ 2 \end{bmatrix}, \quad Z = \begin{bmatrix} 3 \\ 4 \\ -1 \\ 2 \end{bmatrix}, \quad W = \begin{bmatrix} 3 \\ 2 \\ -4 \\ -1 \end{bmatrix}$$

(a) Find scalars a, b, and c such that $W = aX + bY + cZ$.

(b) Show that there do not exist scalars a and b such that $Z = aX + bY$.

(c) Show that if $aX + bY + cZ = O$, then $a = b = c = 0$.

42. Proof Let A, B, and $A + B$ be nonsingular matrices. Prove that $A^{-1} + B^{-1}$ is nonsingular by showing that

$$(A^{-1} + B^{-1})^{-1} = A(A + B)^{-1}B.$$

Finding an *LU*-Factorization of a Matrix **In Exercises 43–46, find an *LU*-factorization of the matrix.**

43. $\begin{bmatrix} 2 & 5 \\ 6 & 14 \end{bmatrix}$ **44.** $\begin{bmatrix} -3 & 1 \\ 12 & 0 \end{bmatrix}$

45. $\begin{bmatrix} 4 & 1 & 0 \\ 0 & 3 & -7 \\ -16 & 11 & 1 \end{bmatrix}$ **46.** $\begin{bmatrix} 1 & 1 & 1 \\ 1 & 2 & 2 \\ 1 & 2 & 3 \end{bmatrix}$

Solving a Linear System Using *LU*-Factorization **In Exercises 47 and 48, use an *LU*-factorization of the coefficient matrix to solve the linear system.**

47. $\begin{aligned} x \quad\quad + z &= 3 \\ 2x + y + 2z &= 7 \\ 3x + 2y + 6z &= 8 \end{aligned}$

48. $\begin{aligned} 2x_1 + x_2 + x_3 - x_4 &= 7 \\ 3x_2 + x_3 - x_4 &= -3 \\ -2x_3 &= 2 \\ 2x_1 + x_2 + x_3 - 2x_4 &= 8 \end{aligned}$

49. Manufacturing A company manufactures tables and chairs at two locations. Matrix C gives the costs of manufacturing at each location.

$$C = \begin{bmatrix} 627 & 681 \\ 135 & 150 \end{bmatrix} \begin{matrix} \text{Tables} \\ \text{Chairs} \end{matrix}$$

with columns labeled Location 1 and Location 2.

(a) Labor accounts for $\frac{2}{3}$ of the cost. Determine the matrix L that gives the labor costs at each location.

(b) Find the matrix M that gives material costs at each location. (Assume there are only labor and material costs.)

50. Manufacturing A corporation has four factories, each of which manufactures sport utility vehicles and pickup trucks. In the matrix

$$A = \begin{bmatrix} 100 & 90 & 70 & 30 \\ 40 & 20 & 60 & 60 \end{bmatrix}$$

a_{ij} represents the number of vehicles of type i produced at factory j in one day. Find the production levels when production increases by 10%.

51. Gasoline Sales Matrix A shows the numbers of liters of 87-octane, 89-octane, and 93-octane gasoline sold at a convenience store over a weekend.

$$A = \begin{bmatrix} 2200 & 3180 & 1210 \\ 2120 & 1590 & 610 \\ 3250 & 3860 & 2040 \end{bmatrix} \begin{matrix} \text{Friday} \\ \text{Saturday} \\ \text{Sunday} \end{matrix}$$

with columns labeled Octane 87, 89, 93.

Matrix B gives the selling prices (in dollars per liter) and the profits (in dollars per liter) for the three grades of gasoline.

$$B = \begin{bmatrix} b_{11} & 0.01 \\ b_{21} & 0.02 \\ b_{31} & 0.03 \end{bmatrix} \begin{matrix} 87 \\ 89 \\ 93 \end{matrix} \text{ Octane}$$

with columns labeled Selling Price and Profit.

(a) Find AB and interpret the result.

(b) Find the convenience store's profit from gasoline sales for the weekend.

52. Final Grades Two midterms and a final exam determine the final grade in a course at a liberal arts college. The matrices below show the grades for six students and two possible grading systems.

	Midterm 1	Midterm 2	Final Exam	
	78	82	80	Student 1
	84	88	85	Student 2
$A =$	92	93	90	Student 3
	88	86	90	Student 4
	74	78	80	Student 5
	96	95	98	Student 6

	Grading System 1	Grading System 2	
	0.25	0.20	Midterm 1
$B =$	0.25	0.20	Midterm 2
	0.50	0.60	Final Exam

(a) Describe the grading systems in matrix B.

(b) Compute the numerical grades for the six students (to the nearest whole number) using the two grading systems.

(c) How many students received an "A" in each grading system? (Assume 90 or greater is an "A.")

Polynomial Function   In Exercises 53 and 54, find $f(A)$ using the definition below.

If $f(x) = a_0 + a_1x + a_2x^2 + \cdots + a_nx^n$ is a polynomial function, then for a square matrix A,

$$f(A) = a_0I + a_1A + a_2A^2 + \cdots + a_nA^n.$$

53. $f(x) = 6 - 7x + x^2, \quad A = \begin{bmatrix} 5 & 4 \\ 1 & 2 \end{bmatrix}$

54. $f(x) = 2 - 3x + x^3, \quad A = \begin{bmatrix} 2 & 1 \\ -1 & 0 \end{bmatrix}$

Stochastic Matrices   In Exercises 55–58, determine whether the matrix is stochastic.

55. $\begin{bmatrix} \frac{12}{25} & \frac{2}{25} \\ \frac{13}{25} & \frac{23}{25} \end{bmatrix}$

56. $\begin{bmatrix} 0.3 & 0.7 \\ 0 & 1 \end{bmatrix}$

57. $\begin{bmatrix} 1 & 0 & 0 \\ 0 & 0.5 & 0.1 \\ 0 & 0.1 & 0.5 \end{bmatrix}$

58. $\begin{bmatrix} 0.3 & 0.4 & 0.1 \\ 0.2 & 0.4 & 0.5 \\ 0.5 & 0.2 & 0.4 \end{bmatrix}$

Finding State Matrices   In Exercises 59–62, use the matrix of transition probabilities P and initial state matrix X_0 to find the state matrices $X_1, X_2,$ and X_3.

59. $P = \begin{bmatrix} \frac{1}{2} & \frac{1}{4} \\ \frac{1}{2} & \frac{3}{4} \end{bmatrix}, \quad X_0 = \begin{bmatrix} \frac{2}{3} \\ \frac{1}{3} \end{bmatrix}$

60. $P = \begin{bmatrix} 0.23 & 0.45 \\ 0.77 & 0.55 \end{bmatrix}, \quad X_0 = \begin{bmatrix} 0.65 \\ 0.35 \end{bmatrix}$

61. $P = \begin{bmatrix} 0.50 & 0.25 & 0 \\ 0.25 & 0.70 & 0.15 \\ 0.25 & 0.05 & 0.85 \end{bmatrix}, \quad X_0 = \begin{bmatrix} 0.5 \\ 0.5 \\ 0 \end{bmatrix}$

62. $P = \begin{bmatrix} \frac{1}{3} & \frac{1}{3} & \frac{2}{3} \\ \frac{1}{3} & 0 & \frac{1}{3} \\ \frac{1}{3} & \frac{2}{3} & 0 \end{bmatrix}, \quad X_0 = \begin{bmatrix} \frac{2}{9} \\ \frac{4}{9} \\ \frac{1}{3} \end{bmatrix}$

63. Caribbean Cruise Three hundred people go on a Caribbean cruise. When the ship stops at a port, each person has a choice of going on shore or not. Seventy percent of the people who go on shore one day will not go on shore the next day. Sixty percent of the people who do not go on shore one day will go on shore the next day. Today, 200 people go on shore. How many people will go on shore (a) tomorrow and (b) the day after tomorrow?

64. Population Migration A country has three regions. Each year, 10% of the residents of Region 1 move to Region 2 and 5% move to Region 3, 15% of the residents of Region 2 move to Region 1 and 5% move to Region 3, and 10% of the residents of Region 3 move to Region 1 and 10% move to Region 2. This year, each region has a population of 100,000. Find the populations of each region (a) in 1 year and (b) in 3 years.

Regular and Steady State Matrix   In Exercises 65–68, determine whether the stochastic matrix P is regular. Then find the steady state matrix $\overline{X}$ of the Markov chain with matrix of transition probabilities P.

65. $P = \begin{bmatrix} 0.8 & 0.5 \\ 0.2 & 0.5 \end{bmatrix}$

66. $P = \begin{bmatrix} 1 & \frac{4}{7} \\ 0 & \frac{3}{7} \end{bmatrix}$

67. $P = \begin{bmatrix} \frac{1}{3} & \frac{1}{6} & 0 \\ \frac{1}{6} & 0 & 0 \\ \frac{1}{2} & \frac{5}{6} & 1 \end{bmatrix}$

68. $P = \begin{bmatrix} 0 & 0 & 0.2 \\ 0.5 & 0.9 & 0 \\ 0.5 & 0.1 & 0.8 \end{bmatrix}$

69. Sales Promotion As a promotional feature, a store conducts a weekly raffle. During any week, 40% of the customers who turn in one or more tickets do not bother to turn in tickets the following week. On the other hand, 30% of the customers who do not turn in tickets will turn in one or more tickets the following week. Find and interpret the steady matrix for this situation.

70. Classified Documents A courtroom has 2000 documents, of which 1250 are classified. Each week, 10% of the classified documents become declassified and 20% are shredded. Also, 20% of the unclassified documents become classified and 5% are shredded. Find and interpret the steady state matrix for this situation.

Absorbing Markov Chains   In Exercises 71 and 72, determine whether the Markov chain with matrix of transition probabilities P is absorbing. Explain.

71. $P = \begin{bmatrix} 0 & 0.4 & 0.1 \\ 0.7 & 0.3 & 0.4 \\ 0.3 & 0.3 & 0.5 \end{bmatrix}$

72. $P = \begin{bmatrix} 1 & 0 & 0.38 \\ 0 & 0.30 & 0 \\ 0 & 0.70 & 0.62 \end{bmatrix}$

True or False?   In Exercises 73–76, determine whether each statement is true or false. If a statement is true, give a reason or cite an appropriate statement from the text. If a statement is false, provide an example that shows the statement is not true in all cases or cite an appropriate statement from the text.

73. (a) Addition of matrices is not commutative.

   (b) The transpose of the sum of matrices is equal to the sum of the transposes of the matrices.

74. (a) If an $n \times n$ matrix A is not symmetric, then A^TA is not symmetric.

   (b) If A and B are nonsingular $n \times n$ matrices, then $A + B$ is a nonsingular matrix.

75. (a) A stochastic matrix can have negative entries.

   (b) A Markov chain that is not regular can have a unique steady state matrix.

76. (a) A regular stochastic matrix can have entries of 0.

   (b) The steady state matrix of an absorbing Markov chain always depends on the initial state matrix.

Encoding a Message   In Exercises 77 and 78, write the uncoded row matrices for the message. Then encode the message using the matrix A.

77. *Message:*   ONE IF BY LAND

   Row Matrix Size: 1×2

   Encoding Matrix: $A = \begin{bmatrix} 5 & 2 \\ 2 & 1 \end{bmatrix}$

78. *Message:*   BEAM ME UP SCOTTY

   Row Matrix Size: 1×3

   Encoding Matrix: $A = \begin{bmatrix} 2 & 1 & 4 \\ 3 & 1 & 3 \\ -2 & -1 & -3 \end{bmatrix}$

Decoding a Message   In Exercises 79–82, use A^{-1} to decode the cryptogram.

79. $A = \begin{bmatrix} 3 & -2 \\ -4 & 3 \end{bmatrix}$,

   $-45\ 34\ 36\ -24\ -43\ 37\ -23\ 22\ -37\ 29\ 57\ -38$
   $-39\ 31$

80. $A = \begin{bmatrix} 1 & 4 \\ -1 & -3 \end{bmatrix}$,

   $11\ 52\ -8\ -9\ -13\ -39\ 5\ 20\ 12\ 56\ 5\ 20\ -2\ 7$
   $9\ 41\ 25\ 100$

81. $A = \begin{bmatrix} 1 & -2 & 2 \\ -1 & 1 & 3 \\ 1 & -1 & -4 \end{bmatrix}$,

   $-2\ 2\ 5\ 39\ -53\ -72\ -6\ -9\ 93\ 4\ -12\ 27\ 31$
   $-49\ -16\ 19\ -24\ -46\ -8\ -7\ 99$

82. $A = \begin{bmatrix} 2 & 0 & 1 \\ 2 & -1 & 0 \\ 1 & 2 & -4 \end{bmatrix}$,

   $66\ 27\ -31\ 37\ 5\ -9\ 61\ 46\ -73\ 46\ -14\ 9\ 94$
   $21\ -49\ 32\ -4\ 12\ 66\ 31\ -53\ 47\ 33\ -67\ 32\ 19$
   $-56\ 43\ -9\ -20\ 68\ 23\ -34$

83. Industrial System   An industrial system has two industries with the input requirements below.

   (a) To produce $1.00 worth of output, Industry A requires $0.20 of its own product and $0.30 of Industry B's product.

   (b) To produce $1.00 worth of output, Industry B requires $0.10 of its own product and $0.50 of Industry A's product.

   Find D, the input-output matrix for this system. Then solve for the output matrix X in the equation $X = DX + E$, where E is the external demand matrix

   $E = \begin{bmatrix} 40,000 \\ 80,000 \end{bmatrix}$.

84. Solving for the Output Matrix   An industrial system with three industries has the input-output matrix D and external demand matrix E below.

$$D = \begin{bmatrix} 0.1 & 0.3 & 0.2 \\ 0.0 & 0.2 & 0.3 \\ 0.4 & 0.1 & 0.1 \end{bmatrix} \text{ and } E = \begin{bmatrix} 3000 \\ 3500 \\ 8500 \end{bmatrix}$$

Solve for the output matrix X in the equation $X = DX + E$.

Finding the Least Squares Regression Line   In Exercises 85–88, find the least squares regression line.

85. $(1, 5), (2, 4), (3, 2)$

86. $(2, 1), (3, 3), (4, 2), (5, 4), (6, 4)$

87. $(1, 1), (1, 3), (1, 2), (1, 4), (2, 5)$

88. $(-2, 4), (-1, 2), (0, 1), (1, -2), (2, -3)$

89. Cellular Phone Subscribers   The table shows the numbers of cellular phone subscribers y (in millions) in the United States from 2008 through 2013. (Source: CTIA–The Wireless Association)

Year	2008	2009	2010	2011	2012	2013
Number, y	270	286	296	316	326	336

   (a) Find the least squares regression line for the data. Let x represent the year, with $x = 8$ corresponding to 2008.

   (b) Use the linear regression capabilities of a graphing utility to find a linear model for the data. How does this model compare with the model obtained in part (a)?

   (c) Use the linear model to create a table of estimated values for y. Compare the estimated values with the actual data.

90. Major League Baseball Salaries   The table shows the average salaries y (in millions of dollars) of Major League Baseball players on opening day of baseball season from 2008 through 2013. (Source: Major League Baseball)

Year	2008	2009	2010	2011	2012	2013
Salary, y	2.93	3.00	3.01	3.10	3.21	3.39

   (a) Find the least squares regression line for the data. Let x represent the year, with $x = 8$ corresponding to 2008.

   (b) Use the linear regression capabilities of a graphing utility to find a linear model for the data. How does this model compare with the model obtained in part (a)?

   (c) Use the linear model to create a table of estimated values for y. Compare the estimated values with the actual data.

2   Projects

1 Exploring Matrix Multiplication

	Test 1	Test 2
Anna	82	98
Bruce	58	70
Chris	76	85
David	84	90

The table shows the first two test scores for Anna, Bruce, Chris, and David. Use the table to create a matrix M to represent the data. Input M into a software program or a graphing utility and use it to answer the questions below.

1. Which test was more difficult? Which was easier? Explain.

2. How would you rank the performances of the four students?

3. Describe the meanings of the matrix products $M\begin{bmatrix} 1 \\ 0 \end{bmatrix}$ and $M\begin{bmatrix} 0 \\ 1 \end{bmatrix}$.

4. Describe the meanings of the matrix products $[1 \ \ 0 \ \ 0 \ \ 0]M$ and $[0 \ \ 0 \ \ 1 \ \ 0]M$.

5. Describe the meanings of the matrix products $M\begin{bmatrix} 1 \\ 1 \end{bmatrix}$ and $\frac{1}{2}M\begin{bmatrix} 1 \\ 1 \end{bmatrix}$.

6. Describe the meanings of the matrix products $[1 \ \ 1 \ \ 1 \ \ 1]M$ and $\frac{1}{4}[1 \ \ 1 \ \ 1 \ \ 1]M$.

7. Describe the meaning of the matrix product $[1 \ \ 1 \ \ 1 \ \ 1]M\begin{bmatrix} 1 \\ 1 \end{bmatrix}$.

8. Use matrix multiplication to find the combined overall average score on both tests.

9. How could you use matrix multiplication to scale the scores on test 1 by a factor of 1.1?

2 Nilpotent Matrices

Let A be a nonzero square matrix. Is it possible that a positive integer k exists such that $A^k = O$? For example, find A^3 for the matrix

$$A = \begin{bmatrix} 0 & 2 & 1 \\ 0 & 0 & 2 \\ 0 & 0 & 0 \end{bmatrix}.$$

A square matrix A is **nilpotent of index k** when $A \neq O$, $A^2 \neq O$, . . . , $A^{k-1} \neq O$, but $A^k = O$. In this project you will explore nilpotent matrices.

1. The matrix in the example above is nilpotent. What is its index?

2. Use a software program or a graphing utility to determine which matrices below are nilpotent and find their indices.

(a) $\begin{bmatrix} 0 & 2 \\ 0 & 0 \end{bmatrix}$ (b) $\begin{bmatrix} 1 & 0 \\ 0 & 1 \end{bmatrix}$ (c) $\begin{bmatrix} 0 & 0 \\ 0 & 1 \end{bmatrix}$

(d) $\begin{bmatrix} 0 & 0 & 0 \\ -1 & 0 & 0 \\ 2 & 1 & 0 \end{bmatrix}$ (e) $\begin{bmatrix} 0 & 1 & 1 \\ 1 & 0 & 1 \\ 1 & 1 & 0 \end{bmatrix}$ (f) $\begin{bmatrix} 0 & 1 & 2 \\ 0 & 0 & 3 \\ 0 & 0 & 0 \end{bmatrix}$

3. Find 3×3 nilpotent matrices of indices 2 and 3.

4. Find 4×4 nilpotent matrices of indices 2, 3, and 4.

5. Find a nilpotent matrix of index 5.

6. Are nilpotent matrices invertible? Prove your answer.

7. When A is nilpotent, what can you say about A^T? Prove your answer.

8. Show that if A is nilpotent, then $I - A$ is invertible.

3 Determinants

Comet Landing (p. 141)

Software Publishing (p. 143)

Engineering and Control (p. 130)

Sudoku (p. 120)

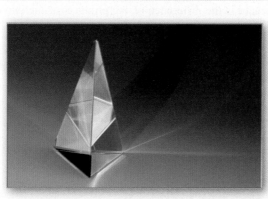

Volume of a Tetrahedron (p. 114)

3.1   The Determinant of a Matrix

■ Find the determinant of a 2 × 2 matrix.

■ Find the minors and cofactors of a matrix.

■ Use expansion by cofactors to find the determinant of a matrix.

■ Find the determinant of a triangular matrix.

THE DETERMINANT OF A 2 × 2 MATRIX

Every *square* matrix can be associated with a real number called its *determinant*. Historically, the use of determinants arose from the recognition of special patterns that occur in the solutions of systems of linear equations. For example, the system

$$a_{11}x_1 + a_{12}x_2 = b_1$$
$$a_{21}x_1 + a_{22}x_2 = b_2$$

has the solution

$$x_1 = \frac{b_1 a_{22} - b_2 a_{12}}{a_{11}a_{22} - a_{21}a_{12}} \quad \text{and} \quad x_2 = \frac{b_2 a_{11} - b_1 a_{21}}{a_{11}a_{22} - a_{21}a_{12}}$$

when $a_{11}a_{22} - a_{21}a_{12} \neq 0$. (See Exercise 53.) Note that both fractions have the same denominator, $a_{11}a_{22} - a_{21}a_{12}$. This quantity is the *determinant* of the coefficient matrix of the system.

REMARK

In this text, det(A) and |A| are used interchangeably to represent the determinant of A. Although vertical bars are also used to denote the absolute value of a real number, the context will show which use is intended. Furthermore, it is common practice to delete the matrix brackets and write

$$\begin{vmatrix} a_{11} & a_{12} \\ a_{21} & a_{22} \end{vmatrix}$$

instead of

$$\begin{Vmatrix} a_{11} & a_{12} \\ a_{21} & a_{22} \end{Vmatrix}.$$

Definition of the Determinant of a 2 × 2 Matrix

The **determinant** of the matrix

$$A = \begin{bmatrix} a_{11} & a_{12} \\ a_{21} & a_{22} \end{bmatrix}$$

is $\det(A) = |A| = a_{11}a_{22} - a_{21}a_{12}$.

The diagram below shows a convenient method for remembering the formula for the determinant of a 2 × 2 matrix.

$$|A| = \begin{vmatrix} a_{11} & a_{12} \\ a_{21} & a_{22} \end{vmatrix} = a_{11}a_{22} - a_{21}a_{12}$$

The determinant is the difference of the products of the two diagonals of the matrix. Note that the order of the products is important.

REMARK

Notice that the determinant of a matrix can be positive, zero, or negative.

EXAMPLE 1   Determinants of Matrices of Order 2

a. For $A = \begin{bmatrix} 2 & -3 \\ 1 & 2 \end{bmatrix}$, $|A| = \begin{vmatrix} 2 & -3 \\ 1 & 2 \end{vmatrix} = 2(2) - 1(-3) = 4 + 3 = 7.$

b. For $B = \begin{bmatrix} 2 & 1 \\ 4 & 2 \end{bmatrix}$, $|B| = \begin{vmatrix} 2 & 1 \\ 4 & 2 \end{vmatrix} = 2(2) - 4(1) = 4 - 4 = 0.$

c. For $C = \begin{bmatrix} 0 & \frac{3}{2} \\ 2 & 4 \end{bmatrix}$, $|C| = \begin{vmatrix} 0 & \frac{3}{2} \\ 2 & 4 \end{vmatrix} = 0(4) - 2(\frac{3}{2}) = 0 - 3 = -3.$

MINORS AND COFACTORS

To define the determinant of a square matrix of order higher than 2, it is convenient to use *minors* and *cofactors*.

Minors and Cofactors of a Square Matrix

If A is a square matrix, then the **minor** M_{ij} of the entry a_{ij} is the determinant of the matrix obtained by deleting the ith row and jth column of A. The **cofactor** C_{ij} of the entry a_{ij} is $C_{ij} = (-1)^{i+j}M_{ij}$.

For example, if A is a 3×3 matrix, then the minors and cofactors of a_{21} and a_{22} are as shown below.

Minor of a_{21}

$$\begin{bmatrix} a_{11} & a_{12} & a_{13} \\ a_{21} & a_{22} & a_{23} \\ a_{31} & a_{32} & a_{33} \end{bmatrix}, \quad M_{21} = \begin{vmatrix} a_{12} & a_{13} \\ a_{32} & a_{33} \end{vmatrix}$$

Delete row 2 and column 1.

Minor of a_{22}

$$\begin{bmatrix} a_{11} & a_{12} & a_{13} \\ a_{21} & a_{22} & a_{23} \\ a_{31} & a_{32} & a_{33} \end{bmatrix}, \quad M_{22} = \begin{vmatrix} a_{11} & a_{13} \\ a_{31} & a_{33} \end{vmatrix}$$

Delete row 2 and column 2.

Cofactor of a_{21}

$$C_{21} = (-1)^{2+1}M_{21} = -M_{21}$$

Cofactor of a_{22}

$$C_{22} = (-1)^{2+2}M_{22} = M_{22}$$

The minors and cofactors of a matrix can differ only in sign. To obtain the cofactors of a matrix, first find the minors and then apply the checkerboard pattern of +'s and −'s shown at the left. Note that *odd* positions (where $i + j$ is odd) have negative signs, and even positions (where $i + j$ is even) have positive signs.

Sign Pattern for Cofactors

$$\begin{bmatrix} + & - & + \\ - & + & - \\ + & - & + \end{bmatrix}$$

3×3 matrix

$$\begin{bmatrix} + & - & + & - \\ - & + & - & + \\ + & - & + & - \\ - & + & - & + \end{bmatrix}$$

4×4 matrix

$$\begin{bmatrix} + & - & + & - & + & \cdots \\ - & + & - & + & - & \cdots \\ + & - & + & - & + & \cdots \\ - & + & - & + & - & \cdots \\ + & - & + & - & + & \cdots \\ \vdots & \vdots & \vdots & \vdots & \vdots \end{bmatrix}$$

$n \times n$ matrix

EXAMPLE 2 Minors and Cofactors of a Matrix

Find all the minors and cofactors of

$$A = \begin{bmatrix} 0 & 2 & 1 \\ 3 & -1 & 2 \\ 4 & 0 & 1 \end{bmatrix}.$$

SOLUTION

To find the minor M_{11}, delete the first row and first column of A and evaluate the determinant of the resulting matrix.

$$\begin{bmatrix} 0 & 2 & 1 \\ 3 & -1 & 2 \\ 4 & 0 & 1 \end{bmatrix}, \quad M_{11} = \begin{vmatrix} -1 & 2 \\ 0 & 1 \end{vmatrix} = -1(1) - 0(2) = -1$$

Verify that the minors are

$$M_{11} = -1 \quad M_{12} = -5 \quad M_{13} = 4$$
$$M_{21} = 2 \quad M_{22} = -4 \quad M_{23} = -8$$
$$M_{31} = 5 \quad M_{32} = -3 \quad M_{33} = -6.$$

Now, to find the cofactors, combine these minors with the checkerboard pattern of signs for a 3×3 matrix shown above.

$$C_{11} = -1 \quad C_{12} = 5 \quad C_{13} = 4$$
$$C_{21} = -2 \quad C_{22} = -4 \quad C_{23} = 8$$
$$C_{31} = 5 \quad C_{32} = 3 \quad C_{33} = -6$$

THE DETERMINANT OF A SQUARE MATRIX

REMARK

The determinant of a matrix of order 1 is simply the entry of the matrix. For example, if $A = [-2]$, then

$$\det(A) = -2.$$

The definition below is **inductive** because it uses the determinant of a square matrix of order $n - 1$ to define the determinant of a square matrix of order n.

Definition of the Determinant of a Square Matrix

If A is a square matrix of order $n \geq 2$, then the determinant of A is the sum of the entries in the first row of A multiplied by their respective cofactors. That is,

$$\det(A) = |A| = \sum_{j=1}^{n} a_{1j}C_{1j} = a_{11}C_{11} + a_{12}C_{12} + \cdots + a_{1n}C_{1n}.$$

Confirm that, for 2×2 matrices, this definition yields

$$|A| = a_{11}a_{22} - a_{21}a_{12}$$

as previously defined.

When you use this definition to evaluate a determinant, you are **expanding by cofactors in the first row.** Example 3 demonstrates this procedure.

EXAMPLE 3   The Determinant of a Matrix of Order 3

Find the determinant of

$$A = \begin{bmatrix} 0 & 2 & 1 \\ 3 & -1 & 2 \\ 4 & 0 & 1 \end{bmatrix}.$$

SOLUTION

This is the same matrix as in Example 2. There you found the cofactors of the entries in the first row to be

$$C_{11} = -1, \quad C_{12} = 5, \quad C_{13} = 4.$$

So, by the definition of a determinant, you have

$$\begin{aligned}
|A| &= a_{11}C_{11} + a_{12}C_{12} + a_{13}C_{13} \qquad \text{First row expansion} \\
&= 0(-1) + 2(5) + 1(4) \\
&= 14.
\end{aligned}$$

Although the determinant is defined as an expansion by the cofactors in the first row, it can be shown that the determinant can be evaluated by expanding in *any* row or column. For instance, you could expand the matrix in Example 3 in the second row to obtain

$$\begin{aligned}
|A| &= a_{21}C_{21} + a_{22}C_{22} + a_{23}C_{23} \qquad \text{Second row expansion} \\
&= 3(-2) + (-1)(-4) + 2(8) \\
&= 14
\end{aligned}$$

or in the first column to obtain

$$\begin{aligned}
|A| &= a_{11}C_{11} + a_{21}C_{21} + a_{31}C_{31} \qquad \text{First column expansion} \\
&= 0(-1) + 3(-2) + 4(5) \\
&= 14.
\end{aligned}$$

Try other possibilities to confirm that the determinant of A can be evaluated by expanding in *any* row or column. The theorem on the next page states this, and is known as Laplace's Expansion of a Determinant, after the French mathematician Pierre Simon de Laplace (1749–1827).

> **THEOREM 3.1   Expansion by Cofactors**
>
> Let A be a square matrix of order n. Then the determinant of A is
>
> $$\det(A) = |A| = \sum_{j=1}^{n} a_{ij}C_{ij} = a_{i1}C_{i1} + a_{i2}C_{i2} + \cdots + a_{in}C_{in} \qquad \text{\textit{i}th row expansion}$$
>
> or
>
> $$\det(A) = |A| = \sum_{i=1}^{n} a_{ij}C_{ij} = a_{1j}C_{1j} + a_{2j}C_{2j} + \cdots + a_{nj}C_{nj}. \qquad \text{\textit{j}th column expansion}$$

When expanding by cofactors, you do not need to find cofactors of zero entries, because zero times its cofactor is zero.

$$a_{ij}C_{ij} = (0)C_{ij}$$
$$= 0$$

The row (or column) containing the most zeros is usually the best choice for expansion by cofactors. The next example demonstrates this.

EXAMPLE 4   **The Determinant of a Matrix of Order 4**

Find the determinant of

$$A = \begin{bmatrix} 1 & -2 & 3 & 0 \\ -1 & 1 & 0 & 2 \\ 0 & 2 & 0 & 3 \\ 3 & 4 & 0 & -2 \end{bmatrix}.$$

SOLUTION

Notice that three of the entries in the third column are zeros. So, to eliminate some of the work in the expansion, use the third column.

$$|A| = 3(C_{13}) + 0(C_{23}) + 0(C_{33}) + 0(C_{43})$$

The cofactors C_{23}, C_{33}, and C_{43} have zero coefficients, so you need only find the cofactor C_{13}. To do this, delete the first row and third column of A and evaluate the determinant of the resulting matrix.

$$C_{13} = (-1)^{1+3} \begin{vmatrix} -1 & 1 & 2 \\ 0 & 2 & 3 \\ 3 & 4 & -2 \end{vmatrix} \qquad \text{Delete 1st row and 3rd column.}$$

$$= \begin{vmatrix} -1 & 1 & 2 \\ 0 & 2 & 3 \\ 3 & 4 & -2 \end{vmatrix} \qquad \text{Simplify.}$$

Expanding by cofactors in the second row yields

$$C_{13} = (0)(-1)^{2+1}\begin{vmatrix} 1 & 2 \\ 4 & -2 \end{vmatrix} + (2)(-1)^{2+2}\begin{vmatrix} -1 & 2 \\ 3 & -2 \end{vmatrix} + (3)(-1)^{2+3}\begin{vmatrix} -1 & 1 \\ 3 & 4 \end{vmatrix}$$

$$= 0 + 2(1)(-4) + 3(-1)(-7)$$
$$= 13.$$

You obtain

$$|A| = 3(13)$$
$$= 39.$$

TECHNOLOGY

Many graphing utilities and software programs can find the determinant of a square matrix. If you use a graphing utility, then you may see something similar to the screen below for Example 4. The **Technology Guide** at *CengageBrain.com* can help you use technology to find a determinant.

```
A
    [[1   -2  3   0 ]
     [-1  1   0   2 ]
     [0   2   0   3 ]
     [3   4   0   -2]]
det A
                    39
```

An alternative method is commonly used to evaluate the determinant of a 3×3 matrix A. To apply this method, copy the first and second columns of A to form fourth and fifth columns. Then obtain the determinant of A by adding (or subtracting) the products of the six diagonals, as shown in the diagram below.

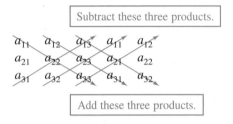

Confirm that the determinant of A is

$$|A| = a_{11}a_{22}a_{33} + a_{12}a_{23}a_{31} + a_{13}a_{21}a_{32} - a_{31}a_{22}a_{13} - a_{32}a_{23}a_{11} - a_{33}a_{21}a_{12}.$$

EXAMPLE 5   The Determinant of a Matrix of Order 3

See LarsonLinearAlgebra.com for an interactive version of this type of example.

Find the determinant of

$$A = \begin{bmatrix} 0 & 2 & 1 \\ 3 & -1 & 2 \\ 4 & -4 & 1 \end{bmatrix}.$$

SOLUTION

Begin by copying the first two columns and then computing the six diagonal products as shown below.

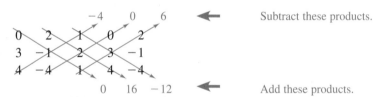

Now, by adding the lower three products and subtracting the upper three products, you can find the determinant of A to be

$$|A| = 0 + 16 + (-12) - (-4) - 0 - 6 = 2.$$

The diagonal process illustrated in Example 5 is valid *only* for matrices of order 3. For matrices of higher order, you must use another method.

LINEAR ALGEBRA APPLIED

Recall that a **tetrahedron** is a polyhedron consisting of four triangular faces. One practical application of determinants is in finding the volume of a tetrahedron in a coordinate plane. If the vertices of a tetrahedron are (x_1, y_1, z_1), (x_2, y_2, z_2), (x_3, y_3, z_3), and (x_4, y_4, z_4), then the volume is

$$\text{Volume} = \pm \tfrac{1}{6} \det \begin{bmatrix} x_1 & y_1 & z_1 & 1 \\ x_2 & y_2 & z_2 & 1 \\ x_3 & y_3 & z_3 & 1 \\ x_4 & y_4 & z_4 & 1 \end{bmatrix}.$$

You will study this and other applications of determinants in Section 3.4.

TRIANGULAR MATRICES

Upper Triangular Matrix

$$\begin{bmatrix} a_{11} & a_{12} & a_{13} & \cdots & a_{1n} \\ 0 & a_{22} & a_{23} & \cdots & a_{2n} \\ 0 & 0 & a_{33} & \cdots & a_{3n} \\ \vdots & \vdots & \vdots & & \vdots \\ 0 & 0 & 0 & \cdots & a_{nn} \end{bmatrix}$$

Lower Triangular Matrix

$$\begin{bmatrix} a_{11} & 0 & 0 & \cdots & 0 \\ a_{21} & a_{22} & 0 & \cdots & 0 \\ a_{31} & a_{32} & a_{33} & \cdots & 0 \\ \vdots & \vdots & \vdots & & \vdots \\ a_{n1} & a_{n2} & a_{n3} & \cdots & a_{nn} \end{bmatrix}$$

Recall from Section 2.4 that a square matrix is *upper triangular* when it has all zero entries below its main diagonal, and *lower triangular* when it has all zero entries above its main diagonal, as shown in the diagram at the left. A matrix that is both upper and lower triangular is a **diagonal matrix.** That is, a diagonal matrix is one in which all entries above and below the main diagonal are zero.

To find the determinant of a triangular matrix, simply form the product of the entries on the main diagonal. It should be easy to see that this procedure is valid for triangular matrices of order 2 or 3. For example, to find the determinant of

$$A = \begin{bmatrix} 2 & 3 & -1 \\ 0 & -1 & 2 \\ 0 & 0 & 3 \end{bmatrix}$$

expand in the third row to obtain

$$|A| = 0(-1)^{3+1}\begin{vmatrix} 3 & -1 \\ -1 & 2 \end{vmatrix} + 0(-1)^{3+2}\begin{vmatrix} 2 & -1 \\ 0 & 2 \end{vmatrix} + 3(-1)^{3+3}\begin{vmatrix} 2 & 3 \\ 0 & -1 \end{vmatrix}$$

$$= 3(1)(-2)$$

$$= -6$$

which is the product of the entries on the main diagonal.

THEOREM 3.2   Determinant of a Triangular Matrix

If A is a triangular matrix of order n, then its determinant is the product of the entries on the main diagonal. That is,

$$\det(A) = |A| = a_{11}a_{22}a_{33} \cdots a_{nn}.$$

PROOF

Use *mathematical induction** to prove this theorem for the case in which A is an upper triangular matrix. The proof of the case in which A is lower triangular is similar. If A has order 1, then $A = [a_{11}]$ and the determinant is $|A| = a_{11}$. Assuming the theorem is true for any upper triangular matrix of order $k - 1$, consider an upper triangular matrix A of order k. Expanding in the kth row, you obtain

$$|A| = 0C_{k1} + 0C_{k2} + \cdots + 0C_{k(k-1)} + a_{kk}C_{kk} = a_{kk}C_{kk}.$$

Now, note that $C_{kk} = (-1)^{2k}M_{kk} = M_{kk}$, where M_{kk} is the determinant of the upper triangular matrix formed by deleting the kth row and kth column of A. This matrix is of order $k - 1$, so apply the induction assumption to write

$$|A| = a_{kk}M_{kk} = a_{kk}(a_{11}a_{22}a_{33} \cdots a_{k-1, k-1}) = a_{11}a_{22}a_{33} \cdots a_{kk}.$$

EXAMPLE 6   The Determinant of a Triangular Matrix

The determinant of the lower triangular matrix

$$A = \begin{bmatrix} 2 & 0 & 0 & 0 \\ 4 & -2 & 0 & 0 \\ -5 & 6 & 1 & 0 \\ 1 & 5 & 3 & 3 \end{bmatrix}$$

is $|A| = (2)(-2)(1)(3) = -12$.

---

*See Appendix for a discussion of mathematical induction.

3.1 Exercises   See CalcChat.com for worked-out solutions to odd-numbered exercises.

The Determinant of a Matrix   **In Exercises 1–12, find the determinant of the matrix.**

1. $[1]$

2. $[-3]$

3. $\begin{bmatrix} 2 & 1 \\ 3 & 4 \end{bmatrix}$

4. $\begin{bmatrix} -3 & 1 \\ 5 & 2 \end{bmatrix}$

5. $\begin{bmatrix} 5 & 2 \\ -6 & 3 \end{bmatrix}$

6. $\begin{bmatrix} 2 & -2 \\ 4 & 3 \end{bmatrix}$

7. $\begin{bmatrix} -7 & 6 \\ \frac{1}{2} & 3 \end{bmatrix}$

8. $\begin{bmatrix} \frac{1}{3} & 5 \\ 4 & -9 \end{bmatrix}$

9. $\begin{bmatrix} 0 & 8 \\ 0 & 4 \end{bmatrix}$

10. $\begin{bmatrix} 2 & -3 \\ -6 & 9 \end{bmatrix}$

11. $\begin{bmatrix} \lambda - 3 & 2 \\ 4 & \lambda - 1 \end{bmatrix}$

12. $\begin{bmatrix} \lambda - 2 & 0 \\ 4 & \lambda - 4 \end{bmatrix}$

Finding the Minors and Cofactors of a Matrix   **In Exercises 13–16, find all (a) minors and (b) cofactors of the matrix.**

13. $\begin{bmatrix} 1 & 2 \\ 3 & 4 \end{bmatrix}$

14. $\begin{bmatrix} -5 & 6 \\ 1 & 0 \end{bmatrix}$

15. $\begin{bmatrix} -3 & 2 & 1 \\ 4 & 5 & 6 \\ 2 & -3 & 1 \end{bmatrix}$

16. $\begin{bmatrix} -3 & 4 & 2 \\ 6 & 3 & 1 \\ 4 & -7 & -8 \end{bmatrix}$

17. Find the determinant of the matrix in Exercise 15 using the method of expansion by cofactors. Use (a) the second row and (b) the second column.

18. Find the determinant of the matrix in Exercise 16 using the method of expansion by cofactors. Use (a) the third row and (b) the first column.

Finding a Determinant   **In Exercises 19–32, use expansion by cofactors to find the determinant of the matrix.**

19. $\begin{bmatrix} 1 & 4 & -2 \\ 3 & 2 & 0 \\ -1 & 4 & 3 \end{bmatrix}$

20. $\begin{bmatrix} 3 & -1 & 2 \\ 4 & 1 & 4 \\ -2 & 0 & 1 \end{bmatrix}$

21. $\begin{bmatrix} 2 & 4 & 6 \\ 0 & 3 & 1 \\ 0 & 0 & -5 \end{bmatrix}$

22. $\begin{bmatrix} -3 & 0 & 0 \\ 7 & 11 & 0 \\ 1 & 2 & 2 \end{bmatrix}$

23. $\begin{bmatrix} -0.4 & 0.4 & 0.3 \\ 0.2 & 0.2 & 0.2 \\ 0.3 & 0.2 & 0.2 \end{bmatrix}$

24. $\begin{bmatrix} 0.1 & 0.2 & 0.3 \\ -0.3 & 0.2 & 0.2 \\ 0.5 & 0.4 & 0.4 \end{bmatrix}$

25. $\begin{bmatrix} x & y & -1 \\ 3 & 2 & 0 \\ 1 & 1 & 1 \end{bmatrix}$

26. $\begin{bmatrix} x & y & 1 \\ -2 & -2 & 1 \\ 1 & 5 & 1 \end{bmatrix}$

27. $\begin{bmatrix} 5 & 3 & 0 & 6 \\ 4 & 6 & 4 & 12 \\ 0 & 2 & -3 & 4 \\ 0 & 1 & -2 & 2 \end{bmatrix}$

28. $\begin{bmatrix} 3 & 0 & 7 & 0 \\ 2 & 6 & 11 & 12 \\ 4 & 1 & -1 & 2 \\ 1 & 5 & 2 & 10 \end{bmatrix}$

29. $\begin{bmatrix} w & x & y & z \\ 21 & -15 & 24 & 30 \\ -10 & 24 & -32 & 18 \\ -40 & 22 & 32 & -35 \end{bmatrix}$

30. $\begin{bmatrix} w & x & y & z \\ 10 & 15 & -25 & 30 \\ -30 & 20 & -15 & -10 \\ 30 & 35 & -25 & -40 \end{bmatrix}$

31. $\begin{bmatrix} 5 & 2 & 0 & 0 & -2 \\ 0 & 1 & 4 & 3 & 2 \\ 0 & 0 & 2 & 6 & 3 \\ 0 & 0 & 3 & 4 & 1 \\ 0 & 0 & 0 & 0 & 2 \end{bmatrix}$

32. $\begin{bmatrix} -4 & 3 & 2 & -1 & -2 \\ 1 & -2 & 7 & -13 & -12 \\ -6 & 2 & -5 & -6 & -7 \\ 0 & 0 & 0 & 0 & 0 \\ 1 & -4 & -2 & 0 & -9 \end{bmatrix}$

Finding a Determinant   **In Exercises 33 and 34, use the method demonstrated in Example 5 to find the determinant of the matrix.**

33. $\begin{bmatrix} 3 & 0 & 4 \\ -2 & 4 & 1 \\ 1 & -3 & 1 \end{bmatrix}$

34. $\begin{bmatrix} 3 & 8 & -7 \\ 0 & -5 & 4 \\ 8 & 1 & 6 \end{bmatrix}$

Finding a Determinant   **In Exercises 35–38, use a software program or a graphing utility to find the determinant of the matrix.**

35. $\begin{bmatrix} 0.1 & 0.6 & -0.3 \\ 0.7 & -0.1 & 0.1 \\ 0.1 & 0.3 & -0.8 \end{bmatrix}$

36. $\begin{bmatrix} 4 & 3 & 2 & 5 \\ 1 & 6 & -1 & 2 \\ -3 & 2 & 4 & 5 \\ 6 & 1 & 3 & -2 \end{bmatrix}$

37. $\begin{bmatrix} 1 & 2 & -1 & 4 \\ 0 & 1 & 2 & -2 \\ 0 & 3 & 2 & -1 \\ 1 & 2 & 0 & -2 \end{bmatrix}$

38. $\begin{bmatrix} 8 & 5 & 1 & -2 & 0 \\ -1 & 0 & 7 & 1 & 6 \\ 0 & 8 & 6 & 5 & -3 \\ 1 & 2 & 5 & -8 & 4 \\ 2 & 6 & -2 & 0 & 6 \end{bmatrix}$

Finding the Determinant of a Triangular Matrix In Exercises 39–42, find the determinant of the triangular matrix.

39. $\begin{bmatrix} -2 & 0 & 0 \\ 4 & 6 & 0 \\ -3 & 7 & 2 \end{bmatrix}$ 40. $\begin{bmatrix} 4 & 0 & 0 \\ 0 & 7 & 0 \\ 0 & 0 & -2 \end{bmatrix}$

41. $\begin{bmatrix} 5 & 8 & -4 & 2 \\ 0 & 0 & 6 & 0 \\ 0 & 0 & 2 & 2 \\ 0 & 0 & 0 & -1 \end{bmatrix}$ 42. $\begin{bmatrix} 4 & 0 & 0 & 0 \\ -1 & \frac{1}{2} & 0 & 0 \\ 3 & 5 & 3 & 0 \\ -8 & 7 & 0 & -2 \end{bmatrix}$

True or False? In Exercises 43 and 44, determine whether each statement is true or false. If a statement is true, give a reason or cite an appropriate statement from the text. If a statement is false, provide an example that shows the statement is not true in all cases or cite an appropriate statement from the text.

43. (a) The determinant of a 2×2 matrix A is $a_{21}a_{12} - a_{11}a_{22}$.
    (b) The determinant of a matrix of order 1 is the entry of the matrix.
    (c) The ij-cofactor of a square matrix A is the matrix obtained by deleting the ith row and jth column of A.

44. (a) To find the determinant of a triangular matrix, add the entries on the main diagonal.
    (b) To find the determinant of a matrix, expand by cofactors in any row or column.
    (c) When expanding by cofactors, you need not evaluate the cofactors of zero entries.

Solving an Equation In Exercises 45–48, solve for x.

45. $\begin{vmatrix} x+3 & 2 \\ 1 & x+2 \end{vmatrix} = 0$ 46. $\begin{vmatrix} x-6 & 3 \\ -2 & x+1 \end{vmatrix} = 0$

47. $\begin{vmatrix} x-1 & 2 \\ 3 & x-2 \end{vmatrix} = 0$ 48. $\begin{vmatrix} x+3 & 1 \\ -4 & x-1 \end{vmatrix} = 0$

Solving an Equation In Exercises 49–52, find the values of λ for which the determinant is zero.

49. $\begin{vmatrix} \lambda+2 & 2 \\ 1 & \lambda \end{vmatrix}$ 50. $\begin{vmatrix} \lambda-5 & 3 \\ 1 & \lambda-5 \end{vmatrix}$

51. $\begin{vmatrix} \lambda & 2 & 0 \\ 0 & \lambda+1 & 2 \\ 0 & 1 & \lambda \end{vmatrix}$ 52. $\begin{vmatrix} \lambda & 0 & 1 \\ 0 & \lambda & 3 \\ 2 & 2 & \lambda-2 \end{vmatrix}$

53. Show that the system of linear equations

$$a_{11}x_1 + a_{12}x_2 = b_1$$
$$a_{21}x_1 + a_{22}x_2 = b_2$$

has the solution

$$x_1 = \frac{b_1a_{22} - b_2a_{12}}{a_{11}a_{22} - a_{21}a_{12}} \quad \text{and} \quad x_2 = \frac{b_2a_{11} - b_1a_{21}}{a_{11}a_{22} - a_{21}a_{12}}$$

when $a_{11}a_{22} - a_{21}a_{12} \neq 0$.

54. **CAPSTONE** For an $n \times n$ matrix A, explain how to find each value.
    (a) The minor M_{ij} of the entry a_{ij}
    (b) The cofactor C_{ij} of the entry a_{ij}
    (c) The determinant of A

Entries Involving Expressions In Exercises 55–62, evaluate the determinant, in which the entries are functions. Determinants of this type occur when changes of variables are made in calculus.

55. $\begin{vmatrix} 6u & -1 \\ -1 & 3v \end{vmatrix}$ 56. $\begin{vmatrix} 3x^2 & -3y^2 \\ 1 & 1 \end{vmatrix}$

57. $\begin{vmatrix} e^{2x} & e^{3x} \\ 2e^{2x} & 3e^{3x} \end{vmatrix}$ 58. $\begin{vmatrix} e^{-x} & xe^{-x} \\ -e^{-x} & (1-x)e^{-x} \end{vmatrix}$

59. $\begin{vmatrix} x & \ln x \\ 1 & 1/x \end{vmatrix}$ 60. $\begin{vmatrix} x & x\ln x \\ 1 & 1+\ln x \end{vmatrix}$

61. $\begin{vmatrix} \cos\theta & -r\sin\theta & 0 \\ \sin\theta & r\cos\theta & 0 \\ 0 & 0 & 1 \end{vmatrix}$ 62. $\begin{vmatrix} 1-v & -u & 0 \\ v(1-w) & u(1-w) & -uv \\ vw & uw & uv \end{vmatrix}$

Verifying an Equation In Exercises 63–68, evaluate the determinants to verify the equation.

63. $\begin{vmatrix} w & x \\ y & z \end{vmatrix} = -\begin{vmatrix} y & z \\ w & x \end{vmatrix}$

64. $\begin{vmatrix} w & cx \\ y & cz \end{vmatrix} = c\begin{vmatrix} w & x \\ y & z \end{vmatrix}$

65. $\begin{vmatrix} w & x \\ y & z \end{vmatrix} = \begin{vmatrix} w & x+cw \\ y & z+cy \end{vmatrix}$

66. $\begin{vmatrix} w & x \\ cw & cx \end{vmatrix} = 0$

67. $\begin{vmatrix} 1 & x & x^2 \\ 1 & y & y^2 \\ 1 & z & z^2 \end{vmatrix} = (y-x)(z-x)(z-y)$

68. $\begin{vmatrix} 1 & 1 & 1 \\ a & b & c \\ a^3 & b^3 & c^3 \end{vmatrix} = (a-b)(b-c)(c-a)(a+b+c)$

69. You are given the equation

$$\begin{vmatrix} x & 0 & c \\ -1 & x & b \\ 0 & -1 & a \end{vmatrix} = ax^2 + bx + c.$$

    (a) Verify the equation.
    (b) Use the equation as a model to find a determinant that is equal to $ax^3 + bx^2 + cx + d$.

70. The determinant of a 2×2 matrix involves two products. The determinant of a 3×3 matrix involves six triple products. Show that the determinant of a 4×4 matrix involves 24 quadruple products.

3.2   Determinants and Elementary Operations

■ Use elementary row operations to evaluate a determinant.

■ Use elementary column operations to evaluate a determinant.

■ Recognize conditions that yield zero determinants.

DETERMINANTS AND ELEMENTARY ROW OPERATIONS

Which of the determinants below is easier to evaluate?

$$|A| = \begin{vmatrix} 1 & -2 & 3 & 1 \\ 4 & -6 & 3 & 2 \\ -2 & 4 & -9 & -3 \\ 3 & -6 & 9 & 2 \end{vmatrix} \quad \text{or} \quad |B| = \begin{vmatrix} 1 & -2 & 3 & 1 \\ 0 & 2 & -9 & -2 \\ 0 & 0 & -3 & -1 \\ 0 & 0 & 0 & -1 \end{vmatrix}$$

Given what you know about the determinant of a triangular matrix, it should be clear that the second determinant is *much* easier to evaluate. Its determinant is simply the product of the entries on the main diagonal. That is, $|B| = (1)(2)(-3)(-1) = 6$. Using expansion by cofactors (the only technique discussed so far) to evaluate the first determinant is messy. For example, when you expand by cofactors in the first row, you have

$$|A| = 1 \begin{vmatrix} -6 & 3 & 2 \\ 4 & -9 & -3 \\ -6 & 9 & 2 \end{vmatrix} + 2 \begin{vmatrix} 4 & 3 & 2 \\ -2 & -9 & -3 \\ 3 & 9 & 2 \end{vmatrix} + 3 \begin{vmatrix} 4 & -6 & 2 \\ -2 & 4 & -3 \\ 3 & -6 & 2 \end{vmatrix} - 1 \begin{vmatrix} 4 & -6 & 3 \\ -2 & 4 & -9 \\ 3 & -6 & 9 \end{vmatrix}.$$

Evaluating the determinants of these four 3×3 matrices produces

$$|A| = (1)(-60) + (2)(39) + (3)(-10) - (1)(-18) = 6.$$

Note that $|A|$ and $|B|$ have the same value. Also note that you can obtain matrix B from matrix A by adding multiples of the first row to the second, third, and fourth rows. (Verify this.) In this section, you will see the effects of elementary row (and column) operations on the value of a determinant.

EXAMPLE 1   **The Effects of Elementary Row Operations on a Determinant**

a. The matrix B is obtained from A by interchanging the rows of A.

$$|A| = \begin{vmatrix} 2 & -3 \\ 1 & 4 \end{vmatrix} = 11 \quad \text{and} \quad |B| = \begin{vmatrix} 1 & 4 \\ 2 & -3 \end{vmatrix} = -11$$

b. The matrix B is obtained from A by adding -2 times the first row of A to the second row of A.

$$|A| = \begin{vmatrix} 1 & -3 \\ 2 & -4 \end{vmatrix} = 2 \quad \text{and} \quad |B| = \begin{vmatrix} 1 & -3 \\ 0 & 2 \end{vmatrix} = 2$$

c. The matrix B is obtained from A by multiplying the first row of A by $\frac{1}{2}$.

$$|A| = \begin{vmatrix} 2 & -8 \\ -2 & 9 \end{vmatrix} = 2 \quad \text{and} \quad |B| = \begin{vmatrix} 1 & -4 \\ -2 & 9 \end{vmatrix} = 1$$

In Example 1, notice that interchanging the two rows of A changes the sign of its determinant, adding -2 times the first row of A to the second row does not change its determinant, and multiplying the first row of A by $\frac{1}{2}$ multiplies its determinant by $\frac{1}{2}$. The next theorem generalizes these observations.

THEOREM 3.3   Elementary Row Operations and Determinants

Let A and B be square matrices.

1. When B is obtained from A by interchanging two rows of A, $\det(B) = -\det(A)$.
2. When B is obtained from A by adding a multiple of a row of A to another row of A, $\det(B) = \det(A)$.
3. When B is obtained from A by multiplying a row of A by a nonzero constant c, $\det(B) = c \det(A)$.

PROOF

The proof of the first property is below. The proofs of the other two properties are left as exercises. (See Exercises 47 and 48.) Assume that A and B are 2×2 matrices

$$A = \begin{bmatrix} a_{11} & a_{12} \\ a_{21} & a_{22} \end{bmatrix} \quad \text{and} \quad B = \begin{bmatrix} a_{21} & a_{22} \\ a_{11} & a_{12} \end{bmatrix}.$$

Then, you have $|A| = a_{11}a_{22} - a_{21}a_{12}$ and $|B| = a_{21}a_{12} - a_{11}a_{22}$. So $|B| = -|A|$. Using mathematical induction, assume the property is true for matrices of order $(n - 1)$. Let A be an $n \times n$ matrix such that B is obtained from A by interchanging two rows of A. Then, to find $|A|$ and $|B|$, expand in a row other than the two interchanged rows. By the induction assumption, the cofactors of B will be the negatives of the cofactors of A because the corresponding $(n - 1) \times (n - 1)$ matrices have two rows interchanged. Finally, $|B| = -|A|$ and the proof is complete. ∎

Theorem 3.3 provides a practical way to evaluate determinants. To find the determinant of a matrix A, you can use elementary row operations to obtain a triangular matrix B that is row-equivalent to A. For each step in the elimination process, use Theorem 3.3 to determine the effect of the elementary row operation on the determinant. Finally, find the determinant of B by multiplying the entries on its main diagonal.

Augustin-Louis Cauchy (1789–1857)
Cauchy's contributions to the study of mathematics were revolutionary, and he is often credited with bringing rigor to modern mathematics. For instance, he was the first to rigorously define limits, continuity, and the convergence of an infinite series. In addition to being known for his work in complex analysis, he contributed to the theories of determinants and differential equations. It is interesting to note that Cauchy's work on determinants preceded Cayley's development of matrices.

EXAMPLE 2   Finding a Determinant Using Elementary Row Operations

Find the determinant of

$$A = \begin{bmatrix} 0 & -7 & 14 \\ 1 & 2 & -2 \\ 0 & 3 & -8 \end{bmatrix}.$$

SOLUTION

Using elementary row operations, rewrite A in triangular form as shown below.

$$\begin{vmatrix} 0 & -7 & 14 \\ 1 & 2 & -2 \\ 0 & 3 & -8 \end{vmatrix} = - \begin{vmatrix} 1 & 2 & -2 \\ 0 & -7 & 14 \\ 0 & 3 & -8 \end{vmatrix}$$ ← Interchange the first two rows.

$$= 7 \begin{vmatrix} 1 & 2 & -2 \\ 0 & 1 & -2 \\ 0 & 3 & -8 \end{vmatrix}$$ ← Factor -7 out of the second row.

$$= 7 \begin{vmatrix} 1 & 2 & -2 \\ 0 & 1 & -2 \\ 0 & 0 & -2 \end{vmatrix}$$ ← Add -3 times the second row to the third row to produce a new third row.

The above matrix is triangular, so the determinant is

$$|A| = 7(1)(1)(-2) = -14.$$

DETERMINANTS AND ELEMENTARY COLUMN OPERATIONS

Although Theorem 3.3 is stated in terms of elementary *row* operations, the therem remains valid when the word "column" replaces the word "row." Operations performed on the columns (rather than on the rows) of a matrix are **elementary column operations,** and two matrices are **column-equivalent** when one can be obtained from the other by elementary column operations. Here are illustrations of the column versions of Theorem 3.3 Properties 1 and 3.

$$\begin{vmatrix} 2 & 1 & -3 \\ 4 & 0 & 1 \\ 0 & 0 & 2 \end{vmatrix} = - \begin{vmatrix} 1 & 2 & -3 \\ 0 & 4 & 1 \\ 0 & 0 & 2 \end{vmatrix}$$

Interchange the first two columns.

$$\begin{vmatrix} 2 & 3 & -5 \\ 4 & 1 & 0 \\ -2 & 4 & -3 \end{vmatrix} = 2 \begin{vmatrix} 1 & 3 & -5 \\ 2 & 1 & 0 \\ -1 & 4 & -3 \end{vmatrix}$$

Factor 2 out of the first column.

In evaluating a determinant, it is occasionally convenient to use elementary column operations, as shown in Example 3.

EXAMPLE 3   **Finding a Determinant Using Elementary Column Operations**

See LarsonLinearAlgebra.com for an interactive version of this type of example.

Find the determinant of $A = \begin{bmatrix} -1 & 2 & 2 \\ 3 & -6 & 4 \\ 5 & -10 & -3 \end{bmatrix}$.

SOLUTION

The first two columns of A are multiples of each other, so you can obtain a column of zeros by adding 2 times the first column to the second column, as shown below.

$$\begin{vmatrix} -1 & 2 & 2 \\ 3 & -6 & 4 \\ 5 & -10 & -3 \end{vmatrix} = \begin{vmatrix} -1 & 0 & 2 \\ 3 & 0 & 4 \\ 5 & 0 & -3 \end{vmatrix}$$

At this point, you do not need to rewrite the matrix in triangular form, because there is an entire column of zeros. Simply conclude that the determinant is zero. The validity of this conclusion follows from Theorem 3.1. Specifically, by expanding by cofactors in the second column, you have

$$|A| = (0)C_{12} + (0)C_{22} + (0)C_{32} = 0.$$

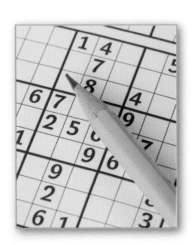

> **LINEAR ALGEBRA APPLIED**
>
> In a Sudoku puzzle, the object is to fill out a partially completed 9×9 grid of boxes with numbers from 1 to 9 so that each column, row, and 3×3 sub-grid contains each number once. For a completed Sudoku grid to be valid, no two rows (or columns) will have the numbers in the same order. If this should happen, then the determinant of the 9×9 matrix formed by the numbers will be zero. This is a direct result of condition 2 of Theorem 3.4 on the next page.

MATRICES AND ZERO DETERMINANTS

Example 3 shows that when two columns of a matrix are scalar multiples of each other, the determinant of the matrix is zero. This is one of three conditions that yield a determinant of zero.

THEOREM 3.4   Conditions That Yield a Zero Determinant

If A is a square matrix and any one of the conditions below is true, then $\det(A) = 0$.

1. An entire row (or an entire column) consists of zeros.
2. Two rows (or columns) are equal.
3. One row (or column) is a multiple of another row (or column).

PROOF

Verify each part of this theorem by using elementary row operations and expansion by cofactors. For instance, if an entire row or column consists of zeros, then each cofactor in the expansion is multiplied by zero. When condition 2 or 3 is true, use elementary row or column operations to create an entire row or column of zeros.

Recognizing the conditions listed in Theorem 3.4 can make evaluating a determinant much easier. For example,

$$\begin{vmatrix} 0 & 0 & 0 \\ 2 & 4 & -5 \\ 3 & -5 & 2 \end{vmatrix} = 0, \quad \begin{vmatrix} 1 & -2 & 4 \\ 0 & 1 & 2 \\ 1 & -2 & 4 \end{vmatrix} = 0, \quad \begin{vmatrix} 1 & 2 & -3 \\ 2 & -1 & -6 \\ -2 & 0 & 6 \end{vmatrix} = 0.$$

| The first row has all zeros. | The first and third rows are the same. | The third column is a multiple of the first column. |

Do not conclude, however, that Theorem 3.4 gives the *only* conditions that produce a zero determinant. This theorem is often used indirectly. That is, you may begin with a matrix that does not satisfy any of the conditions of Theorem 3.4 and, through elementary row or column operations, obtain a matrix that does satisfy one of the conditions. Example 4 demonstrates this.

EXAMPLE 4   **A Matrix with a Zero Determinant**

Find the determinant of

$$A = \begin{bmatrix} 1 & 4 & 1 \\ 2 & -1 & 0 \\ 0 & 18 & 4 \end{bmatrix}.$$

SOLUTION

Adding -2 times the first row to the second row produces

$$|A| = \begin{vmatrix} 1 & 4 & 1 \\ 2 + (-2)(1) & -1 + (-2)(4) & 0 + (-2)(1) \\ 0 & 18 & 4 \end{vmatrix}$$

$$= \begin{vmatrix} 1 & 4 & 1 \\ 0 & -9 & -2 \\ 0 & 18 & 4 \end{vmatrix}.$$

The second and third rows are multiples of each other, so the determinant is zero.

In Example 4, you could have obtained a matrix with a row of all zeros by performing an additional elementary row operation (adding 2 times the second row to the third row). This is true in general. That is, a square matrix has a determinant of zero if and only if it is row- (or column-) equivalent to a matrix that has at least one row (or column) consisting entirely of zeros.

You have now studied two methods for evaluating determinants. Of these, the method of using elementary row operations to reduce the matrix to triangular form is usually faster than cofactor expansion along a row or column. If the matrix is large, then the number of arithmetic operations needed for cofactor expansion can become extremely large. For this reason, most computer and calculator algorithms use the method involving elementary row operations. The table below shows the maximum numbers of additions (plus subtractions) and multiplications (plus divisions) needed for each of these two methods for matrices of orders 3, 5, and 10. (Verify this.)

Order n	Cofactor Expansion		Row Reduction	
	Additions	Multiplications	Additions	Multiplications
3	5	9	8	10
5	119	205	40	44
10	3,628,799	6,235,300	330	339

In fact, the maximum number of additions alone for the cofactor expansion of an $n \times n$ matrix is $n! - 1$. The factorial 30! is approximately equal to 2.65×10^{32}, so even a relatively small 30×30 matrix could require an extremely large number of operations. If a computer could do one trillion operations per second, it could still take more than 22 trillion years to compute the determinant of this matrix using cofactor expansion. Yet, row reduction would take only a fraction of a second.

When evaluating a determinant *by hand*, you sometimes save steps by using elementary row (or column) operations to create a row (or column) having zeros in all but one position and then using cofactor expansion to reduce the order of the matrix by 1. The next two examples illustrate this approach.

EXAMPLE 5   Finding a Determinant

Find the determinant of

$$A = \begin{bmatrix} -3 & 5 & 2 \\ 2 & -4 & -1 \\ -3 & 0 & 6 \end{bmatrix}.$$

SOLUTION

Notice that the matrix A already has one zero in the third row. Create another zero in the third row by adding 2 times the first column to the third column, as shown below.

$$|A| = \begin{vmatrix} -3 & 5 & 2 \\ 2 & -4 & -1 \\ -3 & 0 & 6 \end{vmatrix} = \begin{vmatrix} -3 & 5 & -4 \\ 2 & -4 & 3 \\ -3 & 0 & 0 \end{vmatrix}$$

Expanding by cofactors in the third row produces

$$|A| = \begin{vmatrix} -3 & 5 & -4 \\ 2 & -4 & 3 \\ -3 & 0 & 0 \end{vmatrix} = -3(-1)^4 \begin{vmatrix} 5 & -4 \\ -4 & 3 \end{vmatrix} = -3(1)(-1) = 3.$$

EXAMPLE 6   **Finding a Determinant**

Find the determinant of

$$A = \begin{bmatrix} 2 & 0 & 1 & 3 & -2 \\ -2 & 1 & 3 & 2 & -1 \\ 1 & 0 & -1 & 2 & 3 \\ 3 & -1 & 2 & 4 & -3 \\ 1 & 1 & 3 & 2 & 0 \end{bmatrix}.$$

SOLUTION

The second column of this matrix already has two zeros, so choose it for cofactor expansion. Create two additional zeros in the second column by adding the second row to the fourth row, and then adding -1 times the second row to the fifth row.

$$|A| = \begin{vmatrix} 2 & 0 & 1 & 3 & -2 \\ -2 & 1 & 3 & 2 & -1 \\ 1 & 0 & -1 & 2 & 3 \\ 3 & -1 & 2 & 4 & -3 \\ 1 & 1 & 3 & 2 & 0 \end{vmatrix}$$

$$= \begin{vmatrix} 2 & 0 & 1 & 3 & -2 \\ -2 & 1 & 3 & 2 & -1 \\ 1 & 0 & -1 & 2 & 3 \\ 1 & 0 & 5 & 6 & -4 \\ 3 & 0 & 0 & 0 & 1 \end{vmatrix}$$

$$= (1)(-1)^4 \begin{vmatrix} 2 & 1 & 3 & -2 \\ 1 & -1 & 2 & 3 \\ 1 & 5 & 6 & -4 \\ 3 & 0 & 0 & 1 \end{vmatrix}$$

You have now reduced the problem of finding the determinant of a 5×5 matrix to the problem of finding the determinant of a 4×4 matrix. The fourth row already has two zeros, so choose it for the next cofactor expansion. Add -3 times the fourth column to the first column.

$$|A| = \begin{vmatrix} 2 & 1 & 3 & -2 \\ 1 & -1 & 2 & 3 \\ 1 & 5 & 6 & -4 \\ 3 & 0 & 0 & 1 \end{vmatrix} = \begin{vmatrix} 8 & 1 & 3 & -2 \\ -8 & -1 & 2 & 3 \\ 13 & 5 & 6 & -4 \\ 0 & 0 & 0 & 1 \end{vmatrix}$$

$$= (1)(-1)^8 \begin{vmatrix} 8 & 1 & 3 \\ -8 & -1 & 2 \\ 13 & 5 & 6 \end{vmatrix}$$

Add the second row to the first row and then expand by cofactors in the first row.

$$|A| = \begin{vmatrix} 8 & 1 & 3 \\ -8 & -1 & 2 \\ 13 & 5 & 6 \end{vmatrix} = \begin{vmatrix} 0 & 0 & 5 \\ -8 & -1 & 2 \\ 13 & 5 & 6 \end{vmatrix}$$

$$= 5(-1)^4 \begin{vmatrix} -8 & -1 \\ 13 & 5 \end{vmatrix}$$

$$= 5(1)(-27)$$

$$= -135$$

3.2 Exercises

See CalcChat.com for worked-out solutions to odd-numbered exercises.

Properties of Determinants In Exercises 1–20, determine which property of determinants the equation illustrates.

1. $\begin{vmatrix} 2 & -6 \\ 1 & -3 \end{vmatrix} = 0$

2. $\begin{vmatrix} -4 & 5 \\ 12 & -15 \end{vmatrix} = 0$

3. $\begin{vmatrix} 1 & 4 & 2 \\ 0 & 0 & 0 \\ 5 & 6 & -7 \end{vmatrix} = 0$

4. $\begin{vmatrix} -3 & 2 & 1 \\ 6 & 0 & 0 \\ -3 & 2 & 1 \end{vmatrix} = 0$

5. $\begin{vmatrix} 1 & 3 & 4 \\ -7 & 2 & -5 \\ 6 & 1 & 2 \end{vmatrix} = -\begin{vmatrix} 1 & 4 & 3 \\ -7 & -5 & 2 \\ 6 & 2 & 1 \end{vmatrix}$

6. $\begin{vmatrix} 1 & 3 & 4 & -5 \\ -2 & 2 & 0 & 1 \\ 1 & 6 & 2 & -7 \\ 0 & 5 & 3 & 8 \end{vmatrix} = \begin{vmatrix} 1 & 6 & 2 & -7 \\ -2 & 2 & 0 & 1 \\ 1 & 3 & 4 & -5 \\ 0 & 5 & 3 & 8 \end{vmatrix}$

7. $\begin{vmatrix} 5 & 10 \\ 2 & -7 \end{vmatrix} = 5\begin{vmatrix} 1 & 2 \\ 2 & -7 \end{vmatrix}$

8. $\begin{vmatrix} 9 & 1 \\ 3 & 12 \end{vmatrix} = 3\begin{vmatrix} 3 & 1 \\ 1 & 12 \end{vmatrix}$

9. $\begin{vmatrix} 1 & 8 & -3 \\ 3 & -12 & 6 \\ 7 & 4 & 9 \end{vmatrix} = 12\begin{vmatrix} 1 & 2 & -1 \\ 3 & -3 & 2 \\ 7 & 1 & 3 \end{vmatrix}$

10. $\begin{vmatrix} 1 & 2 & 3 \\ 4 & -8 & 6 \\ 5 & 4 & 12 \end{vmatrix} = 6\begin{vmatrix} 1 & 1 & 1 \\ 4 & -4 & 2 \\ 5 & 2 & 4 \end{vmatrix}$

11. $\begin{vmatrix} -10 & 5 & 5 \\ 35 & -20 & 25 \\ 0 & 15 & 30 \end{vmatrix} = 5^3\begin{vmatrix} -2 & 1 & 1 \\ 7 & -4 & 5 \\ 0 & 3 & 6 \end{vmatrix}$

12. $\begin{vmatrix} 6 & 0 & 0 & 0 \\ 0 & 6 & 0 & 0 \\ 0 & 0 & 6 & 0 \\ 0 & 0 & 0 & 6 \end{vmatrix} = 6^4\begin{vmatrix} 1 & 0 & 0 & 0 \\ 0 & 1 & 0 & 0 \\ 0 & 0 & 1 & 0 \\ 0 & 0 & 0 & 1 \end{vmatrix}$

13. $\begin{vmatrix} 2 & -3 \\ 8 & 7 \end{vmatrix} = \begin{vmatrix} 2 & -3 \\ 0 & 19 \end{vmatrix}$

14. $\begin{vmatrix} 2 & 1 \\ 0 & -1 \end{vmatrix} = \begin{vmatrix} 2 & 1 \\ 4 & 1 \end{vmatrix}$

15. $\begin{vmatrix} 1 & -3 & 2 \\ 5 & 2 & -1 \\ -1 & 0 & 6 \end{vmatrix} = \begin{vmatrix} 1 & -3 & 2 \\ 0 & 17 & -11 \\ -1 & 0 & 6 \end{vmatrix}$

16. $\begin{vmatrix} 3 & 2 & 4 & 11 \\ -2 & 1 & 5 & 6 \\ 5 & -7 & -20 & 15 \\ 4 & -1 & 13 & 12 \end{vmatrix} = \begin{vmatrix} 3 & 2 & -6 & 11 \\ -2 & 1 & 0 & 6 \\ 5 & -7 & 15 & 15 \\ 4 & -1 & 8 & 12 \end{vmatrix}$

17. $\begin{vmatrix} 5 & 4 & 2 \\ 4 & -3 & 4 \\ 7 & 6 & 3 \end{vmatrix} = -\begin{vmatrix} 5 & 4 & 2 \\ -4 & 3 & -4 \\ 7 & 6 & 3 \end{vmatrix}$

18. $\begin{vmatrix} 3 & 2 & -2 \\ -1 & 0 & 3 \\ 4 & 2 & 0 \end{vmatrix} = -\begin{vmatrix} 3 & 2 & -2 \\ 4 & 2 & 0 \\ -1 & 0 & 3 \end{vmatrix}$

19. $\begin{vmatrix} 2 & 1 & -1 & 0 & 4 \\ 1 & 0 & 1 & 3 & 2 \\ 3 & 6 & 1 & -3 & 6 \\ 0 & 4 & 0 & 2 & 0 \\ -1 & 8 & 5 & 3 & 2 \end{vmatrix} = 0$

20. $\begin{vmatrix} 4 & 3 & 1 & 9 & 9 \\ 9 & -1 & 2 & 3 & -3 \\ 3 & 4 & 6 & 9 & 12 \\ 5 & 2 & 0 & 6 & 6 \\ 6 & 0 & 3 & 0 & 0 \end{vmatrix} = 0$

Finding a Determinant In Exercises 21–24, use either elementary row or column operations, or cofactor expansion, to find the determinant by hand. Then use a software program or a graphing utility to verify your answer.

21. $\begin{vmatrix} 1 & 0 & 2 \\ -1 & 1 & 4 \\ 2 & 0 & 3 \end{vmatrix}$

22. $\begin{vmatrix} -1 & 3 & 2 \\ 0 & 2 & 0 \\ 1 & 1 & -1 \end{vmatrix}$

23. $\begin{vmatrix} 5 & 1 & 0 & 1 \\ 1 & 0 & -1 & -1 \\ 2 & 0 & 1 & 2 \\ -1 & 0 & 3 & 1 \end{vmatrix}$

24. $\begin{vmatrix} 3 & 2 & 1 & 1 \\ -1 & 0 & 2 & 0 \\ 4 & 1 & -1 & 0 \\ 3 & 1 & 1 & 0 \end{vmatrix}$

Finding a Determinant In Exercises 25–36, use elementary row or column operations to find the determinant.

25. $\begin{vmatrix} 1 & 7 & -3 \\ 1 & 3 & 1 \\ 4 & 8 & 1 \end{vmatrix}$

26. $\begin{vmatrix} 1 & 1 & 1 \\ 2 & -1 & -2 \\ 1 & -2 & -1 \end{vmatrix}$

27. $\begin{vmatrix} 2 & -1 & -1 \\ 1 & 3 & 2 \\ -6 & 3 & 3 \end{vmatrix}$

28. $\begin{vmatrix} 3 & 0 & 6 \\ 2 & -3 & 4 \\ 1 & -2 & 2 \end{vmatrix}$

29. $\begin{vmatrix} 3 & 2 & -3 \\ 7 & 5 & 1 \\ -1 & 2 & 6 \end{vmatrix}$

30. $\begin{vmatrix} 3 & 8 & -7 \\ 0 & -5 & 4 \\ 6 & 1 & 6 \end{vmatrix}$

31. $\begin{vmatrix} 4 & -7 & 9 & 1 \\ 6 & 2 & 7 & 0 \\ 3 & 6 & -3 & 3 \\ 0 & 7 & 4 & -1 \end{vmatrix}$

32. $\begin{vmatrix} 9 & -4 & 2 & 5 \\ 2 & 7 & 6 & -5 \\ 4 & 1 & -2 & 0 \\ 7 & 3 & 4 & 10 \end{vmatrix}$

33. $\begin{vmatrix} 1 & -2 & 7 & 9 \\ 3 & -4 & 5 & 5 \\ 3 & 6 & 1 & -1 \\ 4 & 5 & 3 & 2 \end{vmatrix}$

34. $\begin{vmatrix} 0 & -4 & 9 & 3 \\ 9 & 2 & -2 & 7 \\ -5 & 7 & 0 & 11 \\ -8 & 0 & 0 & 16 \end{vmatrix}$

35. $\begin{vmatrix} 1 & -1 & 8 & 4 & 2 \\ 2 & 6 & 0 & -4 & 3 \\ 2 & 0 & 2 & 6 & 2 \\ 0 & 2 & 8 & 0 & 0 \\ 0 & 1 & 1 & 2 & 2 \end{vmatrix}$

36. $\begin{vmatrix} 3 & -2 & 4 & 3 & 1 \\ -1 & 0 & 2 & 1 & 0 \\ 5 & -1 & 0 & 3 & 2 \\ 4 & 7 & -8 & 0 & 0 \\ 1 & 2 & 3 & 0 & 2 \end{vmatrix}$

True or False? In Exercises 37 and 38, determine whether each statement is true or false. If a statement is true, give a reason or cite an appropriate statement from the text. If a statement is false, provide an example that shows the statement is not true in all cases or cite an appropriate statement from the text.

37. (a) Interchanging two rows of a square matrix changes the sign of its determinant.

   (b) Multiplying a column of a square matrix by a nonzero constant results in the determinant being multiplied by the same nonzero constant.

   (c) If two rows of a square matrix are equal, then its determinant is 0.

38. (a) Adding a multiple of one column of a square matrix to another column changes only the sign of the determinant.

   (b) Two matrices are column-equivalent when one matrix can be obtained by performing elementary column operations on the other.

   (c) If one row of a square matrix is a multiple of another row, then the determinant is 0.

Finding the Determinant of an Elementary Matrix In Exercises 39–42, find the determinant of the elementary matrix. (Assume $k \neq 0$.)

39. $\begin{bmatrix} 1 & 0 & 0 \\ 0 & k & 0 \\ 0 & 0 & 1 \end{bmatrix}$

40. $\begin{bmatrix} 0 & 0 & 1 \\ 0 & 1 & 0 \\ 1 & 0 & 0 \end{bmatrix}$

41. $\begin{bmatrix} 1 & 0 & 0 \\ k & 1 & 0 \\ 0 & 0 & 1 \end{bmatrix}$

42. $\begin{bmatrix} 1 & 0 & 0 \\ 0 & 1 & 0 \\ 0 & k & 1 \end{bmatrix}$

43. Proof Prove the property.
$$\begin{vmatrix} a_{11} & a_{12} & a_{13} \\ a_{21} & a_{22} & a_{23} \\ a_{31} & a_{32} & a_{33} \end{vmatrix} + \begin{vmatrix} b_{11} & a_{12} & a_{13} \\ b_{21} & a_{22} & a_{23} \\ b_{31} & a_{32} & a_{33} \end{vmatrix} = \begin{vmatrix} (a_{11} + b_{11}) & a_{12} & a_{13} \\ (a_{21} + b_{21}) & a_{22} & a_{23} \\ (a_{31} + b_{31}) & a_{32} & a_{33} \end{vmatrix}$$

44. Proof Prove the property.
$$\begin{vmatrix} 1 + a & 1 & 1 \\ 1 & 1 + b & 1 \\ 1 & 1 & 1 + c \end{vmatrix} = abc\left(1 + \frac{1}{a} + \frac{1}{b} + \frac{1}{c}\right),$$
$a \neq 0, \quad b \neq 0, \quad c \neq 0$

45. Find each determinant.

   (a) $\begin{vmatrix} \cos\theta & \sin\theta \\ -\sin\theta & \cos\theta \end{vmatrix}$ (b) $\begin{vmatrix} \sin\theta & 1 \\ 1 & \sin\theta \end{vmatrix}$

46. CAPSTONE Evaluate each determinant when $a = 1$, $b = 4$, and $c = -3$.

   (a) $\begin{vmatrix} 0 & b & 0 \\ a & 0 & 0 \\ 0 & 0 & c \end{vmatrix}$ (b) $\begin{vmatrix} a & 0 & 1 \\ 0 & c & 0 \\ b & 0 & -16 \end{vmatrix}$

47. Guided Proof Prove Property 2 of Theorem 3.3: When B is obtained from A by adding a multiple of a row of A to another row of A, $\det(B) = \det(A)$.

Getting Started: To prove that the determinant of B is equal to the determinant of A, you need to show that their respective cofactor expansions are equal.

   (i) Begin by letting B be the matrix obtained by adding c times the jth row of A to the ith row of A.

   (ii) Find the determinant of B by expanding in this ith row.

   (iii) Distribute and then group the terms containing a coefficient of c and those not containing a coefficient of c.

   (iv) Show that the sum of the terms not containing a coefficient of c is the determinant of A, and the sum of the terms containing a coefficient of c is equal to 0.

48. Guided Proof Prove Property 3 of Theorem 3.3: When B is obtained from A by multiplying a row of A by a nonzero constant c, $\det(B) = c \det(A)$.

Getting Started: To prove that the determinant of B is equal to c times the determinant of A, you need to show that the determinant of B is equal to c times the cofactor expansion of the determinant of A.

   (i) Begin by letting B be the matrix obtained by multiplying c times the ith row of A.

   (ii) Find the determinant of B by expanding in this ith row.

   (iii) Factor out the common factor c.

   (iv) Show that the result is c times the determinant of A.

3.3   Properties of Determinants

- ■ Find the determinant of a matrix product and a scalar multiple of a matrix.
- ■ Find the determinant of an inverse matrix and recognize equivalent conditions for a nonsingular matrix.
- ■ Find the determinant of the transpose of a matrix.

MATRIX PRODUCTS AND SCALAR MULTIPLES

In this section, you will learn several important properties of determinants. You will begin by considering the determinant of the product of two matrices.

EXAMPLE 1   The Determinant of a Matrix Product

Find $|A|$, $|B|$, and $|AB|$ for the matrices

$$A = \begin{bmatrix} 1 & -2 & 2 \\ 0 & 3 & 2 \\ 1 & 0 & 1 \end{bmatrix} \quad \text{and} \quad B = \begin{bmatrix} 2 & 0 & 1 \\ 0 & -1 & -2 \\ 3 & 1 & -2 \end{bmatrix}.$$

SOLUTION

$|A|$ and $|B|$ have the values

$$|A| = \begin{vmatrix} 1 & -2 & 2 \\ 0 & 3 & 2 \\ 1 & 0 & 1 \end{vmatrix} = -7 \quad \text{and} \quad |B| = \begin{vmatrix} 2 & 0 & 1 \\ 0 & -1 & -2 \\ 3 & 1 & -2 \end{vmatrix} = 11.$$

The matrix product AB is

$$AB = \begin{bmatrix} 1 & -2 & 2 \\ 0 & 3 & 2 \\ 1 & 0 & 1 \end{bmatrix}\begin{bmatrix} 2 & 0 & 1 \\ 0 & -1 & -2 \\ 3 & 1 & -2 \end{bmatrix} = \begin{bmatrix} 8 & 4 & 1 \\ 6 & -1 & -10 \\ 5 & 1 & -1 \end{bmatrix}.$$

Finally,

$$|AB| = \begin{vmatrix} 8 & 4 & 1 \\ 6 & -1 & -10 \\ 5 & 1 & -1 \end{vmatrix} = -77.$$

In Example 1, note that $|AB| = |A||B|$, or $-77 = (-7)(11)$. This is true in general.

REMARK

Theorem 3.5 can be extended to include the product of any finite number of matrices. That is,

$$|A_1 A_2 A_3 \cdots A_k| = |A_1||A_2||A_3| \cdots |A_k|.$$

THEOREM 3.5   Determinant of a Matrix Product

If A and B are square matrices of order n, then $\det(AB) = \det(A)\det(B)$.

PROOF

To begin, observe that if E is an elementary matrix, then, by Theorem 3.3, the next three statements are true. If you obtain E from I by interchanging two rows, then $|E| = -1$. If you obtain E by multiplying a row of I by a nonzero constant c, then $|E| = c$. If you obtain E by adding a multiple of one row of I to another row of I, then $|E| = 1$. Additionally, by Theorem 2.12, if E results from performing an elementary row operation on I and the same elementary row operation is performed on B, then the matrix EB results. It follows that $|EB| = |E||B|$.

This can be generalized to conclude that $|E_k \cdots E_2 E_1 B| = |E_k| \cdots |E_2||E_1||B|$, where E_i is an elementary matrix. Now consider the matrix AB. If A is *nonsingular*, then, by Theorem 2.14, it can be written as the product $A = E_k \cdots E_2 E_1$, and

$$\begin{aligned} |AB| &= |E_k \cdots E_2 E_1 B| \\ &= |E_k| \cdots |E_2||E_1||B| \\ &= |E_k \cdots E_2 E_1||B| \\ &= |A||B|. \end{aligned}$$

If A is *singular,* then A is row-equivalent to a matrix with an entire row of zeros. From Theorem 3.4, $|A| = 0$. Moreover, it follows that AB is also singular. (If AB were nonsingular, then $A[B(AB)^{-1}] = I$ would imply that A is nonsingular.) So, $|AB| = 0$, and you can conclude that $|AB| = |A||B|$.

The next theorem shows the relationship between $|A|$ and $|cA|$.

THEOREM 3.6   Determinant of a Scalar Multiple of a Matrix

If A is a square matrix of order n and c is a scalar, then the determinant of cA is

$$\det(cA) = c^n \det(A).$$

PROOF

This formula can be proven by repeated applications of Property 3 of Theorem 3.3. Factor the scalar c out of each of the n rows of $|cA|$ to obtain $|cA| = c^n|A|$.

EXAMPLE 2   **The Determinant of a Scalar Multiple of a Matrix**

Find the determinant of the matrix.

$$A = \begin{bmatrix} 10 & -20 & 40 \\ 30 & 0 & 50 \\ -20 & -30 & 10 \end{bmatrix}$$

SOLUTION

$$A = 10 \begin{bmatrix} 1 & -2 & 4 \\ 3 & 0 & 5 \\ -2 & -3 & 1 \end{bmatrix} \quad \text{and} \quad \begin{vmatrix} 1 & -2 & 4 \\ 3 & 0 & 5 \\ -2 & -3 & 1 \end{vmatrix} = 5$$

so apply Theorem 3.6 to conclude that

$$|A| = 10^3 \begin{vmatrix} 1 & -2 & 4 \\ 3 & 0 & 5 \\ -2 & -3 & 1 \end{vmatrix} = 1000(5) = 5000.$$

Theorems 3.5 and 3.6 give formulas for the determinants of the product of two matrices and a scalar multiple of a matrix. These theorems do not, however, give a formula for the determinant of the *sum* of two matrices. The sum of the determinants of two matrices usually does not equal the determinant of their sum. That is, in general, $|A| + |B| \neq |A + B|$. For example, if

$$A = \begin{bmatrix} 6 & 2 \\ 2 & 1 \end{bmatrix} \quad \text{and} \quad B = \begin{bmatrix} 3 & 7 \\ 0 & -1 \end{bmatrix}$$

then $|A| = 2$ and $|B| = -3$, but $A + B = \begin{bmatrix} 9 & 9 \\ 2 & 0 \end{bmatrix}$ and $|A + B| = -18$.

DETERMINANTS AND THE INVERSE OF A MATRIX

It can be difficult to tell simply by inspection whether a matrix has an inverse. Can you tell which of the matrices below is invertible?

$$A = \begin{bmatrix} 0 & 2 & -1 \\ 3 & -2 & 1 \\ 3 & 2 & -1 \end{bmatrix} \quad \text{or} \quad B = \begin{bmatrix} 0 & 2 & -1 \\ 3 & -2 & 1 \\ 3 & 2 & 1 \end{bmatrix}$$

The next theorem suggests that determinants are useful for classifying square matrices as invertible or noninvertible.

THEOREM 3.7   Determinant of an Invertible Matrix

A square matrix A is invertible (nonsingular) if and only if $\det(A) \neq 0$.

DISCOVERY

Let

$$A = \begin{bmatrix} 6 & 4 & 1 \\ 0 & 2 & 3 \\ 1 & 1 & 2 \end{bmatrix}.$$

1. Use a software program or a graphing utility to find A^{-1}.

2. Compare $\det(A^{-1})$ with $\det(A)$.

3. Make a conjecture about the determinant of the inverse of a matrix.

PROOF

To prove the theorem in one direction, assume A is invertible. Then $AA^{-1} = I$, and by Theorem 3.5 you can write $|A||A^{-1}| = |I|$. Now, $|I| = 1$, so you know that neither determinant on the left is zero. Specifically, $|A| \neq 0$.

To prove the theorem in the other direction, assume the determinant of A is nonzero. Then, using Gauss-Jordan elimination, find a matrix B, in reduced row-echelon form, that is row-equivalent to A. The matrix B must be the identity matrix I or it must have at least one row that consists entirely of zeros, because B is in reduced row-echelon form. But if B has a row of all zeros, then by Theorem 3.4 you know that $|B| = 0$, which would imply that $|A| = 0$. You assumed that $|A|$ is nonzero, so you can conclude that $B = I$. The matrix A is, therefore, row-equivalent to the identity matrix, and by Theorem 2.15 you know that A is invertible. ∎

EXAMPLE 3   **Classifying Square Matrices as Singular or Nonsingular**

Determine whether each matrix has an inverse.

a. $\begin{bmatrix} 0 & 2 & -1 \\ 3 & -2 & 1 \\ 3 & 2 & -1 \end{bmatrix}$ **b.** $\begin{bmatrix} 0 & 2 & -1 \\ 3 & -2 & 1 \\ 3 & 2 & 1 \end{bmatrix}$

SOLUTION

a. $\begin{vmatrix} 0 & 2 & -1 \\ 3 & -2 & 1 \\ 3 & 2 & -1 \end{vmatrix} = 0$

so this matrix has no inverse (it is singular).

b. $\begin{vmatrix} 0 & 2 & -1 \\ 3 & -2 & 1 \\ 3 & 2 & 1 \end{vmatrix} = -12 \neq 0$

so this matrix has an inverse (it is nonsingular).

The next theorem provides a way to find the determinant of an inverse matrix.

THEOREM 3.8   Determinant of an Inverse Matrix

If A is an $n \times n$ invertible matrix, then $\det(A^{-1}) = \dfrac{1}{\det(A)}$.

PROOF

The matrix A is invertible, so $AA^{-1} = I$, and using Theorem 3.5, $|A||A^{-1}| = |I| = 1$. By Theorem 3.7, you know that $|A| \neq 0$, so you can divide each side by $|A|$ to obtain

$$|A^{-1}| = \frac{1}{|A|}.$$

EXAMPLE 4   **The Determinant of the Inverse of a Matrix**

Find $|A^{-1}|$ for the matrix

$$A = \begin{bmatrix} 1 & 0 & 3 \\ 0 & -1 & 2 \\ 2 & 1 & 0 \end{bmatrix}.$$

REMARK

The inverse of A is

$$A^{-1} = \begin{bmatrix} -\frac{1}{2} & \frac{3}{4} & \frac{3}{4} \\ 1 & -\frac{3}{2} & -\frac{1}{2} \\ \frac{1}{2} & -\frac{1}{4} & -\frac{1}{4} \end{bmatrix}.$$

Evaluate the determinant of this matrix directly. Then compare your answer with that obtained in Example 4.

SOLUTION

One way to solve this problem is to find A^{-1} and then evaluate its determinant. It is easier, however, to apply Theorem 3.8, as shown below. Find the determinant of A,

$$|A| = \begin{vmatrix} 1 & 0 & 3 \\ 0 & -1 & 2 \\ 2 & 1 & 0 \end{vmatrix} = 4$$

and then use the formula $|A^{-1}| = 1/|A|$ to conclude that $|A^{-1}| = \frac{1}{4}$.

Note that Theorem 3.7 provides another equivalent condition that can be added to the list in Theorem 2.15, as shown below.

REMARK

In Section 3.2, you saw that a square matrix A has a determinant of zero when A is row-equivalent to a matrix that has at least one row consisting entirely of zeros. The validity of this statement follows from the equivalence of Statements 4 and 6.

Equivalent Conditions for a Nonsingular Matrix

If A is an $n \times n$ matrix, then the statements below are equivalent.

1. A is invertible.
2. $A\mathbf{x} = \mathbf{b}$ has a unique solution for every $n \times 1$ column matrix $\mathbf{b}$.
3. $A\mathbf{x} = O$ has only the trivial solution.
4. A is row-equivalent to I_n.
5. A can be written as the product of elementary matrices.
6. $\det(A) \neq 0$

EXAMPLE 5   **Systems of Linear Equations**

Which of the systems has a unique solution?

a.
$$\begin{aligned} 2x_2 - x_3 &= -1 \\ 3x_1 - 2x_2 + x_3 &= 4 \\ 3x_1 + 2x_2 - x_3 &= -4 \end{aligned}$$

b.
$$\begin{aligned} 2x_2 - x_3 &= -1 \\ 3x_1 - 2x_2 + x_3 &= 4 \\ 3x_1 + 2x_2 + x_3 &= -4 \end{aligned}$$

SOLUTION

From Example 3, you know that the coefficient matrices for these two systems have the determinants shown below.

a. $\begin{vmatrix} 0 & 2 & -1 \\ 3 & -2 & 1 \\ 3 & 2 & -1 \end{vmatrix} = 0$ b. $\begin{vmatrix} 0 & 2 & -1 \\ 3 & -2 & 1 \\ 3 & 2 & 1 \end{vmatrix} = -12$

Using the preceding list of equivalent conditions, you can conclude that only the second system has a unique solution.

DETERMINANTS AND THE TRANSPOSE OF A MATRIX

The next theorem tells you that the determinant of the transpose of a square matrix is equal to the determinant of the original matrix. This theorem can be proven using mathematical induction and Theorem 3.1, which states that a determinant can be evaluated using cofactor expansion in a row or a column. The details of the proof are left to you. (See Exercise 66.)

THEOREM 3.9   Determinant of a Transpose

If A is a square matrix, then

$$\det(A) = \det(A^T).$$

EXAMPLE 6   **The Determinant of a Transpose**

See LarsonLinearAlgebra.com for an interactive version of this type of example.

Show that $|A| = |A^T|$ for the matrix below.

$$A = \begin{bmatrix} 3 & 1 & -2 \\ 2 & 0 & 0 \\ -4 & -1 & 5 \end{bmatrix}$$

SOLUTION

To find the determinant of A, expand by cofactors in the second *row* to obtain

$$\begin{aligned} |A| &= 2(-1)^3 \begin{vmatrix} 1 & -2 \\ -1 & 5 \end{vmatrix} \\ &= (2)(-1)(3) \\ &= -6. \end{aligned}$$

To find the determinant of

$$A^T = \begin{bmatrix} 3 & 2 & -4 \\ 1 & 0 & -1 \\ -2 & 0 & 5 \end{bmatrix}$$

expand by cofactors in the second *column* to obtain

$$\begin{aligned} |A^T| &= 2(-1)^3 \begin{vmatrix} 1 & -1 \\ -2 & 5 \end{vmatrix} \\ &= (2)(-1)(3) \\ &= -6. \end{aligned}$$

LINEAR ALGEBRA APPLIED

Systems of linear differential equations often arise in engineering and control theory. For a function $f(t)$ that is defined for all positive values of t, the **Laplace transform** of $f(t)$ is

$$F(s) = \int_0^\infty e^{-st} f(t)\, dt$$

provided that the improper integral exists. Laplace transforms and Cramer's Rule, which uses determinants to solve a system of linear equations, can sometimes be used to solve a system of differential equations. You will study Cramer's Rule in the next section.

3.3 Exercises   See CalcChat.com for worked-out solutions to odd-numbered exercises.

The Determinant of a Matrix Product In Exercises 1–6, find (a) $|A|$, (b) $|B|$, (c) AB, and (d) $|AB|$. Then verify that $|A||B| = |AB|$.

1. $A = \begin{bmatrix} -2 & 1 \\ 4 & -2 \end{bmatrix}$, $B = \begin{bmatrix} 1 & 1 \\ 0 & -1 \end{bmatrix}$

2. $A = \begin{bmatrix} 3 & 4 \\ 4 & 3 \end{bmatrix}$, $B = \begin{bmatrix} 2 & -1 \\ 5 & 0 \end{bmatrix}$

3. $A = \begin{bmatrix} -1 & 2 & 1 \\ 1 & 0 & 1 \\ 0 & 1 & 0 \end{bmatrix}$, $B = \begin{bmatrix} -1 & 0 & 0 \\ 0 & 2 & 0 \\ 0 & 0 & 3 \end{bmatrix}$

4. $A = \begin{bmatrix} 2 & 0 & 1 \\ 1 & -1 & 2 \\ 3 & 1 & 0 \end{bmatrix}$, $B = \begin{bmatrix} 2 & -1 & 4 \\ 0 & 1 & 3 \\ 3 & -2 & 1 \end{bmatrix}$

5. $A = \begin{bmatrix} 2 & 0 & 1 & 1 \\ 1 & -1 & 0 & 1 \\ 2 & 3 & 1 & 0 \\ 1 & 2 & 3 & 0 \end{bmatrix}$, $B = \begin{bmatrix} 1 & 0 & -1 & 1 \\ 2 & 1 & 0 & 2 \\ 1 & 1 & -1 & 0 \\ 3 & 2 & 1 & 0 \end{bmatrix}$

6. $A = \begin{bmatrix} 2 & 4 & 7 & 0 \\ 1 & -2 & 1 & 1 \\ 0 & 0 & 2 & 1 \\ 1 & -1 & 1 & 0 \end{bmatrix}$,

$B = \begin{bmatrix} 6 & 1 & -1 & 0 \\ -1 & 2 & 1 & 1 \\ 0 & 0 & 1 & 2 \\ 0 & 0 & 0 & -1 \end{bmatrix}$

The Determinant of a Scalar Multiple of a Matrix In Exercises 7–14, use the fact that $|cA| = c^n|A|$ to evaluate the determinant of the $n \times n$ matrix.

7. $A = \begin{bmatrix} 5 & 15 \\ 10 & -20 \end{bmatrix}$ **8.** $A = \begin{bmatrix} 21 & 7 \\ 28 & -56 \end{bmatrix}$

9. $A = \begin{bmatrix} -3 & 6 & 9 \\ 6 & 9 & 12 \\ 9 & 12 & 15 \end{bmatrix}$ **10.** $A = \begin{bmatrix} 4 & 16 & 0 \\ 12 & -8 & 8 \\ 16 & 20 & -4 \end{bmatrix}$

11. $A = \begin{bmatrix} 2 & -4 & 6 \\ -4 & 6 & -8 \\ 6 & -8 & 10 \end{bmatrix}$ **12.** $A = \begin{bmatrix} 40 & 25 & 10 \\ 30 & 5 & 20 \\ 15 & 35 & 45 \end{bmatrix}$

13. $A = \begin{bmatrix} 5 & 0 & -15 & 0 \\ 0 & 5 & 0 & 0 \\ -10 & 0 & 5 & 0 \\ 0 & -20 & 0 & 5 \end{bmatrix}$

14. $A = \begin{bmatrix} 0 & 16 & -8 & -32 \\ -16 & 8 & -8 & 16 \\ 8 & -24 & 8 & -8 \\ -8 & 32 & 0 & 32 \end{bmatrix}$

The Determinant of a Matrix Sum In Exercises 15–18, find (a) $|A|$, (b) $|B|$, (c) $A + B$, and (d) $|A + B|$. Then verify that $|A| + |B| \neq |A + B|$.

15. $A = \begin{bmatrix} -1 & 1 \\ 2 & 0 \end{bmatrix}$, $B = \begin{bmatrix} 1 & -1 \\ -2 & 0 \end{bmatrix}$

16. $A = \begin{bmatrix} 1 & -2 \\ 1 & 0 \end{bmatrix}$, $B = \begin{bmatrix} 3 & -2 \\ 0 & 0 \end{bmatrix}$

17. $A = \begin{bmatrix} -1 & 1 & 2 \\ 0 & 1 & 1 \\ 1 & 1 & -1 \end{bmatrix}$, $B = \begin{bmatrix} 1 & 0 & 1 \\ -1 & 1 & 2 \\ 0 & 1 & 2 \end{bmatrix}$

18. $A = \begin{bmatrix} 0 & 1 & 2 \\ 1 & -1 & 0 \\ 2 & 1 & 1 \end{bmatrix}$, $B = \begin{bmatrix} 0 & 1 & -1 \\ 2 & 1 & 1 \\ 0 & 1 & 1 \end{bmatrix}$

Classifying Matrices as Singular or Nonsingular In Exercises 19–24, use a determinant to decide whether the matrix is singular or nonsingular.

19. $\begin{bmatrix} 5 & 4 \\ 10 & 8 \end{bmatrix}$ **20.** $\begin{bmatrix} 3 & -6 \\ 4 & 2 \end{bmatrix}$

21. $\begin{bmatrix} \frac{1}{2} & \frac{3}{2} & 2 \\ \frac{2}{3} & -\frac{1}{3} & 0 \\ 1 & 1 & 1 \end{bmatrix}$ **22.** $\begin{bmatrix} 14 & 5 & 7 \\ -15 & 0 & 3 \\ 1 & -5 & -10 \end{bmatrix}$

23. $\begin{bmatrix} 1 & 0 & -8 & 2 \\ 0 & 8 & -1 & 10 \\ 0 & 0 & 0 & 1 \\ 0 & 0 & 0 & 2 \end{bmatrix}$ **24.** $\begin{bmatrix} 0.8 & 0.2 & -0.6 & 0.1 \\ -1.2 & 0.6 & 0.6 & 0 \\ 0.7 & -0.3 & 0.1 & 0 \\ 0.2 & -0.3 & 0.6 & 0 \end{bmatrix}$

The Determinant of the Inverse of a Matrix In Exercises 25–30, find $|A^{-1}|$. Begin by finding A^{-1}, and then evaluate its determinant. Verify your result by finding $|A|$ and then applying the formula from Theorem 3.8, $|A^{-1}| = \dfrac{1}{|A|}$.

25. $A = \begin{bmatrix} 2 & 3 \\ 1 & 4 \end{bmatrix}$ **26.** $A = \begin{bmatrix} 1 & -2 \\ 2 & 2 \end{bmatrix}$

27. $A = \begin{bmatrix} 2 & -2 & 3 \\ 1 & -1 & 2 \\ 3 & 0 & 3 \end{bmatrix}$ **28.** $A = \begin{bmatrix} 1 & 0 & 1 \\ 2 & -1 & 2 \\ 1 & -2 & 3 \end{bmatrix}$

29. $A = \begin{bmatrix} 1 & 0 & -1 & 3 \\ 1 & 0 & 3 & -2 \\ 2 & 0 & 2 & -1 \\ 1 & -3 & 1 & 2 \end{bmatrix}$

30. $A = \begin{bmatrix} 0 & 1 & 0 & 3 \\ 1 & -2 & -3 & 1 \\ 0 & 0 & 2 & -2 \\ 1 & -2 & -4 & 1 \end{bmatrix}$

System of Linear Equations In Exercises 31–36, use the determinant of the coefficient matrix to determine whether the system of linear equations has a unique solution.

31. $x_1 - 3x_2 = 2$
$2x_1 + x_2 = 1$

32. $3x_1 - 4x_2 = 2$
$\frac{2}{3}x_1 - \frac{8}{9}x_2 = 1$

33. $x_1 - x_2 + x_3 = 4$
$2x_1 - x_2 + x_3 = 6$
$3x_1 - 2x_2 + 2x_3 = 0$

34. $x_1 + x_2 - x_3 = 4$
$2x_1 - x_2 + x_3 = 6$
$3x_1 - 2x_2 + 2x_3 = 0$

35. $2x_1 + x_2 + 5x_3 + x_4 = 5$
$x_1 + x_2 - 3x_3 - 4x_4 = -1$
$2x_1 + 2x_2 + 2x_3 - 3x_4 = 2$
$x_1 + 5x_2 - 6x_3 = 3$

36. $x_1 - x_2 - x_3 - x_4 = 0$
$x_1 + x_2 - x_3 - x_4 = 0$
$x_1 + x_2 + x_3 - x_4 = 0$
$x_1 + x_2 + x_3 + x_4 = 6$

Singular Matrices In Exercises 37–42, find the value(s) of k such that A is singular.

37. $A = \begin{bmatrix} k - 1 & 3 \\ 2 & k - 2 \end{bmatrix}$

38. $A = \begin{bmatrix} k - 1 & 2 \\ 2 & k + 2 \end{bmatrix}$

39. $A = \begin{bmatrix} 1 & 0 & 3 \\ 2 & -1 & 0 \\ 4 & 2 & k \end{bmatrix}$

40. $A = \begin{bmatrix} 1 & k & 2 \\ -2 & 0 & -k \\ 3 & 1 & -4 \end{bmatrix}$

41. $A = \begin{bmatrix} 0 & k & 1 \\ k & 1 & k \\ 1 & k & 0 \end{bmatrix}$

42. $A = \begin{bmatrix} k & -3 & -k \\ -2 & k & 1 \\ k & 1 & 0 \end{bmatrix}$

Finding Determinants In Exercises 43–50, find (a) $|A^T|$, (b) $|A^2|$, (c) $|AA^T|$, (d) $|2A|$, and (e) $|A^{-1}|$.

43. $A = \begin{bmatrix} 6 & -11 \\ 4 & -5 \end{bmatrix}$

44. $A = \begin{bmatrix} -4 & 10 \\ 5 & 6 \end{bmatrix}$

45. $A = \begin{bmatrix} 5 & 0 & 0 \\ 1 & -3 & 0 \\ 0 & -1 & 2 \end{bmatrix}$

46. $A = \begin{bmatrix} 1 & 5 & 4 \\ 0 & -6 & 2 \\ 0 & 0 & -3 \end{bmatrix}$

47. $A = \begin{bmatrix} 2 & 0 & 5 \\ 4 & -1 & 6 \\ 3 & 2 & 1 \end{bmatrix}$

48. $A = \begin{bmatrix} 4 & 1 & 9 \\ -1 & 0 & -2 \\ -3 & 3 & 0 \end{bmatrix}$

49. $A = \begin{bmatrix} -3 & 0 & 0 & 0 \\ 0 & 2 & 0 & 0 \\ 0 & 0 & 1 & 0 \\ 0 & 0 & 0 & 5 \end{bmatrix}$

50. $A = \begin{bmatrix} 2 & 0 & 0 & 1 \\ 0 & -3 & 0 & 0 \\ 0 & 0 & 4 & 0 \\ 1 & 0 & 0 & 1 \end{bmatrix}$

Finding Determinants In Exercises 51–56, use a software program or a graphing utility to find (a) $|A|$, (b) $|A^T|$, (c) $|A^2|$, (d) $|2A|$, and (e) $|A^{-1}|$.

51. $A = \begin{bmatrix} 4 & 2 \\ -1 & 5 \end{bmatrix}$

52. $A = \begin{bmatrix} -2 & 4 \\ 6 & 8 \end{bmatrix}$

53. $A = \begin{bmatrix} 3 & 1 & -2 \\ 2 & -1 & 3 \\ -3 & 1 & 2 \end{bmatrix}$

54. $A = \begin{bmatrix} \frac{3}{4} & \frac{2}{3} & -\frac{1}{4} \\ \frac{2}{3} & 1 & \frac{1}{3} \\ -\frac{1}{4} & \frac{1}{3} & \frac{3}{4} \end{bmatrix}$

55. $A = \begin{bmatrix} 4 & -2 & 1 & 5 \\ 3 & 8 & 2 & -1 \\ 6 & 8 & 9 & 2 \\ 2 & 3 & -1 & 0 \end{bmatrix}$

56. $A = \begin{bmatrix} 6 & 5 & 1 & -1 \\ -2 & 4 & 3 & 5 \\ 6 & 1 & -4 & -2 \\ 2 & 2 & 1 & 3 \end{bmatrix}$

57. Let A and B be square matrices of order 4 such that $|A| = -5$ and $|B| = 3$. Find (a) $|A^2|$, (b) $|B^2|$, (c) $|A^3|$, and (d) $|B^4|$.

---

58. CAPSTONE Let A and B be square matrices of order 3 such that $|A| = 4$ and $|B| = 5$.
(a) Find $|AB|$.   (b) Find $|2A|$.
(c) Are A and B singular or nonsingular? Explain.
(d) If A and B are nonsingular, find $|A^{-1}|$ and $|B^{-1}|$.
(e) Find $|(AB)^T|$.

---

59. Proof Let A and B be $n \times n$ matrices such that $AB = I$. Prove that $|A| \neq 0$ and $|B| \neq 0$.

60. Proof Let A and B be $n \times n$ matrices such that AB is singular. Prove that either A or B is singular.

61. Find two 2×2 matrices such that $|A| + |B| = |A + B|$.

62. Verify the equation.

$$\begin{vmatrix} a + b & a & a \\ a & a + b & a \\ a & a & a + b \end{vmatrix} = b^2(3a + b)$$

63. Let A be an $n \times n$ matrix in which the entries of each row sum to zero. Find $|A|$.

64. Illustrate the result of Exercise 63 with the matrix

$$A = \begin{bmatrix} 2 & -1 & -1 \\ -3 & 1 & 2 \\ 0 & -2 & 2 \end{bmatrix}.$$

65. Guided Proof Prove that the determinant of an invertible matrix A is equal to ± 1 when all of the entries of A and A^{-1} are integers.

Getting Started: Denote $\det(A)$ as x and $\det(A^{-1})$ as y. Note that x and y are real numbers. To prove that $\det(A)$ is equal to ± 1, you must show that both x and y are integers such that their product xy is equal to 1.

(i) Use the property for the determinant of a matrix product to show that $xy = 1$.

(ii) Use the definition of a determinant and the fact that the entries of A and A^{-1} are integers to show that both $x = \det(A)$ and $y = \det(A^{-1})$ are integers.

(iii) Conclude that $x = \det(A)$ must be either 1 or -1 because these are the only integer solutions to the equation $xy = 1$.

66. Guided Proof Prove Theorem 3.9: If A is a square matrix, then $\det(A) = \det(A^T)$.

Getting Started: To prove that the determinants of A and A^T are equal, you need to show that their cofactor expansions are equal. The cofactors are $\pm$ determinants of smaller matrices, so you need to use mathematical induction.

(i) Initial step for induction: If A is of order 1, then
$$A = [a_{11}] = A^T$$
so
$$\det(A) = \det(A^T) = a_{11}.$$

(ii) Assume the inductive hypothesis holds for all matrices of order $n - 1$. Let A be a square matrix of order n. Write an expression for the determinant of A by expanding in the first row.

(iii) Write an expression for the determinant of A^T by expanding in the first column.

(iv) Compare the expansions in (ii) and (iii). The entries of the first row of A are the same as the entries of the first column of A^T. Compare cofactors (these are the $\pm$ determinants of smaller matrices that are transposes of one another) and use the inductive hypothesis to conclude that they are equal as well.

67. Writing Let A and P be $n \times n$ matrices, where P is invertible. Does $P^{-1}AP = A$? Illustrate your conclusion with appropriate examples. What can you say about the two determinants $|P^{-1}AP|$ and $|A|$?

68. Writing Let A be an $n \times n$ nonzero matrix satisfying $A^{10} = O$. Explain why A must be singular. What properties of determinants are you using in your argument?

69. Proof A square matrix is **skew-symmetric** when $A^T = -A$. Prove that if A is an $n \times n$ skew-symmetric matrix, then $|A| = (-1)^n|A|$.

70. Proof Let A be a skew-symmetric matrix of odd order. Use the result of Exercise 69 to prove that $|A| = 0$.

True or False? In Exercises 71 and 72, determine whether each statement is true or false. If a statement is true, give a reason or cite an appropriate statement from the text. If a statement is false, provide an example that shows that the statement is not true in all cases or cite an appropriate statement from the text.

71. (a) If A is an $n \times n$ matrix and c is a nonzero scalar, then the determinant of the matrix cA is $nc \cdot \det(A)$.

(b) If A is an invertible matrix, then the determinant of A^{-1} is equal to the reciprocal of the determinant of A.

(c) If A is an invertible $n \times n$ matrix, then $A\mathbf{x} = \mathbf{b}$ has a unique solution for every $\mathbf{b}$.

72. (a) The determinant of the sum of two matrices equals the sum of the determinants of the matrices.

(b) If A and B are square matrices of order n, and $\det(A) = \det(B)$, then $\det(AB) = \det(A^2)$.

(c) If the determinant of an $n \times n$ matrix A is nonzero, then $A\mathbf{x} = O$ has only the trivial solution.

Orthogonal Matrices In Exercises 73–78, determine whether the matrix is orthogonal. An invertible square matrix A is orthogonal when $A^{-1} = A^T$.

73. $\begin{bmatrix} 0 & 1 \\ 1 & 0 \end{bmatrix}$ **74.** $\begin{bmatrix} 1 & 0 \\ 1 & 1 \end{bmatrix}$

75. $\begin{bmatrix} 1 & -1 \\ -1 & -1 \end{bmatrix}$ **76.** $\begin{bmatrix} 1/\sqrt{2} & -1/\sqrt{2} \\ -1/\sqrt{2} & -1/\sqrt{2} \end{bmatrix}$

77. $\begin{bmatrix} 1 & 0 & 0 \\ 0 & 0 & 1 \\ 0 & 1 & 0 \end{bmatrix}$ **78.** $\begin{bmatrix} 1/\sqrt{2} & 0 & -1/\sqrt{2} \\ 0 & 1 & 0 \\ 1/\sqrt{2} & 0 & 1/\sqrt{2} \end{bmatrix}$

79. Proof Prove that the $n \times n$ identity matrix is orthogonal.

80. Proof Prove that if A is an orthogonal matrix, then $|A| = \pm 1$.

Orthogonal Matrices In Exercises 81 and 82, use a graphing utility to determine whether A is orthogonal. Then verify that $|A| = \pm 1$.

81. $A = \begin{bmatrix} \frac{3}{5} & 0 & -\frac{4}{5} \\ 0 & 1 & 0 \\ \frac{4}{5} & 0 & \frac{3}{5} \end{bmatrix}$ **82.** $A = \begin{bmatrix} \frac{2}{3} & -\frac{2}{3} & \frac{1}{3} \\ \frac{2}{3} & \frac{1}{3} & -\frac{2}{3} \\ \frac{1}{3} & \frac{2}{3} & \frac{2}{3} \end{bmatrix}$

83. Proof If A is an idempotent matrix ($A^2 = A$), then prove that the determinant of A is either 0 or 1.

84. Proof Let S be an $n \times n$ singular matrix. Prove that for any $n \times n$ matrix B, the matrix SB is also singular.

3.4 Applications of Determinants

■ Find the adjoint of a matrix and use it to find the inverse of the matrix.

■ Use Cramer's Rule to solve a system of n linear equations in n variables.

■ Use determinants to find area, volume, and the equations of lines and planes.

THE ADJOINT OF A MATRIX

So far in this chapter, you have studied procedures for evaluating, and properties of, determinants. In this section, you will study an explicit formula for the inverse of a nonsingular matrix and use this formula to prove a theorem known as Cramer's Rule. You will also use Cramer's Rule to solve systems of linear equations, and study several other applications of determinants.

Recall from Section 3.1 that the cofactor C_{ij} of a square matrix A is $(-1)^{i+j}$ times the determinant of the matrix obtained by deleting the ith row and jth column of A. The **matrix of cofactors** of A has the form

$$\begin{bmatrix} C_{11} & C_{12} & \cdots & C_{1n} \\ C_{21} & C_{22} & \cdots & C_{2n} \\ \vdots & \vdots & & \vdots \\ C_{n1} & C_{n2} & \cdots & C_{nn} \end{bmatrix}.$$

The transpose of this matrix is the **adjoint** of A and is denoted adj(A). That is,

$$\text{adj}(A) = \begin{bmatrix} C_{11} & C_{21} & \cdots & C_{n1} \\ C_{12} & C_{22} & \cdots & C_{n2} \\ \vdots & \vdots & & \vdots \\ C_{1n} & C_{2n} & \cdots & C_{nn} \end{bmatrix}.$$

| EXAMPLE 1 | Finding the Adjoint of a Square Matrix |

Find the adjoint of $A = \begin{bmatrix} -1 & 3 & 2 \\ 0 & -2 & 1 \\ 1 & 0 & -2 \end{bmatrix}$.

SOLUTION

The cofactor C_{11} is

$$\begin{bmatrix} -1 & 3 & 2 \\ 0 & -2 & 1 \\ 1 & 0 & -2 \end{bmatrix} \quad \rightarrow \quad C_{11} = (-1)^2 \begin{vmatrix} -2 & 1 \\ 0 & -2 \end{vmatrix} = 4.$$

Continuing this process produces the matrix of cofactors of A shown below.

$$\begin{bmatrix} 4 & 1 & 2 \\ 6 & 0 & 3 \\ 7 & 1 & 2 \end{bmatrix}$$

The transpose of this matrix is the adjoint of A. That is, $\text{adj}(A) = \begin{bmatrix} 4 & 6 & 7 \\ 1 & 0 & 1 \\ 2 & 3 & 2 \end{bmatrix}$.

The adjoint of a matrix A can be useful for finding the inverse of A, as shown in the next theorem.

THEOREM 3.10   The Inverse of a Matrix Using Its Adjoint

If A is an $n \times n$ invertible matrix, then $A^{-1} = \dfrac{1}{\det(A)}\text{adj}(A)$.

PROOF

Begin by proving that the product of A and its adjoint is equal to the product of the determinant of A and I_n. Consider the product

$$A[\text{adj}(A)] = \begin{bmatrix} a_{11} & a_{12} & \cdots & a_{1n} \\ a_{21} & a_{22} & \cdots & a_{2n} \\ \vdots & \vdots & & \vdots \\ a_{i1} & a_{i2} & \cdots & a_{in} \\ \vdots & \vdots & & \vdots \\ a_{n1} & a_{n2} & \cdots & a_{nn} \end{bmatrix} \begin{bmatrix} C_{11} & C_{21} & \cdots & C_{j1} & \cdots & C_{n1} \\ C_{12} & C_{22} & \cdots & C_{j2} & \cdots & C_{n2} \\ \vdots & \vdots & & \vdots & & \vdots \\ C_{1n} & C_{2n} & \cdots & C_{jn} & \cdots & C_{nn} \end{bmatrix}.$$

The entry in the ith row and jth column of this product is

$$a_{i1}C_{j1} + a_{i2}C_{j2} + \cdots + a_{in}C_{jn}.$$

If $i = j$, then this sum is simply the cofactor expansion of A in its ith row, which means that the sum is the determinant of A. On the other hand, if $i \neq j$, then the sum is zero. (Verify this.)

$$A[\text{adj}(A)] = \begin{bmatrix} \det(A) & 0 & \cdots & 0 \\ 0 & \det(A) & \cdots & 0 \\ \vdots & \vdots & & \vdots \\ 0 & 0 & \cdots & \det(A) \end{bmatrix} = \det(A)I$$

The matrix A is invertible, so $\det(A) \neq 0$ and you can write

$$\frac{1}{\det(A)}A[\text{adj}(A)] = I \quad \text{or} \quad A\left[\frac{1}{\det(A)}\text{adj}(A)\right] = I.$$

By Theorem 2.7 and the definition of the inverse of a matrix, it follows that

$$\frac{1}{\det(A)}\text{adj}(A) = A^{-1}. \qquad \blacksquare$$

EXAMPLE 2   Using the Adjoint of a Matrix to Find Its Inverse

Use the adjoint of A to find A^{-1}, where $A = \begin{bmatrix} -1 & 3 & 2 \\ 0 & -2 & 1 \\ 1 & 0 & -2 \end{bmatrix}$.

SOLUTION

The determinant of this matrix is 3. Using the adjoint of A (found in Example 1), the inverse of A is

$$A^{-1} = \frac{1}{|A|}\text{adj}(A) = \frac{1}{3}\begin{bmatrix} 4 & 6 & 7 \\ 1 & 0 & 1 \\ 2 & 3 & 2 \end{bmatrix} = \begin{bmatrix} \frac{4}{3} & 2 & \frac{7}{3} \\ \frac{1}{3} & 0 & \frac{1}{3} \\ \frac{2}{3} & 1 & \frac{2}{3} \end{bmatrix}.$$

Check that this matrix is the inverse of A by showing that $AA^{-1} = I = A^{-1}A$. $\blacksquare$

CRAMER'S RULE

Cramer's Rule, named after Gabriel Cramer (1704–1752), uses determinants to solve a system of n linear equations in n variables. This rule applies only to systems with unique solutions. To see how Cramer's Rule works, take another look at the solution described at the beginning of Section 3.1. There, it was pointed out that the system

$$a_{11}x_1 + a_{12}x_2 = b_1$$
$$a_{21}x_1 + a_{22}x_2 = b_2$$

has the solution

$$x_1 = \frac{b_1a_{22} - b_2a_{12}}{a_{11}a_{22} - a_{21}a_{12}} \quad \text{and} \quad x_2 = \frac{b_2a_{11} - b_1a_{21}}{a_{11}a_{22} - a_{21}a_{12}}$$

when $a_{11}a_{22} - a_{21}a_{12} \neq 0$. A determinant can represent each numerator and denominator in this solution, as shown below.

$$x_1 = \frac{\begin{vmatrix} b_1 & a_{12} \\ b_2 & a_{22} \end{vmatrix}}{\begin{vmatrix} a_{11} & a_{12} \\ a_{21} & a_{22} \end{vmatrix}}, \quad x_2 = \frac{\begin{vmatrix} a_{11} & b_1 \\ a_{21} & b_2 \end{vmatrix}}{\begin{vmatrix} a_{11} & a_{12} \\ a_{21} & a_{22} \end{vmatrix}}, \quad a_{11}a_{22} - a_{21}a_{12} \neq 0$$

The denominator for x_1 and x_2 is simply the determinant of the coefficient matrix A of the original system. The numerators for x_1 and x_2 are formed by using the column of constants as replacements for the coefficients of x_1 and x_2 in $|A|$. These two determinants are denoted by $|A_1|$ and $|A_2|$, as shown below.

$$|A_1| = \begin{vmatrix} b_1 & a_{12} \\ b_2 & a_{22} \end{vmatrix} \quad \text{and} \quad |A_2| = \begin{vmatrix} a_{11} & b_1 \\ a_{21} & b_2 \end{vmatrix}$$

You have $x_1 = \frac{|A_1|}{|A|}$ and $x_2 = \frac{|A_2|}{|A|}$. This determinant form of the solution is called **Cramer's Rule.**

EXAMPLE 3   Using Cramer's Rule

Use Cramer's Rule to solve the system of linear equations.

$$4x_1 - 2x_2 = 10$$
$$3x_1 - 5x_2 = 11$$

SOLUTION

First find the determinant of the coefficient matrix.

$$|A| = \begin{vmatrix} 4 & -2 \\ 3 & -5 \end{vmatrix} = -14$$

The determinant is nonzero, so you know the system has a unique solution, and applying Cramer's Rule produces

$$x_1 = \frac{|A_1|}{|A|} = \frac{\begin{vmatrix} 10 & -2 \\ 11 & -5 \end{vmatrix}}{-14} = \frac{-28}{-14} = 2$$

and

$$x_2 = \frac{|A_2|}{|A|} = \frac{\begin{vmatrix} 4 & 10 \\ 3 & 11 \end{vmatrix}}{-14} = \frac{14}{-14} = -1.$$

The solution is $x_1 = 2$ and $x_2 = -1$.

Cramer's Rule generalizes to systems of n linear equations in n variables. The value of each variable is the quotient of two determinants. The denominator is the determinant of the coefficient matrix, and the numerator is the determinant of the matrix formed by replacing the column corresponding to the variable being solved for with the column representing the constants. For example, x_3 in the system

$$
\begin{aligned}
a_{11}x_1 + a_{12}x_2 + a_{13}x_3 &= b_1 \\
a_{21}x_1 + a_{22}x_2 + a_{23}x_3 &= b_2 \\
a_{31}x_1 + a_{32}x_2 + a_{33}x_3 &= b_3
\end{aligned}
\quad \text{is} \quad
x_3 = \frac{|A_3|}{|A|} = \frac{\begin{vmatrix} a_{11} & a_{12} & b_1 \\ a_{21} & a_{22} & b_2 \\ a_{31} & a_{32} & b_3 \end{vmatrix}}{\begin{vmatrix} a_{11} & a_{12} & a_{13} \\ a_{21} & a_{22} & a_{23} \\ a_{31} & a_{32} & a_{33} \end{vmatrix}}.
$$

THEOREM 3.11 Cramer's Rule

If a system of n linear equations in n variables has a coefficient matrix A with a nonzero determinant $|A|$, then the solution of the system is

$$
x_1 = \frac{\det(A_1)}{\det(A)}, \quad x_2 = \frac{\det(A_2)}{\det(A)}, \quad \ldots, \quad x_n = \frac{\det(A_n)}{\det(A)}
$$

where the ith column of A_i is the column of constants in the system of equations.

PROOF

Let the system be represented by $AX = B$. The determinant of A is nonzero, so you can write

$$
X = A^{-1}B = \frac{1}{|A|}\text{adj}(A)B = [x_1 \ x_2 \ \ldots \ x_n]^T.
$$

If the entries of B are $b_1, b_2, \ldots, b_n$, then $x_1 = \frac{1}{|A|}(b_1C_{1i} + b_2C_{2i} + \cdots + b_nC_{ni})$,

but the sum (in parentheses) is precisely the cofactor expansion of A_i, which means that $x_i = |A_i|/|A|$, and the proof is complete.

EXAMPLE 4   Using Cramer's Rule

See LarsonLinearAlgebra.com for an interactive version of this type of example.

Use Cramer's Rule to solve the system of linear equations for x.

$$
\begin{aligned}
-x + 2y - 3z &= 1 \\
2x \qquad + z &= 0 \\
3x - 4y + 4z &= 2
\end{aligned}
$$

SOLUTION

The determinant of the coefficient matrix is $|A| = \begin{vmatrix} -1 & 2 & -3 \\ 2 & 0 & 1 \\ 3 & -4 & 4 \end{vmatrix} = 10$.

The determinant is nonzero, so you know that the solution is unique. Apply Cramer's Rule to solve for x, as shown below.

$$
x = \frac{\begin{vmatrix} 1 & 2 & -3 \\ 0 & 0 & 1 \\ 2 & -4 & 4 \end{vmatrix}}{10} = \frac{(1)(-1)^5 \begin{vmatrix} 1 & 2 \\ 2 & -4 \end{vmatrix}}{10} = \frac{(1)(-1)(-8)}{10} = \frac{4}{5}
$$

REMARK

Apply Cramer's Rule to solve for y and z. You will see that the solution is $y = -\frac{3}{2}$ and $z = -\frac{8}{5}$.

AREA, VOLUME, AND EQUATIONS OF LINES AND PLANES

Determinants have many applications in analytic geometry. One application is in finding the area of a triangle in the xy-plane.

Area of a Triangle in the xy-Plane

The area of a triangle with vertices

$$(x_1, y_1), (x_2, y_2), \text{ and } (x_3, y_3)$$

is

$$\text{Area} = \pm\frac{1}{2}\det\begin{bmatrix} x_1 & y_1 & 1 \\ x_2 & y_2 & 1 \\ x_3 & y_3 & 1 \end{bmatrix}$$

where the sign ($\pm$) is chosen to give a positive area.

PROOF

Prove the case for $y_i > 0$. Assume that $x_1 \leq x_3 \leq x_2$ and that (x_3, y_3) lies above the line segment connecting (x_1, y_1) and (x_2, y_2), as shown in Figure 3.1. Consider the three trapezoids whose vertices are

Trapezoid 1: $(x_1, 0), (x_1, y_1), (x_3, y_3), (x_3, 0)$

Trapezoid 2: $(x_3, 0), (x_3, y_3), (x_2, y_2), (x_2, 0)$

Trapezoid 3: $(x_1, 0), (x_1, y_1), (x_2, y_2), (x_2, 0)$.

The area of the triangle is equal to the sum of the areas of the first two trapezoids minus the area of the third trapezoid. So,

$$\begin{aligned}\text{Area} &= \tfrac{1}{2}(y_1 + y_3)(x_3 - x_1) + \tfrac{1}{2}(y_3 + y_2)(x_2 - x_3) - \tfrac{1}{2}(y_1 + y_2)(x_2 - x_1) \\ &= \tfrac{1}{2}(x_1 y_2 + x_2 y_3 + x_3 y_1 - x_1 y_3 - x_2 y_1 - x_3 y_2) \\ &= \tfrac{1}{2}\begin{vmatrix} x_1 & y_1 & 1 \\ x_2 & y_2 & 1 \\ x_3 & y_3 & 1 \end{vmatrix}.\end{aligned}$$

If the vertices do not occur in the order $x_1 \leq x_3 \leq x_2$ or if the vertex (x_3, y_3) is not above the line segment connecting the other two vertices, then the formula above may yield the negative of the area. So, use $\pm$ and choose the correct sign to give a positive area.

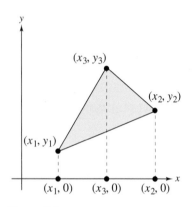

Figure 3.1

EXAMPLE 5   Finding the Area of a Triangle

Find the area of the triangle whose vertices are

$$(1, 1), \quad (2, 2), \quad \text{and} \quad (4, 3).$$

SOLUTION

It is not necessary to know the relative positions of the three vertices. Simply evaluate the determinant

$$\frac{1}{2}\begin{vmatrix} 1 & 1 & 1 \\ 2 & 2 & 1 \\ 4 & 3 & 1 \end{vmatrix} = -\frac{1}{2}$$

and conclude that the area of the triangle is $\frac{1}{2}$ square unit.

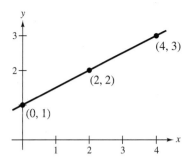

Figure 3.2

If the three points in Example 5 had been on the same line, what would have happened when you applied the area formula? The answer is that the determinant would have been zero. Consider, for example, the three collinear points $(0, 1)$, $(2, 2)$, and $(4, 3)$, as shown in Figure 3.2. The determinant that yields the area of the "triangle" that has these three points as vertices is

$$\frac{1}{2}\begin{vmatrix} 0 & 1 & 1 \\ 2 & 2 & 1 \\ 4 & 3 & 1 \end{vmatrix} = 0.$$

If three points in the xy-plane lie on the same line, then the determinant in the formula for the area of a triangle is zero, as generalized below.

Test for Collinear Points in the xy-Plane

Three points (x_1, y_1), (x_2, y_2), and (x_3, y_3) are collinear if and only if

$$\det \begin{bmatrix} x_1 & y_1 & 1 \\ x_2 & y_2 & 1 \\ x_3 & y_3 & 1 \end{bmatrix} = 0.$$

The test for collinear points can be adapted to another use. That is, when you are given two points in the xy-plane, you can find an equation of the line passing through the two points, as shown below.

Two-Point Form of an Equation of a Line

An equation of the line passing through the distinct points (x_1, y_1) and (x_2, y_2) is

$$\det \begin{bmatrix} x & y & 1 \\ x_1 & y_1 & 1 \\ x_2 & y_2 & 1 \end{bmatrix} = 0.$$

EXAMPLE 6   **Finding an Equation of the Line Passing Through Two Points**

Find an equation of the line passing through the points

$(2, 4)$ and $(-1, 3)$.

SOLUTION

Let $(x_1, y_1) = (2, 4)$ and $(x_2, y_2) = (-1, 3)$. Applying the determinant formula for an equation of a line produces

$$\begin{vmatrix} x & y & 1 \\ 2 & 4 & 1 \\ -1 & 3 & 1 \end{vmatrix} = 0.$$

To evaluate this determinant, expand by cofactors in the first row.

$$x\begin{vmatrix} 4 & 1 \\ 3 & 1 \end{vmatrix} - y\begin{vmatrix} 2 & 1 \\ -1 & 1 \end{vmatrix} + 1\begin{vmatrix} 2 & 4 \\ -1 & 3 \end{vmatrix} = 0$$

$$x(1) - y(3) + 1(10) = 0$$

$$x - 3y + 10 = 0$$

So, an equation of the line is $x - 3y = -10$.

The formula for the area of a triangle in the plane has a straightforward generalization to three-dimensional space, which is presented below without proof.

Volume of a Tetrahedron

The volume of a tetrahedron with vertices (x_1, y_1, z_1), (x_2, y_2, z_2), (x_3, y_3, z_3), and (x_4, y_4, z_4) is

$$\text{Volume} = \pm\tfrac{1}{6}\det\begin{bmatrix} x_1 & y_1 & z_1 & 1 \\ x_2 & y_2 & z_2 & 1 \\ x_3 & y_3 & z_3 & 1 \\ x_4 & y_4 & z_4 & 1 \end{bmatrix}$$

where the sign $(\pm)$ is chosen to give a positive volume.

EXAMPLE 7   Finding the Volume of a Tetrahedron

Find the volume of the tetrahedron shown in Figure 3.3.

SOLUTION

Using the determinant formula for the volume of a tetrahedron produces

$$\frac{1}{6}\begin{vmatrix} 0 & 4 & 1 & 1 \\ 4 & 0 & 0 & 1 \\ 3 & 5 & 2 & 1 \\ 2 & 2 & 5 & 1 \end{vmatrix} = \tfrac{1}{6}(-72) = -12.$$

So, the volume of the tetrahedron is 12 cubic units.

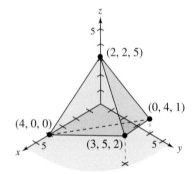

Figure 3.3

If four points in three-dimensional space lie in the same plane, then the determinant in the formula for the volume of a tetrahedron is zero. So, you have the test shown below.

Test for Coplanar Points in Space

Four points (x_1, y_1, z_1), (x_2, y_2, z_2), (x_3, y_3, z_3), and (x_4, y_4, z_4) are coplanar if and only if

$$\det\begin{bmatrix} x_1 & y_1 & z_1 & 1 \\ x_2 & y_2 & z_2 & 1 \\ x_3 & y_3 & z_3 & 1 \\ x_4 & y_4 & z_4 & 1 \end{bmatrix} = 0.$$

An adaptation of this test is the determinant form of an equation of a plane passing through three points in space, as shown below.

Three-Point Form of an Equation of a Plane

An equation of the plane passing through the distinct points (x_1, y_1, z_1), (x_2, y_2, z_2), and (x_3, y_3, z_3) is

$$\det\begin{bmatrix} x & y & z & 1 \\ x_1 & y_1 & z_1 & 1 \\ x_2 & y_2 & z_2 & 1 \\ x_3 & y_3 & z_3 & 1 \end{bmatrix} = 0.$$

EXAMPLE 8

Finding an Equation of the Plane Passing Through Three Points

Find an equation of the plane passing through the points $(0, 1, 0)$, $(-1, 3, 2)$, and $(-2, 0, 1)$.

SOLUTION

Using the determinant form of an equation of a plane produces

$$\begin{vmatrix} x & y & z & 1 \\ 0 & 1 & 0 & 1 \\ -1 & 3 & 2 & 1 \\ -2 & 0 & 1 & 1 \end{vmatrix} = 0.$$

To evaluate this determinant, subtract the fourth column from the second column to obtain

$$\begin{vmatrix} x & y-1 & z & 1 \\ 0 & 0 & 0 & 1 \\ -1 & 2 & 2 & 1 \\ -2 & -1 & 1 & 1 \end{vmatrix} = 0.$$

Expand by cofactors in the second row.

$$x\begin{vmatrix} 2 & 2 \\ -1 & 1 \end{vmatrix} - (y-1)\begin{vmatrix} -1 & 2 \\ -2 & 1 \end{vmatrix} + z\begin{vmatrix} -1 & 2 \\ -2 & -1 \end{vmatrix} = 0$$

$$x(4) - (y-1)(3) + z(5) = 0$$

This produces the equation $4x - 3y + 5z = -3$.

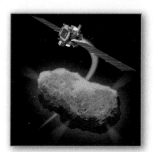

LINEAR ALGEBRA APPLIED

On November 12, 2014, European Space Agency's Rosetta orbiting spacecraft landed the probe Philae on the surface of the comet 67P/Churyumov-Gerasimenko. Comets that orbit the Sun, such as 67P, follow Kepler's First Law of Planetary Motion. This law states that the orbit is an ellipse, with the sun at one focus of the ellipse. The general equation of a conic section, such as an ellipse, is

$$ax^2 + bxy + cy^2 + dx + ey + f = 0.$$

To determine the equation of the comet's orbit, astronomers can find the coordinates of the comet at five different points (x_i, y_i), where $i = 1, 2, 3, 4,$ and 5, substitute these coordinates into the equation

$$\begin{vmatrix} x^2 & xy & y^2 & x & y & 1 \\ x_1^2 & x_1y_1 & y_1^2 & x_1 & y_1 & 1 \\ x_2^2 & x_2y_2 & y_2^2 & x_2 & y_2 & 1 \\ x_3^2 & x_3y_3 & y_3^2 & x_3 & y_3 & 1 \\ x_4^2 & x_4y_4 & y_4^2 & x_4 & y_4 & 1 \\ x_5^2 & x_5x_5 & y_5^2 & x_5 & y_5 & 1 \end{vmatrix} = 0$$

and then expand by cofactors in the first row to find a, b, c, d, e, and f. For example, the coefficient of x^2 is

$$a = \begin{vmatrix} x_1y_1 & y_1^2 & x_1 & y_1 & 1 \\ x_2y_2 & y_2^2 & x_2 & y_2 & 1 \\ x_3y_3 & y_3^2 & x_3 & y_3 & 1 \\ x_4y_4 & y_4^2 & x_4 & y_4 & 1 \\ x_5y_5 & y_5^2 & x_5 & y_5 & 1 \end{vmatrix}.$$

Knowing the equation of 67P's orbit helped astronomers determine the ideal time to release the probe.

3.4   Exercises

See CalcChat.com for worked-out solutions to odd-numbered exercises.

Finding the Adjoint and Inverse of a Matrix In Exercises 1–8, find the adjoint of the matrix A. Then use the adjoint to find the inverse of A (if possible).

1. $A = \begin{bmatrix} 1 & 2 \\ 3 & 4 \end{bmatrix}$ 2. $A = \begin{bmatrix} -1 & 0 \\ 0 & 4 \end{bmatrix}$

3. $A = \begin{bmatrix} 1 & 0 & 0 \\ 0 & 2 & 6 \\ 0 & -4 & -12 \end{bmatrix}$ 4. $A = \begin{bmatrix} 1 & 2 & 3 \\ 0 & 1 & -1 \\ 2 & 2 & 2 \end{bmatrix}$

5. $A = \begin{bmatrix} -3 & -5 & -7 \\ 2 & 4 & 3 \\ 0 & 1 & -1 \end{bmatrix}$ 6. $A = \begin{bmatrix} 0 & 1 & 1 \\ 1 & 2 & 3 \\ -1 & -1 & -2 \end{bmatrix}$

7. $A = \begin{bmatrix} -1 & 2 & 0 & 1 \\ 3 & -1 & 4 & 1 \\ 0 & 0 & 1 & 2 \\ -1 & 1 & 1 & 2 \end{bmatrix}$

8. $A = \begin{bmatrix} 1 & 1 & 1 & 0 \\ 1 & 1 & 0 & 1 \\ 1 & 0 & 1 & 1 \\ 0 & 1 & 1 & 1 \end{bmatrix}$

Using Cramer's Rule   In Exercises 9–22, use Cramer's Rule to solve (if possible) the system of linear equations.

9. $\begin{aligned} x_1 + 2x_2 &= 5 \\ -x_1 + x_2 &= 1 \end{aligned}$ 10. $\begin{aligned} 2x - y &= -10 \\ 3x + 2y &= -1 \end{aligned}$

11. $\begin{aligned} 3x + 4y &= -2 \\ 5x + 3y &= 4 \end{aligned}$ 12. $\begin{aligned} 18x_1 + 12x_2 &= 13 \\ 30x_1 + 24x_2 &= 23 \end{aligned}$

13. $\begin{aligned} 20x + 8y &= 11 \\ 12x - 24y &= 21 \end{aligned}$ 14. $\begin{aligned} 13x - 6y &= 17 \\ 26x - 12y &= 8 \end{aligned}$

15. $\begin{aligned} -0.4x_1 + 0.8x_2 &= 1.6 \\ 2x_1 - 4x_2 &= 5.0 \end{aligned}$ 16. $\begin{aligned} -0.4x_1 + 0.8x_2 &= 1.6 \\ 0.2x_1 + 0.3x_2 &= 0.6 \end{aligned}$

17. $\begin{aligned} 4x - y - z &= 1 \\ 2x + 2y + 3z &= 10 \\ 5x - 2y - 2z &= -1 \end{aligned}$ 18. $\begin{aligned} 4x - 2y + 3z &= -2 \\ 2x + 2y + 5z &= 16 \\ 8x - 5y - 2z &= 4 \end{aligned}$

19. $\begin{aligned} 3x + 4y + 4z &= 11 \\ 4x - 4y + 6z &= 11 \\ 6x - 6y &= 3 \end{aligned}$

20. $\begin{aligned} 14x_1 - 21x_2 - 7x_3 &= -21 \\ -4x_1 + 2x_2 - 2x_3 &= 2 \\ 56x_1 - 21x_2 + 7x_3 &= 7 \end{aligned}$

21. $\begin{aligned} 4x_1 - x_2 + x_3 &= -5 \\ 2x_1 + 2x_2 + 3x_3 &= 10 \\ 5x_1 - 2x_2 + 6x_3 &= 1 \end{aligned}$

22. $\begin{aligned} 2x_1 + 3x_2 + 5x_3 &= 4 \\ 3x_1 + 5x_2 + 9x_3 &= 7 \\ 5x_1 + 9x_2 + 17x_3 &= 13 \end{aligned}$

Using Cramer's Rule In Exercises 23–26, use a software program or a graphing utility and Cramer's Rule to solve (if possible) the system of linear equations.

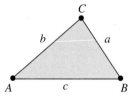

23. $\begin{aligned} \frac{5}{6}x_1 - x_2 &= -20 \\ \frac{4}{3}x_1 - \frac{7}{2}x_2 &= -51 \end{aligned}$

24. $\begin{aligned} -8x_1 + 7x_2 - 10x_3 &= -151 \\ 12x_1 + 3x_2 - 5x_3 &= 86 \\ 15x_1 - 9x_2 + 2x_3 &= 187 \end{aligned}$

25. $\begin{aligned} 3x_1 - 2x_2 + 9x_3 + 4x_4 &= 35 \\ -x_1 - 9x_3 - 6x_4 &= -17 \\ 3x_3 + x_4 &= 5 \\ 2x_1 + 2x_2 + 8x_4 &= -4 \end{aligned}$

26. $\begin{aligned} -x_1 - x_2 + x_4 &= -8 \\ 3x_1 + 5x_2 + 5x_3 &= 24 \\ 2x_3 + x_4 &= -6 \\ -2x_1 - 3x_2 - 3x_3 &= -15 \end{aligned}$

27. Use Cramer's Rule to solve the system of linear equations for x and y.

$$\begin{aligned} kx + (1 - k)y &= 1 \\ (1 - k)x + ky &= 3 \end{aligned}$$

For what value(s) of k will the system be inconsistent?

28. Verify the system of linear equations in $\cos A$, $\cos B$, and $\cos C$ for the triangle shown.

$$\begin{aligned} c \cos B + b \cos C &= a \\ c \cos A + a \cos C &= b \\ b \cos A + a \cos B &= c \end{aligned}$$

Then use Cramer's Rule to solve for $\cos C$, and use the result to verify the Law of Cosines,

$c^2 = a^2 + b^2 - 2ab \cos C.$

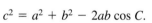

Finding the Area of a Triangle In Exercises 29–32, find the area of the triangle with the given vertices.

29. $(0, 0), (2, 0), (0, 3)$ 30. $(1, 1), (2, 4), (4, 2)$
31. $(-1, 2), (2, 2), (-2, 4)$ 32. $(1, 1), (-1, 1), (0, -2)$

Testing for Collinear Points In Exercises 33–36, determine whether the points are collinear.

33. $(1, 2), (3, 4), (5, 6)$
34. $(-1, 0), (1, 1), (3, 3)$
35. $(-2, 5), (0, -1), (3, -9)$
36. $(-1, -3), (-4, 7), (2, -13)$

Finding an Equation of a Line In Exercises 37–40, find an equation of the line passing through the points.

37. $(0, 0), (3, 4)$

38. $(-4, 7), (2, 4)$

39. $(-2, 3), (-2, -4)$

40. $(1, 4), (3, 4)$

Finding the Volume of a Tetrahedron In Exercises 41–46, find the volume of the tetrahedron with the given vertices.

41. $(1, 0, 0), (0, 1, 0), (0, 0, 1), (1, 1, 1)$

42. $(1, 1, 1), (0, 0, 0), (2, 1, -1), (-1, 1, 2)$

43. $(3, -1, 1), (4, -4, 4), (1, 1, 1), (0, 0, 1)$

44. $(0, 0, 0), (0, 2, 0), (3, 0, 0), (1, 1, 4)$

45. $(-3, -3, -3), (3, -1, -3), (-3, -1, -3), (-2, 3, 2)$

46. $(5, 4, -3), (4, -6, -4), (-6, -6, -5), (0, 0, 10)$

Testing for Coplanar Points In Exercises 47–52, determine whether the points are coplanar.

47. $(-4, 1, 0), (0, 1, 2), (4, 3, -1), (0, 0, 1)$

48. $(1, 2, 3), (-1, 0, 1), (0, -2, -5), (2, 6, 11)$

49. $(0, 0, -1), (0, -1, 0), (1, 1, 0), (2, 1, 2)$

50. $(1, 2, 7), (-3, 6, 6), (4, 4, 2), (3, 3, 4)$

51. $(-3, -2, -1), (2, -1, -2), (-3, -1, -2), (3, 2, 1)$

52. $(1, -5, 9), (-1, -5, 9), (1, -5, -9), (-1, -5, -9)$

Finding an Equation of a Plane In Exercises 53–58, find an equation of the plane passing through the points.

53. $(1, -2, 1), (-1, -1, 7), (2, -1, 3)$

54. $(0, -1, 0), (1, 1, 0), (2, 1, 2)$

55. $(0, 0, 0), (1, -1, 0), (0, 1, -1)$

56. $(1, 2, 7), (4, 4, 2), (3, 3, 4)$

57. $(-4, -4, -4), (4, -1, -4), (-4, -1, -4)$

58. $(3, 2, -2), (3, -2, 2), (-3, -2, -2)$

Using Cramer's Rule In Exercises 59 and 60, determine whether Cramer's Rule is used correctly to solve for the variable. If not, identify the mistake.

59. $\begin{aligned} x + 2y + z &= 2 \\ -x + 3y - 2z &= 4 \\ 4x + y - z &= 6 \end{aligned}$ $y = \dfrac{\begin{vmatrix} 1 & 2 & 1 \\ -1 & 3 & -2 \\ 4 & 1 & -1 \end{vmatrix}}{\begin{vmatrix} 1 & 2 & 1 \\ -1 & 4 & -2 \\ 4 & 6 & -1 \end{vmatrix}}$

60. $\begin{aligned} 5x - 2y + z &= 15 \\ 3x - 3y - z &= -7 \\ 2x - y - 7z &= -3 \end{aligned}$ $x = \dfrac{\begin{vmatrix} 15 & -2 & 1 \\ -7 & -3 & -1 \\ -3 & -1 & -7 \end{vmatrix}}{\begin{vmatrix} 5 & -2 & 1 \\ 3 & -3 & -1 \\ 2 & -1 & -7 \end{vmatrix}}$

61. Software Publishing The table shows the estimated revenues (in billions of dollars) of software publishers in the United States from 2011 through 2013. (Source: U.S. Census Bureau)

Year	Revenues, y
2011	156.8
2012	161.7
2013	177.2

(a) Create a system of linear equations for the data to fit the curve
$$y = at^2 + bt + c$$
where $t = 1$ corresponds to 2011, and y is the revenue.

(b) Use Cramer's Rule to solve the system.

(c) Use a graphing utility to plot the data and graph the polynomial function in the same viewing window.

(d) Briefly describe how well the polynomial function fits the data.

62. CAPSTONE Consider the system of linear equations
$$a_1 x + b_1 y = c_1$$
$$a_2 x + b_2 y = c_2$$
where a_1, b_1, c_1, a_2, b_2, and c_2 represent real numbers. What must be true about the lines represented by the equations when
$$\begin{vmatrix} a_1 & b_1 \\ a_2 & b_2 \end{vmatrix} = 0?$$

63. Proof Prove that if $|A| = 1$ and all entries of A are integers, then all entries of $|A^{-1}|$ must also be integers.

64. Proof Prove that if an $n \times n$ matrix A is not invertible, then $A[\text{adj}(A)]$ is the zero matrix.

Proof In Exercises 65 and 66, prove the formula for a nonsingular $n \times n$ matrix A. Assume $n \geq 2$.

65. $|\text{adj}(A)| = |A|^{n-1}$

66. $\text{adj}[\text{adj}(A)] = |A|^{n-2}A$

67. Illustrate the formula in Exercise 65 using a nonsingular 2×2 matrix A.

68. Illustrate the formula in Exercise 66 using a nonsingular 2×2 matrix A.

69. Proof Prove that if A is an $n \times n$ invertible matrix, then $\text{adj}(A^{-1}) = [\text{adj}(A)]^{-1}$.

70. Illustrate the formula in Exercise 69 using a nonsingular 2×2 matrix A.

3 Review Exercises

See CalcChat.com for worked-out solutions to odd-numbered exercises.

The Determinant of a Matrix **In Exercises 1–18, find the determinant of the matrix.**

1. $\begin{bmatrix} 4 & -1 \\ 2 & 2 \end{bmatrix}$ **2.** $\begin{bmatrix} 0 & -3 \\ 1 & 2 \end{bmatrix}$

3. $\begin{bmatrix} -3 & 1 \\ 6 & -2 \end{bmatrix}$ **4.** $\begin{bmatrix} -2 & 0 \\ 0 & 3 \end{bmatrix}$

5. $\begin{bmatrix} -1 & 3 & -4 \\ 0 & -2 & -1 \\ -1 & -1 & 1 \end{bmatrix}$ **6.** $\begin{bmatrix} 5 & 0 & 2 \\ 0 & -1 & 3 \\ 0 & 0 & 1 \end{bmatrix}$

7. $\begin{bmatrix} -2 & 0 & 0 \\ 0 & -3 & 0 \\ 0 & 0 & -1 \end{bmatrix}$ **8.** $\begin{bmatrix} -15 & 0 & 4 \\ 3 & 0 & -5 \\ 12 & 0 & 6 \end{bmatrix}$

9. $\begin{bmatrix} -3 & 6 & 9 \\ 9 & 12 & -3 \\ 0 & 15 & -6 \end{bmatrix}$ **10.** $\begin{bmatrix} -15 & 0 & 3 \\ 3 & 9 & -6 \\ 12 & -3 & 6 \end{bmatrix}$

11. $\begin{bmatrix} 2 & 0 & -1 & 4 \\ -1 & 2 & 0 & 3 \\ 3 & 0 & 1 & 2 \\ -2 & 0 & 3 & 1 \end{bmatrix}$ **12.** $\begin{bmatrix} 2 & 0 & 0 & 0 \\ -3 & 1 & 0 & 0 \\ 4 & -1 & 3 & 0 \\ 5 & 2 & 1 & -1 \end{bmatrix}$

13. $\begin{bmatrix} -4 & 1 & 2 & 3 \\ 1 & -2 & 1 & 2 \\ 2 & -1 & 3 & 4 \\ 1 & 2 & 2 & -1 \end{bmatrix}$ **14.** $\begin{bmatrix} 3 & -1 & 2 & 1 \\ -2 & 0 & 1 & -3 \\ -1 & 2 & -3 & 4 \\ -2 & 1 & -2 & 1 \end{bmatrix}$

15. $\begin{bmatrix} -1 & 1 & -1 & 0 & 0 \\ 0 & 1 & -1 & 0 & 1 \\ 1 & 0 & 1 & -1 & 0 \\ 0 & -1 & 0 & 1 & -1 \\ 0 & 1 & 1 & -1 & 1 \end{bmatrix}$

16. $\begin{bmatrix} 1 & 2 & -1 & 3 & 4 \\ 2 & 3 & -1 & 2 & -2 \\ 1 & 2 & 0 & 1 & -1 \\ 1 & 0 & 2 & -1 & 0 \\ 0 & -1 & 1 & 0 & 2 \end{bmatrix}$

17. $\begin{bmatrix} -1 & 0 & 0 & 0 & 0 \\ 0 & -1 & 0 & 0 & 0 \\ 0 & 0 & -1 & 0 & 0 \\ 0 & 0 & 0 & -1 & 0 \\ 0 & 0 & 0 & 0 & -1 \end{bmatrix}$

18. $\begin{bmatrix} 0 & 0 & 0 & 0 & 3 \\ 0 & 0 & 0 & 3 & 0 \\ 0 & 0 & 3 & 0 & 0 \\ 0 & 3 & 0 & 0 & 0 \\ 3 & 0 & 0 & 0 & 0 \end{bmatrix}$

Properties of Determinants **In Exercises 19–22, determine which property of determinants the equation illustrates.**

19. $\begin{vmatrix} 4 & -1 \\ 16 & -4 \end{vmatrix} = 0$

20. $\begin{vmatrix} 1 & 2 & -1 \\ 2 & 0 & 3 \\ 4 & -1 & 1 \end{vmatrix} = -\begin{vmatrix} 1 & -1 & 2 \\ 2 & 3 & 0 \\ 4 & 1 & -1 \end{vmatrix}$

21. $\begin{vmatrix} 2 & -4 & 3 & 2 \\ 0 & 4 & 6 & 1 \\ 1 & 8 & 9 & 0 \\ 6 & 12 & -6 & 1 \end{vmatrix} = -12\begin{vmatrix} 2 & 1 & 1 & 2 \\ 0 & -1 & 2 & 1 \\ 1 & -2 & 3 & 0 \\ 6 & -3 & -2 & 1 \end{vmatrix}$

22. $\begin{vmatrix} 1 & 3 & 1 \\ 0 & -1 & 2 \\ 1 & 2 & 1 \end{vmatrix} = \begin{vmatrix} 1 & 3 & 1 \\ 2 & 5 & 4 \\ 1 & 2 & 1 \end{vmatrix}$

The Determinant of a Matrix Product **In Exercises 23 and 24, find (a) $|A|$, (b) $|B|$, (c) AB, and (d) $|AB|$. Then verify that $|A||B| = |AB|$.**

23. $A = \begin{bmatrix} -1 & 2 \\ 0 & 1 \end{bmatrix}$, $B = \begin{bmatrix} 3 & 4 \\ 2 & 1 \end{bmatrix}$

24. $A = \begin{bmatrix} 0 & 1 & 2 \\ 5 & 4 & 3 \\ 7 & 6 & 8 \end{bmatrix}$, $B = \begin{bmatrix} 2 & 1 & 2 \\ 1 & -1 & 0 \\ 0 & 3 & -2 \end{bmatrix}$

Finding Determinants **In Exercises 25 and 26, find (a) $|A^T|$, (b) $|A^3|$, (c) $|A^TA|$, and (d) $|5A|$.**

25. $A = \begin{bmatrix} -3 & 8 \\ 4 & 1 \end{bmatrix}$ **26.** $A = \begin{bmatrix} 3 & 0 & 1 \\ -1 & 0 & 0 \\ 2 & 1 & 2 \end{bmatrix}$

Finding Determinants **In Exercises 27 and 28, find (a) $|A|$ and (b) $|A^{-1}|$.**

27. $A = \begin{bmatrix} 1 & 0 & -4 \\ 0 & 3 & 2 \\ -2 & 7 & 6 \end{bmatrix}$ **28.** $A = \begin{bmatrix} -2 & 1 & 3 \\ 2 & 0 & 4 \\ -1 & 5 & 0 \end{bmatrix}$

The Determinant of the Inverse of a Matrix **In Exercises 29–32, find $|A^{-1}|$. Begin by finding A^{-1}, and then evaluate its determinant. Verify your result by finding $|A|$ and then applying the formula from Theorem 3.8, $|A^{-1}| = \dfrac{1}{|A|}$.**

29. $A = \begin{bmatrix} -2 & 4 \\ 1 & 1 \end{bmatrix}$ **30.** $A = \begin{bmatrix} 10 & 2 \\ -2 & 7 \end{bmatrix}$

31. $A = \begin{bmatrix} 1 & 0 & 1 \\ 2 & -1 & 4 \\ 2 & 6 & 0 \end{bmatrix}$ **32.** $A = \begin{bmatrix} -1 & 1 & 2 \\ 2 & 4 & 8 \\ 1 & -1 & 0 \end{bmatrix}$

Solving a System of Linear Equations   **In Exercises 33–36, solve the system of linear equations by each of the methods listed below.**

(a) Gaussian elimination with back-substitution

(b) Gauss-Jordan elimination

(c) Cramer's Rule

33. $3x_1 + 3x_2 + 5x_3 = 1$
    $3x_1 + 5x_2 + 9x_3 = 2$
    $5x_1 + 9x_2 + 17x_3 = 4$

34. $\quad x_1 + 2x_2 + \ x_3 = 4$
    $-3x_1 + \ x_2 - 2x_3 = 1$
    $\quad 2x_1 + 3x_2 - \ x_3 = 9$

35. $\quad x_1 + 2x_2 - \ x_3 = -7$
    $\quad 2x_1 - 2x_2 - 2x_3 = -8$
    $-x_1 + 3x_2 + 4x_3 = \ 8$

36. $2x_1 + 3x_2 + \ 5x_3 = \ 4$
    $3x_1 + 5x_2 + \ 9x_3 = \ 7$
    $5x_1 + 9x_2 + 13x_3 = 17$

System of Linear Equations   **In Exercises 37–42, use the determinant of the coefficient matrix to determine whether the system of linear equations has a unique solution.**

37. $6x + 5y = \ 0$
    $\ x - \ y = 22$

38. $2x - 5y = 2$
    $3x - 7y = 1$

39. $-x + \ y + 2z = \ 1$
    $2x + 3y + \ z = -2$
    $5x + 4y + 2z = \ 4$

40. $2x + 3y + \ z = \ 10$
    $2x - 3y - 3z = \ 22$
    $8x + 6y \quad\ = -2$

41. $x_1 + 2x_2 + \ 6x_3 = \ 1$
    $2x_1 + 5x_2 + 15x_3 = \ 4$
    $3x_1 + \ x_2 + \ 3x_3 = -6$

42. $x_1 + 5x_2 + 3x_3 \qquad\qquad = 14$
    $4x_1 + 2x_2 + 5x_3 \qquad\qquad = 3$
    $\qquad\qquad 3x_3 + 8x_4 + 6x_5 = 16$
    $2x_1 + 4x_2 \qquad\quad - 2x_5 = 0$
    $2x_1 \qquad - x_3 \qquad\qquad = 0$

43. Let A and B be square matrices of order 4 such that $|A| = 4$ and $|B| = 2$. Find (a) $|BA|$, (b) $|B^2|$, (c) $|2A|$, (d) $|(AB)^T|$, and (e) $|B^{-1}|$.

44. Let A and B be square matrices of order 3 such that $|A| = -2$ and $|B| = 5$. Find (a) $|BA|$, (b) $|B^4|$, (c) $|2A|$, (d) $|(AB)^T|$, and (e) $|B^{-1}|$.

45. **Proof**   Prove the property below.

$$\begin{vmatrix} a_{11} & a_{12} & a_{13} \\ a_{21} & a_{22} & a_{23} \\ a_{31} + c_{31} & a_{32} + c_{32} & a_{33} + c_{33} \end{vmatrix} = \begin{vmatrix} a_{11} & a_{12} & a_{13} \\ a_{21} & a_{22} & a_{23} \\ a_{31} & a_{32} & a_{33} \end{vmatrix}$$

$$+ \begin{vmatrix} a_{11} & a_{12} & a_{13} \\ a_{21} & a_{22} & a_{23} \\ c_{31} & c_{32} & c_{33} \end{vmatrix}$$

46. Illustrate the property in Exercise 45 with A, c_{31}, c_{32}, and c_{33} below.

$$A = \begin{bmatrix} 1 & 0 & 2 \\ 1 & -1 & 2 \\ 2 & 1 & -1 \end{bmatrix}, \quad c_{31} = 3, \quad c_{32} = 0, \quad c_{33} = 1$$

47. Find the determinant of the $n \times n$ matrix.

$$\begin{bmatrix} 1-n & 1 & 1 & \cdots & 1 \\ 1 & 1-n & 1 & \cdots & 1 \\ \vdots & \vdots & \vdots & & \vdots \\ 1 & 1 & 1 & \cdots & 1-n \end{bmatrix}$$

48. Show that

$$\begin{vmatrix} a & 1 & 1 & 1 \\ 1 & a & 1 & 1 \\ 1 & 1 & a & 1 \\ 1 & 1 & 1 & a \end{vmatrix} = (a+3)(a-1)^3.$$

Calculus   **In Exercises 49–54, find the Jacobians of the functions. If x, y, and z are continuous functions of u, v, and w with continuous first partial derivatives, then the Jacobians $J(u, v)$ and $J(u, v, w)$ are**

$$J(u, v) = \begin{vmatrix} \dfrac{\partial x}{\partial u} & \dfrac{\partial x}{\partial v} \\ \dfrac{\partial y}{\partial u} & \dfrac{\partial y}{\partial v} \end{vmatrix} \text{ and } J(u, v, w) = \begin{vmatrix} \dfrac{\partial x}{\partial u} & \dfrac{\partial x}{\partial v} & \dfrac{\partial x}{\partial w} \\ \dfrac{\partial y}{\partial u} & \dfrac{\partial y}{\partial v} & \dfrac{\partial y}{\partial w} \\ \dfrac{\partial z}{\partial u} & \dfrac{\partial z}{\partial v} & \dfrac{\partial z}{\partial w} \end{vmatrix}.$$

49. $x = \frac{1}{2}(v - u), \quad y = \frac{1}{2}(v + u)$

50. $x = au + bv, \quad y = cu + dv$

51. $x = u \cos v, \quad y = u \sin v$

52. $x = e^u \sin v, \quad y = e^u \cos v$

53. $x = \frac{1}{2}(u + v), \quad y = \frac{1}{2}(u - v), \quad z = 2uvw$

54. $x = u - v + w, \quad y = 2uv, \quad z = u + v + w$

55. **Writing**   Compare the various methods for calculating the determinant of a matrix. Which method requires the least amount of computation? Which method do you prefer when the matrix has very few zeros?

56. **Writing**   Use the table on page 122 to compare the numbers of operations involved in calculating the determinant of a 10×10 matrix by cofactor expansion and then by row reduction. Which method would you prefer to use for calculating determinants?

57. **Writing**   Solve the equation for x, if possible.

$$\begin{vmatrix} \cos x & 0 & \sin x \\ \sin x & 0 & \cos x \\ \sin x - \cos x & 1 & \sin x + \cos x \end{vmatrix} = 0$$

58. **Proof**   Prove that if $|A| = |B| \neq 0$, and A and B are of the same size, then there exists a matrix C such that

$$|C| = 1 \quad \text{and} \quad A = CB.$$

Finding the Adjoint of a Matrix **In Exercises 59 and 60, find the adjoint of the matrix.**

59. $\begin{bmatrix} 0 & 1 \\ -2 & 1 \end{bmatrix}$

60. $\begin{bmatrix} 1 & -1 & 1 \\ 0 & 1 & 2 \\ 0 & 0 & -1 \end{bmatrix}$

System of Linear Equations **In Exercises 61–64, use the determinant of the coefficient matrix to determine whether the system of linear equations has a unique solution. If it does, use Cramer's Rule to find the solution.**

61. $0.2x - 0.1y = 0.07$
$0.4x - 0.5y = -0.01$

62. $2x + y = 0.3$
$3x - y = -1.3$

63. $2x_1 + 3x_2 + 3x_3 = 3$
$6x_1 + 6x_2 + 12x_3 = 13$
$12x_1 + 9x_2 - x_3 = 2$

64. $4x_1 + 4x_2 + 4x_3 = 5$
$4x_1 - 2x_2 - 8x_3 = 1$
$8x_1 + 2x_2 - 4x_3 = 6$

Using Cramer's Rule **In Exercises 65 and 66, use a software program or a graphing utility and Cramer's Rule to solve (if possible) the system of linear equations.**

65. $0.2x_1 - 0.6x_2 = 2.4$
$-x_1 + 1.4x_2 = -8.8$

66. $4x_1 - x_2 + x_3 = -5$
$2x_1 + 2x_2 + 3x_3 = 10$
$5x_1 - 2x_2 + 6x_3 = 1$

Finding the Area of a Triangle **In Exercises 67 and 68, use a determinant to find the area of the triangle with the given vertices.**

67. $(1, 0), (5, 0), (5, 8)$

68. $(-4, 0), (4, 0), (0, 6)$

Finding an Equation of a Line **In Exercises 69 and 70, use a determinant to find an equation of the line passing through the points.**

69. $(-4, 0), (4, 4)$

70. $(2, 5), (6, -1)$

Finding an Equation of a Plane **In Exercises 71 and 72, use a determinant to find an equation of the plane passing through the points.**

71. $(0, 0, 0), (1, 0, 3), (0, 3, 4)$

72. $(0, 0, 0), (2, -1, 1), (-3, 2, 5)$

73. Using Cramer's Rule Determine whether Cramer's Rule is used correctly to solve for the variable. If not, identify the mistake.

$x - 4y - z = -1$
$2x - 3y + z = 6$
$x + y - 4z = 1$

$$z = \dfrac{\begin{vmatrix} -1 & -4 & -1 \\ 6 & -3 & 1 \\ 1 & 1 & -4 \end{vmatrix}}{\begin{vmatrix} 1 & -4 & -1 \\ 2 & -3 & 1 \\ 1 & 1 & -4 \end{vmatrix}}$$

74. Health Care Expenditures The table shows annual personal health care expenditures (in billions of dollars) in the United States from 2011 through 2013. (Source: Bureau of Economic Analysis)

Year	2011	2012	2013
Amount, y	1765	1855	1920

(a) Create a system of linear equations for the data to fit the curve

$y = at^2 + bt + c$

where $t = 1$ corresponds to 2011, and y is the amount of the expenditure.

(b) Use Cramer's Rule to solve the system.

(c) Use a graphing utility to plot the data and graph the polynomial function in the same viewing window.

(d) Briefly describe how well the polynomial function fits the data.

True or False? **In Exercises 75–78, determine whether each statement is true or false. If a statement is true, give a reason or cite an appropriate statement from the text. If a statement is false, provide an example that shows the statement is not true in all cases or cite an appropriate statement from the text.**

75. (a) The cofactor C_{22} of a matrix is always a positive number.

(b) If a square matrix B is obtained from A by interchanging two rows, then $\det(B) = \det(A)$.

(c) If one column of a square matrix is a multiple of another column, then the determinant is 0.

(d) If A is a square matrix of order n, then $\det(A) = -\det(A^T)$.

76. (a) If A and B are square matrices of order n such that $\det(AB) = -1$, then both A and B are nonsingular.

(b) If A is a 3×3 matrix with $\det(A) = 5$, then $\det(2A) = 10$.

(c) If A and B are square matrices of order n, then $\det(A + B) = \det(A) + \det(B)$.

77. (a) In Cramer's Rule, the value of x_i is the quotient of two determinants, where the numerator is the determinant of the coefficient matrix.

(b) Three points $(x_1, y_1), (x_2, y_2)$, and (x_3, y_3) are collinear when the determinant of the matrix that has the coordinates as entries in the first two columns and 1's as entries in the third column is nonzero.

78. (a) The matrix of cofactors of a square matrix A is the adjoint of A.

(b) In Cramer's Rule, the denominator is the determinant of the matrix formed by replacing the column corresponding to the variable being solved for with the column representing the constants.

3 Projects

1 Stochastic Matrices

In Section 2.5, you studied a consumer preference model for competing satellite television companies. The matrix of transition probabilities was

$$P = \begin{bmatrix} 0.70 & 0.15 & 0.15 \\ 0.20 & 0.80 & 0.15 \\ 0.10 & 0.05 & 0.70 \end{bmatrix}.$$

When you were given the initial state matrix X_0, you observed that the portions of the total population in the three states (subscribing to Company A, subscribing to Company B, and not subscribing) after 1 year was $X_1 = PX_0$.

$$X_0 = \begin{bmatrix} 0.1500 \\ 0.2000 \\ 0.6500 \end{bmatrix}$$

$$X_1 = PX_0 = \begin{bmatrix} 0.70 & 0.15 & 0.15 \\ 0.20 & 0.80 & 0.15 \\ 0.10 & 0.05 & 0.70 \end{bmatrix}\begin{bmatrix} 0.1500 \\ 0.2000 \\ 0.6500 \end{bmatrix} = \begin{bmatrix} 0.2325 \\ 0.2875 \\ 0.4800 \end{bmatrix}$$

After 15 years, the state matrix had nearly reached a steady state.

$$X_{15} = P^{15}X_0 \approx \begin{bmatrix} 0.3333 \\ 0.4756 \\ 0.1911 \end{bmatrix}$$

That is, for large values of n, the product P^nX approached a limit $\overline{X}$, $P\overline{X} = \overline{X}$.

$P\overline{X} = \overline{X} = 1\overline{X}$, so 1 is an *eigenvalue* of P with corresponding *eigenvector* $\overline{X}$. You will study eigenvalues and eigenvectors in more detail in Chapter 7.

1. Use a software program or a graphing utility to verify the eigenvalues and eigenvectors of P listed below. That is, show that $P\mathbf{x}_i = \lambda_i\mathbf{x}_i$ for $i = 1, 2,$ and 3.

   Eigenvalues: $\lambda_1 = 0.55, \lambda_2 = 0.65, \lambda_3 = 1$

   Eigenvectors: $\mathbf{x}_1 = \begin{bmatrix} 2 \\ -1 \\ -1 \end{bmatrix}, \mathbf{x}_2 = \begin{bmatrix} 0 \\ 1 \\ -1 \end{bmatrix}, \mathbf{x}_3 = \begin{bmatrix} 7 \\ 10 \\ 4 \end{bmatrix}$

2. Let S be the matrix whose columns are the eigenvectors of P. Show that $S^{-1}PS$ is a diagonal matrix D. What are the entries along the diagonal of D?

3. Show that $P^n = (SDS^{-1})^n = SD^nS^{-1}$. Use this result to calculate X_{15} and verify the result above.

2 The Cayley-Hamilton Theorem

The **characteristic polynomial** of a square matrix A is the determinant $|\lambda I - A|$. If the order of A is n, then the characteristic polynomial $p(\lambda)$ is an nth-degree polynomial in the variable λ.

$$p(\lambda) = \det(\lambda I - A) = \lambda^n + c_{n-1}\lambda^{n-1} + \cdots + c_2\lambda^2 + c_1\lambda + c_0$$

The Cayley-Hamilton Theorem asserts that every square matrix satisfies its characteristic polynomial. That is, for the $n \times n$ matrix A, $p(A) = O$, or

$$A^n + c_{n-1}A^{n-1} + \cdots + c_2A^2 + c_1A + c_0I = O.$$

Note that this is a matrix equation. The $n \times n$ zero matrix is on the right, and the coefficient c_0 is multiplied by the $n \times n$ identity matrix I.

1. Verify the Cayley-Hamilton Theorem for the matrix

$$\begin{bmatrix} -2 & 1 \\ 1 & -2 \end{bmatrix}.$$

2. Verify the Cayley-Hamilton Theorem for the matrix

$$\begin{bmatrix} 5 & -1 & 3 \\ -1 & 2 & 4 \\ 1 & 3 & 0 \end{bmatrix}.$$

3. Verify the Cayley-Hamilton Theorem for a general 2×2 matrix A,

$$A = \begin{bmatrix} a & b \\ c & d \end{bmatrix}.$$

4. For a nonsingular $n \times n$ matrix A, show that

$$A^{-1} = \frac{1}{c_0}(-A^{n-1} - c_{n-1}A^{n-2} - \cdots - c_2 A - c_1 I).$$

Use this result to find the inverse of the matrix

$$A = \begin{bmatrix} 2 & 3 \\ 1 & 2 \end{bmatrix}.$$

5. The Cayley-Hamilton Theorem is useful for calculating powers A^n of the square matrix A. For example, the characteristic polynomial of the matrix

$$A = \begin{bmatrix} 5 & -2 \\ 3 & -1 \end{bmatrix}$$

is $p(\lambda) = \lambda^2 - 4\lambda + 1$.

Using the Cayley-Hamilton Theorem,

$$A^2 - 4A + I = O \quad \text{or} \quad A^2 = 4A - I.$$

So, A^2 is written in terms of A and I.

$$A^2 = 4A - I = 4\begin{bmatrix} 5 & -2 \\ 3 & -1 \end{bmatrix} - \begin{bmatrix} 1 & 0 \\ 0 & 1 \end{bmatrix} = \begin{bmatrix} 19 & -8 \\ 12 & -5 \end{bmatrix}$$

Similarly, multiplying both sides of the equation $A^2 = 4A - I$ by A gives A^3 in terms of A^2, A, and I. Moreover, you can write A^3 in terms of A and I by replacing A^2 with $4A - I$, as shown below.

$$A^3 = 4A^2 - A = 4(4A - I) - A = 15A - 4I$$

(a) Write A^4 in terms of A and I.

(b) Find A^5 for the matrix

$$A = \begin{bmatrix} 0 & 0 & 2 \\ 1 & 1 & -2 \\ 2 & 0 & 1 \end{bmatrix}.$$

(*Hint:* Find the characteristic polynomial of A, then use the Cayley-Hamilton Theorem to write A^3 in terms of A^2, A, and I. Inductively write A^5 in terms of A^2, A, and I.)

1–3 Cumulative Test

See CalcChat.com for worked-out solutions
to odd-numbered exercises.

Take this test to review the material in Chapters 1–3. After you are finished, check
your work against the answers in the back of the book.

In Exercises 1 and 2, determine whether the equation is linear in the variables x
and y.

1. $\dfrac{4}{y} - x = 10$

2. $\dfrac{3}{5}x + \dfrac{7}{10}y = 2$

In Exercises 3 and 4, use Gaussian elimination to solve the system of linear
equations.

3.
$$x - 2y = 5$$
$$3x + y = 1$$

4.
$$4x_1 + x_2 - 3x_3 = 11$$
$$2x_1 - 3x_2 + 2x_3 = 9$$
$$x_1 + x_2 + x_3 = -3$$

5. Use a software program or a graphing utility to solve the system of linear equations.

$$0.2x - 2.3y + 1.4z - 0.55w = -110.6$$
$$3.4x + 1.3y - 1.7z + 0.45w = 65.4$$
$$0.5x - 4.9y + 1.1z - 1.6w = -166.2$$
$$0.6x + 2.8y - 3.4z + 0.3w = 189.6$$

6. Find the solution set of the system of linear equations represented by the
augmented matrix.

$$\begin{bmatrix} 0 & 1 & -1 & 0 & 2 \\ 1 & 0 & 2 & -1 & 0 \\ 1 & 2 & 0 & -1 & 4 \end{bmatrix}$$

7. Solve the homogeneous linear system corresponding to the coefficient matrix.

$$\begin{bmatrix} 1 & 2 & 1 & -2 \\ 0 & 0 & 2 & -4 \\ -2 & -4 & 1 & -2 \end{bmatrix}$$

8. Determine the value(s) of k such that the system is consistent.

$$x + 2y - z = 3$$
$$-x - y + z = 2$$
$$-x + y + z = k$$

9. Solve for x and y in the matrix equation $2A - B = I$, given

$$A = \begin{bmatrix} -1 & 1 \\ 2 & 3 \end{bmatrix} \quad \text{and} \quad B = \begin{bmatrix} x & 2 \\ y & 5 \end{bmatrix}.$$

10. Find $A^T A$ for the matrix $A = \begin{bmatrix} 5 & 3 & 1 \\ 2 & 4 & 6 \end{bmatrix}$. Show that this product is symmetric.

In Exercises 11–14, find the inverse of the matrix (if it exists).

11. $\begin{bmatrix} -2 & 3 \\ 4 & 6 \end{bmatrix}$
12. $\begin{bmatrix} -2 & 3 \\ 3 & 6 \end{bmatrix}$
13. $\begin{bmatrix} -1 & 0 & 0 \\ 0 & \frac{1}{2} & 0 \\ 0 & 0 & 3 \end{bmatrix}$
14. $\begin{bmatrix} 1 & 1 & 0 \\ -3 & 6 & 5 \\ 0 & 1 & 0 \end{bmatrix}$

In Exercises 15 and 16, use an inverse matrix to solve the system of linear equations.

15.
$$x + 2y = 0$$
$$3x - 6y = 8$$

16.
$$2x - y = 6$$
$$2x + y = 10$$

17. Find a sequence of elementary matrices whose product is the nonsingular matrix below.

$$\begin{bmatrix} 2 & -4 \\ 1 & 0 \end{bmatrix}$$

18. Find the determinant of the matrix

$$\begin{bmatrix} 4 & 0 & 3 & 2 \\ 0 & 1 & -3 & -5 \\ 0 & 1 & 5 & 1 \\ 1 & 1 & 0 & -3 \end{bmatrix}.$$

19. Find (a) $|A|$, (b) $|B|$, (c) AB, and (d) $|AB|$. Then verify that $|A||B| = |AB|$.

$$A = \begin{bmatrix} 1 & -3 \\ 4 & 2 \end{bmatrix}, \quad B = \begin{bmatrix} -2 & 1 \\ 0 & 5 \end{bmatrix}$$

20. Find (a) $|A|$ and (b) $|A^{-1}|$.

$$A = \begin{bmatrix} 5 & -2 & -3 \\ -1 & 0 & 4 \\ 6 & -8 & 2 \end{bmatrix}$$

21. If $|A| = 7$ and A is of order 4, then find each determinant.

(a) $|3A|$ (b) $|A^T|$ (c) $|A^{-1}|$ (d) $|A^3|$

22. Use the adjoint of

$$A = \begin{bmatrix} 1 & -5 & -1 \\ 0 & -2 & 1 \\ 1 & 0 & 2 \end{bmatrix}$$

to find A^{-1}.

23. Let $\mathbf{x}_1$, $\mathbf{x}_2$, $\mathbf{x}_3$, and $\mathbf{b}$ be the column matrices below.

$$\mathbf{x}_1 = \begin{bmatrix} 1 \\ 0 \\ 1 \end{bmatrix} \quad \mathbf{x}_2 = \begin{bmatrix} 1 \\ 1 \\ 0 \end{bmatrix} \quad \mathbf{x}_3 = \begin{bmatrix} 0 \\ 1 \\ 1 \end{bmatrix} \quad \mathbf{b} = \begin{bmatrix} 1 \\ 2 \\ 3 \end{bmatrix}$$

Find constants a, b, and c such that $a\mathbf{x}_1 + b\mathbf{x}_2 + c\mathbf{x}_3 = \mathbf{b}$.

24. Use a system of linear equations to find the parabola $y = ax^2 + bx + c$ that passes through the points $(-1, 2)$, $(0, 1)$, and $(2, 6)$.

25. Use a determinant to find an equation of the line passing through the points $(1, 4)$ and $(5, -2)$.

26. Use a determinant to find the area of the triangle with vertices $(-2, 2)$, $(8, 2)$, and $(6, -5)$.

27. Determine the currents I_1, I_2, and I_3 for the electrical network shown in the figure at the left.

28. A manufacturer produces three different models of a product and ships them to two warehouses. In the matrix

$$A = \begin{bmatrix} 200 & 300 \\ 600 & 350 \\ 250 & 400 \end{bmatrix}$$

a_{ij} represents the number of units of model i that the manufacturer ships to warehouse j. The matrix

$$B = \begin{bmatrix} 12.50 & 9.00 & 21.50 \end{bmatrix}$$

represents the prices of the three models in dollars per unit. Find the product BA and state what each entry of the matrix represents.

29. Let A, B, and C be three nonzero $n \times n$ matrices such that $AC = BC$. Does it follow that $A = B$? If so, provide a proof. If not, provide a counterexample.

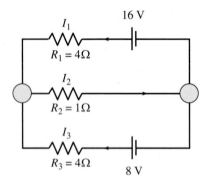

16 V

I_1

$R_1 = 4\Omega$

I_2

$R_2 = 1\Omega$

I_3

$R_3 = 4\Omega$

8 V

Figure for 27

4 Vector Spaces

Satellite Dish (p. 223)

Crystallography (p. 213)

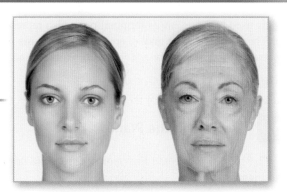

Image Morphing (p. 180)

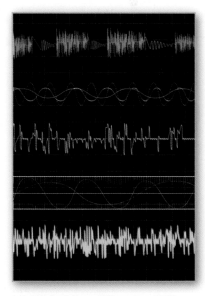

Digital Sampling (p. 172)

Force (p. 157)

4.1 Vectors in R^n

■ Represent a vector as a directed line segment.

■ Perform basic vector operations in R^2 and represent them graphically.

■ Perform basic vector operations in R^n.

■ Write a vector as a linear combination of other vectors.

VECTORS IN THE PLANE

In physics and engineering, a *vector* is characterized by two quantities (length and direction) and is represented by a *directed line segment*. In this chapter you will see that the geometric representation can help you understand the more general definition of a vector.

Geometrically, a **vector in the plane** is represented by a **directed line segment** with its **initial point** at the origin and its **terminal point** at (x_1, x_2), as shown below.

REMARK

The term *vector* derives from the Latin word *vectus*, meaning "to carry." The idea is that if you were to carry something from the origin to the point (x_1, x_2), then the trip could be represented by the directed line segment from $(0, 0)$ to (x_1, x_2). Vectors are represented by lowercase letters set in boldface type (such as **u**, **v**, **w**, and **x**).

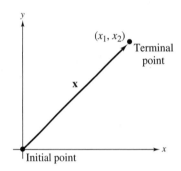

The same **ordered pair** used to represent its terminal point also represents the vector. That is, $\mathbf{x} = (x_1, x_2)$. The coordinates x_1 and x_2 are the **components** of the vector **x**. Two vectors in the plane $\mathbf{u} = (u_1, u_2)$ and $\mathbf{v} = (v_1, v_2)$ are **equal** if and only if

$$u_1 = v_1 \quad \text{and} \quad u_2 = v_2.$$

EXAMPLE 1   **Vectors in the Plane**

a. To represent $\mathbf{u} = (2, 3)$, draw a directed line segment from the origin to the point $(2, 3)$, as shown in Figure 4.1(a).

b. To represent $\mathbf{v} = (-1, 2)$, draw a directed line segment from the origin to the point $(-1, 2)$, as shown in Figure 4.1(b).

a.
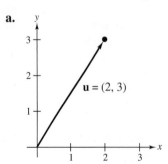

b.
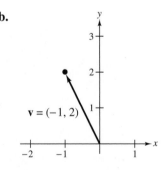

Figure 4.1

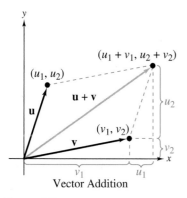

Figure 4.2

VECTOR OPERATIONS

One basic vector operation is **vector addition.** To add two vectors in the plane, add their corresponding components. That is, the **sum** of **u** and **v** is the vector

$$\mathbf{u} + \mathbf{v} = (u_1, u_2) + (v_1, v_2) = (u_1 + v_1, u_2 + v_2).$$

Geometrically, the sum of two vectors in the plane can be represented by the diagonal of a parallelogram having **u** and **v** as its adjacent sides, as shown in Figure 4.2.

In the next example, one of the vectors you will add is the vector $(0, 0)$, the **zero vector.** The zero vector is denoted by **0.**

EXAMPLE 2   **Adding Two Vectors in the Plane**

Find each vector sum $\mathbf{u} + \mathbf{v}$.

a. $\mathbf{u} = (1, 4)$, $\mathbf{v} = (2, -2)$
b. $\mathbf{u} = (3, -2)$, $\mathbf{v} = (-3, 2)$
c. $\mathbf{u} = (2, 1)$, $\mathbf{v} = (0, 0)$

SOLUTION

a. $\mathbf{u} + \mathbf{v} = (1, 4) + (2, -2) = (3, 2)$
b. $\mathbf{u} + \mathbf{v} = (3, -2) + (-3, 2) = (0, 0) = \mathbf{0}$
c. $\mathbf{u} + \mathbf{v} = (2, 1) + (0, 0) = (2, 1)$

Figure 4.3 shows a graphical representation of each sum.

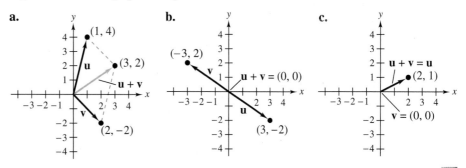

Figure 4.3

Another basic vector operation is **scalar multiplication.** To multiply a vector **v** by a scalar c, multiply each of the components of **v** by c. That is,

$$c\mathbf{v} = c(v_1, v_2) = (cv_1, cv_2).$$

Recall from Chapter 2 that the word *scalar* is used to mean a real number. Historically, this usage arose from the fact that multiplying a vector by a real number changes the "scale" of the vector. For instance, when a vector **v** is multiplied by 2, the resulting vector 2**v** is a vector having the same direction as **v** and twice the length. In general, for a scalar c, the vector $c\mathbf{v}$ will be $|c|$ times as long as **v**. If c is positive, then $c\mathbf{v}$ and **v** have the same direction, and if c is negative, then $c\mathbf{v}$ and **v** have opposite directions. Figure 4.4 shows this.

The product of a vector **v** and the scalar -1 is denoted by

$$-\mathbf{v} = (-1)\mathbf{v}.$$

The vector $-\mathbf{v}$ is the **negative** of **v**. The **difference** of **u** and **v** is

$$\mathbf{u} - \mathbf{v} = \mathbf{u} + (-\mathbf{v}).$$

The vector **v** is **subtracted** from **u** by adding the negative of **v**.

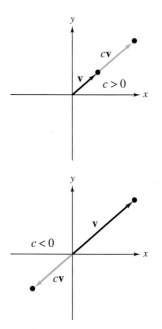

Figure 4.4

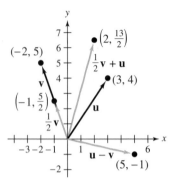

Figure 4.5

EXAMPLE 3   Operations with Vectors in the Plane

Let $\mathbf{v} = (-2, 5)$ and $\mathbf{u} = (3, 4)$. Perform each vector operation.

a. $\frac{1}{2}\mathbf{v}$ **b.** $\mathbf{u} - \mathbf{v}$ **c.** $\frac{1}{2}\mathbf{v} + \mathbf{u}$

SOLUTION

a. $\mathbf{v} = (-2, 5)$, so $\frac{1}{2}\mathbf{v} = \left(\frac{1}{2}(-2), \frac{1}{2}(5)\right) = \left(-1, \frac{5}{2}\right)$.

b. By the definition of vector subtraction, $\mathbf{u} - \mathbf{v} = (3 - (-2), 4 - 5) = (5, -1)$.

c. Using the result of part (a), $\frac{1}{2}\mathbf{v} + \mathbf{u} = \left(-1, \frac{5}{2}\right) + (3, 4) = \left(2, \frac{13}{2}\right)$.

Figure 4.5 shows a graphical representation of these vector operations.

Vector addition and scalar multiplication share many properties with matrix addition and scalar multiplication. The ten properties listed in the next theorem play a fundamental role in linear algebra. In fact, in the next section you will see that it is precisely these ten properties that help define a vector space.

THEOREM 4.1   Properties of Vector Addition and Scalar Multiplication in the Plane

Let $\mathbf{u}$, $\mathbf{v}$, and $\mathbf{w}$ be vectors in the plane, and let c and d be scalars.

1. $\mathbf{u} + \mathbf{v}$ is a vector in the plane.	Closure under addition
2. $\mathbf{u} + \mathbf{v} = \mathbf{v} + \mathbf{u}$	Commutative property of addition
3. $(\mathbf{u} + \mathbf{v}) + \mathbf{w} = \mathbf{u} + (\mathbf{v} + \mathbf{w})$	Associative property of addition
4. $\mathbf{u} + \mathbf{0} = \mathbf{u}$	Additive identity property
5. $\mathbf{u} + (-\mathbf{u}) = \mathbf{0}$	Additive inverse property
6. $c\mathbf{u}$ is a vector in the plane.	Closure under scalar multiplication
7. $c(\mathbf{u} + \mathbf{v}) = c\mathbf{u} + c\mathbf{v}$	Distributive property
8. $(c + d)\mathbf{u} = c\mathbf{u} + d\mathbf{u}$	Distributive property
9. $c(d\mathbf{u}) = (cd)\mathbf{u}$	Associative property of multiplication
10. $1(\mathbf{u}) = \mathbf{u}$	Multiplicative identity property

---

REMARK

Note that the associative property of vector addition allows you to write such expressions as $\mathbf{u} + \mathbf{v} + \mathbf{w}$ without ambiguity, because you obtain the same vector sum regardless of which addition is performed first.

PROOF

The proof of each property is straightforward. For example, to prove the associative property of vector addition, write

$$(\mathbf{u} + \mathbf{v}) + \mathbf{w} = [(u_1, u_2) + (v_1, v_2)] + (w_1, w_2)$$
$$= (u_1 + v_1, u_2 + v_2) + (w_1, w_2)$$
$$= ((u_1 + v_1) + w_1, (u_2 + v_2) + w_2)$$
$$= (u_1 + (v_1 + w_1), u_2 + (v_2 + w_2))$$
$$= (u_1, u_2) + (v_1 + w_1, v_2 + w_2)$$
$$= \mathbf{u} + (\mathbf{v} + \mathbf{w}).$$

Similarly, to prove the right distributive property of scalar multiplication over addition, write

$$(c + d)\mathbf{u} = (c + d)(u_1, u_2)$$
$$= ((c + d)u_1, (c + d)u_2)$$
$$= (cu_1 + du_1, cu_2 + du_2)$$
$$= (cu_1, cu_2) + (du_1, du_2)$$
$$= c\mathbf{u} + d\mathbf{u}.$$

The proofs of the other eight properties are left as an exercise. (See Exercise 63.)

VECTORS IN R^n

The discussion of vectors in the plane can be extended to a discussion of vectors in n-space. An **ordered n-tuple** represents a vector in n-space. For instance, an ordered triple has the form (x_1, x_2, x_3), an ordered quadruple has the form (x_1, x_2, x_3, x_4), and a general ordered n-tuple has the form $(x_1, x_2, x_3, \ldots, x_n)$. The set of all n-tuples is **n-space** and is denoted by R^n.

$R^1 = $ 1-space $= $ set of all real numbers
$R^2 = $ 2-space $= $ set of all ordered pairs of real numbers
$R^3 = $ 3-space $= $ set of all ordered triples of real numbers
$\vdots$
$R^n = $ n-space $= $ set of all ordered n-tuples of real numbers

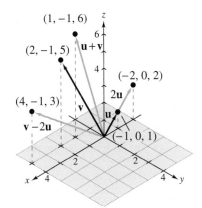

(1, −1, 6)
(2, −1, 5) **u + v**
(−2, 0, 2)
2u
(4, −1, 3) **v** **u**
v − 2u (−1, 0, 1)

Figure 4.6

An n-tuple $(x_1, x_2, x_3, \ldots, x_n)$ can be viewed as a **point** in R^n with the x_i's as its coordinates, or as a **vector** $x = (x_1, x_2, x_3, \ldots, x_n)$ with the x_i's as its components. As with vectors in the plane (or R^2), two vectors in R^n are **equal** if and only if corresponding components are equal. [In the case of $n = 2$ or $n = 3$, the familiar (x, y) or (x, y, z) notation is used occasionally.]

The sum of two vectors in R^n and the scalar multiple of a vector in R^n are the **standard operations in R^n** and are defined below.

Definitions of Vector Addition and Scalar Multiplication in R^n

Let $\mathbf{u} = (u_1, u_2, u_3, \ldots, u_n)$ and $\mathbf{v} = (v_1, v_2, v_3, \ldots, v_n)$ be vectors in R^n and let c be a real number. The sum of $\mathbf{u}$ and $\mathbf{v}$ is the vector

$$\mathbf{u} + \mathbf{v} = (u_1 + v_1, u_2 + v_2, u_3 + v_3, \ldots, u_n + v_n)$$

and the **scalar multiple** of $\mathbf{u}$ by c is the vector

$$c\mathbf{u} = (cu_1, cu_2, cu_3, \ldots, cu_n).$$

As with 2-space, the **negative** of a vector in R^n is

$$-\mathbf{u} = (-u_1, -u_2, -u_3, \ldots, -u_n)$$

and the **difference** of two vectors in R^n is

$$\mathbf{u} - \mathbf{v} = (u_1 - v_1, u_2 - v_2, u_3 - v_3, \ldots, u_n - v_n).$$

The **zero vector** in R^n is denoted by $\mathbf{0} = (0, 0, 0, \ldots, 0)$.

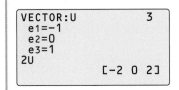
EXAMPLE 4 **Vector Operations in R^3**

See LarsonLinearAlgebra.com for an interactive version of this type of example.

Let $\mathbf{u} = (-1, 0, 1)$ and $\mathbf{v} = (2, -1, 5)$ in R^3. Perform each vector operation.

a. $\mathbf{u} + \mathbf{v}$ **b.** $2\mathbf{u}$ **c.** $\mathbf{v} - 2\mathbf{u}$

SOLUTION

a. To add two vectors, add their corresponding components.

$$\mathbf{u} + \mathbf{v} = (-1, 0, 1) + (2, -1, 5) = (1, -1, 6)$$

b. To multiply a vector by a scalar, multiply each component by the scalar.

$$2\mathbf{u} = 2(-1, 0, 1) = (-2, 0, 2)$$

c. Using the result of part (b), $\mathbf{v} - 2\mathbf{u} = (2, -1, 5) - (-2, 0, 2) = (4, -1, 3)$.

Figure 4.6 shows a graphical representation of these vector operations in R^3.

The properties of vector addition and scalar multiplication for vectors in R^n listed below are similar to those listed in Theorem 4.1 for vectors in R^2. Their proofs, based on the definitions of vector addition and scalar multiplication in R^n, are left as an exercise. (See Exercise 64.)

William Rowan Hamilton
(1805–1865)
Hamilton is considered to be Ireland's most famous mathematician. In 1828, he published an impressive work on optics entitled *A Theory of Systems of Rays.* In it, Hamilton included some of his own methods for working with systems of linear equations. He also introduced the notion of the characteristic equation of a matrix (see Section 7.1). Hamilton's work led to the development of modern vector notation. We still use his **i, j,** and **k** notation for the standard unit vectors in R^3 (see Section 5.1).

THEOREM 4.2   Properties of Vector Addition and Scalar Multiplication in R^n

Let **u, v,** and **w** be vectors in R^n, and let c and d be scalars.

1. $\mathbf{u} + \mathbf{v}$ is a vector in R^n.	Closure under addition
2. $\mathbf{u} + \mathbf{v} = \mathbf{v} + \mathbf{u}$	Commutative property of addition
3. $(\mathbf{u} + \mathbf{v}) + \mathbf{w} = \mathbf{u} + (\mathbf{v} + \mathbf{w})$	Associative property of addition
4. $\mathbf{u} + \mathbf{0} = \mathbf{u}$	Additive identity property
5. $\mathbf{u} + (-\mathbf{u}) = \mathbf{0}$	Additive inverse property
6. $c\mathbf{u}$ is a vector in R^n.	Closure under scalar multiplication
7. $c(\mathbf{u} + \mathbf{v}) = c\mathbf{u} + c\mathbf{v}$	Distributive property
8. $(c + d)\mathbf{u} = c\mathbf{u} + d\mathbf{u}$	Distributive property
9. $c(d\mathbf{u}) = (cd)\mathbf{u}$	Associative property of multiplication
10. $1(\mathbf{u}) = \mathbf{u}$	Multiplicative identity property

---

Using the ten properties from Theorem 4.2, you can perform algebraic manipulations with vectors in R^n in much the same way as you do with real numbers, as demonstrated in the next example.

EXAMPLE 5   **Vector Operations in R^4**

Let $\mathbf{u} = (2, -1, 5, 0)$, $\mathbf{v} = (4, 3, 1, -1)$, and $\mathbf{w} = (-6, 2, 0, 3)$ be vectors in R^4. Find **x** using each equation.

a. $\mathbf{x} = 2\mathbf{u} - (\mathbf{v} + 3\mathbf{w})$

b. $3(\mathbf{x} + \mathbf{w}) = 2\mathbf{u} - \mathbf{v} + \mathbf{x}$

SOLUTION

a. Using the properties listed in Theorem 4.2, you have

$$\mathbf{x} = 2\mathbf{u} - (\mathbf{v} + 3\mathbf{w})$$
$$= 2\mathbf{u} - \mathbf{v} - 3\mathbf{w}$$
$$= 2(2, -1, 5, 0) - (4, 3, 1, -1) - 3(-6, 2, 0, 3)$$
$$= (4, -2, 10, 0) - (4, 3, 1, -1) - (-18, 6, 0, 9)$$
$$= (4 - 4 + 18, -2 - 3 - 6, 10 - 1 - 0, 0 + 1 - 9)$$
$$= (18, -11, 9, -8).$$

b. Begin by solving for **x.**

$$3(\mathbf{x} + \mathbf{w}) = 2\mathbf{u} - \mathbf{v} + \mathbf{x}$$
$$3\mathbf{x} + 3\mathbf{w} = 2\mathbf{u} - \mathbf{v} + \mathbf{x}$$
$$3\mathbf{x} - \mathbf{x} = 2\mathbf{u} - \mathbf{v} - 3\mathbf{w}$$
$$2\mathbf{x} = 2\mathbf{u} - \mathbf{v} - 3\mathbf{w}$$
$$\mathbf{x} = \tfrac{1}{2}(2\mathbf{u} - \mathbf{v} - 3\mathbf{w})$$

Using the result of part (a),

$$\mathbf{x} = \tfrac{1}{2}(18, -11, 9, -8)$$
$$= \left(9, -\tfrac{11}{2}, \tfrac{9}{2}, -4\right).$$

The zero vector $\mathbf{0}$ in R^n is the **additive identity** in R^n. Similarly, the vector $-\mathbf{v}$ is the **additive inverse** of $\mathbf{v}$. The next theorem summarizes several important properties of the additive identity and additive inverse in R^n.

THEOREM 4.3 Properties of Additive Identity and Additive Inverse

Let $\mathbf{v}$ be a vector in R^n, and let c be a scalar. Then the properties below are true.

1. The additive identity is unique. That is, if $\mathbf{v} + \mathbf{u} = \mathbf{v}$, then $\mathbf{u} = \mathbf{0}$.
2. The additive inverse of $\mathbf{v}$ is unique. That is, if $\mathbf{v} + \mathbf{u} = \mathbf{0}$, then $\mathbf{u} = -\mathbf{v}$.
3. $0\mathbf{v} = \mathbf{0}$
4. $c\mathbf{0} = \mathbf{0}$
5. If $c\mathbf{v} = \mathbf{0}$, then $c = 0$ or $\mathbf{v} = \mathbf{0}$.
6. $-(-\mathbf{v}) = \mathbf{v}$

PROOF

To prove the first property, assume $\mathbf{v} + \mathbf{u} = \mathbf{v}$. Then Theorem 4.2 justifies the steps below.

$\mathbf{v} + \mathbf{u} = \mathbf{v}$	Given
$(\mathbf{v} + \mathbf{u}) + (-\mathbf{v}) = \mathbf{v} + (-\mathbf{v})$	Add $-\mathbf{v}$ to both sides.
$(\mathbf{v} + \mathbf{u}) + (-\mathbf{v}) = \mathbf{0}$	Additive inverse
$(\mathbf{u} + \mathbf{v}) + (-\mathbf{v}) = \mathbf{0}$	Commutative property
$\mathbf{u} + [\mathbf{v} + (-\mathbf{v})] = \mathbf{0}$	Associative property
$\mathbf{u} + \mathbf{0} = \mathbf{0}$	Additive inverse
$\mathbf{u} = \mathbf{0}$	Additive identity

To prove the second property, assume $\mathbf{v} + \mathbf{u} = \mathbf{0}$, and again use Theorem 4.2 to justify the steps below.

$\mathbf{v} + \mathbf{u} = \mathbf{0}$	Given
$(-\mathbf{v}) + (\mathbf{v} + \mathbf{u}) = (-\mathbf{v}) + \mathbf{0}$	Add $-\mathbf{v}$ to both sides.
$(-\mathbf{v}) + (\mathbf{v} + \mathbf{u}) = -\mathbf{v}$	Additive identity
$[(-\mathbf{v}) + \mathbf{v}] + \mathbf{u} = -\mathbf{v}$	Associative property
$\mathbf{0} + \mathbf{u} = -\mathbf{v}$	Additive inverse
$\mathbf{u} + \mathbf{0} = -\mathbf{v}$	Commutative property
$\mathbf{u} = -\mathbf{v}$	Additive identity

As you gain experience in reading and writing proofs involving vector algebra, you will not need to list as many steps as shown above. For now, however, it is a good idea to list as many steps as possible. The proofs of the other four properties are left as exercises. (See Exercises 65–68.)

LINEAR ALGEBRA APPLIED

Vectors have a wide variety of applications in engineering and the physical sciences. For example, to determine the amount of force required to pull an object up a ramp that has an angle of elevation θ, use the figure at the right.

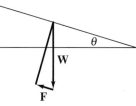

In the figure, the vector labeled $\mathbf{W}$ represents the weight of the object, and the vector labeled $\mathbf{F}$ represents the required force. Using similar triangles and some trigonometry, the required force is $\mathbf{F} = \mathbf{W} \sin \theta$. (Verify this.)

LINEAR COMBINATIONS OF VECTORS

An important type of problem in linear algebra involves writing one vector $\mathbf{x}$ as the sum of scalar multiples of other vectors $\mathbf{v}_1, \mathbf{v}_2, \ldots,$ and $\mathbf{v}_n$. That is, for scalars $c_1, c_2, \ldots, c_n$,

$$\mathbf{x} = c_1\mathbf{v}_1 + c_2\mathbf{v}_2 + \cdots + c_n\mathbf{v}_n.$$

The vector $\mathbf{x}$ is called a **linear combination** of the vectors $\mathbf{v}_1, \mathbf{v}_2, \ldots,$ and $\mathbf{v}_n$.

See LarsonLinearAlgebra.com for an interactive version of this type of exercise.

DISCOVERY

1. Is the vector (1, 1) a linear combination of the vectors (1, 2) and (−2, −4)? Graph these vectors and explain your answer geometrically.

2. Similarly, determine whether the vector (1, 1) is a linear combination of the vectors (1, 2) and (2, 1).

3. What is the geometric significance of questions 1 and 2?

4. Is every vector in R^2 a linear combination of the vectors (1, −2) and (−2, 1)? Give a geometric explanation for your answer.

EXAMPLE 6   Writing a Vector as a Linear Combination of Other Vectors

Let $\mathbf{x} = (-1, -2, -2)$, $\mathbf{u} = (0, 1, 4)$, $\mathbf{v} = (-1, 1, 2)$, and $\mathbf{w} = (3, 1, 2)$ in R^3. Find scalars a, b, and c such that

$$\mathbf{x} = a\mathbf{u} + b\mathbf{v} + c\mathbf{w}.$$

SOLUTION

Write

$$\overbrace{(-1, -2, -2)}^{\mathbf{x}} = a\overbrace{(0, 1, 4)}^{\mathbf{u}} + b\overbrace{(-1, 1, 2)}^{\mathbf{v}} + c\overbrace{(3, 1, 2)}^{\mathbf{w}}$$
$$= (-b + 3c, a + b + c, 4a + 2b + 2c)$$

and equate corresponding components so that they form the system of three linear equations in a, b, and c shown below.

$$\begin{aligned} -b + 3c &= -1 &&\text{Equation from first component} \\ a + b + c &= -2 &&\text{Equation from second component} \\ 4a + 2b + 2c &= -2 &&\text{Equation from third component} \end{aligned}$$

Solve for a, b, and c to get $a = 1$, $b = -2$, and $c = -1$. As a linear combination of $\mathbf{u}$, $\mathbf{v}$, and $\mathbf{w}$,

$$\mathbf{x} = \mathbf{u} - 2\mathbf{v} - \mathbf{w}.$$

Use vector addition and scalar multiplication to check this result.

You will often find it useful to represent a vector $\mathbf{u} = (u_1, u_2, \ldots, u_n)$ in R^n as either a $1 \times n$ row matrix (row vector) or an $n \times 1$ column matrix (column vector). This approach is valid because the matrix operations of addition and scalar multiplication give the same results as the corresponding vector operations. That is, the matrix sums

$$\mathbf{u} + \mathbf{v} = [u_1 \ u_2 \ \ldots \ u_n] + [v_1 \ v_2 \ \ldots \ v_n]$$
$$= [u_1 + v_1 \ u_2 + v_2 \ \ldots \ u_n + v_n]$$

and

$$\mathbf{u} + \mathbf{v} = \begin{bmatrix} u_1 \\ u_2 \\ \vdots \\ u_n \end{bmatrix} + \begin{bmatrix} v_1 \\ v_2 \\ \vdots \\ v_n \end{bmatrix} = \begin{bmatrix} u_1 + v_1 \\ u_2 + v_2 \\ \vdots \\ u_n + v_n \end{bmatrix}$$

yield the same results as the vector operation of addition,

$$\mathbf{u} + \mathbf{v} = (u_1, u_2, \ldots, u_n) + (v_1, v_2, \ldots, v_n)$$
$$= (u_1 + v_1, u_2 + v_2, \ldots, u_n + v_n).$$

The same argument applies to scalar multiplication. The only difference in each set of notations is how the components (entries) are displayed.

4.1 Exercises
See CalcChat.com for worked-out solutions to odd-numbered exercises.

Finding the Component Form of a Vector In Exercises 1 and 2, find the component form of the vector.

1.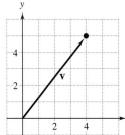
2.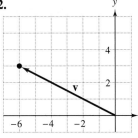

Representing a Vector In Exercises 3–6, use a directed line segment to represent the vector.

3. $u = (2, -4)$
4. $v = (-2, 3)$
5. $u = (-3, -4)$
6. $v = (-2, -5)$

Finding the Sum of Two Vectors In Exercises 7–10, find the sum of the vectors and illustrate the sum geometrically.

7. $u = (1, 3)$, $v = (2, -2)$
8. $u = (-1, 4)$, $v = (4, -3)$
9. $u = (2, -3)$, $v = (-3, -1)$
10. $u = (4, -2)$, $v = (-2, -3)$

Vector Operations In Exercises 11–16, find the vector v and illustrate the specified vector operations geometrically, where $u = (-2, 3)$ and $w = (-3, -2)$.

11. $v = \frac{3}{2}u$
12. $v = u + w$
13. $v = u + 2w$
14. $v = -u + w$
15. $v = \frac{1}{2}(3u + w)$
16. $v = u - 2w$

17. For the vector $v = (2, 1)$, sketch (a) $2v$, (b) $-3v$, and (c) $\frac{1}{2}v$.
18. For the vector $v = (3, -2)$, sketch (a) $4v$, (b) $-\frac{1}{2}v$, and (c) $0v$.

Vector Operations In Exercises 19–24, let $u = (1, 2, 3)$, $v = (2, 2, -1)$, and $w = (4, 0, -4)$.

19. Find $u - v$ and $v - u$.
20. Find $u - v + 2w$.
21. Find $2u + 4v - w$.
22. Find $5u - 3v - \frac{1}{2}w$.
23. Find z, where $3u - 4z = w$.
24. Find z, where $2u + v - w + 3z = 0$.
25. For the vector $v = (1, 2, 2)$, sketch (a) $2v$, (b) $-v$, and (c) $\frac{1}{2}v$.
26. For the vector $v = (2, 0, 1)$, sketch (a) $-v$, (b) $2v$, and (c) $\frac{1}{2}v$.
27. Determine whether each vector is a scalar multiple of $z = (3, 2, -5)$.
   (a) $v = \left(\frac{9}{2}, 3, -\frac{15}{2}\right)$
   (b) $w = (9, -6, -15)$

28. Determine whether each vector is a scalar multiple of $z = \left(\frac{1}{2}, -\frac{2}{3}, \frac{3}{4}\right)$.
   (a) $u = (6, -4, 9)$
   (b) $v = \left(-1, \frac{4}{3}, -\frac{3}{2}\right)$

Vector Operations In Exercises 29–32, find (a) $u - v$, (b) $2(u + 3v)$, and (c) $2v - u$.

29. $u = (4, 0, -3, 5)$, $v = (0, 2, 5, 4)$
30. $u = (0, 4, 3, 4, 4)$, $v = (6, 8, -3, 3, -5)$
31. $u = (-7, 0, 0, 0, 9)$, $v = (2, -3, -2, 3, 3)$
32. $u = (6, -5, 4, 3)$, $v = \left(-2, \frac{5}{3}, -\frac{4}{3}, -1\right)$

Vector Operations In Exercises 33 and 34, use a graphing utility to perform each operation where $u = (1, 2, -3, 1)$, $v = (0, 2, -1, -2)$, and $w = (2, -2, 1, 3)$.

33. (a) $u + 2v$
    (b) $w - 3u$
    (c) $4v + \frac{1}{2}u - w$
34. (a) $v + 3w$
    (b) $2w - \frac{1}{2}u$
    (c) $\frac{1}{2}(4v - 3u + w)$

Solving a Vector Equation In Exercises 35–38, solve for w, where $u = (1, -1, 0, 1)$ and $v = (0, 2, 3, -1)$.

35. $3w = u - 2v$
36. $w + u = -v$
37. $\frac{1}{2}w = 2u + 3v$
38. $w + 3v = -2u$

Solving a Vector Equation In Exercises 39 and 40, find w such that $2u + v - 3w = 0$.

39. $u = (0, 2, 7, 5)$, $v = (-3, 1, 4, -8)$
40. $u = (-6, 0, 2, 0)$, $v = (5, -3, 0, 1)$

Writing a Linear Combination In Exercises 41–46, write v as a linear combination of u and w, if possible, where $u = (1, 2)$ and $w = (1, -1)$.

41. $v = (2, 1)$
42. $v = (0, 3)$
43. $v = (3, 3)$
44. $v = (1, -1)$
45. $v = (-1, -2)$
46. $v = (1, -4)$

Writing a Linear Combination In Exercises 47–50, write v as a linear combination of u_1, u_2, and u_3, if possible.

47. $v = (10, 1, 4)$, $u_1 = (2, 3, 5)$, $u_2 = (1, 2, 4)$, $u_3 = (-2, 2, 3)$
48. $v = (-1, 7, 2)$, $u_1 = (1, 3, 5)$, $u_2 = (2, -1, 3)$, $u_3 = (-3, 2, -4)$
49. $v = (0, 5, 3, 0)$, $u_1 = (1, 1, 2, 2)$, $u_2 = (2, 3, 5, 6)$, $u_3 = (-3, 1, -4, 2)$
50. $v = (7, 2, 5, -3)$, $u_1 = (2, 1, 1, 2)$, $u_2 = (-3, 3, 4, -5)$, $u_3 = (-6, 3, 1, 2)$

Writing a Linear Combination In Exercises 51 and 52, write the third column of the matrix as a linear combination of the first two columns, if possible.

51. $\begin{bmatrix} 1 & 2 & 3 \\ 7 & 8 & 9 \\ 4 & 5 & 6 \end{bmatrix}$ 52. $\begin{bmatrix} 1 & 2 & 3 \\ 7 & 8 & 9 \\ 4 & 5 & 7 \end{bmatrix}$

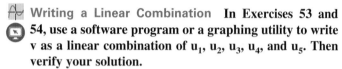 **Writing a Linear Combination** In Exercises 53 and 54, use a software program or a graphing utility to write v as a linear combination of u_1, u_2, u_3, u_4, and u_5. Then verify your solution.

53. $\mathbf{v} = (5, 3, -11, 11, 9)$
$\mathbf{u}_1 = (1, 2, -3, 4, -1)$
$\mathbf{u}_2 = (1, 2, 0, 2, 1)$
$\mathbf{u}_3 = (0, 1, 1, 1, -4)$
$\mathbf{u}_4 = (2, 1, -1, 2, 1)$
$\mathbf{u}_5 = (0, 2, 2, -1, -1)$

54. $\mathbf{v} = (5, 8, 7, -2, 4)$
$\mathbf{u}_1 = (1, 1, -1, 2, 1)$
$\mathbf{u}_2 = (2, 1, 2, -1, 1)$
$\mathbf{u}_3 = (1, 2, 0, 1, 2)$
$\mathbf{u}_4 = (0, 2, 0, 1, -4)$
$\mathbf{u}_5 = (1, 1, 2, -1, 2)$

Writing a Linear Combination In Exercises 55 and 56, the zero vector $\mathbf{0} = (0, 0, 0)$ can be written as a linear combination of the vectors v_1, v_2, and v_3 because $\mathbf{0} = 0\mathbf{v}_1 + 0\mathbf{v}_2 + 0\mathbf{v}_3$. This is the *trivial* solution. Find a *nontrivial* way of writing 0 as a linear combination of the three vectors, if possible.

55. $\mathbf{v}_1 = (1, 0, 1)$, $\mathbf{v}_2 = (-1, 1, 2)$, $\mathbf{v}_3 = (0, 1, 4)$
56. $\mathbf{v}_1 = (1, 0, 1)$, $\mathbf{v}_2 = (-1, 1, 2)$, $\mathbf{v}_3 = (0, 1, 3)$

True or False? In Exercises 57 and 58, determine whether each statement is true or false. If a statement is true, give a reason or cite an appropriate statement from the text. If a statement is false, provide an example that shows the statement is not true in all cases or cite an appropriate statement from the text.

57. (a) Two vectors in R^n are equal if and only if their corresponding components are equal.
   (b) The vector $-\mathbf{v}$ is the additive identity of $\mathbf{v}$.

58. (a) To subtract two vectors in R^n, subtract their corresponding components.
   (b) The zero vector $\mathbf{0}$ in R^n is the additive inverse of a vector.

59. **Writing** Let $A\mathbf{x} = \mathbf{b}$ be a system of m linear equations in n variables. Designate the columns of A as $\mathbf{a}_1$, $\mathbf{a}_2, \ldots, \mathbf{a}_n$. When $\mathbf{b}$ is a linear combination of these n column vectors, explain why this implies that the linear system is consistent. What can you conclude about the linear system when $\mathbf{b}$ is not a linear combination of the columns of A?

60. **Writing** How could you describe vector subtraction geometrically? What is the relationship between vector subtraction and the basic vector operations of addition and scalar multiplication?

61. Illustrate properties 1–10 of Theorem 4.2 for $\mathbf{u} = (2, -1, 3, 6)$, $\mathbf{v} = (1, 4, 0, 1)$, $\mathbf{w} = (3, 0, 2, 0)$, $c = 5$, and $d = -2$.

62. **CAPSTONE** Consider the vectors $\mathbf{u} = (3, -4)$ and $\mathbf{v} = (9, 1)$.
   (a) Use directed line segments to represent each vector graphically.
   (b) Find $\mathbf{u} + \mathbf{v}$.
   (c) Find $2\mathbf{v} - \mathbf{u}$.
   (d) Write $\mathbf{w} = (39, 0)$ as a linear combination of $\mathbf{u}$ and $\mathbf{v}$.

63. **Proof** Complete the proof of Theorem 4.1.

64. **Proof** Prove each property of vector addition and scalar multiplication from Theorem 4.2.

Proof In Exercises 65–68, complete the proofs of the remaining properties of Theorem 4.3 by supplying the justification for each step. Use the properties of vector addition and scalar multiplication from Theorem 4.2.

65. Property 3: $0\mathbf{v} = \mathbf{0}$

$$0\mathbf{v} = (0 + 0)\mathbf{v} \qquad \text{a. _____}$$
$$0\mathbf{v} = 0\mathbf{v} + 0\mathbf{v} \qquad \text{b. _____}$$
$$0\mathbf{v} + (-0\mathbf{v}) = (0\mathbf{v} + 0\mathbf{v}) + (-0\mathbf{v}) \qquad \text{c. _____}$$
$$\mathbf{0} = 0\mathbf{v} + (0\mathbf{v} + (-0\mathbf{v})) \qquad \text{d. _____}$$
$$\mathbf{0} = 0\mathbf{v} + \mathbf{0} \qquad \text{e. _____}$$
$$\mathbf{0} = 0\mathbf{v} \qquad \text{f. _____}$$

66. Property 4: $c\mathbf{0} = \mathbf{0}$

$$c\mathbf{0} = c(\mathbf{0} + \mathbf{0}) \qquad \text{a. _____}$$
$$c\mathbf{0} = c\mathbf{0} + c\mathbf{0} \qquad \text{b. _____}$$
$$c\mathbf{0} + (-c\mathbf{0}) = (c\mathbf{0} + c\mathbf{0}) + (-c\mathbf{0}) \qquad \text{c. _____}$$
$$\mathbf{0} = c\mathbf{0} + (c\mathbf{0} + (-c\mathbf{0})) \qquad \text{d. _____}$$
$$\mathbf{0} = c\mathbf{0} + \mathbf{0} \qquad \text{e. _____}$$
$$\mathbf{0} = c\mathbf{0} \qquad \text{f. _____}$$

67. Property 5: If $c\mathbf{v} = \mathbf{0}$, then $c = 0$ or $\mathbf{v} = \mathbf{0}$. If $c = 0$, then you are done. If $c \neq 0$, then c^{-1} exists, and you have

$$c^{-1}(c\mathbf{v}) = c^{-1}\mathbf{0} \qquad \text{a. _____}$$
$$(c^{-1}c)\mathbf{v} = \mathbf{0} \qquad \text{b. _____}$$
$$1\mathbf{v} = \mathbf{0} \qquad \text{c. _____}$$
$$\mathbf{v} = \mathbf{0}. \qquad \text{d. _____}$$

68. Property 6: $-(-\mathbf{v}) = \mathbf{v}$

$$-(-\mathbf{v}) + (-\mathbf{v}) = \mathbf{0} \text{ and } \mathbf{v} + (-\mathbf{v}) = \mathbf{0} \qquad \text{a. _____}$$
$$-(-\mathbf{v}) + (-\mathbf{v}) = \mathbf{v} + (-\mathbf{v}) \qquad \text{b. _____}$$
$$-(-\mathbf{v}) + (-\mathbf{v}) + \mathbf{v} = \mathbf{v} + (-\mathbf{v}) + \mathbf{v} \qquad \text{c. _____}$$
$$-(-\mathbf{v}) + ((-\mathbf{v}) + \mathbf{v}) = \mathbf{v} + ((-\mathbf{v}) + \mathbf{v}) \qquad \text{d. _____}$$
$$-(-\mathbf{v}) + \mathbf{0} = \mathbf{v} + \mathbf{0} \qquad \text{e. _____}$$
$$-(-\mathbf{v}) = \mathbf{v} \qquad \text{f. _____}$$

4.2   Vector Spaces

■ Define a vector space and recognize some important vector spaces.

■ Show that a given set is not a vector space.

DEFINITION OF A VECTOR SPACE

Theorem 4.2 lists ten properties of vector addition and scalar multiplication in R^n. Suitable definitions of addition and scalar multiplication reveal that many other mathematical quantities (such as matrices, polynomials, and functions) also share these ten properties. *Any* set that satisfies these properties (or **axioms**) is called a **vector space,** and the objects in the set are **vectors.**

It is important to realize that the definition of a vector space below is precisely that—a *definition*. You do not need to prove anything because you are simply listing the axioms required of vector spaces. This type of definition is an **abstraction** because you are abstracting a collection of properties from a particular setting, R^n, to form the axioms for a more general setting.

Definition of a Vector Space

Let V be a set on which two operations (**vector addition** and **scalar multiplication**) are defined. If the listed axioms are satisfied for every **u**, **v**, and **w** in V and every scalar (real number) c and d, then V is a **vector space.**

Addition:

1. **u** + **v** is in V.	Closure under addition
2. **u** + **v** = **v** + **u**	Commutative property
3. **u** + (**v** + **w**) = (**u** + **v**) + **w**	Associative property
4. V has a **zero vector 0** such that for every **u** in V, **u** + **0** = **u**.	Additive identity
5. For every **u** in V, there is a vector in V denoted by −**u** such that **u** + (−**u**) = **0**.	Additive inverse

Scalar Multiplication:

6. c**u** is in V.	Closure under scalar multiplication
7. c(**u** + **v**) = c**u** + c**v**	Distributive property
8. $(c + d)$**u** = c**u** + d**u**	Distributive property
9. $c(d$**u**$) = (cd)$**u**	Associative property
10. $1($**u**$) =$ **u**	Scalar identity

It is important to realize that a vector space consists of four entities: a set of vectors, a set of scalars, and two operations. When you refer to a vector space V, be sure that all four entities are clearly stated or understood. Unless stated otherwise, assume that the set of scalars is the set of real numbers.

The first two examples of vector spaces should not be surprising. They are, in fact, the models used to form the ten vector space axioms.

EXAMPLE 1   *R^2 with the Standard Operations Is a Vector Space*

The set of all ordered pairs of real numbers R^2 with the standard operations is a vector space. To verify this, look back at Theorem 4.1. Vectors in this space have the form

$$\mathbf{v} = (v_1, v_2).$$

REMARK

From Example 2 you can conclude that R^1, the set of real numbers (with the usual operations of addition and multiplication), is a vector space.

EXAMPLE 2   **R^n with the Standard Operations Is a Vector Space**

The set of all ordered n-tuples of real numbers R^n with the standard operations is a vector space. Theorem 4.2 verifies this. Vectors in this space are of the form

$$\mathbf{v} = (v_1, v_2, v_3, \ldots, v_n).$$

The next three examples describe vector spaces in which the basic set V does not consist of ordered n-tuples. Each example describes the set V and defines the two vector operations. To show that the set is a vector space, you must verify all ten axioms.

EXAMPLE 3   **The Vector Space of All 2×3 Matrices**

Show that the set of all 2×3 matrices with the operations of matrix addition and scalar multiplication is a vector space.

SOLUTION

If A and B are 2×3 matrices and c is a scalar, then $A + B$ and cA are also 2×3 matrices. The set is, therefore, closed under matrix addition and scalar multiplication. Moreover, the other eight vector space axioms follow directly from Theorems 2.1 and 2.2 (see Section 2.2). So, the set is a vector space. Vectors in this space have the form

$$\mathbf{a} = A = \begin{bmatrix} a_{11} & a_{12} & a_{13} \\ a_{21} & a_{22} & a_{23} \end{bmatrix}.$$

REMARK

In the same way you are able to show that the set of all 2×3 matrices is a vector space, you can show that the set of all $m \times n$ matrices, denoted by $M_{m,n}$, is a vector space.

EXAMPLE 4   **The Vector Space of All Polynomials of Degree 2 or Less**

Let P_2 be the set of all polynomials of the form $p(x) = a_0 + a_1 x + a_2 x^2$, where a_0, a_1, and a_2 are real numbers. The *sum* of two polynomials $p(x) = a_0 + a_1 x + a_2 x^2$ and $q(x) = b_0 + b_1 x + b_2 x^2$ is defined in the usual way,

$$p(x) + q(x) = (a_0 + b_0) + (a_1 + b_1)x + (a_2 + b_2)x^2$$

and the *scalar multiple* of $p(x)$ by the scalar c is defined by

$$cp(x) = ca_0 + ca_1 x + ca_2 x^2.$$

Show that P_2 is a vector space.

SOLUTION

Verification of each of the ten vector space axioms is a straightforward application of the properties of real numbers. For example, the set of real numbers is closed under addition, so it follows that $a_0 + b_0$, $a_1 + b_1$, and $a_2 + b_2$ are real numbers, and

$$p(x) + q(x) = (a_0 + b_0) + (a_1 + b_1)x + (a_2 + b_2)x^2$$

is in the set P_2 because it is a polynomial of degree 2 or less. So, P_2 is closed under addition. To verify the commutative property of addition, write

$$\begin{aligned} p(x) + q(x) &= (a_0 + a_1 x + a_2 x^2) + (b_0 + b_1 x + b_2 x^2) \\ &= (a_0 + b_0) + (a_1 + b_1)x + (a_2 + b_2)x^2 \\ &= (b_0 + a_0) + (b_1 + a_1)x + (b_2 + a_2)x^2 \\ &= (b_0 + b_1 x + b_2 x^2) + (a_0 + a_1 x + a_2 x^2) \\ &= q(x) + p(x). \end{aligned}$$

REMARK

Even though the zero polynomial $\mathbf{0}(x) = 0$ has no degree, P_2 is often called the set of all polynomials of degree 2 *or less*.

Can you see where the commutative property of addition of real numbers was used? The zero vector in this space is the zero polynomial $\mathbf{0}(x) = 0 + 0x + 0x^2$. Verify the other vector space axioms to show that P_2 is a vector space.

P_n is defined as the set of all polynomials of degree n or less (together with the zero polynomial). The procedure used to verify that P_2 is a vector space can be extended to show that P_n, with the usual operations of polynomial addition and scalar multiplication, is a vector space.

EXAMPLE 5   The Vector Space of Continuous Functions (Calculus)

See LarsonLinearAlgebra.com for an interactive version of this type of example.

Let $C(-\infty, \infty)$ be the set of all real-valued continuous functions defined on the entire real line. This set consists of all polynomial functions and all other continuous functions on the entire real line. For example, $f(x) = \sin x$ and $g(x) = e^x$ are members of this set.

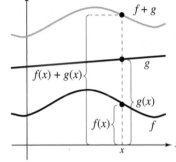

Addition is defined by

$$(f + g)(x) = f(x) + g(x)$$

as shown at the right. Scalar multiplication is defined by

$$(cf)(x) = c[f(x)].$$

Show that $C(-\infty, \infty)$ is a vector space.

SOLUTION

To verify that the set $C(-\infty, \infty)$ is closed under addition and scalar multiplication, use a result from calculus—the sum of two continuous functions is continuous and the product of a scalar and a continuous function is continuous. To verify that the set $C(-\infty, \infty)$ has an additive identity, consider the function f_0 that has a value of zero for all x, that is,

$$f_0(x) = 0, \quad x \text{ is any real number.}$$

This function is continuous on the entire real line (its graph is simply the line $y = 0$), which means that it is in the set $C(-\infty, \infty)$. Moreover, if f is any other function that is continuous on the entire real line, then

$$(f + f_0)(x) = f(x) + f_0(x) = f(x) + 0 = f(x).$$

This shows that f_0 is the additive identity in $C(-\infty, \infty)$. The verification of the other vector space axioms is left to you.

The summary below lists some important vector spaces frequently referenced in the remainder of this text. The operations are the standard operations in each case.

Summary of Important Vector Spaces

$$R = \text{set of all real numbers}$$
$$R^2 = \text{set of all ordered pairs}$$
$$R^3 = \text{set of all ordered triples}$$
$$R^n = \text{set of all } n\text{-tuples}$$
$$C(-\infty, \infty) = \text{set of all continuous functions defined on the real number line}$$
$$C[a, b] = \text{set of all continuous functions defined on a closed interval } [a, b],$$
$$\text{where } a \neq b$$
$$P = \text{set of all polynomials}$$
$$P_n = \text{set of all polynomials of degree} \leq n \text{ (together with the zero polynomial)}$$
$$M_{m,n} = \text{set of all } m \times n \text{ matrices}$$
$$M_{n,n} = \text{set of all } n \times n \text{ square matrices}$$

You have seen the versatility of the concept of a vector space. For instance, a vector can be a real number, an *n*-tuple, a matrix, a polynomial, a continuous function, and so on. But what is the purpose of this abstraction, and why bother to define it? There are several reasons, but the most important reason applies to efficiency. Once a theorem has been proved for an abstract vector space, you need not give separate proofs for *n*-tuples, matrices, polynomials, or other forms. Simply point out that the theorem is true for any vector space, regardless of the form the vectors have. Theorem 4.4 illustrates this process.

THEOREM 4.4   Properties of Scalar Multiplication

Let $\mathbf{v}$ be any element of a vector space V, and let c be any scalar. Then the properties below are true.

1. $0\mathbf{v} = \mathbf{0}$ **2.** $c\mathbf{0} = \mathbf{0}$

3. If $c\mathbf{v} = \mathbf{0}$, then $c = 0$ or $\mathbf{v} = \mathbf{0}$.   **4.** $(-1)\mathbf{v} = -\mathbf{v}$

PROOF

To prove these properties, use the appropriate vector space axioms. For example, to prove the second property, note from axiom 4 that $\mathbf{0} = \mathbf{0} + \mathbf{0}$. This allows you to write the steps below.

$$c\mathbf{0} = c(\mathbf{0} + \mathbf{0}) \qquad \text{Additive identity}$$
$$c\mathbf{0} = c\mathbf{0} + c\mathbf{0} \qquad \text{Left distributive property}$$
$$c\mathbf{0} + (-c\mathbf{0}) = (c\mathbf{0} + c\mathbf{0}) + (-c\mathbf{0}) \qquad \text{Add } -c\mathbf{0} \text{ to both sides.}$$
$$c\mathbf{0} + (-c\mathbf{0}) = c\mathbf{0} + [c\mathbf{0} + (-c\mathbf{0})] \qquad \text{Associative property}$$
$$\mathbf{0} = c\mathbf{0} + \mathbf{0} \qquad \text{Additive inverse}$$
$$\mathbf{0} = c\mathbf{0} \qquad \text{Additive identity}$$

To prove the third property, let $c\mathbf{v} = \mathbf{0}$. To show that this implies either $c = 0$ or $\mathbf{v} = \mathbf{0}$, assume that $c \neq 0$. (When $c = 0$, you have nothing more to prove.) Now, $c \neq 0$, so you can use the reciprocal $1/c$ to show that $\mathbf{v} = \mathbf{0}$, as shown below.

$$\mathbf{v} = 1\mathbf{v} = \left(\frac{1}{c}\right)(c)\mathbf{v} = \frac{1}{c}(c\mathbf{v}) = \frac{1}{c}(\mathbf{0}) = \mathbf{0}$$

Note that the last step uses Property 2 (the one you just proved). The proofs of the first and fourth properties are left as exercises. (See Exercises 51 and 52.)

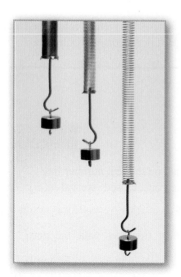

LINEAR ALGEBRA APPLIED

In a mass-spring system, motion is assumed to occur in only the vertical direction. That is, the system has one *degree of freedom*. When the mass is pulled downward and then released, the system will oscillate. If there are no forces present to slow or stop the oscillation, then the system is undamped and will oscillate indefinitely. Applying Newton's Second Law of Motion to the mass yields the second order differential equation

$$x'' + \omega^2 x = 0$$

where x is the displacement at time t, and ω is a fixed constant called the *natural frequency* of the system. The general solution of this differential equation is

$$x(t) = a_1 \sin \omega t + a_2 \cos \omega t$$

where a_1 and a_2 are arbitrary constants. (Verify this.) In Exercise 45, you are asked to show that the set of all functions $x(t)$ is a vector space.

SETS THAT ARE NOT VECTOR SPACES

The remaining examples in this section describe some sets (with operations) that *do not* form vector spaces. To show that a set is not a vector space, you need only find one axiom that is not satisfied.

EXAMPLE 6   The Set of Integers Is Not a Vector Space

The set of all integers (with the standard operations) does not form a vector space because it is not closed under scalar multiplication. For example,

$$\tfrac{1}{2}(1) = \tfrac{1}{2}.$$

Scalar Integer Noninteger

In Example 4, it was shown that the set of all polynomials of degree 2 or less forms a vector space. You will now see that the set of all polynomials whose degree is exactly 2 does not form a vector space.

EXAMPLE 7   The Set of Second-Degree Polynomials Is Not a Vector Space

The set of all second-degree polynomials is not a vector space because it is not closed under addition. To see this, consider the second-degree polynomials $p(x) = x^2$ and $q(x) = 1 + x - x^2$, whose sum is the first-degree polynomial $p(x) + q(x) = 1 + x$.

The sets in Examples 6 and 7 are not vector spaces because they fail one or both closure axioms. In the next example, you will look at a set that passes both tests for closure but still fails to be a vector space.

EXAMPLE 8   A Set That Is Not a Vector Space

Let $V = R^2$, the set of all ordered pairs of real numbers, with the standard operation of addition and the *nonstandard* definition of scalar multiplication listed below.

$$c(x_1, x_2) = (cx_1, 0)$$

Show that V is not a vector space.

SOLUTION

In this example, the operation of scalar multiplication is nonstandard. For instance, the product of the scalar 2 and the ordered pair $(3, 4)$ does not equal $(6, 8)$. Instead, the second component of the product is 0,

$$2(3, 4) = (2 \cdot 3, 0) = (6, 0).$$

This example is interesting because it satisfies the first nine axioms of the definition of a vector space (show this). In attempting to verify the tenth axiom, the nonstandard definition of scalar multiplication gives you

$$1(1, 1) = (1, 0) \neq (1, 1).$$

The tenth axiom is not satisfied and the set (together with the two operations) is not a vector space.

Do not be confused by the notation used for scalar multiplication in Example 8. In writing $c(x_1, x_2) = (cx_1, 0)$, the scalar multiple of (x_1, x_2) by c is *defined* to be $(cx_1, 0)$ in this example.

4.2 Exercises

See CalcChat.com for worked-out solutions to odd-numbered exercises.

Describing the Additive Identity In Exercises 1–6, describe the zero vector (the additive identity) of the vector space.

1. R^4

2. $C[-1, 0]$

3. $M_{4,3}$

4. $M_{5,1}$

5. P_3

6. $M_{2,2}$

Describing the Additive Inverse In Exercises 7–12, describe the additive inverse of a vector in the vector space.

7. R^3

8. $C(-\infty, \infty)$

9. $M_{2,3}$

10. $M_{1,4}$

11. P_4

12. $M_{5,5}$

Testing for a Vector Space In Exercises 13–36, determine whether the set, together with the standard operations, is a vector space. If it is not, identify at least one of the ten vector space axioms that fails.

13. $M_{4,6}$

14. $M_{1,1}$

15. The set of all third-degree polynomials

16. The set of all fifth-degree polynomials

17. The set of all first-degree polynomial functions ax, $a \neq 0$, whose graphs pass through the origin

18. The set of all first-degree polynomial functions $ax + b$, $a, b \neq 0$, whose graphs *do not* pass through the origin

19. The set of all polynomials of degree four or less

20. The set of all quadratic functions whose graphs pass through the origin

21. The set
$$\{(x, y): x \geq 0, y \text{ is a real number}\}$$

22. The set
$$\{(x, y): x \geq 0, y \geq 0\}$$

23. The set
$$\{(x, x): x \text{ is a real number}\}$$

24. The set
$$\left\{\left(x, \tfrac{1}{2}x\right): x \text{ is a real number}\right\}$$

25. The set of all 2×2 matrices of the form
$$\begin{bmatrix} a & b \\ c & 0 \end{bmatrix}$$

26. The set of all 2×2 matrices of the form
$$\begin{bmatrix} a & b \\ c & 1 \end{bmatrix}$$

27. The set of all 3×3 matrices of the form
$$\begin{bmatrix} 0 & a & b \\ c & 0 & d \\ e & f & 0 \end{bmatrix}$$

28. The set of all 3×3 matrices of the form
$$\begin{bmatrix} 1 & a & b \\ c & 1 & d \\ e & f & 1 \end{bmatrix}$$

29. The set of all 4×4 matrices of the form
$$\begin{bmatrix} 0 & a & b & c \\ a & 0 & b & c \\ a & b & 0 & c \\ a & b & c & 1 \end{bmatrix}$$

30. The set of all 4×4 matrices of the form
$$\begin{bmatrix} 0 & a & b & c \\ a & 0 & b & c \\ a & b & 0 & c \\ a & b & c & 0 \end{bmatrix}$$

31. The set of all 2×2 singular matrices

32. The set of all 2×2 nonsingular matrices

33. The set of all 2×2 diagonal matrices

34. The set of all 3×3 upper triangular matrices

35. $C[0, 1]$, the set of all continuous functions defined on the interval $[0, 1]$

36. $C[-1, 1]$, the set of all continuous functions defined on the interval $[-1, 1]$

37. Let V be the set of all positive real numbers. Determine whether V is a vector space with the operations shown below.

$x + y = xy$ Addition

$cx = x^c$ Scalar multiplication

If it is, verify each vector space axiom; if it is not, state all vector space axioms that fail.

38. Determine whether the set R^2 with the operations

$$(x_1, y_1) + (x_2, y_2) = (x_1 x_2, y_1 y_2)$$

and

$$c(x_1, y_1) = (cx_1, cy_1)$$

is a vector space. If it is, verify each vector space axiom; if it is not, state all vector space axioms that fail.

39. **Proof** Prove in full detail that the set $\{(x, 2x): x \text{ is a real number}\}$, with the standard operations in R^2, is a vector space.

40. Proof Prove in full detail that $M_{2,2}$, with the standard operations, is a vector space.

41. Rather than use the standard definitions of addition and scalar multiplication in R^2, let these two operations be defined as shown below.

(a) $(x_1, y_1) + (x_2, y_2) = (x_1 + x_2, y_1 + y_2)$
$$c(x, y) = (cx, y)$$

(b) $(x_1, y_1) + (x_2, y_2) = (x_1, 0)$
$$c(x, y) = (cx, cy)$$

(c) $(x_1, y_1) + (x_2, y_2) = (x_1 + x_2, y_1 + y_2)$
$$c(x, y) = \left(\sqrt{cx}, \sqrt{cy}\right)$$

With each of these new definitions, is R^2 a vector space? Justify your answers.

42. Rather than use the standard definitions of addition and scalar multiplication in R^3, let these two operations be defined as shown below.

(a) $(x_1, y_1, z_1) + (x_2, y_2, z_2)$
$$= (x_1 + x_2, y_1 + y_2, z_1 + z_2)$$
$$c(x, y, z) = (cx, cy, 0)$$

(b) $(x_1, y_1, z_1) + (x_2, y_2, z_2) = (0, 0, 0)$
$$c(x, y, z) = (cx, cy, cz)$$

(c) $(x_1, y_1, z_1) + (x_2, y_2, z_2)$
$$= (x_1 + x_2 + 1, y_1 + y_2 + 1, z_1 + z_2 + 1)$$
$$c(x, y, z) = (cx, cy, cz)$$

(d) $(x_1, y_1, z_1) + (x_2, y_2, z_2)$
$$= (x_1 + x_2 + 1, y_1 + y_2 + 1, z_1 + z_2 + 1)$$
$$c(x, y, z) = (cx + c - 1, cy + c - 1, cz + c - 1)$$

With each of these new definitions, is R^3 a vector space? Justify your answers.

43. Prove that in a given vector space V, the zero vector is unique.

44. Prove that in a given vector space V, the additive inverse of a vector is unique.

45. Mass-Spring System The mass in a mass-spring system (see figure) is pulled downward and then released, causing the system to oscillate according to

$$x(t) = a_1 \sin \omega t + a_2 \cos \omega t$$

where x is the displacement at time t, a_1 and a_2 are arbitrary constants, and ω is a fixed constant. Show that the set of all functions $x(t)$ is a vector space.

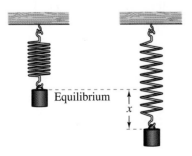

Equilibrium

x

46. CAPSTONE

(a) Describe the conditions under which a set may be classified as a vector space.

(b) Give an example of a set that is a vector space and an example of a set that is not a vector space.

47. Proof Complete the proof of the cancellation property of vector addition by justifying each step.

Prove that if **u**, **v**, and **w** are vectors in a vector space V such that **u** + **w** = **v** + **w**, then **u** = **v**.

$$\mathbf{u} + \mathbf{w} = \mathbf{v} + \mathbf{w}$$
$$(\mathbf{u} + \mathbf{w}) + (-\mathbf{w}) = (\mathbf{v} + \mathbf{w}) + (-\mathbf{w}) \qquad \text{a.} \underline{\hspace{1cm}}$$
$$\mathbf{u} + (\mathbf{w} + (-\mathbf{w})) = \mathbf{v} + (\mathbf{w} + (-\mathbf{w})) \qquad \text{b.} \underline{\hspace{1cm}}$$
$$\mathbf{u} + \mathbf{0} = \mathbf{v} + \mathbf{0} \qquad \text{c.} \underline{\hspace{1cm}}$$
$$\mathbf{u} = \mathbf{v} \qquad \text{d.} \underline{\hspace{1cm}}$$

48. Let R^∞ be the set of all infinite sequences of real numbers, with the operations

$$\mathbf{u} + \mathbf{v} = (u_1, u_2, u_3, \dots) + (v_1, v_2, v_3, \dots)$$
$$= (u_1 + v_1, u_2 + v_2, u_3 + v_3, \dots)$$

and

$$c\mathbf{u} = c(u_1, u_2, u_3, \dots)$$
$$= (cu_1, cu_2, cu_3, \dots).$$

Determine whether R^∞ is a vector space. If it is, verify each vector space axiom; if it is not, state all vector space axioms that fail.

True or False? In Exercises 49 and 50, determine whether each statement is true or false. If a statement is true, give a reason or cite an appropriate statement from the text. If a statement is false, provide an example that shows the statement is not true in all cases or cite an appropriate statement from the text.

49. (a) A vector space consists of four entities: a set of vectors, a set of scalars, and two operations.

(b) The set of all integers with the standard operations is a vector space.

(c) The set of all ordered triples (x, y, z) of real numbers, where $y \geq 0$, with the standard operations on R^3 is a vector space.

50. (a) To show that a set is not a vector space, it is sufficient to show that just one axiom is not satisfied.

(b) The set of all first-degree polynomials with the standard operations is a vector space.

(c) The set of all pairs of real numbers of the form $(0, y)$, with the standard operations on R^2, is a vector space.

51. Proof Prove Property 1 of Theorem 4.4.

52. Proof Prove Property 4 of Theorem 4.4.

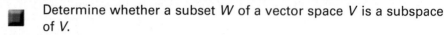

4.3   Subspaces of Vector Spaces

■ Determine whether a subset W of a vector space V is a subspace of V.

■ Determine subspaces of R^n.

SUBSPACES

In many applications in linear algebra, vector spaces occur as **subspaces** of larger spaces. For instance, you will see that the solution set of a homogeneous system of linear equations in n variables is a subspace of R^n. (See Theorem 4.16.)

A nonempty subset of a vector space is a subspace when it is a vector space with the *same* operations defined in the original vector space, as stated in the next definition.

REMARK

Note that if W is a subspace of V, then it must be closed under the operations inherited from V.

> ### Definition of a Subspace of a Vector Space
>
> A nonempty subset W of a vector space V is a **subspace** of V when W is a vector space under the operations of addition and scalar multiplication defined in V.

EXAMPLE 1   A Subspace of R^3

Show that the set $W = \{(x_1, 0, x_3): x_1 \text{ and } x_3 \text{ are real numbers}\}$ is a subspace of R^3 with the standard operations.

SOLUTION

The set W is nonempty because it contains the zero vector $(0, 0, 0)$.

Graphically, the set W can be interpreted as the xz-plane, as shown in Figure 4.7. The set W is closed under addition because the sum of any two vectors in the xz-plane must also lie in the xz-plane. That is, if $(x_1, 0, x_3)$ and $(y_1, 0, y_3)$ are in W, then their sum $(x_1 + y_1, 0, x_3 + y_3)$ is also in W. Similarly, to see that W is closed under scalar multiplication, let $(x_1, 0, x_3)$ be in W and let c be a scalar. Then $c(x_1, 0, x_3) = (cx_1, 0, cx_3)$ has zero as its second component and must be in W. The verifications of the other eight vector space axioms are left to you.

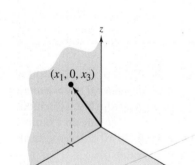

Figure 4.7

To establish that a set W is a vector space, you must verify all ten vector space axioms. If W is a nonempty subset of a larger vector space V (and the operations defined on W are the *same* as those defined on V), however, then most of the ten properties are *inherited* from the larger space and need no verification. The next theorem states that it is sufficient to test for closure in order to establish that a nonempty subset of a vector space is a subspace.

> ### THEOREM 4.5   Test for a Subspace
>
> If W is a nonempty subset of a vector space V, then W is a subspace of V if and only if the two closure conditions listed below hold.
>
> 1. If $\mathbf{u}$ and $\mathbf{v}$ are in W, then $\mathbf{u} + \mathbf{v}$ is in W.
> 2. If $\mathbf{u}$ is in W and c is any scalar, then $c\mathbf{u}$ is in W.

PROOF

The proof of this theorem in one direction is straightforward. That is, if W is a subspace of V, then W is a vector space and must be closed under addition and scalar multiplication.

To prove the theorem in the other direction, assume that W is closed under addition and scalar multiplication. Note that if $\mathbf{u}$, $\mathbf{v}$, and $\mathbf{w}$ are in W, then they are also in V. Consequently, vector space axioms 2, 3, 7, 8, 9, and 10 are satisfied automatically. W is closed under addition and scalar multiplication, so it follows that for any $\mathbf{v}$ in W and scalar $c = 0$, $c\mathbf{v} = \mathbf{0}$, and $(-1)\mathbf{v} = -\mathbf{v}$ both lie in W, which satisfies axioms 4 and 5. ■

REMARK

Note that if W is a subspace of a vector space V, then both W and V must have the same zero vector $\mathbf{0}$. (In Exercise 55, you are asked to prove this.)

A subspace of a vector space is also a vector space, so it must contain the zero vector. In fact, the simplest subspace of a vector space V is the one consisting of only the zero vector, $W = \{\mathbf{0}\}$. This subspace is the **zero subspace**. Another subspace of V is V itself. Every vector space contains these two trivial subspaces, and subspaces other than these are called **proper** (or nontrivial) subspaces.

EXAMPLE 2 A Subspace of $M_{2,2}$

Let W be the set of all 2×2 symmetric matrices. Show that W is a subspace of the vector space $M_{2,2}$, with the standard operations of matrix addition and scalar multiplication.

SOLUTION

Recall that a square matrix is *symmetric* when it is equal to its own transpose. The set $M_{2,2}$ is a vector space, so you only need to show that W (a subset of $M_{2,2}$) satisfies the conditions of Theorem 4.5. Begin by observing that W is *nonempty*. W is closed under addition because for matrices A_1 and A_2 in W, $A_1 = A_1^T$ and $A_2 = A_2^T$, which implies that

$$(A_1 + A_2)^T = A_1^T + A_2^T = A_1 + A_2.$$

So, if A_1 and A_2 are symmetric matrices of order 2, then so is $A_1 + A_2$. Similarly, W is closed under scalar multiplication because $A = A^T$ implies that $(cA)^T = cA^T = cA$. If A is a symmetric matrix of order 2, then so is cA. ■

The result of Example 2 can be generalized. That is, for any positive integer n, the set of symmetric matrices of order n is a subspace of the vector space $M_{n,n}$ with the standard operations. The next example describes a subset of $M_{n,n}$ that is not a subspace.

EXAMPLE 3 The Set of Singular Matrices Is Not a Subspace of $M_{n,n}$

Let W be the set of singular matrices of order 2. Show that W is not a subspace of $M_{2,2}$ with the standard operations.

SOLUTION

By Theorem 4.5, to show that a subset W is not a subspace, show that W is empty, W is not closed under addition, or W is not closed under scalar multiplication. In this example, W is nonempty and closed under scalar multiplication, but it is not closed under addition. To see this, let A and B be

$$A = \begin{bmatrix} 1 & 0 \\ 0 & 0 \end{bmatrix} \quad \text{and} \quad B = \begin{bmatrix} 0 & 0 \\ 0 & 1 \end{bmatrix}.$$

Then A and B are both singular (noninvertible), but their sum

$$A + B = \begin{bmatrix} 1 & 0 \\ 0 & 1 \end{bmatrix}$$

is nonsingular (invertible). So W is not closed under addition, and by Theorem 4.5, W is not a subspace of $M_{2,2}$. ■

EXAMPLE 4   A Subset of R^2 That Is Not a Subspace

Show that $W = \{(x_1, x_2): x_1 \geq 0 \text{ and } x_2 \geq 0\}$, with the standard operations, is not a subspace of R^2.

SOLUTION

This set is nonempty and closed under addition. It is not, however, closed under scalar multiplication. To see this, note that $(1, 1)$ is in W, but the scalar multiple $(-1)(1, 1) = (-1, -1)$ is not in W. So W is not a subspace of R^2.

You will often encounter sequences of nested subspaces. For example, consider the vector spaces $P_0, P_1, P_2, P_3, \ldots, P_n$, where P_k is the set of all polynomials of degree less than or equal to k, with the standard operations. You can write $P_0 \subset P_1 \subset P_2 \subset P_3 \subset \cdots \subset P_n$. If $j \leq k$, then P_j is a subspace of P_k. (In Exercise 45, you are asked to show this.) Example 5 describes another nesting of subspaces.

EXAMPLE 5   Subspaces of Functions (Calculus)

Let W_5 be the vector space of all functions defined on $[0, 1]$, and let W_1, W_2, W_3, and W_4 be defined as shown below.

W_1 = set of all polynomial functions that are defined on $[0, 1]$
W_2 = set of all functions that are differentiable on $[0, 1]$
W_3 = set of all functions that are continuous on $[0, 1]$
W_4 = set of all functions that are integrable on $[0, 1]$

Show that $W_1 \subset W_2 \subset W_3 \subset W_4 \subset W_5$ and that W_i is a subspace of W_j for $i \leq j$.

SOLUTION

From calculus you know that every polynomial function is differentiable on $[0, 1]$. So, $W_1 \subset W_2$. Moreover, every differentiable function is continuous, every continuous function is integrable, and every integrable function is a function, which means that $W_2 \subset W_3 \subset W_4 \subset W_5$. So, you have $W_1 \subset W_2 \subset W_3 \subset W_4 \subset W_5$, as shown in Figure 4.8. It is left to you to show that W_i is a subspace of W_j for $i \leq j$. (See Exercise 46.)

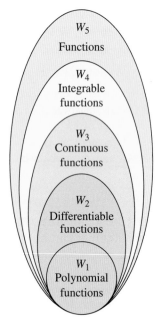

Figure 4.8

As implied in Example 5, if U, V, and W are vector spaces such that W is a subspace of V and V is a subspace of U, then W is also a subspace of U. The next theorem states that the intersection of two subspaces is also a subspace, as shown in Figure 4.9.

THEOREM 4.6   The Intersection of Two Subspaces Is a Subspace

If V and W are both subspaces of a vector space U, then the intersection of V and W (denoted by $V \cap W$) is also a subspace of U.

REMARK

Theorem 4.6 states that the *intersection* of two subspaces is a subspace. In Exercise 56 you are asked to show that the *union* of two subspaces is not necessarily a subspace.

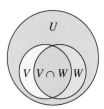

Figure 4.9 The intersection of two subspaces is a subspace.

PROOF

V and W are both subspaces of U, so both contain $\mathbf{0}$, and $V \cap W$ is nonempty. To show that $V \cap W$ is closed under addition, let $\mathbf{v}_1$ and $\mathbf{v}_2$ be any two vectors in $V \cap W$. V and W are both subspaces of U, which means that both are closed under addition. Both $\mathbf{v}_1$ and $\mathbf{v}_2$ are in V, so their sum $\mathbf{v}_1 + \mathbf{v}_2$ must be in V. Similarly, $\mathbf{v}_1 + \mathbf{v}_2$ is in W because both $\mathbf{v}_1$ and $\mathbf{v}_2$ are also in W. But this implies that $\mathbf{v}_1 + \mathbf{v}_2$ is in $V \cap W$, and it follows that $V \cap W$ is closed under addition. It is left to you to show (by a similar argument) that $V \cap W$ is closed under scalar multiplication. (See Exercise 59.)

SUBSPACES OF R^n

R^n is a convenient source for examples of vector spaces, so the remainder of this section is devoted to looking at subspaces of R^n.

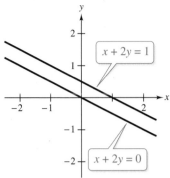

Figure 4.10

EXAMPLE 6 Determining Subspaces of R^2

Determine whether each subset is a subspace of R^2.

a. The set of points on the line $x + 2y = 0$

b. The set of points on the line $x + 2y = 1$

SOLUTION

a. Solving for x, a point in R^2 is on the line $x + 2y = 0$ if and only if it has the form $(-2t, t)$, where t is any real number. (See Figure 4.10.)

To show that this set is closed under addition, let $\mathbf{v}_1 = (-2t_1, t_1)$ and $\mathbf{v}_2 = (-2t_2, t_2)$ be any two points on the line. Then you have

$$\mathbf{v}_1 + \mathbf{v}_2 = (-2t_1, t_1) + (-2t_2, t_2) = (-2(t_1 + t_2), t_1 + t_2) = (-2t_3, t_3)$$

where $t_3 = t_1 + t_2$. $\mathbf{v}_1 + \mathbf{v}_2$ lies on the line, and the set is closed under addition. In a similar way, you can show that the set is closed under scalar multiplication. So, this set is a subspace of R^2.

b. This subset of R^2 is *not* a subspace of R^2 because every subspace must contain the zero vector $(0, 0)$, which is not on the line $x + 2y = 1$. (See Figure 4.10.) ◼

Of the two lines in Example 6, the one that is a subspace of R^2 is the one that passes through the origin. This is characteristic of subspaces of R^2. That is, if W is a subset of R^2, then it is a subspace if and only if it has one of the forms listed below.

1. W consists of the *single point* $(0, 0)$.

2. W consists of all points on a *line* that passes through the origin.

3. W consists of all of R^2.

Figure 4.11 shows these three possibilities graphically.

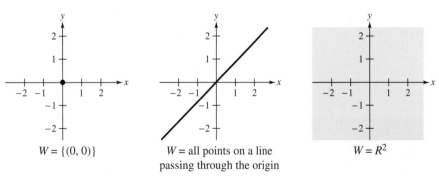

Figure 4.11

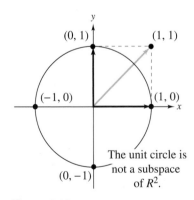

Figure 4.12

REMARK

Another way to tell that the subset shown in Figure 4.12 is not a subspace of R^2 is by noting that it does not contain the zero vector (the origin).

EXAMPLE 7 A Subset of R^2 That Is Not a Subspace

Show that the subset of R^2 consisting of all points on $x^2 + y^2 = 1$ is not a subspace.

SOLUTION

This subset of R^2 is *not* a subspace because the points $(1, 0)$ and $(0, 1)$ are in the subset, but their sum $(1, 1)$ is not. (See Figure 4.12.) So, this subset is not closed under addition. ◼

EXAMPLE 8   **Determining Subspaces of R^3**

See LarsonLinearAlgebra.com for an interactive version of this type of example.

Determine whether each subset is a subspace of R^3.

a. $W = \{(x_1, x_2, 1): x_1 \text{ and } x_2 \text{ are real numbers}\}$

b. $W = \{(x_1, x_1 + x_3, x_3): x_1 \text{ and } x_3 \text{ are real numbers}\}$

SOLUTION

a. The zero vector $\mathbf{0} = (0, 0, 0)$ is not in W, so W is *not* a subspace of R^3.

b. This set is nonempty because it contains the zero vector $(0, 0, 0)$. Let

$$\mathbf{v} = (v_1, v_1 + v_3, v_3) \quad \text{and} \quad \mathbf{u} = (u_1, u_1 + u_3, u_3)$$

be two vectors in W, and let c be any real number. W is closed under addition because

$$\begin{aligned}
\mathbf{v} + \mathbf{u} &= (v_1 + u_1, v_1 + v_3 + u_1 + u_3, v_3 + u_3) \\
&= (v_1 + u_1, (v_1 + u_1) + (v_3 + u_3), v_3 + u_3) \\
&= (x_1, x_1 + x_3, x_3)
\end{aligned}$$

where $x_1 = v_1 + u_1$ and $x_3 = v_3 + u_3$, which means that $\mathbf{v} + \mathbf{u}$ is in W. Similarly, W is closed under scalar multiplication because

$$\begin{aligned}
c\mathbf{v} &= (cv_1, c(v_1 + v_3), cv_3) \\
&= (cv_1, cv_1 + cv_3, cv_3) \\
&= (x_1, x_1 + x_3, x_3)
\end{aligned}$$

where $x_1 = cv_1$ and $x_3 = cv_3$, which means that $c\mathbf{v}$ is in W. So, W is a subspace of R^3.

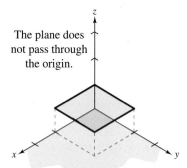

The plane does not pass through the origin.

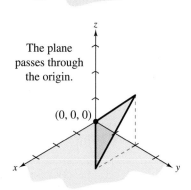

The plane passes through the origin.

$(0, 0, 0)$

Figure 4.13

In Example 8, note that the graph of each subset is a plane in R^3, but the only subset that is a *subspace* is the one represented by a plane that passes through the origin. (See Figure 4.13.) You can show that a subset W of R^3 is a subspace of R^3 if and only if it has one of the forms listed below.

1. W consists of the *single point* $(0, 0, 0)$.

2. W consists of all points on a *line* that passes through the origin.

3. W consists of all points in a *plane* that passes through the origin.

4. W consists of all of R^3.

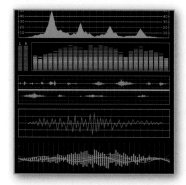

LINEAR ALGEBRA APPLIED

Digital signal processing depends on sampling, which converts continuous signals into discrete sequences that can be used by digital devices. Traditionally, sampling is uniform and pointwise, and is obtained from a single vector space. Then, the resulting sequence is reconstructed into a continuous-domain signal. Such a process, however, can involve a significant reduction in information, which could result in a low-quality reconstructed signal. In applications such as radar, geophysics, and wireless communications, researchers have determined situations in which sampling from a *union* of vector subspaces can be more appropriate. *(Source: Sampling Signals from a Union of Subspaces—A New Perspective for the Extension of This Theory, Lu, Y.M. and Do, M.N., IEEE Signal Processing Magazine)*

4.3 Exercises

See CalcChat.com for worked-out solutions to odd-numbered exercises.

Verifying Subspaces In Exercises 1–6, verify that W is a subspace of V. In each case, assume that V has the standard operations.

1. $W = \{(x_1, x_2, x_3, 0): x_1, x_2, \text{ and } x_3 \text{ are real numbers}\}$
$V = R^4$

2. $\{(x, y, 4x - 5y): x \text{ and } y \text{ are real numbers}\}$
$V = R^3$

3. W is the set of all 2×2 matrices of the form
$$\begin{bmatrix} 0 & a \\ b & 0 \end{bmatrix}.$$
$V = M_{2,2}$

4. W is the set of all 3×2 matrices of the form
$$\begin{bmatrix} a & b \\ a - 2b & 0 \\ 0 & c \end{bmatrix}.$$
$V = M_{3,2}$

5. Calculus W is the set of all functions that are continuous on $[-1, 1]$. V is the set of all functions that are integrable on $[-1, 1]$.

6. Calculus W is the set of all functions that are differentiable on $[-1, 1]$. V is the set of all functions that are continuous on $[-1, 1]$.

Subsets That Are Not Subspaces In Exercises 7–20, W is not a subspace of the vector space. Verify this by giving a specific example that violates the test for a vector subspace (Theorem 4.5).

7. W is the set of all vectors in R^3 whose third component is -1.

8. W is the set of all vectors in R^2 whose first component is 2.

9. W is the set of all vectors in R^2 whose components are rational numbers.

10. W is the set of all vectors in R^2 whose components are integers.

11. W is the set of all nonnegative functions in $C(-\infty, \infty)$.

12. W is the set of all linear functions $ax + b, a \neq 0$, in $C(-\infty, \infty)$.

13. W is the set of all vectors in R^3 whose components are nonnegative.

14. W is the set of all vectors in R^3 whose components are Pythagorean triples.

15. W is the set of all matrices in $M_{3,3}$ of the form
$$\begin{bmatrix} 1 & a & b \\ c & 1 & d \\ e & f & 0 \end{bmatrix}.$$

16. W is the set of all matrices in $M_{3,1}$ of the form
$$\begin{bmatrix} \sqrt{a} & 0 & 3a \end{bmatrix}^T.$$

17. W is the set of all matrices in $M_{n,n}$ with determinants equal to 1.

18. W is the set of all matrices in $M_{n,n}$ such that $A^2 = A$.

19. W is the set of all vectors in R^2 whose second component is the cube of the first.

20. W is the set of all vectors in R^2 whose second component is the square of the first.

Determining Subspaces of $C(-\infty, \infty)$ In Exercises 21–28, determine whether the subset of $C(-\infty, \infty)$ is a subspace of $C(-\infty, \infty)$ with the standard operations. Justify your answer.

21. The set of all positive functions: $f(x) > 0$

22. The set of all negative functions: $f(x) < 0$

23. The set of all even functions: $f(-x) = f(x)$

24. The set of all odd functions: $f(-x) = -f(x)$

25. The set of all constant functions: $f(x) = c$

26. The set of all exponential functions $f(x) = a^x$, where $a > 0$

27. The set of all functions such that $f(0) = 0$

28. The set of all functions such that $f(0) = 1$

Determining Subspaces of $M_{n,n}$ In Exercises 29–36, determine whether the subset of $M_{n,n}$ is a subspace of $M_{n,n}$ with the standard operations. Justify your answer.

29. The set of all $n \times n$ upper triangular matrices

30. The set of all $n \times n$ diagonal matrices

31. The set of all $n \times n$ matrices with integer entries

32. The set of all $n \times n$ matrices A that commute with a given matrix B; that is, $AB = BA$

33. The set of all $n \times n$ singular matrices

34. The set of all $n \times n$ invertible matrices

35. The set of all $n \times n$ matrices whose entries sum to zero

36. The set of all $n \times n$ matrices whose trace is nonzero (Recall that the **trace** of a matrix is the sum of the main diagonal entries of the matrix.)

Determining Subspaces of R^3 In Exercises 37–42, determine whether the set W is a subspace of R^3 with the standard operations. Justify your answer.

37. $W = \{(0, x_2, x_3): x_2 \text{ and } x_3 \text{ are real numbers}\}$

38. $W = \{(x_1, x_2, 4): x_1 \text{ and } x_2 \text{ are real numbers}\}$

39. $W = \{(a, a - 3b, b): a \text{ and } b \text{ are real numbers}\}$

40. $W = \{(s, t, s + t): s \text{ and } t \text{ are real numbers}\}$

41. $W = \{(x_1, x_2, x_1 x_2): x_1 \text{ and } x_2 \text{ are real numbers}\}$

42. $W = \{(x_1, 1/x_1, x_3): x_1 \text{ and } x_3 \text{ are real numbers}, x_1 \neq 0\}$

True or False? **In Exercises 43 and 44, determine whether each statement is true or false. If a statement is true, give a reason or cite an appropriate statement from the text. If a statement is false, provide an example that shows the statement is not true in all cases or cite an appropriate statement from the text.**

43. (a) If W is a subspace of a vector space V, then it has closure under scalar multiplication as defined in V.

   (b) If V and W are both subspaces of a vector space U, then the intersection of V and W is also a subspace.

   (c) If U, V, and W are vector spaces such that W is a subspace of V and U is a subspace of V, then $W = U$.

44. (a) Every vector space V contains two proper subspaces that are the zero subspace and itself.

   (b) If W is a subspace of R^2, then W must contain the vector $(0, 0)$.

   (c) If W is a subspace of a vector space V, then it has closure under addition as defined in V.

   (d) If W is a subspace of a vector space V, then W is also a vector space.

45. Consider the vector spaces

   $$P_0, P_1, P_2, \ldots, P_n$$

   where P_k is the set of all polynomials of degree less than or equal to k, with the standard operations. Show that if $j \le k$, then P_j is a subspace of P_k.

46. **Calculus** Let W_1, W_2, W_3, W_4, and W_5 be defined as in Example 5. Show that W_i is a subspace of W_j for $i \le j$.

47. **Calculus** Let $F(-\infty, \infty)$ be the vector space of real-valued functions defined on the entire real line. Show that each set is a subspace of $F(-\infty, \infty)$.

   (a) $C(-\infty, \infty)$

   (b) The set of all differentiable functions f defined on the real number line

   (c) The set of all differentiable functions f defined on the real number line that satisfy the differential equation

   $$f' - 3f = 0$$

48. **Calculus** Determine whether the set

   $$S = \left\{ f \in C[0, 1] : \int_0^1 f(x)\,dx = 0 \right\}$$

   is a subspace of $C[0, 1]$. Prove your answer.

49. Let W be the subset of R^3 consisting of all points on a line that passes through the origin. Such a line can be represented by the parametric equations

   $$x = at, \quad y = bt, \quad \text{and} \quad z = ct.$$

   Use these equations to show that W is a subspace of R^3.

50. **CAPSTONE** Explain why it is sufficient to test for closure to establish that a nonempty subset of a vector space is a subspace.

51. **Guided Proof** Prove that a nonempty set W is a subspace of a vector space V if and only if $a\mathbf{x} + b\mathbf{y}$ is an element of W for all scalars a and b and all vectors $\mathbf{x}$ and $\mathbf{y}$ in W.

   Getting Started: In one direction, assume W is a subspace, and show by using closure axioms that $a\mathbf{x} + b\mathbf{y}$ is an element of W. In the other direction, assume $a\mathbf{x} + b\mathbf{y}$ is an element of W for all scalars a and b and all vectors $\mathbf{x}$ and $\mathbf{y}$ in W, and verify that W is closed under addition and scalar multiplication.

   (i) If W is a subspace of V, then use scalar multiplication closure to show that $a\mathbf{x}$ and $b\mathbf{y}$ are in W. Now use additive closure to get the desired result.

   (ii) Conversely, assume $a\mathbf{x} + b\mathbf{y}$ is in W. By cleverly assigning specific values to a and b, show that W is closed under addition and scalar multiplication.

52. Let $\mathbf{x}$, $\mathbf{y}$, and $\mathbf{z}$ be vectors in a vector space V. Show that the set of all linear combinations of $\mathbf{x}$, $\mathbf{y}$, and $\mathbf{z}$,

   $$W = \{a\mathbf{x} + b\mathbf{y} + c\mathbf{z} : a, b, \text{ and } c \text{ are scalars}\}$$

   is a subspace of V. This subspace is the **span** of $\{\mathbf{x}, \mathbf{y}, \mathbf{z}\}$.

53. **Proof** Let A be a fixed 2×3 matrix. Prove that the set

   $$W = \left\{ \mathbf{x} \in R^3 : A\mathbf{x} = \begin{bmatrix} 1 \\ 2 \end{bmatrix} \right\}$$

   is not a subspace of R^3.

54. **Proof** Let A be a fixed $m \times n$ matrix. Prove that the set

   $$W = \{\mathbf{x} \in R^n : A\mathbf{x} = \mathbf{0}\}$$

   is a subspace of R^n.

55. **Proof** Let W be a subspace of the vector space V. Prove that the zero vector in V is also the zero vector in W.

56. Give an example showing that the union of two subspaces of a vector space V is not necessarily a subspace of V.

57. **Proof** Let A and B be fixed 2×2 matrices. Prove that the set

   $$W = \{X : XAB = BAX\}$$

   is a subspace of $M_{2,2}$.

58. **Proof** Let V and W be two subspaces of a vector space U.

   (a) Prove that the set

   $$V + W = \{\mathbf{u} : \mathbf{u} = \mathbf{v} + \mathbf{w}, \mathbf{v} \in V \text{ and } \mathbf{w} \in W\}$$

   is a subspace of U.

   (b) Describe $V + W$ when V and W are the subspaces of $U = R^2$:

   $V = \{(x, 0) : x \text{ is a real number}\}$ and $W = \{(0, y) : y$ is a real number$\}$.

59. **Proof** Complete the proof of Theorem 4.6 by showing that the intersection of two subspaces of a vector space is closed under scalar multiplication.

4.4 Spanning Sets and Linear Independence

- Write a linear combination of a set of vectors in a vector space V.
- Determine whether a set S of vectors in a vector space V is a spanning set of V.
- Determine whether a set of vectors in a vector space V is linearly independent.

LINEAR COMBINATIONS OF VECTORS IN A VECTOR SPACE

This section begins to develop procedures for representing each vector in a vector space as a **linear combination** of a select number of vectors in the space.

Definition of a Linear Combination of Vectors

A vector $\mathbf{v}$ in a vector space V is a **linear combination** of the vectors $\mathbf{u}_1$, $\mathbf{u}_2, \ldots, \mathbf{u}_k$ in V when $\mathbf{v}$ can be written in the form

$$\mathbf{v} = c_1 \mathbf{u}_1 + c_2 \mathbf{u}_2 + \cdots + c_k \mathbf{u}_k$$

where $c_1, c_2, \ldots, c_k$ are scalars.

Often, one or more of the vectors in a set can be written as linear combinations of other vectors in the set. Examples 1 and 2 illustrate this possibility.

EXAMPLE 1   **Examples of Linear Combinations**

a. For the set of vectors in R^3

$$S = \{\overset{\mathbf{v}_1}{(1, 3, 1)}, \overset{\mathbf{v}_2}{(0, 1, 2)}, \overset{\mathbf{v}_3}{(1, 0, -5)}\}$$

$\mathbf{v}_1$ is a linear combination of $\mathbf{v}_2$ and $\mathbf{v}_3$ because

$$\mathbf{v}_1 = 3\mathbf{v}_2 + \mathbf{v}_3 = 3(0, 1, 2) + (1, 0, -5) = (1, 3, 1).$$

b. For the set of vectors in $M_{2,2}$

$$S = \left\{ \overset{\mathbf{v}_1}{\begin{bmatrix} 0 & 8 \\ 2 & 1 \end{bmatrix}}, \overset{\mathbf{v}_2}{\begin{bmatrix} 0 & 2 \\ 1 & 0 \end{bmatrix}}, \overset{\mathbf{v}_3}{\begin{bmatrix} -1 & 3 \\ 1 & 2 \end{bmatrix}}, \overset{\mathbf{v}_4}{\begin{bmatrix} -2 & 0 \\ 1 & 3 \end{bmatrix}} \right\}$$

$\mathbf{v}_1$ is a linear combination of $\mathbf{v}_2$, $\mathbf{v}_3$, and $\mathbf{v}_4$ because

$$
\begin{aligned}
\mathbf{v}_1 &= \mathbf{v}_2 + 2\mathbf{v}_3 - \mathbf{v}_4 \\
&= \begin{bmatrix} 0 & 2 \\ 1 & 0 \end{bmatrix} + 2\begin{bmatrix} -1 & 3 \\ 1 & 2 \end{bmatrix} - \begin{bmatrix} -2 & 0 \\ 1 & 3 \end{bmatrix} \\
&= \begin{bmatrix} 0 & 8 \\ 2 & 1 \end{bmatrix}.
\end{aligned}
$$

In Example 1, it is relatively easy to verify that one of the vectors in the set S is a linear combination of the other vectors because the coefficients to form the linear combination are given. Example 2 demonstrates a procedure for finding the coefficients.

EXAMPLE 2   **Finding a Linear Combination**

Write the vector $\mathbf{w} = (1, 1, 1)$ as a linear combination of vectors in the set

$$S = \{\underset{\mathbf{v}_1}{(1, 2, 3)}, \underset{\mathbf{v}_2}{(0, 1, 2)}, \underset{\mathbf{v}_3}{(-1, 0, 1)}\}.$$

SOLUTION

Find scalars c_1, c_2, and c_3 such that

$$\begin{aligned}
(1, 1, 1) &= c_1(1, 2, 3) + c_2(0, 1, 2) + c_3(-1, 0, 1) \\
&= (c_1, 2c_1, 3c_1) + (0, c_2, 2c_2) + (-c_3, 0, c_3) \\
&= (c_1 - c_3, 2c_1 + c_2, 3c_1 + 2c_2 + c_3).
\end{aligned}$$

Equating corresponding components yields the system of linear equations below.

$$\begin{aligned}
c_1 \quad\quad - c_3 &= 1 \\
2c_1 + c_2 \quad\quad &= 1 \\
3c_1 + 2c_2 + c_3 &= 1
\end{aligned}$$

Using Gauss-Jordan elimination, the augmented matrix of this system row reduces to

$$\begin{bmatrix} 1 & 0 & -1 & 1 \\ 0 & 1 & 2 & -1 \\ 0 & 0 & 0 & 0 \end{bmatrix}.$$

So, this system has infinitely many solutions, each of the form

$$c_1 = 1 + t, \quad c_2 = -1 - 2t, \quad c_3 = t.$$

To obtain one solution, you could let $t = 1$. Then $c_3 = 1$, $c_2 = -3$, and $c_1 = 2$, and you have

$$\mathbf{w} = 2\mathbf{v}_1 - 3\mathbf{v}_2 + \mathbf{v}_3.$$

(Verify this.) Other choices for t would yield different ways to write $\mathbf{w}$ as a linear combination of $\mathbf{v}_1$, $\mathbf{v}_2$, and $\mathbf{v}_3$.

EXAMPLE 3   **Finding a Linear Combination**

If possible, write the vector

$$\mathbf{w} = (1, -2, 2)$$

as a linear combination of vectors in the set S in Example 2.

SOLUTION

Following the procedure from Example 2 results in the system

$$\begin{aligned}
c_1 \quad\quad - c_3 &= 1 \\
2c_1 + c_2 \quad\quad &= -2 \\
3c_1 + 2c_2 + c_3 &= 2.
\end{aligned}$$

The augmented matrix of this system row reduces to

$$\begin{bmatrix} 1 & 0 & -1 & 0 \\ 0 & 1 & 2 & 0 \\ 0 & 0 & 0 & 1 \end{bmatrix}.$$

From the third row you can conclude that the system of equations is inconsistent, which means that there is no solution. Consequently, $\mathbf{w}$ *cannot* be written as a linear combination of $\mathbf{v}_1$, $\mathbf{v}_2$, and $\mathbf{v}_3$.

SPANNING SETS

If every vector in a vector space can be written as a linear combination of vectors in a set S, then S is a **spanning set** of the vector space.

Definition of a Spanning Set of a Vector Space

Let $S = \{v_1, v_2, \ldots, v_k\}$ be a subset of a vector space V. The set S is a **spanning set** of V when *every* vector in V can be written as a linear combination of vectors in S. In such cases it is said that S **spans** V.

EXAMPLE 4   Examples of Spanning Sets

a. The set $S = \{(1, 0, 0), (0, 1, 0), (0, 0, 1)\}$ spans R^3 because any vector $\mathbf{u} = (u_1, u_2, u_3)$ in R^3 can be written as

$$\mathbf{u} = u_1(1, 0, 0) + u_2(0, 1, 0) + u_3(0, 0, 1) = (u_1, u_2, u_3).$$

b. The set $S = \{1, x, x^2\}$ spans P_2 because any polynomial function $p(x) = a + bx + cx^2$ in P_2 can be written as

$$p(x) = a(1) + b(x) + c(x^2)$$
$$= a + bx + cx^2.$$

The spanning sets in Example 4 are called the **standard spanning sets** of R^3 and P_2, respectively. (You will learn more about standard spanning sets in the next section.) In the next example, you will look at a nonstandard spanning set of R^3.

EXAMPLE 5   A Spanning Set of R^3

Show that the set $S = \{(1, 2, 3), (0, 1, 2), (-2, 0, 1)\}$ spans R^3.

SOLUTION

Let $\mathbf{u} = (u_1, u_2, u_3)$ be *any* vector in R^3. Find scalars $c_1, c_2,$ and c_3 such that

$$(u_1, u_2, u_3) = c_1(1, 2, 3) + c_2(0, 1, 2) + c_3(-2, 0, 1)$$
$$= (c_1 - 2c_3, 2c_1 + c_2, 3c_1 + 2c_2 + c_3).$$

This vector equation produces the system

$$\begin{aligned} c_1 \qquad\quad - 2c_3 &= u_1 \\ 2c_1 + c_2 \qquad &= u_2 \\ 3c_1 + 2c_2 + c_3 &= u_3. \end{aligned}$$

REMARK

The coefficient matrix of the system in Example 3,

$$\begin{bmatrix} 1 & 0 & -1 \\ 2 & 1 & 0 \\ 3 & 2 & 1 \end{bmatrix}$$

has a determinant of zero. (Verify this.)

The coefficient matrix of this system has a nonzero determinant (verify that it is equal to -1), and it follows from the list of equivalent conditions in Section 3.3 that the system has a unique solution. So, any vector in R^3 can be written as a linear combination of the vectors in S, and you can conclude that the set S spans R^3.

EXAMPLE 6   A Set That Does Not Span R^3

From Example 3 you know that the set

$$S = \{(1, 2, 3), (0, 1, 2), (-1, 0, 1)\}$$

does not span R^3 because $\mathbf{w} = (1, -2, 2)$ is in R^3 and cannot be expressed as a linear combination of the vectors in S.

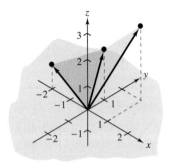

$S_1 = \{(1, 2, 3), (0, 1, 2), (-2, 0, 1)\}$
The vectors in S_1 do not lie
in a common plane.

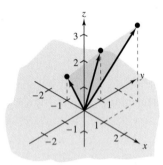

$S_2 = \{(1, 2, 3), (0, 1, 2), (-1, 0, 1)\}$
The vectors in S_2 lie in a
common plane.

Figure 4.14

Comparing the sets of vectors in Examples 5 and 6, note that the sets are the same except for a seemingly insignificant difference in the third vector.

$$S_1 = \{(1, 2, 3), (0, 1, 2), (-2, 0, 1)\} \qquad \text{Example 5}$$
$$S_2 = \{(1, 2, 3), (0, 1, 2), (-1, 0, 1)\} \qquad \text{Example 6}$$

The difference, however, is significant, because the set S_1 spans R^3 whereas the set S_2 does not. The reason for this difference can be seen in Figure 4.14. The vectors in S_2 lie in a common plane; the vectors in S_1 do not.

Although the set S_2 does not span all of R^3, it does span a subspace of R^3—namely, the plane in which the three vectors of S_2 lie. This subspace is the **span of S_2,** as stated in the next definition.

Definition of the Span of a Set

If $S = \{\mathbf{v}_1, \mathbf{v}_2, \ldots, \mathbf{v}_k\}$ is a set of vectors in a vector space V, then the **span of S** is the set of all linear combinations of the vectors in S,

$$\text{span}(S) = \{c_1\mathbf{v}_1 + c_2\mathbf{v}_2 + \cdots + c_k\mathbf{v}_k : c_1, c_2, \ldots, c_k \text{ are real numbers}\}.$$

The span of S is denoted by

$$\text{span}(S) \quad \text{or} \quad \text{span}\{\mathbf{v}_1, \mathbf{v}_2, \ldots, \mathbf{v}_k\}.$$

When $\text{span}(S) = V$, it is said that V is **spanned** by $\{\mathbf{v}_1, \mathbf{v}_2, \ldots, \mathbf{v}_k\}$, or that S **spans** V.

The next theorem tells you that the span of any finite nonempty subset of a vector space V is a subspace of V.

THEOREM 4.7   Span(S) Is a Subspace of V

If $S = \{\mathbf{v}_1, \mathbf{v}_2, \ldots, \mathbf{v}_k\}$ is a set of vectors in a vector space V, then $\text{span}(S)$ is a subspace of V. Moreover, $\text{span}(S)$ is the smallest subspace of V that contains S, in the sense that every other subspace of V that contains S must contain $\text{span}(S)$.

PROOF

To show that $\text{span}(S)$, the set of all linear combinations of $\mathbf{v}_1, \mathbf{v}_2, \ldots, \mathbf{v}_k$, is a subspace of V, show that it is closed under addition and scalar multiplication. Consider any two vectors $\mathbf{u}$ and $\mathbf{v}$ in $\text{span}(S)$,

$$\mathbf{u} = c_1\mathbf{v}_1 + c_2\mathbf{v}_2 + \cdots + c_k\mathbf{v}_k$$
$$\mathbf{v} = d_1\mathbf{v}_1 + d_2\mathbf{v}_2 + \cdots + d_k\mathbf{v}_k$$

where

$$c_1, c_2, \ldots, c_k \quad \text{and} \quad d_1, d_2, \ldots, d_k$$

are scalars. Then

$$\mathbf{u} + \mathbf{v} = (c_1 + d_1)\mathbf{v}_1 + (c_2 + d_2)\mathbf{v}_2 + \cdots + (c_k + d_k)\mathbf{v}_k$$

and

$$c\mathbf{u} = (cc_1)\mathbf{v}_1 + (cc_2)\mathbf{v}_2 + \cdots + (cc_k)\mathbf{v}_k$$

which means that $\mathbf{u} + \mathbf{v}$ and $c\mathbf{u}$ are also in $\text{span}(S)$ because they can be written as linear combinations of vectors in S. So, $\text{span}(S)$ is a subspace of V. It is left to you to prove that $\text{span}(S)$ is the smallest subspace of V that contains S. (See Exercise 59.) ∎

LINEAR DEPENDENCE AND LINEAR INDEPENDENCE

For a set of vectors

$$S = \{\mathbf{v}_1, \mathbf{v}_2, \ldots, \mathbf{v}_k\}$$

in a vector space V, the vector equation

$$c_1\mathbf{v}_1 + c_2\mathbf{v}_2 + \cdots + c_k\mathbf{v}_k = \mathbf{0}$$

always has the trivial solution

$$c_1 = 0, c_2 = 0, \ldots, c_k = 0.$$

Sometimes, however, there are also nontrivial solutions. For instance, in Example 1(a) you saw that in the set

$$\overset{\mathbf{v}_1}{} \quad \overset{\mathbf{v}_2}{} \quad \overset{\mathbf{v}_3}{}$$
$$S = \{(1, 3, 1), (0, 1, 2), (1, 0, -5)\}$$

the vector $\mathbf{v}_1$ can be written as a linear combination of the other two vectors, as shown below.

$$\mathbf{v}_1 = 3\mathbf{v}_2 + \mathbf{v}_3$$

So, the vector equation

$$c_1\mathbf{v}_1 + c_2\mathbf{v}_2 + c_3\mathbf{v}_3 = \mathbf{0}$$

has a nontrivial solution in which the coefficients are *not all zero:*

$$c_1 = 1, \quad c_2 = -3, \quad c_3 = -1.$$

When a nontrivial solution exists, the set S is **linearly dependent.** Had the only solution been the trivial one ($c_1 = c_2 = c_3 = 0$), then the set S would have been **linearly independent.** This concept is essential to the study of linear algebra.

Definition of Linear Dependence and Linear Independence

A set of vectors $S = \{\mathbf{v}_1, \mathbf{v}_2, \ldots, \mathbf{v}_k\}$ in a vector space V is **linearly independent** when the vector equation

$$c_1\mathbf{v}_1 + c_2\mathbf{v}_2 + \cdots + c_k\mathbf{v}_k = \mathbf{0}$$

has only the trivial solution

$$c_1 = 0, c_2 = 0, \ldots, c_k = 0.$$

If there are also nontrivial solutions, then S is **linearly dependent.**

EXAMPLE 7   **Examples of Linearly Dependent Sets**

a. The set $S = \{(1, 2), (2, 4)\}$ in R^2 is linearly dependent because

$$-2(1, 2) + (2, 4) = (0, 0).$$

b. The set $S = \{(1, 0), (0, 1), (-2, 5)\}$ in R^2 is linearly dependent because

$$2(1, 0) - 5(0, 1) + (-2, 5) = (0, 0).$$

c. The set $S = \{(0, 0), (1, 2)\}$ in R^2 is linearly dependent because

$$1(0, 0) + 0(1, 2) = (0, 0).$$

The next example demonstrates a test to determine whether a set of vectors is linearly independent or linearly dependent.

EXAMPLE 8   **Testing for Linear Independence**

See LarsonLinearAlgebra.com for an interactive version of this type of example.

Determine whether the set of vectors in R^3 is linearly independent or linearly dependent.

$$S = \{\mathbf{v}_1, \mathbf{v}_2, \mathbf{v}_3\} = \{(1, 2, 3), (0, 1, 2), (-2, 0, 1)\}$$

SOLUTION

To test for linear independence or linear dependence, form the vector equation

$$c_1\mathbf{v}_1 + c_2\mathbf{v}_2 + c_3\mathbf{v}_3 = \mathbf{0}.$$

If the only solution of this equation is $c_1 = c_2 = c_3 = 0$, then the set S is linearly independent. Otherwise, S is linearly dependent. Expanding this equation, you have

$$c_1(1, 2, 3) + c_2(0, 1, 2) + c_3(-2, 0, 1) = (0, 0, 0)$$
$$(c_1 - 2c_3, 2c_1 + c_2, 3c_1 + 2c_2 + c_3) = (0, 0, 0)$$

which yields the homogeneous system of linear equations in c_1, c_2, and c_3 below.

$$\begin{aligned} c_1 \qquad\quad - 2c_3 &= 0 \\ 2c_1 + \ c_2 \qquad\quad &= 0 \\ 3c_1 + 2c_2 + \ c_3 &= 0 \end{aligned}$$

The augmented matrix of this system reduces by Gauss-Jordan elimination as shown.

$$\begin{bmatrix} 1 & 0 & -2 & 0 \\ 2 & 1 & 0 & 0 \\ 3 & 2 & 1 & 0 \end{bmatrix} \longrightarrow \begin{bmatrix} 1 & 0 & 0 & 0 \\ 0 & 1 & 0 & 0 \\ 0 & 0 & 1 & 0 \end{bmatrix}$$

This implies that the only solution is the trivial solution $c_1 = c_2 = c_3 = 0$. So, S is linearly independent.

The steps in Example 8 are summarized below.

Testing for Linear Independence and Linear Dependence

Let $S = \{\mathbf{v}_1, \mathbf{v}_2, \ldots, \mathbf{v}_k\}$ be a set of vectors in a vector space V. To determine whether S is linearly independent or linearly dependent, use the steps below.

1. From the vector equation $c_1\mathbf{v}_1 + c_2\mathbf{v}_2 + \cdots + c_k\mathbf{v}_k = \mathbf{0}$, write a system of linear equations in the variables $c_1, c_2, \ldots,$ and c_k.
2. Determine whether the system has a unique solution.
3. If the system has only the trivial solution, $c_1 = 0, c_2 = 0, \ldots, c_k = 0$, then the set S is linearly independent. If the system also has nontrivial solutions, then S is linearly dependent.

LINEAR ALGEBRA APPLIED

Image morphing is the process of transforming one image into another by generating a sequence of synthetic intermediate images. Morphing has a wide variety of applications, such as movie special effects, age progression software, and simulating wound healing and cosmetic surgery results. Morphing an image uses a process called warping, in which a piece of an image is distorted. The mathematics behind warping and morphing can include forming a linear combination of the vectors that bound a triangular piece of an image, and performing an *affine transformation* to form new vectors and a distorted image piece.

dundanim/Shutterstock.com

EXAMPLE 9   Testing for Linear Independence

Determine whether the set of vectors in P_2 is linearly independent or linearly dependent.

$$\overset{v_1}{}\quad\overset{v_2}{}\quad\overset{v_3}{}$$
$$S = \{1 + x - 2x^2, 2 + 5x - x^2, x + x^2\}$$

SOLUTION

Expanding the equation $c_1v_1 + c_2v_2 + c_3v_3 = \mathbf{0}$ produces

$$c_1(1 + x - 2x^2) + c_2(2 + 5x - x^2) + c_3(x + x^2) = 0 + 0x + 0x^2$$
$$(c_1 + 2c_2) + (c_1 + 5c_2 + c_3)x + (-2c_1 - c_2 + c_3)x^2 = 0 + 0x + 0x^2.$$

Equating corresponding coefficients of powers of x yields the homogeneous system of linear equations in c_1, c_2, and c_3 below.

$$\begin{aligned} c_1 + 2c_2 &= 0 \\ c_1 + 5c_2 + c_3 &= 0 \\ -2c_1 - c_2 + c_3 &= 0 \end{aligned}$$

The augmented matrix of this system reduces by Gaussian elimination as shown below.

$$\begin{bmatrix} 1 & 2 & 0 & 0 \\ 1 & 5 & 1 & 0 \\ -2 & -1 & 1 & 0 \end{bmatrix} \rightarrow \begin{bmatrix} 1 & 2 & 0 & 0 \\ 0 & 1 & \frac{1}{3} & 0 \\ 0 & 0 & 0 & 0 \end{bmatrix}$$

This implies that the system has infinitely many solutions. So, the system must have nontrivial solutions, and you can conclude that the set S is linearly dependent.
    One nontrivial solution is

$$c_1 = 2, \quad c_2 = -1, \quad \text{and} \quad c_3 = 3$$

which yields the nontrivial linear combination

$$(2)(1 + x - 2x^2) + (-1)(2 + 5x - x^2) + (3)(x + x^2) = 0.$$

EXAMPLE 10   Testing for Linear Independence

Determine whether the set of vectors in $M_{2,2}$ is linearly independent or linearly dependent.

$$S = \left\{ \overset{v_1}{\begin{bmatrix} 2 & 1 \\ 0 & 1 \end{bmatrix}}, \overset{v_2}{\begin{bmatrix} 3 & 0 \\ 2 & 1 \end{bmatrix}}, \overset{v_3}{\begin{bmatrix} 1 & 0 \\ 2 & 0 \end{bmatrix}} \right\}$$

SOLUTION

From the equation $c_1v_1 + c_2v_2 + c_3v_3 = \mathbf{0}$, you have

$$c_1\begin{bmatrix} 2 & 1 \\ 0 & 1 \end{bmatrix} + c_2\begin{bmatrix} 3 & 0 \\ 2 & 1 \end{bmatrix} + c_3\begin{bmatrix} 1 & 0 \\ 2 & 0 \end{bmatrix} = \begin{bmatrix} 0 & 0 \\ 0 & 0 \end{bmatrix}$$

which produces the system of linear equations in c_1, c_2, and c_3 below.

$$\begin{aligned} 2c_1 + 3c_2 + c_3 &= 0 \\ c_1 &= 0 \\ 2c_2 + 2c_3 &= 0 \\ c_1 + c_2 &= 0 \end{aligned}$$

Use Gaussian elimination to show that the system has only the trivial solution, which means that the set S is linearly independent.

EXAMPLE 11   **Testing for Linear Independence**

Determine whether the set of vectors in $M_{4,1}$ is linearly independent or linearly dependent.

$$S = \{\mathbf{v}_1, \mathbf{v}_2, \mathbf{v}_3, \mathbf{v}_4\} = \left\{ \begin{bmatrix} 1 \\ 0 \\ -1 \\ 0 \end{bmatrix}, \begin{bmatrix} 1 \\ 1 \\ 0 \\ 2 \end{bmatrix}, \begin{bmatrix} 0 \\ 3 \\ 1 \\ -2 \end{bmatrix}, \begin{bmatrix} 0 \\ 1 \\ -1 \\ 2 \end{bmatrix} \right\}$$

SOLUTION

From the equation $c_1\mathbf{v}_1 + c_2\mathbf{v}_2 + c_3\mathbf{v}_3 + c_4\mathbf{v}_4 = \mathbf{0}$, you obtain

$$c_1 \begin{bmatrix} 1 \\ 0 \\ -1 \\ 0 \end{bmatrix} + c_2 \begin{bmatrix} 1 \\ 1 \\ 0 \\ 2 \end{bmatrix} + c_3 \begin{bmatrix} 0 \\ 3 \\ 1 \\ -2 \end{bmatrix} + c_4 \begin{bmatrix} 0 \\ 1 \\ -1 \\ 2 \end{bmatrix} = \begin{bmatrix} 0 \\ 0 \\ 0 \\ 0 \end{bmatrix}.$$

This equation produces the system of linear equations in c_1, c_2, c_3, and c_4 below.

$$\begin{aligned} c_1 + c_2 \qquad\qquad\quad &= 0 \\ c_2 + 3c_3 + c_4 &= 0 \\ -c_1 \qquad + c_3 - c_4 &= 0 \\ 2c_2 - 2c_3 + 2c_4 &= 0 \end{aligned}$$

Use Gaussian elimination to show that the system has only the trivial solution, which means that the set S is linearly independent.

If a set of vectors is linearly dependent, then by definition the equation $c_1\mathbf{v}_1 + c_2\mathbf{v}_2 + \cdots + c_k\mathbf{v}_k = \mathbf{0}$ has a nontrivial solution (a solution for which not all the c_i's are zero). For instance, if $c_1 \neq 0$, then you can solve this equation for $\mathbf{v}_1$ and write $\mathbf{v}_1$ as a linear combination of the other vectors $\mathbf{v}_2, \mathbf{v}_3, \ldots,$ and $\mathbf{v}_k$. In other words, the vector $\mathbf{v}_1$ *depends* on the other vectors in the set. This property is characteristic of a linearly dependent set.

THEOREM 4.8   A Property of Linearly Dependent Sets

A set $S = \{\mathbf{v}_1, \mathbf{v}_2, \ldots, \mathbf{v}_k\}$, $k \geq 2$, is linearly dependent if and only if at least one of the vectors $\mathbf{v}_i$ can be written as a linear combination of the other vectors in S.

PROOF

To prove the theorem in one direction, assume S is a linearly dependent set. Then there exist scalars $c_1, c_2, c_3, \ldots, c_k$ (not all zero) such that

$$c_1\mathbf{v}_1 + c_2\mathbf{v}_2 + c_3\mathbf{v}_3 + \cdots + c_k\mathbf{v}_k = \mathbf{0}.$$

One of the coefficients must be nonzero, so no generality is lost by assuming $c_1 \neq 0$. Then solving for $\mathbf{v}_1$ as a linear combination of the other vectors produces

$$c_1\mathbf{v}_1 = -c_2\mathbf{v}_2 - c_3\mathbf{v}_3 - \cdots - c_k\mathbf{v}_k$$

$$\mathbf{v}_1 = -\frac{c_2}{c_1}\mathbf{v}_2 - \frac{c_3}{c_1}\mathbf{v}_3 - \cdots - \frac{c_k}{c_1}\mathbf{v}_k.$$

Conversely, assume the vector $\mathbf{v}_1$ in S is a linear combination of the other vectors. That is,

$$\mathbf{v}_1 = c_2\mathbf{v}_2 + c_3\mathbf{v}_3 + \cdots + c_k\mathbf{v}_k.$$

Then the equation $-\mathbf{v}_1 + c_2\mathbf{v}_2 + c_3\mathbf{v}_3 + \cdots + c_k\mathbf{v}_k = \mathbf{0}$ has at least one coefficient, -1, that is nonzero, and you can conclude that S is linearly dependent.

EXAMPLE 12 | **Writing a Vector as a Linear Combination of Other Vectors**

In Example 9, you determined that the set

$$S = \{\overset{\mathbf{v}_1}{1 + x - 2x^2}, \overset{\mathbf{v}_2}{2 + 5x - x^2}, \overset{\mathbf{v}_3}{x + x^2}\}$$

is linearly dependent. Show that one of the vectors in this set can be written as a linear combination of the other two.

SOLUTION

In Example 9, the equation $c_1\mathbf{v}_1 + c_2\mathbf{v}_2 + c_3\mathbf{v}_3 = \mathbf{0}$ produced the system

$$
\begin{aligned}
c_1 + 2c_2 &= 0 \\
c_1 + 5c_2 + c_3 &= 0 \\
-2c_1 - c_2 + c_3 &= 0.
\end{aligned}
$$

This system has infinitely many solutions represented by $c_3 = 3t$, $c_2 = -t$, and $c_1 = 2t$. Letting $t = 1$ results in the equation $2\mathbf{v}_1 - \mathbf{v}_2 + 3\mathbf{v}_3 = \mathbf{0}$. So, $\mathbf{v}_2$ can be written as a linear combination of $\mathbf{v}_1$ and $\mathbf{v}_3$, as shown below.

$$\mathbf{v}_2 = 2\mathbf{v}_1 + 3\mathbf{v}_3$$

A check yields

$$2 + 5x - x^2 = 2(1 + x - 2x^2) + 3(x + x^2) = 2 + 5x - x^2.$$

Theorem 4.8 has a practical corollary that provides a simple test for determining whether *two* vectors are linearly dependent. In Exercise 77 you are asked to prove this corollary.

THEOREM 4.8   Corollary

Two vectors $\mathbf{u}$ and $\mathbf{v}$ in a vector space V are linearly dependent if and only if one is a scalar multiple of the other.

EXAMPLE 13 | **Testing for Linear Dependence of Two Vectors**

a. The set $S = \{\mathbf{v}_1, \mathbf{v}_2\} = \{(1, 2, 0), (-2, 2, 1)\}$ is linearly independent because $\mathbf{v}_1$ and $\mathbf{v}_2$ are not scalar multiples of each other, as shown in Figure 4.15(a).

b. The set $S = \{\mathbf{v}_1, \mathbf{v}_2\} = \{(4, -4, -2), (-2, 2, 1)\}$ is linearly dependent because $\mathbf{v}_1 = -2\mathbf{v}_2$, as shown in Figure 4.15(b).

a.

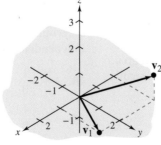

$S = \{(1, 2, 0), (-2, 2, 1)\}$
The set S is linearly independent.

b.

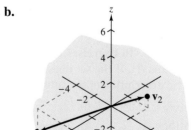

$S = \{(4, -4, -2), (-2, 2, 1)\}$
The set S is linearly dependent.

Figure 4.15

4.4 Exercises

See CalcChat.com for worked-out solutions to odd-numbered exercises.

Linear Combinations **In Exercises 1–4, write each vector as a linear combination of the vectors in S (if possible).**

1. $S = \{(2, -1, 3), (5, 0, 4)\}$
   (a) $\mathbf{z} = (-1, -2, 2)$ (b) $\mathbf{v} = \left(8, -\frac{1}{4}, \frac{27}{4}\right)$
   (c) $\mathbf{w} = (1, -8, 12)$ (d) $\mathbf{u} = (1, 1, -1)$

2. $S = \{(1, 2, -2), (2, -1, 1)\}$
   (a) $\mathbf{z} = (-4, -3, 3)$ (b) $\mathbf{v} = (-2, -6, 6)$
   (c) $\mathbf{w} = (-1, -22, 22)$ (d) $\mathbf{u} = (1, -5, -5)$

3. $S = \{(2, 0, 7), (2, 4, 5), (2, -12, 13)\}$
   (a) $\mathbf{u} = (-1, 5, -6)$ (b) $\mathbf{v} = (-3, 15, 18)$
   (c) $\mathbf{w} = \left(\frac{1}{3}, \frac{4}{3}, \frac{1}{2}\right)$ (d) $\mathbf{z} = (2, 20, -3)$

4. $S = \{(6, -7, 8, 6), (4, 6, -4, 1)\}$
   (a) $\mathbf{u} = (2, 19, -16, -4)$ (b) $\mathbf{v} = \left(\frac{49}{2}, \frac{99}{4}, -14, \frac{19}{2}\right)$
   (c) $\mathbf{w} = \left(-4, -14, \frac{27}{2}, \frac{53}{8}\right)$ (d) $\mathbf{z} = \left(8, 4, -1, \frac{17}{4}\right)$

Linear Combinations **In Exercises 5–8, for the matrices**
$$A = \begin{bmatrix} 2 & -3 \\ 4 & 1 \end{bmatrix} \text{ and } B = \begin{bmatrix} 0 & 5 \\ 1 & -2 \end{bmatrix}$$
in $M_{2,2}$, determine whether the given matrix is a linear combination of A and B.

5. $\begin{bmatrix} 6 & -19 \\ 10 & 7 \end{bmatrix}$ 6. $\begin{bmatrix} 6 & 2 \\ 9 & 11 \end{bmatrix}$

7. $\begin{bmatrix} -2 & 23 \\ 0 & -9 \end{bmatrix}$ 8. $\begin{bmatrix} 0 & 0 \\ 0 & 0 \end{bmatrix}$

Spanning Sets **In Exercises 9–18, determine whether the set S spans R^2. If the set does not span R^2, then give a geometric description of the subspace that it does span.**

9. $S = \{(2, 1), (-1, 2)\}$ 10. $S = \{(-1, 1), (3, 1)\}$
11. $S = \{(5, 0), (5, -4)\}$ 12. $S = \{(2, 0), (0, 1)\}$
13. $S = \{(-3, 5)\}$ 14. $S = \{(1, 1)\}$
15. $S = \{(-1, 2), (2, -4)\}$ 16. $S = \{(0, 2), (1, 4)\}$
17. $S = \{(1, 3), (-2, -6), (4, 12)\}$
18. $S = \{(-1, 2), (2, -1), (1, 1)\}$

Spanning Sets **In Exercises 19–24, determine whether the set S spans R^3. If the set does not span R^3, then give a geometric description of the subspace that it does span.**

19. $S = \{(4, 7, 3), (-1, 2, 6), (2, -3, 5)\}$
20. $S = \{(5, 6, 5), (2, 1, -5), (0, -4, 1)\}$
21. $S = \{(-2, 5, 0), (4, 6, 3)\}$
22. $S = \{(1, 0, 1), (1, 1, 0), (0, 1, 1)\}$
23. $S = \{(1, -2, 0), (0, 0, 1), (-1, 2, 0)\}$
24. $S = \{(1, 0, 3), (2, 0, -1), (4, 0, 5), (2, 0, 6)\}$

25. Determine whether the set $S = \{1, x^2, 2 + x^2\}$ spans P_2.

26. Determine whether the set
$$S = \{-2x + x^2, 8 + x^3, -x^2 + x^3, -4 + x^2\}$$
spans P_3.

Testing for Linear Independence **In Exercises 27–40, determine whether the set S is linearly independent or linearly dependent.**

27. $S = \{(-2, 2), (3, 5)\}$ 28. $S = \{(3, -6), (-1, 2)\}$
29. $S = \{(0, 0), (1, -1)\}$
30. $S = \{(1, 0), (1, 1), (2, -1)\}$
31. $S = \{(1, -4, 1), (6, 3, 2)\}$
32. $S = \{(6, 2, 1), (-1, 3, 2)\}$
33. $S = \{(-2, 1, 3), (2, 9, -3), (2, 3, -3)\}$
34. $S = \{(1, 1, 1), (2, 2, 2), (3, 3, 3)\}$
35. $S = \left\{\left(\frac{3}{4}, \frac{5}{2}, \frac{3}{2}\right), \left(3, 4, \frac{7}{2}\right), \left(-\frac{3}{2}, 6, 2\right)\right\}$
36. $S = \{(-4, -3, 4), (1, -2, 3), (6, 0, 0)\}$
37. $S = \{(1, 0, 0), (0, 4, 0), (0, 0, -6), (1, 5, -3)\}$
38. $S = \{(4, -3, 6, 2), (1, 8, 3, 1), (3, -2, -1, 0)\}$
39. $S = \{(0, 0, 0, 1), (0, 0, 1, 1), (0, 1, 1, 1), (1, 1, 1, 1)\}$
40. $S = \{(4, 1, 2, 3), (3, 2, 1, 4), (1, 5, 5, 9), (1, 3, 9, 7)\}$

Testing for Linear Independence **In Exercises 41–48, determine whether the set of vectors in P_2 is linearly independent or linearly dependent.**

41. $S = \{2 - x, 2x - x^2, 6 - 5x + x^2\}$
42. $S = \{-1 + x^2, 5 + 2x\}$
43. $S = \{1 + 3x + x^2, -1 + x + 2x^2, 4x\}$
44. $S = \{x^2, 1 + x^2\}$
45. $S = \{-x + x^2, -5 + x, -5 + x^2\}$
46. $S = \{-2 - x, 2 + 3x + x^2, 6 + 5x + x^2\}$
47. $S = \{7 - 3x + 4x^2, 6 + 2x - x^2, 1 - 8x + 5x^2\}$
48. $S = \{7 - 4x + 4x^2, 6 + 2x - 3x^2, 20 - 6x + 5x^2\}$

Testing for Linear Independence **In Exercises 49–52, determine whether the set of vectors in $M_{2,2}$ is linearly independent or linearly dependent.**

49. $A = \begin{bmatrix} 1 & 0 \\ 0 & -2 \end{bmatrix}, B = \begin{bmatrix} 0 & 1 \\ 1 & 0 \end{bmatrix}, C = \begin{bmatrix} -2 & 1 \\ 1 & 4 \end{bmatrix}$

50. $A = \begin{bmatrix} 1 & 0 \\ 0 & 1 \end{bmatrix}, B = \begin{bmatrix} 0 & 1 \\ 0 & 0 \end{bmatrix}, C = \begin{bmatrix} 0 & 0 \\ 1 & 0 \end{bmatrix}$

51. $A = \begin{bmatrix} 1 & -1 \\ 4 & 5 \end{bmatrix}, B = \begin{bmatrix} 4 & 3 \\ -2 & 3 \end{bmatrix}, C = \begin{bmatrix} 1 & -8 \\ 22 & 23 \end{bmatrix}$

52. $A = \begin{bmatrix} 2 & 0 \\ -3 & 1 \end{bmatrix}, B = \begin{bmatrix} -4 & -1 \\ 0 & 5 \end{bmatrix}, C = \begin{bmatrix} -8 & -3 \\ -6 & 17 \end{bmatrix}$

Showing Linear Dependence In Exercises 53–56, show that the set is linearly dependent by finding a nontrivial linear combination of vectors in the set whose sum is the zero vector. Then express one of the vectors in the set as a linear combination of the other vectors in the set.

53. $S = \{(3, 4), (-1, 1), (2, 0)\}$
54. $S = \{(2, 4), (-1, -2), (0, 6)\}$
55. $S = \{(1, 1, 1), (1, 1, 0), (0, 1, 1), (0, 0, 1)\}$
56. $S = \{(1, 2, 3, 4), (1, 0, 1, 2), (1, 4, 5, 6)\}$

57. For which values of t is each set linearly independent?
   (a) $S = \{(t, 1, 1), (1, t, 1), (1, 1, t)\}$
   (b) $S = \{(t, 1, 1), (1, 0, 1), (1, 1, 3t)\}$
58. For which values of t is each set linearly independent?
   (a) $S = \{(t, 0, 0), (0, 1, 0), (0, 0, 1)\}$
   (b) $S = \{(t, t, t), (t, 1, 0), (t, 0, 1)\}$
59. **Proof** Complete the proof of Theorem 4.7.

60. **CAPSTONE** By inspection, determine why each of the sets is linearly dependent.
   (a) $S = \{(1, -2), (2, 3), (-2, 4)\}$
   (b) $S = \{(1, -6, 2), (2, -12, 4)\}$
   (c) $S = \{(0, 0), (1, 0)\}$

Spanning the Same Subspace In Exercises 61 and 62, show that the sets S_1 and S_2 span the same subspace of R^3.
61. $S_1 = \{(1, 2, -1), (0, 1, 1), (2, 5, -1)\}$
   $S_2 = \{(-2, -6, 0), (1, 1, -2)\}$
62. $S_1 = \{(0, 0, 1), (0, 1, 1), (2, 1, 1)\}$
   $S_2 = \{(1, 1, 1), (1, 1, 2), (2, 1, 1)\}$

True or False? In Exercises 63 and 64, determine whether each statement is true or false. If a statement is true, give a reason or cite an appropriate statement from the text. If a statement is false, provide an example that shows the statement is not true in all cases or cite an appropriate statement from the text.

63. (a) A set of vectors $S = \{v_1, v_2, \ldots, v_k\}$ in a vector space is linearly dependent when the vector equation $c_1v_1 + c_2v_2 + \cdots + c_kv_k = 0$ has only the trivial solution.
   (b) The set $S = \{(1, 0, 0, 0), (0, -1, 0, 0), (0, 0, 1, 0), (0, 0, 0, 1)\}$ spans R^4.
64. (a) A set $S = \{v_1, v_2, \ldots, v_k\}$, $k \geq 2$, is linearly independent if and only if at least one of the vectors v_i can be written as a linear combination of the other vectors in S.
   (b) If a subset S spans a vector space V, then every vector in V can be written as a linear combination of the vectors in S.

Proof In Exercises 65 and 66, prove that the set of vectors is linearly independent and spans R^3.
65. $B = \{(1, 1, 1), (1, 1, 0), (1, 0, 0)\}$
66. $B = \{(1, 2, 3), (3, 2, 1), (0, 0, 1)\}$

67. **Guided Proof** Prove that a nonempty subset of a finite set of linearly independent vectors is linearly independent.

   Getting Started: You need to show that a subset of a linearly independent set of vectors cannot be linearly dependent.
   (i) Assume S is a set of linearly independent vectors. Let T be a subset of S.
   (ii) If T is linearly dependent, then there exist constants not all zero satisfying the vector equation $c_1v_1 + c_2v_2 + \cdots + c_kv_k = 0$.
   (iii) Use this fact to derive a contradiction and conclude that T is linearly independent.

68. **Proof** Prove that if S_1 is a nonempty subset of the finite set S_2, and S_1 is linearly dependent, then so is S_2.
69. **Proof** Prove that any set of vectors containing the zero vector is linearly dependent.
70. **Proof** When the set of vectors $\{u_1, u_2, \ldots, u_n\}$ is linearly independent and the set $\{u_1, u_2, \ldots, u_n, v\}$ is linearly dependent, prove that v is a linear combination of the u_i's.
71. **Proof** Let $\{v_1, v_2, \ldots, v_k\}$ be a linearly independent set of vectors in a vector space V. Delete the vector v_k from this set and prove that the set $\{v_1, v_2, \ldots, v_{k-1}\}$ cannot span V.
72. **Proof** When V is spanned by $\{v_1, v_2, \ldots, v_k\}$ and one of these vectors can be written as a linear combination of the other $k - 1$ vectors, prove that the span of these $k - 1$ vectors is also V.
73. **Proof** Let $S = \{u, v\}$ be a linearly independent set. Prove that the set $\{u + v, u - v\}$ is linearly independent.
74. Let u, v, and w be any three vectors from a vector space V. Determine whether the set of vectors $\{v - u, w - v, u - w\}$ is linearly independent or linearly dependent.
75. **Proof** Let A be a nonsingular matrix of order 3. Prove that if $\{v_1, v_2, v_3\}$ is a linearly independent set in $M_{3,1}$, then the set $\{Av_1, Av_2, Av_3\}$ is also linearly independent. Explain, by means of an example, why this is not true when A is singular.
76. Let $f_1(x) = 3x$ and $f_2(x) = |x|$. Graph both functions on the interval $-2 \leq x \leq 2$. Show that these functions are linearly dependent in the vector space $C[0, 1]$, but linearly independent in $C[-1, 1]$.
77. **Proof** Prove the corollary to Theorem 4.8: Two vectors u and v are linearly dependent if and only if one is a scalar multiple of the other.

4.5   Basis and Dimension

■ Recognize bases in the vector spaces R^n, P_n, and $M_{m,n}$.

■ Find the dimension of a vector space.

BASIS FOR A VECTOR SPACE

REMARK

This definition tells you that a basis has two features. A basis S must have *enough vectors* to span V, but *not so many vectors* that one of them could be written as a linear combination of the other vectors in S.

In this section, you will continue your study of spanning sets. In particular, you will look at spanning sets in a vector space that are both linearly independent *and* span the entire space. Such a set forms a **basis** for the vector space. (The plural of *basis* is *bases*.)

Definition of Basis

A set of vectors $S = \{\mathbf{v}_1, \mathbf{v}_2, \ldots, \mathbf{v}_n\}$ in a vector space V is a **basis** for V when the conditions below are true.

1. S spans V.   **2.** S is linearly independent.

This definition does not imply that every vector space has a basis consisting of a finite number of vectors. This text, however, restricts the discussion to such bases. Moreover, if a vector space V has a basis with a finite number of vectors, then V is **finite dimensional.** Otherwise, V is **infinite dimensional.** [The vector space P of *all* polynomials is infinite dimensional, as is the vector space $C(-\infty, \infty)$ of all continuous functions defined on the real line.] The vector space $V = \{\mathbf{0}\}$, consisting of the zero vector alone, is finite dimensional.

EXAMPLE 1   **The Standard Basis for R^3**

Show that the set below is a basis for R^3.

$$S = \{(1, 0, 0), (0, 1, 0), (0, 0, 1)\}$$

SOLUTION

Example 4(a) in Section 4.4 showed that S spans R^3. Furthermore, S is linearly independent because the vector equation

$$c_1(1, 0, 0) + c_2(0, 1, 0) + c_3(0, 0, 1) = (0, 0, 0)$$

has only the trivial solution

$$c_1 = c_2 = c_3 = 0.$$

(Verify this.) So, S is a basis for R^3. (See Figure 4.16.)

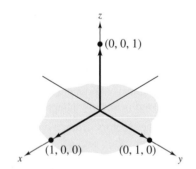

Figure 4.16

The basis

$$S = \{(1, 0, 0), (0, 1, 0), (0, 0, 1)\}$$

is the **standard basis** for R^3. This can be generalized to n-space. That is, the vectors

$$\mathbf{e}_1 = (1, 0, \ldots, 0)$$
$$\mathbf{e}_2 = (0, 1, \ldots, 0)$$
$$\vdots$$
$$\mathbf{e}_n = (0, 0, \ldots, 1)$$

form the **standard basis** for R^n.

The next two examples describe nonstandard bases for R^2 and R^3.

EXAMPLE 2 A Nonstandard Basis for R^2

Show that the set

$$v_1 \quad v_2$$
$$S = \{(1, 1), (1, -1)\}$$

is a basis for R^2.

SOLUTION

According to the definition of a basis for a vector space, you must show that S spans R^2 and S is linearly independent.

To verify that S spans R^2, let

$$\mathbf{x} = (x_1, x_2)$$

represent an arbitrary vector in R^2. To show that $\mathbf{x}$ can be written as a linear combination of $\mathbf{v}_1$ and $\mathbf{v}_2$, consider the equation

$$c_1\mathbf{v}_1 + c_2\mathbf{v}_2 = \mathbf{x}$$
$$c_1(1, 1) + c_2(1, -1) = (x_1, x_2)$$
$$(c_1 + c_2, c_1 - c_2) = (x_1, x_2).$$

Equating corresponding components yields the system of linear equations below.

$$c_1 + c_2 = x_1$$
$$c_1 - c_2 = x_2$$

The coefficient matrix of this system has a nonzero determinant, which means that the system has a unique solution. So, S spans R^2.

One way to show that S is linearly independent is to let $(x_1, x_2) = (0, 0)$ in the above system, yielding the homogeneous system

$$c_1 + c_2 = 0$$
$$c_1 - c_2 = 0.$$

This system has only the trivial solution

$$c_1 = c_2 = 0.$$

So, S is linearly independent. An alternative way to show that S is linearly independent is to note that

$$\mathbf{v}_1 = (1, 1) \quad \text{and} \quad \mathbf{v}_2 = (1, -1)$$

are not scalar multiples of each other. This means, by the corollary to Theorem 4.8, that $S = \{\mathbf{v}_1, \mathbf{v}_2\}$ is linearly independent.

You can conclude that S is a basis for R^2 because it is a spanning set for R^2 and it is linearly independent.

EXAMPLE 3 A Nonstandard Basis for R^3

See LarsonLinearAlgebra.com for an interactive version of this type of example.

From Examples 5 and 8 in the preceding section, you know that

$$S = \{(1, 2, 3), (0, 1, 2), (-2, 0, 1)\}$$

spans R^3 and is linearly independent. So, S is a basis for R^3.

EXAMPLE 4   A Basis for Polynomials

Show that the vector space P_3 has the basis

$$S = \{1, x, x^2, x^3\}.$$

SOLUTION

It is clear that S spans P_3 because the span of S consists of all polynomials of the form

$$a_0 + a_1x + a_2x^2 + a_3x^3, \quad a_0, a_1, a_2, \text{ and } a_3 \text{ are real numbers}$$

which is precisely the form of all polynomials in P_3.

To verify the linear independence of S, recall that the zero vector $\mathbf{0}$ in P_3 is the polynomial $\mathbf{0}(x) = 0$ for all x. The test for linear independence yields the equation

$$a_0 + a_1x + a_2x^2 + a_3x^3 = \mathbf{0}(x) = 0, \quad \text{for all } x.$$

This third-degree polynomial is *identically equal to zero*. From algebra you know that for a polynomial to be identically equal to zero, all of its coefficients must be zero; that is,

$$a_0 = a_1 = a_2 = a_3 = 0.$$

So, S is linearly independent and is a basis for P_3.

REMARK

The basis $S = \{1, x, x^2, x^3\}$ is the **standard basis** for P_3. Similarly, the **standard basis** for P_n is

$$S = \{1, x, x^2, \ldots, x^n\}.$$

EXAMPLE 5   A Basis for $M_{2,2}$

The set

$$S = \left\{ \begin{bmatrix} 1 & 0 \\ 0 & 0 \end{bmatrix}, \begin{bmatrix} 0 & 1 \\ 0 & 0 \end{bmatrix}, \begin{bmatrix} 0 & 0 \\ 1 & 0 \end{bmatrix}, \begin{bmatrix} 0 & 0 \\ 0 & 1 \end{bmatrix} \right\}.$$

is a basis for $M_{2,2}$. This set is the **standard basis** for $M_{2,2}$. In a similar manner, the standard basis for the vector space $M_{m,n}$ consists of the mn distinct $m \times n$ matrices having a single entry equal to 1 and all the other entries equal to 0.

THEOREM 4.9   Uniqueness of Basis Representation

If $S = \{\mathbf{v}_1, \mathbf{v}_2, \ldots, \mathbf{v}_n\}$ is a basis for a vector space V, then every vector in V can be written in one and only one way as a linear combination of vectors in S.

PROOF

The existence portion of the proof is straightforward. That is, S spans V, so you know that an arbitrary vector $\mathbf{u}$ in V can be expressed as $\mathbf{u} = c_1\mathbf{v}_1 + c_2\mathbf{v}_2 + \cdots + c_n\mathbf{v}_n$.

To prove uniqueness (that a vector can be represented in only one way), assume $\mathbf{u}$ has another representation

$$\mathbf{u} = b_1\mathbf{v}_1 + b_2\mathbf{v}_2 + \cdots + b_n\mathbf{v}_n.$$

Subtracting the second representation from the first produces

$$\mathbf{u} - \mathbf{u} = (c_1 - b_1)\mathbf{v}_1 + (c_2 - b_2)\mathbf{v}_2 + \cdots + (c_n - b_n)\mathbf{v}_n = \mathbf{0}.$$

S is linearly independent, however, so the only solution to this equation is the trivial solution

$$c_1 - b_1 = 0, \quad c_2 - b_2 = 0, \quad \ldots, \quad c_n - b_n = 0$$

which means that $c_i = b_i$ for all $i = 1, 2, \ldots, n$, and $\mathbf{u}$ has only one representation for the basis S.

EXAMPLE 6   Uniqueness of Basis Representation

Let $\mathbf{u} = \{u_1, u_2, u_3\}$ be any vector in R^3. Show that the equation $\mathbf{u} = c_1\mathbf{v}_1 + c_2\mathbf{v}_2 + c_3\mathbf{v}_3$ has a unique solution for the basis $S = \{\mathbf{v}_1, \mathbf{v}_2, \mathbf{v}_3\} = \{(1, 2, 3), (0, 1, 2), (-2, 0, 1)\}$.

SOLUTION

From the equation

$$(u_1, u_2, u_3) = c_1(1, 2, 3) + c_2(0, 1, 2) + c_3(-2, 0, 1)$$
$$= (c_1 - 2c_3, 2c_1 + c_2, 3c_1 + 2c_2 + c_3)$$

you obtain the system of linear equations below.

$$
\begin{array}{r}
c_1 \qquad\; - 2c_3 = u_1 \\
2c_1 + c_2 \qquad\;\; = u_2 \\
3c_1 + 2c_2 + c_3 = u_3
\end{array}
\qquad
\underbrace{\begin{bmatrix} 1 & 0 & -2 \\ 2 & 1 & 0 \\ 3 & 2 & 1 \end{bmatrix}}_{A}
\underbrace{\begin{bmatrix} c_1 \\ c_2 \\ c_3 \end{bmatrix}}_{\mathbf{c}}
=
\underbrace{\begin{bmatrix} u_1 \\ u_2 \\ u_3 \end{bmatrix}}_{\mathbf{u}}
$$

The matrix A is invertible, so you know this system has a unique solution, $\mathbf{c} = A^{-1}\mathbf{u}$. Verify by finding A^{-1} that

$$c_1 = -u_1 + 4u_2 - 2u_3$$
$$c_2 = 2u_1 - 7u_2 + 4u_3$$
$$c_3 = -u_1 + 2u_2 - u_3.$$

For example, $\mathbf{u} = (1, 0, 0)$ can be represented uniquely as $-\mathbf{v}_1 + 2\mathbf{v}_2 - \mathbf{v}_3$.

You will now study two important theorems concerning bases.

THEOREM 4.10   Bases and Linear Dependence

If $S = \{\mathbf{v}_1, \mathbf{v}_2, \ldots, \mathbf{v}_n\}$ is a basis for a vector space V, then every set containing more than n vectors in V is linearly dependent.

PROOF

Let $S_1 = \{\mathbf{u}_1, \mathbf{u}_2, \ldots, \mathbf{u}_m\}$ be any set of m vectors in V, where $m > n$. To show that S_1 is linearly *dependent,* you need to find scalars $k_1, k_2, \ldots, k_m$ (not all zero) such that

$$k_1\mathbf{u}_1 + k_2\mathbf{u}_2 + \cdots + k_m\mathbf{u}_m = \mathbf{0}. \qquad \text{Equation 1}$$

S is a basis for V, so each $\mathbf{u}_i$ can be represented as a linear combination of vectors in S:

$$\mathbf{u}_1 = c_{11}\mathbf{v}_1 + c_{21}\mathbf{v}_2 + \cdots + c_{n1}\mathbf{v}_n$$
$$\mathbf{u}_2 = c_{12}\mathbf{v}_1 + c_{22}\mathbf{v}_2 + \cdots + c_{n2}\mathbf{v}_n$$
$$\vdots \qquad \vdots \qquad \vdots \qquad\qquad \vdots$$
$$\mathbf{u}_m = c_{1m}\mathbf{v}_1 + c_{2m}\mathbf{v}_2 + \cdots + c_{nm}\mathbf{v}_n.$$

Substituting into Equation 1 and regrouping terms produces

$$d_1\mathbf{v}_1 + d_2\mathbf{v}_2 + \cdots + d_n\mathbf{v}_n = \mathbf{0}$$

where $d_i = c_{i1}k_1 + c_{i2}k_2 + \cdots + c_{im}k_m$. The $\mathbf{v}_i$'s form a linearly independent set, so each $d_i = 0$, and you obtain the system of equations below.

$$c_{11}k_1 + c_{12}k_2 + \cdots + c_{1m}k_m = 0$$
$$c_{21}k_1 + c_{22}k_2 + \cdots + c_{2m}k_m = 0$$
$$\vdots \qquad \vdots \qquad\quad \vdots \qquad \vdots$$
$$c_{n1}k_1 + c_{n2}k_2 + \cdots + c_{nm}k_m = 0$$

But this homogeneous system has fewer equations than variables $k_1, k_2, \ldots, k_m$, and from Theorem 1.1, it has *nontrivial* solutions. Consequently, S_1 is linearly dependent.

EXAMPLE 7   Linearly Dependent Sets in R^3 and P_3

a. R^3 has a basis consisting of three vectors, so the set

$$S = \{(1, 2, -1), (1, 1, 0), (2, 3, 0), (5, 9, -1)\}$$

must be linearly dependent.

b. P_3 has a basis consisting of four vectors, so the set

$$S = \{1, 1 + x, 1 - x, 1 + x + x^2, 1 - x + x^2\}$$

must be linearly dependent.

R^n has the standard basis consisting of n vectors, so it follows from Theorem 4.10 that every set of vectors in R^n containing more than n vectors must be linearly dependent. The next theorem states another significant consequence of Theorem 4.10.

THEOREM 4.11   Number of Vectors in a Basis

If a vector space V has one basis with n vectors, then every basis for V has n vectors.

PROOF

Let $S_1 = \{\mathbf{v}_1, \mathbf{v}_2, \ldots, \mathbf{v}_n\}$ be a basis for V, and let $S_2 = \{\mathbf{u}_1, \mathbf{u}_2, \ldots, \mathbf{u}_m\}$ be any other basis for V. Theorem 4.10 implies that $m \leq n$, because S_1 is a basis and S_2 is linearly independent. Similarly, $n \leq m$ because S_1 is linearly independent and S_2 is a basis. Consequently, $n = m$.

EXAMPLE 8   Spanning Sets and Bases

Use Theorem 4.11 to explain why each statement is true.

a. The set $S_1 = \{(3, 2, 1), (7, -1, 4)\}$ is not a basis for R^3.

b. The set $S_2 = \{2 + x, x^2, -1 + x^3, 1 + 3x, 3 - 2x + x^2\}$ is not a basis for P_3.

SOLUTION

a. The standard basis for R^3, $S = \{(1, 0, 0), (0, 1, 0), (0, 0, 1)\}$, has three vectors, and S_1 has only two vectors. By Theorem 4.11, S_1 cannot be a basis for R^3.

b. The standard basis for P_3, $S = \{1, x, x^2, x^3\}$, has four vectors. By Theorem 4.11, the set S_2 has too many vectors to be a basis for P_3.

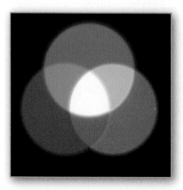

LINEAR ALGEBRA APPLIED

The RGB color model uses combinations of red ($\mathbf{r}$), green ($\mathbf{g}$), and blue ($\mathbf{b}$), known as the *primary additive colors,* to create all other colors in a system. Using the standard basis for R^3, where $\mathbf{r} = (1, 0, 0)$, $\mathbf{g} = (0, 1, 0)$, and $\mathbf{b} = (0, 0, 1)$, any visible color can be represented as a linear combination $c_1\mathbf{r} + c_2\mathbf{g} + c_3\mathbf{b}$ of the primary additive colors. The coefficients c_i are values between 0 and a specified maximum a, inclusive. When $c_1 = c_2 = c_3$, the color is *grayscale*, with $c_i = 0$ representing black and $c_i = a$ representing white. The RGB color model is commonly used in computers, smart phones, televisions, and other electronics with a color display.

THE DIMENSION OF A VECTOR SPACE

By Theorem 4.11, if a vector space V has a basis consisting of n vectors, then every other basis for the space also has n vectors. This number n is the **dimension** of V.

> **Definition of the Dimension of a Vector Space**
>
> If a vector space V has a basis consisting of n vectors, then the number n is the **dimension** of V, denoted by $\dim(V) = n$. When V consists of the zero vector alone, the dimension of V is defined as zero.

This definition allows you to state the dimensions of familiar vector spaces. In each example listed below, the dimension is simply the number of vectors in the standard basis.

1. The dimension of R^n with the standard operations is n.

2. The dimension of P_n with the standard operations is $n + 1$.

3. The dimension of $M_{m,n}$ with the standard operations is mn.

If W is a subspace of a vector space V that has dimension n, then it can be shown that the dimension of W is less than or equal to n. (See Exercise 83.) The next three examples show a technique for finding the dimension of a subspace. Basically, you determine the dimension by finding a set of linearly independent vectors that spans the subspace. This set is a basis for the subspace, and the dimension of the subspace is the number of vectors in the basis.

EXAMPLE 9   Finding Dimensions of Subspaces

Find the dimension of each subspace of R^3.

a. $W = \{(d, c - d, c):\ c \text{ and } d \text{ are real numbers}\}$
b. $W = \{(2b, b, 0):\ b \text{ is a real number}\}$

SOLUTION

a. By writing the representative vector $(d, c - d, c)$ as

$$(d, c - d, c) = (0, c, c) + (d, -d, 0) = c(0, 1, 1) + d(1, -1, 0)$$

you can see that W is spanned by the set $S = \{(0, 1, 1), (1, -1, 0)\}$. Using the techniques described in the preceding section, you can show that this set is linearly independent. So, S is a basis for W, and W is a two-dimensional subspace of R^3.

b. By writing the representative vector $(2b, b, 0)$ as $b(2, 1, 0)$, you can see that W is spanned by the set $S = \{(2, 1, 0)\}$. So, W is a one-dimensional subspace of R^3.

REMARK

In Example 9(a), the subspace W is the plane in R^3 determined by the vectors $(0, 1, 1)$ and $(1, -1, 0)$. In Example 9(b), the subspace is the line determined by the vector $(2, 1, 0)$.

EXAMPLE 10   Finding the Dimension of a Subspace

Find the dimension of the subspace W of R^4 spanned by

$$S = \{\mathbf{v}_1, \mathbf{v}_2, \mathbf{v}_3\} = \{(-1, 2, 5, 0), (3, 0, 1, -2), (-5, 4, 9, 2)\}.$$

SOLUTION

Although W is spanned by the set S, S is not a basis for W because S is a linearly dependent set. Specifically, $\mathbf{v}_3$ can be written as $\mathbf{v}_3 = 2\mathbf{v}_1 - \mathbf{v}_2$. This means that W is spanned by the set $S_1 = \{\mathbf{v}_1, \mathbf{v}_2\}$. Moreover, S_1 is linearly independent because neither vector is a scalar multiple of the other, and you can conclude that the dimension of W is 2.

EXAMPLE 11   **Finding the Dimension of a Subspace**

Let W be the subspace of all symmetric matrices in $M_{2,2}$. What is the dimension of W?

SOLUTION

Every 2×2 symmetric matrix has the form

$$A = \begin{bmatrix} a & b \\ b & c \end{bmatrix} = a\begin{bmatrix} 1 & 0 \\ 0 & 0 \end{bmatrix} + b\begin{bmatrix} 0 & 1 \\ 1 & 0 \end{bmatrix} + c\begin{bmatrix} 0 & 0 \\ 0 & 1 \end{bmatrix}.$$

So, the set

$$S = \left\{ \begin{bmatrix} 1 & 0 \\ 0 & 0 \end{bmatrix}, \begin{bmatrix} 0 & 1 \\ 1 & 0 \end{bmatrix}, \begin{bmatrix} 0 & 0 \\ 0 & 1 \end{bmatrix} \right\}$$

spans W. Moreover, S can be shown to be linearly independent, and you can conclude that the dimension of W is 3.

Usually, to conclude that a set $S = \{\mathbf{v}_1, \mathbf{v}_2, \ldots, \mathbf{v}_n\}$ is a basis for a vector space V, you must show that S satisfies two conditions: S spans V and is linearly independent. If V is known to have a dimension of n, however, then the next theorem tells you that you do not need to check both conditions. Either one will suffice. The proof is left as an exercise. (See Exercise 82.)

THEOREM 4.12   Basis Tests in an n-Dimensional Space

Let V be a vector space of dimension n.

1. If $S = \{\mathbf{v}_1, \mathbf{v}_2, \ldots, \mathbf{v}_n\}$ is a linearly independent set of vectors in V, then S is a basis for V.
2. If $S = \{\mathbf{v}_1, \mathbf{v}_2, \ldots, \mathbf{v}_n\}$ spans V, then S is a basis for V.

EXAMPLE 12   **Testing for a Basis in an n-Dimensional Space**

Show that the set of vectors is a basis for $M_{5,1}$.

$$S = \left\{ \underset{\mathbf{v}_1}{\begin{bmatrix} 1 \\ 2 \\ -1 \\ 3 \\ 4 \end{bmatrix}}, \underset{\mathbf{v}_2}{\begin{bmatrix} 0 \\ 1 \\ 3 \\ -2 \\ 3 \end{bmatrix}}, \underset{\mathbf{v}_3}{\begin{bmatrix} 0 \\ 0 \\ 2 \\ -1 \\ 5 \end{bmatrix}}, \underset{\mathbf{v}_4}{\begin{bmatrix} 0 \\ 0 \\ 0 \\ 2 \\ -3 \end{bmatrix}}, \underset{\mathbf{v}_5}{\begin{bmatrix} 0 \\ 0 \\ 0 \\ 0 \\ -2 \end{bmatrix}} \right\}$$

SOLUTION

S has five vectors and the dimension of $M_{5,1}$ is 5, so apply Theorem 4.12 to verify that S is a basis by showing either that S is linearly independent or that S spans $M_{5,1}$. To show that S is linearly independent, form the vector equation $c_1\mathbf{v}_1 + c_2\mathbf{v}_2 + c_3\mathbf{v}_3 + c_4\mathbf{v}_4 + c_5\mathbf{v}_5 = \mathbf{0}$, which yields the linear system below.

$$\begin{aligned} c_1 \qquad\qquad\qquad\qquad &= 0 \\ 2c_1 + c_2 \qquad\qquad\qquad &= 0 \\ -c_1 + 3c_2 + 2c_3 \qquad\qquad &= 0 \\ 3c_1 - 2c_2 - c_3 + 2c_4 \qquad &= 0 \\ 4c_1 + 3c_2 + 5c_3 - 3c_4 - 2c_5 &= 0 \end{aligned}$$

This system has only the trivial solution, so S is linearly independent. By Theorem 4.12, S is a basis for $M_{5,1}$.

4.5 Exercises

See CalcChat.com for worked-out solutions to odd-numbered exercises.

Writing the Standard Basis In Exercises 1–6, write the standard basis for the vector space.

1. R^6
2. R^4
3. $M_{3,3}$
4. $M_{4,1}$
5. P_4
6. P_2

Explaining Why a Set Is Not a Basis In Exercises 7–14, explain why S is not a basis for R^2.

7. $S = \{(-4, 5), (0, 0)\}$
8. $S = \{(2, 3), (6, 9)\}$
9. $S = \{(-3, 2)\}$
10. $S = \{(5, -7)\}$
11. $S = \{(1, 2), (1, 0), (0, 1)\}$
12. $S = \{(-1, 2), (1, -2), (2, 4)\}$
13. $S = \{(6, -5), (12, -10)\}$
14. $S = \{(4, -3), (8, -6)\}$

Explaining Why a Set Is Not a Basis In Exercises 15–22, explain why S is not a basis for R^3.

15. $S = \{(1, 3, 0), (4, 1, 2), (-2, 5, -2)\}$
16. $S = \{(2, 1, -2), (-2, -1, 2), (4, 2, -4)\}$
17. $S = \{(7, 0, 3), (8, -4, 1)\}$
18. $S = \{(1, 1, 2), (0, 2, 1)\}$
19. $S = \{(0, 0, 0), (1, 0, 0), (0, 1, 0)\}$
20. $S = \{(-1, 0, 0), (0, 0, 1), (1, 0, 0)\}$
21. $S = \{(1, 1, 1), (0, 1, 1), (1, 0, 1), (0, 0, 0)\}$
22. $S = \{(6, 4, 1), (3, -5, 1), (8, 13, 6), (0, 6, 9)\}$

Explaining Why a Set Is Not a Basis In Exercises 23–30, explain why S is not a basis for P_2.

23. $S = \{1, 2x, -4 + x^2, 5x\}$
24. $S = \{2, x, 3 + x, 3x^2\}$
25. $S = \{-x, 4x^2\}$
26. $S = \{-1, 11x\}$
27. $S = \{1 + x^2, 1 - x^2\}$
28. $S = \{1 - 2x + x^2, 3 - 6x + 3x^2, -2 + 4x - 2x^2\}$
29. $S = \{1 - x, 1 - x^2, -1 - 2x + 3x^2\}$
30. $S = \{-3 + 6x, 3x^2, 1 - 2x - x^2\}$

Explaining Why a Set Is Not a Basis In Exercises 31–34, explain why S is not a basis for $M_{2,2}$.

31. $S = \left\{ \begin{bmatrix} 1 & 0 \\ 0 & 1 \end{bmatrix}, \begin{bmatrix} 0 & 1 \\ 1 & 0 \end{bmatrix} \right\}$

32. $S = \left\{ \begin{bmatrix} 1 & 1 \\ 0 & 0 \end{bmatrix}, \begin{bmatrix} 0 & 1 \\ 1 & 0 \end{bmatrix}, \begin{bmatrix} -1 & 0 \\ 1 & 0 \end{bmatrix}, \begin{bmatrix} 0 & 0 \\ 0 & 1 \end{bmatrix} \right\}$

33. $S = \left\{ \begin{bmatrix} 1 & 0 \\ 0 & 0 \end{bmatrix}, \begin{bmatrix} 0 & 1 \\ 1 & 0 \end{bmatrix}, \begin{bmatrix} 1 & 0 \\ 0 & 1 \end{bmatrix}, \begin{bmatrix} 8 & -4 \\ -4 & 3 \end{bmatrix} \right\}$

34. $S = \left\{ \begin{bmatrix} 1 & 0 \\ 0 & 1 \end{bmatrix}, \begin{bmatrix} 0 & 1 \\ 1 & 0 \end{bmatrix}, \begin{bmatrix} 1 & 1 \\ 0 & 0 \end{bmatrix} \right\}$

Determining Whether a Set Is a Basis In Exercises 35–38, determine whether the set $\{v_1, v_2\}$ is a basis for R^2.

35.
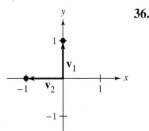

36.
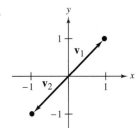

37.
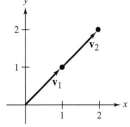

38.
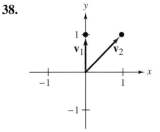

Determining Whether a Set Is a Basis In Exercises 39–46, determine whether S is a basis for the given vector space.

39. $S = \{(4, -3), (5, 2)\}$ for R^2
40. $S = \{(1, 2), (1, -1)\}$ for R^2
41. $S = \{(1, 5, 3), (0, 1, 2), (0, 0, 6)\}$ for R^3
42. $S = \{(2, 1, 0), (0, -1, 1)\}$ for R^3
43. $S = \{(0, 3, -2), (4, 0, 3), (-8, 15, -16)\}$ for R^3
44. $S = \{(0, 0, 0), (1, 5, 6), (6, 2, 1)\}$ for R^3
45. $S = \{(-1, 2, 0, 0), (2, 0, -1, 0), (3, 0, 0, 4), (0, 0, 5, 0)\}$ for R^4
46. $S = \{(1, 0, 0, 1), (0, 2, 0, 2), (1, 0, 1, 0), (0, 2, 2, 0)\}$ for R^4

Determining Whether a Set Is a Basis In Exercises 47–50, determine whether S is a basis for P_3.

47. $S = \{1 - 2t^2 + t^3, -4 + t^2, 2t + t^3, 5t\}$
48. $S = \{4t - t^2, 5 + t^3, 5 + 3t, -3t^2 + 2t^3\}$
49. $S = \{4 - t, t^3, 6t^2, 3t + t^3, -1 + 4t\}$
50. $S = \{-1 + t^3, 2t^2, 3 + t, 5 + 2t + 2t^2 + t^3\}$

Determining Whether a Set Is a Basis In Exercises 51 and 52, determine whether S is a basis for $M_{2,2}$.

51. $S = \left\{ \begin{bmatrix} 2 & 0 \\ 0 & 3 \end{bmatrix}, \begin{bmatrix} 1 & 4 \\ 0 & 1 \end{bmatrix}, \begin{bmatrix} 0 & 1 \\ 3 & 2 \end{bmatrix}, \begin{bmatrix} 0 & 1 \\ 2 & 0 \end{bmatrix} \right\}$

52. $S = \left\{ \begin{bmatrix} 1 & 2 \\ -5 & 4 \end{bmatrix}, \begin{bmatrix} 2 & -7 \\ 6 & 2 \end{bmatrix}, \begin{bmatrix} 4 & -9 \\ 11 & 12 \end{bmatrix}, \begin{bmatrix} 12 & -16 \\ 17 & 42 \end{bmatrix} \right\}$

Determining Whether a Set Is a Basis **In Exercises 53–56, determine whether S is a basis for R^3. If it is, write u = $(8, 3, 8)$ as a linear combination of the vectors in S.**

53. $S = \{(4, 3, 2), (0, 3, 2), (0, 0, 2)\}$
54. $S = \{(1, 0, 0), (1, 1, 0), (1, 1, 1)\}$
55. $S = \{(0, 0, 0), (1, 3, 4), (6, 1, -2)\}$
56. $S = \left\{\left(\frac{2}{3}, \frac{5}{2}, 1\right), \left(1, \frac{3}{2}, 0\right), (2, 12, 6)\right\}$

Finding the Dimension of a Vector Space **In Exercises 57–64, find the dimension of the vector space.**

57. R^6 58. R

59. P_7 60. P_4

61. $M_{2,3}$ 62. $M_{3,2}$

63. R^{3m} 64. $P_{2m-1}, m \geq 1$

65. Find a basis for the vector space of all 3×3 diagonal matrices. What is the dimension of this vector space?

66. Find a basis for the vector space of all 3×3 symmetric matrices. What is the dimension of this vector space?

67. Find all subsets of the set

$$S = \{(1, 0), (0, 1), (1, 1)\}$$

that form a basis for R^2.

68. Find all subsets of the set

$$S = \{(1, 3, -2), (-4, 1, 1), (-2, 7, -3), (2, 1, 1)\}$$

that form a basis for R^3.

69. Find a basis for R^2 that includes the vector $(2, 2)$.

70. Find a basis for R^3 that includes the vectors $(1, 0, 2)$ and $(0, 1, 1)$.

Geometric Description, Basis, and Dimension **In Exercises 71 and 72, (a) give a geometric description of, (b) find a basis for, and (c) find the dimension of the subspace W of R^2.**

71. $W = \{(2t, t): t \text{ is a real number}\}$
72. $W = \{(0, t): t \text{ is a real number}\}$

Geometric Description, Basis, and Dimension **In Exercises 73 and 74, (a) give a geometric description of, (b) find a basis for, and (c) find the dimension of the subspace W of R^3.**

73. $W = \{(2t, t, -t): t \text{ is a real number}\}$
74. $W = \{(2s - t, s, t): s \text{ and } t \text{ are real numbers}\}$

Basis and Dimension **In Exercises 75–78, find (a) a basis for and (b) the dimension of the subspace W of R^4.**

75. $W = \{(2s - t, s, t, s): s \text{ and } t \text{ are real numbers}\}$
76. $W = \{(5t, -3t, t, t): t \text{ is a real number}\}$
77. $W = \{(0, 6t, t, -t): t \text{ is a real number}\}$
78. $W = \{(s + 4t, t, s, 2s - t): s \text{ and } t \text{ are real numbers}\}$

True or False? **In Exercises 79 and 80, determine whether each statement is true or false. If a statement is true, give a reason or cite an appropriate statement from the text. If a statement is false, provide an example that shows the statement is not true in all cases or cite an appropriate statement from the text.**

79. (a) If $\dim(V) = n$, then there exists a set of $n - 1$ vectors in V that span V.

   (b) If $\dim(V) = n$, then there exists a set of $n + 1$ vectors in V that span V.

80. (a) If $\dim(V) = n$, then any set of $n + 1$ vectors in V must be linearly dependent.

   (b) If $\dim(V) = n$, then any set of $n - 1$ vectors in V must be linearly independent.

81. **Proof** Prove that if $S = \{\mathbf{v}_1, \mathbf{v}_2, \ldots, \mathbf{v}_n\}$ is a basis for a vector space V and c is a nonzero scalar, then the set $S_1 = \{c\mathbf{v}_1, c\mathbf{v}_2, \ldots, c\mathbf{v}_n\}$ is also a basis for V.

82. **Proof** Prove Theorem 4.12.

83. **Proof** Prove that if W is a subspace of a finite dimensional vector space V, then $\dim(W) \leq \dim(V)$.

---

84. **CAPSTONE**

(a) A set S_1 consists of two vectors of the form $\mathbf{u} = (u_1, u_2, u_3)$. Explain why S_1 is not a basis for R^3.

(b) A set S_2 consists of four vectors of the form $\mathbf{u} = (u_1, u_2, u_3)$. Explain why S_2 is not a basis for R^3.

(c) A set S_3 consists of three vectors of the form $\mathbf{u} = (u_1, u_2, u_3)$. Determine the conditions under which S_3 is a basis for R^3.

---

85. **Proof** Let S be a linearly independent set of vectors from a finite dimensional vector space V. Prove that there exists a basis for V containing S.

86. **Guided Proof** Let S be a spanning set for a finite dimensional vector space V. Prove that there exists a subset S' of S that forms a basis for V.

   Getting Started: S is a spanning set, but it may not be a basis because it may be linearly dependent. You need to remove extra vectors so that a subset S' is a spanning set and is also linearly independent.

   (i) If S is a linearly independent set, then you are done. If not, remove some vector $\mathbf{v}$ from S that is a linear combination of the other vectors in S. Call this set S_1.

   (ii) If S_1 is a linearly independent set, then you are done. If not, then continue to remove dependent vectors until you produce a linearly independent subset S'.

   (iii) Conclude that this subset is the minimal spanning set S'.

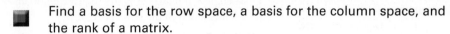

4.6 Rank of a Matrix and Systems of Linear Equations

■ Find a basis for the row space, a basis for the column space, and the rank of a matrix.

■ Find the nullspace of a matrix.

■ Find the solution of a consistent system $A\mathbf{x} = \mathbf{b}$ in the form $\mathbf{x}_p + \mathbf{x}_h$.

ROW SPACE, COLUMN SPACE, AND RANK OF A MATRIX

In this section, you will investigate the vector space spanned by the row vectors (or column vectors) of a matrix. Then you will see how such vector spaces relate to solutions of systems of linear equations.

For an $m \times n$ matrix A, recall that the n-tuples corresponding to the rows of A are the row vectors of A.

Row Vectors of A

$$A = \begin{bmatrix} a_{11} & a_{12} & \cdots & a_{1n} \\ a_{21} & a_{22} & \cdots & a_{2n} \\ \vdots & \vdots & & \vdots \\ a_{m1} & a_{m2} & \cdots & a_{mn} \end{bmatrix} \qquad \begin{matrix} (a_{11}, a_{12}, \ldots, a_{1n}) \\ (a_{21}, a_{22}, \ldots, a_{2n}) \\ \vdots \\ (a_{m1}, a_{m2}, \ldots, a_{mn}) \end{matrix}$$

Similarly, the $m \times 1$ matrices corresponding to the columns of A are the column vectors of A.

Column Vectors of A

$$A = \begin{bmatrix} a_{11} & a_{12} & \cdots & a_{1n} \\ a_{21} & a_{22} & \cdots & a_{2n} \\ \vdots & \vdots & & \vdots \\ a_{m1} & a_{m2} & \cdots & a_{mn} \end{bmatrix} \qquad \begin{bmatrix} a_{11} \\ a_{21} \\ \vdots \\ a_{m1} \end{bmatrix} \begin{bmatrix} a_{12} \\ a_{22} \\ \vdots \\ a_{m2} \end{bmatrix} \cdots \begin{bmatrix} a_{1n} \\ a_{2n} \\ \vdots \\ a_{mn} \end{bmatrix}$$

EXAMPLE 1   **Row Vectors and Column Vectors**

For the matrix $A = \begin{bmatrix} 0 & 1 & -1 \\ -2 & 3 & 4 \end{bmatrix}$, the row vectors are $(0, 1, -1)$ and $(-2, 3, 4)$

and the column vectors are $\begin{bmatrix} 0 \\ -2 \end{bmatrix}$, $\begin{bmatrix} 1 \\ 3 \end{bmatrix}$, and $\begin{bmatrix} -1 \\ 4 \end{bmatrix}$. ■

In Example 1, note that for an $m \times n$ matrix A, the row vectors are vectors in R^n and the column vectors are vectors in R^m. This leads to the definitions of the **row space** and **column space** of a matrix listed below.

Definitions of Row Space and Column Space of a Matrix

Let A be an $m \times n$ matrix.

1. The **row space** of A is the subspace of R^n spanned by the row vectors of A.
2. The **column space** of A is the subspace of R^m spanned by the column vectors of A.

Recall that two matrices are row-equivalent when one can be obtained from the other by elementary row operations. The next theorem tells you that row-equivalent matrices have the same row space.

THEOREM 4.13   Row-Equivalent Matrices Have the Same Row Space

If an $m \times n$ matrix A is row-equivalent to an $m \times n$ matrix B, then the row space of A is equal to the row space of B.

PROOF

The rows of B can be obtained from the rows of A by elementary row operations (scalar multiplication and addition), so it follows that the row vectors of B can be written as linear combinations of the row vectors of A. The row vectors of B lie in the row space of A, and the subspace spanned by the row vectors of B is contained in the row space of A. But it is also true that the rows of A can be obtained from the rows of B by elementary row operations. So, the two row spaces are subspaces of each other, making them equal.

If a matrix B is in row-echelon form, then its nonzero row vectors form a linearly independent set. (Verify this.) Consequently, they form a basis for the row space of B, and by Theorem 4.13 they also form a basis for the row space of A. The next theorem states this important result.

THEOREM 4.14   Basis for the Row Space of a Matrix

If a matrix A is row-equivalent to a matrix B in row-echelon form, then the nonzero row vectors of B form a basis for the row space of A.

EXAMPLE 2   Finding a Basis for a Row Space

Find a basis for the row space of

$$A = \begin{bmatrix} 1 & 3 & 1 & 3 \\ 0 & 1 & 1 & 0 \\ -3 & 0 & 6 & -1 \\ 3 & 4 & -2 & 1 \\ 2 & 0 & -4 & -2 \end{bmatrix}.$$

SOLUTION

Using elementary row operations, rewrite A in row-echelon form as shown below.

$$B = \begin{bmatrix} 1 & 3 & 1 & 3 \\ 0 & 1 & 1 & 0 \\ 0 & 0 & 0 & 1 \\ 0 & 0 & 0 & 0 \\ 0 & 0 & 0 & 0 \end{bmatrix} \begin{matrix} \mathbf{w}_1 \\ \mathbf{w}_2 \\ \mathbf{w}_3 \\ \\ \end{matrix}$$

By Theorem 4.14, the nonzero row vectors of B, $\mathbf{w}_1 = (1, 3, 1, 3)$, $\mathbf{w}_2 = (0, 1, 1, 0)$, and $\mathbf{w}_3 = (0, 0, 0, 1)$, form a basis for the row space of A.

The technique used in Example 2 to find a basis for the row space of a matrix can be used to find a basis for the subspace spanned by the set $S = \{\mathbf{v}_1, \mathbf{v}_2, \ldots, \mathbf{v}_k\}$ in R^n. Use the vectors in S to form the rows of a matrix A, then use elementary row operations to rewrite A in row-echelon form. The nonzero rows of this matrix will then form a basis for the subspace spanned by S. Example 3 demonstrates this process.

<table>
<tr><td>EXAMPLE 3</td><td>**Finding a Basis for a Subspace**</td></tr>
</table>

Find a basis for the subspace of R^3 spanned by

$$S = \{\mathbf{v}_1, \mathbf{v}_2, \mathbf{v}_3\} = \{(-1, 2, 5), (3, 0, 3), (5, 1, 8)\}.$$

SOLUTION

Use $\mathbf{v}_1$, $\mathbf{v}_2$, and $\mathbf{v}_3$ to form the rows of a matrix A. Then write A in row-echelon form.

$$A = \begin{bmatrix} -1 & 2 & 5 \\ 3 & 0 & 3 \\ 5 & 1 & 8 \end{bmatrix} \begin{matrix} \mathbf{v}_1 \\ \mathbf{v}_2 \\ \mathbf{v}_3 \end{matrix} \longrightarrow B = \begin{bmatrix} 1 & -2 & -5 \\ 0 & 1 & 3 \\ 0 & 0 & 0 \end{bmatrix} \begin{matrix} \mathbf{w}_1 \\ \mathbf{w}_2 \\ \end{matrix}$$

The nonzero row vectors of B, $\mathbf{w}_1 = (1, -2, -5)$ and $\mathbf{w}_2 = (0, 1, 3)$, form a basis for the row space of A. That is, they form a basis for the subspace spanned by $S = \{\mathbf{v}_1, \mathbf{v}_2, \mathbf{v}_3\}$.

To find a basis for the column space of a matrix A, you have two options. On the one hand, you could use the fact that the column space of A is equal to the row space of A^T and apply the technique of Example 2 to the matrix A^T. On the other hand, observe that although row operations can change the column space of a matrix, they do not change the dependency relationships among columns. (You are asked to prove this in Exercise 80.) For example, consider the row-equivalent matrices A and B from Example 2.

$$A = \begin{bmatrix} 1 & 3 & 1 & 3 \\ 0 & 1 & 1 & 0 \\ -3 & 0 & 6 & -1 \\ 3 & 4 & -2 & 1 \\ 2 & 0 & -4 & -2 \end{bmatrix} \quad B = \begin{bmatrix} 1 & 3 & 1 & 3 \\ 0 & 1 & 1 & 0 \\ 0 & 0 & 0 & 1 \\ 0 & 0 & 0 & 0 \\ 0 & 0 & 0 & 0 \end{bmatrix}$$
$$\quad\;\; \mathbf{a}_1 \;\; \mathbf{a}_2 \;\; \mathbf{a}_3 \;\; \mathbf{a}_4 \qquad\qquad\quad \mathbf{b}_1 \;\; \mathbf{b}_2 \;\; \mathbf{b}_3 \;\; \mathbf{b}_4$$

Notice that columns 1, 2, and 3 of matrix B satisfy $\mathbf{b}_3 = -2\mathbf{b}_1 + \mathbf{b}_2$, and the corresponding columns of matrix A satisfy $\mathbf{a}_3 = -2\mathbf{a}_1 + \mathbf{a}_2$. Similarly, the column vectors $\mathbf{b}_1$, $\mathbf{b}_2$, and $\mathbf{b}_4$ of matrix B are linearly independent, as are the corresponding columns of matrix A.

The next two examples show how to find a basis for the column space of a matrix using these methods.

<table>
<tr><td>EXAMPLE 4</td><td>**Finding a Basis for the Column Space of a Matrix (Method 1)**</td></tr>
</table>

Find a basis for the column space of matrix A from Example 2 by finding a basis for the row space of A^T.

SOLUTION

Write the transpose of A and use elementary row operations to write A^T in row-echelon form.

$$A^T = \begin{bmatrix} 1 & 0 & -3 & 3 & 2 \\ 3 & 1 & 0 & 4 & 0 \\ 1 & 1 & 6 & -2 & -4 \\ 3 & 0 & -1 & 1 & -2 \end{bmatrix} \longrightarrow \begin{bmatrix} 1 & 0 & -3 & 3 & 2 \\ 0 & 1 & 9 & -5 & -6 \\ 0 & 0 & 1 & -1 & -1 \\ 0 & 0 & 0 & 0 & 0 \end{bmatrix} \begin{matrix} \mathbf{w}_1 \\ \mathbf{w}_2 \\ \mathbf{w}_3 \\ \end{matrix}$$

So, $\mathbf{w}_1 = (1, 0, -3, 3, 2)$, $\mathbf{w}_2 = (0, 1, 9, -5, -6)$, and $\mathbf{w}_3 = (0, 0, 1, -1, -1)$ form a basis for the row space of A^T. This is equivalent to saying that the column vectors $[1 \quad 0 \quad -3 \quad 3 \quad 2]^T$, $[0 \quad 1 \quad 9 \quad -5 \quad -6]^T$, and $[0 \quad 0 \quad 1 \quad -1 \quad -1]^T$ form a basis for the column space of A.

EXAMPLE 5 **Finding a Basis for the Column Space of a Matrix (Method 2)**

Find a basis for the column space of matrix A from Example 2 by using the dependency relationships among columns.

SOLUTION

In Example 2, row operations were used on the original matrix A to obtain its row-echelon form B. As mentioned earlier, in matrix B, the first, second, and fourth column vectors are linearly independent (these columns have the leading 1's), as are the corresponding columns of matrix A. So, a basis for the column space of A consists of the vectors

REMARK

Notice that the row-echelon form B tells you which columns of A form the basis for the column space. You do not use the column vectors of B to form the basis.

$$\begin{bmatrix} 1 \\ 0 \\ -3 \\ 3 \\ 2 \end{bmatrix}, \begin{bmatrix} 3 \\ 1 \\ 0 \\ 4 \\ 0 \end{bmatrix}, \quad \text{and} \quad \begin{bmatrix} 3 \\ 0 \\ -1 \\ 1 \\ -2 \end{bmatrix}.$$

Notice that the basis for the column space obtained in Example 5 is different than that obtained in Example 4. Verify that these bases both span the column space of A by writing the columns of A as linear combinations of the vectors in each basis.

Also notice in Examples 2, 4, and 5 that both the row space and the column space of A have a dimension of 3 (because there are *three* vectors in both bases). The next theorem generalizes this.

THEOREM 4.15   Row and Column Spaces Have Equal Dimensions

The row space and column space of an $m \times n$ matrix A have the same dimension.

PROOF

Let $\mathbf{v}_1, \mathbf{v}_2, \ldots, \mathbf{v}_m$ be the row vectors and $\mathbf{u}_1, \mathbf{u}_2, \ldots, \mathbf{u}_n$ be the column vectors of

$$A = \begin{bmatrix} a_{11} & a_{12} & \cdots & a_{1n} \\ a_{21} & a_{22} & \cdots & a_{2n} \\ \vdots & \vdots & & \vdots \\ a_{m1} & a_{m2} & \cdots & a_{mn} \end{bmatrix}.$$

Assume the row space of A has dimension r and basis $S = \{\mathbf{b}_1, \mathbf{b}_2, \ldots, \mathbf{b}_r\}$, where $\mathbf{b}_i = (b_{i1}, b_{i2}, \ldots, b_{in})$. Using this basis, write the row vectors of A as

$$\begin{aligned} \mathbf{v}_1 &= c_{11}\mathbf{b}_1 + c_{12}\mathbf{b}_2 + \cdots + c_{1r}\mathbf{b}_r \\ \mathbf{v}_2 &= c_{21}\mathbf{b}_1 + c_{22}\mathbf{b}_2 + \cdots + c_{2r}\mathbf{b}_r \\ &\vdots \\ \mathbf{v}_m &= c_{m1}\mathbf{b}_1 + c_{m2}\mathbf{b}_2 + \cdots + c_{mr}\mathbf{b}_r. \end{aligned}$$

Rewrite this system of vector equations as shown below.

$$\begin{aligned} (a_{11}, a_{12}, \ldots, a_{1n}) &= c_{11}(b_{11}, b_{12}, \ldots, b_{1n}) + c_{12}(b_{21}, b_{22}, \ldots, b_{2n}) + \cdots + c_{1r}(b_{r1}, b_{r2}, \ldots, b_{rn}) \\ (a_{21}, a_{22}, \ldots, a_{2n}) &= c_{21}(b_{11}, b_{12}, \ldots, b_{1n}) + c_{22}(b_{21}, b_{22}, \ldots, b_{2n}) + \cdots + c_{2r}(b_{r1}, b_{r2}, \ldots, b_{rn}) \\ &\vdots \\ (a_{m1}, a_{m2}, \ldots, a_{mn}) &= c_{m1}(b_{11}, b_{12}, \ldots, b_{1n}) + c_{m2}(b_{21}, b_{22}, \ldots, b_{2n}) + \cdots + c_{mr}(b_{r1}, b_{r2}, \ldots, b_{rn}) \end{aligned}$$

Now, take only entries corresponding to the first column of matrix A to obtain the system of scalar equations shown on the next page.

$$a_{11} = c_{11}b_{11} + c_{12}b_{21} + \cdots + c_{1r}b_{r1}$$
$$a_{21} = c_{21}b_{11} + c_{22}b_{21} + \cdots + c_{2r}b_{r1}$$
$$\vdots$$
$$a_{m1} = c_{m1}b_{11} + c_{m2}b_{21} + \cdots + c_{mr}b_{r1}$$

Similarly, for the entries of the jth column, you can obtain the system below.

$$a_{1j} = c_{11}b_{1j} + c_{12}b_{2j} + \cdots + c_{1r}b_{rj}$$
$$a_{2j} = c_{21}b_{1j} + c_{22}b_{2j} + \cdots + c_{2r}b_{rj}$$
$$\vdots$$
$$a_{mj} = c_{m1}b_{1j} + c_{m2}b_{2j} + \cdots + c_{mr}b_{rj}$$

Now, let the vectors

$$\mathbf{c}_i = \begin{bmatrix} c_{1i} & c_{2i} & \ldots & c_{mi} \end{bmatrix}^T.$$

Then the system for the jth column can be rewritten in a vector form as

$$\mathbf{u}_j = b_{1j}\mathbf{c}_1 + b_{2j}\mathbf{c}_2 + \cdots + b_{rj}\mathbf{c}_r.$$

Put all column vectors together to obtain

$$\mathbf{u}_1 = \begin{bmatrix} a_{11} & a_{12} & \ldots & a_{m1} \end{bmatrix}^T = b_{11}\mathbf{c}_1 + b_{21}\mathbf{c}_2 + \cdots + b_{r1}\mathbf{c}_r$$
$$\mathbf{u}_2 = \begin{bmatrix} a_{12} & a_{22} & \ldots & a_{m2} \end{bmatrix}^T = b_{12}\mathbf{c}_1 + b_{22}\mathbf{c}_2 + \cdots + b_{r2}\mathbf{c}_r$$
$$\vdots$$
$$\mathbf{u}_n = \begin{bmatrix} a_{1n} & a_{2n} & \ldots & a_{mn} \end{bmatrix}^T = b_{1n}\mathbf{c}_1 + b_{2n}\mathbf{c}_2 + \cdots + b_{rn}\mathbf{c}_r.$$

Each column vector of A is a linear combination of r vectors, so you know that the dimension of the column space of A is less than or equal to r (the dimension of the row space of A). That is,

$$\dim(\text{column space of } A) \leq \dim(\text{row space of } A).$$

Repeating this procedure for A^T, you can conclude that the dimension of the column space of A^T is less than or equal to the dimension of the row space of A^T. But this implies that the dimension of the row space of A is less than or equal to the dimension of the column space of A. That is,

$$\dim(\text{row space of } A) \leq \dim(\text{column space of } A).$$

So, the two dimensions must be equal.

REMARK

Some texts distinguish between the *row rank* and the *column rank* of a matrix, but these ranks are equal (Theorem 4.15). So, this text will not distinguish between them.

The dimension of the row (or column) space of a matrix is the **rank** of the matrix.

Definition of the Rank of a Matrix

The dimension of the row (or column) space of a matrix A is the **rank** of A and is denoted by rank(A).

EXAMPLE 6   Finding the Rank of a Matrix

To find the rank of the matrix A below, convert to a matrix B in row-echelon form as shown.

$$A = \begin{bmatrix} 1 & -2 & 0 & 1 \\ 2 & 1 & 5 & -3 \\ 0 & 1 & 3 & 5 \end{bmatrix} \quad \rightarrow \quad B = \begin{bmatrix} 1 & -2 & 0 & 1 \\ 0 & 1 & 1 & -1 \\ 0 & 0 & 1 & 3 \end{bmatrix}$$

The matrix B has three nonzero rows, so the rank of A is 3.

THE NULLSPACE OF A MATRIX

Row and column spaces and rank have some important applications to systems of linear equations. Consider first the homogeneous linear system $A\mathbf{x} = \mathbf{0}$, where A is an $m \times n$ matrix, $\mathbf{x} = [x_1 \ x_2 \ \ldots \ x_n]^T$ is the column vector of variables, and $\mathbf{0} = [0 \ 0 \ \ldots \ 0]^T$ is the zero vector in R^m. The next theorem tells you that the set of all solutions of this homogeneous system is a subspace of R^n.

REMARK

The nullspace of A is also called the **solution space** of the system $A\mathbf{x} = \mathbf{0}$.

THEOREM 4.16   Solutions of a Homogeneous System

If A is an $m \times n$ matrix, then the set of all solutions of the homogeneous system of linear equations $A\mathbf{x} = \mathbf{0}$ is a subspace of R^n called the **nullspace** of A and is denoted by $N(A)$. So,

$$N(A) = \{\mathbf{x} \in R^n \colon A\mathbf{x} = \mathbf{0}\}.$$

The dimension of the nullspace of A is the **nullity** of A.

PROOF

The size of A is $m \times n$, so you know that $\mathbf{x}$ has size $n \times 1$, and the set of all solutions of the system is a *subset* of R^n. This set is clearly nonempty, because $A\mathbf{0} = \mathbf{0}$. Verify that it is a subspace by showing that it is closed under the operations of addition and scalar multiplication. Let $\mathbf{x}_1$ and $\mathbf{x}_2$ be two solution vectors of the system $A\mathbf{x} = \mathbf{0}$, and let c be a scalar. Both $A\mathbf{x}_1 = \mathbf{0}$ and $A\mathbf{x}_2 = \mathbf{0}$, so you know that

$$A(\mathbf{x}_1 + \mathbf{x}_2) = A\mathbf{x}_1 + A\mathbf{x}_2 = \mathbf{0} + \mathbf{0} = \mathbf{0} \qquad \text{Addition}$$

and

$$A(c\mathbf{x}_1) = c(A\mathbf{x}_1) = c(\mathbf{0}) = \mathbf{0}. \qquad \text{Scalar multiplication}$$

So, both $(\mathbf{x}_1 + \mathbf{x}_2)$ and $c\mathbf{x}_1$ are solutions of $A\mathbf{x} = \mathbf{0}$, and you can conclude that the set of all solutions forms a subspace of R^n.

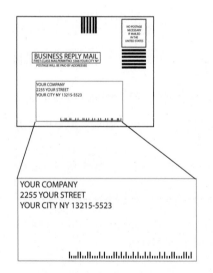

LINEAR ALGEBRA APPLIED

The U.S. Postal Service uses barcodes to represent such information as ZIP codes and delivery addresses. The ZIP + 4 barcode shown at the left starts with a long bar, then has a sequence of short and long bars to represent each digit in the ZIP + 4 code, an additional digit for error checking, and then the code ends with a long bar. The code for the digits is shown below.

0 = ‖‖ııı   1 = ııı‖‖   2 = ıı‖ıl   3 = ıı‖‖ı   4 = ıluı

5 = ılılı   6 = ı‖ılı   7 = lıılı   8 = lılılı   9 = lıluı

The error checking digit is such that when it is summed with the digits in the ZIP + 4 code, the result is a multiple of 10. (Verify this, as well as whether the ZIP + 4 code shown is coded correctly.) More sophisticated barcodes will also include error correcting digit(s). In an analogous way, matrices can be used to check for errors in transmitted messages. Information in the form of column vectors can be multiplied by an error detection matrix. When the resulting product is in the nullspace of the error detection matrix, no error in transmission exists. Otherwise, an error exists somewhere in the message. If the error detection matrix also has error correction, then the resulting matrix product will also tell where the error is occurring.

EXAMPLE 7   **Finding the Nullspace of a Matrix**

Find the nullspace of the matrix.

$$A = \begin{bmatrix} 1 & 2 & -2 & 1 \\ 3 & 6 & -5 & 4 \\ 1 & 2 & 0 & 3 \end{bmatrix}$$

SOLUTION

The nullspace of A is the solution space of the homogeneous system

$$A\mathbf{x} = \mathbf{0}.$$

To solve this system, you could write the augmented matrix $\begin{bmatrix} A & \mathbf{0} \end{bmatrix}$ in reduced row-echelon form. However, the last column of the augmented matrix consists entirely of zeros and will not change as you perform row operations, so it is sufficient to find the reduced row-echelon form of A.

$$A = \begin{bmatrix} 1 & 2 & -2 & 1 \\ 3 & 6 & -5 & 4 \\ 1 & 2 & 0 & 3 \end{bmatrix} \longrightarrow \begin{bmatrix} 1 & 2 & 0 & 3 \\ 0 & 0 & 1 & 1 \\ 0 & 0 & 0 & 0 \end{bmatrix}$$

The system of equations corresponding to the reduced row-echelon form is

$$x_1 + 2x_2 + 3x_4 = 0$$
$$x_3 + x_4 = 0.$$

Choose x_2 and x_4 as free variables to represent the solutions in parametric form.

$$x_1 = -2s - 3t, \quad x_2 = s, \quad x_3 = -t, \quad x_4 = t$$

This means that the solution space of $A\mathbf{x} = \mathbf{0}$ consists of all solution vectors of the form

$$\mathbf{x} = \begin{bmatrix} x_1 \\ x_2 \\ x_3 \\ x_4 \end{bmatrix} = \begin{bmatrix} -2s - 3t \\ s \\ -t \\ t \end{bmatrix} = s\begin{bmatrix} -2 \\ 1 \\ 0 \\ 0 \end{bmatrix} + t\begin{bmatrix} -3 \\ 0 \\ -1 \\ 1 \end{bmatrix}.$$

So, a basis for the nullspace of A consists of the vectors

$$\begin{bmatrix} -2 \\ 1 \\ 0 \\ 0 \end{bmatrix} \quad \text{and} \quad \begin{bmatrix} -3 \\ 0 \\ -1 \\ 1 \end{bmatrix}.$$

In other words, these two vectors are solutions of $A\mathbf{x} = \mathbf{0}$, and all linear combinations of these two vectors are also solutions.

REMARK

Although Example 7 shows that the basis spans the solution set, it does not show that the vectors in the basis are linearly independent. When you solve homogeneous systems from the reduced row-echelon form, the spanning set is always linearly independent. Verify this for the basis found in Example 7.

In Example 7, matrix A has four columns. Furthermore, the rank of A is 2, and the dimension of the nullspace is 2. So,

Number of columns = rank + nullity.

One way to see this is to look at the reduced row-echelon form of A.

$$\begin{bmatrix} 1 & 2 & 0 & 3 \\ 0 & 0 & 1 & 1 \\ 0 & 0 & 0 & 0 \end{bmatrix}$$

The columns with the leading 1's (columns 1 and 3) determine the rank of the matrix. The other columns (2 and 4) determine the nullity of the matrix because they correspond to the free variables. The next theorem generalizes this relationship.

THEOREM 4.17   Dimension of the Solution Space

If A is an $m \times n$ matrix of rank r, then the dimension of the solution space of $A\mathbf{x} = \mathbf{0}$ is $n - r$. That is, $n = \text{rank}(A) + \text{nullity}(A)$.

PROOF

A has rank r, so you know it is row-equivalent to a reduced row-echelon matrix B with r nonzero rows. No generality is lost by assuming that the upper left corner of B has the form of the $r \times r$ identity matrix I_r. Moreover, the zero rows of B contribute nothing to the solution, so discard them to form the $r \times n$ matrix B', where B' is the augmented matrix $[I_r \quad C]$. The matrix C has $n - r$ columns corresponding to the variables x_{r+1}, $x_{r+2}, \dots, x_n$, and the solution space of $A\mathbf{x} = \mathbf{0}$ can be represented by the system

$$
\begin{aligned}
x_1 + & & c_{11}x_{r+1} + c_{12}x_{r+2} + \cdots + c_{1,\,n-r}x_n &= 0 \\
& x_2 + & c_{21}x_{r+1} + c_{22}x_{r+2} + \cdots + c_{2,\,n-r}x_n &= 0 \\
& & \quad\vdots \qquad\qquad \vdots \qquad\qquad\qquad \vdots \\
& x_r + & c_{r1}x_{r+1} + c_{r2}x_{r+2} + \cdots + c_{r,\,n-r}x_n &= 0.
\end{aligned}
$$

Solving for the first r variables in terms of the last $n - r$ variables produces $n - r$ vectors in the basis for the solution space, so the solution space has dimension $n - r$. ∎

Example 8 illustrates this theorem and further explores the column space of a matrix.

EXAMPLE 8   Rank, Nullity of a Matrix, and Basis for the Column Space

See LarsonLinearAlgebra.com for an interactive version of this type of example.

Let the column vectors of the matrix A be denoted by $\mathbf{a}_1, \mathbf{a}_2, \mathbf{a}_3, \mathbf{a}_4$, and $\mathbf{a}_5$. Find (a) the rank and nullity of A, and (b) a subset of the column vectors of A that forms a basis for the column space of A.

$$
A = \begin{bmatrix} 1 & 0 & -2 & 1 & 0 \\ 0 & -1 & -3 & 1 & 3 \\ -2 & -1 & 1 & -1 & 3 \\ 0 & 3 & 9 & 0 & -12 \end{bmatrix}
$$
$$
\;\;\;\;\;\mathbf{a}_1 \;\; \mathbf{a}_2 \;\; \mathbf{a}_3 \;\; \mathbf{a}_4 \;\; \mathbf{a}_5
$$

SOLUTION

Let B be the reduced row-echelon form of A.

$$
A = \begin{bmatrix} 1 & 0 & -2 & 1 & 0 \\ 0 & -1 & -3 & 1 & 3 \\ -2 & -1 & 1 & -1 & 3 \\ 0 & 3 & 9 & 0 & -12 \end{bmatrix} \longrightarrow B = \begin{bmatrix} 1 & 0 & -2 & 0 & 1 \\ 0 & 1 & 3 & 0 & -4 \\ 0 & 0 & 0 & 1 & -1 \\ 0 & 0 & 0 & 0 & 0 \end{bmatrix}
$$

a. B has three nonzero rows, so the rank of A is 3. Also, the number of columns of A is $n = 5$, which implies that the nullity of A is $n - \text{rank} = 5 - 3 = 2$.

b. The first, second, and fourth column vectors of B are linearly independent, so the corresponding column vectors of A,

$$
\mathbf{a}_1 = \begin{bmatrix} 1 \\ 0 \\ -2 \\ 0 \end{bmatrix}, \quad \mathbf{a}_2 = \begin{bmatrix} 0 \\ -1 \\ -1 \\ 3 \end{bmatrix}, \quad \text{and} \quad \mathbf{a}_4 = \begin{bmatrix} 1 \\ 1 \\ -1 \\ 0 \end{bmatrix}
$$

form a basis for the column space of A.

SOLUTIONS OF SYSTEMS OF LINEAR EQUATIONS

You now know that the set of all solution vectors of the *homogeneous* linear system $A\mathbf{x} = \mathbf{0}$ is a subspace. The set of all solution vectors of the *nonhomogeneous* system $A\mathbf{x} = \mathbf{b}$, where $\mathbf{b} \neq \mathbf{0}$, is *not* a subspace because the zero vector is never a solution of a nonhomogeneous system. There is a relationship, however, between the sets of solutions of the two systems $A\mathbf{x} = \mathbf{0}$ and $A\mathbf{x} = \mathbf{b}$. Specifically, if $\mathbf{x}_p$ is a *particular* solution of the nonhomogeneous system $A\mathbf{x} = \mathbf{b}$, then *every* solution of this system can be written in the form $\mathbf{x} = \mathbf{x}_p + \mathbf{x}_h$, where $\mathbf{x}_h$ is a solution of the corresponding homogeneous system $A\mathbf{x} = \mathbf{0}$. The next theorem states this important concept.

THEOREM 4.18   Solutions of a Nonhomogeneous Linear System

If $\mathbf{x}_p$ is a particular solution of the nonhomogeneous system $A\mathbf{x} = \mathbf{b}$, then every solution of this system can be written in the form $\mathbf{x} = \mathbf{x}_p + \mathbf{x}_h$, where $\mathbf{x}_h$ is a solution of the corresponding homogeneous system $A\mathbf{x} = \mathbf{0}$.

PROOF

Let $\mathbf{x}$ be any solution of $A\mathbf{x} = \mathbf{b}$. Then $(\mathbf{x} - \mathbf{x}_p)$ is a solution of the homogeneous system $A\mathbf{x} = \mathbf{0}$, because

$$A(\mathbf{x} - \mathbf{x}_p) = A\mathbf{x} - A\mathbf{x}_p = \mathbf{b} - \mathbf{b} = \mathbf{0}.$$

Letting $\mathbf{x}_h = \mathbf{x} - \mathbf{x}_p$, you have $\mathbf{x} = \mathbf{x}_p + \mathbf{x}_h$.

EXAMPLE 9   **Finding the Solution Set of a Nonhomogeneous System**

Find the set of all solution vectors of the system of linear equations.

$$\begin{array}{rrrcr}
x_1 & & -\,2x_3 & +\,x_4 & = & 5 \\
3x_1 & +\,x_2 & -\,5x_3 & & = & 8 \\
x_1 & +\,2x_2 & & -\,5x_4 & = & -9
\end{array}$$

SOLUTION

The augmented matrix for the system $A\mathbf{x} = \mathbf{b}$ reduces as shown below.

$$\begin{bmatrix} 1 & 0 & -2 & 1 & 5 \\ 3 & 1 & -5 & 0 & 8 \\ 1 & 2 & 0 & -5 & -9 \end{bmatrix} \longrightarrow \begin{bmatrix} 1 & 0 & -2 & 1 & 5 \\ 0 & 1 & 1 & -3 & -7 \\ 0 & 0 & 0 & 0 & 0 \end{bmatrix}$$

The system of linear equations corresponding to the reduced row-echelon matrix is

$$\begin{array}{rrrcr}
x_1 & -\,2x_3 & +\,x_4 & = & 5 \\
x_2 & +\,x_3 & -\,3x_4 & = & -7.
\end{array}$$

Letting $x_3 = s$ and $x_4 = t$, write a representative solution vector of $A\mathbf{x} = \mathbf{b}$ as shown below.

$$\mathbf{x} = \begin{bmatrix} x_1 \\ x_2 \\ x_3 \\ x_4 \end{bmatrix} = \begin{bmatrix} 5 + 2s - t \\ -7 - s + 3t \\ 0 + s + 0t \\ 0 + 0s + t \end{bmatrix} = \begin{bmatrix} 5 \\ -7 \\ 0 \\ 0 \end{bmatrix} + s\begin{bmatrix} 2 \\ -1 \\ 1 \\ 0 \end{bmatrix} + t\begin{bmatrix} -1 \\ 3 \\ 0 \\ 1 \end{bmatrix} = \mathbf{x}_p + s\mathbf{u}_1 + t\mathbf{u}_2$$

$\mathbf{x}_p$ is a *particular* solution vector of $A\mathbf{x} = \mathbf{b}$, and $\mathbf{x}_h = s\mathbf{u}_1 + t\mathbf{u}_2$ represents an arbitrary vector in the solution space of $A\mathbf{x} = \mathbf{0}$.

The next theorem describes how the column space of a matrix can be used to determine whether a system of linear equations is consistent.

THEOREM 4.19   Solutions of a System of Linear Equations

The system $A\mathbf{x} = \mathbf{b}$ is consistent if and only if $\mathbf{b}$ is in the column space of A.

PROOF

For the system $A\mathbf{x} = \mathbf{b}$, let A, $\mathbf{x}$, and $\mathbf{b}$ be the $m \times n$ coefficient matrix, the $n \times 1$ column matrix of variables, and the $m \times 1$ right-hand side, respectively. Then

$$A\mathbf{x} = \begin{bmatrix} a_{11} & a_{12} & \cdots & a_{1n} \\ a_{21} & a_{22} & \cdots & a_{2n} \\ \vdots & \vdots & & \vdots \\ a_{m1} & a_{m2} & \cdots & a_{mn} \end{bmatrix} \begin{bmatrix} x_1 \\ x_2 \\ \vdots \\ x_n \end{bmatrix} = x_1 \begin{bmatrix} a_{11} \\ a_{21} \\ \vdots \\ a_{m1} \end{bmatrix} + x_2 \begin{bmatrix} a_{12} \\ a_{22} \\ \vdots \\ a_{m2} \end{bmatrix} + \cdots + x_n \begin{bmatrix} a_{1n} \\ a_{2n} \\ \vdots \\ a_{mn} \end{bmatrix}.$$

So, $A\mathbf{x} = \mathbf{b}$ if and only if $\mathbf{b} = [b_1 \ b_2 \ \cdots \ b_m]^T$ is a linear combination of the columns of A. That is, the system is consistent if and only if $\mathbf{b}$ is in the subspace of R^m spanned by the columns of A. ∎

EXAMPLE 10   **Consistency of a System of Linear Equations**

Consider the system of linear equations

$$\begin{aligned} x_1 + x_2 - x_3 &= -1 \\ x_1 + x_3 &= 3 \\ 3x_1 + 2x_2 - x_3 &= 1. \end{aligned}$$

The augmented matrix for the system is

$$[A \ \ \mathbf{b}] = \begin{bmatrix} 1 & 1 & -1 & -1 \\ 1 & 0 & 1 & 3 \\ 3 & 2 & -1 & 1 \end{bmatrix}.$$
$$\quad\; \mathbf{a}_1 \quad \mathbf{a}_2 \quad \mathbf{a}_3 \quad\ \mathbf{b}$$

Notice that $\mathbf{b} = 2\mathbf{a}_1 - 2\mathbf{a}_2 + \mathbf{a}_3$. So, $\mathbf{b}$ is in the column space of A, and the system of linear equations is consistent. ∎

REMARK

The reduced row-echelon form of $[A \ \ \mathbf{b}]$ is

$$\begin{bmatrix} 1 & 0 & 1 & 3 \\ 0 & 1 & -2 & -4 \\ 0 & 0 & 0 & 0 \end{bmatrix}$$

(verify this). So, there are infinitely many ways to write $\mathbf{b}$ as a linear combination of the columns of A.

The summary below presents several major results involving systems of linear equations, matrices, determinants, and vector spaces.

Summary of Equivalent Conditions for Square Matrices

If A is an $n \times n$ matrix, then the conditions below are equivalent.

1. A is invertible.
2. $A\mathbf{x} = \mathbf{b}$ has a unique solution for any $n \times 1$ matrix $\mathbf{b}$.
3. $A\mathbf{x} = \mathbf{0}$ has only the trivial solution.
4. A is row-equivalent to I_n.
5. $|A| \neq 0$
6. Rank$(A) = n$
7. The n row vectors of A are linearly independent.
8. The n column vectors of A are linearly independent.

4.6 Exercises

Row Vectors and Column Vectors **In Exercises 1–4,** write (a) the row vectors and (b) the column vectors of the matrix.

1. $\begin{bmatrix} 0 & -2 \\ 1 & -3 \end{bmatrix}$

2. $\begin{bmatrix} 6 & 5 & -1 \end{bmatrix}$

3. $\begin{bmatrix} 4 & 3 & 1 \\ 1 & -4 & 0 \end{bmatrix}$

4. $\begin{bmatrix} 0 & 3 & -4 \\ 4 & 0 & -1 \\ -6 & 1 & 1 \end{bmatrix}$

Finding a Basis for a Row Space and Rank **In Exercises 5–12, find (a) a basis for the row space and (b) the rank of the matrix.**

5. $\begin{bmatrix} 1 & 0 \\ 0 & 2 \end{bmatrix}$

6. $\begin{bmatrix} 0 & 1 & -2 \end{bmatrix}$

7. $\begin{bmatrix} 1 & -3 & 2 \\ 4 & 2 & 1 \end{bmatrix}$

8. $\begin{bmatrix} 2 & 5 \\ -2 & -5 \\ -6 & -15 \end{bmatrix}$

9. $\begin{bmatrix} 1 & 6 & 18 \\ 7 & 40 & 116 \\ -3 & -12 & -27 \end{bmatrix}$

10. $\begin{bmatrix} 2 & -3 & 1 \\ 5 & 10 & 6 \\ 8 & -7 & 5 \end{bmatrix}$

11. $\begin{bmatrix} -2 & -4 & 4 & 5 \\ 3 & 6 & -6 & -4 \\ -2 & -4 & 4 & 9 \end{bmatrix}$

12. $\begin{bmatrix} 4 & 0 & 2 & 3 & 1 \\ 2 & -1 & 2 & 0 & 1 \\ 5 & 2 & 2 & 1 & -1 \\ 4 & 0 & 2 & 2 & 1 \\ 2 & -2 & 0 & 0 & 1 \end{bmatrix}$

Finding a Basis for a Subspace **In Exercises 13–16, find a basis for the subspace of R^3 spanned by S.**

13. $S = \{(1, 2, 4), (-1, 3, 4), (2, 3, 1)\}$

14. $S = \{(2, 3, -1), (1, 3, -9), (0, 1, 5)\}$

15. $S = \{(4, 4, 8), (1, 1, 2), (1, 1, 1)\}$

16. $S = \{(1, 2, 2), (-1, 0, 0), (1, 1, 1)\}$

Finding a Basis for a Subspace **In Exercises 17–20, find a basis for the subspace of R^4 spanned by S.**

17. $S = \{(2, 9, -2, 53), (-3, 2, 3, -2), (8, -3, -8, 17), (0, -3, 0, 15)\}$

18. $S = \{(6, -3, 6, 34), (3, -2, 3, 19), (8, 3, -9, 6), (-2, 0, 6, -5)\}$

19. $S = \{(-3, 2, 5, 28), (-6, 1, -8, -1), (14, -10, 12, -10), (0, 5, 12, 50)\}$

20. $S = \{(2, 5, -3, -2), (-2, -3, 2, -5), (1, 3, -2, 2), (-1, -5, 3, 5)\}$

Finding a Basis for a Column Space and Rank **In Exercises 21–26, find (a) a basis for the column space and (b) the rank of the matrix.**

21. $\begin{bmatrix} 2 & 4 \\ 1 & 6 \end{bmatrix}$

22. $\begin{bmatrix} 1 & 2 & 3 \end{bmatrix}$

23. $\begin{bmatrix} 1 & 2 & 4 \\ -1 & 2 & 1 \end{bmatrix}$

24. $\begin{bmatrix} 4 & 20 & 31 \\ 6 & -5 & -6 \\ 2 & -11 & -16 \end{bmatrix}$

25. $\begin{bmatrix} 2 & 4 & -3 & -6 \\ 7 & 14 & -6 & -3 \\ -2 & -4 & 1 & -2 \\ 2 & 4 & -2 & -2 \end{bmatrix}$

26. $\begin{bmatrix} 2 & 4 & -2 & 1 & 1 \\ 2 & 5 & 4 & -2 & 2 \\ 4 & 3 & 1 & 1 & 2 \\ 2 & -4 & 2 & -1 & 1 \\ 0 & 1 & 4 & 2 & -1 \end{bmatrix}$

Finding the Nullspace of a Matrix **In Exercises 27–40, find the nullspace of the matrix.**

27. $A = \begin{bmatrix} 2 & -1 \\ -6 & 3 \end{bmatrix}$

28. $A = \begin{bmatrix} 2 & -1 \\ 1 & 3 \end{bmatrix}$

29. $A = \begin{bmatrix} 1 & 2 & 3 \end{bmatrix}$

30. $A = \begin{bmatrix} 1 & 4 & 2 \end{bmatrix}$

31. $A = \begin{bmatrix} 1 & 2 & 3 \\ 0 & 1 & 0 \end{bmatrix}$

32. $A = \begin{bmatrix} 1 & 4 & 2 \\ 0 & 0 & 1 \end{bmatrix}$

33. $A = \begin{bmatrix} 1 & 2 & -3 \\ 2 & -1 & 4 \\ 4 & 3 & -2 \end{bmatrix}$

34. $A = \begin{bmatrix} 3 & -6 & 21 \\ -2 & 4 & -14 \\ 1 & -2 & 7 \end{bmatrix}$

35. $A = \begin{bmatrix} 5 & 2 \\ 3 & -1 \\ 2 & 1 \end{bmatrix}$

36. $A = \begin{bmatrix} -16 & 1 \\ 48 & -3 \\ -80 & 5 \end{bmatrix}$

37. $A = \begin{bmatrix} 1 & 3 & -2 & 4 \\ 0 & 1 & -1 & 2 \\ -2 & -6 & 4 & -8 \end{bmatrix}$

38. $A = \begin{bmatrix} 1 & 4 & 2 & 1 \\ 0 & 1 & 1 & -1 \\ -2 & -8 & -4 & -2 \end{bmatrix}$

39. $A = \begin{bmatrix} 2 & 6 & 3 & 1 \\ 2 & 1 & 0 & -2 \\ 3 & -2 & 1 & 1 \\ 0 & 6 & 2 & 0 \end{bmatrix}$

40. $A = \begin{bmatrix} 1 & 4 & 2 & 1 \\ 2 & -1 & 1 & 1 \\ 4 & 2 & 1 & 1 \\ 0 & 4 & 2 & 0 \end{bmatrix}$

Rank, Nullity, Bases, and Linear Independence In Exercises 41 and 42, use the fact that matrices A and B are row-equivalent.

(a) Find the rank and nullity of A.

(b) Find a basis for the nullspace of A.

(c) Find a basis for the row space of A.

(d) Find a basis for the column space of A.

(e) Determine whether the rows of A are linearly independent.

(f) Let the columns of A be denoted by $a_1, a_2, a_3, a_4,$ and a_5. Determine whether each set is linearly independent.

   (i) $\{a_1, a_2, a_4\}$ (ii) $\{a_1, a_2, a_3\}$ (iii) $\{a_1, a_3, a_5\}$

41. $A = \begin{bmatrix} 1 & 2 & 1 & 0 & 0 \\ 2 & 5 & 1 & 1 & 0 \\ 3 & 7 & 2 & 2 & -2 \\ 4 & 9 & 3 & -1 & 4 \end{bmatrix}$

$B = \begin{bmatrix} 1 & 0 & 3 & 0 & -4 \\ 0 & 1 & -1 & 0 & 2 \\ 0 & 0 & 0 & 1 & -2 \\ 0 & 0 & 0 & 0 & 0 \end{bmatrix}$

42. $A = \begin{bmatrix} -2 & -5 & 8 & 0 & -17 \\ 1 & 3 & -5 & 1 & 5 \\ 3 & 11 & -19 & 7 & 1 \\ 1 & 7 & -13 & 5 & -3 \end{bmatrix}$

$B = \begin{bmatrix} 1 & 0 & 1 & 0 & 1 \\ 0 & 1 & -2 & 0 & 3 \\ 0 & 0 & 0 & 1 & -5 \\ 0 & 0 & 0 & 0 & 0 \end{bmatrix}$

Finding a Basis and Dimension In Exercises 43–48, find (a) a basis for and (b) the dimension of the solution space of the homogeneous system of linear equations.

43. $\begin{aligned} -x + y + z &= 0 \\ 3x - y &= 0 \\ 2x - 4y - 5z &= 0 \end{aligned}$

44. $\begin{aligned} x - 2y + 3z &= 0 \\ -3x + 6y - 9z &= 0 \end{aligned}$

45. $\begin{aligned} 3x_1 + 3x_2 + 15x_3 + 11x_4 &= 0 \\ x_1 - 3x_2 + x_3 + x_4 &= 0 \\ 2x_1 + 3x_2 + 11x_3 + 8x_4 &= 0 \end{aligned}$

46. $\begin{aligned} 2x_1 + 2x_2 + 4x_3 - 2x_4 &= 0 \\ x_1 + 2x_2 + x_3 + 2x_4 &= 0 \\ -x_1 + x_2 + 4x_3 - 2x_4 &= 0 \end{aligned}$

47. $\begin{aligned} 9x_1 - 4x_2 - 2x_3 - 20x_4 &= 0 \\ 12x_1 - 6x_2 - 4x_3 - 29x_4 &= 0 \\ 3x_1 - 2x_2 - 7x_4 &= 0 \\ 3x_1 - 2x_2 - x_3 - 8x_4 &= 0 \end{aligned}$

48. $\begin{aligned} x_1 + 3x_2 + 2x_3 + 22x_4 + 13x_5 &= 0 \\ x_1 + x_3 - 2x_4 + x_5 &= 0 \\ 3x_1 + 6x_2 + 5x_3 + 42x_4 + 27x_5 &= 0 \end{aligned}$

Nonhomogeneous System In Exercises 49–56, determine whether the nonhomogeneous system $Ax = b$ is consistent. If it is, write the solution in the form $x = x_p + x_h$, where x_p is a particular solution of $Ax = b$ and x_h is a solution of $Ax = 0$.

49. $\begin{aligned} x - 4y &= 17 \\ 3x - 12y &= 51 \\ -2x + 8y &= -34 \end{aligned}$

50. $\begin{aligned} x + 2y - 4z &= -1 \\ -3x - 6y + 12z &= 3 \end{aligned}$

51. $\begin{aligned} x + 3y + 10z &= 18 \\ -2x + 7y + 32z &= 29 \\ -x + 3y + 14z &= 12 \\ x + y + 2z &= 8 \end{aligned}$

52. $\begin{aligned} 2x - 4y + 5z &= 8 \\ -7x + 14y + 4z &= -28 \\ 3x - 6y + z &= 12 \end{aligned}$

53. $\begin{aligned} 3x - 8y + 4z &= 19 \\ -6y + 2z + 4w &= 5 \\ 5x + 22z + w &= 29 \\ x - 2y + 2z &= 8 \end{aligned}$

54. $\begin{aligned} 3w - 2x + 16y - 2z &= -7 \\ -w + 5x - 14y + 18z &= 29 \\ 3w - x + 14y + 2z &= 1 \end{aligned}$

55. $\begin{aligned} x_1 + 2x_2 + x_3 + x_4 + 5x_5 &= 0 \\ -5x_1 - 10x_2 + 3x_3 + 3x_4 + 55x_5 &= -8 \\ x_1 + 2x_2 + 2x_3 - 3x_4 - 5x_5 &= 14 \\ -x_1 - 2x_2 + x_3 + x_4 + 15x_5 &= -2 \end{aligned}$

56. $\begin{aligned} 5x_1 - 4x_2 + 12x_3 - 33x_4 + 14x_5 &= -4 \\ -2x_1 + x_2 - 6x_3 + 12x_4 - 8x_5 &= 1 \\ 2x_1 - x_2 + 6x_3 - 12x_4 + 8x_5 &= -1 \end{aligned}$

Consistency of $Ax = b$ In Exercises 57–62, determine whether b is in the column space of A. If it is, write b as a linear combination of the column vectors of A.

57. $A = \begin{bmatrix} -1 & 2 \\ 4 & 0 \end{bmatrix}$, $b = \begin{bmatrix} 3 \\ 4 \end{bmatrix}$

58. $A = \begin{bmatrix} -1 & 2 \\ 2 & -4 \end{bmatrix}$, $b = \begin{bmatrix} 2 \\ 4 \end{bmatrix}$

59. $A = \begin{bmatrix} 1 & 3 & 2 \\ -1 & 1 & 2 \\ 0 & 1 & 1 \end{bmatrix}$, $b = \begin{bmatrix} 1 \\ 1 \\ 0 \end{bmatrix}$

60. $A = \begin{bmatrix} 1 & 3 & 0 \\ -1 & 1 & 0 \\ 2 & 0 & 1 \end{bmatrix}$, $b = \begin{bmatrix} 1 \\ 2 \\ -3 \end{bmatrix}$

61. $A = \begin{bmatrix} -1 & -1 & 1 \\ 1 & 0 & 1 \\ -3 & -2 & 1 \end{bmatrix}$, $b = \begin{bmatrix} 0 \\ 3 \\ -3 \end{bmatrix}$

62. $A = \begin{bmatrix} 5 & 4 & 4 \\ -3 & 1 & -2 \\ 1 & 0 & 8 \end{bmatrix}$, $b = \begin{bmatrix} -9 \\ 11 \\ -25 \end{bmatrix}$

63. Proof Prove that if A is not square, then either the row vectors of A or the column vectors of A form a linearly dependent set.

64. Give an example showing that the rank of the product of two matrices can be less than the rank of either matrix.

65. Give examples of matrices A and B of the same size such that

(a) $\text{rank}(A + B) < \text{rank}(A)$ and $\text{rank}(A + B) < \text{rank}(B)$

(b) $\text{rank}(A + B) = \text{rank}(A)$ and $\text{rank}(A + B) = \text{rank}(B)$

(c) $\text{rank}(A + B) > \text{rank}(A)$ and $\text{rank}(A + B) > \text{rank}(B)$.

66. Proof Prove that the nonzero row vectors of a matrix in row-echelon form are linearly independent.

67. Let A be an $m \times n$ matrix (where $m < n$) whose rank is r.

(a) What is the largest value r can be?

(b) How many vectors are in a basis for the row space of A?

(c) How many vectors are in a basis for the column space of A?

(d) Which vector space R^k has the row space as a subspace?

(e) Which vector space R^k has the column space as a subspace?

68. Show that the three points (x_1, y_1), (x_2, y_2), and (x_3, y_3) in a plane are collinear if and only if the matrix

$$\begin{bmatrix} x_1 & y_1 & 1 \\ x_2 & y_2 & 1 \\ x_3 & y_3 & 1 \end{bmatrix}$$

has rank less than 3.

69. Consider an $m \times n$ matrix A and an $n \times p$ matrix B. Show that the row vectors of AB are in the row space of B and the column vectors of AB are in the column space of A.

70. Find the rank of the matrix

$$\begin{bmatrix} 1 & 2 & 3 & \cdots & n \\ n+1 & n+2 & n+3 & \cdots & 2n \\ 2n+1 & 2n+2 & 2n+3 & \cdots & 3n \\ \vdots & \vdots & \vdots & & \vdots \\ n^2-n+1 & n^2-n+2 & n^2-n+3 & \cdots & n^2 \end{bmatrix}$$

for $n = 2, 3,$ and 4. Can you find a pattern in these ranks?

71. Proof Prove each property of the system of linear equations in n variables $A\mathbf{x} = \mathbf{b}$.

(a) If $\text{rank}(A) = \text{rank}([A \ \mathbf{b}]) = n$, then the system has a unique solution.

(b) If $\text{rank}(A) = \text{rank}([A \ \mathbf{b}]) < n$, then the system has infinitely many solutions.

(c) If $\text{rank}(A) < \text{rank}([A \ \mathbf{b}])$, then the system is inconsistent.

72. Proof Let A be an $m \times n$ matrix. Prove that $N(A) \subset N(A^T A)$.

True or False? In Exercises 73–76, determine whether each statement is true or false. If a statement is true, give a reason or cite an appropriate statement from the text. If a statement is false, provide an example that shows the statement is not true in all cases or cite an appropriate statement from the text.

73. (a) The nullspace of a matrix A is the solution space of the homogeneous system $A\mathbf{x} = \mathbf{0}$.

(b) The dimension of the nullspace of a matrix A is the nullity of A.

74. (a) If an $m \times n$ matrix A is row-equivalent to an $m \times n$ matrix B, then the row space of A is equivalent to the row space of B.

(b) If A is an $m \times n$ matrix of rank r, then the dimension of the solution space of $A\mathbf{x} = \mathbf{0}$ is $m - r$.

75. (a) If an $m \times n$ matrix B can be obtained from elementary row operations on an $m \times n$ matrix A, then the column space of B is equal to the column space of A.

(b) The system of linear equations $A\mathbf{x} = \mathbf{b}$ is inconsistent if and only if $\mathbf{b}$ is in the column space of A.

76. (a) The column space of a matrix A is equal to the row space of A^T.

(b) The row space of a matrix A is equal to the column space of A^T.

77. Let A and B be square matrices of order n satisfying $A\mathbf{x} = B\mathbf{x}$ for all $\mathbf{x}$ in R^n.

(a) Find the rank and nullity of $A - B$.

(b) Show that A and B must be identical.

---

78. CAPSTONE The dimension of the row space of a 3×5 matrix A is 2.

(a) What is the dimension of the column space of A?

(b) What is the rank of A?

(c) What is the nullity of A?

(d) What is the dimension of the solution space of the homogeneous system $A\mathbf{x} = \mathbf{0}$?

---

79. Proof Let A be an $m \times n$ matrix.

(a) Prove that the system of linear equations $A\mathbf{x} = \mathbf{b}$ is consistent for all column vectors $\mathbf{b}$ if and only if the rank of A is m.

(b) Prove that the homogeneous system of linear equations $A\mathbf{x} = \mathbf{0}$ has only the trivial solution if and only if the columns of A are linearly independent.

80. Proof Prove that row operations do not change the dependency relationships among the columns of an $m \times n$ matrix.

81. Writing Explain why the row vectors of a 4×3 matrix form a linearly dependent set. (Assume all matrix entries are distinct.)

4.7   Coordinates and Change of Basis

■ Find a coordinate matrix relative to a basis in R^n.

■ Find the transition matrix from the basis B to the basis B' in R^n.

■ Represent coordinates in general n-dimensional spaces.

COORDINATE REPRESENTATION IN R^n

In Theorem 4.9, you saw that if B is a basis for a vector space V, then every vector $\mathbf{x}$ in V can be expressed in one and only one way as a linear combination of vectors in B. The coefficients in the linear combination are the **coordinates of x relative to B.** In the context of coordinates, the order of the vectors in the basis is important, so this will sometimes be emphasized by referring to the basis B as an *ordered* basis.

Coordinate Representation Relative to a Basis

Let $B = \{\mathbf{v}_1, \mathbf{v}_2, \ldots, \mathbf{v}_n\}$ be an ordered basis for a vector space V and let $\mathbf{x}$ be a vector in V such that

$$\mathbf{x} = c_1\mathbf{v}_1 + c_2\mathbf{v}_2 + \cdots + c_n\mathbf{v}_n.$$

The scalars $c_1, c_2, \ldots, c_n$ are the **coordinates of x relative to the basis B.** The **coordinate matrix** (or **coordinate vector**) of $\mathbf{x}$ relative to B is the column matrix in R^n whose components are the coordinates of $\mathbf{x}$.

$$[\mathbf{x}]_B = \begin{bmatrix} c_1 \\ c_2 \\ \vdots \\ c_n \end{bmatrix}$$

In R^n, column notation is used for the coordinate matrix. For the vector $\mathbf{x} = (x_1, x_2, \ldots, x_n)$, the x_i's are the coordinates of $\mathbf{x}$ *relative to the standard basis S for R^n.* So, you have

$$[\mathbf{x}]_S = \begin{bmatrix} x_1 \\ x_2 \\ \vdots \\ x_n \end{bmatrix}.$$

EXAMPLE 1   Coordinates and Components in R^n

Find the coordinate matrix of $\mathbf{x} = (-2, 1, 3)$ in R^3 relative to the standard basis

$$S = \{(1, 0, 0), (0, 1, 0), (0, 0, 1)\}.$$

SOLUTION

The vector $\mathbf{x}$ can be written as $\mathbf{x} = (-2, 1, 3) = -2(1, 0, 0) + 1(0, 1, 0) + 3(0, 0, 1)$, so the coordinate matrix of $\mathbf{x}$ relative to the standard basis is simply

$$[\mathbf{x}]_S = \begin{bmatrix} -2 \\ 1 \\ 3 \end{bmatrix}.$$

The components of $\mathbf{x}$ are the same as its coordinates relative to the standard basis.

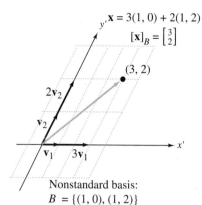

$\mathbf{x} = 3(1, 0) + 2(1, 2)$

$[\mathbf{x}]_B = \begin{bmatrix} 3 \\ 2 \end{bmatrix}$

$(3, 2)$

$2\mathbf{v}_2$

$\mathbf{v}_2$

$\mathbf{v}_1 \quad 3\mathbf{v}_1$

Nonstandard basis:
$B = \{(1, 0), (1, 2)\}$

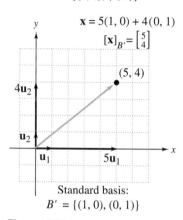

$\mathbf{x} = 5(1, 0) + 4(0, 1)$

$[\mathbf{x}]_{B'} = \begin{bmatrix} 5 \\ 4 \end{bmatrix}$

$(5, 4)$

$4\mathbf{u}_2$

$\mathbf{u}_2$

$\mathbf{u}_1 \quad 5\mathbf{u}_1$

Standard basis:
$B' = \{(1, 0), (0, 1)\}$

Figure 4.17

REMARK

It would be incorrect to write the coordinate matrix as

$$\mathbf{x} = \begin{bmatrix} 5 \\ -8 \\ -2 \end{bmatrix}.$$

Do you see why?

EXAMPLE 2 **Finding a Coordinate Matrix Relative to a Standard Basis**

The coordinate matrix of $\mathbf{x}$ in R^2 relative to the (nonstandard) ordered basis $B = \{\mathbf{v}_1, \mathbf{v}_2\} = \{(1, 0), (1, 2)\}$ is

$$[\mathbf{x}]_B = \begin{bmatrix} 3 \\ 2 \end{bmatrix}.$$

Find the coordinate matrix of $\mathbf{x}$ relative to the standard basis $B' = \{\mathbf{u}_1, \mathbf{u}_2\} = \{(1, 0), (0, 1)\}$.

SOLUTION

The coordinate matrix of $\mathbf{x}$ relative to B is $[\mathbf{x}]_B = \begin{bmatrix} 3 \\ 2 \end{bmatrix}$, so

$$\mathbf{x} = 3\mathbf{v}_1 + 2\mathbf{v}_2 = 3(1, 0) + 2(1, 2) = (5, 4) = 5(1, 0) + 4(0, 1).$$

It follows that the coordinate matrix of $\mathbf{x}$ relative to B' is

$$[\mathbf{x}]_{B'} = \begin{bmatrix} 5 \\ 4 \end{bmatrix}.$$

Figure 4.17 compares these two coordinate representations.

Example 2 shows that the procedure for finding the coordinate matrix relative to a *standard* basis is straightforward. It is more difficult, however, to find the coordinate matrix relative to a *nonstandard* basis. Here is an example.

EXAMPLE 3 **Finding a Coordinate Matrix Relative to a Nonstandard Basis**

Find the coordinate matrix of $\mathbf{x} = (1, 2, -1)$ in R^3 relative to the (nonstandard) basis

$$B' = \{\mathbf{u}_1, \mathbf{u}_2, \mathbf{u}_3\} = \{(1, 0, 1), (0, -1, 2), (2, 3, -5)\}.$$

SOLUTION

Begin by writing $\mathbf{x}$ as a linear combination of $\mathbf{u}_1$, $\mathbf{u}_2$, and $\mathbf{u}_3$.

$$\mathbf{x} = c_1\mathbf{u}_1 + c_2\mathbf{u}_2 + c_3\mathbf{u}_3$$
$$(1, 2, -1) = c_1(1, 0, 1) + c_2(0, -1, 2) + c_3(2, 3, -5)$$

Equating corresponding components produces the system of linear equations and corresponding matrix equation below.

$$\begin{array}{rcl} c_1 \phantom{{}+ 2c_2} + 2c_3 &=& 1 \\ -c_2 + 3c_3 &=& 2 \\ c_1 + 2c_2 - 5c_3 &=& -1 \end{array}$$

$$\begin{bmatrix} 1 & 0 & 2 \\ 0 & -1 & 3 \\ 1 & 2 & -5 \end{bmatrix} \begin{bmatrix} c_1 \\ c_2 \\ c_3 \end{bmatrix} = \begin{bmatrix} 1 \\ 2 \\ -1 \end{bmatrix}$$

The solution of this system is $c_1 = 5$, $c_2 = -8$, and $c_3 = -2$. So,

$$\mathbf{x} = 5(1, 0, 1) + (-8)(0, -1, 2) + (-2)(2, 3, -5)$$

and the coordinate matrix of $\mathbf{x}$ relative to B' is

$$[\mathbf{x}]_{B'} = \begin{bmatrix} 5 \\ -8 \\ -2 \end{bmatrix}.$$

CHANGE OF BASIS IN R^n

The procedure demonstrated in Examples 2 and 3 is called a **change of basis.** That is, you were given the coordinates of a vector relative to a basis B and were asked to find the coordinates relative to another basis B'.

For instance, if in Example 3 you let B be the standard basis, then the problem of finding the coordinate matrix of $\mathbf{x} = (1, 2, -1)$ relative to the basis B' becomes one of solving for c_1, c_2, and c_3 in the matrix equation

$$\underset{P}{\begin{bmatrix} 1 & 0 & 2 \\ 0 & -1 & 3 \\ 1 & 2 & -5 \end{bmatrix}} \underset{[\mathbf{x}]_{B'}}{\begin{bmatrix} c_1 \\ c_2 \\ c_3 \end{bmatrix}} = \underset{[\mathbf{x}]_{B}}{\begin{bmatrix} 1 \\ 2 \\ -1 \end{bmatrix}}.$$

The matrix P is the *transition matrix from B' to B,* where $[\mathbf{x}]_{B'}$ is the coordinate matrix of $\mathbf{x}$ relative to B', and $[\mathbf{x}]_B$ is the coordinate matrix of $\mathbf{x}$ relative to B. Multiplication by the transition matrix P changes a coordinate matrix relative to B' into a coordinate matrix relative to B. That is,

$$P[\mathbf{x}]_{B'} = [\mathbf{x}]_B. \qquad \text{Change of basis from } B' \text{ to } B$$

To perform a change of basis from B to B', use the matrix P^{-1} (the *transition matrix from B to B'*) and write

$$[\mathbf{x}]_{B'} = P^{-1}[\mathbf{x}]_B. \qquad \text{Change of basis from } B \text{ to } B'$$

So, the change of basis problem in Example 3 can be represented by the matrix equation

$$\begin{bmatrix} c_1 \\ c_2 \\ c_3 \end{bmatrix} = \underset{P^{-1}}{\begin{bmatrix} -1 & 4 & 2 \\ 3 & -7 & -3 \\ 1 & -2 & -1 \end{bmatrix}} \underset{[\mathbf{x}]_B}{\begin{bmatrix} 1 \\ 2 \\ -1 \end{bmatrix}} = \underset{[\mathbf{x}]_{B'}}{\begin{bmatrix} 5 \\ -8 \\ -2 \end{bmatrix}}.$$

Generalizing this discussion, assume that

$$B = \{\mathbf{v}_1, \mathbf{v}_2, \dots, \mathbf{v}_n\} \quad \text{and} \quad B' = \{\mathbf{u}_1, \mathbf{u}_2, \dots, \mathbf{u}_n\}$$

are two ordered bases for R^n. If $\mathbf{x}$ is a vector in R^n and

$$[\mathbf{x}]_B = \begin{bmatrix} c_1 \\ c_2 \\ \vdots \\ c_n \end{bmatrix} \quad \text{and} \quad [\mathbf{x}]_{B'} = \begin{bmatrix} d_1 \\ d_2 \\ \vdots \\ d_n \end{bmatrix}$$

are the coordinate matrices of $\mathbf{x}$ relative to B and B', then the **transition matrix P from B' to B** is the matrix P such that

$$[\mathbf{x}]_B = P[\mathbf{x}]_{B'}.$$

The next theorem tells you that the transition matrix P is invertible and its inverse is the **transition matrix from B to B'**. That is,

$$[\mathbf{x}]_{B'} = P^{-1}[\mathbf{x}]_B.$$

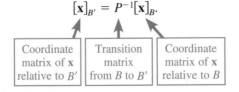

Coordinate matrix of $\mathbf{x}$ relative to B'	Transition matrix from B to B'	Coordinate matrix of $\mathbf{x}$ relative to B

THEOREM 4.20   The Inverse of a Transition Matrix

If P is the transition matrix from a basis B' to a basis B in R^n, then P is invertible and the transition matrix from B to B' is P^{-1}.

Before proving Theorem 4.20, it is necessary to look at and prove a preliminary lemma.

LEMMA

Let $B = \{\mathbf{v}_1, \mathbf{v}_2, \ldots, \mathbf{v}_n\}$ and $B' = \{\mathbf{u}_1, \mathbf{u}_2, \ldots, \mathbf{u}_n\}$ be two bases for a vector space V. If

$$
\begin{aligned}
\mathbf{v}_1 &= c_{11}\mathbf{u}_1 + c_{21}\mathbf{u}_2 + \cdots + c_{n1}\mathbf{u}_n \\
\mathbf{v}_2 &= c_{12}\mathbf{u}_1 + c_{22}\mathbf{u}_2 + \cdots + c_{n2}\mathbf{u}_n \\
&\ \vdots \\
\mathbf{v}_n &= c_{1n}\mathbf{u}_1 + c_{2n}\mathbf{u}_2 + \cdots + c_{nn}\mathbf{u}_n
\end{aligned}
$$

then the transition matrix from B to B' is

$$
Q = \begin{bmatrix} c_{11} & c_{12} & \cdots & c_{1n} \\ c_{21} & c_{22} & \cdots & c_{2n} \\ \vdots & \vdots & & \vdots \\ c_{n1} & c_{n2} & \cdots & c_{nn} \end{bmatrix}.
$$

PROOF (OF LEMMA)

Let $\mathbf{v} = d_1\mathbf{v}_1 + d_2\mathbf{v}_2 + \cdots + d_n\mathbf{v}_n$ be an arbitrary vector in V. The coordinate matrix of $\mathbf{v}$ with respect to the basis B is

$$
[\mathbf{v}]_B = \begin{bmatrix} d_1 \\ d_2 \\ \vdots \\ d_n \end{bmatrix}.
$$

Then you have

$$
Q[\mathbf{v}]_B = \begin{bmatrix} c_{11} & c_{12} & \cdots & c_{1n} \\ c_{21} & c_{22} & \cdots & c_{2n} \\ \vdots & \vdots & & \vdots \\ c_{n1} & c_{n2} & \cdots & c_{nn} \end{bmatrix}\begin{bmatrix} d_1 \\ d_2 \\ \vdots \\ d_n \end{bmatrix} = \begin{bmatrix} c_{11}d_1 + c_{12}d_2 + \cdots + c_{1n}d_n \\ c_{21}d_1 + c_{22}d_2 + \cdots + c_{2n}d_n \\ \vdots & \vdots & \vdots \\ c_{n1}d_1 + c_{n2}d_2 + \cdots + c_{nn}d_n \end{bmatrix}.
$$

On the other hand,

$$
\begin{aligned}
\mathbf{v} &= d_1\mathbf{v}_1 + d_2\mathbf{v}_2 + \cdots + d_n\mathbf{v}_n \\
&= d_1(c_{11}\mathbf{u}_1 + c_{21}\mathbf{u}_2 + \cdots + c_{n1}\mathbf{u}_n) + d_2(c_{12}\mathbf{u}_1 + c_{22}\mathbf{u}_2 + \cdots + c_{n2}\mathbf{u}_n) + \cdots \\
&\quad + d_n(c_{1n}\mathbf{u}_1 + c_{2n}\mathbf{u}_2 + \cdots + c_{nn}\mathbf{u}_n) \\
&= (d_1c_{11} + d_2c_{12} + \cdots + d_nc_{1n})\mathbf{u}_1 + (d_1c_{21} + d_2c_{22} + \cdots + d_nc_{2n})\mathbf{u}_2 + \cdots \\
&\quad + (d_1c_{n1} + d_2c_{n2} + \cdots + d_nc_{nn})\mathbf{u}_n
\end{aligned}
$$

which implies

$$
[\mathbf{v}]_{B'} = \begin{bmatrix} c_{11}d_1 + c_{12}d_2 + \cdots + c_{1n}d_n \\ c_{21}d_1 + c_{22}d_2 + \cdots + c_{2n}d_n \\ \vdots & \vdots & \vdots \\ c_{n1}d_1 + c_{n2}d_2 + \cdots + c_{nn}d_n \end{bmatrix}.
$$

So, $Q[\mathbf{v}]_B = [\mathbf{v}]_{B'}$ and you can conclude that Q is the transition matrix from B to B'.

PROOF (OF THEOREM 4.20)

From the preceding lemma, let Q be the transition matrix from B to B'. Then $[\mathbf{v}]_B = P[\mathbf{v}]_{B'}$ and $[\mathbf{v}]_{B'} = Q[\mathbf{v}]_B$, which implies that $[\mathbf{v}]_B = PQ[\mathbf{v}]_B$ for every vector $\mathbf{v}$ in R^n. From this it follows that $PQ = I$. So, P is invertible and P^{-1} is equal to Q, the transition matrix from B to B'.

REMARK

Verify that the transition matrix P^{-1} from B to B' is $(B')^{-1}B$. Also verify that the transition matrix P from B' to B is $B^{-1}B'$. You can use these relationships to check the results obtained by Gauss-Jordan elimination.

Gauss-Jordan elimination can be used to find the transition matrix P^{-1}. First define two matrices B and B' whose columns correspond to the vectors in B and B'. That is,

$$B = \begin{bmatrix} v_{11} & v_{12} & \cdots & v_{1n} \\ v_{21} & v_{22} & \cdots & v_{2n} \\ \vdots & \vdots & & \vdots \\ v_{n1} & v_{n2} & \cdots & v_{nn} \end{bmatrix} \quad \text{and} \quad B' = \begin{bmatrix} u_{11} & u_{12} & \cdots & u_{1n} \\ u_{21} & u_{22} & \cdots & u_{2n} \\ \vdots & \vdots & & \vdots \\ u_{n1} & u_{n2} & \cdots & u_{nn} \end{bmatrix}.$$
$$ \mathbf{v}_1 \;\; \mathbf{v}_2 \;\;\;\;\;\; \mathbf{v}_n \phantom{\text{and} \quad B' = \;\;} \mathbf{u}_1 \;\; \mathbf{u}_2 \;\;\;\;\;\;\; \mathbf{u}_n$$

Then, by reducing the $n \times 2n$ matrix $\begin{bmatrix} B' & B \end{bmatrix}$ so that the identity matrix I_n occurs in place of B', you obtain the matrix $\begin{bmatrix} I_n & P^{-1} \end{bmatrix}$. The next theorem states this procedure formally.

THEOREM 4.21   Transition Matrix from B to B'

Let

$$B = \{\mathbf{v}_1, \mathbf{v}_2, \ldots, \mathbf{v}_n\} \quad \text{and} \quad B' = \{\mathbf{u}_1, \mathbf{u}_2, \ldots, \mathbf{u}_n\}$$

be two bases for R^n. Then the transition matrix P^{-1} from B to B' can be found by using Gauss-Jordan elimination on the $n \times 2n$ matrix $\begin{bmatrix} B' & B \end{bmatrix}$, as shown below.

$$\begin{bmatrix} B' & B \end{bmatrix} \quad \longrightarrow \quad \begin{bmatrix} I_n & P^{-1} \end{bmatrix}$$

PROOF

To begin, let

$$\begin{aligned} \mathbf{v}_1 &= c_{11}\mathbf{u}_1 + c_{21}\mathbf{u}_2 + \cdots + c_{n1}\mathbf{u}_n \\ \mathbf{v}_2 &= c_{12}\mathbf{u}_1 + c_{22}\mathbf{u}_2 + \cdots + c_{n2}\mathbf{u}_n \\ &\;\;\vdots \\ \mathbf{v}_n &= c_{1n}\mathbf{u}_1 + c_{2n}\mathbf{u}_2 + \cdots + c_{nn}\mathbf{u}_n \end{aligned}$$

which implies that

$$c_{1i}\begin{bmatrix} u_{11} \\ u_{21} \\ \vdots \\ u_{n1} \end{bmatrix} + c_{2i}\begin{bmatrix} u_{12} \\ u_{22} \\ \vdots \\ u_{n2} \end{bmatrix} + \cdots + c_{ni}\begin{bmatrix} u_{1n} \\ u_{2n} \\ \vdots \\ u_{nn} \end{bmatrix} = \begin{bmatrix} v_{1i} \\ v_{2i} \\ \vdots \\ v_{ni} \end{bmatrix}$$

for $i = 1, 2, \ldots, n$. From these vector equations, write the n systems of linear equations

$$\begin{aligned} u_{11}c_{1i} + u_{12}c_{2i} + \cdots + u_{1n}c_{ni} &= v_{1i} \\ u_{21}c_{1i} + u_{22}c_{2i} + \cdots + u_{2n}c_{ni} &= v_{2i} \\ &\;\;\vdots \\ u_{n1}c_{1i} + u_{n2}c_{2i} + \cdots + u_{nn}c_{ni} &= v_{ni} \end{aligned}$$

for $i = 1, 2, \ldots, n$. Each of the n systems has the same coefficient matrix, so you can reduce all n systems simultaneously using the augmented matrix below.

$$\begin{bmatrix} u_{11} & u_{12} & \cdots & u_{1n} & v_{11} & v_{12} & \cdots & v_{1n} \\ u_{21} & u_{22} & \cdots & u_{2n} & v_{21} & v_{22} & \cdots & v_{2n} \\ \vdots & \vdots & & \vdots & \vdots & \vdots & & \vdots \\ u_{n1} & u_{n2} & \cdots & u_{nn} & v_{n1} & v_{n2} & \cdots & v_{nn} \end{bmatrix}$$
$$ B' B$$

Applying Gauss-Jordan elimination to this matrix produces

$$\begin{bmatrix} 1 & 0 & \cdots & 0 & c_{11} & c_{12} & \cdots & c_{1n} \\ 0 & 1 & \cdots & 0 & c_{21} & c_{22} & \cdots & c_{2n} \\ \vdots & \vdots & & \vdots & \vdots & \vdots & & \vdots \\ 0 & 0 & \cdots & 1 & c_{n1} & c_{n2} & \cdots & c_{nn} \end{bmatrix}.$$

By the lemma following Theorem 4.20, however, the right-hand side of this matrix is $Q = P^{-1}$, which implies that the matrix has the form $\begin{bmatrix} I & P^{-1} \end{bmatrix}$, which proves the theorem. ◼

In the next example, you will apply this procedure to the change of basis problem from Example 3.

EXAMPLE 4   **Finding a Transition Matrix**

See LarsonLinearAlgebra.com for an interactive version of this type of example.

Find the transition matrix from B to B' for the bases for R^3 below.

$$B = \{(1, 0, 0), (0, 1, 0), (0, 0, 1)\} \quad \text{and} \quad B' = \{(1, 0, 1), (0, -1, 2), (2, 3, -5)\}$$

SOLUTION

First use the vectors in the two bases to form the matrices B and B'.

$$B = \begin{bmatrix} 1 & 0 & 0 \\ 0 & 1 & 0 \\ 0 & 0 & 1 \end{bmatrix} \quad \text{and} \quad B' = \begin{bmatrix} 1 & 0 & 2 \\ 0 & -1 & 3 \\ 1 & 2 & -5 \end{bmatrix}$$

Then form the matrix $\begin{bmatrix} B' & B \end{bmatrix}$ and use Gauss-Jordan elimination to rewrite $\begin{bmatrix} B' & B \end{bmatrix}$ as $\begin{bmatrix} I_3 & P^{-1} \end{bmatrix}$.

$$\begin{bmatrix} 1 & 0 & 2 & 1 & 0 & 0 \\ 0 & -1 & 3 & 0 & 1 & 0 \\ 1 & 2 & -5 & 0 & 0 & 1 \end{bmatrix} \Longrightarrow \begin{bmatrix} 1 & 0 & 0 & -1 & 4 & 2 \\ 0 & 1 & 0 & 3 & -7 & -3 \\ 0 & 0 & 1 & 1 & -2 & -1 \end{bmatrix}$$

From this, you can conclude that the transition matrix from B to B' is

$$P^{-1} = \begin{bmatrix} -1 & 4 & 2 \\ 3 & -7 & -3 \\ 1 & -2 & -1 \end{bmatrix}.$$

Multiply P^{-1} by the coordinate matrix of $\mathbf{x} = \begin{bmatrix} 1 & 2 & -1 \end{bmatrix}^T$ to see that the result is the same as that obtained in Example 3. ◼

DISCOVERY

1. Let $B = \{(1, 0), (1, 2)\}$ and $B' = \{(1, 0), (0, 1)\}$. Form the matrix $\begin{bmatrix} B' & B \end{bmatrix}$.

2. Make a conjecture about the necessity of using Gauss-Jordan elimination to obtain the transition matrix P^{-1} when the change of basis is from a nonstandard basis to a standard basis.

LINEAR ALGEBRA APPLIED

Crystallography is the science of atomic and molecular structure. In a crystal, atoms are in a repeating pattern called a *lattice*. The simplest repeating unit in a lattice is a *unit cell*. Crystallographers can use bases and coordinate matrices in R^3 to designate the locations of atoms in a unit cell. For example, the figure below shows the unit cell known as *end-centered monoclinic*.

One possible coordinate matrix for the top end-centered (blue) atom is $[\mathbf{x}]_{B'} = \begin{bmatrix} \frac{1}{2} & \frac{1}{2} & 1 \end{bmatrix}^T$.

Note that when B is the standard basis, as in Example 4, the process of changing $[B'\ \ B]$ to $[I_n\ \ P^{-1}]$ becomes

$$[B'\ \ I_n]\ \ \longrightarrow\ \ [I_n\ \ P^{-1}].$$

But this is the same process that was used to find inverse matrices in Section 2.3. In other words, if B is the standard basis for R^n, then the transition matrix from B to B' is

$$P^{-1} = (B')^{-1}.\qquad \text{Standard basis to nonstandard basis}$$

The process is even simpler when B' is the standard basis, because the matrix $[B'\ \ B]$ is already in the form

$$[I_n\ \ B] = [I_n\ \ P^{-1}].$$

In this case, the transition matrix is simply

$$P^{-1} = B.\qquad \text{Nonstandard basis to standard basis}$$

For instance, the transition matrix in Example 2 from $B = \{(1,0),(1,2)\}$ to $B' = \{(1,0),(0,1)\}$ is

$$P^{-1} = B = \begin{bmatrix} 1 & 1 \\ 0 & 2 \end{bmatrix}.$$

EXAMPLE 5 **Finding a Transition Matrix**

Find the transition matrix from B to B' for the bases for R^2 below.

$$B = \{(-3,2),(4,-2)\} \quad\text{and}\quad B' = \{(-1,2),(2,-2)\}$$

SOLUTION

Begin by forming the matrix

$$[B'\ \ B] = \begin{bmatrix} -1 & 2 & -3 & 4 \\ 2 & -2 & 2 & -2 \end{bmatrix}$$

and use Gauss-Jordan elimination to obtain the transition matrix P^{-1} from B to B':

$$[I_2\ \ P^{-1}] = \begin{bmatrix} 1 & 0 & -1 & 2 \\ 0 & 1 & -2 & 3 \end{bmatrix}.$$

So, you have

$$P^{-1} = \begin{bmatrix} -1 & 2 \\ -2 & 3 \end{bmatrix}.$$

In Example 5, if you had found the transition matrix from B' to B (rather than from B to B'), then you would have obtained

$$[B\ \ B'] = \begin{bmatrix} -3 & 4 & -1 & 2 \\ 2 & -2 & 2 & -2 \end{bmatrix}$$

which reduces to

$$[I_2\ \ P] = \begin{bmatrix} 1 & 0 & 3 & -2 \\ 0 & 1 & 2 & -1 \end{bmatrix}.$$

The transition matrix from B' to B is

$$P = \begin{bmatrix} 3 & -2 \\ 2 & -1 \end{bmatrix}.$$

Verify that this is the inverse of the transition matrix found in Example 5 by multiplying PP^{-1} to obtain I_2.

COORDINATE REPRESENTATION IN GENERAL n-DIMENSIONAL SPACES

One benefit of coordinate representation is that it enables you to represent vectors in any n-dimensional space using the same notation used in R^n. For instance, in Example 6, note that the coordinate matrix of a vector in P_3 is a vector in R^4.

EXAMPLE 6   Coordinate Representation in P_3

Find the coordinate matrix of

$$p = 4 - 2x^2 + 3x^3$$

relative to the standard basis for P_3,

$$S = \{1, x, x^2, x^3\}.$$

SOLUTION

Write p as a linear combination of the basis vectors (in the given order).

$$p = 4(1) + 0(x) + (-2)(x^2) + 3(x^3)$$

So, the coordinate matrix of p relative to S is

$$[p]_S = \begin{bmatrix} 4 \\ 0 \\ -2 \\ 3 \end{bmatrix}.$$

In the next example, the coordinate matrix of a vector in $M_{3,1}$ is a vector in R^3.

EXAMPLE 7   Coordinate Representation in $M_{3,1}$

Find the coordinate matrix of

$$X = \begin{bmatrix} -1 \\ 4 \\ 3 \end{bmatrix}$$

relative to the standard basis for $M_{3,1}$,

$$S = \left\{ \begin{bmatrix} 1 \\ 0 \\ 0 \end{bmatrix}, \begin{bmatrix} 0 \\ 1 \\ 0 \end{bmatrix}, \begin{bmatrix} 0 \\ 0 \\ 1 \end{bmatrix} \right\}.$$

SOLUTION

X can be written as

$$X = \begin{bmatrix} -1 \\ 4 \\ 3 \end{bmatrix} = (-1)\begin{bmatrix} 1 \\ 0 \\ 0 \end{bmatrix} + 4\begin{bmatrix} 0 \\ 1 \\ 0 \end{bmatrix} + 3\begin{bmatrix} 0 \\ 0 \\ 1 \end{bmatrix}$$

so the coordinate matrix of X relative to S is

$$[X]_S = \begin{bmatrix} -1 \\ 4 \\ 3 \end{bmatrix}.$$

REMARK

In Section 6.2 you will learn more about the use of R^n to represent an arbitrary n-dimensional vector space.

Theorems 4.20 and 4.21 can be generalized to cover arbitrary n-dimensional spaces. This text, however, does not cover the generalizations of these theorems.

4.7 Exercises   See CalcChat.com for worked-out solutions to odd-numbered exercises.

Finding a Coordinate Matrix   In Exercises 1–4, find the coordinate matrix of x in R^n relative to the standard basis.

1. $\mathbf{x} = (5, -2)$

2. $\mathbf{x} = (1, -3, 0)$

3. $\mathbf{x} = (7, -4, -1, 2)$

4. $\mathbf{x} = (-6, 12, -4, 9, -8)$

Finding a Coordinate Matrix   In Exercises 5–10, given the coordinate matrix of x relative to a (nonstandard) basis B for R^n, find the coordinate matrix of x relative to the standard basis.

5. $B = \{(2, -1), (0, 1)\}$,

$[\mathbf{x}]_B = \begin{bmatrix} 4 \\ 1 \end{bmatrix}$

6. $B = \{(-2, 3), (3, -2)\}$,

$[\mathbf{x}]_B = \begin{bmatrix} -1 \\ 4 \end{bmatrix}$

7. $B = \{(1, 0, 1), (1, 1, 0), (0, 1, 1)\}$,

$[\mathbf{x}]_B = \begin{bmatrix} 2 \\ 3 \\ 1 \end{bmatrix}$

8. $B = \left\{ \left(\frac{3}{4}, \frac{5}{2}, \frac{3}{2}\right), \left(3, 4, \frac{7}{2}\right), \left(-\frac{3}{2}, 6, 2\right) \right\}$,

$[\mathbf{x}]_B = \begin{bmatrix} 2 \\ 0 \\ 4 \end{bmatrix}$

9. $B = \{(0, 0, 0, 1), (0, 0, 1, 1), (0, 1, 1, 1), (1, 1, 1, 1)\}$,

$[\mathbf{x}]_B = \begin{bmatrix} 1 \\ -2 \\ 3 \\ -1 \end{bmatrix}$

10. $B = \{(4, 0, 7, 3), (0, 5, -1, -1), (-3, 4, 2, 1), (0, 1, 5, 0)\}$,

$[\mathbf{x}]_B = \begin{bmatrix} -2 \\ 3 \\ 4 \\ 1 \end{bmatrix}$

Finding a Coordinate Matrix   In Exercises 11–16, find the coordinate matrix of x in R^n relative to the basis B'.

11. $B' = \{(4, 0), (0, 3)\}$, $\mathbf{x} = (12, 6)$

12. $B' = \{(-5, 6), (3, -2)\}$, $\mathbf{x} = (-17, 22)$

13. $B' = \{(8, 11, 0), (7, 0, 10), (1, 4, 6)\}$, $\mathbf{x} = (3, 19, 2)$

14. $B' = \left\{ \left(\frac{3}{2}, 4, 1\right), \left(\frac{3}{4}, \frac{5}{2}, 0\right), \left(1, \frac{1}{2}, 2\right) \right\}$, $\mathbf{x} = \left(3, -\frac{1}{2}, 8\right)$

15. $B' = \{(4, 3, 3), (-11, 0, 11), (0, 9, 2)\}$,

$\mathbf{x} = (11, 18, -7)$

16. $B' = \{(9, -3, 15, 4), (3, 0, 0, 1), (0, -5, 6, 8), (3, -4, 2, -3)\}$,

$\mathbf{x} = (0, -20, 7, 15)$

Finding a Transition Matrix   In Exercises 17–24, find the transition matrix from B to B'.

17. $B = \{(1, 0), (0, 1)\}$, $B' = \{(2, 4), (1, 3)\}$

18. $B = \{(1, 0), (0, 1)\}$, $B' = \{(1, 1), (5, 6)\}$

19. $B = \{(2, 4), (-1, 3)\}$, $B' = \{(1, 0), (0, 1)\}$

20. $B = \{(1, 1), (1, 0)\}$, $B' = \{(1, 0), (0, 1)\}$

21. $B = \{(-1, 0, 0), (0, 1, 0), (0, 0, -1)\}$,

$B' = \{(0, 0, 2), (1, 4, 0), (5, 0, 2)\}$

22. $B = \{(1, 0, 0), (0, 1, 0), (0, 0, 1)\}$,

$B' = \{(1, 3, -1), (2, 7, -4), (2, 9, -7)\}$

23. $B = \{(3, 4, 0), (-2, -1, 1), (1, 0, -3)\}$,

$B' = \{(1, 0, 0), (0, 1, 0), (0, 0, 1)\}$

24. $B = \{(1, 3, 2), (2, -1, 2), (5, 6, 1)\}$,

$B' = \{(1, 0, 0), (0, 1, 0), (0, 0, 1)\}$

Finding a Transition Matrix   In Exercises 25–36, use a software program or a graphing utility to find the transition matrix from B to B'.

25. $B = \{(2, 5), (1, 2)\}$, $B' = \{(2, 1), (-1, 2)\}$

26. $B = \{(-2, 1), (3, 2)\}$, $B' = \{(1, 2), (-1, 0)\}$

27. $B = \{(-3, 4), (3, -5)\}$, $B' = \{(-5, -6), (7, -8)\}$

28. $B = \{(2, -2), (-2, -2)\}$, $B' = \{(3, -3), (-3, -3)\}$

29. $B = \{(1, 0, 0), (0, 1, 0), (0, 0, 1)\}$,

$B' = \{(1, 3, 3), (1, 5, 6), (1, 4, 5)\}$

30. $B = \{(1, 0, 0), (0, 1, 0), (0, 0, 1)\}$,

$B' = \{(2, -1, 4), (0, 2, 1), (-3, 2, 1)\}$

31. $B = \{(1, 2, 4), (-1, 2, 0), (2, 4, 0)\}$,

$B' = \{(0, 2, 1), (-2, 1, 0), (1, 1, 1)\}$

32. $B = \{(3, 2, 1), (1, 1, 2), (1, 2, 0)\}$,

$B' = \{(1, 1, -1), (0, 1, 2), (-1, 4, 0)\}$

33. $B = \{(1, 0, 0, 0), (0, 1, 0, 0), (0, 0, 1, 0), (0, 0, 0, 1)\}$,

$B' = \{(1, 3, 2, -1), (-2, -5, -5, 4), (-1, -2, -2, 4), (-2, -3, -5, 11)\}$

34. $B = \{(1, 0, 0, 0), (0, 1, 0, 0), (0, 0, 1, 0), (0, 0, 0, 1)\}$,

$B' = \{(1, 1, 1, 1), (0, 1, 1, 1), (0, 0, 1, 1), (0, 0, 0, 1)\}$

35. $B = \{(1, 0, 0, 0, 0), (0, 1, 0, 0, 0), (0, 0, 1, 0, 0), (0, 0, 0, 1, 0), (0, 0, 0, 0, 1)\}$,

$B' = \{(1, 2, 4, -1, 2), (-2, -3, 4, 2, 1), (0, 1, 2, -2, 1), (0, 1, 2, 2, 1), (1, -1, 0, 1, 2)\}$

36. $B = \{(1, 0, 0, 0, 0), (0, 1, 0, 0, 0), (0, 0, 1, 0, 0), (0, 0, 0, 1, 0), (0, 0, 0, 0, 1)\}$,

$B' = \{(2, 4, -2, 1, 0), (3, -1, 0, 1, 2), (0, 0, -2, 4, 5), (2, -1, 2, 1, 1), (0, 1, 2, -3, 1)\}$

Finding Transition and Coordinate Matrices
In Exercises 37–40, (a) find the transition matrix from B to B', (b) find the transition matrix from B' to B, (c) verify that the two transition matrices are inverses of each other, and (d) find the coordinate matrix $[\mathbf{x}]_B$, given the coordinate matrix $[\mathbf{x}]_{B'}$.

37. $B = \{(1, 3), (-2, -2)\}$, $B' = \{(-12, 0), (-4, 4)\}$,
$$[\mathbf{x}]_{B'} = \begin{bmatrix} -1 \\ 3 \end{bmatrix}$$

38. $B = \{(2, -2), (6, 3)\}$, $B' = \{(1, 1), (32, 31)\}$,
$$[\mathbf{x}]_{B'} = \begin{bmatrix} 2 \\ -1 \end{bmatrix}$$

39. $B = \{(1, 0, 2), (0, 1, 3), (1, 1, 1)\}$,
$B' = \{(2, 1, 1), (1, 0, 0), (0, 2, 1)\}$,
$$[\mathbf{x}]_{B'} = \begin{bmatrix} 1 \\ 2 \\ -1 \end{bmatrix}$$

40. $B = \{(1, 1, 1), (1, -1, 1), (0, 0, 1)\}$,
$B' = \{(2, 2, 0), (0, 1, 1), (1, 0, 1)\}$,
$$[\mathbf{x}]_{B'} = \begin{bmatrix} 2 \\ 3 \\ 1 \end{bmatrix}$$

Finding Transition and Coordinate Matrices
In Exercises 41–44, use a software program or a graphing utility to (a) find the transition matrix from B to B', (b) find the transition matrix from B' to B, (c) verify that the two transition matrices are inverses of each other, and (d) find the coordinate matrix $[\mathbf{x}]_B$, given the coordinate matrix $[\mathbf{x}]_{B'}$.

41. $B = \{(4, 2, -4), (6, -5, -6), (2, -1, 8)\}$,
$B' = \{(1, 0, 4), (4, 2, 8), (2, 5, -2)\}$,
$[\mathbf{x}]_{B'} = \begin{bmatrix} 1 & -1 & 2 \end{bmatrix}^T$

42. $B = \{(1, 3, 4), (2, -5, 2), (-4, 2, -6)\}$,
$B' = \{(1, 2, -2), (4, 1, -4), (-2, 5, 8)\}$,
$[\mathbf{x}]_{B'} = \begin{bmatrix} -1 & 0 & 2 \end{bmatrix}^T$

43. $B = \{(2, 0, -1), (0, -1, 3), (1, -3, -2)\}$,
$B' = \{(0, -1, -3), (-1, 3, -2), (-3, -2, 0)\}$,
$[\mathbf{x}]_{B'} = \begin{bmatrix} 4 & -3 & -2 \end{bmatrix}^T$

44. $B = \{(1, -1, 9), (-9, 1, 1), (1, 9, -1)\}$,
$B' = \{(3, 0, 3), (-3, 3, 0), (0, -3, 3)\}$,
$[\mathbf{x}]_{B'} = \begin{bmatrix} -5 & -4 & 1 \end{bmatrix}^T$

Coordinate Representation in P_3 In Exercises 45–48, find the coordinate matrix of p relative to the standard basis for P_3.

45. $p = 1 + 5x - 2x^2 + x^3$ **46.** $p = -2 - 3x + 4x^3$
47. $p = 13 + 114x + 3x^2$
48. $p = 4 + 11x + x^2 + 2x^3$

Coordinate Representation in $M_{3,1}$ In Exercises 49–52, find the coordinate matrix of X relative to the standard basis for $M_{3,1}$.

49. $X = \begin{bmatrix} 0 \\ 3 \\ 2 \end{bmatrix}$ **50.** $X = \begin{bmatrix} 2 \\ -1 \\ 4 \end{bmatrix}$

51. $X = \begin{bmatrix} 1 \\ 2 \\ -1 \end{bmatrix}$ **52.** $X = \begin{bmatrix} 1 \\ 0 \\ -4 \end{bmatrix}$

53. Writing Is it possible for a transition matrix to equal the identity matrix? Explain.

54. CAPSTONE Let B and B' be two bases for R^n.
(a) When $B = I_n$, write the transition matrix from B to B' in terms of B'.
(b) When $B' = I_n$, write the transition matrix from B to B' in terms of B.
(c) When $B = I_n$, write the transition matrix from B' to B in terms of B'.
(d) When $B' = I_n$, write the transition matrix from B' to B in terms of B.

True or False? In Exercises 55 and 56, determine whether each statement is true or false. If a statement is true, give a reason or cite an appropriate statement from the text. If a statement is false, provide an example that shows the statement is not true in all cases or cite an appropriate statement from the text.

55. (a) If P is the transition matrix from a basis B to B', then the equation $P[\mathbf{x}]_{B'} = [\mathbf{x}]_B$ represents the change of basis from B to B'.
(b) If B is the standard basis in R^n, then the transition matrix from B to B' is $P^{-1} = (B')^{-1}$.
(c) For any 4×1 matrix X, the coordinate matrix $[X]_S$ relative to the standard basis for $M_{4,1}$ is equal to X itself.

56. (a) If P is the transition matrix from a basis B' to B, then P^{-1} is the transition matrix from B to B'.
(b) To perform the change of basis from a nonstandard basis B' to the standard basis B, the transition matrix P^{-1} is simply B'.
(c) The coordinate matrix of $p = -3 + x + 5x^2$ relative to the standard basis for P_2 is $[p]_S = \begin{bmatrix} 5 & 1 & -3 \end{bmatrix}^T$.

57. Let P be the transition matrix from B'' to B', and let Q be the transition matrix from B' to B. What is the transition matrix from B'' to B?

58. Let P be the transition matrix from B'' to B', and let Q be the transition matrix from B' to B. What is the transition matrix from B to B''?

4.8   Applications of Vector Spaces

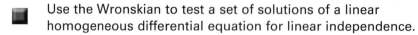

■ Use the Wronskian to test a set of solutions of a linear homogeneous differential equation for linear independence.

■ Identify and sketch the graph of a conic section and perform a rotation of axes.

LINEAR DIFFERENTIAL EQUATIONS (CALCULUS)

A **linear differential equation of order n** is of the form

$$y^{(n)} + g_{n-1}(x)y^{(n-1)} + \cdots + g_1(x)y' + g_0(x)y = f(x)$$

where $g_0, g_1, \ldots, g_{n-1}$ and f are functions of x with a common domain. If $f(x) = 0$, then the equation is **homogeneous.** Otherwise it is **nonhomogeneous.** A function y is a **solution** of the linear differential equation if the equation is satisfied when y and its first n derivatives are substituted into the equation.

EXAMPLE 1   A Second-Order Linear Differential Equation

Show that both $y = e^x$ and $y_2 = e^{-x}$ are solutions of the second-order linear differential equation $y'' - y = 0$.

SOLUTION

For the function $y_1 = e^x$, you have $y_1' = e^x$ and $y_1'' = e^x$. So,

$$y_1'' - y_1 = e^x - e^x = 0$$

which means that $y_1 = e^x$ is a solution of the differential equation. Similarly, for $y_2 = e^{-x}$, you have

$$y_2' = -e^{-x} \quad \text{and} \quad y_2'' = e^{-x}.$$

This implies that

$$y_2'' - y_2 = e^{-x} - e^{-x} = 0.$$

So, $y_2 = e^{-x}$ is also a solution of the linear differential equation.

There are two important observations you can make about Example 1. The first is that in the vector space $C''(-\infty, \infty)$ of all twice differentiable functions defined on the entire real line, the two solutions $y_1 = e^x$ and $y_2 = e^{-x}$ are *linearly independent*. This means that the only solution of

$$C_1 y_1 + C_2 y_2 = 0$$

that is valid for all x is $C_1 = C_2 = 0$. The second observation is that every *linear combination* of y_1 and y_2 is also a solution of the linear differential equation. To see this, let $y = C_1 y_1 + C_2 y_2$. Then

$$y = C_1 e^x + C_2 e^{-x}$$
$$y' = C_1 e^x - C_2 e^{-x}$$
$$y'' = C_1 e^x + C_2 e^{-x}.$$

Substituting into the differential equation $y'' - y = 0$ produces

$$y'' - y = (C_1 e^x + C_2 e^{-x}) - (C_1 e^x + C_2 e^{-x}) = 0.$$

So, $y = C_1 e^x + C_2 e^{-x}$ is a solution.

The next theorem, which is stated without proof, generalizes these observations.

REMARK

The solution

$$y = C_1 y_1 + C_2 y_2 + \cdots + C_n y_n$$

is the **general solution** of the differential equation.

Solutions of a Linear Homogeneous Differential Equation

Every nth-order linear homogeneous differential equation

$$y^{(n)} + g_{n-1}(x)y^{(n-1)} + \cdots + g_1(x)y' + g_0(x)y = 0$$

has n linearly independent solutions. Moreover, if $\{y_1, y_2, \ldots, y_n\}$ is a set of linearly independent solutions, then every solution is of the form

$$y = C_1 y_1 + C_2 y_2 + \cdots + C_n y_n$$

where $C_1, C_2, \ldots, C_n$ are real numbers.

In light of the preceding theorem, you can see the importance of being able to determine whether a set of solutions is linearly independent. Before describing a way of testing for linear independence, consider the definition below.

REMARK

The Wronskian of a set of functions is named after the Polish mathematician Josef Maria Wronski (1778–1853).

Definition of the Wronskian of a Set of Functions

Let $\{y_1, y_2, \ldots, y_n\}$ be a set of functions, each of which has $n-1$ derivatives on an interval I. The determinant

$$W(y_1, y_2, \ldots, y_n) = \begin{vmatrix} y_1 & y_2 & \cdots & y_n \\ y_1' & y_2' & \cdots & y_n' \\ \vdots & \vdots & & \vdots \\ y_1^{(n-1)} & y_2^{(n-1)} & \cdots & y_n^{(n-1)} \end{vmatrix}$$

is the **Wronskian** of the set of functions.

EXAMPLE 2   Finding the Wronskian of a Set of Functions

a. The Wronskian of the set $\{1 - x, 1 + x, 2 - x\}$ is

$$W = \begin{vmatrix} 1-x & 1+x & 2-x \\ -1 & 1 & -1 \\ 0 & 0 & 0 \end{vmatrix} = 0.$$

b. The Wronskian of the set $\{x, x^2, x^3\}$ is

$$W = \begin{vmatrix} x & x^2 & x^3 \\ 1 & 2x & 3x^2 \\ 0 & 2 & 6x \end{vmatrix} = 2x^3.$$

The Wronskian in part (a) of Example 2 is **identically equal to zero,** because it is zero for any value of x. The Wronskian in part (b) is not identically equal to zero because values of x exist for which this Wronskian is nonzero.

The next theorem shows how the Wronskian of a set of functions can be used to test for linear independence.

REMARK

This test does *not* apply to an arbitrary set of functions. Each of the functions $y_1, y_2, \ldots,$ and y_n must be a solution of the same linear homogeneous differential equation of order n.

Wronskian Test for Linear Independence

Let $\{y_1, y_2, \ldots, y_n\}$ be a set of n solutions of an nth-order linear homogeneous differential equation. This set is linearly independent if and only if the Wronskian is not identically equal to zero.

The proof of this theorem for the case where $n = 2$ is left as an exercise. (See Exercise 40.)

EXAMPLE 3   **Testing a Set of Solutions for Linear Independence**

Determine whether $\{1, \cos x, \sin x\}$ is a set of linearly independent solutions of the linear homogeneous differential equation

$$y''' + y' = 0.$$

SOLUTION

Begin by observing that each of the functions is a solution of $y''' + y' = 0$. (Check this.) Next, testing for linear independence produces the Wronskian of the three functions, as shown below.

$$W = \begin{vmatrix} 1 & \cos x & \sin x \\ 0 & -\sin x & \cos x \\ 0 & -\cos x & -\sin x \end{vmatrix}$$

$$= \sin^2 x + \cos^2 x$$

$$= 1$$

The Wronskian W is not identically equal to zero, so the set

$$\{1, \cos x, \sin x\}$$

is linearly independent. Moreover, this set consists of three linearly independent solutions of a third-order linear homogeneous differential equation, so the general solution is

$$y = C_1 + C_2 \cos x + C_3 \sin x$$

where C_1, C_2, and C_3 are real numbers.

EXAMPLE 4   **Testing a Set of Solutions for Linear Independence**

See LarsonLinearAlgebra.com for an interactive version of this type of example.

Determine whether $\{e^x, xe^x, (x + 1)e^x\}$ is a set of linearly independent solutions of the linear homogeneous differential equation

$$y''' - 3y'' + 3y' - y = 0.$$

SOLUTION

As in Example 3, begin by verifying that each of the functions is a solution of $y''' - 3y'' + 3y' - y = 0$. (This verification is left to you.) Testing for linear independence produces the Wronskian of the three functions, as shown below.

$$W = \begin{vmatrix} e^x & xe^x & (x + 1)e^x \\ e^x & (x + 1)e^x & (x + 2)e^x \\ e^x & (x + 2)e^x & (x + 3)e^x \end{vmatrix} = 0$$

So, the set $\{e^x, xe^x, (x + 1)e^x\}$ is linearly dependent.

In Example 4, the Wronskian is used to determine that the set $\{e^x, xe^x, (x + 1)e^x\}$ is linearly dependent. Another way to determine the linear dependence of this set is to observe that the third function is a linear combination of the first two. That is,

$$(x + 1)e^x = e^x + xe^x.$$

Verify that a different set, $\{e^x, xe^x, x^2e^x\}$, forms a linearly independent set of solutions of the differential equation

$$y''' - 3y'' + 3y' - y = 0.$$

CONIC SECTIONS AND ROTATION

Every conic section in the xy-plane has an equation that can be written in the form

$$ax^2 + bxy + cy^2 + dx + ey + f = 0.$$

Identifying the graph of this equation is fairly simple as long as b, the coefficient of the xy-term, is zero. When b is zero, the conic axes are parallel to the coordinate axes, and the identification is accomplished by writing the equation in standard (completed square) form. The standard forms of the equations of the four basic conics are given in the summary below. For circles, ellipses, and hyperbolas, the point (h, k) is the center. For parabolas, the point (h, k) is the vertex.

Standard Forms of Equations of Conics

Circle (r = radius): $(x - h)^2 + (y - k)^2 = r^2$

Ellipse (2α = major axis length, 2β = minor axis length):

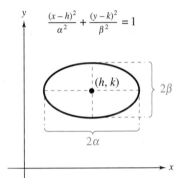

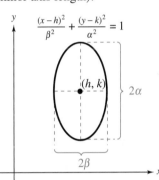

Hyperbola (2α = transverse axis length, 2β = conjugate axis length):

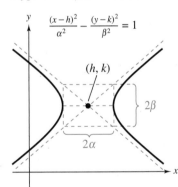

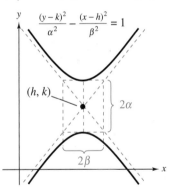

Parabola (p = directed distance from vertex to focus):

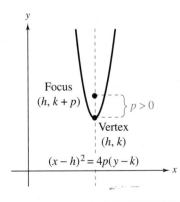

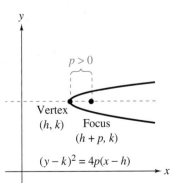

EXAMPLE 5   **Identifying Conic Sections**

a. The standard form of $x^2 - 2x + 4y - 3 = 0$ is

$$(x - 1)^2 = 4(-1)(y - 1).$$

The graph of this equation is a parabola with the vertex at $(h, k) = (1, 1)$. The axis of the parabola is vertical. The directed distance p from the vertex to the focus is $p = -1$, so the focus is the point $(1, 0)$. Finally, the focus lies below the vertex, so the parabola opens downward, as shown in Figure 4.18(a).

b. The standard form of $x^2 + 4y^2 + 6x - 8y + 9 = 0$ is

$$\frac{(x + 3)^2}{4} + \frac{(y - 1)^2}{1} = 1.$$

The graph of this equation is an ellipse with its center at $(h, k) = (-3, 1)$. The major axis is horizontal, and its length is $2\alpha = 4$. The length of the minor axis is $2\beta = 2$. The vertices of this ellipse occur at $(-5, 1)$ and $(-1, 1)$, and the endpoints of the minor axis occur at $(-3, 2)$ and $(-3, 0)$, as shown in Figure 4.18(b).

a.

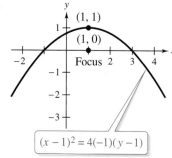

b.

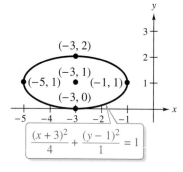

Figure 4.18

Note that the equations of the conics in Example 5 have no xy-term, so the axes of the graphs of these conics are parallel to the coordinate axes. For second-degree equations that have an xy-term, the axes of the graphs of the corresponding conics are not parallel to the coordinate axes. In such cases, it is helpful to *rotate* the standard axes to form a new x'-axis and y'-axis. The required rotation angle θ (measured counterclockwise) can be found using the equation $\cot 2\theta = (a - c)/b$. Then, the standard basis for R^2,

$$B = \{(1, 0), (0, 1)\}$$

rotates to form the new basis

$$B' = \{(\cos \theta, \sin \theta), (-\sin \theta, \cos \theta)\}$$

as shown below.

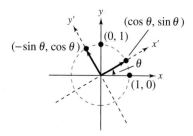

To find the coordinates of a point (x, y) relative to this new basis, you can use a transition matrix, as demonstrated in Example 6.

EXAMPLE 6   **A Transition Matrix for Rotation in R^2**

Find the coordinates of a point (x, y) in R^2 relative to the basis

$$B' = \{(\cos \theta, \sin \theta), (-\sin \theta, \cos \theta)\}.$$

SOLUTION

By Theorem 4.21 you have

$$[B' \quad B] = \begin{bmatrix} \cos \theta & -\sin \theta & 1 & 0 \\ \sin \theta & \cos \theta & 0 & 1 \end{bmatrix}.$$

B is the standard basis for R^2, so P^{-1} is represented by $(B')^{-1}$. You can use the formula given in Section 2.3 (page 66) for the inverse of a 2×2 matrix to find $(B')^{-1}$. This results in

$$[I \quad P^{-1}] = \begin{bmatrix} 1 & 0 & \cos \theta & \sin \theta \\ 0 & 1 & -\sin \theta & \cos \theta \end{bmatrix}.$$

By letting (x', y') be the coordinates of (x, y) relative to B', you can use the transition matrix P^{-1} as shown below.

$$\begin{bmatrix} \cos \theta & \sin \theta \\ -\sin \theta & \cos \theta \end{bmatrix} \begin{bmatrix} x \\ y \end{bmatrix} = \begin{bmatrix} x' \\ y' \end{bmatrix}$$

The x'- and y'-coordinates are $x' = x \cos \theta + y \sin \theta$ and $y' = -x \sin \theta + y \cos \theta$. ∎

The last two equations in Example 6 give the $x'y'$-coordinates in terms of the xy-coordinates. To perform a rotation of axes for a general second-degree equation, it is helpful to express the xy-coordinates in terms of the $x'y'$-coordinates. To do this, solve the last two equations in Example 6 for x and y to obtain

$$x = x' \cos \theta - y' \sin \theta \quad \text{and} \quad y = x' \sin \theta + y' \cos \theta.$$

Substituting these expressions for x and y into the given second-degree equation produces a second-degree equation in x' and y' that has no $x'y'$-term.

Rotation of Axes

The general second-degree equation $ax^2 + bxy + cy^2 + dx + ey + f = 0$ can be written in the form

$$a'(x')^2 + c'(y')^2 + d'x' + e'y' + f' = 0$$

by rotating the coordinate axes counterclockwise through the angle θ, where θ is found using the equation $\cot 2\theta = \dfrac{a - c}{b}$. The coefficients of the new equation are obtained from the substitutions

$$x = x' \cos \theta - y' \sin \theta \quad \text{and} \quad y = x' \sin \theta + y' \cos \theta.$$

The proof of the above result is left to you. (See Exercise 80.)

LINEAR ALGEBRA APPLIED

A satellite dish is an antenna that is designed to transmit signals to or receive signals from a communications satellite. A standard satellite dish consists of a bowl-shaped surface and a *feed horn* that is aimed toward the surface. The bowl-shaped surface is typically in the shape of a rotated elliptic paraboloid. (See Section 7.4.) The cross section of the surface is typically in the shape of a rotated parabola.

Example 7 demonstrates how to identify the graph of a second-degree equation by rotating the coordinate axes.

EXAMPLE 7   Rotation of a Conic Section

Perform a rotation of axes to eliminate the xy-term in

$$5x^2 - 6xy + 5y^2 + 14\sqrt{2}x - 2\sqrt{2}y + 18 = 0$$

and sketch the graph of the resulting equation in the $x'y'$-plane.

SOLUTION

Find the angle of rotation θ using

$$\cot 2\theta = \frac{a - c}{b} = \frac{5 - 5}{-6} = 0.$$

This implies that $\theta = \pi/4$. So,

$$\sin \theta = \frac{1}{\sqrt{2}} \quad \text{and} \quad \cos \theta = \frac{1}{\sqrt{2}}.$$

By substituting

$$x = x' \cos \theta - y' \sin \theta = \frac{1}{\sqrt{2}}(x' - y')$$

and

$$y = x' \sin \theta + y' \cos \theta = \frac{1}{\sqrt{2}}(x' + y')$$

into the original equation and simplifying, verify that you obtain

$$(x')^2 + 4(y')^2 + 6x' - 8y' + 9 = 0.$$

Finally, by completing the square, the standard form of this equation is

$$\frac{(x' + 3)^2}{2^2} + \frac{(y' - 1)^2}{1^2} = \frac{(x' + 3)^2}{4} + \frac{(y' - 1)^2}{1} = 1$$

which is the equation of an ellipse, as shown in Figure 4.19.

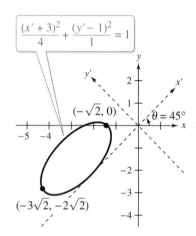

$$\frac{(x' + 3)^2}{4} + \frac{(y' - 1)^2}{1} = 1$$

$(-\sqrt{2}, 0)$

$\theta = 45°$

$(-3\sqrt{2}, -2\sqrt{2})$

Figure 4.19

In Example 7, the new (rotated) basis for R^2 is

$$B' = \left\{ \left(\frac{1}{\sqrt{2}}, \frac{1}{\sqrt{2}} \right), \left(-\frac{1}{\sqrt{2}}, \frac{1}{\sqrt{2}} \right) \right\}$$

and the coordinates of the vertices of the ellipse relative to B' are

$$\begin{bmatrix} -5 \\ 1 \end{bmatrix} \quad \text{and} \quad \begin{bmatrix} -1 \\ 1 \end{bmatrix}.$$

To find the coordinates of the vertices relative to the standard basis $B = \{(1, 0), (0, 1)\}$, use the equations

$$x = \frac{1}{\sqrt{2}}(x' - y')$$

and

$$y = \frac{1}{\sqrt{2}}(x' + y')$$

to obtain $\left(-3\sqrt{2}, -2\sqrt{2} \right)$ and $\left(-\sqrt{2}, 0 \right)$, as shown in Figure 4.19.

4.8 Exercises

See CalcChat.com for worked-out solutions to odd-numbered exercises.

Determining Solutions of a Differential Equation
In Exercises 1–12, determine which functions are solutions of the linear differential equation.

1. $y'' + y = 0$

   (a) e^x (b) $\sin x$

   (c) $\cos x$ (d) $\sin x - \cos x$

2. $y''' + y = 0$

   (a) e^x (b) e^{-x} (c) e^{-2x} (d) $2e^{-x}$

3. $y''' + y'' + y' + y = 0$

   (a) x (b) e^x (c) e^{-x} (d) xe^{-x}

4. $y'' - 6y' + 9y = 0$

   (a) e^{3x} (b) xe^{3x}

   (c) x^2e^{3x} (d) $(x + 3)e^{3x}$

5. $y^{(4)} + y''' - 2y'' = 0$

   (a) 1 (b) x (c) x^2 (d) e^x

6. $y^{(4)} - 16y = 0$

   (a) $3\cos x$ (b) $3\cos 2x$

   (c) e^{-2x} (d) $3e^{2x} - 4\sin 2x$

7. $x^2y'' - 2y = 0$

   (a) $\dfrac{1}{x^2}$ (b) x^2 (c) e^{x^2} (d) e^{-x^2}

8. $y' + (2x - 1)y = 0$

   (a) e^{x-x^2} (b) $2e^{x-x^2}$ (c) $3e^{x-x^2}$ (d) $4e^{x-x^2}$

9. $xy' - 2y = 0$

   (a) $\sqrt{x}$ (b) x (c) x^2 (d) x^3

10. $xy'' + 2y' = 0$

   (a) x (b) $\dfrac{1}{x}$ (c) xe^x (d) xe^{-x}

11. $y'' - y' - 12y = 0$

   (a) e^{-4x} (b) e^{4x} (c) e^{-3x} (d) e^{3x}

12. $y' - 2xy = 0$

   (a) $3e^{x^2}$ (b) xe^{x^2} (c) x^2e^x (d) xe^{-x}

Finding the Wronskian for a Set of Functions
In Exercises 13–26, find the Wronskian for the set of functions.

13. $\{x, -\sin x\}$ **14.** $\{e^{3x}, \sin 2x\}$

15. $\{e^x, e^{-x}\}$ **16.** $\{e^{x^2}, e^{-x^2}\}$

17. $\{x, \sin x, \cos x\}$ **18.** $\{x, -\sin x, \cos x\}$

19. $\{e^{-x}, xe^{-x}, (x + 3)e^{-x}\}$ **20.** $\{x, e^{-x}, e^x\}$

21. $\{1, e^x, e^{2x}\}$ **22.** $\{x^2, e^{x^2}, x^2e^x\}$

23. $\{1, x, x^2, x^3\}$ **24.** $\{x, x^2, e^x, e^{-x}\}$

25. $\{1, x, \cos x, e^{-x}\}$ **26.** $\{x, e^x, \sin x, \cos x\}$

Showing Linear Independence **In Exercises 27–30, show that the set of solutions of a second-order linear homogeneous differential equation is linearly independent.**

27. $\{e^{ax}, e^{bx}\}, a \neq b$ **28.** $\{e^{ax}, xe^{ax}\}$

29. $\{\cos ax, \sin ax\}, a \neq 0$

30. $\{e^{ax}\cos bx, e^{ax}\sin bx\}, b \neq 0$

Testing for Linear Independence **In Exercises 31–38, (a) verify that each solution satisfies the differential equation, (b) test the set of solutions for linear independence, and (c) if the set is linearly independent, then write the general solution of the differential equation.**

Differential Equation	Solutions
31. $y'' + 16y = 0$	$\{\sin 4x, \cos 4x\}$
32. $y'' - 4y' + 5y = 0$	$\{e^{2x}\sin x, e^{2x}\cos x\}$
33. $y''' + 4y'' + 4y' = 0$	$\{e^{-2x}, xe^{-2x}, (2x + 1)e^{-2x}\}$
34. $y''' + 4y' = 0$	$\{1, 2\cos 2x, 2 + \cos 2x\}$
35. $y''' + 4y' = 0$	$\{1, \sin 2x, \cos 2x\}$
36. $y''' + 3y'' + 3y' + y = 0$	$\{e^{-x}, xe^{-x}, x^2e^{-x}\}$
37. $y''' + 3y'' + 3y' + y = 0$	$\{e^{-x}, xe^{-x}, e^{-x} + xe^{-x}\}$
38. $y^{(4)} - 2y''' + y'' = 0$	$\{1, x, e^x, xe^x\}$

39. Pendulum Consider a pendulum of length L that swings by the force of gravity only.

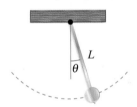

For small values of $\theta = \theta(t)$, the motion of the pendulum can be approximated by the differential equation

$$\frac{d^2\theta}{dt^2} + \frac{g}{L}\theta = 0$$

where g is the acceleration due to gravity.

(a) Verify that

$$\left\{\sin\sqrt{\frac{g}{L}}t, \cos\sqrt{\frac{g}{L}}t\right\}$$

is a set of linearly independent solutions of the differential equation.

(b) Find the general solution of the differential equation and show that it can be written in the form

$$\theta(t) = A\cos\left[\sqrt{\frac{g}{L}}(t + \phi)\right].$$

40. Proof Let $\{y_1, y_2\}$ be a set of solutions of a second-order linear homogeneous differential equation. Prove that this set is linearly independent if and only if the Wronskian is not identically equal to zero.

41. Writing Is the sum of two solutions of a nonhomogeneous linear differential equation also a solution? Explain.

42. Writing Is the scalar multiple of a solution of a nonhomogeneous linear differential equation also a solution? Explain.

Identifying and Graphing a Conic Section
In Exercises 43–58, identify and sketch the graph of the conic section.

43. $y^2 + x = 0$
44. $x^2 - 6y = 0$
45. $x^2 + 4y^2 - 16 = 0$
46. $5x^2 + 3y^2 - 15 = 0$
47. $\dfrac{x^2}{9} - \dfrac{y^2}{16} - 1 = 0$
48. $\dfrac{x^2}{36} - \dfrac{y^2}{49} = 1$
49. $x^2 + 4x + 6y - 2 = 0$ **50.** $y^2 - 6y - 4x + 21 = 0$
51. $16x^2 + 36y^2 - 64x - 36y + 73 = 0$
52. $4x^2 + y^2 - 8x + 3 = 0$
53. $9x^2 - y^2 + 54x + 10y + 55 = 0$
54. $4y^2 - 2x^2 - 4y - 8x - 15 = 0$
55. $x^2 + 4y^2 + 4x + 32y + 64 = 0$
56. $4y^2 + 4x^2 - 24x + 35 = 0$
57. $2x^2 - y^2 + 4x + 10y - 22 = 0$
58. $y^2 + 8x + 6y + 25 = 0$

Matching a Graph with an Equation In Exercises 59–62, match the graph with its equation. [The graphs are labeled (a), (b), (c), and (d).]

(a)
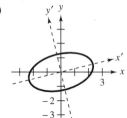

(b)
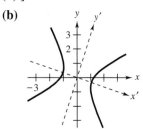

(c)
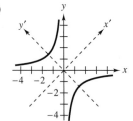

(d)
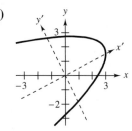

59. $xy + 2 = 0$
60. $-2x^2 + 3xy + 2y^2 + 3 = 0$
61. $x^2 - xy + 3y^2 - 5 = 0$
62. $x^2 - 4xy + 4y^2 + 10x - 30 = 0$

Rotation of a Conic Section In Exercises 63–74, perform a rotation of axes to eliminate the xy-term, and sketch the graph of the conic.

63. $xy + 1 = 0$ **64.** $xy - 8x - 4y = 0$
65. $4x^2 + 2xy + 4y^2 - 15 = 0$
66. $x^2 + 2xy + y^2 - 8x + 8y = 0$
67. $2x^2 - 3xy - 2y^2 + 10 = 0$
68. $5x^2 - 2xy + 5y^2 - 24 = 0$
69. $9x^2 + 24xy + 16y^2 + 90x - 130y = 0$
70. $5x^2 - 6xy + 5y^2 - 12 = 0$
71. $7x^2 - 6\sqrt{3}xy + 13y^2 - 64 = 0$
72. $7x^2 - 2\sqrt{3}xy + 5y^2 = 16$
73. $3x^2 - 2\sqrt{3}xy + y^2 + 2x + 2\sqrt{3}y = 0$
74. $x^2 + 2\sqrt{3}xy + 3y^2 - 2\sqrt{3}x + 2y + 16 = 0$

Rotation of a Degenerate Conic Section In Exercises 75–78, perform a rotation of axes to eliminate the xy-term, and sketch the graph of the "degenerate" conic.

75. $x^2 - 2xy + y^2 = 0$ **76.** $5x^2 - 2xy + 5y^2 = 0$
77. $x^2 + 2xy + y^2 - 1 = 0$ **78.** $x^2 - 10xy + y^2 = 0$

79. Proof Prove that a rotation of $\theta = \pi/4$ will eliminate the xy-term from the equation
$$ax^2 + bxy + ay^2 + dx + ey + f = 0.$$

80. Proof Prove that a rotation of θ, where $\cot 2\theta = (a - c)/b$, will eliminate the xy-term from the equation
$$ax^2 + bxy + cy^2 + dx + ey + f = 0.$$

81. Proof For the equation $ax^2 + bxy + cy^2 = 0$, define the matrix A as
$$A = \begin{bmatrix} a & b/2 \\ b/2 & c \end{bmatrix}.$$
(a) Prove that if $|A| = 0$, then the graph of $ax^2 + bxy + cy^2 = 0$ is a line.
(b) Prove that if $|A| \neq 0$, then the graph of $ax^2 + bxy + cy^2 = 0$ is two intersecting lines.

82. CAPSTONE
(a) Explain how to use the Wronskian to test a set of solutions of a linear homogeneous differential equation for linear independence.
(b) Explain how to eliminate the xy-term when it appears in the general equation of a conic section.

83. Use your school's library, the Internet, or some other reference source to find real-life applications of (a) linear differential equations and (b) rotation of conic sections that are different than those discussed in this section.

4   Review Exercises

See CalcChat.com for worked-out solutions to odd-numbered exercises.

Vector Operations In Exercises 1–4, find (a) $\mathbf{u} + \mathbf{v}$, (b) $2\mathbf{v}$, (c) $\mathbf{u} - \mathbf{v}$, and (d) $3\mathbf{u} - 2\mathbf{v}$.

1. $\mathbf{u} = (1, -2, -3)$, $\mathbf{v} = (3, 1, 0)$
2. $\mathbf{u} = (-1, 2, 1)$, $\mathbf{v} = (0, 1, 1)$
3. $\mathbf{u} = (3, -1, 2, 3)$, $\mathbf{v} = (0, 2, 2, 1)$
4. $\mathbf{u} = (0, 1, -1, 2)$, $\mathbf{v} = (1, 0, 0, 2)$

Solving a Vector Equation In Exercises 5–8, solve for x, where $\mathbf{u} = (1, -1, 2)$, $\mathbf{v} = (0, 2, 3)$, and $\mathbf{w} = (0, 1, 1)$.

5. $2\mathbf{x} - \mathbf{u} + 3\mathbf{v} + \mathbf{w} = \mathbf{0}$ 6. $3\mathbf{x} + 2\mathbf{u} - \mathbf{v} + 2\mathbf{w} = \mathbf{0}$
7. $5\mathbf{u} - 2\mathbf{x} = 3\mathbf{v} + \mathbf{w}$ 8. $3\mathbf{u} + 2\mathbf{x} = \mathbf{w} - \mathbf{v}$

Writing a Linear Combination In Exercises 9–12, write v as a linear combination of $\mathbf{u}_1$, $\mathbf{u}_2$, and $\mathbf{u}_3$, if possible.

9. $\mathbf{v} = (3, 0, -6)$, $\mathbf{u}_1 = (1, -1, 2)$, $\mathbf{u}_2 = (2, 4, -2)$, $\mathbf{u}_3 = (1, 2, -4)$
10. $\mathbf{v} = (4, 4, 5)$, $\mathbf{u}_1 = (1, 2, 3)$, $\mathbf{u}_2 = (-2, 0, 1)$, $\mathbf{u}_3 = (1, 0, 0)$
11. $\mathbf{v} = (1, 2, 3, 5)$, $\mathbf{u}_1 = (1, 2, 3, 4)$, $\mathbf{u}_2 = (-1, -2, -3, 4)$, $\mathbf{u}_3 = (0, 0, 1, 1)$
12. $\mathbf{v} = (4, -13, -5, -4)$, $\mathbf{u}_1 = (1, -2, 1, 1)$, $\mathbf{u}_2 = (-1, 2, 3, 2)$, $\mathbf{u}_3 = (0, -1, -1, -1)$

Describing the Zero Vector and the Additive Inverse In Exercises 13–16, describe the zero vector and the additive inverse of a vector in the vector space.

13. $M_{4,2}$ 14. P_8
15. R^5 16. $M_{2,3}$

Determining Subspaces In Exercises 17–24, determine whether W is a subspace of the vector space V.

17. $W = \{(x, y): x = 2y\}$, $V = R^2$
18. $W = \{(x, y): x - y = 1\}$, $V = R^2$
19. $W = \{(x, y): y = ax, a \text{ is an integer}\}$, $V = R^2$
20. $W = \{(x, y): y = ax^2\}$, $V = R^2$
21. $W = \{(x, 2x, 3x): x \text{ is a real number}\}$, $V = R^3$
22. $W = \{(x, y, z): x \ge 0\}$, $V = R^3$
23. $W = \{f: f(0) = -1\}$, $V = C[-1, 1]$
24. $W = \{f: f(-1) = 0\}$, $V = C[-1, 1]$

25. Which of the subsets of R^3 is a subspace of R^3?
    (a) $W = \{(x_1, x_2, x_3): x_1^2 + x_2^2 + x_3^2 = 0\}$
    (b) $W = \{(x_1, x_2, x_3): x_1^2 + x_2^2 + x_3^2 = 1\}$
26. Which of the subsets of R^3 is a subspace of R^3?
    (a) $W = \{(x_1, x_2, x_3): x_1 + x_2 + x_3 = 0\}$
    (b) $W = \{(x_1, x_2, x_3): x_1 + x_2 + x_3 = 1\}$

Spanning Sets, Linear Independence, and Bases In Exercises 27–32, determine whether the set (a) spans R^3, (b) is linearly independent, and (c) is a basis for R^3.

27. $S = \{(1, -5, 4), (11, 6, -1), (2, 3, 5)\}$
28. $S = \{(4, 0, 1), (0, -3, 2), (5, 10, 0)\}$
29. $S = \{(-\frac{1}{2}, \frac{3}{4}, -1), (5, 2, 3), (-4, 6, -8)\}$
30. $S = \{(2, 0, 1), (2, -1, 1), (4, 2, 0)\}$
31. $S = \{(1, 0, 0), (0, 1, 0), (0, 0, 1), (-1, 2, -3)\}$
32. $S = \{(1, 0, 0), (0, 1, 0), (0, 0, 1), (2, -1, 0)\}$

33. Determine whether
    $$S = \{1 - t, 2t + 3t^2, t^2 - 2t^3, 2 + t^3\}$$
    is a basis for P_3.

34. Determine whether $S = \{1, t, 1 + t^2\}$ is a basis for P_2.

Determining Whether a Set Is a Basis In Exercises 35 and 36, determine whether the set is a basis for $M_{2,2}$.

35. $S = \left\{ \begin{bmatrix} -2 & 3 \\ 1 & 0 \end{bmatrix}, \begin{bmatrix} 2 & 0 \\ -4 & 0 \end{bmatrix}, \begin{bmatrix} 1 & 3 \\ -1 & 1 \end{bmatrix}, \begin{bmatrix} 1 & 0 \\ 2 & 1 \end{bmatrix} \right\}$

36. $S = \left\{ \begin{bmatrix} 1 & 0 \\ 0 & 1 \end{bmatrix}, \begin{bmatrix} -1 & 0 \\ 1 & 1 \end{bmatrix}, \begin{bmatrix} 2 & 1 \\ 1 & 0 \end{bmatrix}, \begin{bmatrix} 1 & 1 \\ 0 & 1 \end{bmatrix} \right\}$

Finding the Nullspace, Nullity, and Rank of a Matrix In Exercises 37–42, find (a) the nullspace, (b) the nullity, and (c) the rank of the matrix A. Then verify that rank(A) + nullity$(A) = n$, where n is the number of columns of A.

37. $A = \begin{bmatrix} -4 & 3 \\ 12 & -9 \end{bmatrix}$

38. $A = \begin{bmatrix} 1 & 4 \\ 3 & 2 \end{bmatrix}$

39. $A = \begin{bmatrix} 2 & -3 & -6 & -4 \\ 1 & 5 & -3 & 11 \\ 2 & 7 & -6 & 16 \end{bmatrix}$

40. $A = \begin{bmatrix} 1 & 0 & -2 & 0 \\ 4 & -2 & 4 & -2 \\ -2 & 0 & 1 & 3 \end{bmatrix}$

41. $A = \begin{bmatrix} 1 & 3 & 2 \\ 4 & -1 & -18 \\ -1 & 3 & 10 \\ 1 & 2 & 0 \end{bmatrix}$

42. $A = \begin{bmatrix} 1 & 2 & 1 & 2 \\ 1 & 4 & 0 & 3 \\ -2 & 3 & 0 & 2 \\ 1 & 2 & 6 & 1 \end{bmatrix}$

Finding a Basis for a Row Space and Rank In Exercises 43–46, find (a) a basis for the row space and (b) the rank of the matrix.

43. $\begin{bmatrix} 1 & 2 \\ -4 & 3 \\ 6 & 1 \end{bmatrix}$ 44. $\begin{bmatrix} 2 & -1 & 4 \\ 1 & 5 & 6 \\ 1 & 16 & 14 \end{bmatrix}$

45. $\begin{bmatrix} 7 & 0 & 2 \\ 4 & 1 & 6 \\ -1 & 16 & 14 \end{bmatrix}$ 46. $\begin{bmatrix} 1 & 2 & 0 \\ -1 & 4 & 1 \\ 0 & 1 & 3 \end{bmatrix}$

Finding a Basis and Dimension In Exercises 47–50, find (a) a basis for and (b) the dimension of the solution space of the homogeneous system of linear equations.

47. $2x_1 + 4x_2 + 3x_3 - 6x_4 = 0$
    $x_1 + 2x_2 + 2x_3 - 5x_4 = 0$
    $3x_1 + 6x_2 + 5x_3 - 11x_4 = 0$

48. $16x_1 + 24x_2 + 8x_3 - 32x_4 = 0$
    $4x_1 + 6x_2 + 2x_3 - 8x_4 = 0$
    $2x_1 + 3x_2 + x_3 - 4x_4 = 0$

49. $x_1 - 3x_2 + x_3 + x_4 = 0$
    $2x_1 + x_2 - x_3 + 2x_4 = 0$
    $x_1 + 4x_2 - 2x_3 + x_4 = 0$
    $5x_1 - 8x_2 + 2x_3 + 5x_4 = 0$

50. $-x_1 + 2x_2 - x_3 + 2x_4 = 0$
    $-2x_1 + 2x_2 + x_3 + 4x_4 = 0$
    $3x_1 + 2x_2 + 2x_3 + 5x_4 = 0$
    $-3x_1 + 8x_2 + 5x_3 + 17x_4 = 0$

Finding a Coordinate Matrix In Exercises 51–56, given the coordinate matrix of x relative to a (nonstandard) basis B for R^n, find the coordinate matrix of x relative to the standard basis.

51. $B = \{(1, 1), (-1, 1)\}$, $[\mathbf{x}]_B = [3 \quad 5]^T$
52. $B = \{(2, 0), (3, 3)\}$, $[\mathbf{x}]_B = [1 \quad 1]^T$
53. $B = \{(\frac{1}{2}, \frac{1}{2}), (1, 0)\}$, $[\mathbf{x}]_B = [\frac{1}{2} \quad \frac{1}{2}]^T$
54. $B = \{(2, 4), (-1, 1)\}$, $[\mathbf{x}]_B = [4 \quad -7]^T$
55. $B = \{(1, 0, 0), (1, 1, 0), (0, 1, 1)\}$,
    $[\mathbf{x}]_B = [2 \quad 0 \quad -1]^T$
56. $B = \{(1, 0, 1), (0, 1, 0), (0, 1, 1)\}$, $[\mathbf{x}]_B = [4 \quad 0 \quad 2]^T$

Finding a Coordinate Matrix In Exercises 57–62, find the coordinate matrix of x in R^n relative to the basis B'.

57. $B' = \{(5, 0), (0, -8)\}$, $\mathbf{x} = (2, 2)$
58. $B' = \{(2, 2), (0, -1)\}$, $\mathbf{x} = (-1, 2)$
59. $B' = \{(1, 2, 3), (1, 2, 0), (0, -6, 2)\}$, $\mathbf{x} = (3, -3, 0)$
60. $B' = \{(1, 0, 0), (0, 1, 0), (1, 1, 1)\}$, $\mathbf{x} = (4, -2, 9)$
61. $B' = \{(9, -3, 15, 4), (-3, 0, 0, -1), (0, -5, 6, 8),$
    $(-3, 4, -2, 3)\}$, $\mathbf{x} = (21, -5, 43, 14)$
62. $B' = \{(1, -1, 2, 1), (1, 1, -4, 3), (1, 2, 0, 3),$
    $(1, 2, -2, 0)\}$, $\mathbf{x} = (5, 3, -6, 2)$

Finding a Transition Matrix In Exercises 63–68, find the transition matrix from B to B'.

63. $B = \{(1, -1), (3, 1)\}$, $B' = \{(1, 0), (0, 1)\}$
64. $B = \{(1, -1), (3, 1)\}$, $B' = \{(1, 2), (-1, 0)\}$
65. $B = \{(1, 0, 0), (0, 1, 0), (0, 0, 1)\}$,
    $B' = \{(0, 0, 1), (0, 1, 0), (1, 0, 0)\}$
66. $B = \{(1, 1, 1), (1, 1, 0), (1, 0, 0)\}$,
    $B' = \{(1, 2, 3), (0, 1, 0), (1, 0, 1)\}$
67. $B = \{(1, 1, 2), (2, 3, 4), (3, 3, 3)\}$,
    $B' = \{(7, -1, -1), (-3, 1, 0), (-3, 0, 1)\}$
68. $B = \{(1, 1, 1), (3, 4, 3), (3, 3, 4)\}$,
    $B' = \{(1, -1, \frac{2}{3}), (-2, 1, 0), (1, 0, -\frac{1}{3})\}$

Finding Transition and Coordinate Matrices In Exercises 69–72, (a) find the transition matrix from B to B', (b) find the transition matrix from B' to B, (c) verify that the two transition matrices are inverses of each other, and (d) find the coordinate matrix $[\mathbf{x}]_{B'}$, given the coordinate matrix $[\mathbf{x}]_B$.

69. $B = \{(-2, 1), (1, -1)\}$, $B' = \{(0, 2), (1, 1)\}$,
    $[\mathbf{x}]_B = [6 \quad -6]^T$
70. $B = \{(1, 0), (1, -1)\}$, $B' = \{(1, 1), (1, -1)\}$,
    $[\mathbf{x}]_B = [2 \quad -2]^T$
71. $B = \{(1, 0, 0), (1, 1, 0), (1, 1, 1)\}$,
    $B' = \{(0, 0, 1), (0, 1, 1), (1, 1, 1)\}$,
    $[\mathbf{x}]_B = [-1 \quad 2 \quad -3]^T$
72. $B = \{(1, 1, -1), (1, 1, 0), (1, -1, 0)\}$,
    $B' = \{(1, -1, 2), (2, 2, -1), (2, 2, 2)\}$,
    $[\mathbf{x}]_B = [2 \quad 2 \quad -1]^T$

73. Let W be the subspace of P_3 [the set of all polynomials $p(x)$ of degree 3 or less] such that $p(0) = 0$, and let U be the subspace of P_3 such that $p(1) = 0$. Find a basis for W, a basis for U, and a basis for their intersection $W \cap U$.

74. **Calculus** Let $V = C'(-\infty, \infty)$, the vector space of all continuously differentiable functions on the real line.
    (a) Prove that $W = \{f : f' = 4f\}$ is a subspace of V.
    (b) Prove that $U = \{f : f' = f + 1\}$ is not a subspace of V.

75. **Writing** Let $B = \{p_1(x), p_2(x), \ldots, p_n(x), p_{n+1}(x)\}$ be a basis for P_n. Must B contain a polynomial of each degree 0, 1, 2, . . . , n? Explain.

76. **Proof** Let A and B be $n \times n$ square matrices with $A \neq O$ and $B \neq O$. Prove that if A is symmetric and B is skew-symmetric ($B^T = -B$), then $\{A, B\}$ is a linearly independent set.

77. **Proof** Let $V = P_5$ and consider the set W of all polynomials of the form $(x^3 + x)p(x)$, where $p(x)$ is in P_2. Is W a subspace of V? Prove your answer.

78. Let $\mathbf{v}_1$, $\mathbf{v}_2$, and $\mathbf{v}_3$ be three linearly independent vectors in a vector space V. Is the set $\{\mathbf{v}_1 - 2\mathbf{v}_2, 2\mathbf{v}_2 - 3\mathbf{v}_3, 3\mathbf{v}_3 - \mathbf{v}_1\}$ linearly dependent or linearly independent? Explain.

79. Proof Let A be an $n \times n$ square matrix. Prove that the row vectors of A are linearly dependent if and only if the column vectors of A are linearly dependent.

80. Proof Let A be an $n \times n$ square matrix, and let λ be a scalar. Prove that the set
$$S = \{\mathbf{x}: A\mathbf{x} = \lambda\mathbf{x}\}$$
is a subspace of R^n. Determine the dimension of S when $\lambda = 3$ and
$$A = \begin{bmatrix} 3 & 1 & 0 \\ 0 & 3 & 0 \\ 0 & 0 & 1 \end{bmatrix}.$$

81. Let $f(x) = x$ and $g(x) = |x|$.

(a) Show that f and g are linearly independent in $C[-1, 1]$.

(b) Show that f and g are linearly dependent in $C[0, 1]$.

82. Describe how the domain of a set of functions can influence whether the set is linearly independent or dependent.

True or False? In Exercises 83–86, determine whether each statement is true or false. If a statement is true, give a reason or cite an appropriate statement from the text. If a statement is false, provide an example that shows the statement is not true in all cases or cite an appropriate statement from the text.

83. (a) The standard operations in R^n are vector addition and scalar multiplication.

(b) The additive inverse of a vector is not unique.

(c) A vector space consists of four entities: a set of vectors, a set of scalars, and two operations.

84. (a) The set $W = \{(0, x^2, x^3): x^2 \text{ and } x^3 \text{ are real numbers}\}$ is a subspace of R^3.

(b) A set of vectors S in a vector space V is a basis for V when S spans V and S is linearly independent.

(c) If A is an invertible $n \times n$ matrix, then the n row vectors of A are linearly dependent.

85. (a) The set of all n-tuples is n-space and is denoted by R^n.

(b) The additive identity of a vector space is not unique.

(c) Once a theorem has been proved for an abstract vector space, you need not give separate proofs for n-tuples, matrices, and polynomials.

86. (a) The set of points on the line $x + y = 0$ is a subspace of R^2.

(b) Elementary row operations preserve the column space of the matrix A.

Determining Solutions of a Differential Equation In Exercises 87–90, determine which functions are solutions of the linear differential equation.

87. $y'' - y' - 6y = 0$
 (a) e^{3x} (b) e^{2x} (c) e^{-3x} (d) e^{-2x}

88. $y^{(4)} - y = 0$
 (a) e^x (b) e^{-x} (c) $\cos x$ (d) $\sin x$

89. $y' + 2y = 0$
 (a) e^{-2x} (b) xe^{-2x} (c) x^2e^{-x} (d) $2xe^{-2x}$

90. $y'' + 25y = 0$
 (a) $\sin 5x + \cos 5x$ (b) $5\sin x + 5\cos x$
 (c) $\sin 5x$ (d) $\cos 5x$

Finding the Wronskian for a Set of Functions In Exercises 91–94, find the Wronskian for the set of functions.

91. $\{1, x, e^x\}$ **92.** $\{2, x^2, 3 + x\}$
93. $\{1, \sin 2x, \cos 2x\}$ **94.** $\{x, \sin^2 x, \cos^2 x\}$

Testing for Linear Independence In Exercises 95–98, (a) verify that each solution satisfies the differential equation, (b) test the set of solutions for linear independence, and (c) if the set is linearly independent, then write the general solution of the differential equation.

Differential Equation	Solutions
95. $y'' + 6y' + 9y = 0$	$\{e^{-3x}, xe^{-3x}\}$
96. $y'' + 6y' + 9y = 0$	$\{e^{-3x}, 3e^{-3x}\}$
97. $y''' - 6y'' + 11y' - 6y = 0$	$\{e^x, e^{2x}, e^x - e^{2x}\}$
98. $y'' + 9y = 0$	$\{\sin 3x, \cos 3x\}$

Identifying and Graphing a Conic Section In Exercises 99–106, identify and sketch the graph of the conic section.

99. $x^2 + y^2 + 4x - 2y - 11 = 0$
100. $9x^2 + 9y^2 + 18x - 18y + 14 = 0$
101. $x^2 - y^2 + 2x - 3 = 0$
102. $4x^2 - y^2 + 8x - 6y + 4 = 0$
103. $2x^2 - 20x - y + 46 = 0$
104. $y^2 - 4x - 4 = 0$
105. $4x^2 + y^2 + 32x + 4y + 63 = 0$
106. $16x^2 + 25y^2 - 32x - 50y + 16 = 0$

Rotation of a Conic Section In Exercises 107–110, perform a rotation of axes to eliminate the xy-term, and sketch the graph of the conic.

107. $xy = 3$
108. $9x^2 + 4xy + 9y^2 - 20 = 0$
109. $16x^2 - 24xy + 9y^2 - 60x - 80y + 100 = 0$
110. $x^2 + 2xy + y^2 + \sqrt{2}x - \sqrt{2}y = 0$

4   Projects

1   Solutions of Linear Systems

Write a paragraph to answer the question. Do not perform any calculations, but instead base your explanations on appropriate properties from the text.

1. One solution of the homogeneous linear system

$$2x + y + 4z + \ w = 0$$
$$4x + y \quad\quad - \ w = 0$$
$$\quad\ 2y + \ z - 9w = 0$$

is $x = -1, y = 5, z = -1$, and $w = 1$. Explain why $x = 2, y = -10, z = 2$, and $w = -2$ is also a solution.

2. The vectors $\mathbf{x}_1$ and $\mathbf{x}_2$ are solutions of the homogeneous linear system $A\mathbf{x} = \mathbf{0}$. Explain why the vector $5\mathbf{x}_1 - 2\mathbf{x}_2$ is also a solution.

3. Consider the two systems represented by the augmented matrices.

$$\begin{bmatrix} 2 & -1 & 5 & 0 \\ 0 & 1 & 1 & 4 \\ 1 & -2 & 1 & -6 \end{bmatrix} \quad \begin{bmatrix} 2 & -1 & 5 & 0 \\ 0 & 1 & 1 & -8 \\ 1 & -2 & 1 & 12 \end{bmatrix}$$

If the first system is consistent, then why is the second system also consistent?

4. The vectors $\mathbf{x}_1$ and $\mathbf{x}_2$ are solutions of the linear system $A\mathbf{x} = \mathbf{b}$. Is the vector $5\mathbf{x}_1 - 2\mathbf{x}_2$ also a solution? Why or why not?

5. The linear systems $A\mathbf{x} = \mathbf{b}_1$ and $A\mathbf{x} = \mathbf{b}_2$ are consistent. Is the system $A\mathbf{x} = \mathbf{b}_1 + \mathbf{b}_2$ necessarily consistent? Why or why not?

2   Direct Sum

In this project, you will explore the **sum** and **direct sum** of subspaces. In Exercise 58 in Section 4.3, you proved that for two subspaces U and W of a vector space V, the sum $U + W$ of the subspaces, defined as $U + W = \{\mathbf{u} + \mathbf{w} : \mathbf{u} \in U, \mathbf{w} \in W\}$, is also a subspace of V.

1. Consider the subspaces of $V = R^3$ below.

$$U = \{(x + y, x, y) : x, y \in R\}$$
$$W = \{(x, x, 0) : x \in R\}$$
$$Z = \{(x, x, x) : x \in R\}$$

Find $U + W$, $U + Z$, and $W + Z$.

2. If U and W are subspaces of V such that $V = U + W$ and $U \cap W = \{\mathbf{0}\}$, then prove that every vector in V has a *unique* representation of the form $\mathbf{u} + \mathbf{w}$, where $\mathbf{u}$ is in U and $\mathbf{w}$ is in W. V is called the **direct sum** of U and W, and is written as

$$V = U \oplus W. \qquad \text{Direct sum}$$

Which of the sums in part (1) are direct sums?

3. Let $V = U \oplus W$, and let $\{\mathbf{u}_1, \mathbf{u}_2, \ldots, \mathbf{u}_k\}$ be a basis for the subspace U and $\{\mathbf{w}_1, \mathbf{w}_2, \ldots, \mathbf{w}_m\}$ be a basis for the subspace W. Prove that the set $\{\mathbf{u}_1, \ldots, \mathbf{u}_k, \mathbf{w}_1, \ldots, \mathbf{w}_m\}$ is a basis for V.

4. Consider the subspaces $U = \{(y, 0, x) : x, y \in R\}$ and $W = \{(x, y, 0) : x, y \in R\}$ of $V = R^3$. Show that $R^3 = U + W$. Is R^3 the *direct* sum of U and W? What are the dimensions of U, W, $U \cap W$, and $U + W$? Formulate a conjecture that relates the dimensions of U, W, $U \cap W$, and $U + W$.

5. Do there exist two two-dimensional subspaces of R^3 whose intersection is the zero vector? Why or why not?

5 Inner Product Spaces

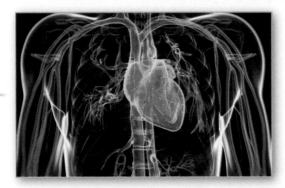

Revenue (p. 266)

Torque (p. 277)

Heart Rhythm Analysis (p. 255)

Work (p. 248)

Electric/Magnetic Flux (p. 240)

5.1   Length and Dot Product in R^n

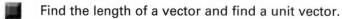

■ Find the length of a vector and find a unit vector.

■ Find the distance between two vectors.

■ Find a dot product and the angle between two vectors, determine orthogonality, and verify the Cauchy-Schwarz Inequality, the triangle inequality, and the Pythagorean Theorem.

■ Use a matrix product to represent a dot product.

VECTOR LENGTH AND UNIT VECTORS

Section 4.1 mentioned that vectors can be characterized by two quantities, *length* and *direction*. This section defines these and other geometric properties (such as distance and angle) of vectors in R^n. Section 5.2 extends these ideas to general vector spaces.

You will begin by reviewing the definition of the length of a vector in R^2. If $\mathbf{v} = (v_1, v_2)$ is a vector in R^2, then the *length*, or *norm*, of $\mathbf{v}$, denoted by $\|\mathbf{v}\|$, is the length of the hypotenuse of a right triangle whose legs have lengths of $|v_1|$ and $|v_2|$, as shown in Figure 5.1. Applying the Pythagorean Theorem produces

$$\|\mathbf{v}\|^2 = |v_1|^2 + |v_2|^2 = v_1^2 + v_2^2$$
$$\|\mathbf{v}\| = \sqrt{v_1^2 + v_2^2}.$$

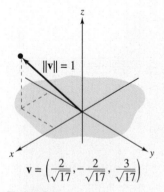

$$\|\mathbf{v}\| = \sqrt{v_1^2 + v_2^2}$$

Figure 5.1

Using R^2 as a model, the length of a vector in R^n is defined below.

Definition of the Length of a Vector in R^n

The **length,** or **norm,** of a vector $\mathbf{v} = (v_1, v_2, \ldots, v_n)$ in R^n is

$$\|\mathbf{v}\| = \sqrt{v_1^2 + v_2^2 + \cdots + v_n^2}.$$

The length of a vector is also called its **magnitude.** If $\|\mathbf{v}\| = 1$, then the vector $\mathbf{v}$ is a **unit vector.**

This definition shows that the length of a vector cannot be negative. That is, $\|\mathbf{v}\| \geq 0$. Moreover, $\|\mathbf{v}\| = 0$ if and only if $\mathbf{v}$ is the zero vector $\mathbf{0}$.

EXAMPLE 1   The Length of a Vector in R^n

a. In R^5, the length of $\mathbf{v} = (0, -2, 1, 4, -2)$ is

$$\|\mathbf{v}\| = \sqrt{0^2 + (-2)^2 + 1^2 + 4^2 + (-2)^2} = \sqrt{25} = 5.$$

b. In R^3, the length of $\mathbf{v} = \left(2/\sqrt{17}, -2/\sqrt{17}, 3/\sqrt{17}\right)$ is

$$\|\mathbf{v}\| = \sqrt{\left(2/\sqrt{17}\right)^2 + \left(-2/\sqrt{17}\right)^2 + \left(3\sqrt{17}\right)^2} = \sqrt{17/17} = 1.$$

The length of $\mathbf{v}$ is 1, so $\mathbf{v}$ is a unit vector, as shown in Figure 5.2.

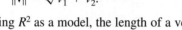

$$\mathbf{v} = \left(\frac{2}{\sqrt{17}}, -\frac{2}{\sqrt{17}}, \frac{3}{\sqrt{17}}\right)$$

Figure 5.2

Each vector in the standard basis for R^n has length 1 and is a **standard unit vector** in R^n. It is common to denote the standard unit vectors in R^2 and R^3 as

$$\{\mathbf{i}, \mathbf{j}\} = \{(1, 0), (0, 1)\} \quad \text{and} \quad \{\mathbf{i}, \mathbf{j}, \mathbf{k}\} = \{(1, 0, 0), (0, 1, 0), (0, 0, 1)\}.$$

Two nonzero vectors $\mathbf{u}$ and $\mathbf{v}$ in R^n are **parallel** when one is a scalar multiple of the other—that is, $\mathbf{u} = c\mathbf{v}$. Moreover, if $c > 0$, then $\mathbf{u}$ and $\mathbf{v}$ have the **same direction,** and if $c < 0$, then $\mathbf{u}$ and $\mathbf{v}$ have **opposite directions.** The next theorem gives a formula for finding the length of a scalar multiple of a vector.

THEOREM 5.1   Length of a Scalar Multiple

Let $\mathbf{v}$ be a vector in R^n and let c be a scalar. Then

$$\|c\mathbf{v}\| = |c|\,\|\mathbf{v}\|$$

where $|c|$ is the absolute value of c.

PROOF

$c\mathbf{v} = (cv_1, cv_2, \ldots, cv_n)$, so it follows that

$$
\begin{aligned}
\|c\mathbf{v}\| &= \|(cv_1, cv_2, \ldots, cv_n)\| \\
&= \sqrt{(cv_1)^2 + (cv_2)^2 + \cdots + (cv_n)^2} \\
&= |c|\sqrt{v_1^2 + v_2^2 + \cdots + v_n^2} \\
&= |c|\,\|\mathbf{v}\|.
\end{aligned}
$$

One important use of Theorem 5.1 is in finding a unit vector having the same direction as a given vector. Theorem 5.2 provides a procedure for doing this.

THEOREM 5.2   Unit Vector in the Direction of v

If $\mathbf{v}$ is a nonzero vector in R^n, then the vector

$$\mathbf{u} = \frac{\mathbf{v}}{\|\mathbf{v}\|}$$

has length 1 and has the same direction as $\mathbf{v}$. This vector $\mathbf{u}$ is the **unit vector in the direction of v.**

PROOF

$\mathbf{v} \neq \mathbf{0}$, so you know that $\|\mathbf{v}\| \neq 0$. You also know that $1/\|\mathbf{v}\|$ is positive, so you can write $\mathbf{u}$ as a positive scalar multiple of $\mathbf{v}$.

$$\mathbf{u} = \left(\frac{1}{\|\mathbf{v}\|}\right)\mathbf{v}$$

It follows that $\mathbf{u}$ has the same direction as $\mathbf{v}$, and $\mathbf{u}$ has length 1 because

$$\|\mathbf{u}\| = \left\|\frac{\mathbf{v}}{\|\mathbf{v}\|}\right\| = \frac{1}{\|\mathbf{v}\|}\|\mathbf{v}\| = 1.$$

The process of finding the unit vector in the direction of $\mathbf{v}$ is called **normalizing** the vector $\mathbf{v}$. The next example demonstrates this procedure.

EXAMPLE 2   Finding a Unit Vector

Find the unit vector in the direction of $\mathbf{v} = (3, -1, 2)$, and verify that this vector has length 1.

SOLUTION

The unit vector in the direction of $\mathbf{v}$ is

$$\frac{\mathbf{v}}{\|\mathbf{v}\|} = \frac{(3, -1, 2)}{\sqrt{3^2 + (-1)^2 + 2^2}} = \frac{1}{\sqrt{14}}(3, -1, 2) = \left(\frac{3}{\sqrt{14}}, -\frac{1}{\sqrt{14}}, \frac{2}{\sqrt{14}}\right)$$

which is a unit vector because

$$\sqrt{\left(\frac{3}{\sqrt{14}}\right)^2 + \left(-\frac{1}{\sqrt{14}}\right)^2 + \left(\frac{2}{\sqrt{14}}\right)^2} = \sqrt{\frac{14}{14}} = 1. \text{ (See Figure 5.3.)}$$

TECHNOLOGY

You can use a graphing utility or software program to find the length of a vector $\mathbf{v}$, the length of a scalar multiple $c\mathbf{v}$ of a vector, or a unit vector in the direction of $\mathbf{v}$. For instance, if you use a graphing utility to verify the result of Example 2, then you may see something similar to the screen below.

```
VECTOR:V        3
 e1=3
 e2=-1
 e3=2
unitV V
   [.8018 -.2673 .5345]
```

Note that $\dfrac{3}{\sqrt{14}} \approx 0.8018$,

$-\dfrac{1}{\sqrt{14}} \approx -0.2673$, and

$\dfrac{2}{\sqrt{14}} \approx 0.5345$.

See LarsonLinearAlgebra.com for an interactive example.

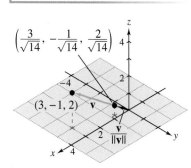

Figure 5.3

DISTANCE BETWEEN TWO VECTORS IN R^n

To define the **distance between two vectors** in R^n, R^2 will be used as a model. The Distance Formula from analytic geometry states that the distance d between two points in R^2, (u_1, u_2) and (v_1, v_2), is

$$d = \sqrt{(u_1 - v_1)^2 + (u_2 - v_2)^2}.$$

In vector terminology, this distance can be viewed as the length of $\mathbf{u} - \mathbf{v}$, where $\mathbf{u} = (u_1, u_2)$ and $\mathbf{v} = (v_1, v_2)$, as shown below. That is,

$$\|\mathbf{u} - \mathbf{v}\| = \sqrt{(u_1 - v_1)^2 + (u_2 - v_2)^2}.$$

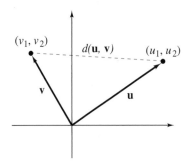

$$d(\mathbf{u}, \mathbf{v}) = \|\mathbf{u} - \mathbf{v}\| = \sqrt{(u_1 - v_1)^2 + (u_2 - v_2)^2}$$

This leads to the next definition.

Definition of Distance Between Two Vectors

The distance between two vectors $\mathbf{u}$ and $\mathbf{v}$ in R^n is

$$d(\mathbf{u}, \mathbf{v}) = \|\mathbf{u} - \mathbf{v}\|.$$

Verify the three properties of distance listed below.

1. $d(\mathbf{u}, \mathbf{v}) \geq 0$
2. $d(\mathbf{u}, \mathbf{v}) = 0$ if and only if $\mathbf{u} = \mathbf{v}$.
3. $d(\mathbf{u}, \mathbf{v}) = d(\mathbf{v}, \mathbf{u})$

EXAMPLE 3   Finding the Distance Between Two Vectors

a. The distance between $\mathbf{u} = (-1, -4)$ and $\mathbf{v} = (2, 3)$ is

$$d(\mathbf{u}, \mathbf{v}) = \|\mathbf{u} - \mathbf{v}\| = \|(-1 - 2, -4 - 3)\| = \sqrt{(-3)^2 + (-7)^2} = \sqrt{58}.$$

b. The distance between $\mathbf{u} = (0, 2, 2)$ and $\mathbf{v} = (2, 0, 1)$ is

$$d(\mathbf{u}, \mathbf{v}) = \|\mathbf{u} - \mathbf{v}\| = \|(0 - 2, 2 - 0, 2 - 1)\| = \sqrt{(-2)^2 + 2^2 + 1^2} = 3.$$

c. The distance between $\mathbf{u} = (3, -1, 0, -3)$ and $\mathbf{v} = (4, 0, 1, 2)$ is

$$d(\mathbf{u}, \mathbf{v}) = \|\mathbf{u} - \mathbf{v}\|$$
$$= \|(3 - 4, -1 - 0, 0 - 1, -3 - 2)\|$$
$$= \sqrt{(-1)^2 + (-1)^2 + (-1)^2 + (-5)^2}$$
$$= \sqrt{28}$$
$$= 2\sqrt{7}.$$

photos provided by Jacobs, Konrad/Oberwolfach Photo Collection

DOT PRODUCT AND THE ANGLE BETWEEN TWO VECTORS

To find the angle θ $(0 \le \theta \le \pi)$ between two nonzero vectors $\mathbf{u} = (u_1, u_2)$ and $\mathbf{v} = (v_1, v_2)$ in R^2, apply the Law of Cosines to the triangle shown to obtain

$$\|\mathbf{v} - \mathbf{u}\|^2 = \|\mathbf{u}\|^2 + \|\mathbf{v}\|^2 - 2\|\mathbf{u}\| \|\mathbf{v}\| \cos \theta.$$

Expanding and solving for $\cos \theta$ yields

$$\cos \theta = \frac{u_1 v_1 + u_2 v_2}{\|\mathbf{u}\| \|\mathbf{v}\|}.$$

The numerator of the quotient above is the **dot product** of $\mathbf{u}$ and $\mathbf{v}$ and is denoted by

$$\mathbf{u} \cdot \mathbf{v} = u_1 v_1 + u_2 v_2.$$

The definition below generalizes the dot product to R^n.

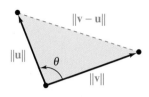

Angle Between Two Vectors

Definition of Dot Product in R^n

The **dot product** of $\mathbf{u} = (u_1, u_2, \ldots, u_n)$ and $\mathbf{v} = (v_1, v_2, \ldots, v_n)$ is the *scalar* quantity

$$\mathbf{u} \cdot \mathbf{v} = u_1 v_1 + u_2 v_2 + \cdots + u_n v_n.$$

EXAMPLE 4 **Finding the Dot Product of Two Vectors**

The dot product of $\mathbf{u} = (1, 2, 0, -3)$ and $\mathbf{v} = (3, -2, 4, 2)$ is

$$\mathbf{u} \cdot \mathbf{v} = (1)(3) + (2)(-2) + (0)(4) + (-3)(2) = -7.$$

THEOREM 5.3 Properties of the Dot Product

If $\mathbf{u}$, $\mathbf{v}$, and $\mathbf{w}$ are vectors in R^n and c is a scalar, then the properties listed below are true.

1. $\mathbf{u} \cdot \mathbf{v} = \mathbf{v} \cdot \mathbf{u}$
2. $\mathbf{u} \cdot (\mathbf{v} + \mathbf{w}) = \mathbf{u} \cdot \mathbf{v} + \mathbf{u} \cdot \mathbf{w}$
3. $c(\mathbf{u} \cdot \mathbf{v}) = (c\mathbf{u}) \cdot \mathbf{v} = \mathbf{u} \cdot (c\mathbf{v})$
4. $\mathbf{v} \cdot \mathbf{v} = \|\mathbf{v}\|^2$
5. $\mathbf{v} \cdot \mathbf{v} \ge 0$, and $\mathbf{v} \cdot \mathbf{v} = 0$ if and only if $\mathbf{v} = \mathbf{0}$.

PROOF

The proofs of these properties follow from the definition of dot product. For example, to prove the first property, write

$$\begin{aligned} \mathbf{u} \cdot \mathbf{v} &= u_1 v_1 + u_2 v_2 + \cdots + u_n v_n \\ &= v_1 u_1 + v_2 u_2 + \cdots + v_n u_n \\ &= \mathbf{v} \cdot \mathbf{u}. \end{aligned}$$

In Section 4.1, R^n was defined to be the set of all ordered n-tuples of real numbers. When R^n is combined with the standard operations of vector addition, scalar multiplication, vector length, and the dot product, the resulting vector space is **Euclidean n-space.** In the remainder of this text, unless stated otherwise, assume that R^n has the standard Euclidean operations.

EXAMPLE 5	**Finding Dot Products**

Let $\mathbf{u} = (2, -2)$, $\mathbf{v} = (5, 8)$, and $\mathbf{w} = (-4, 3)$. Find each quantity.

a. $\mathbf{u} \cdot \mathbf{v}$

b. $(\mathbf{u} \cdot \mathbf{v})\mathbf{w}$

c. $\mathbf{u} \cdot (2\mathbf{v})$

d. $\|\mathbf{w}\|^2$

e. $\mathbf{u} \cdot (\mathbf{v} - 2\mathbf{w})$

SOLUTION

a. By definition, you have

$$\mathbf{u} \cdot \mathbf{v} = 2(5) + (-2)(8) = -6.$$

b. Using the result in part (a), you have

$$(\mathbf{u} \cdot \mathbf{v})\mathbf{w} = -6\mathbf{w} = -6(-4, 3) = (24, -18).$$

c. By Property 3 of Theorem 5.3, you have

$$\mathbf{u} \cdot (2\mathbf{v}) = 2(\mathbf{u} \cdot \mathbf{v}) = 2(-6) = -12.$$

d. By Property 4 of Theorem 5.3, you have

$$\|\mathbf{w}\|^2 = \mathbf{w} \cdot \mathbf{w} = (-4)(-4) + (3)(3) = 25.$$

e. $2\mathbf{w} = (-8, 6)$, so you have

$$\mathbf{v} - 2\mathbf{w} = (5 - (-8), 8 - 6) = (13, 2).$$

Consequently,

$$\mathbf{u} \cdot (\mathbf{v} - 2\mathbf{w}) = 2(13) + (-2)(2) = 26 - 4 = 22.$$

EXAMPLE 6	**Using Properties of the Dot Product**

Consider two vectors $\mathbf{u}$ and $\mathbf{v}$ in R^n such that $\mathbf{u} \cdot \mathbf{u} = 39$, $\mathbf{u} \cdot \mathbf{v} = -3$, and $\mathbf{v} \cdot \mathbf{v} = 79$. Evaluate $(\mathbf{u} + 2\mathbf{v}) \cdot (3\mathbf{u} + \mathbf{v})$.

SOLUTION

Using Theorem 5.3, rewrite the dot product as

$$\begin{aligned}
(\mathbf{u} + 2\mathbf{v}) \cdot (3\mathbf{u} + \mathbf{v}) &= \mathbf{u} \cdot (3\mathbf{u} + \mathbf{v}) + (2\mathbf{v}) \cdot (3\mathbf{u} + \mathbf{v}) \\
&= \mathbf{u} \cdot (3\mathbf{u}) + \mathbf{u} \cdot \mathbf{v} + (2\mathbf{v}) \cdot (3\mathbf{u}) + (2\mathbf{v}) \cdot \mathbf{v} \\
&= 3(\mathbf{u} \cdot \mathbf{u}) + \mathbf{u} \cdot \mathbf{v} + 6(\mathbf{v} \cdot \mathbf{u}) + 2(\mathbf{v} \cdot \mathbf{v}) \\
&= 3(\mathbf{u} \cdot \mathbf{u}) + 7(\mathbf{u} \cdot \mathbf{v}) + 2(\mathbf{v} \cdot \mathbf{v}) \\
&= 3(39) + 7(-3) + 2(79) \\
&= 254.
\end{aligned}$$

To define the angle θ between two nonzero vectors $\mathbf{u}$ and $\mathbf{v}$ in R^n, use the formula in R^2

$$\cos \theta = \frac{\mathbf{u} \cdot \mathbf{v}}{\|\mathbf{u}\| \|\mathbf{v}\|}.$$

For such a definition to make sense, however, the absolute value of the right-hand side of this formula cannot exceed 1. This fact comes from a famous theorem named after the French mathematician Augustin-Louis Cauchy (1789–1857) and the German mathematician Hermann Schwarz (1843–1921).

DISCOVERY

1. Let $\mathbf{u} = (1, 1)$ and $\mathbf{v} = (-4, -3)$. Calculate $\mathbf{u} \cdot \mathbf{v}$ and $\|\mathbf{u}\| \|\mathbf{v}\|$.

2. Repeat with other choices for $\mathbf{u}$ and $\mathbf{v}$.

3. Formulate a conjecture about the relationship between the dot product of two vectors and the product of their lengths.

> **THEOREM 5.4   The Cauchy-Schwarz Inequality**
>
> If **u** and **v** are vectors in R^n, then
>
> $$|\mathbf{u} \cdot \mathbf{v}| \leq \|\mathbf{u}\| \|\mathbf{v}\|$$
>
> where $|\mathbf{u} \cdot \mathbf{v}|$ denotes the *absolute value* of $\mathbf{u} \cdot \mathbf{v}$.

PROOF

Case 1. If $\mathbf{u} = \mathbf{0}$, then it follows that $|\mathbf{u} \cdot \mathbf{v}| = |\mathbf{0} \cdot \mathbf{v}| = 0$ and $\|\mathbf{u}\| \|\mathbf{v}\| = 0\|\mathbf{v}\| = 0$. So, the theorem is true when $\mathbf{u} = \mathbf{0}$.

Case 2. When $\mathbf{u} \neq \mathbf{0}$. let t be any real number and consider the vector $t\mathbf{u} + \mathbf{v}$. The product $(t\mathbf{u} + \mathbf{v}) \cdot (t\mathbf{u} + \mathbf{v})$ is nonnegative, so it follows that

$$(t\mathbf{u} + \mathbf{v}) \cdot (t\mathbf{u} + \mathbf{v}) = t^2(\mathbf{u} \cdot \mathbf{u}) + 2t(\mathbf{u} \cdot \mathbf{v}) + \mathbf{v} \cdot \mathbf{v} \geq 0.$$

Now, let $a = \mathbf{u} \cdot \mathbf{u}$, $b = 2(\mathbf{u} \cdot \mathbf{v})$, and $c = \mathbf{v} \cdot \mathbf{v}$ to obtain the quadratic inequality $at^2 + bt + c \geq 0$. This quadratic is never negative, so it has either no real roots or a single repeated real root. But by the Quadratic Formula, this implies that the discriminant, $b^2 - 4ac$, is less than or equal to zero.

$$b^2 - 4ac \leq 0$$
$$b^2 \leq 4ac$$
$$4(\mathbf{u} \cdot \mathbf{v})^2 \leq 4(\mathbf{u} \cdot \mathbf{u})(\mathbf{v} \cdot \mathbf{v})$$
$$(\mathbf{u} \cdot \mathbf{v})^2 \leq (\mathbf{u} \cdot \mathbf{u})(\mathbf{v} \cdot \mathbf{v})$$

Taking the square roots of both sides produces

$$|\mathbf{u} \cdot \mathbf{v}| \leq \sqrt{\mathbf{u} \cdot \mathbf{u}}\sqrt{\mathbf{v} \cdot \mathbf{v}} = \|\mathbf{u}\| \|\mathbf{v}\|.$$

EXAMPLE 7   **Verifying the Cauchy-Schwarz Inequality**

Verify the Cauchy-Schwarz Inequality for $\mathbf{u} = (1, -1, 3)$ and $\mathbf{v} = (2, 0, -1)$.

SOLUTION

$\mathbf{u} \cdot \mathbf{v} = -1$, $\mathbf{u} \cdot \mathbf{u} = 11$, and $\mathbf{v} \cdot \mathbf{v} = 5$, so you have

$$|\mathbf{u} \cdot \mathbf{v}| = |-1| = 1$$

and

$$\|\mathbf{u}\| \|\mathbf{v}\| = \sqrt{\mathbf{u} \cdot \mathbf{u}}\sqrt{\mathbf{v} \cdot \mathbf{v}}$$
$$= \sqrt{11}\sqrt{5}$$
$$= \sqrt{55}.$$

The inequality $|\mathbf{u} \cdot \mathbf{v}| \leq \|\mathbf{u}\| \|\mathbf{v}\|$ holds, because $1 \leq \sqrt{55}$.

The Cauchy-Schwarz Inequality allows the definition of the angle between two nonzero vectors to be extended to R^n.

REMARK

The angle between the zero vector and another vector is not defined.

> **Definition of the Angle Between Two Vectors in R^n**
>
> The **angle** θ between two nonzero vectors in R^n can be found using
>
> $$\cos \theta = \frac{\mathbf{u} \cdot \mathbf{v}}{\|\mathbf{u}\| \|\mathbf{v}\|}, \quad 0 \leq \theta \leq \pi.$$

EXAMPLE 8   Finding the Angle Between Two Vectors

See LarsonLinearAlgebra.com for an interactive version of this type of example.

The angle between $\mathbf{u} = (-4, 0, 2, -2)$ and $\mathbf{v} = (2, 0, -1, 1)$ is

$$\cos \theta = \frac{\mathbf{u} \cdot \mathbf{v}}{\|\mathbf{u}\| \|\mathbf{v}\|} = \frac{-12}{\sqrt{24}\sqrt{6}} = -\frac{12}{\sqrt{144}} = -1.$$

Consequently, $\theta = \pi$. It makes sense that $\mathbf{u}$ and $\mathbf{v}$ should have opposite directions, because $\mathbf{u} = -2\mathbf{v}$.

Note that $\|\mathbf{u}\|$ and $\|\mathbf{v}\|$ are always positive, so $\mathbf{u} \cdot \mathbf{v}$ and $\cos \theta$ will always have the same sign. Moreover, the cosine is positive in the first quadrant and negative in the second quadrant, so the sign of the dot product of two vectors can be used to determine, for instance, whether the angle between them is acute or obtuse.

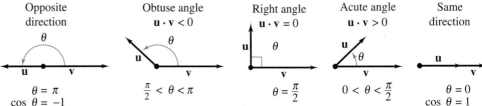

Opposite direction $\theta = \pi$, $\cos \theta = -1$

Obtuse angle $\mathbf{u} \cdot \mathbf{v} < 0$, $\frac{\pi}{2} < \theta < \pi$

Right angle $\mathbf{u} \cdot \mathbf{v} = 0$, $\theta = \frac{\pi}{2}$

Acute angle $\mathbf{u} \cdot \mathbf{v} > 0$, $0 < \theta < \frac{\pi}{2}$

Same direction $\theta = 0$, $\cos \theta = 1$

REMARK

Even though the angle between the zero vector and another vector is not defined, it is convenient to extend the definition of orthogonality to include the zero vector. In other words, the vector $\mathbf{0}$ is said to be orthogonal to every vector.

Note from the above that two nonzero vectors meet at a right angle if and only if their dot product is zero. Two such vectors are **orthogonal** (or perpendicular).

Definition of Orthogonal Vectors

Two vectors $\mathbf{u}$ and $\mathbf{v}$ in R^n are **orthogonal** when

$$\mathbf{u} \cdot \mathbf{v} = 0.$$

EXAMPLE 9   Orthogonal Vectors in R^n

a. The vectors $\mathbf{u} = (1, 0, 0)$ and $\mathbf{v} = (0, 1, 0)$ are orthogonal because

$$\mathbf{u} \cdot \mathbf{v} = (1)(0) + (0)(1) + (0)(0) = 0.$$

b. The vectors $\mathbf{u} = (3, 2, -1, 4)$ and $\mathbf{v} = (1, -1, 1, 0)$ are orthogonal because

$$\mathbf{u} \cdot \mathbf{v} = (3)(1) + (2)(-1) + (-1)(1) + (4)(0) = 0.$$

EXAMPLE 10   Finding Orthogonal Vectors

Determine all vectors in R^2 that are orthogonal to $\mathbf{u} = (4, 2)$.

SOLUTION

Let $\mathbf{v} = (v_1, v_2)$ be orthogonal to $\mathbf{u}$. Then

$$\mathbf{u} \cdot \mathbf{v} = (4, 2) \cdot (v_1, v_2) = 4v_1 + 2v_2 = 0$$

which implies that $2v_2 = -4v_1$ and $v_2 = -2v_1$. So, every vector that is orthogonal to $(4, 2)$ is of the form

$$\mathbf{v} = (t, -2t) = t(1, -2)$$

where t is a real number. (See Figure 5.4.)

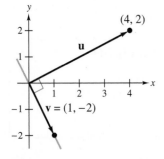

Figure 5.4

a.

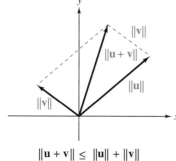

$$\|\mathbf{u} + \mathbf{v}\| \leq \|\mathbf{u}\| + \|\mathbf{v}\|$$

b.

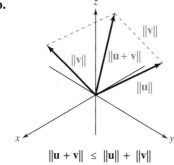

$$\|\mathbf{u} + \mathbf{v}\| \leq \|\mathbf{u}\| + \|\mathbf{v}\|$$

Figure 5.5

REMARK

Equality occurs in the triangle inequality if and only if the vectors **u** and **v** have the same direction. (See Exercise 86.)

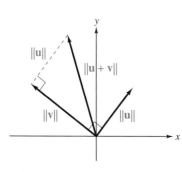

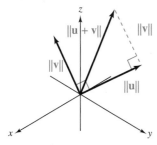

Figure 5.6

The Cauchy-Schwarz Inequality can be used to prove another well-known inequality called the **triangle inequality** (Theorem 5.5 below). The name "triangle inequality" is derived from the interpretation of the theorem in R^2, illustrated for the vectors **u** and **v** in Figure 5.5(a). When you consider

$$\|\mathbf{u}\| \quad \text{and} \quad \|\mathbf{v}\|$$

to be the lengths of two sides of a triangle, the length of the third side is

$$\|\mathbf{u} + \mathbf{v}\|.$$

Moreover, the length of any side of a triangle cannot be greater than the sum of the lengths of the other two sides, so you have

$$\|\mathbf{u} + \mathbf{v}\| \leq \|\mathbf{u}\| + \|\mathbf{v}\|.$$

Figure 5.5(b) illustrates the triangle inequality for the vectors **u** and **v** in R^3. The theorem below generalizes these results to R^n.

THEOREM 5.5   The Triangle Inequality

If **u** and **v** are vectors in R^n, then

$$\|\mathbf{u} + \mathbf{v}\| \leq \|\mathbf{u}\| + \|\mathbf{v}\|.$$

PROOF

Using the properties of the dot product, you have

$$
\begin{aligned}
\|\mathbf{u} + \mathbf{v}\|^2 &= (\mathbf{u} + \mathbf{v}) \cdot (\mathbf{u} + \mathbf{v}) \\
&= \mathbf{u} \cdot (\mathbf{u} + \mathbf{v}) + \mathbf{v} \cdot (\mathbf{u} + \mathbf{v}) \\
&= \mathbf{u} \cdot \mathbf{u} + 2(\mathbf{u} \cdot \mathbf{v}) + \mathbf{v} \cdot \mathbf{v} \\
&= \|\mathbf{u}\|^2 + 2(\mathbf{u} \cdot \mathbf{v}) + \|\mathbf{v}\|^2 \\
&\leq \|\mathbf{u}\|^2 + 2|\mathbf{u} \cdot \mathbf{v}| + \|\mathbf{v}\|^2.
\end{aligned}
$$

Now, by the Cauchy-Schwarz Inequality, $|\mathbf{u} \cdot \mathbf{v}| \leq \|\mathbf{u}\|\|\mathbf{v}\|$, and

$$
\begin{aligned}
\|\mathbf{u} + \mathbf{v}\|^2 &\leq \|\mathbf{u}\|^2 + 2|\mathbf{u} \cdot \mathbf{v}| + \|\mathbf{v}\|^2 \\
&\leq \|\mathbf{u}\|^2 + 2\|\mathbf{u}\|\|\mathbf{v}\| + \|\mathbf{v}\|^2 \\
&= \left(\|\mathbf{u}\| + \|\mathbf{v}\|\right)^2.
\end{aligned}
$$

Both $\|\mathbf{u} + \mathbf{v}\|$ and $\left(\|\mathbf{u}\| + \|\mathbf{v}\|\right)$ are nonnegative, so taking the square roots of both sides yields

$$\|\mathbf{u} + \mathbf{v}\| \leq \|\mathbf{u}\| + \|\mathbf{v}\|.$$

From the proof of the triangle inequality, you have

$$\|\mathbf{u} + \mathbf{v}\|^2 = \|\mathbf{u}\|^2 + 2(\mathbf{u} \cdot \mathbf{v}) + \|\mathbf{v}\|^2.$$

If **u** and **v** are orthogonal, then $\mathbf{u} \cdot \mathbf{v} = 0$, and you have the extension of the **Pythagorean Theorem** to R^n, shown below.

THEOREM 5.6   The Pythagorean Theorem

If **u** and **v** are vectors in R^n, then **u** and **v** are orthogonal if and only if

$$\|\mathbf{u} + \mathbf{v}\|^2 = \|\mathbf{u}\|^2 + \|\mathbf{v}\|^2.$$

Figure 5.6 illustrates this relationship graphically for R^2 and R^3.

THE DOT PRODUCT AND MATRIX MULTIPLICATION

It is often useful to represent a vector in R^n as an $n \times 1$ column matrix. In this notation, the dot product of two vectors

$$\mathbf{u} = \begin{bmatrix} u_1 \\ u_2 \\ \vdots \\ u_n \end{bmatrix} \quad \text{and} \quad \mathbf{v} = \begin{bmatrix} v_1 \\ v_2 \\ \vdots \\ v_n \end{bmatrix}$$

can be represented as the matrix product of the transpose of $\mathbf{u}$ multiplied by $\mathbf{v}$.

$$\mathbf{u} \cdot \mathbf{v} = \mathbf{u}^T \mathbf{v} = \begin{bmatrix} u_1 & u_2 & \dots & u_n \end{bmatrix} \begin{bmatrix} v_1 \\ v_2 \\ \vdots \\ v_n \end{bmatrix} = \begin{bmatrix} u_1 v_1 + u_2 v_2 + \dots + u_n v_n \end{bmatrix}$$

EXAMPLE 11   **Using Matrix Multiplication to Find Dot Products**

a. The dot product of the vectors

$$\mathbf{u} = \begin{bmatrix} 2 \\ 0 \end{bmatrix} \quad \text{and} \quad \mathbf{v} = \begin{bmatrix} 3 \\ 1 \end{bmatrix}$$

is $\mathbf{u} \cdot \mathbf{v} = \mathbf{u}^T \mathbf{v} = \begin{bmatrix} 2 & 0 \end{bmatrix} \begin{bmatrix} 3 \\ 1 \end{bmatrix} = [(2)(3) + (0)(1)] = 6.$

b. The dot product of the vectors

$$\mathbf{u} = \begin{bmatrix} 1 \\ 2 \\ -1 \end{bmatrix} \quad \text{and} \quad \mathbf{v} = \begin{bmatrix} 3 \\ -2 \\ 4 \end{bmatrix}$$

is $\mathbf{u} \cdot \mathbf{v} = \mathbf{u}^T \mathbf{v} = \begin{bmatrix} 1 & 2 & -1 \end{bmatrix} \begin{bmatrix} 3 \\ -2 \\ 4 \end{bmatrix} = [(1)(3) + (2)(-2) + (-1)(4)] = -5.$

Many of the properties of the dot product are direct consequences of the corresponding properties of matrix multiplication. In Exercise 87, you are asked to use the properties of matrix multiplication to prove the first three properties of Theorem 5.3.

LINEAR ALGEBRA APPLIED

Electrical engineers can use the dot product to calculate electric or magnetic *flux*, which is a measure of the strength of the electric or magnetic field penetrating a surface. Consider an arbitrarily shaped surface with an element of area dA, normal (perpendicular) vector $d\mathbf{A}$, electric field vector $\mathbf{E}$, and magnetic field vector $\mathbf{B}$. The electric flux Φ_e can be found using the surface integral $\Phi_e = \int \mathbf{E} \cdot d\mathbf{A}$ and the magnetic flux Φ_m can be found using the surface integral $\Phi_m = \int \mathbf{B} \cdot d\mathbf{A}$. It is interesting to note that for a closed surface that surrounds an electrical charge, the net electric flux is proportional to the charge, but the net magnetic flux is zero. This is because electric fields initiate at positive charges and terminate at negative charges, but magnetic fields form closed loops, so they do not initiate or terminate at any point. This means that the magnetic field entering a closed surface must equal the magnetic field leaving the closed surface.

5.1 Exercises

See CalcChat.com for worked-out solutions to odd-numbered exercises.

Finding the Length of a Vector In Exercises 1–4, find the length of the vector.

1. $\mathbf{v} = (4, 3)$
2. $\mathbf{v} = (0, 1)$
3. $\mathbf{v} = (5, -3, -4)$
4. $\mathbf{v} = (2, 0, -5, 5)$

Finding the Length of a Vector In Exercises 5–8, find (a) $\|\mathbf{u}\|$, (b) $\|\mathbf{v}\|$, and (c) $\|\mathbf{u} + \mathbf{v}\|$.

5. $\mathbf{u} = \left(-1, \frac{1}{4}\right)$, $\mathbf{v} = \left(4, -\frac{1}{8}\right)$
6. $\mathbf{u} = \left(1, \frac{1}{2}\right)$, $\mathbf{v} = \left(2, -\frac{1}{2}\right)$
7. $\mathbf{u} = (3, 1, 3)$, $\mathbf{v} = (0, -1, 1)$
8. $\mathbf{u} = (0, 1, -1, 2)$, $\mathbf{v} = (1, 1, 3, 0)$

Finding a Unit Vector In Exercises 9–12, find a unit vector (a) in the direction of u and (b) in the direction opposite that of u. Verify that each vector has length 1.

9. $\mathbf{u} = (-5, 12)$
10. $\mathbf{u} = (2, -2)$
11. $\mathbf{u} = (3, 2, -5)$
12. $\mathbf{u} = (-1, 3, 4)$

Finding a Vector In Exercises 13–16, find the vector v with the given length and the same direction as u.

13. $\|\mathbf{v}\| = 4$, $\mathbf{u} = (1, 1)$
14. $\|\mathbf{v}\| = 4$, $\mathbf{u} = (-1, 1)$
15. $\|\mathbf{v}\| = 5$, $\mathbf{u} = \left(\sqrt{5}, 5, 0\right)$
16. $\|\mathbf{v}\| = 3$, $\mathbf{u} = (0, 2, 1, -1)$

17. Consider the vector $\mathbf{v} = (-1, 3, 0, 4)$. Find $\mathbf{u}$ such that

   (a) $\mathbf{u}$ has the same direction as $\mathbf{v}$ and one-half its length.

   (b) $\mathbf{u}$ has the direction opposite that of $\mathbf{v}$ and twice its length.

18. For what values of c is $\|c(1, 2, 3)\| = 1$?

Finding the Distance Between Two Vectors In Exercises 19–22, find the distance between u and v.

19. $\mathbf{u} = (1, -1)$, $\mathbf{v} = (-1, 1)$
20. $\mathbf{u} = (-1, 2, 5)$, $\mathbf{v} = (3, 0, -1)$
21. $\mathbf{u} = (1, 2, 0)$, $\mathbf{v} = (-1, 4, 1)$
22. $\mathbf{u} = (0, 1, -1, 2)$, $\mathbf{v} = (1, 1, 2, 2)$

Finding Dot Products In Exercises 23–26, find (a) $\mathbf{u} \cdot \mathbf{v}$, (b) $\mathbf{v} \cdot \mathbf{v}$, (c) $\|\mathbf{u}\|^2$, (d) $(\mathbf{u} \cdot \mathbf{v})\mathbf{v}$, and (e) $\mathbf{u} \cdot (5\mathbf{v})$.

23. $\mathbf{u} = (3, 4)$, $\mathbf{v} = (2, -3)$
24. $\mathbf{u} = (-1, 2)$, $\mathbf{v} = (2, -2)$
25. $\mathbf{u} = (2, -2, 1)$, $\mathbf{v} = (2, -1, -6)$
26. $\mathbf{u} = (4, 0, -3, 5)$, $\mathbf{v} = (0, 2, 5, 4)$

27. Find $(\mathbf{u} + \mathbf{v}) \cdot (2\mathbf{u} - \mathbf{v})$ when $\mathbf{u} \cdot \mathbf{u} = 4$, $\mathbf{u} \cdot \mathbf{v} = -5$, and $\mathbf{v} \cdot \mathbf{v} = 10$.

28. Find $(3\mathbf{u} - \mathbf{v}) \cdot (\mathbf{u} - 3\mathbf{v})$ when $\mathbf{u} \cdot \mathbf{u} = 8$, $\mathbf{u} \cdot \mathbf{v} = 7$, and $\mathbf{v} \cdot \mathbf{v} = 6$.

Finding Lengths, Unit Vectors, and Dot Products In Exercises 29–34, use a software program or a graphing utility to find (a) the lengths of u and v, (b) a unit vector in the direction of v, (c) a unit vector in the direction opposite that of u, (d) $\mathbf{u} \cdot \mathbf{v}$, (e) $\mathbf{u} \cdot \mathbf{u}$, and (f) $\mathbf{v} \cdot \mathbf{v}$.

29. $\mathbf{u} = \left(1, \frac{1}{8}, \frac{2}{5}\right)$, $\mathbf{v} = \left(0, \frac{1}{4}, \frac{1}{5}\right)$
30. $\mathbf{u} = \left(-1, \frac{1}{2}, \frac{1}{4}\right)$, $\mathbf{v} = \left(0, \frac{1}{4}, -\frac{1}{2}\right)$
31. $\mathbf{u} = \left(0, 1, \sqrt{2}\right)$, $\mathbf{v} = \left(-1, \sqrt{2}, -1\right)$
32. $\mathbf{u} = \left(-1, \sqrt{3}, 2\right)$, $\mathbf{v} = \left(\sqrt{2}, -1, -\sqrt{2}\right)$
33. $\mathbf{u} = \left(2, \sqrt{3}, \sqrt{2}, \sqrt{3}\right)$, $\mathbf{v} = \left(-2, \sqrt{2}, -\sqrt{3}, -\sqrt{2}\right)$
34. $\mathbf{u} = \left(1, \sqrt{2}, -1, \sqrt{2}\right)$, $\mathbf{v} = \left(1, -\frac{1}{\sqrt{2}}, 1, -\frac{1}{\sqrt{2}}\right)$

Verifying the Cauchy-Schwarz Inequality In Exercises 35–38, verify the Cauchy-Schwarz Inequality for the vectors.

35. $\mathbf{u} = (6, 8)$, $\mathbf{v} = (3, -2)$
36. $\mathbf{u} = (-1, 0)$, $\mathbf{v} = (1, 1)$
37. $\mathbf{u} = (1, 1, -2)$, $\mathbf{v} = (1, -3, -2)$
38. $\mathbf{u} = (1, -1, 0)$, $\mathbf{v} = (0, 1, -1)$

Finding the Angle Between Two Vectors In Exercises 39–46, find the angle θ between the vectors.

39. $\mathbf{u} = (3, 1)$, $\mathbf{v} = (-2, 4)$
40. $\mathbf{u} = (-4, 1)$, $\mathbf{v} = (5, 0)$
41. $\mathbf{u} = \left(\cos \frac{\pi}{6}, \sin \frac{\pi}{6}\right)$, $\mathbf{v} = \left(\cos \frac{3\pi}{4}, \sin \frac{3\pi}{4}\right)$
42. $\mathbf{u} = \left(\cos \frac{\pi}{3}, \sin \frac{\pi}{3}\right)$, $\mathbf{v} = \left(\cos \frac{\pi}{4}, \sin \frac{\pi}{4}\right)$
43. $\mathbf{u} = (1, 1, 1)$, $\mathbf{v} = (2, 1, -1)$
44. $\mathbf{u} = (2, 3, 1)$, $\mathbf{v} = (-3, 2, 0)$
45. $\mathbf{u} = (0, 1, 0, 1)$, $\mathbf{v} = (3, 3, 3, 3)$
46. $\mathbf{u} = (1, -1, 0, 1)$, $\mathbf{v} = (-1, 2, -1, 0)$

Determining a Relationship Between Two Vectors In Exercises 47–54, determine whether u and v are orthogonal, parallel, or neither.

47. $\mathbf{u} = (2, 18)$, $\mathbf{v} = \left(\frac{3}{2}, -\frac{1}{6}\right)$
48. $\mathbf{u} = (4, 3)$, $\mathbf{v} = \left(\frac{1}{2}, -\frac{2}{3}\right)$
49. $\mathbf{u} = \left(-\frac{1}{3}, \frac{2}{3}\right)$, $\mathbf{v} = (2, -4)$
50. $\mathbf{u} = (1, -1)$, $\mathbf{v} = (0, -1)$
51. $\mathbf{u} = (0, 1, 0)$, $\mathbf{v} = (1, -2, 0)$
52. $\mathbf{u} = (0, 3, -4)$, $\mathbf{v} = (1, -8, -6)$
53. $\mathbf{u} = (-2, 5, 1, 0)$, $\mathbf{v} = \left(\frac{1}{4}, -\frac{5}{4}, 0, 1\right)$
54. $\mathbf{u} = \left(4, \frac{3}{2}, -1, \frac{1}{2}\right)$, $\mathbf{v} = \left(-2, -\frac{3}{4}, \frac{1}{2}, -\frac{1}{4}\right)$

Finding Orthogonal Vectors   In Exercises 55–58, determine all vectors v that are orthogonal to u.

55. $\mathbf{u} = (0, 5)$ **56.** $\mathbf{u} = (11, 2)$

57. $\mathbf{u} = (2, -1, 1)$ **58.** $\mathbf{u} = (4, -1, 0)$

Verifying the Triangle Inequality   In Exercises 59–62, verify the triangle inequality for the vectors u and v.

59. $\mathbf{u} = (4, 0)$, $\mathbf{v} = (1, 1)$ **60.** $\mathbf{u} = (-1, 1)$, $\mathbf{v} = (2, 0)$

61. $\mathbf{u} = (1, 1, 1)$, $\mathbf{v} = (0, 1, -2)$

62. $\mathbf{u} = (1, -1, 0)$, $\mathbf{v} = (0, 1, 2)$

Verifying the Pythagorean Theorem   In Exercises 63–66, verify the Pythagorean Theorem for the vectors u and v.

63. $\mathbf{u} = (1, -1)$, $\mathbf{v} = (1, 1)$

64. $\mathbf{u} = (3, -2)$, $\mathbf{v} = (4, 6)$

65. $\mathbf{u} = (3, 4, -2)$, $\mathbf{v} = (4, -3, 0)$

66. $\mathbf{u} = (4, 1, -5)$, $\mathbf{v} = (2, -3, 1)$

67. Rework Exercise 23 using matrix multiplication.

68. Rework Exercise 24 using matrix multiplication.

69. Rework Exercise 25 using matrix multiplication.

70. Rework Exercise 26 using matrix multiplication.

Writing   In Exercises 71 and 72, determine whether the vectors are orthogonal, parallel, or neither. Explain.

71. $\mathbf{u} = (\cos\theta, \sin\theta, -1)$, $\mathbf{v} = (\sin\theta, -\cos\theta, 0)$

72. $\mathbf{u} = (-\sin\theta, \cos\theta, 1)$, $\mathbf{v} = (\sin\theta, -\cos\theta, 0)$

True or False?   In Exercises 73 and 74, determine whether each statement is true or false. If a statement is true, give a reason or cite an appropriate statement from the text. If a statement is false, provide an example that shows the statement is not true in all cases or cite an appropriate statement from the text.

73. (a) The length or norm of a vector is
$$\|\mathbf{v}\| = |v_1 + v_2 + v_3 + \cdots + v_n|.$$
    (b) The dot product of two vectors $\mathbf{u}$ and $\mathbf{v}$ is another vector represented by
$$\mathbf{u} \cdot \mathbf{v} = (u_1v_1, u_2v_2, u_3v_3, \ldots, u_nv_n).$$

74. (a) If $\mathbf{v}$ is a nonzero vector in R^n, then the unit vector in the direction of $\mathbf{v}$ is $\mathbf{u} = \|\mathbf{v}\|/\mathbf{v}$.

    (b) If $\mathbf{u} \cdot \mathbf{v} < 0$, then the angle θ between $\mathbf{u}$ and $\mathbf{v}$ is acute.

Writing   In Exercises 75 and 76, explain why each expression involving dot product(s) is meaningless. Assume that u and v are vectors in R^n, and that c is a scalar.

75. (a) $(\mathbf{u} \cdot \mathbf{v}) - \mathbf{v}$ (b) $\mathbf{u} + (\mathbf{u} \cdot \mathbf{v})$

76. (a) $(\mathbf{u} \cdot \mathbf{v}) \cdot \mathbf{u}$ (b) $c \cdot (\mathbf{u} \cdot \mathbf{v})$

Orthogonal Vectors   In Exercises 77 and 78, let $\mathbf{v} = (v_1, v_2)$ be a vector in R^2. Show that $(v_2, -v_1)$ is orthogonal to v, and use this fact to find two unit vectors orthogonal to the given vector.

77. $\mathbf{v} = (12, 5)$ **78.** $\mathbf{v} = (8, 15)$

79. Revenue The vector $\mathbf{u} = (3140, 2750)$ gives the numbers of hamburgers and hot dogs, respectively, sold at a fast-food stand in one month. The vector $\mathbf{v} = (2.25, 1.75)$ gives the prices (in dollars) of the food items. Find the dot product $\mathbf{u} \cdot \mathbf{v}$ and interpret the result in the context of the problem.

80. Revenue The vector $\mathbf{u} = (4600, 4290, 5250)$ gives the numbers of units of three models of cellular phones manufactured. The vector $\mathbf{v} = (499.99, 199.99, 99.99)$ gives the prices in dollars of the three models of cellular phones, respectively. Find the dot product $\mathbf{u} \cdot \mathbf{v}$ and interpret the result in the context of the problem.

81. Find the angle between the diagonal of a cube and one of its edges.

82. Find the angle between the diagonal of a cube and the diagonal of one of its sides.

83. Guided Proof Prove that if $\mathbf{u}$ is orthogonal to $\mathbf{v}$ and $\mathbf{w}$, then $\mathbf{u}$ is orthogonal to $c\mathbf{v} + d\mathbf{w}$ for any scalars c and d.

Getting Started: To prove that $\mathbf{u}$ is orthogonal to $c\mathbf{v} + d\mathbf{w}$, you need to show that the dot product of $\mathbf{u}$ and $c\mathbf{v} + d\mathbf{w}$ is 0.

(i) Rewrite the dot product of $\mathbf{u}$ and $c\mathbf{v} + d\mathbf{w}$ as a linear combination of $(\mathbf{u} \cdot \mathbf{v})$ and $(\mathbf{u} \cdot \mathbf{w})$ using Properties 2 and 3 of Theorem 5.3.

(ii) Use the fact that $\mathbf{u}$ is orthogonal to $\mathbf{v}$ and $\mathbf{w}$, and the result of part (i), to lead to the conclusion that $\mathbf{u}$ is orthogonal to $c\mathbf{v} + d\mathbf{w}$.

84. Proof Prove that if $\mathbf{u}$ and $\mathbf{v}$ are vectors in R^n, then $\mathbf{u} \cdot \mathbf{v} = \frac{1}{4}\|\mathbf{u} + \mathbf{v}\|^2 - \frac{1}{4}\|\mathbf{u} - \mathbf{v}\|^2$.

85. Proof Prove that the vectors $\mathbf{u} = (\cos\theta, -\sin\theta)$ and $\mathbf{v} = (\sin\theta, \cos\theta)$ are orthogonal unit vectors for any value of θ. Graph $\mathbf{u}$ and $\mathbf{v}$ for $\theta = \pi/3$.

86. Proof Prove that $\|\mathbf{u} + \mathbf{v}\| = \|\mathbf{u}\| + \|\mathbf{v}\|$ if and only if $\mathbf{u}$ and $\mathbf{v}$ have the same direction.

87. Proof Use the properties of matrix multiplication to prove the first three properties of Theorem 5.3.

88. CAPSTONE What do you know about θ, the angle between two nonzero vectors $\mathbf{u}$ and $\mathbf{v}$, under each condition?
    (a) $\mathbf{u} \cdot \mathbf{v} = 0$ (b) $\mathbf{u} \cdot \mathbf{v} > 0$ (c) $\mathbf{u} \cdot \mathbf{v} < 0$

89. Writing Let $\mathbf{x}$ be a solution to the $m \times n$ homogeneous linear system of equations $A\mathbf{x} = \mathbf{0}$. Explain why $\mathbf{x}$ is orthogonal to the row vectors of A.

5.2 Inner Product Spaces

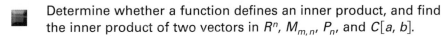 Determine whether a function defines an inner product, and find the inner product of two vectors in R^n, $M_{m,n}$, P_n, and $C[a, b]$.

Find an orthogonal projection of a vector onto another vector in an inner product space.

INNER PRODUCTS

In Section 5.1, the concepts of length, distance, and angle were extended from R^2 to R^n. This section extends these concepts one step further—to general vector spaces—by using the idea of an **inner product** of two vectors.

You are already familiar with one example of an inner product: the *dot product* in R^n. The dot product, called the **Euclidean inner product,** is only one of several inner products that can be defined on R^n. To distinguish between the standard inner product and other possible inner products, use the notation below.

$\mathbf{u} \cdot \mathbf{v} =$ dot product (Euclidean inner product for R^n)

$\langle \mathbf{u}, \mathbf{v} \rangle =$ general inner product for a vector space V

A general inner product is defined in much the same way that a general vector space is defined—that is, in order for a function to qualify as an inner product, it must satisfy a set of axioms. The axioms below parallel Properties 1, 2, 3, and 5 of the dot product given in Theorem 5.3.

Definition of Inner Product

Let $\mathbf{u}$, $\mathbf{v}$, and $\mathbf{w}$ be vectors in a vector space V, and let c be any scalar. An **inner product** on V is a function that associates a real number $\langle \mathbf{u}, \mathbf{v} \rangle$ with each pair of vectors $\mathbf{u}$ and $\mathbf{v}$ and satisfies the axioms listed below.

1. $\langle \mathbf{u}, \mathbf{v} \rangle = \langle \mathbf{v}, \mathbf{u} \rangle$
2. $\langle \mathbf{u}, \mathbf{v} + \mathbf{w} \rangle = \langle \mathbf{u}, \mathbf{v} \rangle + \langle \mathbf{u}, \mathbf{w} \rangle$
3. $c\langle \mathbf{u}, \mathbf{v} \rangle = \langle c\mathbf{u}, \mathbf{v} \rangle$
4. $\langle \mathbf{v}, \mathbf{v} \rangle \geq 0$, and $\langle \mathbf{v}, \mathbf{v} \rangle = 0$ if and only if $\mathbf{v} = \mathbf{0}$.

A vector space V with an inner product is an **inner product space.** Whenever an inner product space is referred to, assume that the set of scalars is the set of real numbers.

EXAMPLE 1   The Euclidean Inner Product for R^n

Show that the dot product in R^n satisfies the four axioms of an inner product.

SOLUTION

In R^n, the dot product of two vectors $\mathbf{u} = (u_1, u_2, \ldots, u_n)$ and $\mathbf{v} = (v_1, v_2, \ldots, v_n)$ is

$\mathbf{u} \cdot \mathbf{v} = u_1 v_1 + u_2 v_2 + \cdots + u_n v_n.$

By Theorem 5.3, you know that this dot product satisfies the required four axioms, which verifies that it is an inner product on R^n.

The Euclidean inner product is not the only inner product that can be defined on R^n. Example 2 illustrates a different inner product. To show that a function is an inner product, you must show that it satisfies the four inner product axioms.

EXAMPLE 2   **A Different Inner Product for R^2**

Show that the function below defines an inner product on R^2, where $\mathbf{u} = (u_1, u_2)$ and $\mathbf{v} = (v_1, v_2)$.

$$\langle \mathbf{u}, \mathbf{v} \rangle = u_1 v_1 + 2u_2 v_2$$

SOLUTION

1. The product of real numbers is commutative, so

$$\langle \mathbf{u}, \mathbf{v} \rangle = u_1 v_1 + 2u_2 v_2 = v_1 u_1 + 2v_2 u_2 = \langle \mathbf{v}, \mathbf{u} \rangle.$$

2. Let $\mathbf{w} = (w_1, w_2)$ Then

$$\begin{aligned} \langle \mathbf{u}, \mathbf{v} + \mathbf{w} \rangle &= u_1(v_1 + w_1) + 2u_2(v_2 + w_2) \\ &= u_1 v_1 + u_1 w_1 + 2u_2 v_2 + 2u_2 w_2 \\ &= (u_1 v_1 + 2u_2 v_2) + (u_1 w_1 + 2u_2 w_2) \\ &= \langle \mathbf{u}, \mathbf{v} \rangle + \langle \mathbf{u}, \mathbf{w} \rangle. \end{aligned}$$

3. If c is any scalar, then

$$c\langle \mathbf{u}, \mathbf{v} \rangle = c(u_1 v_1 + 2u_2 v_2) = (cu_1)v_1 + 2(cu_2)v_2 = \langle c\mathbf{u}, \mathbf{v} \rangle.$$

4. The square of a real number is nonnegative, so

$$\langle \mathbf{v}, \mathbf{v} \rangle = v_1^2 + 2v_2^2 \geq 0.$$

Moreover, this expression is equal to zero if and only if $\mathbf{v} = \mathbf{0}$ (that is, if and only if $v_1 = v_2 = 0$).

Example 2 can be generalized. The function

$$\langle \mathbf{u}, \mathbf{v} \rangle = c_1 u_1 v_1 + c_2 u_2 v_2 + \cdots + c_n u_n v_n, \quad c_i > 0$$

is an inner product on R^n. (In Exercise 89, you are asked to prove this.) The positive constants $c_1, \ldots, c_n$ are **weights.** If any c_i is negative or 0, then this function does not define an inner product.

EXAMPLE 3   **A Function That Is Not an Inner Product**

Show that the function below is not an inner product on R^3, where $\mathbf{u} = (u_1, u_2, u_3)$ and $\mathbf{v} = (v_1, v_2, v_3)$.

$$\langle \mathbf{u}, \mathbf{v} \rangle = u_1 v_1 - 2u_2 v_2 + u_3 v_3$$

SOLUTION

Observe that Axiom 4 is not satisfied. For example, let $\mathbf{v} = (1, 2, 1)$. Then $\langle \mathbf{v}, \mathbf{v} \rangle = (1)(1) - 2(2)(2) + (1)(1) = -6$, which is less than zero.

EXAMPLE 4   **An Inner Product on $M_{2,2}$**

Let $A = \begin{bmatrix} a_{11} & a_{12} \\ a_{21} & a_{22} \end{bmatrix}$ and $B = \begin{bmatrix} b_{11} & b_{12} \\ b_{21} & b_{22} \end{bmatrix}$ be matrices in the vector space $M_{2,2}$. The function

$$\langle A, B \rangle = a_{11}b_{11} + a_{12}b_{12} + a_{21}b_{21} + a_{22}b_{22}$$

is an inner product on $M_{2,2}$. The verification of the four inner product axioms is left to you. (See Exercise 27.)

You obtain the inner product in the next example from calculus. The verification of the inner product properties depends on the properties of the definite integral.

REMARK

Remember that a and b must be distinct, otherwise

$$\int_a^b f(x)g(x)\,dx$$

is zero regardless of which functions f and g you use.

EXAMPLE 5   **An Inner Product Defined by a Definite Integral (Calculus)**

Let f and g be real-valued continuous functions in the vector space $C[a, b]$. Show that

$$\langle f, g \rangle = \int_a^b f(x)g(x)\,dx$$

defines an inner product on $C[a, b]$.

SOLUTION

Use familiar properties from calculus to verify the four parts of the definition.

1. $\langle f, g \rangle = \displaystyle\int_a^b f(x)g(x)\,dx = \int_a^b g(x)f(x)\,dx = \langle g, f \rangle$

2. $\langle f, g + h \rangle = \displaystyle\int_a^b f(x)[g(x) + h(x)]\,dx = \int_a^b [f(x)g(x) + f(x)h(x)]\,dx$

$$= \int_a^b f(x)g(x)\,dx + \int_a^b f(x)h(x)\,dx = \langle f, g \rangle + \langle f, h \rangle$$

3. $c\langle f, g \rangle = c\displaystyle\int_a^b f(x)g(x)\,dx = \int_a^b cf(x)g(x)\,dx = \langle cf, g \rangle$

4. $[f(x)]^2 \geq 0$ for all x, so you know from calculus that

$$\langle f, f \rangle = \int_a^b [f(x)]^2\,dx \geq 0$$

with

$$\langle f, f \rangle = \int_a^b [f(x)]^2\,dx = 0$$

if and only if f is the zero function in $C[a, b]$.

The next theorem lists some properties of inner products.

THEOREM 5.7   Properties of Inner Products

Let $\mathbf{u}$, $\mathbf{v}$, and $\mathbf{w}$ be vectors in an inner product space V, and let c be any real number.

1. $\langle \mathbf{0}, \mathbf{v} \rangle = \langle \mathbf{v}, \mathbf{0} \rangle = 0$
2. $\langle \mathbf{u} + \mathbf{v}, \mathbf{w} \rangle = \langle \mathbf{u}, \mathbf{w} \rangle + \langle \mathbf{v}, \mathbf{w} \rangle$
3. $\langle \mathbf{u}, c\mathbf{v} \rangle = c\langle \mathbf{u}, \mathbf{v} \rangle$

PROOF

The proof of the first property is given here. The proofs of the other two properties are left as exercises. (See Exercises 91 and 92.) From the definition of an inner product, you know $\langle \mathbf{0}, \mathbf{v} \rangle = \langle \mathbf{v}, \mathbf{0} \rangle$, so you only need to show one of these to be zero. Using the fact that $0(\mathbf{v}) = \mathbf{0}$,

$$\langle \mathbf{0}, \mathbf{v} \rangle = \langle 0(\mathbf{v}), \mathbf{v} \rangle$$
$$= 0\langle \mathbf{v}, \mathbf{v} \rangle$$
$$= 0.$$

The definitions of length (or norm), distance, and angle for general inner product spaces closely parallel those for Euclidean n-space.

Definitions of Length, Distance, and Angle

Let $\mathbf{u}$ and $\mathbf{v}$ be vectors in an inner product space V.

1. The **length** (or **norm**) of $\mathbf{u}$ is $\|\mathbf{u}\| = \sqrt{\langle \mathbf{u}, \mathbf{u} \rangle}$.
2. The **distance** between $\mathbf{u}$ and $\mathbf{v}$ is $d(\mathbf{u}, \mathbf{v}) = \|\mathbf{u} - \mathbf{v}\|$.
3. The **angle** between two nonzero vectors $\mathbf{u}$ and $\mathbf{v}$ can be found using

$$\cos\theta = \frac{\langle \mathbf{u}, \mathbf{v} \rangle}{\|\mathbf{u}\| \, \|\mathbf{v}\|}, \quad 0 \le \theta \le \pi.$$

4. $\mathbf{u}$ and $\mathbf{v}$ are **orthogonal** when $\langle \mathbf{u}, \mathbf{v} \rangle = 0$.

If $\|\mathbf{u}\| = 1$, then $\mathbf{u}$ is a **unit vector**. Moreover, if $\mathbf{v}$ is any nonzero vector in an inner product space V, then the vector $\mathbf{u} = \mathbf{v}/\|\mathbf{v}\|$ is the **unit vector in the direction of v.**

Note that the definition of the angle θ between $\mathbf{u}$ and $\mathbf{v}$ presumes that

$$-1 \le \frac{\langle \mathbf{u}, \mathbf{v} \rangle}{\|\mathbf{u}\| \, \|\mathbf{v}\|} \le 1$$

for a general inner product (as with Euclidean n-space), which follows from the Cauchy-Schwarz Inequality given later in Theorem 5.8.

EXAMPLE 6   Finding Inner Products

For polynomials $p = a_0 + a_1 x + \cdots + a_n x^n$ and $q = b_0 + b_1 x + \cdots + b_n x^n$ in the vector space P_n, the function $\langle p, q \rangle = a_0 b_0 + a_1 b_1 + \cdots + a_n b_n$ is an inner product. (In Exercise 34, you are asked to show this.) Let $p(x) = 1 - 2x^2$, $q(x) = 4 - 2x + x^2$, and $r(x) = x + 2x^2$ be polynomials in P_2, and find each quantity.

a. $\langle p, q \rangle$ **b.** $\langle q, r \rangle$ **c.** $\|q\|$ **d.** $d(p, q)$

SOLUTION

a. The inner product of p and q is

$$\langle p, q \rangle = a_0 b_0 + a_1 b_1 + a_2 b_2 = (1)(4) + (0)(-2) + (-2)(1) = 2.$$

b. The inner product of q and r is $\langle q, r \rangle = (4)(0) + (-2)(1) + (1)(2) = 0$. Notice that the vectors q and r are orthogonal.

c. The length of q is $\|q\| = \sqrt{\langle q, q \rangle} = \sqrt{4^2 + (-2)^2 + 1^2} = \sqrt{21}$.

d. The distance between p and q is

$$\begin{aligned}
d(p, q) &= \|p - q\| \\
&= \|(1 - 2x^2) - (4 - 2x + x^2)\| \\
&= \|-3 + 2x - 3x^2\| \\
&= \sqrt{(-3)^2 + 2^2 + (-3)^2} \\
&= \sqrt{22}.
\end{aligned}$$

Orthogonality depends on the inner product. That is, two vectors may be orthogonal with respect to one inner product but not to another. Rework parts (a) and (b) of Example 6 using the inner product $\langle p, q \rangle = a_0 b_0 + a_1 b_1 + 2 a_2 b_2$. With this inner product, p and q are orthogonal, but q and r are not.

EXAMPLE 7   **Using the Inner Product on $C[0, 1]$ (Calculus)**

Use the inner product defined in Example 5 and the functions $f(x) = x$ and $g(x) = x^2$ in $C[0, 1]$ to find each quantity.

a. $\|f\|$ **b.** $d(f, g)$

SOLUTION

a. $f(x) = x$, so you have

$$\|f\|^2 = \langle f, f \rangle = \int_0^1 (x)(x)\, dx = \int_0^1 x^2\, dx = \left[\frac{x^3}{3}\right]_0^1 = \frac{1}{3}.$$

So, $\|f\| = \dfrac{1}{\sqrt{3}}$.

b. To find $d(f, g)$, write

$$[d(f, g)]^2 = \langle f - g, f - g \rangle$$

$$= \int_0^1 [f(x) - g(x)]^2 dx = \int_0^1 [x - x^2]^2\, dx$$

$$= \int_0^1 [x^2 - 2x^3 + x^4]\, dx = \left[\frac{x^3}{3} - \frac{x^4}{2} + \frac{x^5}{5}\right]_0^1 = \frac{1}{30}.$$

So, $d(f, g) = \dfrac{1}{\sqrt{30}}$.

In Example 7, the distance between the functions $f(x) = x$ and $g(x) = x^2$ in $C[0, 1]$ is $1/\sqrt{30} \approx 0.183$. In practice, the distance between a pair of vectors is not as useful as the *relative* distance(s) between more than one pair. For example, the distance between $g(x) = x^2$ and $h(x) = x^2 + 1$ in $C[0, 1]$ is 1. (Verify this.) From the figures below, it seems reasonable to say that f and g are closer than g and h.

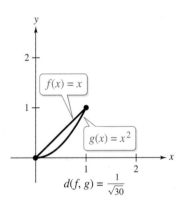

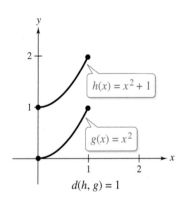

The properties of length and distance listed for R^n in the preceding section also hold for general inner product spaces. For instance, if $\mathbf{u}$ and $\mathbf{v}$ are vectors in an inner product space, then the properties listed below are true.

Properties of Length	*Properties of Distance*
1. $\|\mathbf{u}\| \geq 0$	1. $d(\mathbf{u}, \mathbf{v}) \geq 0$
2. $\|\mathbf{u}\| = 0$ if and only if $\mathbf{u} = \mathbf{0}$.	2. $d(\mathbf{u}, \mathbf{v}) = 0$ if and only if $\mathbf{u} = \mathbf{v}$.
3. $\|c\mathbf{u}\| = \|c\| \|\mathbf{u}\|$	3. $d(\mathbf{u}, \mathbf{v}) = d(\mathbf{v}, \mathbf{u})$

Theorem 5.8 lists the general inner product space versions of the Cauchy-Schwarz Inequality, the triangle inequality, and the Pythagorean Theorem.

THEOREM 5.8

Let **u** and **v** be vectors in an inner product space V.

1. Cauchy-Schwarz Inequality: $|\langle \mathbf{u}, \mathbf{v} \rangle| \leq \|\mathbf{u}\| \|\mathbf{v}\|$
2. Triangle inequality: $\|\mathbf{u} + \mathbf{v}\| \leq \|\mathbf{u}\| + \|\mathbf{v}\|$
3. Pythagorean Theorem: **u** and **v** are orthogonal if and only if

$$\|\mathbf{u} + \mathbf{v}\|^2 = \|\mathbf{u}\|^2 + \|\mathbf{v}\|^2.$$

The proof of each part of Theorem 5.8 parallels the proofs of Theorems 5.4, 5.5, and 5.6, respectively. Simply substitute $\langle \mathbf{u}, \mathbf{v} \rangle$ for the Euclidean inner product $\mathbf{u} \cdot \mathbf{v}$ in each proof.

EXAMPLE 8   An Example of the Cauchy-Schwarz Inequality (Calculus)

Let $f(x) = 1$ and $g(x) = x$ be functions in the vector space $C[0, 1]$, with the inner product defined in Example 5. Verify that $|\langle f, g \rangle| \leq \|f\| \|g\|$.

SOLUTION

For the left side of this inequality, you have

$$\langle f, g \rangle = \int_0^1 f(x)g(x)\, dx = \int_0^1 x\, dx = \left.\frac{x^2}{2}\right]_0^1 = \frac{1}{2}.$$

For the right side of the inequality, you have

$$\|f\|^2 = \int_0^1 f(x)f(x)\, dx = \int_0^1 dx = \left. x \right]_0^1 = 1$$

and

$$\|g\|^2 = \int_0^1 g(x)g(x)\, dx = \int_0^1 x^2\, dx = \left.\frac{x^3}{3}\right]_0^1 = \frac{1}{3}.$$

So,

$$\|f\| \|g\| = \sqrt{(1)\left(\frac{1}{3}\right)} = \frac{1}{\sqrt{3}} \approx 0.577, \text{ and } |\langle f, g \rangle| \leq \|f\| \|g\|.$$

LINEAR ALGEBRA APPLIED

The concept of work is important for determining the energy needed to perform various jobs. If a constant force **F** acts at an angle θ with the line of motion of an object to move the object from point A to point B (see figure below), then the work W done by the force is

$$W = (\cos \theta)\|\mathbf{F}\| \|\overrightarrow{AB}\|$$
$$= \mathbf{F} \cdot \overrightarrow{AB}$$

where $\overrightarrow{AB}$ represents the directed line segment from A to B. The quantity $(\cos \theta)\|\mathbf{F}\|$ is the length of the *orthogonal projection* of **F** onto $\overrightarrow{AB}$. Orthogonal projections are discussed on the next page.

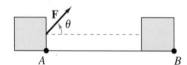

ORTHOGONAL PROJECTIONS IN INNER PRODUCT SPACES

Let $\mathbf{u}$ and $\mathbf{v}$ be vectors in R^2. If $\mathbf{v}$ is nonzero, then $\mathbf{u}$ can be orthogonally projected onto $\mathbf{v}$, as shown in Figure 5.7. This projection is denoted by $\text{proj}_{\mathbf{v}}\mathbf{u}$ and is a scalar multiple of $\mathbf{v}$, so you can write $\text{proj}_{\mathbf{v}}\mathbf{u} = a\mathbf{v}$. If $a > 0$, as shown in Figure 5.7(a), then $\cos\theta > 0$ and the length of $\text{proj}_{\mathbf{v}}\mathbf{u}$ is

$$\|a\mathbf{v}\| = |a|\,\|\mathbf{v}\| = a\|\mathbf{v}\| = \|\mathbf{u}\|\cos\theta = \frac{\|\mathbf{u}\|\,\|\mathbf{v}\|\cos\theta}{\|\mathbf{v}\|} = \frac{\mathbf{u}\cdot\mathbf{v}}{\|\mathbf{v}\|}$$

which implies that $a = (\mathbf{u}\cdot\mathbf{v})/\|\mathbf{v}\|^2 = (\mathbf{u}\cdot\mathbf{v})/(\mathbf{v}\cdot\mathbf{v})$. So,

$$\text{proj}_{\mathbf{v}}\mathbf{u} = \frac{\mathbf{u}\cdot\mathbf{v}}{\mathbf{v}\cdot\mathbf{v}}\mathbf{v}.$$

If $a < 0$, as shown in Figure 5.7(b), then the orthogonal projection of $\mathbf{u}$ onto $\mathbf{v}$ can be found using the same formula. (Verify this.)

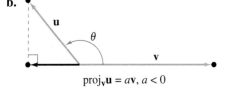

a. $\text{proj}_{\mathbf{v}}\mathbf{u} = a\mathbf{v}, \ a > 0$ **b.** $\text{proj}_{\mathbf{v}}\mathbf{u} = a\mathbf{v}, \ a < 0$

Figure 5.7

EXAMPLE 9   Finding the Orthogonal Projection of u onto v

In R^2, the orthogonal projection of $\mathbf{u} = (4, 2)$ onto $\mathbf{v} = (3, 4)$ is

$$\text{proj}_{\mathbf{v}}\mathbf{u} = \frac{\mathbf{u}\cdot\mathbf{v}}{\mathbf{v}\cdot\mathbf{v}}\mathbf{v} = \frac{(4, 2)\cdot(3, 4)}{(3, 4)\cdot(3, 4)}(3, 4) = \frac{20}{25}(3, 4) = \left(\frac{12}{5}, \frac{16}{5}\right)$$

as shown in Figure 5.8.

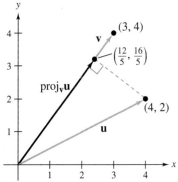

Figure 5.8

REMARK

If $\mathbf{v}$ is a unit vector, then $\langle \mathbf{v}, \mathbf{v} \rangle = \|\mathbf{v}\|^2 = 1$, and the formula for the orthogonal projection of $\mathbf{u}$ onto $\mathbf{v}$ takes the simpler form

$$\text{proj}_{\mathbf{v}}\mathbf{u} = \langle \mathbf{u}, \mathbf{v} \rangle\mathbf{v}.$$

An orthogonal projection in a general inner product space is defined below.

Definition of Orthogonal Projection

Let $\mathbf{u}$ and $\mathbf{v}$ be vectors in an inner product space V, such that $\mathbf{v} \neq \mathbf{0}$. Then the **orthogonal projection** of $\mathbf{u}$ onto $\mathbf{v}$ is

$$\text{proj}_{\mathbf{v}}\mathbf{u} = \frac{\langle \mathbf{u}, \mathbf{v} \rangle}{\langle \mathbf{v}, \mathbf{v} \rangle}\mathbf{v}.$$

EXAMPLE 10   Finding an Orthogonal Projection in R^3

See LarsonLinearAlgebra.com for an interactive version of this type of example.

Use the Euclidean inner product in R^3 to find the orthogonal projection of $\mathbf{u} = (6, 2, 4)$ onto $\mathbf{v} = (1, 2, 0)$.

SOLUTION

$\mathbf{u}\cdot\mathbf{v} = 10$ and $\mathbf{v}\cdot\mathbf{v} = 5$, so the orthogonal projection of $\mathbf{u}$ onto $\mathbf{v}$ is

$$\text{proj}_{\mathbf{v}}\mathbf{u} = \frac{\mathbf{u}\cdot\mathbf{v}}{\mathbf{v}\cdot\mathbf{v}}\mathbf{v} = \frac{10}{5}(1, 2, 0) = 2(1, 2, 0) = (2, 4, 0)$$

as shown in Figure 5.9.

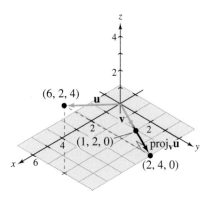

Figure 5.9

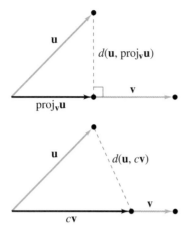

Figure 5.10

Verify in Example 10 that $\mathbf{u} - \text{proj}_\mathbf{v}\mathbf{u} = (6, 2, 4) - (2, 4, 0) = (4, -2, 4)$ is orthogonal to $\mathbf{v} = (1, 2, 0)$. This is true in general. If $\mathbf{u}$ and $\mathbf{v}$ are nonzero vectors in an inner product space, then $\mathbf{u} - \text{proj}_\mathbf{v}\mathbf{u}$ is orthogonal to $\mathbf{v}$. (In Exercise 90, you are asked to prove this.)

An important property of orthogonal projections used in mathematical modeling (see Section 5.4) is given in the next theorem. It states that, of all possible scalar multiples of a vector $\mathbf{v}$, the orthogonal projection of $\mathbf{u}$ onto $\mathbf{v}$ is the one closest to $\mathbf{u}$, as shown in Figure 5.10. For instance, in Example 10, this theorem implies that, of all the scalar multiples of the vector $\mathbf{v} = (1, 2, 0)$, the vector $\text{proj}_\mathbf{v}\mathbf{u} = (2, 4, 0)$ is closest to $\mathbf{u} = (6, 2, 4)$. You are asked to prove this explicitly in Exercise 101.

THEOREM 5.9   Orthogonal Projection and Distance

Let $\mathbf{u}$ and $\mathbf{v}$ be two vectors in an inner product space V, such that $\mathbf{v} \neq \mathbf{0}$. Then

$$d(\mathbf{u}, \text{proj}_\mathbf{v}\mathbf{u}) < d(\mathbf{u}, c\mathbf{v}), \quad c \neq \frac{\langle \mathbf{u}, \mathbf{v} \rangle}{\langle \mathbf{v}, \mathbf{v} \rangle}.$$

PROOF

Let $b = \langle \mathbf{u}, \mathbf{v} \rangle / \langle \mathbf{v}, \mathbf{v} \rangle$. Then

$$\|\mathbf{u} - c\mathbf{v}\|^2 = \|(\mathbf{u} - b\mathbf{v}) + (b - c)\mathbf{v}\|^2$$

where $(\mathbf{u} - b\mathbf{v})$ and $(b - c)\mathbf{v}$ are orthogonal. Verify this by using the inner product axioms to show that $\langle (\mathbf{u} - b\mathbf{v}), (b - c)\mathbf{v} \rangle = 0$. Now, by the Pythagorean Theorem,

$$\|(\mathbf{u} - b\mathbf{v}) + (b - c)\mathbf{v}\|^2 = \|\mathbf{u} - b\mathbf{v}\|^2 + \|(b - c)\mathbf{v}\|^2$$

which implies that

$$\|\mathbf{u} - c\mathbf{v}\|^2 = \|\mathbf{u} - b\mathbf{v}\|^2 + (b - c)^2\|\mathbf{v}\|^2.$$

$b \neq c$ and $\mathbf{v} \neq \mathbf{0}$, so you know that $(b - c)^2\|\mathbf{v}\|^2 > 0$. This means that

$$\|\mathbf{u} - b\mathbf{v}\|^2 < \|\mathbf{u} - c\mathbf{v}\|^2$$

and it follows that $d(\mathbf{u}, b\mathbf{v}) < d(\mathbf{u}, c\mathbf{v})$.

The next example discusses an orthogonal projection in the inner product space $C[a, b]$.

EXAMPLE 11   **Finding an Orthogonal Projection in $C[a, b]$ (Calculus)**

Let $f(x) = 1$ and $g(x) = x$ be functions in $C[0, 1]$. Use the inner product on $C[a, b]$ defined in Example 5,

$$\langle f, g \rangle = \int_a^b f(x)g(x)\, dx$$

to find the orthogonal projection of f onto g.

SOLUTION

From Example 8, you know that

$$\langle f, g \rangle = \frac{1}{2} \quad \text{and} \quad \langle g, g \rangle = \|g\|^2 = \frac{1}{3}.$$

So, the orthogonal projection of f onto g is

$$\text{proj}_g f = \frac{\langle f, g \rangle}{\langle g, g \rangle} g = \frac{1/2}{1/3}x = \frac{3}{2}x.$$

5.2 Exercises

See CalcChat.com for worked-out solutions to odd-numbered exercises.

Showing That a Function Is an Inner Product In Exercises 1–4, show that the function defines an inner product on R^2, where $\mathbf{u} = (u_1, u_2)$ and $\mathbf{v} = (v_1, v_2)$.

1. $\langle \mathbf{u}, \mathbf{v} \rangle = 3u_1v_1 + u_2v_2$ 2. $\langle \mathbf{u}, \mathbf{v} \rangle = u_1v_1 + 9u_2v_2$
3. $\langle \mathbf{u}, \mathbf{v} \rangle = \frac{1}{2}u_1v_1 + \frac{1}{4}u_2v_2$
4. $\langle \mathbf{u}, \mathbf{v} \rangle = 2u_1v_2 + u_2v_1 + u_1v_2 + 2u_2v_2$

Showing That a Function Is an Inner Product In Exercises 5–8, show that the function defines an inner product on R^3, where $\mathbf{u} = (u_1, u_2, u_3)$ and $\mathbf{v} = (v_1, v_2, v_3)$.

5. $\langle \mathbf{u}, \mathbf{v} \rangle = 2u_1v_1 + 3u_2v_2 + u_3v_3$
6. $\langle \mathbf{u}, \mathbf{v} \rangle = u_1v_1 + 2u_2v_2 + u_3v_3$
7. $\langle \mathbf{u}, \mathbf{v} \rangle = 4u_1v_1 + 3u_2v_2 + 2u_3v_3$
8. $\langle \mathbf{u}, \mathbf{v} \rangle = \frac{1}{2}u_1v_1 + \frac{1}{4}u_2v_2 + \frac{1}{2}u_3v_3$

Showing That a Function Is Not an Inner Product In Exercises 9–12, show that the function *does not* define an inner product on R^3, where $\mathbf{u} = (u_1, u_2)$ and $\mathbf{v} = (v_1, v_2)$.

9. $\langle \mathbf{u}, \mathbf{v} \rangle = u_1v_1$ 10. $\langle \mathbf{u}, \mathbf{v} \rangle = u_1v_1 - 6u_2v_2$
11. $\langle \mathbf{u}, \mathbf{v} \rangle = u_1^2v_1^2 - u_2^2v_2^2$ 12. $\langle \mathbf{u}, \mathbf{v} \rangle = 3u_1v_2 - u_2v_1$

Showing That a Function Is Not an Inner Product In Exercises 13–16, show that the function *does not* define an inner product on R^3, where $\mathbf{u} = (u_1, u_2, u_3)$ and $\mathbf{v} = (v_1, v_2, v_3)$.

13. $\langle \mathbf{u}, \mathbf{v} \rangle = -u_1u_2u_3$
14. $\langle \mathbf{u}, \mathbf{v} \rangle = u_1v_1 - u_2v_2 - u_3v_3$
15. $\langle \mathbf{u}, \mathbf{v} \rangle = u_1^2v_1^2 + u_2^2v_2^2 + u_3^2v_2^2$
16. $\langle \mathbf{u}, \mathbf{v} \rangle = 2u_1u_2 + 3v_1v_2 + u_3v_3$

Finding Inner Product, Length, and Distance In Exercises 17–26, find (a) $\langle \mathbf{u}, \mathbf{v} \rangle$, (b) $\|\mathbf{u}\|$, (c) $\|\mathbf{v}\|$, and (d) $d(\mathbf{u}, \mathbf{v})$ for the given inner product defined on R^n.

17. $\mathbf{u} = (3, 4), \quad \mathbf{v} = (5, -12), \quad \langle \mathbf{u}, \mathbf{v} \rangle = \mathbf{u} \cdot \mathbf{v}$
18. $\mathbf{u} = (-1, 1), \quad \mathbf{v} = (6, 8), \quad \langle \mathbf{u}, \mathbf{v} \rangle = \mathbf{u} \cdot \mathbf{v}$
19. $\mathbf{u} = (-4, 3), \quad \mathbf{v} = (0, 5), \quad \langle \mathbf{u}, \mathbf{v} \rangle = 3u_1v_1 + u_2v_2$
20. $\mathbf{u} = (0, -6), \quad \mathbf{v} = (-1, 1), \quad \langle \mathbf{u}, \mathbf{v} \rangle = u_1v_1 + 2u_2v_2$
21. $\mathbf{u} = (0, 7, 2), \quad \mathbf{v} = (9, -3, -2), \quad \langle \mathbf{u}, \mathbf{v} \rangle = \mathbf{u} \cdot \mathbf{v}$
22. $\mathbf{u} = (0, 1, 2), \quad \mathbf{v} = (1, 2, 0), \quad \langle \mathbf{u}, \mathbf{v} \rangle = \mathbf{u} \cdot \mathbf{v}$
23. $\mathbf{u} = (8, 0, -8), \quad \mathbf{v} = (8, 3, 16),$
    $\langle \mathbf{u}, \mathbf{v} \rangle = 2u_1v_1 + 3u_2v_2 + u_3v_3$
24. $\mathbf{u} = (1, 1, 1), \quad \mathbf{v} = (2, 5, 2),$
    $\langle \mathbf{u}, \mathbf{v} \rangle = u_1v_1 + 2u_2v_2 + u_3v_3$
25. $\mathbf{u} = (-1, 2, 0, 1), \quad \mathbf{v} = (0, 1, 2, 2), \quad \langle \mathbf{u}, \mathbf{v} \rangle = \mathbf{u} \cdot \mathbf{v}$
26. $\mathbf{u} = (1, -1, 2, 0), \quad \mathbf{v} = (2, 1, 0, -1),$
    $\langle \mathbf{u}, \mathbf{v} \rangle = \mathbf{u} \cdot \mathbf{v}$

Showing That a Function Is an Inner Product In Exercises 27 and 28, let

$$A = \begin{bmatrix} a_{11} & a_{12} \\ a_{21} & a_{22} \end{bmatrix} \quad \text{and} \quad B = \begin{bmatrix} b_{11} & b_{12} \\ b_{21} & b_{22} \end{bmatrix}$$

be matrices in the vector space $M_{2,2}$. Show that the function defines an inner product on $M_{2,2}$.

27. $\langle A, B \rangle = a_{11}b_{11} + a_{12}b_{12} + a_{21}b_{21} + a_{22}b_{22}$
28. $\langle A, B \rangle = 2a_{11}b_{11} + a_{12}b_{12} + a_{21}b_{21} + 2a_{22}b_{22}$

Finding Inner Product, Length, and Distance In Exercises 29–32, find (a) $\langle A, B \rangle$, (b) $\|A\|$, (c) $\|B\|$, and (d) $d(A, B)$ for the matrices in $M_{2,2}$ using the inner product $\langle A, B \rangle = 2a_{11}b_{11} + a_{12}b_{12} + a_{21}b_{21} + 2a_{22}b_{22}$.

29. $A = \begin{bmatrix} 2 & -4 \\ -3 & 1 \end{bmatrix}, \quad B = \begin{bmatrix} -2 & 1 \\ 1 & 0 \end{bmatrix}$

30. $A = \begin{bmatrix} 1 & 0 \\ 0 & 1 \end{bmatrix}, \quad B = \begin{bmatrix} 0 & 1 \\ 1 & 0 \end{bmatrix}$

31. $A = \begin{bmatrix} 1 & -1 \\ 2 & 4 \end{bmatrix}, \quad B = \begin{bmatrix} 0 & 1 \\ -2 & 0 \end{bmatrix}$

32. $A = \begin{bmatrix} 1 & 0 \\ 0 & -1 \end{bmatrix}, \quad B = \begin{bmatrix} 1 & 1 \\ 0 & -1 \end{bmatrix}$

Showing That a Function Is an Inner Product In Exercises 33 and 34, show that the function defines an inner product for polynomials $p(x) = a_0 + a_1x + \cdots + a_nx^n$ and $q(x) = b_0 + b_1x + \cdots + b_nx^n$.

33. $\langle p, q \rangle = a_0b_0 + 2a_1b_1 + a_2b_2$ in P_2
34. $\langle p, q \rangle = a_0b_0 + a_1b_1 + \cdots + a_nb_n$ in P_n

Finding Inner Product, Length, and Distance In Exercises 35–38, find (a) $\langle p, q \rangle$, (b) $\|p\|$, (c) $\|q\|$, and (d) $d(p, q)$ for the polynomials in P_2 using the inner product $\langle p, q \rangle = a_0b_0 + a_1b_1 + a_2b_2$.

35. $p(x) = 1 - x + 3x^2, \quad q(x) = x - x^2$
36. $p(x) = 1 + x + \frac{1}{2}x^2, \quad q(x) = 1 + 2x^2$
37. $p(x) = 1 + x^2, \quad q(x) = 1 - x^2$
38. $p(x) = 1 - 3x + x^2, \quad q(x) = -x + 2x^2$

Calculus In Exercises 39–42, use the functions f and g in $C[-1, 1]$ to find (a) $\langle f, g \rangle$, (b) $\|f\|$, (c) $\|g\|$, and (d) $d(f, g)$ for the inner product

$$\langle f, g \rangle = \int_{-1}^{1} f(x)g(x)\, dx.$$

39. $f(x) = 1, \quad g(x) = 4x^2 - 1$
40. $f(x) = -x, \quad g(x) = x^2 - x + 2$
41. $f(x) = x, \quad g(x) = e^x$
42. $f(x) = x, \quad g(x) = e^{-x}$

Finding the Angle Between Two Vectors   In Exercises 43–52, find the angle θ between the vectors.

43. $\mathbf{u} = (3, 4)$, $\mathbf{v} = (5, -12)$, $\langle \mathbf{u}, \mathbf{v} \rangle = \mathbf{u} \cdot \mathbf{v}$

44. $\mathbf{u} = (3, -1)$, $\mathbf{v} = \left(\frac{1}{3}, 1\right)$, $\langle \mathbf{u}, \mathbf{v} \rangle = \mathbf{u} \cdot \mathbf{v}$

45. $\mathbf{u} = (-4, 3)$, $\mathbf{v} = (0, 5)$, $\langle \mathbf{u}, \mathbf{v} \rangle = 3u_1v_1 + u_2v_2$

46. $\mathbf{u} = \left(\frac{1}{4}, -1\right)$, $\mathbf{v} = (2, 1)$,

$\langle \mathbf{u}, \mathbf{v} \rangle = 2u_1v_1 + u_2v_2$

47. $\mathbf{u} = (1, 1, 1)$, $\mathbf{v} = (2, -2, 2)$,

$\langle \mathbf{u}, \mathbf{v} \rangle = u_1v_1 + 2u_2v_2 + u_3v_3$

48. $\mathbf{u} = (0, 1, -2)$, $\mathbf{v} = (3, -2, 1)$, $\langle \mathbf{u}, \mathbf{v} \rangle = \mathbf{u} \cdot \mathbf{v}$

49. $p(x) = 1 - x + x^2$, $q(x) = 1 + x + x^2$,

$\langle p, q \rangle = a_0b_0 + a_1b_1 + a_2b_2$

50. $p(x) = 1 + x^2$, $q(x) = x - x^2$,

$\langle p, q \rangle = a_0b_0 + 2a_1b_1 + a_2b_2$

51. **Calculus** $f(x) = x$, $g(x) = x^2$,

$\langle f, g \rangle = \displaystyle\int_{-1}^{1} f(x)g(x)\, dx$

52. **Calculus** $f(x) = 1$, $g(x) = x^2$,

$\langle f, g \rangle = \displaystyle\int_{-1}^{1} f(x)g(x)\, dx$

Verifying Inequalities   In Exercises 53–64, verify (a) the Cauchy-Schwarz Inequality and (b) the triangle inequality for the given vectors and inner products.

53. $\mathbf{u} = (5, 12)$, $\mathbf{v} = (3, 4)$, $\langle \mathbf{u}, \mathbf{v} \rangle = \mathbf{u} \cdot \mathbf{v}$

54. $\mathbf{u} = (-1, 1)$, $\mathbf{v} = (1, -1)$, $\langle \mathbf{u}, \mathbf{v} \rangle = \mathbf{u} \cdot \mathbf{v}$

55. $\mathbf{u} = (0, 1, 5)$, $\mathbf{v} = (-4, 3, 3)$, $\langle \mathbf{u}, \mathbf{v} \rangle = \mathbf{u} \cdot \mathbf{v}$

56. $\mathbf{u} = (1, 0, 2)$, $\mathbf{v} = (1, 2, 0)$, $\langle \mathbf{u}, \mathbf{v} \rangle = \mathbf{u} \cdot \mathbf{v}$

57. $p(x) = 2x$, $q(x) = 1 + 3x^2$,

$\langle p, q \rangle = a_0b_0 + a_1b_1 + a_2b_2$

58. $p(x) = x$, $q(x) = 1 - x^2$,

$\langle p, q \rangle = a_0b_0 + 2a_1b_1 + a_2b_2$

59. $A = \begin{bmatrix} 0 & 3 \\ 2 & 1 \end{bmatrix}$, $B = \begin{bmatrix} -3 & 1 \\ 4 & 3 \end{bmatrix}$,

$\langle A, B \rangle = a_{11}b_{11} + a_{12}b_{12} + a_{21}b_{21} + a_{22}b_{22}$

60. $A = \begin{bmatrix} 0 & 1 \\ 2 & -1 \end{bmatrix}$, $B = \begin{bmatrix} 1 & 1 \\ 2 & -2 \end{bmatrix}$,

$\langle A, B \rangle = a_{11}b_{11} + a_{12}b_{12} + a_{21}b_{21} + a_{22}b_{22}$

61. **Calculus** $f(x) = \sin x$, $g(x) = \cos x$,

$\langle f, g \rangle = \displaystyle\int_{0}^{\pi/4} f(x)g(x)\, dx$

62. **Calculus** $f(x) = x$, $g(x) = \cos \pi x$,

$\langle f, g \rangle = \displaystyle\int_{0}^{2} f(x)g(x)\, dx$

63. **Calculus** $f(x) = x$, $g(x) = e^x$,

$\langle f, g \rangle = \displaystyle\int_{0}^{1} f(x)g(x)\, dx$

64. **Calculus** $f(x) = x$, $g(x) = e^{-x}$,

$\langle f, g \rangle = \displaystyle\int_{0}^{1} f(x)g(x)\, dx$

Calculus   In Exercises 65–68, show that f and g are orthogonal in the inner product space $C[a, b]$ with the inner product

$$\langle f, g \rangle = \int_{a}^{b} f(x)g(x)\, dx.$$

65. $C[-\pi/2, \pi/2]$, $f(x) = \cos x$, $g(x) = \sin x$

66. $C[-1, 1]$, $f(x) = x$, $g(x) = \frac{1}{2}(3x^2 - 1)$

67. $C[-1, 1]$, $f(x) = x$, $g(x) = \frac{1}{2}(5x^3 - 3x)$

68. $C[0, \pi]$, $f(x) = 1$, $g(x) = \cos(2nx)$,

$n = 1, 2, 3, \ldots$

Finding and Graphing Orthogonal Projections in R^2 In Exercises 69–72, (a) find $\text{proj}_\mathbf{v}\mathbf{u}$, (b) find $\text{proj}_\mathbf{u}\mathbf{v}$, and (c) sketch a graph of both $\text{proj}_\mathbf{v}\mathbf{u}$ and $\text{proj}_\mathbf{u}\mathbf{v}$. Use the Euclidean inner product.

69. $\mathbf{u} = (1, 2)$, $\mathbf{v} = (2, 1)$

70. $\mathbf{u} = (-3, -1)$, $\mathbf{v} = (6, 3)$

71. $\mathbf{u} = (-1, 3)$, $\mathbf{v} = (4, 4)$

72. $\mathbf{u} = (2, -2)$, $\mathbf{v} = (3, 1)$

Finding Orthogonal Projections   In Exercises 73–76, find (a) $\text{proj}_\mathbf{v}\mathbf{u}$ and (b) $\text{proj}_\mathbf{u}\mathbf{v}$. Use the Euclidean inner product.

73. $\mathbf{u} = (5, -3, 1)$, $\mathbf{v} = (1, -1, 0)$

74. $\mathbf{u} = (1, 2, -1)$, $\mathbf{v} = (-1, 2, -1)$

75. $\mathbf{u} = (0, 1, 3, -6)$, $\mathbf{v} = (-1, 1, 2, 2)$

76. $\mathbf{u} = (-1, 4, -2, 3)$, $\mathbf{v} = (2, -1, 2, -1)$

Calculus   In Exercises 77–84, find the orthogonal projection of f onto g. Use the inner product in $C[a, b]$

$$\langle f, g \rangle = \int_{a}^{b} f(x)g(x)\, dx.$$

77. $C[-1, 1]$, $f(x) = x$, $g(x) = 1$

78. $C[-1, 1]$, $f(x) = x^3 - x$, $g(x) = 2x - 1$

79. $C[0, 1]$, $f(x) = x$, $g(x) = e^x$

80. $C[0, 1]$, $f(x) = x$, $g(x) = e^{-x}$

81. $C[-\pi, \pi]$, $f(x) = \sin x$, $g(x) = \cos x$

82. $C[-\pi, \pi]$, $f(x) = \sin 2x$, $g(x) = \cos 2x$

83. $C[-\pi, \pi]$, $f(x) = x$, $g(x) = \sin 2x$

84. $C[-\pi, \pi]$, $f(x) = x$, $g(x) = \cos 2x$

True or False? **In Exercises 85 and 86, determine whether each statement is true or false. If a statement is true, give a reason or cite an appropriate statement from the text. If a statement is false, provide an example that shows the statement is not true in all cases or cite an appropriate statement from the text.**

85. (a) The dot product is the only inner product that can be defined in R^n.

    (b) A nonzero vector in an inner product can have a norm of zero.

86. (a) The norm of the vector $\mathbf{u}$ is the angle between $\mathbf{u}$ and the positive x-axis.

    (b) The angle θ between a vector $\mathbf{v}$ and the projection of $\mathbf{u}$ onto $\mathbf{v}$ is obtuse when the scalar $a < 0$ and acute when $a > 0$, where $a\mathbf{v} = \text{proj}_{\mathbf{v}}\mathbf{u}$.

87. Let $\mathbf{u} = (4, 2)$ and $\mathbf{v} = (2, -2)$ be vectors in R^2 with the inner product $\langle \mathbf{u}, \mathbf{v} \rangle = u_1 v_1 + 2u_2 v_2$.

    (a) Show that $\mathbf{u}$ and $\mathbf{v}$ are orthogonal.

    (b) Sketch $\mathbf{u}$ and $\mathbf{v}$. Are they orthogonal in the Euclidean sense?

88. **Proof** Prove that
$$\|\mathbf{u} + \mathbf{v}\|^2 + \|\mathbf{u} - \mathbf{v}\|^2 = 2\|\mathbf{u}\|^2 + 2\|\mathbf{v}\|^2$$
for any vectors $\mathbf{u}$ and $\mathbf{v}$ in an inner product space V.

89. **Proof** Prove that the function is an inner product on R^n.
$$\langle \mathbf{u}, \mathbf{v} \rangle = c_1 u_1 v_1 + c_2 u_2 v_2 + \cdots + c_n u_n v_n, \quad c_i > 0$$

90. **Proof** Let $\mathbf{u}$ and $\mathbf{v}$ be nonzero vectors in an inner product space V. Prove that $\mathbf{u} - \text{proj}_{\mathbf{v}}\mathbf{u}$ is orthogonal to $\mathbf{v}$.

91. **Proof** Prove Property 2 of Theorem 5.7: If $\mathbf{u}$, $\mathbf{v}$, and $\mathbf{w}$ are vectors in an inner product space V, then $\langle \mathbf{u} + \mathbf{v}, \mathbf{w} \rangle = \langle \mathbf{u}, \mathbf{w} \rangle + \langle \mathbf{v}, \mathbf{w} \rangle$.

92. **Proof** Prove Property 3 of Theorem 5.7: If $\mathbf{u}$ and $\mathbf{v}$ are vectors in an inner product space V and c is any real number, then $\langle \mathbf{u}, c\mathbf{v} \rangle = c\langle \mathbf{u}, \mathbf{v} \rangle$.

93. **Guided Proof** Let W be a subspace of the inner product space V. Prove that the set
$$W^{\perp} = \{\mathbf{v} \in V : \langle \mathbf{v}, \mathbf{w} \rangle = 0 \text{ for all } \mathbf{w} \in W\}$$
is a subspace of V.

    Getting Started: To prove that $W^{\perp}$ is a subspace of V, you must show that $W^{\perp}$ is nonempty and that the closure conditions for a subspace hold (Theorem 4.5).

    (i) Find a vector in $W^{\perp}$ to conclude that it is nonempty.

    (ii) To show the closure of $W^{\perp}$ under addition, you need to show that $\langle \mathbf{v}_1 + \mathbf{v}_2, \mathbf{w} \rangle = 0$ for all $\mathbf{w} \in W$ and for any $\mathbf{v}_1, \mathbf{v}_2 \in W^{\perp}$. Use the properties of inner products and the fact that $\langle \mathbf{v}_1, \mathbf{w} \rangle$ and $\langle \mathbf{v}_2, \mathbf{w} \rangle$ are both zero to show this.

    (iii) To show closure under multiplication by a scalar, proceed as in part (ii). Use the properties of inner products and the condition of belonging to $W^{\perp}$.

94. Use the result of Exercise 93 to find $W^{\perp}$ when W is the span of $(1, 2, 3)$ in $V = R^3$.

95. **Guided Proof** Let $\langle \mathbf{u}, \mathbf{v} \rangle$ be the Euclidean inner product on R^n. Use the fact that $\langle \mathbf{u}, \mathbf{v} \rangle = \mathbf{u}^T \mathbf{v}$ to prove that for any $n \times n$ matrix A,

    (a) $\langle A^T A \mathbf{u}, \mathbf{v} \rangle = \langle \mathbf{u}, A\mathbf{v} \rangle$

    and

    (b) $\langle A^T A \mathbf{u}, \mathbf{u} \rangle = \|A\mathbf{u}\|^2$.

    Getting Started: To prove (a) and (b), make use of both the properties of transposes (Theorem 2.6) and the properties of the dot product (Theorem 5.3).

    (i) To prove part (a), make repeated use of the property $\langle \mathbf{u}, \mathbf{v} \rangle = \mathbf{u}^T \mathbf{v}$ and Property 4 of Theorem 2.6.

    (ii) To prove part (b), make use of the property $\langle \mathbf{u}, \mathbf{v} \rangle = \mathbf{u}^T \mathbf{v}$, Property 4 of Theorem 2.6, and Property 4 of Theorem 5.3.

96. **CAPSTONE**

    (a) Explain how to determine whether a function defines an inner product.

    (b) Let $\mathbf{u}$ and $\mathbf{v}$ be vectors in an inner product space V, such that $\mathbf{v} \neq \mathbf{0}$. Explain how to find the orthogonal projection of $\mathbf{u}$ onto $\mathbf{v}$.

Finding Inner Product Weights **In Exercises 97–100, find c_1 and c_2 for the inner product of R^2,**
$$\langle \mathbf{u}, \mathbf{v} \rangle = c_1 u_1 v_1 + c_2 u_2 v_2$$
such that the graph represents a unit circle as shown.

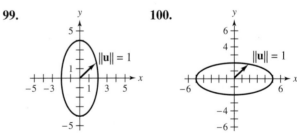

97.

98.

99.

100.

101. Consider the vectors
$$\mathbf{u} = (6, 2, 4) \text{ and } \mathbf{v} = (1, 2, 0)$$
from Example 10. Without using Theorem 5.9, show that among all the scalar multiples $c\mathbf{v}$ of the vector $\mathbf{v}$, the projection of $\mathbf{u}$ onto $\mathbf{v}$ is the vector closest to $\mathbf{u}$—that is, show that $d(\mathbf{u}, \text{proj}_{\mathbf{v}}\mathbf{u})$ is a minimum.

5.3 Orthonormal Bases: Gram-Schmidt Process

■ Show that a set of vectors is orthogonal and forms an orthonormal basis, and represent a vector relative to an orthonormal basis.

■ Apply the Gram-Schmidt orthonormalization process.

ORTHOGONAL AND ORTHONORMAL SETS

You saw in Section 4.7 that a vector space can have many different bases. While studying that section, you may have noticed that some bases are more convenient than others. For example, R^3 has the basis $B = \{(1, 0, 0), (0, 1, 0), (0, 0, 1)\}$. This set is the *standard* basis for R^3 because it has important characteristics that are particularly useful. One important characteristic is that the three vectors in the basis are *mutually orthogonal*. That is,

$$(1, 0, 0) \cdot (0, 1, 0) = 0$$
$$(1, 0, 0) \cdot (0, 0, 1) = 0$$
$$(0, 1, 0) \cdot (0, 0, 1) = 0.$$

A second important characteristic is that each vector in the basis is a *unit* vector. (Verify this by inspection.)

This section identifies some advantages of using bases consisting of mutually orthogonal unit vectors and develops a procedure for constructing such bases, known as the *Gram-Schmidt orthonormalization process*.

Definitions of Orthogonal and Orthonormal Sets

A set S of vectors in an inner product space V is **orthogonal** when every pair of vectors in S is orthogonal. If, in addition, each vector in the set is a unit vector, then S is **orthonormal.**

For $S = \{\mathbf{v}_1, \mathbf{v}_2, \ldots, \mathbf{v}_n\}$, this definition has the form below.

Orthogonal	*Orthonormal*
1. $\langle \mathbf{v}_i, \mathbf{v}_j \rangle = 0, \; i \neq j$	**1.** $\langle \mathbf{v}_i, \mathbf{v}_j \rangle = 0, \; i \neq j$
	2. $\|\mathbf{v}_i\| = 1, \; i = 1, 2, \ldots, n$

If S is a *basis,* then it is an **orthogonal basis** or an **orthonormal basis,** respectively.

The standard basis for R^n is orthonormal, but it is not the only orthonormal basis for R^n. For example, a nonstandard orthonormal basis for R^3 can be formed by rotating the standard basis about the z-axis, resulting in

$$B = \{(\cos \theta, \sin \theta, 0), (-\sin \theta, \cos \theta, 0), (0, 0, 1)\}$$

as shown below. Verify that the dot product of any two distinct vectors in B is zero, and that each vector in B is a unit vector.

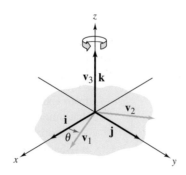

Example 1 describes another nonstandard orthonormal basis for R^3.

EXAMPLE 1   **A Nonstandard Orthonormal Basis for R^3**

Show that the set is an orthonormal basis for R^3.

$$S = \{\mathbf{v}_1, \mathbf{v}_2, \mathbf{v}_3\} = \left\{ \left(\frac{1}{\sqrt{2}}, \frac{1}{\sqrt{2}}, 0 \right), \left(-\frac{\sqrt{2}}{6}, \frac{\sqrt{2}}{6}, \frac{2\sqrt{2}}{3} \right), \left(\frac{2}{3}, -\frac{2}{3}, \frac{1}{3} \right) \right\}$$

SOLUTION

First show that the three vectors are mutually orthogonal.

$$\mathbf{v}_1 \cdot \mathbf{v}_2 = -\frac{1}{6} + \frac{1}{6} + 0 = 0$$

$$\mathbf{v}_1 \cdot \mathbf{v}_3 = \frac{2}{3\sqrt{2}} - \frac{2}{3\sqrt{2}} + 0 = 0$$

$$\mathbf{v}_2 \cdot \mathbf{v}_3 = -\frac{\sqrt{2}}{9} - \frac{\sqrt{2}}{9} + \frac{2\sqrt{2}}{9} = 0$$

Now, each vector is of length 1 because

$$\|\mathbf{v}_1\| = \sqrt{\mathbf{v}_1 \cdot \mathbf{v}_1} = \sqrt{\tfrac{1}{2} + \tfrac{1}{2} + 0} = 1$$

$$\|\mathbf{v}_2\| = \sqrt{\mathbf{v}_2 \cdot \mathbf{v}_2} = \sqrt{\tfrac{1}{18} + \tfrac{1}{18} + \tfrac{8}{9}} = 1$$

$$\|\mathbf{v}_3\| = \sqrt{\mathbf{v}_3 \cdot \mathbf{v}_3} = \sqrt{\tfrac{4}{9} + \tfrac{4}{9} + \tfrac{1}{9}} = 1.$$

So, S is an orthonormal set. The three vectors do not lie in the same plane (see Figure 5.11), so you know that they span R^3. By Theorem 4.12, they form a (nonstandard) orthonormal basis for R^3.

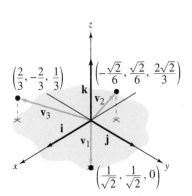

Figure 5.11

EXAMPLE 2   **An Orthonormal Basis for P_3**

In P_3, with the inner product

$$\langle p, q \rangle = a_0 b_0 + a_1 b_1 + a_2 b_2 + a_3 b_3$$

the standard basis $B = \{1, x, x^2, x^3\}$ is orthonormal. The verification of this is left as an exercise. (See Exercise 17.)

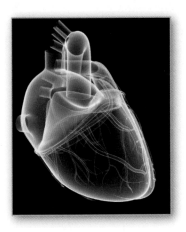

LINEAR ALGEBRA APPLIED

Time-frequency analysis of irregular physiological signals, such as beat-to-beat cardiac rhythm variations (also known as heart rate variability or HRV), can be difficult. This is because the structure of a signal can include multiple periodic, nonperiodic, and pseudo-periodic components. Researchers have proposed and validated a simplified HRV analysis method called orthonormal-basis partitioning and time-frequency representation (OPTR). This method can detect both abrupt and slow changes in the HRV signal's structure, divide a nonstationary HRV signal into segments that are "less nonstationary," and determine patterns in the HRV. The researchers found that although it had poor time resolution with signals that changed gradually, the OPTR method accurately represented multicomponent and abrupt changes in both real-life and simulated HRV signals.
(Source: Orthonormal-Basis Partitioning and Time-Frequency Representation of Cardiac Rhythm Dynamics, Aysin, Benhur, et al, IEEE Transactions on Biomedical Engineering, 52, no. 5)

The orthogonal set in the next example is used to construct Fourier approximations of continuous functions. (See Section 5.5.)

EXAMPLE 3   An Orthogonal Set in $C[0, 2\pi]$ (Calculus)

In $C[0, 2\pi]$, with the inner product

$$\langle f, g \rangle = \int_0^{2\pi} f(x)g(x)\, dx$$

show that the set $S = \{1, \sin x, \cos x, \sin 2x, \cos 2x, \ldots, \sin nx, \cos nx\}$ is orthogonal.

SOLUTION

To show that this set is orthogonal, verify the inner products listed below, where m and n are positive integers.

$$\langle 1, \sin nx \rangle = \int_0^{2\pi} \sin nx\, dx = 0$$

$$\langle 1, \cos nx \rangle = \int_0^{2\pi} \cos nx\, dx = 0$$

$$\langle \sin mx, \cos nx \rangle = \int_0^{2\pi} \sin mx \cos nx\, dx = 0$$

$$\langle \sin mx, \sin nx \rangle = \int_0^{2\pi} \sin mx \sin nx\, dx = 0, \quad m \neq n$$

$$\langle \cos mx, \cos nx \rangle = \int_0^{2\pi} \cos mx \cos nx\, dx = 0, \quad m \neq n$$

One of these inner products is verified below, and the others are left to you. If $m \neq n$, then use the formula for rewriting a product of trigonometric functions as a sum to obtain

$$\int_0^{2\pi} \sin mx \cos nx\, dx = \frac{1}{2}\int_0^{2\pi} [\sin(m + n)x + \sin(m - n)x]\, dx = 0.$$

If $m = n$, then

$$\int_0^{2\pi} \sin nx \cos nx\, dx = \frac{1}{2n}\left[\sin^2 nx \right]_0^{2\pi} = 0.$$

Jean-Baptiste Joseph Fourier (1768–1830)

Fourier was born in Auxerre, France. He is credited as a significant contributor to the field of education for scientists, mathematicians, and engineers. His research led to important results pertaining to eigenvalues (Section 7.1), differential equations, and what would later become known as Fourier series (representations of functions using trigonometric series). His work forced mathematicians to reconsider the accepted, but narrow, definition of a function.

The set S in Example 3 is orthogonal but not orthonormal. An orthonormal set can be formed, however, by normalizing each vector in S. That is,

$$\|1\|^2 = \int_0^{2\pi} dx = 2\pi$$

$$\|\sin nx\|^2 = \int_0^{2\pi} \sin^2 nx\, dx = \pi$$

$$\|\cos nx\|^2 = \int_0^{2\pi} \cos^2 nx\, dx = \pi$$

so it follows that the set

$$\left\{ \frac{1}{\sqrt{2\pi}}, \frac{1}{\sqrt{\pi}}\sin x, \frac{1}{\sqrt{\pi}}\cos x, \ldots, \frac{1}{\sqrt{\pi}}\sin nx, \frac{1}{\sqrt{\pi}}\cos nx \right\}$$

is orthonormal.

Each set in Examples 1, 2, and 3 is linearly independent. This is a characteristic of any orthogonal set of nonzero vectors, as stated in the next theorem.

THEOREM 5.10   Orthogonal Sets Are Linearly Independent

If $S = \{\mathbf{v}_1, \mathbf{v}_2, \ldots, \mathbf{v}_n\}$ is an orthogonal set of *nonzero* vectors in an inner product space V, then S is linearly independent.

PROOF

You need to show that the vector equation

$$c_1\mathbf{v}_1 + c_2\mathbf{v}_2 + \cdots + c_n\mathbf{v}_n = \mathbf{0}$$

implies $c_1 = c_2 = \cdots = c_n = 0$. To do this, form the inner product of both sides of the equation with each vector in S. That is, for each i,

$$\langle (c_1\mathbf{v}_1 + c_2\mathbf{v}_2 + \cdots + c_i\mathbf{v}_i + \cdots + c_n\mathbf{v}_n), \mathbf{v}_i \rangle = \langle \mathbf{0}, \mathbf{v}_i \rangle$$

$$c_1\langle \mathbf{v}_1, \mathbf{v}_i \rangle + c_2\langle \mathbf{v}_2, \mathbf{v}_i \rangle + \cdots + c_i\langle \mathbf{v}_i, \mathbf{v}_i \rangle + \cdots + c_n\langle \mathbf{v}_n, \mathbf{v}_i \rangle = 0.$$

Now, S is orthogonal, so $\langle \mathbf{v}_i, \mathbf{v}_j \rangle = 0$ for $j \neq i$, and the equation reduces to

$$c_i\langle \mathbf{v}_i, \mathbf{v}_i \rangle = 0.$$

But each vector in S is nonzero, so you know that

$$\langle \mathbf{v}_i, \mathbf{v}_i \rangle = \|\mathbf{v}_i\|^2 \neq 0.$$

This means that every c_i must be zero and the set must be linearly independent.

As a consequence of Theorems 4.12 and 5.10, you have the corollary below.

THEOREM 5.10   Corollary

If V is an inner product space of dimension n, then any orthogonal set of n nonzero vectors is a basis for V.

EXAMPLE 4   **Using Orthogonality to Test for a Basis**

Show that the set S below is a basis for R^4.

$$S = \{\underset{\mathbf{v}_1}{(2, 3, 2, -2)}, \underset{\mathbf{v}_2}{(1, 0, 0, 1)}, \underset{\mathbf{v}_3}{(-1, 0, 2, 1)}, \underset{\mathbf{v}_4}{(-1, 2, -1, 1)}\}$$

SOLUTION

The set S has four nonzero vectors. By the corollary to Theorem 5.10, you can show that S is a basis for R^4 by showing that it is an orthogonal set.

$$\mathbf{v}_1 \cdot \mathbf{v}_2 = 2 + 0 + 0 - 2 = 0$$
$$\mathbf{v}_1 \cdot \mathbf{v}_3 = -2 + 0 + 4 - 2 = 0$$
$$\mathbf{v}_1 \cdot \mathbf{v}_4 = -2 + 6 - 2 - 2 = 0$$
$$\mathbf{v}_2 \cdot \mathbf{v}_3 = -1 + 0 + 0 + 1 = 0$$
$$\mathbf{v}_2 \cdot \mathbf{v}_4 = -1 + 0 + 0 + 1 = 0$$
$$\mathbf{v}_3 \cdot \mathbf{v}_4 = 1 + 0 - 2 + 1 = 0$$

S is orthogonal, and by the corollary to Theorem 5.10, it is a basis for R^4.

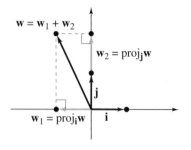

$$\mathbf{w} = \mathbf{w}_1 + \mathbf{w}_2 = c_1\mathbf{i} + c_2\mathbf{j}$$

Figure 5.12

Section 4.7 discusses a technique for finding a coordinate representation relative to a nonstandard basis. When the basis is *orthonormal*, this procedure can be streamlined.

Before looking at this procedure, consider an example in R^2. Figure 5.12 shows an orthonormal basis for R^2, $\mathbf{i} = (1, 0)$ and $\mathbf{j} = (0, 1)$. Any vector $\mathbf{w}$ in R^2 can be represented as $\mathbf{w} = \mathbf{w}_1 + \mathbf{w}_2$, where $\mathbf{w}_1 = \text{proj}_\mathbf{i}\mathbf{w}$ and $\mathbf{w}_2 = \text{proj}_\mathbf{j}\mathbf{w}$. The vectors $\mathbf{i}$ and $\mathbf{j}$ are unit vectors, so it follows that $\mathbf{w}_1 = (\mathbf{w} \cdot \mathbf{i})\mathbf{i}$ and $\mathbf{w}_2 = (\mathbf{w} \cdot \mathbf{j})\mathbf{j}$. Consequently,

$$\mathbf{w} = \mathbf{w}_1 + \mathbf{w}_2 = (\mathbf{w} \cdot \mathbf{i})\mathbf{i} + (\mathbf{w} \cdot \mathbf{j})\mathbf{j} = c_1\mathbf{i} + c_2\mathbf{j}$$

which shows that the coefficients c_1 and c_2 are simply the dot products of $\mathbf{w}$ with the respective basis vectors. The next theorem generalizes this.

THEOREM 5.11   Coordinates Relative to an Orthonormal Basis

If $B = \{\mathbf{v}_1, \mathbf{v}_2, \ldots, \mathbf{v}_n\}$ is an orthonormal basis for an inner product space V, then the coordinate representation of a vector $\mathbf{w}$ relative to B is

$$\mathbf{w} = \langle \mathbf{w}, \mathbf{v}_1 \rangle \mathbf{v}_1 + \langle \mathbf{w}, \mathbf{v}_2 \rangle \mathbf{v}_2 + \cdots + \langle \mathbf{w}, \mathbf{v}_n \rangle \mathbf{v}_n.$$

PROOF

B is a basis for V, so there must exist unique scalars $c_1, c_2, \ldots, c_n$ such that

$$\mathbf{w} = c_1\mathbf{v}_1 + c_2\mathbf{v}_2 + \cdots + c_n\mathbf{v}_n.$$

Taking the inner product (with $\mathbf{v}_i$) of both sides of this equation, you have

$$\langle \mathbf{w}, \mathbf{v}_i \rangle = \langle (c_1\mathbf{v}_1 + c_2\mathbf{v}_2 + \cdots + c_n\mathbf{v}_n), \mathbf{v}_i \rangle$$
$$= c_1\langle \mathbf{v}_1, \mathbf{v}_i \rangle + c_2\langle \mathbf{v}_2, \mathbf{v}_i \rangle + \cdots + c_n\langle \mathbf{v}_n, \mathbf{v}_i \rangle$$

and by the orthogonality of B, this equation reduces to

$$\langle \mathbf{w}, \mathbf{v}_i \rangle = c_i\langle \mathbf{v}_i, \mathbf{v}_i \rangle.$$

B is orthonormal, so you have $\langle \mathbf{v}_i, \mathbf{v}_i \rangle = \|\mathbf{v}_i\|^2 = 1$, and it follows that $\langle \mathbf{w}, \mathbf{v}_i \rangle = c_i$. ■

In Theorem 5.11, the coordinates of $\mathbf{w}$ relative to the *orthonormal* basis B are called the **Fourier coefficients** of $\mathbf{w}$ relative to B, after Jean-Baptiste Joseph Fourier. The corresponding coordinate matrix of $\mathbf{w}$ relative to B is

$$[\mathbf{w}]_B = [c_1 \ c_2 \ldots c_n]^T = [\langle \mathbf{w}, \mathbf{v}_1 \rangle \ \langle \mathbf{w}, \mathbf{v}_2 \rangle \ \ldots \ \langle \mathbf{w}, \mathbf{v}_n \rangle]^T.$$

EXAMPLE 5   **Representing Vectors Relative to an Orthonormal Basis**

Find the coordinate matrix of $\mathbf{w} = (5, -5, 2)$ relative to the orthonormal basis B for R^3 below.

$$B = \left\{ \begin{matrix} \mathbf{v}_1 \\ \left(\frac{3}{5}, \frac{4}{5}, 0\right), \end{matrix} \begin{matrix} \mathbf{v}_2 \\ \left(-\frac{4}{5}, \frac{3}{5}, 0\right), \end{matrix} \begin{matrix} \mathbf{v}_3 \\ (0, 0, 1) \end{matrix} \right\}$$

SOLUTION

B is orthonormal (verify this), so use Theorem 5.11 to find the coordinates.

$$\mathbf{w} \cdot \mathbf{v}_1 = (5, -5, 2) \cdot \left(\tfrac{3}{5}, \tfrac{4}{5}, 0\right) = -1$$
$$\mathbf{w} \cdot \mathbf{v}_2 = (5, -5, 2) \cdot \left(-\tfrac{4}{5}, \tfrac{3}{5}, 0\right) = -7$$
$$\mathbf{w} \cdot \mathbf{v}_3 = (5, -5, 2) \cdot (0, 0, 1) = 2$$

So, the coordinate matrix relative to B is $[\mathbf{w}]_B = [-1 \ \ -7 \ \ 2]^T$. ■

GRAM-SCHMIDT ORTHONORMALIZATION PROCESS

Having seen one of the advantages of orthonormal bases (the straightforwardness of coordinate representation), you will now look at a procedure for finding such a basis. This procedure is called the **Gram-Schmidt orthonormalization process,** after the Danish mathematician Jorgen Pederson Gram (1850–1916) and the German mathematician Erhardt Schmidt (1876–1959). It has three steps.

1. Begin with a basis for the inner product space. It need not be orthogonal nor consist of unit vectors.

2. Convert the basis to an orthogonal basis.

3. Normalize each vector in the orthogonal basis to form an orthonormal basis.

THEOREM 5.12 Gram-Schmidt Orthonormalization Process

1. Let $B = \{\mathbf{v}_1, \mathbf{v}_2, \ldots, \mathbf{v}_n\}$ be a basis for an inner product space V.
2. Let $B' = \{\mathbf{w}_1, \mathbf{w}_2, \ldots, \mathbf{w}_n\}$, where

$$\mathbf{w}_1 = \mathbf{v}_1$$

$$\mathbf{w}_2 = \mathbf{v}_2 - \frac{\langle \mathbf{v}_2, \mathbf{w}_1 \rangle}{\langle \mathbf{w}_1, \mathbf{w}_1 \rangle} \mathbf{w}_1$$

$$\mathbf{w}_3 = \mathbf{v}_3 - \frac{\langle \mathbf{v}_3, \mathbf{w}_1 \rangle}{\langle \mathbf{w}_1, \mathbf{w}_1 \rangle} \mathbf{w}_1 - \frac{\langle \mathbf{v}_3, \mathbf{w}_2 \rangle}{\langle \mathbf{w}_2, \mathbf{w}_2 \rangle} \mathbf{w}_2$$

$$\vdots$$

$$\mathbf{w}_n = \mathbf{v}_n - \frac{\langle \mathbf{v}_n, \mathbf{w}_1 \rangle}{\langle \mathbf{w}_1, \mathbf{w}_1 \rangle} \mathbf{w}_1 - \frac{\langle \mathbf{v}_n, \mathbf{w}_2 \rangle}{\langle \mathbf{w}_2, \mathbf{w}_2 \rangle} \mathbf{w}_2 - \cdots - \frac{\langle \mathbf{v}_n, \mathbf{w}_{n-1} \rangle}{\langle \mathbf{w}_{n-1}, \mathbf{w}_{n-1} \rangle} \mathbf{w}_{n-1}.$$

Then B' is an *orthogonal* basis for V.

3. Let $\mathbf{u}_i = \dfrac{\mathbf{w}_i}{\|\mathbf{w}_i\|}$. Then $B'' = \{\mathbf{u}_1, \mathbf{u}_2, \ldots, \mathbf{u}_n\}$ is an *orthonormal* basis for V. Also, $\mathrm{span}\{\mathbf{v}_1, \mathbf{v}_2, \ldots, \mathbf{v}_k\} = \mathrm{span}\{\mathbf{u}_1, \mathbf{u}_2, \ldots, \mathbf{u}_k\}$ for $k = 1, 2, \ldots, n$.

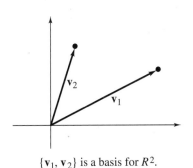

$\{\mathbf{v}_1, \mathbf{v}_2\}$ is a basis for R^2.

Figure 5.13

Rather than give a proof of this theorem, it is more instructive to discuss a special case for which you can use a geometric model. Let $\{\mathbf{v}_1, \mathbf{v}_2\}$ be a basis for R^2, as shown in Figure 5.13. To determine an orthogonal basis for R^2, first choose one of the original vectors, say $\mathbf{v}_1$, and call it $\mathbf{w}_1$. Now you want to find a second vector orthogonal to $\mathbf{w}_1$. The figure below shows that $\mathbf{v}_2 - \mathrm{proj}_{\mathbf{v}_1}\mathbf{v}_2$ has this property.

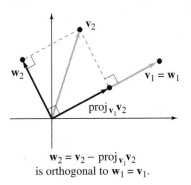

$\mathbf{w}_2 = \mathbf{v}_2 - \mathrm{proj}_{\mathbf{v}_1}\mathbf{v}_2$
is orthogonal to $\mathbf{w}_1 = \mathbf{v}_1$.

By letting $\mathbf{w}_1 = \mathbf{v}_1$ and $\mathbf{w}_2 = \mathbf{v}_2 - \mathrm{proj}_{\mathbf{v}_1}\mathbf{v}_2 = \mathbf{v}_2 - \dfrac{\mathbf{v}_2 \cdot \mathbf{w}_1}{\mathbf{w}_1 \cdot \mathbf{w}_1} \mathbf{w}_1$, you can conclude that the set $\{\mathbf{w}_1, \mathbf{w}_2\}$ is orthogonal. By the corollary to Theorem 5.10, it is a basis for R^2. Finally, by normalizing $\mathbf{w}_1$ and $\mathbf{w}_2$, you obtain the orthonormal basis for R^2 below.

$$\{\mathbf{u}_1, \mathbf{u}_2\} = \left\{ \frac{\mathbf{w}_1}{\|\mathbf{w}_1\|}, \frac{\mathbf{w}_2}{\|\mathbf{w}_2\|} \right\}$$

REMARK

An orthonormal set derived by the Gram-Schmidt orthonormalization process depends on the order of the vectors in the basis. For instance, rework Example 6 with the original basis ordered as $\{\mathbf{v}_2, \mathbf{v}_1\}$ rather than $\{\mathbf{v}_1, \mathbf{v}_2\}$.

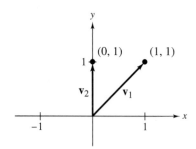

Given basis: $B = \{\mathbf{v}_1, \mathbf{v}_2\}$

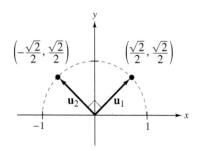

Orthonormal basis: $B'' = \{\mathbf{u}_1, \mathbf{u}_2\}$

Figure 5.14

EXAMPLE 6   Applying the Gram-Schmidt Orthonormalization Process

Apply the Gram-Schmidt orthonormalization process to the basis B for R^2 below.

$$\overset{\mathbf{v}_1}{} \quad \overset{\mathbf{v}_2}{}$$
$$B = \{(1, 1), (0, 1)\}$$

SOLUTION

The Gram-Schmidt orthonormalization process produces

$$\mathbf{w}_1 = \mathbf{v}_1 = (1, 1)$$
$$\mathbf{w}_2 = \mathbf{v}_2 - \frac{\mathbf{v}_2 \cdot \mathbf{w}_1}{\mathbf{w}_1 \cdot \mathbf{w}_1}\mathbf{w}_1 = (0, 1) - \frac{1}{2}(1, 1) = \left(-\frac{1}{2}, \frac{1}{2}\right).$$

The set $B' = \{\mathbf{w}_1, \mathbf{w}_2\}$ is an orthogonal basis for R^2. By normalizing each vector in B', you obtain

$$\mathbf{u}_1 = \frac{\mathbf{w}_1}{\|\mathbf{w}_1\|} = \frac{1}{\sqrt{2}}(1, 1) = \left(\frac{\sqrt{2}}{2}, \frac{\sqrt{2}}{2}\right)$$
$$\mathbf{u}_2 = \frac{\mathbf{w}_2}{\|\mathbf{w}_2\|} = \frac{1}{1/\sqrt{2}}\left(-\frac{1}{2}, \frac{1}{2}\right) = \sqrt{2}\left(-\frac{1}{2}, \frac{1}{2}\right) = \left(-\frac{\sqrt{2}}{2}, \frac{\sqrt{2}}{2}\right).$$

So, $B'' = \{\mathbf{u}_1, \mathbf{u}_2\}$ is an orthonormal basis for R^2. See Figure 5.14.

EXAMPLE 7   Applying the Gram-Schmidt Orthonormalization Process

Apply the Gram-Schmidt orthonormalization process to the basis B for R^3 below.

$$\overset{\mathbf{v}_1}{} \quad \overset{\mathbf{v}_2}{} \quad \overset{\mathbf{v}_3}{}$$
$$B = \{(1, 1, 0), (1, 2, 0), (0, 1, 2)\}$$

SOLUTION

Applying the Gram-Schmidt orthonormalization process produces

$$\mathbf{w}_1 = \mathbf{v}_1 = (1, 1, 0)$$
$$\mathbf{w}_2 = \mathbf{v}_2 - \frac{\mathbf{v}_2 \cdot \mathbf{w}_1}{\mathbf{w}_1 \cdot \mathbf{w}_1}\mathbf{w}_1 = (1, 2, 0) - \frac{3}{2}(1, 1, 0) = \left(-\frac{1}{2}, \frac{1}{2}, 0\right)$$
$$\mathbf{w}_3 = \mathbf{v}_3 - \frac{\mathbf{v}_3 \cdot \mathbf{w}_1}{\mathbf{w}_1 \cdot \mathbf{w}_1}\mathbf{w}_1 - \frac{\mathbf{v}_3 \cdot \mathbf{w}_2}{\mathbf{w}_2 \cdot \mathbf{w}_2}\mathbf{w}_2$$
$$= (0, 1, 2) - \frac{1}{2}(1, 1, 0) - \frac{1/2}{1/2}\left(-\frac{1}{2}, \frac{1}{2}, 0\right)$$
$$= (0, 0, 2).$$

The set $B' = \{\mathbf{w}_1, \mathbf{w}_2, \mathbf{w}_3\}$ is an orthogonal basis for R^3. Normalizing each vector in B' produces

$$\mathbf{u}_1 = \frac{\mathbf{w}_1}{\|\mathbf{w}_1\|} = \frac{1}{\sqrt{2}}(1, 1, 0) = \left(\frac{\sqrt{2}}{2}, \frac{\sqrt{2}}{2}, 0\right)$$
$$\mathbf{u}_2 = \frac{\mathbf{w}_2}{\|\mathbf{w}_2\|} = \frac{1}{1/\sqrt{2}}\left(-\frac{1}{2}, \frac{1}{2}, 0\right) = \left(-\frac{\sqrt{2}}{2}, \frac{\sqrt{2}}{2}, 0\right)$$
$$\mathbf{u}_3 = \frac{\mathbf{w}_3}{\|\mathbf{w}_3\|} = \frac{1}{2}(0, 0, 2) = (0, 0, 1).$$

So, $B'' = \{\mathbf{u}_1, \mathbf{u}_2, \mathbf{u}_3\}$ is an orthonormal basis for R^3.

Examples 6 and 7 apply the Gram-Schmidt orthonormalization process to bases for R^2 and R^3. The process works equally well for a subspace of an inner product space. The next example demonstrates.

EXAMPLE 8 Applying the Gram-Schmidt Orthonormalization Process

See LarsonLinearAlgebra.com for an interactive version of this type of example.

The vectors

$$\mathbf{v}_1 = (0, 1, 0) \quad \text{and} \quad \mathbf{v}_2 = (1, 1, 1)$$

span a plane in R^3. Find an orthonormal basis for this subspace.

SOLUTION

Applying the Gram-Schmidt orthonormalization process produces

$$\mathbf{w}_1 = \mathbf{v}_1 = (0, 1, 0)$$

$$\mathbf{w}_2 = \mathbf{v}_2 - \frac{\mathbf{v}_2 \cdot \mathbf{w}_1}{\mathbf{w}_1 \cdot \mathbf{w}_1}\mathbf{w}_1 = (1, 1, 1) - \frac{1}{1}(0, 1, 0) = (1, 0, 1)$$

Normalizing $\mathbf{w}_1$ and $\mathbf{w}_2$ produces the orthonormal set

$$\mathbf{u}_1 = \frac{\mathbf{w}_1}{\|\mathbf{w}_1\|} = (0, 1, 0)$$

$$\mathbf{u}_2 = \frac{\mathbf{w}_2}{\|\mathbf{w}_2\|} = \frac{1}{\sqrt{2}}(1, 0, 1) = \left(\frac{\sqrt{2}}{2}, 0, \frac{\sqrt{2}}{2}\right).$$

See Figure 5.15.

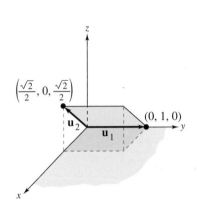

Figure 5.15

EXAMPLE 9 Applying the Gram-Schmidt Orthonormalization Process (Calculus)

Apply the Gram-Schmidt orthonormalization process to the basis $B = \{1, x, x^2\}$ in P_2, using the inner product

$$\langle p, q \rangle = \int_{-1}^{1} p(x)q(x)\, dx.$$

SOLUTION

Let $B = \{1, x, x^2\} = \{\mathbf{v}_1, \mathbf{v}_2, \mathbf{v}_3\}$. Then you have

$$\mathbf{w}_1 = \mathbf{v}_1 = 1$$

$$\mathbf{w}_2 = \mathbf{v}_2 - \frac{\langle \mathbf{v}_2, \mathbf{w}_1 \rangle}{\langle \mathbf{w}_1, \mathbf{w}_1 \rangle}\mathbf{w}_1 = x - \frac{0}{2}(1) = x$$

$$\mathbf{w}_3 = \mathbf{v}_3 - \frac{\langle \mathbf{v}_3, \mathbf{w}_1 \rangle}{\langle \mathbf{w}_1, \mathbf{w}_1 \rangle}\mathbf{w}_1 - \frac{\langle \mathbf{v}_3, \mathbf{w}_2 \rangle}{\langle \mathbf{w}_2, \mathbf{w}_2 \rangle}\mathbf{w}_2 = x^2 - \frac{2/3}{2}(1) - \frac{0}{2/3}(x) = x^2 - \frac{1}{3}.$$

Now, by normalizing $B' = \{\mathbf{w}_1, \mathbf{w}_2, \mathbf{w}_3\}$, you have

$$\mathbf{u}_1 = \frac{\mathbf{w}_1}{\|\mathbf{w}_1\|} = \frac{1}{\sqrt{2}}(1) = \frac{1}{\sqrt{2}}$$

$$\mathbf{u}_2 = \frac{\mathbf{w}_2}{\|\mathbf{w}_2\|} = \frac{1}{\sqrt{2/3}}(x) = \frac{\sqrt{3}}{\sqrt{2}}x$$

$$\mathbf{u}_3 = \frac{\mathbf{w}_3}{\|\mathbf{w}_3\|} = \frac{1}{\sqrt{8/45}}\left(x^2 - \frac{1}{3}\right) = \frac{\sqrt{5}}{2\sqrt{2}}(3x^2 - 1).$$

In Exercises 43–48, you are asked to verify these calculations.

REMARK

The polynomials $\mathbf{u}_1$, $\mathbf{u}_2$, and $\mathbf{u}_3$ in Example 9 are called the first three **normalized Legendre polynomials,** after the French mathematician Adrien-Marie Legendre (1752–1833).

The computations in the Gram-Schmidt orthonormalization process are sometimes simpler when you normalize each vector $\mathbf{w}_i$ *before* you use it to determine the next vector. This **alternative form of the Gram-Schmidt orthonormalization process** has the steps listed below.

$$\mathbf{u}_1 = \frac{\mathbf{w}_1}{\|\mathbf{w}_1\|} = \frac{\mathbf{v}_1}{\|\mathbf{v}_1\|}$$

$$\mathbf{u}_2 = \frac{\mathbf{w}_2}{\|\mathbf{w}_2\|}, \text{ where } \mathbf{w}_2 = \mathbf{v}_2 - \langle \mathbf{v}_2, \mathbf{u}_1 \rangle \mathbf{u}_1$$

$$\mathbf{u}_3 = \frac{\mathbf{w}_3}{\|\mathbf{w}_3\|}, \text{ where } \mathbf{w}_3 = \mathbf{v}_3 - \langle \mathbf{v}_3, \mathbf{u}_1 \rangle \mathbf{u}_1 - \langle \mathbf{v}_3, \mathbf{u}_2 \rangle \mathbf{u}_2$$

$$\vdots$$

$$\mathbf{u}_n = \frac{\mathbf{w}_n}{\|\mathbf{w}_n\|}, \text{ where } \mathbf{w}_n = \mathbf{v}_n - \langle \mathbf{v}_n, \mathbf{u}_1 \rangle \mathbf{u}_1 - \cdots - \langle \mathbf{v}_n, \mathbf{u}_{n-1} \rangle \mathbf{u}_{n-1}$$

EXAMPLE 10 **Alternative Form of the Gram-Schmidt Orthonormalization Process**

Find an orthonormal basis for the solution space of the homogeneous linear system.

$$x_1 + x_2 \qquad\quad + 7x_4 = 0$$
$$2x_1 + x_2 + 2x_3 + 6x_4 = 0$$

SOLUTION

The augmented matrix for this system reduces as shown below.

$$\begin{bmatrix} 1 & 1 & 0 & 7 & 0 \\ 2 & 1 & 2 & 6 & 0 \end{bmatrix} \rightarrow \begin{bmatrix} 1 & 0 & 2 & -1 & 0 \\ 0 & 1 & -2 & 8 & 0 \end{bmatrix}$$

If you let $x_3 = s$ and $x_4 = t$, then each solution of the system has the form

$$\begin{bmatrix} x_1 \\ x_2 \\ x_3 \\ x_4 \end{bmatrix} = \begin{bmatrix} -2s + t \\ 2s - 8t \\ s \\ t \end{bmatrix} = s \begin{bmatrix} -2 \\ 2 \\ 1 \\ 0 \end{bmatrix} + t \begin{bmatrix} 1 \\ -8 \\ 0 \\ 1 \end{bmatrix}.$$

So, one basis for the solution space is

$$B = \{\mathbf{v}_1, \mathbf{v}_2\} = \{(-2, 2, 1, 0), (1, -8, 0, 1)\}.$$

To find an orthonormal basis $B' = \{\mathbf{u}_1, \mathbf{u}_2\}$, use the alternative form of the Gram-Schmidt orthonormalization process, as shown below.

$$\mathbf{u}_1 = \frac{\mathbf{v}_1}{\|\mathbf{v}_1\|}$$
$$= \left(-\frac{2}{3}, \frac{2}{3}, \frac{1}{3}, 0\right)$$

$$\mathbf{w}_2 = \mathbf{v}_2 - \langle \mathbf{v}_2, \mathbf{u}_1 \rangle \mathbf{u}_1$$
$$= (1, -8, 0, 1) - \left[(1, -8, 0, 1) \cdot \left(-\frac{2}{3}, \frac{2}{3}, \frac{1}{3}, 0\right)\right]\left(-\frac{2}{3}, \frac{2}{3}, \frac{1}{3}, 0\right)$$
$$= (-3, -4, 2, 1)$$

$$\mathbf{u}_2 = \frac{\mathbf{w}_2}{\|\mathbf{w}_2\|}$$
$$= \left(-\frac{3}{\sqrt{30}}, -\frac{4}{\sqrt{30}}, \frac{2}{\sqrt{30}}, \frac{1}{\sqrt{30}}\right)$$

5.3 Exercises

See CalcChat.com for worked-out solutions to odd-numbered exercises.

Orthogonal and Orthonormal Sets In Exercises 1–12, (a) determine whether the set of vectors in R^n is orthogonal, (b) if the set is orthogonal, then determine whether it is also orthonormal, and (c) determine whether the set is a basis for R^n.

1. $\{(2, -4), (2, 1)\}$ **2.** $\{(-3, 5), (4, 0)\}$

3. $\left\{\left(\frac{3}{5}, \frac{4}{5}\right), \left(-\frac{4}{5}, \frac{3}{5}\right)\right\}$ **4.** $\left\{(2, 1), \left(\frac{1}{3}, -\frac{2}{3}\right)\right\}$

5. $\{(4, -1, 1), (-1, 0, 4), (-4, -17, -1)\}$

6. $\{(2, -4, 2), (0, 2, 4), (-10, -4, 2)\}$

7. $\left\{\left(\frac{\sqrt{2}}{3}, 0, -\frac{\sqrt{2}}{6}\right), \left(0, \frac{2\sqrt{5}}{5}, -\frac{\sqrt{5}}{5}\right), \left(\frac{\sqrt{5}}{5}, 0, \frac{1}{2}\right)\right\}$

8. $\left\{\left(\frac{\sqrt{2}}{2}, 0, \frac{\sqrt{2}}{2}\right), \left(-\frac{\sqrt{6}}{6}, \frac{\sqrt{6}}{3}, \frac{\sqrt{6}}{6}\right), \left(\frac{\sqrt{3}}{3}, \frac{\sqrt{3}}{3}, -\frac{\sqrt{3}}{3}\right)\right\}$

9. $\{(2, 5, -3), (4, 2, 6)\}$

10. $\{(-6, 3, 2, 1), (2, 0, 6, 0)\}$

11. $\left\{\left(\frac{\sqrt{2}}{2}, 0, 0, \frac{\sqrt{2}}{2}\right), \left(0, \frac{\sqrt{2}}{2}, \frac{\sqrt{2}}{2}, 0\right), \left(-\frac{1}{2}, \frac{1}{2}, -\frac{1}{2}, \frac{1}{2}\right)\right\}$

12. $\left\{\left(\frac{\sqrt{10}}{10}, 0, 0, \frac{3\sqrt{10}}{10}\right), (0, 0, 1, 0), (0, 1, 0, 0),\right.$ $\left.\left(-\frac{3\sqrt{10}}{10}, 0, 0, \frac{\sqrt{10}}{10}\right)\right\}$

Normalizing an Orthogonal Set In Exercises 13–16, (a) show that the set of vectors in R^n is orthogonal, and (b) normalize the set to produce an orthonormal set.

13. $\{(-1, 3), (12, 4)\}$ **14.** $\{(2, -5), (10, 4)\}$

15. $\left\{\left(\sqrt{3}, \sqrt{3}, \sqrt{3}\right), \left(-\sqrt{2}, 0, \sqrt{2}\right)\right\}$

16. $\left\{\left(\frac{6}{13}, -\frac{2}{13}, \frac{3}{13}\right), \left(\frac{2}{13}, \frac{6}{13}, 0\right)\right\}$

17. Complete Example 2 by verifying that $\{1, x, x^2, x^3\}$ is an orthonormal basis for P_3 with the inner product $\langle p, q \rangle = a_0 b_0 + a_1 b_1 + a_2 b_2 + a_3 b_3$.

18. Verify that $\{(\sin\theta, \cos\theta), (\cos\theta, -\sin\theta)\}$ is an orthonormal basis for R^2.

Finding a Coordinate Matrix In Exercises 19–24, find the coordinate matrix of w relative to the orthonormal basis B in R^n.

19. $\mathbf{w} = (1, 2)$, $B = \left\{\left(-\frac{2\sqrt{13}}{13}, \frac{3\sqrt{13}}{13}\right), \left(\frac{3\sqrt{13}}{13}, \frac{2\sqrt{13}}{13}\right)\right\}$

20. $\mathbf{w} = (4, -3)$, $B = \left\{\left(\frac{\sqrt{3}}{3}, \frac{\sqrt{6}}{3}\right), \left(-\frac{\sqrt{6}}{3}, \frac{\sqrt{3}}{3}\right)\right\}$

21. $\mathbf{w} = (2, -2, 1)$, $B = \left\{\left(\frac{\sqrt{10}}{10}, 0, \frac{3\sqrt{10}}{10}\right), (0, 1, 0), \left(-\frac{3\sqrt{10}}{10}, 0, \frac{\sqrt{10}}{10}\right)\right\}$

22. $\mathbf{w} = (3, -5, 11)$, $B = \{(1, 0, 0), (0, 1, 0), (0, 0, 1)\}$

23. $\mathbf{w} = (5, 10, 15)$, $B = \left\{\left(\frac{3}{5}, \frac{4}{5}, 0\right), \left(-\frac{4}{5}, \frac{3}{5}, 0\right), (0, 0, 1)\right\}$

24. $\mathbf{w} = (2, -1, 4, 3)$, $B = \left\{\left(\frac{5}{13}, 0, \frac{12}{13}, 0\right), (0, 1, 0, 0), \left(-\frac{12}{13}, 0, \frac{5}{13}, 0\right), (0, 0, 0, 1)\right\}$

Applying the Gram-Schmidt Process In Exercises 25–34, apply the Gram-Schmidt orthonormalization process to transform the given basis for R^n into an orthonormal basis. Use the vectors in the order in which they are given.

25. $B = \{(3, 4), (1, 0)\}$ **26.** $B = \{(-1, 2), (1, 0)\}$

27. $B = \{(0, 1), (2, 5)\}$ **28.** $B = \{(4, -3), (3, 2)\}$

29. $B = \{(2, 1, -2), (1, 2, 2), (2, -2, 1)\}$

30. $B = \{(1, 0, 0), (1, 1, 1), (1, 1, -1)\}$

31. $B = \{(4, -3, 0), (1, 2, 0), (0, 0, 4)\}$

32. $B = \{(0, 1, 2), (2, 0, 0), (1, 1, 1)\}$

33. $B = \{(0, 1, 1), (1, 1, 0), (1, 0, 1)\}$

34. $B = \{(3, 4, 0, 0), (-1, 1, 0, 0), (2, 1, 0, -1), (0, 1, 1, 0)\}$

Applying the Gram-Schmidt Process In Exercises 35–40, apply the Gram-Schmidt orthonormalization process to transform the given basis for a subspace of R^n into an orthonormal basis for the subspace. Use the vectors in the order in which they are given.

35. $B = \{(-8, 3, 5)\}$ **36.** $B = \{(2, -9, 6)\}$

37. $B = \{(3, 4, 0), (2, 0, 0)\}$

38. $B = \{(1, 3, 0), (3, 0, -3)\}$

39. $B = \{(1, 2, -1, 0), (2, 2, 0, 1), (1, 1, -1, 0)\}$

40. $B = \{(7, 24, 0, 0), (0, 0, 1, 1), (0, 0, 1, -2)\}$

41. Use the inner product $\langle \mathbf{u}, \mathbf{v} \rangle = 2u_1 v_1 + u_2 v_2$ in R^2 and the Gram-Schmidt orthonormalization process to transform $\{(2, -1), (-2, 10)\}$ into an orthonormal basis.

42. **Writing** Explain why the result of Exercise 41 is not an orthonormal basis when you use the Euclidean inner product on R^2.

Calculus In Exercises 43–48, let $B = \{1, x, x^2\}$ be a basis for P_2 with the inner product

$$\langle p, q \rangle = \int_{-1}^{1} p(x)q(x)\,dx.$$

Complete Example 9 by verifying the inner products.

43. $\langle x, 1 \rangle = 0$ **44.** $\langle 1, 1 \rangle = 2$

45. $\langle x^2, 1 \rangle = \frac{2}{3}$ **46.** $\langle x^2, x \rangle = 0$

47. $\langle x, x \rangle = \frac{2}{3}$ **48.** $\left\langle x^2 - \frac{1}{3}, x^2 - \frac{1}{3} \right\rangle = \frac{8}{45}$

Applying the Alternative Form of the Gram-Schmidt Process In Exercises 49–54, apply the alternative form of the Gram-Schmidt orthonormalization process to find an orthonormal basis for the solution space of the homogeneous linear system.

49. $x_1 - 2x_2 + x_3 = 0$ **50.** $x_1 + 3x_2 - 3x_3 = 0$

51. $x_1 - x_2 + x_3 + x_4 = 0$
$x_1 - 2x_2 + x_3 + x_4 = 0$

52. $x_1 + x_2 - x_3 - x_4 = 0$
$2x_1 + x_2 - 2x_3 - 2x_4 = 0$

53. $2x_1 + x_2 - 6x_3 + 2x_4 = 0$
$x_1 + 2x_2 - 3x_3 + 4x_4 = 0$
$x_1 + x_2 - 3x_3 + 2x_4 = 0$

54. $-x_1 + x_2 - x_3 + x_4 - x_5 = 0$
$2x_1 - x_2 + 2x_3 - x_4 + 2x_5 = 0$

True or False? In Exercises 55 and 56, determine whether each statement is true or false. If a statement is true, give a reason or cite an appropriate statement from the text. If a statement is false, provide an example that shows the statement is not true in all cases or cite an appropriate statement from the text.

55. (a) A set S of vectors in an inner product space V is orthogonal when every pair of vectors in S is orthogonal.

(b) An orthonormal basis derived by the Gram-Schmidt orthonormalization process does not depend on the order of the vectors in the basis.

56. (a) A set S of vectors in an inner product space V is orthonormal when every vector is a unit vector and each pair of vectors is orthogonal.

(b) If a set of nonzero vectors S in an inner product space V is orthogonal, then S is linearly independent.

Orthonormal Sets in P_2 In Exercises 57–62, let $p(x) = a_0 + a_1x + a_2x^2$ and $q(x) = b_0 + b_1x + b_2x^2$ be vectors in P_2 with $\langle p, q \rangle = a_0b_0 + a_1b_1 + a_2b_2$. Determine whether the polynomials form an orthonormal set, and if not, apply the Gram-Schmidt orthonormalization process to form an orthonormal set.

57. $\{1, x, x^2\}$ **58.** $\{x^2, 2x + x^2, 1 + 2x + x^2\}$

59. $\{-1 + x^2, -1 + x\}$ **60.** $\left\{\dfrac{5x + 12x^2}{13}, \dfrac{12x - 5x^2}{13}, 1\right\}$

61. $\left\{\dfrac{1 + x^2}{\sqrt{2}}, \dfrac{-1 + x + x^2}{\sqrt{3}}\right\}$

62. $\left\{\sqrt{2}(-1 + x^2), \sqrt{2}(2 + x + x^2)\right\}$

63. Proof Let $\{\mathbf{u}_1, \mathbf{u}_2, \ldots, \mathbf{u}_n\}$ be an orthonormal basis for R^n. Prove that
$$\|\mathbf{v}\|^2 = |\mathbf{v} \cdot \mathbf{u}_1|^2 + |\mathbf{v} \cdot \mathbf{u}_2|^2 + \cdots + |\mathbf{v} \cdot \mathbf{u}_n|^2$$
for any vector $\mathbf{v}$ in R^n. This equation is **Parseval's equality.**

64. Guided Proof Prove that if $\mathbf{w}$ is orthogonal to each vector in $S = \{\mathbf{v}_1, \mathbf{v}_2, \ldots, \mathbf{v}_n\}$, then $\mathbf{w}$ is orthogonal to every linear combination of vectors in S.

Getting Started: To prove that $\mathbf{w}$ is orthogonal to every linear combination of vectors in S, you need to show that their inner product is 0.

(i) Write $\mathbf{v}$ as a linear combination of vectors, with arbitrary scalars $c_1, \ldots, c_n$, in S.

(ii) Form the inner product of $\mathbf{w}$ and $\mathbf{v}$.

(iii) Use the properties of inner products to rewrite the inner product $\langle \mathbf{w}, \mathbf{v} \rangle$ as a linear combination of the inner products $\langle \mathbf{w}, \mathbf{v}_i \rangle$, $i = 1, \ldots, n$.

(iv) Use the fact that $\mathbf{w}$ is orthogonal to each vector in S to lead to the conclusion that $\mathbf{w}$ is orthogonal to $\mathbf{v}$.

65. Proof Let P be an $n \times n$ matrix. Prove that the three conditions are equivalent.

(a) $P^{-1} = P^T$. (Such a matrix is *orthogonal*.)

(b) The row vectors of P form an orthonormal basis for R^n.

(c) The column vectors of P form an orthonormal basis for R^n.

66. Proof Let W be a subspace of R^n. Prove that the intersection of W and $W^\perp$ is $\{\mathbf{0}\}$, where $W^\perp$ is the subspace of R^n given by
$$W^\perp = \{\mathbf{v}: \mathbf{w} \cdot \mathbf{v} = 0 \text{ for every } \mathbf{w} \text{ in } W\}.$$

Fundamental Subspaces In Exercises 67 and 68, find bases for the four fundamental subspaces of the matrix A listed below.

$N(A) =$ **nullspace of A** $N(A^T) =$ **nullspace of A^T**
$R(A) =$ **column space of A** $R(A^T) =$ **column space of A^T**

Then show that $N(A) = R(A^T)^\perp$ and $N(A^T) = R(A)^\perp$.

67. $\begin{bmatrix} 1 & 1 & -1 \\ 0 & 2 & 1 \\ 1 & 3 & 0 \end{bmatrix}$ **68.** $\begin{bmatrix} 0 & 1 & -1 \\ 0 & -2 & 2 \\ 0 & -1 & 1 \end{bmatrix}$

69. Let A be an $m \times n$ matrix and let $N(A)$, $N(A^T)$, $R(A)$, and $R(A^T)$ be the subspaces in Exercises 67 and 68.

(a) Explain why $R(A^T)$ is the same as the row space of A.

(b) Prove that $N(A) \subset R(A^T)^\perp$.

(c) Prove that $N(A) = R(A^T)^\perp$.

(d) Prove that $N(A^T) = R(A)^\perp$.

70. CAPSTONE Let B be a basis for an inner product space V. Explain how to apply the Gram-Schmidt orthonormalization process to form an orthonormal basis B' for V.

71. Find an orthonormal basis for R^4 that includes the vectors
$$\mathbf{v}_1 = \left(\frac{1}{\sqrt{2}}, 0, \frac{1}{\sqrt{2}}, 0\right) \quad \text{and} \quad \mathbf{v}_2 = \left(0, -\frac{1}{\sqrt{2}}, 0, \frac{1}{\sqrt{2}}\right).$$

5.4   Mathematical Models and Least Squares Analysis

■ Define the least squares problem.

■ Find the orthogonal complement of a subspace and the projection of a vector onto a subspace.

■ Find the four fundamental subspaces of a matrix.

■ Solve a least squares problem.

■ Use least squares for mathematical modeling.

THE LEAST SQUARES PROBLEM

In this section, you will study *inconsistent* systems of linear equations and learn how to find the "best possible solution" of such a system. The necessity of "solving" inconsistent systems arises in the computation of least squares regression lines, as illustrated in Example 1.

EXAMPLE 1   Least Squares Regression Line

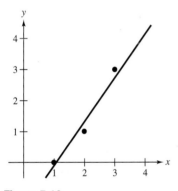

Figure 5.16

Let $(1, 0)$, $(2, 1)$, and $(3, 3)$ be three points in R^2, as shown in Figure 5.16. How can you find the line $y = c_0 + c_1 x$ that "best fits" these points? One way is to note that if the three points were collinear, then the system of equations below would be consistent.

$$
\begin{aligned}
c_0 + c_1 &= 0 \\
c_0 + 2c_1 &= 1 \\
c_0 + 3c_1 &= 3
\end{aligned}
$$

This system can be written in the matrix form $Ax = b$, where

$$
A = \begin{bmatrix} 1 & 1 \\ 1 & 2 \\ 1 & 3 \end{bmatrix}, \quad b = \begin{bmatrix} 0 \\ 1 \\ 3 \end{bmatrix}, \quad \text{and} \quad x = \begin{bmatrix} c_0 \\ c_1 \end{bmatrix}.
$$

The points are not collinear, however, so the system is inconsistent. Although it is impossible to find x such that $Ax = b$, you can look for an x that *minimizes* the norm of the error $\|Ax - b\|$. The solution $x = \begin{bmatrix} c_0 & c_1 \end{bmatrix}^T$ of this minimization problem results in the **least squares regression line** $y = c_0 + c_1 x$.

In Section 2.6, you briefly studied the least squares regression line and how to calculate it using matrices. Now you will combine the ideas of orthogonality and projection to develop this concept in more generality. To begin, consider the linear system $Ax = b$, where A is an $m \times n$ matrix and b is a column vector in R^m. You know how to use Gaussian elimination with back-substitution to solve for x when the system is consistent. When the system is inconsistent, however, it is still useful to find the "best possible" solution; that is, the vector x for which the difference between Ax and b is smallest. One way to define "best possible" is to require that the norm of $Ax - b$ be minimized. This definition is the heart of the **least squares problem.**

REMARK

The term **least squares** comes from the fact that minimizing $\|Ax - b\|$ is equivalent to minimizing $\|Ax - b\|^2$, which is a sum of squares.

Least Squares Problem

Given an $m \times n$ matrix A and a vector b in R^m, the **least squares problem** is to find x in R^n such that $\|Ax - b\|^2$ is minimized.

ORTHOGONAL SUBSPACES

To solve the least squares problem, you first need to develop the concept of orthogonal subspaces. Two subspaces of R^n are **orthogonal** when the vectors in each subspace are orthogonal to the vectors in the other subspace.

Definition of Orthogonal Subspaces

The subspaces S_1 and S_2 of R^n are **orthogonal** when $\mathbf{v}_1 \cdot \mathbf{v}_2 = 0$ for all $\mathbf{v}_1$ in S_1 and all $\mathbf{v}_2$ in S_2.

EXAMPLE 2   **Orthogonal Subspaces**

The subspaces

$$S_1 = \text{span}\left\{ \begin{bmatrix} 1 \\ 0 \\ 1 \end{bmatrix}, \begin{bmatrix} 1 \\ 1 \\ 0 \end{bmatrix} \right\} \quad \text{and} \quad S_2 = \text{span}\left\{ \begin{bmatrix} -1 \\ 1 \\ 1 \end{bmatrix} \right\}$$

are orthogonal because the dot product of any vector in S_1 and any vector in S_2 is zero.

Notice in Example 2 that the zero vector is the only vector common to both S_1 and S_2. This is true in general. If S_1 and S_2 are orthogonal subspaces of R^n, then their intersection consists of only the zero vector. You are asked to prove this in Exercise 45.

Given a subspace S of R^n, the set of all vectors orthogonal to every vector in S is the **orthogonal complement** of S, as stated in the next definition.

Definition of Orthogonal Complement

If S is a subspace of R^n, then the **orthogonal complement** of S is the set $S^\perp = \{\mathbf{u} \in R^n : \mathbf{v} \cdot \mathbf{u} = 0 \text{ for all vectors } \mathbf{v} \in S\}$.

The orthogonal complement of the trivial subspace $\{\mathbf{0}\}$ is all of R^n, and, conversely, the orthogonal complement of R^n is the trivial subspace $\{\mathbf{0}\}$. In Example 2, the subspace S_1 is the orthogonal complement of S_2, and the subspace S_2 is the orthogonal complement of S_1. The orthogonal complement of a subspace of R^n is itself a subspace of R^n (see Exercise 46). You can find the orthogonal complement of a subspace of R^n by finding the nullspace of a matrix, as illustrated in Example 3.

LINEAR ALGEBRA APPLIED

The least squares problem has a wide variety of real-life applications. To illustrate, Examples 9 and 10 and Exercises 39, 40, and 41 are all least squares analysis problems, and they involve such diverse subject matter as world population, astronomy, master's degrees awarded, company revenues, and galloping speeds of animals. In each of these applications, you are given a set of data and you are asked to come up with mathematical model(s) for the data. For example, in Exercise 40, you are given the annual revenues from 2008 through 2013 for General Dynamics Corporation. You are asked to find the least squares regression quadratic and cubic polynomials for the data, to predict the revenue for the year 2018, and to decide which of the models appears to be more accurate for predicting future revenues.

EXAMPLE 3   **Finding the Orthogonal Complement**

Find the orthogonal complement of the subspace S of R^4 spanned by the two column vectors $\mathbf{v}_1$ and $\mathbf{v}_2$ of the matrix A.

$$A = \begin{bmatrix} 1 & 0 \\ 2 & 0 \\ 1 & 0 \\ 0 & 1 \end{bmatrix}$$

$\qquad\;\; \mathbf{v}_1 \quad \mathbf{v}_2$

SOLUTION

A vector $\mathbf{u} \in R^4$ is in the orthogonal complement of S when its dot product with each of the columns of A, $\mathbf{v}_1$ and $\mathbf{v}_2$, is zero. So, the orthogonal complement of S consists of all the vectors $\mathbf{u}$ such that $A^T\mathbf{u} = \mathbf{0}$.

$$A^T\mathbf{u} = \mathbf{0}$$

$$\begin{bmatrix} 1 & 2 & 1 & 0 \\ 0 & 0 & 0 & 1 \end{bmatrix} \begin{bmatrix} x_1 \\ x_2 \\ x_3 \\ x_4 \end{bmatrix} = \begin{bmatrix} 0 \\ 0 \end{bmatrix}$$

That is, the orthogonal complement of S is the nullspace of the matrix A^T:

$$S^\perp = N(A^T).$$

Using the techniques for solving homogeneous linear systems, you can find that a basis for the orthogonal complement consists of the vectors

$$\mathbf{u}_1 = \begin{bmatrix} -2 & 1 & 0 & 0 \end{bmatrix}^T \quad \text{and} \quad \mathbf{u}_2 = \begin{bmatrix} -1 & 0 & 1 & 0 \end{bmatrix}^T.$$

Notice that R^4 in Example 3 is split into two subspaces, $S = \text{span}\{\mathbf{v}_1, \mathbf{v}_2\}$ and $S^\perp = \text{span}\{\mathbf{u}_1, \mathbf{u}_2\}$. In fact, the four vectors $\mathbf{v}_1$, $\mathbf{v}_2$, $\mathbf{u}_1$, and $\mathbf{u}_2$ form a basis for R^4. Each vector in R^4 can be *uniquely* written as a sum of a vector from S and a vector from $S^\perp$. The next definition generalizes this concept.

Definition of Direct Sum

Let S_1 and S_2 be two subspaces of R^n. If each vector $\mathbf{x} \in R^n$ can be uniquely written as a sum of a vector $\mathbf{s}_1$ from S_1 and a vector $\mathbf{s}_2$ from S_2, $\mathbf{x} = \mathbf{s}_1 + \mathbf{s}_2$, then R^n is the **direct sum** of S_1 and S_2 and you can write $R^n = S_1 \oplus S_2$.

EXAMPLE 4   **Direct Sum**

a. From Example 2, R^3 is the direct sum of the subspaces

$$S_1 = \text{span}\left\{ \begin{bmatrix} 1 \\ 0 \\ 1 \end{bmatrix}, \begin{bmatrix} 1 \\ 1 \\ 0 \end{bmatrix} \right\} \quad \text{and} \quad S_2 = \text{span}\left\{ \begin{bmatrix} -1 \\ 1 \\ 1 \end{bmatrix} \right\}.$$

b. From Example 3, you can see that $R^4 = S \oplus S^\perp$, where

$$S = \text{span}\left\{ \begin{bmatrix} 1 \\ 2 \\ 1 \\ 0 \end{bmatrix}, \begin{bmatrix} 0 \\ 0 \\ 0 \\ 1 \end{bmatrix} \right\} \quad \text{and} \quad S^\perp = \text{span}\left\{ \begin{bmatrix} -2 \\ 1 \\ 0 \\ 0 \end{bmatrix}, \begin{bmatrix} -1 \\ 0 \\ 1 \\ 0 \end{bmatrix} \right\}.$$

The next theorem lists some important facts about orthogonal complements and direct sums.

THEOREM 5.13　Properties of Orthogonal Subspaces

Let S be a subspace of R^n. Then the properties listed below are true.

1. $\dim(S) + \dim(S^\perp) = n$
2. $R^n = S \oplus S^\perp$
3. $(S^\perp)^\perp = S$

PROOF

1. If $S = R^n$ or $S = \{0\}$, then Property 1 is trivial. So let $\{v_1, v_2, \ldots, v_t\}$ be a basis for S, $0 < t < n$. Let A be the $n \times t$ matrix whose columns are the basis vectors v_i. Then $S = R(A)$ (the column space of A), which implies that $S^\perp = N(A^T)$, where A^T is a $t \times n$ matrix of rank t (see Section 5.3, Exercise 69). The dimension of $N(A^T)$ is $n - t$, so you have shown that

$$\dim(S) + \dim(S^\perp) = t + (n - t) = n.$$

2. If $S = R^n$ or $S = \{0\}$, then Property 2 is trivial. So let $\{v_1, v_2, \ldots, v_t\}$ be a basis for S and let $\{v_{t+1}, v_{t+2}, \ldots, v_n\}$ be a basis for $S^\perp$. The set $\{v_1, v_2, \ldots, v_t, v_{t+1}, \ldots, v_n\}$ is linearly independent and forms a basis for R^n. (Verify this.) Let $x \in R^n$, $x = c_1 v_1 + \cdots + c_t v_t + c_{t+1} v_{t+1} + \cdots + c_n v_n$. If you write $v = c_1 v_1 + \cdots + c_t v_t$ and $w = c_{t+1} v_{t+1} + \cdots + c_n v_n$, then you have expressed an arbitrary vector x as the sum of a vector from S and a vector from $S^\perp$, $x = v + w$.

   To show the uniqueness of this representation, assume $x = v + w = \hat{v} + \hat{w}$ (where $\hat{v}$ is in S and $\hat{w}$ is in $S^\perp$). This implies that $\hat{v} - v = w - \hat{w}$. The two vectors $\hat{v} - v$ and $w - \hat{w}$ are in both S and $S^\perp$, and $S \cap S^\perp = \{0\}$. So, you must have $\hat{v} = v$ and $w = \hat{w}$.

3. Let $v \in S$. Then $v \cdot u = 0$ for all $u \in S^\perp$, which implies that $v \in (S^\perp)^\perp$. On the other hand, let $v \in (S^\perp)^\perp$. Then $v \in R^n = S \oplus S^\perp$, so you can write v as the unique sum of a vector s from S and a vector w from $S^\perp$, $v = s + w$. The vector w is in $S^\perp$, so it is orthogonal to every vector in S, and also to v. So,

$$0 = w \cdot v = w \cdot (s + w) = w \cdot s + w \cdot w = w \cdot w.$$

This implies that $w = 0$ and $v = s + w = s \in S$.

You studied the projection of one vector onto another in Section 5.2. This is now generalized to projections of a vector v onto a subspace S. $R^n = S \oplus S^\perp$, so every vector v in R^n can be uniquely written as a sum of a vector from S and a vector from $S^\perp$:

$$v = v_1 + v_2, \quad v_1 \in S, \quad v_2 \in S^\perp.$$

The vector v_1 is the **projection** of v onto the subspace S, and is denoted by $v_1 = \text{proj}_S v$. So, $v_2 = v - v_1 = v - \text{proj}_S v$, which implies that the vector $v - \text{proj}_S v$ is orthogonal to the subspace S.

Given a subspace S of R^n, you can apply the Gram-Schmidt orthonormalization process to find an orthonormal basis for S. You can then find the projection of a vector v onto S using the next theorem. (You are asked to prove this theorem in Exercise 47.)

THEOREM 5.14　Projection onto a Subspace

If $\{u_1, u_2, \ldots, u_t\}$ is an orthonormal basis for the subspace S of R^n, and $v \in R^n$, then

$$\text{proj}_S v = (v \cdot u_1)u_1 + (v \cdot u_2)u_2 + \cdots + (v \cdot u_t)u_t.$$

EXAMPLE 5	Projection Onto a Subspace

Find the projection of the vector $\mathbf{v} = \begin{bmatrix} 1 \\ 1 \\ 3 \end{bmatrix}$ onto the subspace S of R^3 spanned by the vectors

$$\mathbf{w}_1 = \begin{bmatrix} 0 \\ 3 \\ 1 \end{bmatrix} \quad \text{and} \quad \mathbf{w}_2 = \begin{bmatrix} 2 \\ 0 \\ 0 \end{bmatrix}.$$

SOLUTION

By normalizing $\mathbf{w}_1$ and $\mathbf{w}_2$ you obtain an orthonormal basis for S.

$$\{\mathbf{u}_1, \mathbf{u}_2\} = \left\{ \frac{1}{\sqrt{10}} \mathbf{w}_1, \frac{1}{2} \mathbf{w}_2 \right\} = \left\{ \begin{bmatrix} 0 \\ \dfrac{3}{\sqrt{10}} \\ \dfrac{1}{\sqrt{10}} \end{bmatrix}, \begin{bmatrix} 1 \\ 0 \\ 0 \end{bmatrix} \right\}$$

Use Theorem 5.14 to find the projection of $\mathbf{v}$ onto S.

$$\text{proj}_S\mathbf{v} = (\mathbf{v} \cdot \mathbf{u}_1)\mathbf{u}_1 + (\mathbf{v} \cdot \mathbf{u}_2)\mathbf{u}_2 = \frac{6}{\sqrt{10}} \begin{bmatrix} 0 \\ \dfrac{3}{\sqrt{10}} \\ \dfrac{1}{\sqrt{10}} \end{bmatrix} + 1 \begin{bmatrix} 1 \\ 0 \\ 0 \end{bmatrix} = \begin{bmatrix} 1 \\ \dfrac{9}{5} \\ \dfrac{3}{5} \end{bmatrix}$$

Figure 5.17 illustrates the projection of $\mathbf{v}$ onto the plane S.

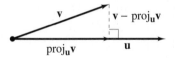

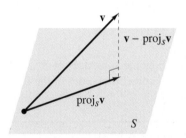

Figure 5.17

Theorem 5.9 states that among all the scalar multiples of a vector $\mathbf{u}$, the orthogonal projection of $\mathbf{v}$ onto $\mathbf{u}$ is the one closest to $\mathbf{v}$. Example 5 suggests that this property is also true for projections onto subspaces. That is, among all the vectors in the subspace S, the vector $\text{proj}_S\mathbf{v}$ is the closest vector to $\mathbf{v}$. Figure 5.18 illustrates these two results.

Figure 5.18

THEOREM 5.15   Orthogonal Projection and Distance

Let S be a subspace of R^n and let $\mathbf{v} \in R^n$. Then, for all $\mathbf{u} \in S$, $\mathbf{u} \neq \text{proj}_S\mathbf{v}$,

$$\|\mathbf{v} - \text{proj}_S\mathbf{v}\| < \|\mathbf{v} - \mathbf{u}\|.$$

PROOF

Let $\mathbf{u} \in S$, $\mathbf{u} \neq \text{proj}_S\mathbf{v}$. By adding and subtracting the same quantity $\text{proj}_S\mathbf{v}$ to and from the vector $\mathbf{v} - \mathbf{u}$, you obtain

$$\mathbf{v} - \mathbf{u} = (\mathbf{v} - \text{proj}_S\mathbf{v}) + (\text{proj}_S\mathbf{v} - \mathbf{u}).$$

Observe that $(\text{proj}_S\mathbf{v} - \mathbf{u})$ is in S and $(\mathbf{v} - \text{proj}_S\mathbf{v})$ is orthogonal to S. So, $(\mathbf{v} - \text{proj}_S\mathbf{v})$ and $(\text{proj}_S\mathbf{v} - \mathbf{u})$ are orthogonal vectors, and you can use the Pythagorean Theorem (Theorem 5.6) to obtain

$$\|\mathbf{v} - \mathbf{u}\|^2 = \|\mathbf{v} - \text{proj}_S\mathbf{v}\|^2 + \|\text{proj}_S\mathbf{v} - \mathbf{u}\|^2.$$

$\mathbf{u} \neq \text{proj}_S\mathbf{v}$, so $\|\text{proj}_S\mathbf{v} - \mathbf{u}\|^2$ is positive, and you have

$$\|\mathbf{v} - \text{proj}_S\mathbf{v}\| < \|\mathbf{v} - \mathbf{u}\|.$$

FUNDAMENTAL SUBSPACES OF A MATRIX

Recall that if A is an $m \times n$ matrix, then the column space of A is a subspace of R^m consisting of all vectors of the form $A\mathbf{x}$, $\mathbf{x} \in R^n$. The four **fundamental subspaces** of the matrix A are listed below (see Exercises 67 and 68 in Section 5.3).

$N(A) = $ nullspace of A $\qquad\qquad N(A^T) = $ nullspace of A^T

$R(A) = $ column space of A $\qquad\quad R(A^T) = $ column space of A^T

These subspaces play a crucial role in the solution of the least squares problem.

EXAMPLE 6   **Fundamental Subspaces**

Find the four fundamental subspaces of the matrix

$$A = \begin{bmatrix} 1 & 2 & 0 \\ 0 & 0 & 1 \\ 0 & 0 & 0 \\ 0 & 0 & 0 \end{bmatrix}.$$

SOLUTION

The column space of A is simply the span of the first and third columns, because the second column is a scalar multiple of the first column. The column space of A^T is equal to the row space of A, which is spanned by the first two rows. The nullspace of A is a solution space of the homogeneous system $A\mathbf{x} = \mathbf{0}$. Finally, the nullspace of A^T is a solution space of the homogeneous system whose coefficient matrix is A^T. A summary of these results is shown below.

$$R(A) = \text{span}\left\{ \begin{bmatrix} 1 \\ 0 \\ 0 \\ 0 \end{bmatrix}, \begin{bmatrix} 0 \\ 1 \\ 0 \\ 0 \end{bmatrix} \right\} \qquad R(A^T) = \text{span}\left\{ \begin{bmatrix} 1 \\ 2 \\ 0 \end{bmatrix}, \begin{bmatrix} 0 \\ 0 \\ 1 \end{bmatrix} \right\}$$

$$N(A) = \text{span}\left\{ \begin{bmatrix} -2 \\ 1 \\ 0 \end{bmatrix} \right\} \qquad N(A^T) = \text{span}\left\{ \begin{bmatrix} 0 \\ 0 \\ 1 \\ 0 \end{bmatrix}, \begin{bmatrix} 0 \\ 0 \\ 0 \\ 1 \end{bmatrix} \right\}$$

In Example 6, observe that $R(A)$ and $N(A^T)$ are orthogonal subspaces of R^4, and $R(A^T)$ and $N(A)$ are orthogonal subspaces of R^3. These and other properties of the four fundamental subspaces are stated in the next theorem.

THEOREM 5.16   Fundamental Subspaces of a Matrix

If A is an $m \times n$ matrix, then

1. $R(A)$ and $N(A^T)$ are orthogonal subspaces of R^m.
2. $R(A^T)$ and $N(A)$ are orthogonal subspaces of R^n.
3. $R(A) \oplus N(A^T) = R^m$.
4. $R(A^T) \oplus N(A) = R^n$.

PROOF

To prove Property 1, let $\mathbf{v} \in R(A)$ and $\mathbf{u} \in N(A^T)$. The column space of A is equal to the row space of A^T, so $A^T \mathbf{u} = \mathbf{0}$ implies $\mathbf{u} \cdot \mathbf{v} = 0$. Property 2 follows from applying Property 1 to A^T.

To prove Property 3, observe that $R(A)^{\perp} = N(A^T)$ and $R^m = R(A) \oplus R(A)^{\perp}$. So, $R^m = R(A) \oplus N(A^T)$. A similar argument applied to $R(A^T)$ proves Property 4.

SOLVING THE LEAST SQUARES PROBLEM

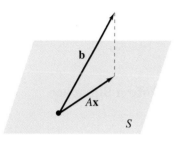

Figure 5.19

You have now developed all the tools needed to solve the least squares problem. Recall that you are attempting to find a vector $\mathbf{x}$ that minimizes $\|A\mathbf{x} - \mathbf{b}\|$, where A is an $m \times n$ matrix and $\mathbf{b}$ is a vector in R^m. Let S be the column space of A: $S = R(A)$. Assume that $\mathbf{b}$ is not in S, because otherwise the system $A\mathbf{x} = \mathbf{b}$ would be consistent. You are looking for a vector $A\mathbf{x}$ in S that is as close as possible to $\mathbf{b}$, as illustrated in Figure 5.19.

From Theorem 5.15, you know that the desired vector is the projection of $\mathbf{b}$ onto S. So, $A\mathbf{x} = \text{proj}_S\mathbf{b}$ and $A\mathbf{x} - \mathbf{b} = \text{proj}_S\mathbf{b} - \mathbf{b}$ is orthogonal to $S = R(A)$. However, this implies that $A\mathbf{x} - \mathbf{b}$ is in $R(A)^\perp$, which equals $N(A^T)$. This is the crucial observation: $A\mathbf{x} - \mathbf{b}$ is in the nullspace of A^T. So, you have

$$A^T(A\mathbf{x} - \mathbf{b}) = \mathbf{0}$$
$$A^TA\mathbf{x} - A^T\mathbf{b} = \mathbf{0}$$
$$A^TA\mathbf{x} = A^T\mathbf{b}.$$

The solution of the least squares problem comes down to solving the $n \times n$ linear system of equations $A^TA\mathbf{x} = A^T\mathbf{b}$. These equations are the **normal equations** of the least squares problem $A\mathbf{x} = \mathbf{b}$.

EXAMPLE 7   Finding the Least Squares Solution

See LarsonLinearAlgebra.com for an interactive version of this type of example.

Find the solution of the least squares problem

$$A\mathbf{x} = \mathbf{b}$$

$$\begin{bmatrix} 1 & 1 \\ 1 & 2 \\ 1 & 3 \end{bmatrix} \begin{bmatrix} c_0 \\ c_1 \end{bmatrix} = \begin{bmatrix} 0 \\ 1 \\ 3 \end{bmatrix}$$

from Example 1.

SOLUTION

Begin by finding the matrix products below.

$$A^TA = \begin{bmatrix} 1 & 1 & 1 \\ 1 & 2 & 3 \end{bmatrix} \begin{bmatrix} 1 & 1 \\ 1 & 2 \\ 1 & 3 \end{bmatrix} = \begin{bmatrix} 3 & 6 \\ 6 & 14 \end{bmatrix}$$

$$A^T\mathbf{b} = \begin{bmatrix} 1 & 1 & 1 \\ 1 & 2 & 3 \end{bmatrix} \begin{bmatrix} 0 \\ 1 \\ 3 \end{bmatrix} = \begin{bmatrix} 4 \\ 11 \end{bmatrix}$$

The normal equations are represented by the system

$$A^TA\mathbf{x} = A^T\mathbf{b}$$

$$\begin{bmatrix} 3 & 6 \\ 6 & 14 \end{bmatrix} \begin{bmatrix} c_0 \\ c_1 \end{bmatrix} = \begin{bmatrix} 4 \\ 11 \end{bmatrix}.$$

The solution of this system of equations is

$$\mathbf{x} = \begin{bmatrix} -\frac{5}{3} \\ \frac{3}{2} \end{bmatrix}$$

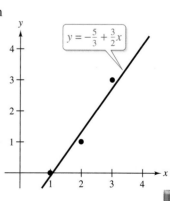

which implies that the least squares regression line for the data is $y = -\frac{5}{3} + \frac{3}{2}x$, as shown in the figure.

For an $m \times n$ matrix A, the normal equations form an $n \times n$ system of linear equations. This system is always consistent, but it may have infinitely many solutions. It can be shown, however, that there is a unique solution when the rank of A is n.

The next example illustrates how to solve the projection problem from Example 5 using normal equations.

EXAMPLE 8   Orthogonal Projection Onto a Subspace

Find the orthogonal projection of the vector

$$\mathbf{b} = \begin{bmatrix} 1 \\ 1 \\ 3 \end{bmatrix}$$

onto the column space S of the matrix

$$A = \begin{bmatrix} 0 & 2 \\ 3 & 0 \\ 1 & 0 \end{bmatrix}.$$

SOLUTION

To find the orthogonal projection of $\mathbf{b}$ onto S, first solve the least squares problem

$$A\mathbf{x} = \mathbf{b}.$$

As in Example 7, find the matrix products A^TA and $A^T\mathbf{b}$.

$$A^TA = \begin{bmatrix} 0 & 3 & 1 \\ 2 & 0 & 0 \end{bmatrix} \begin{bmatrix} 0 & 2 \\ 3 & 0 \\ 1 & 0 \end{bmatrix}$$

$$= \begin{bmatrix} 10 & 0 \\ 0 & 4 \end{bmatrix}$$

$$A^T\mathbf{b} = \begin{bmatrix} 0 & 3 & 1 \\ 2 & 0 & 0 \end{bmatrix} \begin{bmatrix} 1 \\ 1 \\ 3 \end{bmatrix}$$

$$= \begin{bmatrix} 6 \\ 2 \end{bmatrix}$$

The normal equations are represented by the system

$$A^TA\mathbf{x} = A^T\mathbf{b}$$

$$\begin{bmatrix} 10 & 0 \\ 0 & 4 \end{bmatrix} \begin{bmatrix} x_1 \\ x_2 \end{bmatrix} = \begin{bmatrix} 6 \\ 2 \end{bmatrix}.$$

The solution of these equations is

$$\mathbf{x} = \begin{bmatrix} x_1 \\ x_2 \end{bmatrix} = \begin{bmatrix} \frac{3}{5} \\ \frac{1}{2} \end{bmatrix}.$$

Finally, the projection of $\mathbf{b}$ onto S is

$$A\mathbf{x} = \begin{bmatrix} 0 & 2 \\ 3 & 0 \\ 1 & 0 \end{bmatrix} \begin{bmatrix} \frac{3}{5} \\ \frac{1}{2} \end{bmatrix} = \begin{bmatrix} 1 \\ \frac{9}{5} \\ \frac{3}{5} \end{bmatrix}$$

which agrees with the solution obtained in Example 5.

MATHEMATICAL MODELING

Least squares problems play a fundamental role in mathematical modeling of real-life phenomena. The next example shows how to model the world population using a least squares quadratic polynomial.

EXAMPLE 9   **World Population**

The table shows the world population (in billions) for six different years. (Source: U.S. Census Bureau)

Year	1985	1990	1995	2000	2005	2010
Population, y	4.9	5.3	5.7	6.1	6.5	6.9

Let $x = 5$ represent the year 1985. Find the least squares regression quadratic polynomial $y = c_0 + c_1 x + c_2 x^2$ for the data and use the model to estimate the population for the year 2020.

SOLUTION

By substituting the data points $(5, 4.9)$, $(10, 5.3)$, $(15, 5.7)$, $(20, 6.1)$, $(25, 6.5)$, and $(30, 6.9)$ into the quadratic polynomial $y = c_0 + c_1 x + c_2 x^2$, you obtain the system of linear equations below.

$$c_0 + 5c_1 + 25c_2 = 4.9$$
$$c_0 + 10c_1 + 100c_2 = 5.3$$
$$c_0 + 15c_1 + 225c_2 = 5.7$$
$$c_0 + 20c_1 + 400c_2 = 6.1$$
$$c_0 + 25c_1 + 625c_2 = 6.5$$
$$c_0 + 30c_1 + 900c_2 = 6.9$$

This produces the least squares problem

$$A\mathbf{x} = \mathbf{b}$$

$$\begin{bmatrix} 1 & 5 & 25 \\ 1 & 10 & 100 \\ 1 & 15 & 225 \\ 1 & 20 & 400 \\ 1 & 25 & 625 \\ 1 & 30 & 900 \end{bmatrix} \begin{bmatrix} c_0 \\ c_1 \\ c_2 \end{bmatrix} = \begin{bmatrix} 4.9 \\ 5.3 \\ 5.7 \\ 6.1 \\ 6.5 \\ 6.9 \end{bmatrix}$$

The normal equations are represented by the system

$$A^T A \mathbf{x} = A^T \mathbf{b}$$

$$\begin{bmatrix} 6 & 105 & 2275 \\ 105 & 2275 & 55{,}125 \\ 2275 & 55{,}125 & 1{,}421{,}875 \end{bmatrix} \begin{bmatrix} c_0 \\ c_1 \\ c_2 \end{bmatrix} = \begin{bmatrix} 35.4 \\ 654.5 \\ 14{,}647.5 \end{bmatrix}$$

and their solution is $\mathbf{x} = \begin{bmatrix} c_0 \\ c_1 \\ c_2 \end{bmatrix} = \begin{bmatrix} 4.5 \\ 0.08 \\ 0 \end{bmatrix}$.

Note that $c_2 = 0$. So, the least squares polynomial is the *linear* polynomial $y = 4.5 + 0.08x$. Evaluating this polynomial at $x = 40$ gives the estimate of the world population for the year 2020: $y = 4.5 + 0.08(40) = 7.7$ billion.

Least squares models can arise in many other contexts. Section 5.5 explores some applications of least squares models to approximations of functions. The next example uses data from Section 1.3 to find a nonlinear relationship between the period of a planet and its mean distance from the Sun.

EXAMPLE 10   Application to Astronomy

The table shows the mean distances x and the periods y of the six planets that are closest to the Sun. The mean distances are in astronomical units and the periods are in years. Find a model for the data.

Planet	Mercury	Venus	Earth	Mars	Jupiter	Saturn
Distance, x	0.387	0.723	1.000	1.524	5.203	9.537
Period, y	0.241	0.615	1.000	1.881	11.862	29.457

SOLUTION

When you plot the data as given, they do not lie in a straight line. By taking the natural logarithm of each coordinate, however, you obtain points of the form $(\ln x, \ln y)$, as shown below.

Planet	Mercury	Venus	Earth	Mars	Jupiter	Saturn
$\ln x$	-0.949	-0.324	0.0	0.421	1.649	2.255
$\ln y$	-1.423	-0.486	0.0	0.632	2.473	3.383

A plot of the transformed points suggests that the least squares regression line would be a good fit.

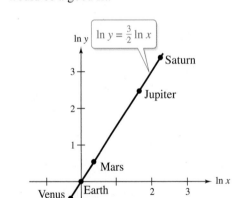

Use the techniques of this section on the system

$$c_0 - 0.949c_1 = -1.423$$
$$c_0 - 0.324c_1 = -0.486$$
$$c_0 \qquad\quad = \quad 0.0$$
$$c_0 + 0.421c_1 = \quad 0.632$$
$$c_0 + 1.649c_1 = \quad 2.473$$
$$c_0 + 2.255c_1 = \quad 3.383$$

to verify that the equation of the line is

$$\ln y = \tfrac{3}{2}\ln x \quad \text{or} \quad y = x^{3/2}.$$

TECHNOLOGY

You can use a graphing utility or software program to verify the result of Example 10. For instance, using the data in the first table, a graphing utility gives a power regression model of $y \approx 1.00029x^{1.49972}$. The **Technology Guide** at *CengageBrain.com* can help you use technology to model data.

5.4 Exercises

See CalcChat.com for worked-out solutions to odd-numbered exercises.

Least Squares Regression Line In Exercises 1–4, determine whether the points are collinear. If so, find the line $y = c_0 + c_1 x$ that fits the points.

1. $(0, 1), (1, 3), (2, 5)$ **2.** $(0, 0), (3, 1), (4, 2)$

3. $(-2, 0), (0, 2), (2, 2)$ **4.** $(-1, 5), (1, -1), (1, -4)$

Orthogonal Subspaces In Exercises 5–8, determine whether the subspaces are orthogonal.

5. $S_1 = \text{span}\left\{\begin{bmatrix} 3 \\ 2 \\ -2 \end{bmatrix}, \begin{bmatrix} 0 \\ 1 \\ 0 \end{bmatrix}\right\}$ $S_2 = \text{span}\left\{\begin{bmatrix} 2 \\ -3 \\ 0 \end{bmatrix}\right\}$

6. $S_1 = \text{span}\left\{\begin{bmatrix} -3 \\ 0 \\ 1 \end{bmatrix}\right\}$ $S_2 = \text{span}\left\{\begin{bmatrix} 2 \\ 1 \\ 6 \end{bmatrix}, \begin{bmatrix} 0 \\ 1 \\ 0 \end{bmatrix}\right\}$

7. $S_1 = \text{span}\left\{\begin{bmatrix} 1 \\ 1 \\ 1 \\ 1 \end{bmatrix}\right\}$ $S_2 = \text{span}\left\{\begin{bmatrix} -1 \\ 1 \\ -1 \\ 1 \end{bmatrix}, \begin{bmatrix} 0 \\ 2 \\ -2 \\ 0 \end{bmatrix}\right\}$

8. $S_1 = \text{span}\left\{\begin{bmatrix} 0 \\ 0 \\ 2 \\ 1 \end{bmatrix}, \begin{bmatrix} 0 \\ 0 \\ 1 \\ -2 \end{bmatrix}\right\}$ $S_2 = \text{span}\left\{\begin{bmatrix} 3 \\ 2 \\ 0 \\ 0 \end{bmatrix}, \begin{bmatrix} 0 \\ 1 \\ -2 \\ 2 \end{bmatrix}\right\}$

Finding the Orthogonal Complement and Direct Sum In Exercises 9–14, (a) find the orthogonal complement $S^\perp$, and (b) find the direct sum $S \oplus S^\perp$.

9. $S = \text{span}\left\{\begin{bmatrix} 0 \\ 1 \\ 0 \end{bmatrix}, \begin{bmatrix} 2 \\ 0 \\ 1 \end{bmatrix}\right\}$ **10.** $S = \text{span}\left\{\begin{bmatrix} 0 \\ -2 \\ 1 \end{bmatrix}\right\}$

11. $S = \text{span}\left\{\begin{bmatrix} 0 \\ 1 \\ -1 \\ 1 \end{bmatrix}\right\}$

12. $S = \text{span}\left\{\begin{bmatrix} 0 \\ 1 \\ -1 \\ 1 \\ -1 \end{bmatrix}, \begin{bmatrix} 0 \\ 1 \\ 0 \\ 2 \\ -1 \end{bmatrix}, \begin{bmatrix} 2 \\ 0 \\ 1 \\ 0 \\ 2 \end{bmatrix}\right\}$

13. S is the subspace of R^3 consisting of the xz-plane.

14. S is the subspace of R^5 consisting of all vectors whose third and fourth components are zero.

15. Find the orthogonal complement of the solution of Exercise 11(a).

16. Find the orthogonal complement of the solution of Exercise 12(a).

Projection Onto a Subspace In Exercises 17–20, find the projection of the vector v onto the subspace S.

17. $S = \text{span}\left\{\begin{bmatrix} 0 \\ 0 \\ -1 \\ 1 \end{bmatrix}, \begin{bmatrix} 0 \\ 1 \\ 1 \\ 1 \end{bmatrix}\right\}$, $\mathbf{v} = \begin{bmatrix} 1 \\ 0 \\ 1 \\ 1 \end{bmatrix}$

18. $S = \text{span}\left\{\begin{bmatrix} -1 \\ 2 \\ 0 \\ 0 \end{bmatrix}, \begin{bmatrix} 0 \\ 0 \\ 1 \\ 0 \end{bmatrix}, \begin{bmatrix} 0 \\ 0 \\ 0 \\ 1 \end{bmatrix}\right\}$, $\mathbf{v} = \begin{bmatrix} 1 \\ 1 \\ 1 \\ 1 \end{bmatrix}$

19. $S = \text{span}\left\{\begin{bmatrix} 1 \\ 0 \\ 1 \end{bmatrix}, \begin{bmatrix} 0 \\ 1 \\ 1 \end{bmatrix}\right\}$, $\mathbf{v} = \begin{bmatrix} 2 \\ 3 \\ 4 \end{bmatrix}$

20. $S = \text{span}\left\{\begin{bmatrix} 1 \\ 1 \\ 1 \\ 1 \end{bmatrix}, \begin{bmatrix} 0 \\ 1 \\ -1 \\ 0 \end{bmatrix}, \begin{bmatrix} 0 \\ 1 \\ 1 \\ 0 \end{bmatrix}\right\}$, $\mathbf{v} = \begin{bmatrix} 1 \\ 2 \\ 3 \\ 4 \end{bmatrix}$

Fundamental Subspaces In Exercises 21–24, find bases for the four fundamental subspaces of the matrix A.

21. $A = \begin{bmatrix} 1 & 2 & 3 \\ 0 & 1 & 0 \end{bmatrix}$ **22.** $A = \begin{bmatrix} 0 & -1 & 1 \\ 1 & 2 & 0 \\ 1 & 1 & 1 \end{bmatrix}$

23. $A = \begin{bmatrix} 1 & 0 & 0 \\ 0 & 1 & 1 \\ 1 & 1 & 1 \\ 1 & 2 & 2 \end{bmatrix}$ **24.** $A = \begin{bmatrix} 1 & 0 & -1 \\ 0 & -1 & 1 \\ 1 & 1 & 0 \\ 1 & 0 & 1 \end{bmatrix}$

Finding the Least Squares Solution In Exercises 25–28, find the least squares solution of the system $A\mathbf{x} = \mathbf{b}$.

25. $A = \begin{bmatrix} 2 & 1 \\ 1 & 2 \\ 1 & 1 \end{bmatrix}$ $\mathbf{b} = \begin{bmatrix} 2 \\ 0 \\ -3 \end{bmatrix}$

26. $A = \begin{bmatrix} 1 & -1 & 1 \\ 1 & 1 & 1 \\ 0 & 1 & 1 \\ 1 & 0 & 1 \end{bmatrix}$ $\mathbf{b} = \begin{bmatrix} 2 \\ 1 \\ 0 \\ 2 \end{bmatrix}$

27. $A = \begin{bmatrix} 1 & 0 & 1 \\ 1 & 1 & 1 \\ 0 & 1 & 1 \\ 1 & 1 & 0 \end{bmatrix}$ $\mathbf{b} = \begin{bmatrix} 4 \\ -1 \\ 0 \\ 1 \end{bmatrix}$

28. $A = \begin{bmatrix} 0 & 2 & 1 \\ 1 & 1 & -1 \\ 2 & 1 & 0 \\ 1 & 1 & 1 \\ 0 & 2 & -1 \end{bmatrix}$ $\mathbf{b} = \begin{bmatrix} 1 \\ 0 \\ 1 \\ -1 \\ 0 \end{bmatrix}$

Orthogonal Projection Onto a Subspace **In Exercises 29 and 30, use the method of Example 8 to find the orthogonal projection of $b = \begin{bmatrix} 2 & -2 & 1 \end{bmatrix}^T$ onto the column space of the matrix A.**

29. $A = \begin{bmatrix} 1 & 2 \\ 0 & 1 \\ 1 & 1 \end{bmatrix}$ 30. $A = \begin{bmatrix} 0 & 2 \\ 1 & 1 \\ 1 & 3 \end{bmatrix}$

Finding the Least Squares Regression Line **In Exercises 31–34, find the least squares regression line for the data points. Graph the points and the line on the same set of axes.**

31. $(-1, 1), (1, 0), (3, -3)$

32. $(1, 1), (2, 3), (4, 5)$

33. $(-2, 1), (-1, 2), (0, 1), (1, 2), (2, 1)$

34. $(-2, 0), (-1, 2), (0, 3), (1, 5), (2, 6)$

Finding the Least Squares Quadratic Polynomial **In Exercises 35–38, find the least squares regression quadratic polynomial for the data points.**

35. $(0, 0), (2, 2), (3, 6), (4, 12)$

36. $(0, 2), \left(1, \frac{3}{2}\right), \left(2, \frac{5}{2}\right), (3, 4)$

37. $(-2, 0), (-1, 0), (0, 1), (1, 2), (2, 5)$

38. $(-2, 6), (-1, 5), \left(0, \frac{7}{2}\right), (1, 2), (2, -1)$

39. **Master's Degrees** The table shows the numbers of master's degrees y (in thousands) conferred in the United States from 2009 through 2012. Find the least squares regression line for the data. Then use the model to predict the number of degrees conferred in 2019. Let t represent the year, with $t = 9$ corresponding to 2009. (Source: U.S. National Center for Education Statistics)

Year	2009	2010	2011	2012
Master's Degrees, y	662.1	693.0	730.6	754.2

40. **Revenue** The table shows the revenues y (in billions of dollars) for General Dynamics Corporation from 2008 through 2013. Find the least squares regression quadratic and cubic polynomials for the data. Then use each model to predict the revenue in 2018. Let t represent the year, with $t = 8$ corresponding to 2008. Which model appears to be more accurate for predicting future revenues? Explain. (Source: General Dynamics Corporation)

Year	2008	2009	2010
Revenue, y	29.3	32.0	32.5

Year	2011	2012	2013
Revenue, y	32.7	31.7	31.2

41. **Galloping Speeds of Animals** Four-legged animals run with two different types of motion: trotting and galloping. An animal that is trotting has at least one foot on the ground at all times, whereas an animal that is galloping has all four feet off the ground at some point in its stride. The number of strides per minute at which an animal breaks from a trot to a gallop depends on the weight of the animal. Use the table and the method of Example 10 to find an equation that relates an animal's weight x (in kilograms) and its lowest galloping speed y (in strides per minute).

Weight, x	11.3	15.9	22.7
Galloping Speed, y	191.5	182.7	173.8

Weight, x	34.0	226.8	453.6
Galloping Speed, y	164.2	125.9	114.2

42. **CAPSTONE** Explain how orthogonality, orthogonal complements, the projection of a vector, and fundamental subspaces are used to find the solution of a least squares problem.

True or False? **In Exercises 43 and 44, determine whether each statement is true or false. If a statement is true, give a reason or cite an appropriate statement from the text. If a statement is false, provide an example that shows the statement is not true in all cases or cite an appropriate statement from the text.**

43. (a) The orthogonal complement of R^n is the empty set.

(b) If each vector $\mathbf{v} \in R^n$ can be uniquely written as a sum of a vector $\mathbf{s}_1$ from S_1 and a vector $\mathbf{s}_2$ from S_2, then R^n is the direct sum of S_1 and S_2.

44. (a) If A is an $m \times n$ matrix, then $R(A)$ and $N(A^T)$ are orthogonal subspaces of R^n.

(b) The set of all vectors orthogonal to every vector in a subspace S is the orthogonal complement of S.

(c) Given an $m \times n$ matrix A and a vector $\mathbf{b}$ in R^m, the least squares problem is to find $\mathbf{x}$ in R^n such that $\|A\mathbf{x} - \mathbf{b}\|^2$ is minimized.

45. **Proof** Prove that if S_1 and S_2 are orthogonal subspaces of R^n, then their intersection consists of only the zero vector.

46. **Proof** Prove that the orthogonal complement of a subspace of R^n is itself a subspace of R^n.

47. **Proof** Prove Theorem 5.14.

48. **Proof** Prove that if S_1 and S_2 are subspaces of R^n and if

$$R^n = S_1 \oplus S_2$$

then

$$S_1 \cap S_2 = \{\mathbf{0}\}.$$

5.5 Applications of Inner Product Spaces

■ Find the cross product of two vectors in R^3.

■ Find the linear or quadratic least squares approximation of a function.

■ Find the nth-order Fourier approximation of a function.

THE CROSS PRODUCT OF TWO VECTORS IN R^3

Here you will look at a vector product that yields a vector in R^3 orthogonal to two vectors. This vector product is called the **cross product,** and it is most conveniently defined and calculated with vectors written in standard unit vector form

$$\mathbf{v} = (v_1, v_2, v_3) = v_1\mathbf{i} + v_2\mathbf{j} + v_3\mathbf{k}.$$

REMARK

The cross product is defined only for vectors in R^3. The cross product of two vectors in R^n, $n \neq 3$, is not defined here.

Definition of the Cross Product of Two Vectors

Let $\mathbf{u} = u_1\mathbf{i} + u_2\mathbf{j} + u_3\mathbf{k}$ and $\mathbf{v} = v_1\mathbf{i} + v_2\mathbf{j} + v_3\mathbf{k}$ be vectors in R^3. The **cross product** of $\mathbf{u}$ and $\mathbf{v}$ is the vector

$$\mathbf{u} \times \mathbf{v} = (u_2v_3 - u_3v_2)\mathbf{i} - (u_1v_3 - u_3v_1)\mathbf{j} + (u_1v_2 - u_2v_1)\mathbf{k}.$$

A convenient way to remember the formula for the cross product $\mathbf{u} \times \mathbf{v}$ is to use the determinant form below.

$$\mathbf{u} \times \mathbf{v} = \begin{vmatrix} \mathbf{i} & \mathbf{j} & \mathbf{k} \\ u_1 & u_2 & u_3 \\ v_1 & v_2 & v_3 \end{vmatrix} \qquad \longleftarrow \quad \text{Components of } \mathbf{u} \\ \qquad\qquad\qquad\quad \longleftarrow \quad \text{Components of } \mathbf{v}$$

Technically this is not a determinant because it represents a vector and not a real number. Nevertheless, it is useful because it can help you remember the cross product formula. Using cofactor expansion in the first row produces

$$\mathbf{u} \times \mathbf{v} = \begin{vmatrix} u_2 & u_3 \\ v_2 & v_3 \end{vmatrix}\mathbf{i} - \begin{vmatrix} u_1 & u_3 \\ v_1 & v_3 \end{vmatrix}\mathbf{j} + \begin{vmatrix} u_1 & u_2 \\ v_1 & v_2 \end{vmatrix}\mathbf{k}$$

$$= (u_2v_3 - u_3v_2)\mathbf{i} - (u_1v_3 - u_3v_1)\mathbf{j} + (u_1v_2 - u_2v_1)\mathbf{k}$$

which yields the formula in the definition. Be sure to note that the $\mathbf{j}$-component is preceded by a minus sign.

LINEAR ALGEBRA APPLIED

In physics, the cross product can be used to measure *torque*—the moment $\mathbf{M}$ of a force $\mathbf{F}$ about a point A, as shown in the figure below. When the point of application of the force is B, the moment of $\mathbf{F}$ about A is

$$\mathbf{M} = \overrightarrow{AB} \times \mathbf{F}$$

where $\overrightarrow{AB}$ represents the vector whose initial point is A and whose terminal point is B. The magnitude of the moment $\mathbf{M}$ measures the tendency of $\overrightarrow{AB}$ to rotate counterclockwise about an axis directed along the vector $\mathbf{M}$.

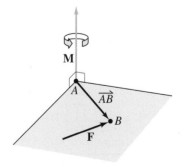

EXAMPLE 1   **Finding the Cross Product of Two Vectors**

Let $\mathbf{u} = \mathbf{i} - 2\mathbf{j} + \mathbf{k}$ and $\mathbf{v} = 3\mathbf{i} + \mathbf{j} - 2\mathbf{k}$. Find each cross product.

a. $\mathbf{u} \times \mathbf{v}$ **b.** $\mathbf{v} \times \mathbf{u}$ **c.** $\mathbf{v} \times \mathbf{v}$

SOLUTION

a. $\mathbf{u} \times \mathbf{v} = \begin{vmatrix} \mathbf{i} & \mathbf{j} & \mathbf{k} \\ 1 & -2 & 1 \\ 3 & 1 & -2 \end{vmatrix}$

$= \begin{vmatrix} -2 & 1 \\ 1 & -2 \end{vmatrix}\mathbf{i} - \begin{vmatrix} 1 & 1 \\ 3 & -2 \end{vmatrix}\mathbf{j} + \begin{vmatrix} 1 & -2 \\ 3 & 1 \end{vmatrix}\mathbf{k}$

$= 3\mathbf{i} + 5\mathbf{j} + 7\mathbf{k}$

b. $\mathbf{v} \times \mathbf{u} = \begin{vmatrix} \mathbf{i} & \mathbf{j} & \mathbf{k} \\ 3 & 1 & -2 \\ 1 & -2 & 1 \end{vmatrix}$

$= \begin{vmatrix} 1 & -2 \\ -2 & 1 \end{vmatrix}\mathbf{i} - \begin{vmatrix} 3 & -2 \\ 1 & 1 \end{vmatrix}\mathbf{j} + \begin{vmatrix} 3 & 1 \\ 1 & -2 \end{vmatrix}\mathbf{k}$

$= -3\mathbf{i} - 5\mathbf{j} - 7\mathbf{k}$

Note that this result is the negative of that in part (a).

c. $\mathbf{v} \times \mathbf{v} = \begin{vmatrix} \mathbf{i} & \mathbf{j} & \mathbf{k} \\ 3 & 1 & -2 \\ 3 & 1 & -2 \end{vmatrix}$

$= \begin{vmatrix} 1 & -2 \\ 1 & -2 \end{vmatrix}\mathbf{i} - \begin{vmatrix} 3 & -2 \\ 3 & -2 \end{vmatrix}\mathbf{j} + \begin{vmatrix} 3 & 1 \\ 3 & 1 \end{vmatrix}\mathbf{k}$

$= 0\mathbf{i} + 0\mathbf{j} + 0\mathbf{k} = \mathbf{0}$

The results obtained in Example 1 suggest some interesting *algebraic* properties of the cross product. For instance,

$$\mathbf{u} \times \mathbf{v} = -(\mathbf{v} \times \mathbf{u}) \quad \text{and} \quad \mathbf{v} \times \mathbf{v} = \mathbf{0}.$$

Theorem 5.17 states these properties along with several others.

THEOREM 5.17   Algebraic Properties of the Cross Product

If $\mathbf{u}$, $\mathbf{v}$, and $\mathbf{w}$ are vectors in R^3 and c is a scalar, then the properties listed below are true.

1. $\mathbf{u} \times \mathbf{v} = -(\mathbf{v} \times \mathbf{u})$
2. $\mathbf{u} \times (\mathbf{v} + \mathbf{w}) = (\mathbf{u} \times \mathbf{v}) + (\mathbf{u} \times \mathbf{w})$
3. $c(\mathbf{u} \times \mathbf{v}) = c\mathbf{u} \times \mathbf{v} = \mathbf{u} \times c\mathbf{v}$
4. $\mathbf{u} \times \mathbf{0} = \mathbf{0} \times \mathbf{u} = \mathbf{0}$
5. $\mathbf{u} \times \mathbf{u} = \mathbf{0}$
6. $\mathbf{u} \cdot (\mathbf{v} \times \mathbf{w}) = (\mathbf{u} \times \mathbf{v}) \cdot \mathbf{w}$

PROOF

The proof of the first property is given here. The proofs of the other properties are left to you. (See Exercises 55–59.) Let $\mathbf{u}$ and $\mathbf{v}$ be

$$\mathbf{u} = u_1\mathbf{i} + u_2\mathbf{j} + u_3\mathbf{k}$$

and

$$\mathbf{v} = v_1\mathbf{i} + v_2\mathbf{j} + v_3\mathbf{k}.$$

Then $\mathbf{u} \times \mathbf{v}$ is

$$\mathbf{u} \times \mathbf{v} = \begin{vmatrix} \mathbf{i} & \mathbf{j} & \mathbf{k} \\ u_1 & u_2 & u_3 \\ v_1 & v_2 & v_3 \end{vmatrix}$$

$$= (u_2 v_3 - u_3 v_2)\mathbf{i} - (u_1 v_3 - u_3 v_1)\mathbf{j} + (u_1 v_2 - u_2 v_1)\mathbf{k}$$

and $\mathbf{v} \times \mathbf{u}$ is

$$\mathbf{v} \times \mathbf{u} = \begin{vmatrix} \mathbf{i} & \mathbf{j} & \mathbf{k} \\ v_1 & v_2 & v_3 \\ u_1 & u_2 & u_3 \end{vmatrix}$$

$$= (v_2 u_3 - v_3 u_2)\mathbf{i} - (v_1 u_3 - v_3 u_1)\mathbf{j} + (v_1 u_2 - v_2 u_1)\mathbf{k}$$

$$= -(u_2 v_3 - u_3 v_2)\mathbf{i} + (u_1 v_3 - u_3 v_1)\mathbf{j} - (u_1 v_2 - u_2 v_1)\mathbf{k}$$

$$= -(\mathbf{v} \times \mathbf{u}).$$

Property 1 of Theorem 5.17 tells you that the vectors $\mathbf{u} \times \mathbf{v}$ and $\mathbf{v} \times \mathbf{u}$ have equal lengths but opposite directions. The geometric implication of this will be discussed after establishing some geometric properties of the cross product of two vectors.

THEOREM 5.18 Geometric Properties of the Cross Product

If $\mathbf{u}$ and $\mathbf{v}$ are nonzero vectors in R^3, then the properties listed below are true.

1. $\mathbf{u} \times \mathbf{v}$ is orthogonal to both $\mathbf{u}$ and $\mathbf{v}$.
2. The angle θ between $\mathbf{u}$ and $\mathbf{v}$ is found using $\|\mathbf{u} \times \mathbf{v}\| = \|\mathbf{u}\| \|\mathbf{v}\| \sin \theta$.
3. $\mathbf{u}$ and $\mathbf{v}$ are parallel if and only if $\mathbf{u} \times \mathbf{v} = \mathbf{0}$.
4. The parallelogram having $\mathbf{u}$ and $\mathbf{v}$ as adjacent sides has an area of $\|\mathbf{u} \times \mathbf{v}\|$.

PROOF

The proof of Property 4 is presented here. The proofs of the other properties are left to you. (See Exercises 63–65.) Let $\mathbf{u}$ and $\mathbf{v}$ represent adjacent sides of a parallelogram, as shown in Figure 5.20. By Property 2, the area of the parallelogram is

$$\overset{\text{Base}}{\overbrace{\phantom{\|\mathbf{u}\|}}} \overset{\text{Height}}{\overbrace{\phantom{\|\mathbf{v}\| \sin \theta}}}$$

$$\text{Area} = \|\mathbf{u}\| \|\mathbf{v}\| \sin \theta = \|\mathbf{u} \times \mathbf{v}\|.$$

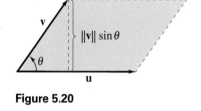

Figure 5.20

Property 1 states that the vector $\mathbf{u} \times \mathbf{v}$ is orthogonal to both $\mathbf{u}$ and $\mathbf{v}$. This implies that $\mathbf{u} \times \mathbf{v}$ (and $\mathbf{v} \times \mathbf{u}$) is orthogonal to the plane determined by $\mathbf{u}$ and $\mathbf{v}$. One way to remember the orientation of the vectors $\mathbf{u}$, $\mathbf{v}$, and $\mathbf{u} \times \mathbf{v}$ is to compare them with the unit vectors $\mathbf{i}$, $\mathbf{j}$, and $\mathbf{k}$, as shown below. The three vectors $\mathbf{u}$, $\mathbf{v}$, and $\mathbf{u} \times \mathbf{v}$ form a *right-handed system*, whereas the three vectors $\mathbf{u}$, $\mathbf{v}$, and $\mathbf{v} \times \mathbf{u}$ form a *left-handed system*.

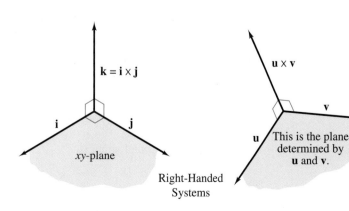

Right-Handed Systems

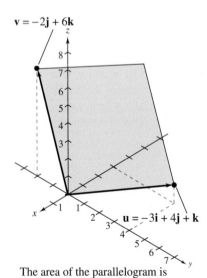

Figure 5.21

| EXAMPLE 2 | Finding a Vector Orthogonal to Two Given Vectors |

See LarsonLinearAlgebra.com for an interactive version of this type of example.

Find a unit vector orthogonal to both

$$\mathbf{u} = \mathbf{i} - 4\mathbf{j} + \mathbf{k}$$

and

$$\mathbf{v} = 2\mathbf{i} + 3\mathbf{j}.$$

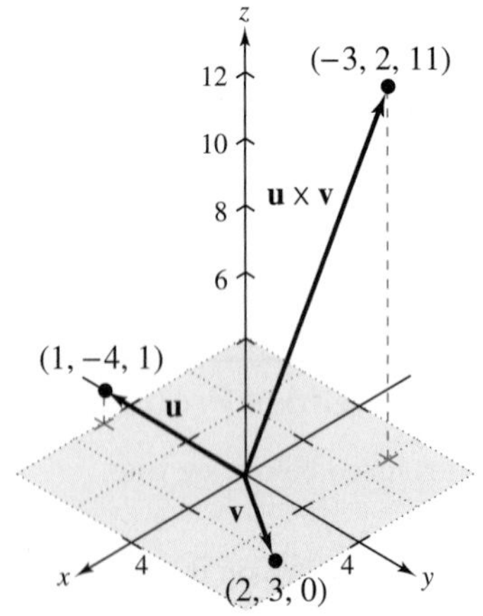

SOLUTION

From Property 1 of Theorem 5.18, you know that the cross product

$$\mathbf{u} \times \mathbf{v} = \begin{vmatrix} \mathbf{i} & \mathbf{j} & \mathbf{k} \\ 1 & -4 & 1 \\ 2 & 3 & 0 \end{vmatrix} = -3\mathbf{i} + 2\mathbf{j} + 11\mathbf{k}$$

is orthogonal to both $\mathbf{u}$ and $\mathbf{v}$, as shown in Figure 5.21. Then, by dividing by the length of $\mathbf{u} \times \mathbf{v}$,

$$\|\mathbf{u} \times \mathbf{v}\| = \sqrt{(-3)^2 + 2^2 + 11^2} = \sqrt{134}$$

you obtain the unit vector

$$\frac{\mathbf{u} \times \mathbf{v}}{\|\mathbf{u} \times \mathbf{v}\|} = -\frac{3}{\sqrt{134}}\mathbf{i} + \frac{2}{\sqrt{134}}\mathbf{j} + \frac{11}{\sqrt{134}}\mathbf{k}$$

which is orthogonal to both $\mathbf{u}$ and $\mathbf{v}$, because

$$\left(-\frac{3}{\sqrt{134}}, \frac{2}{\sqrt{134}}, \frac{11}{\sqrt{134}}\right) \cdot (1, -4, 1) = 0$$

and

$$\left(-\frac{3}{\sqrt{134}}, \frac{2}{\sqrt{134}}, \frac{11}{\sqrt{134}}\right) \cdot (2, 3, 0) = 0.$$

| EXAMPLE 3 | Finding the Area of a Parallelogram |

$\mathbf{v} = -2\mathbf{j} + 6\mathbf{k}$

$\mathbf{u} = -3\mathbf{i} + 4\mathbf{j} + \mathbf{k}$

The area of the parallelogram is
$\|\mathbf{u} \times \mathbf{v}\| = \sqrt{1036}.$

Figure 5.22

Find the area of the parallelogram that has

$$\mathbf{u} = -3\mathbf{i} + 4\mathbf{j} + \mathbf{k}$$

and

$$\mathbf{v} = -2\mathbf{j} + 6\mathbf{k}$$

as adjacent sides, as shown in Figure 5.22.

SOLUTION

From Property 4 of Theorem 5.18, you know that the area of this parallelogram is $\|\mathbf{u} \times \mathbf{v}\|$. The cross product is

$$\mathbf{u} \times \mathbf{v} = \begin{vmatrix} \mathbf{i} & \mathbf{j} & \mathbf{k} \\ -3 & 4 & 1 \\ 0 & -2 & 6 \end{vmatrix} = 26\mathbf{i} + 18\mathbf{j} + 6\mathbf{k}.$$

So, the area of the parallelogram is

$$\|\mathbf{u} \times \mathbf{v}\| = \sqrt{26^2 + 18^2 + 6^2} = \sqrt{1036} \approx 32.19 \text{ square units.}$$

LEAST SQUARES APPROXIMATIONS (CALCULUS)

Many problems in the physical sciences and engineering involve an approximation of a function f by another function g. If f is in $C[a, b]$ (the inner product space of all continuous functions on $[a, b]$), then g is usually chosen from a subspace W of $C[a, b]$. For example, to approximate the function

$$f(x) = e^x, 0 \leq x \leq 1$$

you could choose one of the forms of g listed below.

1. $g(x) = a_0 + a_1 x, \quad 0 \leq x \leq 1$ Linear

2. $g(x) = a_0 + a_1 x + a_2 x^2, \quad 0 \leq x \leq 1$ Quadratic

3. $g(x) = a_0 + a_1 \cos x + a_2 \sin x, \quad 0 \leq x \leq 1$ Trigonometric

Before discussing ways of finding the function g, you must define how one function can "best" approximate another function. One natural way would require the area bounded by the graphs of f and g on the interval $[a, b]$,

$$\text{Area} = \int_a^b |f(x) - g(x)| \, dx$$

to be a minimum with respect to other functions in the subspace W, as shown below.

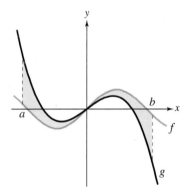

Integrands involving absolute value are often difficult to evaluate, however, so it is more common to square the integrand to obtain

$$\int_a^b [f(x) - g(x)]^2 \, dx.$$

With this criterion, the function g is the **least squares approximation** of f with respect to the inner product space W.

Definition of Least Squares Approximation

Let f be continuous on $[a, b]$, and let W be a subspace of $C[a, b]$. A function g in W is the **least squares approximation** of f with respect to W when the value of

$$I = \int_a^b [f(x) - g(x)]^2 \, dx$$

is a minimum with respect to all other functions in W.

Note that if the subspace W in this definition is the entire space $C[a, b]$, then $g(x) = f(x)$, which implies that $I = 0$.

EXAMPLE 4   **Finding a Least Squares Approximation**

Find the least squares approximation $g(x) = a_0 + a_1 x$ of

$$f(x) = e^x, \quad 0 \le x \le 1.$$

SOLUTION

For this approximation, you need to find the constants a_0 and a_1 that minimize the value of

$$I = \int_0^1 [f(x) - g(x)]^2 \, dx$$

$$= \int_0^1 (e^x - a_0 - a_1 x)^2 \, dx.$$

Evaluating this integral, you have

$$I = \int_0^1 (e^x - a_0 - a_1 x)^2 \, dx$$

$$= \int_0^1 (e^{2x} - 2a_0 e^x - 2a_1 x e^x + a_0^2 + 2a_0 a_1 x + a_1^2 x^2) \, dx$$

$$= \left[\frac{1}{2} e^{2x} - 2a_0 e^x - 2a_1 e^x (x - 1) + a_0^2 x + a_0 a_1 x^2 + a_1^2 \frac{x^3}{3} \right]_0^1$$

$$= \frac{1}{2}(e^2 - 1) - 2a_0(e - 1) - 2a_1 + a_0^2 + a_0 a_1 + \frac{1}{3} a_1^2.$$

Now, considering I to be a function of the variables a_0 and a_1, use calculus to determine the values of a_0 and a_1 that minimize I. Specifically, by setting the partial derivatives

$$\frac{\partial I}{\partial a_0} = 2a_0 - 2e + 2 + a_1$$

$$\frac{\partial I}{\partial a_1} = a_0 + \frac{2}{3} a_1 - 2$$

equal to zero, you obtain the two linear equations in a_0 and a_1 below.

$$2a_0 + \quad a_1 = 2(e - 1)$$
$$3a_0 + 2a_1 = 6$$

The solution of this system is

$$a_0 = 4e - 10 \approx 0.873 \quad \text{and} \quad a_1 = 18 - 6e \approx 1.690.$$

(Verify this.) So, the best *linear approximation* of $f(x) = e^x$ on the interval $[0, 1]$ is

$$g(x) = 4e - 10 + (18 - 6e)x \approx 0.873 + 1.690x.$$

Figure 5.23 shows the graphs of f and g on $[0, 1]$. ■

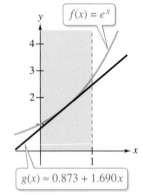

$f(x) = e^x$

$g(x) \approx 0.873 + 1.690x$

Figure 5.23

Of course, whether the approximation obtained in Example 4 is the best approximation depends on the definition of the best approximation. For instance, if the definition of the best approximation had been the *Taylor polynomial of degree 1* centered at 0.5, then the approximating function g would have been

$$g(x) = f(0.5) + f'(0.5)(x - 0.5)$$
$$= e^{0.5} + e^{0.5}(x - 0.5)$$
$$\approx 0.824 + 1.649x.$$

Moreover, the function g obtained in Example 4 is only the best *linear* approximation of f (according to the least squares criterion). In Example 5 you will find the best *quadratic* approximation.

EXAMPLE 5   **Finding a Least Squares Approximation**

Find the least squares approximation $g(x) = a_0 + a_1x + a_2x^2$ of $f(x) = e^x, 0 \leq x \leq 1$.

SOLUTION

For this approximation you need to find the values of a_0, a_1, and a_2 that minimize the value of

$$
\begin{aligned}
I &= \int_0^1 [f(x) - g(x)]^2 \, dx \\
&= \int_0^1 (e^x - a_0 - a_1x - a_2x^2)^2 \, dx \\
&= \frac{1}{2}(e^2 - 1) + 2a_0(1 - e) + 2a_2(2 - e) \\
&\quad + a_0^2 + a_0a_1 + \frac{2}{3}a_0a_2 + \frac{1}{2}a_1a_2 + \frac{1}{3}a_1^2 + \frac{1}{5}a_2^2 - 2a_1.
\end{aligned}
$$

Setting the partial derivatives of I (with respect to a_0, a_1, and a_2) equal to zero produces the system of linear equations below.

$$
\begin{aligned}
6a_0 + 3a_1 + 2a_2 &= 6(e - 1) \\
6a_0 + 4a_1 + 3a_2 &= 12 \\
20a_0 + 15a_1 + 12a_2 &= 60(e - 2)
\end{aligned}
$$

(Verify this.) The solution of this system is

$$
\begin{aligned}
a_0 &= -105 + 39e \approx 1.013 \\
a_1 &= 588 - 216e \approx 0.851 \\
a_2 &= -570 + 210e \approx 0.839.
\end{aligned}
$$

(Verify this.) So, the approximating function g is $g(x) \approx 1.013 + 0.851x + 0.839x^2$. Figure 5.24 shows the graphs of f and g on $[0, 1]$.

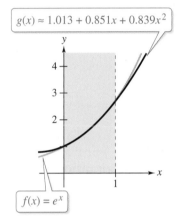

$g(x) \approx 1.013 + 0.851x + 0.839x^2$

$f(x) = e^x$

Figure 5.24

The integral I given in the definition of the least squares approximation can be expressed in vector form. To do this, use the inner product defined in Example 5 in Section 5.2:

$$
\langle f, g \rangle = \int_a^b f(x)g(x) \, dx.
$$

With this inner product you have

$$
I = \int_a^b [f(x) - g(x)]^2 \, dx = \langle f - g, f - g \rangle = \|f - g\|^2.
$$

This means that the least squares approximating function g is the function that minimizes $\|f - g\|^2$ or, equivalently, minimizes $\|f - g\|$. In other words, the least squares approximation of a function f is the function g (in the subspace W) closest to f in terms of the inner product $\langle f, g \rangle$. The next theorem gives you a way of determining the function g.

THEOREM 5.19   Least Squares Approximation

Let f be continuous on $[a, b]$, and let W be a finite-dimensional subspace of $C[a, b]$. The least squares approximating function of f with respect to W is

$$
g = \langle f, \mathbf{w}_1 \rangle \mathbf{w}_1 + \langle f, \mathbf{w}_2 \rangle \mathbf{w}_2 + \cdots + \langle f, \mathbf{w}_n \rangle \mathbf{w}_n
$$

where $B = \{\mathbf{w}_1, \mathbf{w}_2, \ldots, \mathbf{w}_n\}$ is an orthonormal basis for W.

PROOF

To show that g is the least squares approximating function of f, prove that the inequality $\|f - g\| \le \|f - \mathbf{w}\|$ is true for any vector $\mathbf{w}$ in W. Writing $f - g$ as

$$f - g = f - \langle f, \mathbf{w}_1 \rangle \mathbf{w}_1 - \langle f, \mathbf{w}_2 \rangle \mathbf{w}_2 - \cdots - \langle f, \mathbf{w}_n \rangle \mathbf{w}_n$$

shows that $f - g$ is orthogonal to each $\mathbf{w}_i$, which in turn implies that it is orthogonal to each vector in W. In particular, $f - g$ is orthogonal to $g - \mathbf{w}$. This allows you to apply the Pythagorean Theorem to the vector sum $f - \mathbf{w} = (f - g) + (g - \mathbf{w})$ to conclude that $\|f - \mathbf{w}\|^2 = \|f - g\|^2 + \|g - \mathbf{w}\|^2$. So, it follows that $\|f - g\|^2 \le \|f - \mathbf{w}\|^2$, which then implies that $\|f - g\| \le \|f - \mathbf{w}\|$. ∎

Now observe how Theorem 5.19 can be used to produce the least squares approximation obtained in Example 4. First apply the Gram-Schmidt orthonormalization process to the standard basis $\{1, x\}$ to obtain the orthonormal basis $B = \{1, \sqrt{3}(2x - 1)\}$. (Verify this.) Then, by Theorem 5.19, the least squares approximation of e^x in the subspace of all linear functions is

$$g(x) = \langle e^x, 1 \rangle (1) + \left\langle e^x, \sqrt{3}(2x - 1) \right\rangle \sqrt{3}(2x - 1)$$
$$= \int_0^1 e^x \, dx + \sqrt{3}(2x - 1) \int_0^1 \sqrt{3}e^x(2x - 1) \, dx$$
$$= \int_0^1 e^x \, dx + 3(2x - 1) \int_0^1 e^x(2x - 1) \, dx$$
$$= 4e - 10 + (18 - 6e)x$$

which agrees with the result obtained in Example 4.

EXAMPLE 6   **Finding a Least Squares Approximation**

Find the least squares approximation of $f(x) = \sin x$, $0 \le x \le \pi$, with respect to the subspace W of polynomial functions of degree 2 or less.

SOLUTION

To use Theorem 5.19, apply the Gram-Schmidt orthonormalization process to the standard basis for W, $\{1, x, x^2\}$, to obtain the orthonormal basis

$$B = \{\mathbf{w}_1, \mathbf{w}_2, \mathbf{w}_3\} = \left\{ \frac{1}{\sqrt{\pi}}, \frac{\sqrt{3}}{\pi\sqrt{\pi}}(2x - \pi), \frac{\sqrt{5}}{\pi^2\sqrt{\pi}}(6x^2 - 6\pi x + \pi^2) \right\}.$$

(Verify this.) The least squares approximating function g is

$$g(x) = \langle f, \mathbf{w}_1 \rangle \mathbf{w}_1 + \langle f, \mathbf{w}_2 \rangle \mathbf{w}_2 + \langle f, \mathbf{w}_3 \rangle \mathbf{w}_3$$

and you have

$$\langle f, \mathbf{w}_1 \rangle = \frac{1}{\sqrt{\pi}} \int_0^\pi \sin x \, dx = \frac{2}{\sqrt{\pi}}$$

$$\langle f, \mathbf{w}_2 \rangle = \frac{\sqrt{3}}{\pi\sqrt{\pi}} \int_0^\pi \sin x(2x - \pi) \, dx = 0$$

$$\langle f, \mathbf{w}_3 \rangle = \frac{\sqrt{5}}{\pi^2\sqrt{\pi}} \int_0^\pi \sin x(6x^2 - 6\pi x + \pi^2) \, dx = \frac{2\sqrt{5}}{\pi^2\sqrt{\pi}}(\pi^2 - 12).$$

So, g is

$$g(x) = \frac{2}{\pi} + \frac{10(\pi^2 - 12)}{\pi^5}(6x^2 - 6\pi x + \pi^2) \approx -0.4177x^2 + 1.3122x - 0.0505.$$

Figure 5.25 shows the graphs of f and g.

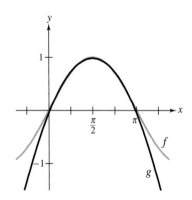

Figure 5.25

5.5 Applications of Inner Product Spaces 285

FOURIER APPROXIMATIONS (CALCULUS)

You will now look at a special type of least squares approximation called a **Fourier approximation.** For this approximation, consider functions of the form

$$g(x) = \frac{a_0}{2} + a_1 \cos x + \cdots + a_n \cos nx + b_1 \sin x + \cdots + b_n \sin nx$$

in the subspace W of

$$C[0, 2\pi]$$

spanned by the basis

$$S = \{1, \cos x, \cos 2x, \ldots, \cos nx, \sin x, \sin 2x, \ldots, \sin nx\}.$$

These $2n + 1$ vectors are orthogonal in the inner product space $C[0, 2\pi]$ because

$$\langle f, g \rangle = \int_0^{2\pi} f(x)g(x)\, dx$$
$$= 0, \quad f \neq g$$

as demonstrated in Example 3 in Section 5.3. Moreover, by normalizing each function in this basis, you obtain the orthonormal basis

$$B = \{\mathbf{w}_0, \mathbf{w}_1, \ldots, \mathbf{w}_n, \mathbf{w}_{n+1}, \ldots, \mathbf{w}_{2n}\}$$
$$= \left\{ \frac{1}{\sqrt{2\pi}}, \frac{1}{\sqrt{\pi}} \cos x, \ldots, \frac{1}{\sqrt{\pi}} \cos nx, \frac{1}{\sqrt{\pi}} \sin x, \ldots, \frac{1}{\sqrt{\pi}} \sin nx \right\}.$$

With this orthonormal basis, you can apply Theorem 5.19 to write

$$g(x) = \langle f, \mathbf{w}_0 \rangle \mathbf{w}_0 + \langle f, \mathbf{w}_1 \rangle \mathbf{w}_1 + \cdots + \langle f, \mathbf{w}_{2n} \rangle \mathbf{w}_{2n}.$$

The coefficients

$$a_0, a_1, \ldots, a_n, b_1, \ldots, b_n$$

for $g(x)$ in the equation

$$g(x) = \frac{a_0}{2} + a_1 \cos x + \cdots + a_n \cos nx + b_1 \sin x + \cdots + b_n \sin nx$$

are found using the integrals below.

$$a_0 = \langle f, \mathbf{w}_0 \rangle \frac{2}{\sqrt{2\pi}} = \frac{2}{\sqrt{2\pi}} \int_0^{2\pi} f(x) \frac{1}{\sqrt{2\pi}} dx = \frac{1}{\pi} \int_0^{2\pi} f(x)\, dx$$

$$a_1 = \langle f, \mathbf{w}_1 \rangle \frac{1}{\sqrt{\pi}} = \frac{1}{\sqrt{\pi}} \int_0^{2\pi} f(x) \frac{1}{\sqrt{\pi}} \cos x\, dx = \frac{1}{\pi} \int_0^{2\pi} f(x) \cos x\, dx$$

$$\vdots$$

$$a_n = \langle f, \mathbf{w}_n \rangle \frac{1}{\sqrt{\pi}} = \frac{1}{\sqrt{\pi}} \int_0^{2\pi} f(x) \frac{1}{\sqrt{\pi}} \cos nx\, dx = \frac{1}{\pi} \int_0^{2\pi} f(x) \cos nx\, dx$$

$$b_1 = \langle f, \mathbf{w}_{n+1} \rangle \frac{1}{\sqrt{\pi}} = \frac{1}{\sqrt{\pi}} \int_0^{2\pi} f(x) \frac{1}{\sqrt{\pi}} \sin x\, dx = \frac{1}{\pi} \int_0^{2\pi} f(x) \sin x\, dx$$

$$\vdots$$

$$b_n = \langle f, \mathbf{w}_{2n} \rangle \frac{1}{\sqrt{\pi}} = \frac{1}{\sqrt{\pi}} \int_0^{2\pi} f(x) \frac{1}{\sqrt{\pi}} \sin nx\, dx = \frac{1}{\pi} \int_0^{2\pi} f(x) \sin nx\, dx$$

The function $g(x)$ is the ***n*th-order Fourier approximation** of f on the interval $[0, 2\pi]$. Like Fourier coefficients, this function is named after the French mathematician Jean-Baptiste Joseph Fourier. This brings you to Theorem 5.20.

THEOREM 5.20   Fourier Approximation

On the interval $[0, 2\pi]$, the least squares approximation of a continuous function f with respect to the vector space spanned by

$$\{1, \cos x, \ldots, \cos nx, \sin x, \ldots, \sin nx\}$$

is

$$g(x) = \frac{a_0}{2} + a_1 \cos x + \cdots + a_n \cos nx + b_1 \sin x + \cdots + b_n \sin nx$$

where the **Fourier coefficients** $a_0, a_1, \ldots, a_n, b_1, \ldots, b_n$ are

$$a_0 = \frac{1}{\pi} \int_0^{2\pi} f(x)\, dx$$

$$a_j = \frac{1}{\pi} \int_0^{2\pi} f(x) \cos jx\, dx, \quad j = 1, 2, \ldots, n$$

$$b_j = \frac{1}{\pi} \int_0^{2\pi} f(x) \sin jx\, dx, \quad j = 1, 2, \ldots, n.$$

EXAMPLE 7   Finding a Fourier Approximation

Find the third-order Fourier approximation of $f(x) = x, \, 0 \le x \le 2\pi$.

SOLUTION

Using Theorem 5.20, you have

$$g(x) = \frac{a_0}{2} + a_1 \cos x + a_2 \cos 2x + a_3 \cos 3x + b_1 \sin x + b_2 \sin 2x + b_3 \sin 3x$$

where

$$a_0 = \frac{1}{\pi} \int_0^{2\pi} x\, dx = \frac{1}{\pi} 2\pi^2 = 2\pi$$

$$a_j = \frac{1}{\pi} \int_0^{2\pi} x \cos jx\, dx = \left[\frac{1}{\pi j^2} \cos jx + \frac{x}{\pi j} \sin jx \right]_0^{2\pi} = 0$$

$$b_j = \frac{1}{\pi} \int_0^{2\pi} x \sin jx\, dx = \left[\frac{1}{\pi j^2} \sin jx - \frac{x}{\pi j} \cos jx \right]_0^{2\pi} = -\frac{2}{j}.$$

This implies that $a_0 = 2\pi$, $a_1 = 0$, $a_2 = 0$, $a_3 = 0$, $b_1 = -2$, $b_2 = -\frac{2}{2} = -1$, and $b_3 = -\frac{2}{3}$. So, you have

$$g(x) = \frac{2\pi}{2} - 2 \sin x - \sin 2x - \frac{2}{3} \sin 3x$$

$$= \pi - 2 \sin x - \sin 2x - \frac{2}{3} \sin 3x.$$

The figure at the right compares the graphs of f and g.

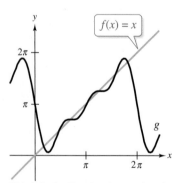

Third-Order Fourier Approximation

In Example 7, the pattern for the Fourier coefficients is $a_0 = 2\pi$, $a_1 = a_2 = \cdots = a_n = 0$, and

$$b_1 = -\frac{2}{1}, b_2 = -\frac{2}{2}, \ldots, b_n = -\frac{2}{n}.$$

The nth-order Fourier approximation of $f(x) = x$ is

$$g(x) = \pi - 2\left(\sin x + \frac{1}{2}\sin 2x + \frac{1}{3}\sin 3x + \cdots + \frac{1}{n}\sin nx\right).$$

As n increases, the Fourier approximation improves. For example, the figures below show the fourth- and fifth-order Fourier approximations of $f(x) = x, 0 \le x \le 2\pi$.

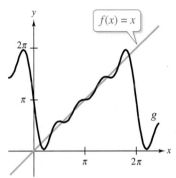

Fourth-Order Fourier Approximation

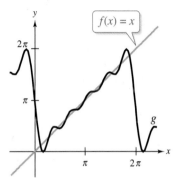

Fifth-Order Fourier Approximation

In advanced courses it is shown that as $n \to \infty$, the approximation error $\|f - g\|$ approaches zero. The infinite *series* for $g(x)$ is a *Fourier series*.

EXAMPLE 8   Finding a Fourier Approximation

Find the fourth-order Fourier approximation of $f(x) = |x - \pi|, 0 \le x \le 2\pi$.

SOLUTION

Using Theorem 5.20, find the Fourier coefficients as shown below.

$$a_0 = \frac{1}{\pi}\int_0^{2\pi} |x - \pi|\, dx = \pi$$

$$a_j = \frac{1}{\pi}\int_0^{2\pi} |x - \pi|\cos jx\, dx$$

$$= \frac{2}{\pi}\int_0^{\pi} (\pi - x)\cos jx\, dx$$

$$= \frac{2}{\pi j^2}(1 - \cos j\pi)$$

$$b_j = \frac{1}{\pi}\int_0^{2\pi} |x - \pi|\sin jx\, dx$$

$$= 0$$

So, $a_0 = \pi$, $a_1 = 4/\pi$, $a_2 = 0$, $a_3 = 4/(9\pi)$, $a_4 = 0$, $b_1 = 0$, $b_2 = 0$, $b_3 = 0$, and $b_4 = 0$, which means that the fourth-order Fourier approximation of f is

$$g(x) = \frac{\pi}{2} + \frac{4}{\pi}\cos x + \frac{4}{9\pi}\cos 3x.$$

Figure 5.26 compares the graphs of f and g.

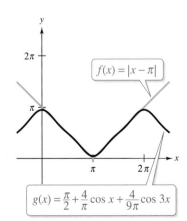

Figure 5.26

5.5   Exercises

See CalcChat.com for worked-out solutions to odd-numbered exercises.

Finding the Cross Product   In Exercises 1–6, find the cross product of the unit vectors [where i = (1, 0, 0), j = (0, 1, 0), and k = (0, 0, 1)]. Sketch your result.

1. $j \times i$
2. $i \times j$
3. $j \times k$
4. $k \times j$
5. $i \times k$
6. $k \times i$

Finding the Cross Product   In Exercises 7–14, find (a) u × v, (b) v × u, and (c) v × v.

7. $u = i - j,\quad v = j + k$
8. $u = 2i + k,\quad v = i + 3k$
9. $u = i + 2j - k,\quad v = i + j + 2k$
10. $u = i - j - k,\quad v = 2i + 2j + 2k$
11. $u = (-1, -1, 1),\quad v = (-1, 1, -1)$
12. $u = (3, -3, -3),\quad v = (3, -3, 3)$
13. $u = (3, -2, 4),\quad v = (1, 5, -3)$
14. $u = (-2, 9, -3),\quad v = (4, 6, -5)$

Finding the Cross Product   In Exercises 15–26, find u × v and show that it is orthogonal to both u and v.

15. $u = (0, 1, -2),\quad v = (1, -1, 0)$
16. $u = (-1, 1, 2),\quad v = (0, 1, -1)$
17. $u = (12, -3, 1),\quad v = (-2, 5, 1)$
18. $u = (-2, 1, 1),\quad v = (4, 2, 0)$
19. $u = (2, -3, 1),\quad v = (1, -2, 1)$
20. $u = (4, 1, 0),\quad v = (3, 2, -2)$
21. $u = j + 6k,\quad v = 2i - k$
22. $u = 2i - j + k,\quad v = 3i - j$
23. $u = i + j + k,\quad v = 2i + j - k$
24. $u = i - 2j + k,\quad v = -i + 3j - 2k$
25. $u = 3i + 2j + 4k,\quad v = 4i + 5j + 6k$
26. $u = -5i + 19j - 12k,\quad v = 5i - 19j + 12k$

Finding the Cross Product   In Exercises 27–34, use a graphing utility to find u × v, and then show that it is orthogonal to both u and v.

27. $u = (1, 2, -1),\quad v = (2, 1, 2)$
28. $u = (1, 2, -3),\quad v = (-1, 1, 2)$
29. $u = (0, 1, -1),\quad v = (1, 2, 0)$
30. $u = (2, 0, -1),\quad v = (-1, 0, -4)$
31. $u = -2i + j - k,\quad v = -i + 2j - k$
32. $u = 3i - j + k,\quad v = 2i + j - k$
33. $u = 2i + j - k,\quad v = i - j + 2k$
34. $u = 4i + 2j,\quad v = i - 4k$

Using the Cross Product   In Exercises 35–42, find a unit vector orthogonal to both u and v.

35. $u = (-4, 3, -2)$
    $v = (-1, 1, 0)$
36. $u = (2, -1, 3)$
    $v = (1, 0, -2)$
37. $u = 3i + j$
    $v = j + k$
38. $u = i + 2j$
    $v = i - 3k$
39. $u = -3i + 2j - 5k$
    $v = \frac{1}{2}i - \frac{3}{4}j + \frac{1}{10}k$
40. $u = 7i - 14j + 5k$
    $v = 14i + 28j - 15k$
41. $u = -i - j + k$
    $v = i - j - k$
42. $u = i - 2j + 2k$
    $v = 2i - j - 2k$

Finding the Area of a Parallelogram   In Exercises 43–46, find the area of the parallelogram that has the vectors as adjacent sides.

43. $u = j,\quad v = j + k$
44. $u = i - j + k,\quad v = i + k$
45. $u = (3, 2, -1),\quad v = (1, 2, 3)$
46. $u = (2, -1, 0),\quad v = (-1, 2, 0)$

Geometric Application of the Cross Product   In Exercises 47 and 48, verify that the points are the vertices of a parallelogram, and then find its area.

47. $(1, 1, 1), (2, 3, 4), (6, 5, 2), (7, 7, 5)$
48. $(1, -2, 0), (4, 0, 3), (-1, 0, 0), (2, 2, 3)$

Finding the Area of a Triangle   In Exercises 49 and 50, find the area of the triangle with the given vertices. Use the fact that the area A of the triangle having u and v as adjacent sides is $A = \frac{1}{2}\|u \times v\|$.

49. $(3, 5, 7), (5, 5, 0), (-4, 0, 4)$
50. $(2, -3, 4), (0, 1, 2), (-1, 2, 0)$

Triple Scalar Product   In Exercises 51–54, find u · (v × w). This quantity is called the triple scalar product of u, v, and w.

51. $u = i,\quad v = j,\quad w = k$
52. $u = -i,\quad v = -j,\quad w = k$
53. $u = (3, 3, 3),\quad v = (1, 2, 0),\quad w = (0, -1, 0)$
54. $u = (2, 0, 1),\quad v = (0, 3, 0),\quad w = (0, 0, 1)$

55. **Proof**   Prove that $u \times (v + w) = (u \times v) + (u \times w)$.
56. **Proof**   Prove that $c(u \times v) = cu \times v = u \times cv$.
57. **Proof**   Prove that $u \times 0 = 0 \times u = 0$.
58. **Proof**   Prove that $u \times u = 0$.
59. **Proof**   Prove that $u \cdot (v \times w) = (u \times v) \cdot w$.
60. **Proof**   Prove **Lagrange's Identity:**
$$\|u \times v\|^2 = \|u\|^2\|v\|^2 - (u \cdot v)^2.$$

61. Volume of a Parallelepiped Show that the volume V of a parallelepiped having $\mathbf{u}, \mathbf{v},$ and $\mathbf{w}$ as adjacent edges is $V = |\mathbf{u} \cdot (\mathbf{v} \times \mathbf{w})|$.

62. Finding the Volume of a Parallelepiped Use the result of Exercise 61 to find the volume of each parallelepiped.

(a) $\mathbf{u} = \mathbf{i} + \mathbf{j}$
$\mathbf{v} = \mathbf{j} + \mathbf{k}$
$\mathbf{w} = \mathbf{i} + 2\mathbf{k}$

(b) $\mathbf{u} = \mathbf{i} + \mathbf{j}$
$\mathbf{v} = \mathbf{j} + \mathbf{k}$
$\mathbf{w} = \mathbf{i} + \mathbf{k}$

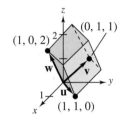 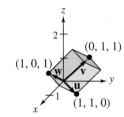

(c) $\mathbf{u} = (0, 2, 2)$
$\mathbf{v} = (0, 0, -2)$
$\mathbf{w} = (3, 0, 2)$

(d) $\mathbf{u} = (1, 2, -1)$
$\mathbf{v} = (-1, 2, 2)$
$\mathbf{w} = (2, 0, 1)$

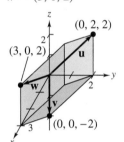 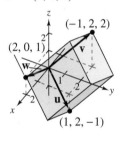

63. Proof Prove that $\mathbf{u} \times \mathbf{v}$ is orthogonal to both $\mathbf{u}$ and $\mathbf{v}$.

64. Proof Prove that the angle θ between $\mathbf{u}$ and $\mathbf{v}$ is found using $\|\mathbf{u} \times \mathbf{v}\| = \|\mathbf{u}\| \|\mathbf{v}\| \sin \theta$.

65. Proof Prove that $\mathbf{u}$ and $\mathbf{v}$ are parallel if and only if $\mathbf{u} \times \mathbf{v} = \mathbf{0}$.

66. Proof

(a) Prove that
$$\mathbf{u} \times (\mathbf{v} \times \mathbf{w}) = (\mathbf{u} \cdot \mathbf{w})\mathbf{v} - (\mathbf{u} \cdot \mathbf{v})\mathbf{w}.$$

(b) Find an example for which
$$\mathbf{u} \times (\mathbf{v} \times \mathbf{w}) \neq (\mathbf{u} \times \mathbf{v}) \times \mathbf{w}.$$

Finding a Least Squares Approximation **In Exercises 67–72, (a) find the least squares approximation $g(x) = a_0 + a_1 x$ of the function f, and (b) use a graphing utility to graph f and g in the same viewing window.**

67. $f(x) = x^2, \quad 0 \le x \le 1$

68. $f(x) = \sqrt{x}, \quad 1 \le x \le 4$

69. $f(x) = e^{2x}, \quad 0 \le x \le 1$

70. $f(x) = e^{-2x}, \quad 0 \le x \le 1$

71. $f(x) = \cos x, \quad 0 \le x \le \pi$

72. $f(x) = \sin x, \quad 0 \le x \le \pi/2$

Finding a Least Squares Approximation **In Exercises 73–76, (a) find the least squares approximation $g(x) = a_0 + a_1 x + a_2 x^2$ of the function f, and (b) use a graphing utility to graph f and g in the same viewing window.**

73. $f(x) = x^3, \quad 0 \le x \le 1$ 74. $f(x) = \sqrt{x}, \quad 1 \le x \le 4$

75. $f(x) = \sin x, \quad -\pi/2 \le x \le \pi/2$

76. $f(x) = \cos x, \quad -\pi/2 \le x \le \pi/2$

Finding a Fourier Approximation **In Exercises 77–88, find the Fourier approximation with the specified order of the function on the interval $[0, 2\pi]$.**

77. $f(x) = \pi - x, \quad$ third order

78. $f(x) = \pi - x, \quad$ fourth order

79. $f(x) = (x - \pi)^2, \quad$ third order

80. $f(x) = (x - \pi)^2, \quad$ fourth order

81. $f(x) = e^{-x}, \quad$ first order

82. $f(x) = e^{-x}, \quad$ second order

83. $f(x) = e^{-2x}, \quad$ first order

84. $f(x) = e^{-2x}, \quad$ second order

85. $f(x) = 1 + x, \quad$ third order

86. $f(x) = 1 + x, \quad$ fourth order

87. $f(x) = 2 \sin x \cos x, \quad$ fourth order

88. $f(x) = \sin^2 x, \quad$ fourth order

89. Use the results of Exercises 77 and 78 to find the nth-order Fourier approximation of $f(x) = \pi - x$ on the interval $[0, 2\pi]$.

90. Use the results of Exercises 79 and 80 to find the nth-order Fourier approximation of $f(x) = (x - \pi)^2$ on the interval $[0, 2\pi]$.

91. Use the results of Exercises 81 and 82 to find the nth-order Fourier approximation of $f(x) = e^{-x}$ on the interval $[0, 2\pi]$.

92. CAPSTONE

(a) Explain how to find the cross product of two vectors in R^3.

(b) Explain how to find the least squares approximation of a function $f \in C[a, b]$ with respect to a subspace W of $C[a, b]$.

(c) Explain how to find the nth-order Fourier approximation on the interval $[0, 2\pi]$ of a continuous function f with respect to the vector space spanned by $\{1, \cos x, \ldots, \cos nx, \sin x, \ldots, \sin nx\}$.

93. Use your school's library, the Internet, or some other reference source to find real-life applications of approximations of functions.

5 Review Exercises

See CalcChat.com for worked-out solutions to odd-numbered exercises.

Finding Lengths, Dot Product, and Distance   In Exercises 1–8, find (a) $\|\mathbf{u}\|$, (b) $\|\mathbf{v}\|$, (c) $\mathbf{u} \cdot \mathbf{v}$, and (d) $d(\mathbf{u}, \mathbf{v})$.

1. $\mathbf{u} = (1, 4)$, $\mathbf{v} = (2, 1)$

2. $\mathbf{u} = (-1, 2)$, $\mathbf{v} = (2, 3)$

3. $\mathbf{u} = (2, 1, 1)$, $\mathbf{v} = (3, 2, -1)$

4. $\mathbf{u} = (-3, 2, -2)$, $\mathbf{v} = (1, 3, 5)$

5. $\mathbf{u} = (1, -2, 0, 1)$, $\mathbf{v} = (1, 1, -1, 0)$

6. $\mathbf{u} = (1, -2, 2, 0)$, $\mathbf{v} = (2, -1, 0, 2)$

7. $\mathbf{u} = (0, 1, -1, 1, 2)$, $\mathbf{v} = (0, 1, -2, 1, 1)$

8. $\mathbf{u} = (1, -1, 0, 1, 1)$, $\mathbf{v} = (0, 1, -2, 2, 1)$

Finding Length and a Unit Vector   In Exercises 9–12, find $\|\mathbf{v}\|$ and find a unit vector in the direction of v.

9. $\mathbf{v} = (5, 3, -2)$ 10. $\mathbf{v} = (-1, -4, 1)$

11. $\mathbf{v} = (-1, 1, 2)$ 12. $\mathbf{v} = (0, 2, -1)$

13. Consider the vector $\mathbf{v} = (8, 8, 6)$. Find $\mathbf{u}$ such that

    (a) $\mathbf{u}$ has the same direction as $\mathbf{v}$ and one-half its length.

    (b) $\mathbf{u}$ has the direction opposite that of $\mathbf{v}$ and one-fourth its length.

    (c) $\mathbf{u}$ has the direction opposite that of $\mathbf{v}$ and twice its length.

14. For what values of c is $\|c(2, 2, -1)\| = 3$?

Finding the Angle Between Two Vectors   In Exercises 15–20, find the angle θ between the two vectors.

15. $\mathbf{u} = (3, 3)$, $\mathbf{v} = (-2, 2)$

16. $\mathbf{u} = (1, -1)$, $\mathbf{v} = (0, 1)$

17. $\mathbf{u} = \left(\cos \dfrac{3\pi}{4}, \sin \dfrac{3\pi}{4}\right)$, $\mathbf{v} = \left(\cos \dfrac{2\pi}{3}, \sin \dfrac{2\pi}{3}\right)$

18. $\mathbf{u} = \left(\cos \dfrac{\pi}{6}, \sin \dfrac{\pi}{6}\right)$, $\mathbf{v} = \left(\cos \dfrac{5\pi}{6}, \sin \dfrac{5\pi}{6}\right)$

19. $\mathbf{u} = (10, -5, 15)$, $\mathbf{v} = (-2, 1, -3)$

20. $\mathbf{u} = (0, 4, 0, -1)$, $\mathbf{v} = (1, 1, 3, -3)$

Finding Orthogonal Vectors   In Exercises 21–24, determine all vectors v that are orthogonal to u.

21. $\mathbf{u} = (0, -4, 3)$

22. $\mathbf{u} = (1, -2, 1)$

23. $\mathbf{u} = (2, -1, 1, 2)$

24. $\mathbf{u} = (0, 1, 2, -1)$

25. For $\mathbf{u} = \left(4, -\frac{3}{2}, -1\right)$ and $\mathbf{v} = \left(\frac{1}{2}, 3, 1\right)$, (a) find the inner product represented by $\langle \mathbf{u}, \mathbf{v}\rangle = u_1v_1 + 2u_2v_2 + 3u_3v_3$, and (b) use this inner product to find the distance between $\mathbf{u}$ and $\mathbf{v}$.

26. For $\mathbf{u} = \left(0, 3, \frac{1}{3}\right)$ and $\mathbf{v} = \left(\frac{4}{3}, 1, -3\right)$, (a) find the inner product represented by $\langle \mathbf{u}, \mathbf{v}\rangle = 2u_1v_1 + u_2v_2 + 2u_3v_3$ and (b) use this inner product to find the distance between $\mathbf{u}$ and $\mathbf{v}$.

27. Verify the triangle inequality and the Cauchy-Schwarz Inequality for $\mathbf{u}$ and $\mathbf{v}$ from Exercise 25. (Use the inner product given in Exercise 25.)

28. Verify the triangle inequality and the Cauchy-Schwarz Inequality for $\mathbf{u}$ and $\mathbf{v}$ from Exercise 26. (Use the inner product given in Exercise 26.)

Calculus   In Exercises 29 and 30, (a) find the inner product, (b) determine whether the vectors are orthogonal, and (c) verify the Cauchy-Schwarz Inequality for the vectors.

29. $f(x) = x$, $g(x) = \dfrac{1}{x^2 + 1}$, $\langle f, g\rangle = \displaystyle\int_{-1}^{1} f(x)g(x)\, dx$

30. $f(x) = x$, $g(x) = 4x^2$, $\langle f, g\rangle = \displaystyle\int_{0}^{1} f(x)g(x)\, dx$

Finding an Orthogonal Projection   In Exercises 31–36, find $\text{proj}_{\mathbf{v}}\mathbf{u}$.

31. $\mathbf{u} = (2, 4)$, $\mathbf{v} = (1, -5)$

32. $\mathbf{u} = (2, 3)$, $\mathbf{v} = (0, 4)$

33. $\mathbf{u} = (2, 5)$, $\mathbf{v} = (0, 5)$

34. $\mathbf{u} = (2, -1)$, $\mathbf{v} = (7, 6)$

35. $\mathbf{u} = (0, -1, 2)$, $\mathbf{v} = (3, 2, 4)$

36. $\mathbf{u} = (-1, 3, 1)$, $\mathbf{v} = (4, 0, 5)$

Applying the Gram-Schmidt Process   In Exercises 37–40, apply the Gram-Schmidt orthonormalization process to transform the given basis for R^n into an orthonormal basis. Use the Euclidean inner product for R^n and use the vectors in the order in which they are given.

37. $B = \{(1, 1), (0, 2)\}$

38. $B = \{(3, 4), (1, 2)\}$

39. $B = \{(0, 3, 4), (1, 0, 0), (1, 1, 0)\}$

40. $B = \{(0, 0, 2), (0, 1, 1), (1, 1, 1)\}$

41. Let $B = \{(0, 2, -2), (1, 0, -2)\}$ be a basis for a subspace of R^3, and consider $\mathbf{x} = (-1, 4, -2)$, a vector in the subspace.

    (a) Write $\mathbf{x}$ as a linear combination of the vectors in B. That is, find the coordinates of $\mathbf{x}$ relative to B.

    (b) Apply the Gram-Schmidt orthonormalization process to transform B into an orthonormal set B'.

    (c) Write $\mathbf{x}$ as a linear combination of the vectors in B'. That is, find the coordinates of $\mathbf{x}$ relative to B'.

42. Repeat Exercise 41 for $B = \{(-1, 2, 2), (1, 0, 0)\}$ and $\mathbf{x} = (-3, 4, 4)$.

Calculus In Exercises 43–46, let f and g be functions in the vector space $C[a, b]$ with inner product

$$\langle f, g \rangle = \int_a^b f(x)g(x)\, dx.$$

43. Show that $f(x) = \sin x$ and $g(x) = \cos x$ are orthogonal in $C[0, \pi]$.

44. Show that $f(x) = \sqrt{1 - x^2}$ and $g(x) = 2x\sqrt{1 - x^2}$ are orthogonal in $C[-1, 1]$.

45. Let $f(x) = x$ and $g(x) = x^3$ be vectors in $C[0, 1]$.

(a) Find $\langle f, g \rangle$.

(b) Find $\|g\|$.

(c) Find $d(f, g)$.

(d) Orthonormalize the set $B = \{f, g\}$.

46. Let $f(x) = x + 2$ and $g(x) = 15x - 8$ be vectors in $C[0, 1]$.

(a) Find $\langle f, g \rangle$.

(b) Find $\langle -4f, g \rangle$.

(c) Find $\|f\|$.

(d) Orthonormalize the set $B = \{f, g\}$.

47. Find an orthonormal basis for the subspace of Euclidean 3-space below.

$$W = \{(x_1, x_2, x_3): x_1 + x_2 + x_3 = 0\}$$

48. Find an orthonormal basis for the solution space of the homogeneous system of linear equations.

$$x + y - z + \ w = 0$$
$$2x - y + z + 2w = 0$$

49. Proof Prove that if $\mathbf{u}$, $\mathbf{v}$, and $\mathbf{w}$ are vectors in R^n, then $(\mathbf{u} + \mathbf{v}) \cdot \mathbf{w} = \mathbf{u} \cdot \mathbf{w} + \mathbf{v} \cdot \mathbf{w}$.

50. Proof Prove that if $\mathbf{u}$ and $\mathbf{v}$ are vectors in R^n, then $\|\mathbf{u} + \mathbf{v}\|^2 + \|\mathbf{u} - \mathbf{v}\|^2 = 2\|\mathbf{u}\|^2 + 2\|\mathbf{v}\|^2$.

51. Proof Prove that if $\mathbf{u}$ and $\mathbf{v}$ are vectors in an inner product space such that $\|\mathbf{u}\| \leq 1$ and $\|\mathbf{v}\| \leq 1$, then $|\langle \mathbf{u}, \mathbf{v} \rangle| \leq 1$.

52. Proof Prove that if $\mathbf{u}$ and $\mathbf{v}$ are vectors in an inner product space V, then

$$\big| \|\mathbf{u}\| - \|\mathbf{v}\| \big| \leq \|\mathbf{u} \pm \mathbf{v}\|.$$

53. Proof Let V be an m-dimensional subspace of R^n such that $m < n$. Prove that any vector $\mathbf{u}$ in R^n can be uniquely written in the form $\mathbf{u} = \mathbf{v} + \mathbf{w}$, where $\mathbf{v}$ is in V and $\mathbf{w}$ is orthogonal to every vector in V.

54. Let V be the two-dimensional subspace of R^4 spanned by $(0, 1, 0, 1)$ and $(0, 2, 0, 0)$. Write the vector $\mathbf{u} = (1, 1, 1, 1)$ in the form $\mathbf{u} = \mathbf{v} + \mathbf{w}$, where $\mathbf{v}$ is in V and $\mathbf{w}$ is orthogonal to every vector in V.

55. Proof Let $\{\mathbf{u}_1, \mathbf{u}_2, \ldots, \mathbf{u}_m\}$ be an orthonormal subset of R^n, and let $\mathbf{v}$ be any vector in R^n. Prove that

$$\|\mathbf{v}\|^2 \geq \sum_{i=1}^m (\mathbf{v} \cdot \mathbf{u}_i)^2.$$

(This inequality is called **Bessel's Inequality**.)

56. Proof Let $\{x_1, x_2, \ldots, x_n\}$ be a set of real numbers. Use the Cauchy-Schwarz Inequality to prove that

$$(x_1 + x_2 + \cdots + x_n)^2 \leq n(x_1^2 + x_2^2 + \cdots + x_n^2).$$

57. Proof Let $\mathbf{u}$ and $\mathbf{v}$ be vectors in an inner product space V. Prove that $\|\mathbf{u} + \mathbf{v}\| = \|\mathbf{u} - \mathbf{v}\|$ if and only if $\mathbf{u}$ and $\mathbf{v}$ are orthogonal.

58. Writing Let $\{\mathbf{u}_1, \mathbf{u}_2, \ldots, \mathbf{u}_n\}$ be a dependent set of vectors in an inner product space V. Describe the result of applying the Gram-Schmidt orthonormalization process to this set.

59. Find the orthogonal complement $S^\perp$ of the subspace S of R^3 spanned by the two column vectors of the matrix

$$A = \begin{bmatrix} 1 & 2 \\ 2 & 1 \\ 0 & -1 \end{bmatrix}.$$

60. Find the projection of the vector $\mathbf{v} = \begin{bmatrix} 1 & 0 & -2 \end{bmatrix}^T$ onto the subspace

$$S = \text{span}\left\{ \begin{bmatrix} 0 \\ -1 \\ 1 \end{bmatrix}, \begin{bmatrix} 0 \\ 1 \\ 1 \end{bmatrix} \right\}.$$

61. Find bases for the four fundamental subspaces of the matrix

$$A = \begin{bmatrix} 0 & 1 & 0 \\ 0 & -3 & 0 \\ 1 & 0 & 1 \end{bmatrix}.$$

62. Find the least squares regression line for the set of data points

$$\{(-2, 2), (-1, 1), (0, 1), (1, 3)\}.$$

Graph the points and the line on the same set of axes.

63. Revenue The table shows the revenues y (in billions of dollars) for Google, Incorporated from 2006 through 2013. Find the least squares regression cubic polynomial for the data. Then use the model to predict the revenue in 2018. Let t represent the year, with $t = 6$ corresponding to 2006. (Source: Google, Incorporated)

Year	2006	2007	2008	2009
Revenue, y	10.6	16.6	21.8	23.7

Year	2010	2011	2012	2013
Revenue, y	29.3	37.9	50.2	59.8

64. Petroleum Production The table shows the North American petroleum productions y (in millions of barrels per day) from 2006 through 2013. Find the least squares regression linear and quadratic polynomials for the data. Then use the model to predict the petroleum production in 2018. Let t represent the year, with $t = 6$ corresponding to 2006. Which model appears to be more accurate for predicting future petroleum productions? Explain. (Source: U.S. Energy Information Administration)

Year	2006	2007	2008	2009
Petroleum Production, y	15.3	15.4	15.1	15.4

Year	2010	2011	2012	2013
Petroleum Production, y	16.1	16.7	17.9	19.3

Finding the Cross Product In Exercises 65–68, find $\mathbf{u} \times \mathbf{v}$ and show that it is orthogonal to both $\mathbf{u}$ and $\mathbf{v}$.

65. $\mathbf{u} = (1, 1, 0), \quad \mathbf{v} = (0, 3, 0)$

66. $\mathbf{u} = (1, -1, 1), \quad \mathbf{v} = (0, 1, 1)$

67. $\mathbf{u} = \mathbf{j} + 6\mathbf{k}, \quad \mathbf{v} = \mathbf{i} - 2\mathbf{j} + \mathbf{k}$

68. $\mathbf{u} = 2\mathbf{i} - \mathbf{k}, \quad \mathbf{v} = \mathbf{i} + \mathbf{j} - \mathbf{k}$

Finding the Volume of a Parallelepiped In Exercises 69–72, find the volume V of the parallelepiped that has $\mathbf{u}, \mathbf{v},$ and $\mathbf{w}$ as adjacent edges using the formula $V = |\mathbf{u} \cdot (\mathbf{v} \times \mathbf{w})|$.

69. $\mathbf{u} = (1, 0, 0)$
$\mathbf{v} = (0, 0, 1)$
$\mathbf{w} = (0, 1, 0)$

70. $\mathbf{u} = (1, 2, 1)$
$\mathbf{v} = (-1, -1, 0)$
$\mathbf{w} = (3, 4, -1)$

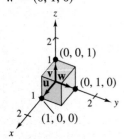

71. $\mathbf{u} = -2\mathbf{i} + \mathbf{j}$
$\mathbf{v} = 3\mathbf{i} - 2\mathbf{j} + \mathbf{k}$
$\mathbf{w} = 2\mathbf{i} - 3\mathbf{j} - 2\mathbf{k}$

72. $\mathbf{u} = \mathbf{i} + \mathbf{j} + 3\mathbf{k}$
$\mathbf{v} = 3\mathbf{j} + 3\mathbf{k}$
$\mathbf{w} = 3\mathbf{i} + 3\mathbf{k}$

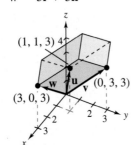

73. Find the area of the parallelogram that has
$\mathbf{u} = (1, 3, 0) \quad \text{and} \quad \mathbf{v} = (-1, 0, 2)$
as adjacent sides.

74. Proof Prove that
$\|\mathbf{u} \times \mathbf{v}\| = \|\mathbf{u}\| \, \|\mathbf{v}\|$
if and only if $\mathbf{u}$ and $\mathbf{v}$ are orthogonal.

Finding a Least Squares Approximation In Exercises 75–78, (a) find the least squares approximation $g(x) = a_0 + a_1 x$ of the function f, and (b) use a graphing utility to graph f and g in the same viewing window.

75. $f(x) = x^3, \ -1 \le x \le 1$

76. $f(x) = x^3, \ 0 \le x \le 2$

77. $f(x) = \sin 2x, \ 0 \le x \le \pi/2$

78. $f(x) = \sin x \cos x, \ 0 \le x \le \pi$

Finding a Least Squares Approximation In Exercises 79 and 80, (a) find the least squares approximation $g(x) = a_0 + a_1 x + a_2 x^2$ of the function f, and (b) use a graphing utility to graph f and g in the same viewing window.

79. $f(x) = \sqrt{x}, \ 0 \le x \le 1$ **80.** $f(x) = \dfrac{1}{x}, \ 1 \le x \le 2$

Finding a Fourier Approximation In Exercises 81 and 82, find the Fourier approximation with the specified order of the function on the interval $[-\pi, \pi]$.

81. $f(x) = x^2$, first order

82. $f(x) = x$, second order

True or False? In Exercises 83 and 84, determine whether each statement is true or false. If a statement is true, give a reason or cite an appropriate statement from the text. If a statement is false, provide an example that shows the statement is not true in all cases or cite an appropriate statement from the text.

83. (a) The cross product of two nonzero vectors in R^3 yields a vector orthogonal to the two vectors that produced it.

(b) The cross product of two nonzero vectors in R^3 is commutative.

(c) The least squares approximation of a function f is the function g (in the subspace W) closest to f in terms of the inner product $\langle f, g \rangle$.

84. (a) The vectors $\mathbf{u} \times \mathbf{v}$ and $\mathbf{v} \times \mathbf{u}$ in R^3 have equal lengths but opposite directions.

(b) If $\mathbf{u}$ and $\mathbf{v}$ are two nonzero vectors in R^3, then $\mathbf{u}$ and $\mathbf{v}$ are parallel if and only if $\mathbf{u} \times \mathbf{v} = \mathbf{0}$.

(c) A special type of least squares approximation, the Fourier approximation, is spanned by the basis $S = \{1, \cos x, \cos 2x, \ldots, \cos nx, \sin x, \sin 2x, \ldots, \sin nx\}$.

5 Projects

1 The *QR*-Factorization

The Gram-Schmidt orthonormalization process leads to an important factorization of matrices called the **QR-factorization.** If A is an $m \times n$ matrix of rank n, then A can be expressed as the product $A = QR$ of an $m \times n$ matrix Q and an $n \times n$ matrix R, where Q has orthonormal columns and R is upper triangular.

The columns of A can be considered a basis for a subspace of R^m, and the columns of Q are the result of applying the Gram-Schmidt orthonormalization process to this set of column vectors.

Recall that Example 7, Section 5.3, used the Gram-Schmidt orthonormalization process on the column vectors $\mathbf{v}_1$, $\mathbf{v}_2$ and $\mathbf{v}_3$ of the matrix

$$A = \begin{bmatrix} 1 & 1 & 0 \\ 1 & 2 & 1 \\ 0 & 0 & 2 \end{bmatrix}$$

to produce an orthonormal basis for R^3, which is labeled here as $\mathbf{q}_1$, $\mathbf{q}_2$, $\mathbf{q}_3$.

$$\mathbf{q}_1 = \left(\sqrt{2}/2, \sqrt{2}/2, 0 \right), \quad \mathbf{q}_2 = \left(-\sqrt{2}/2, \sqrt{2}/2, 0 \right), \quad \mathbf{q}_3 = (0, 0, 1)$$

These vectors form the columns of the matrix Q.

$$Q = \begin{bmatrix} \sqrt{2}/2 & -\sqrt{2}/2 & 0 \\ \sqrt{2}/2 & \sqrt{2}/2 & 0 \\ 0 & 0 & 1 \end{bmatrix}$$

The upper triangular matrix R is

$$R = \begin{bmatrix} \mathbf{v}_1 \cdot \mathbf{q}_1 & \mathbf{v}_2 \cdot \mathbf{q}_1 & \mathbf{v}_3 \cdot \mathbf{q}_1 \\ 0 & \mathbf{v}_2 \cdot \mathbf{q}_2 & \mathbf{v}_3 \cdot \mathbf{q}_2 \\ 0 & 0 & \mathbf{v}_3 \cdot \mathbf{q}_3 \end{bmatrix} = \begin{bmatrix} \sqrt{2} & 3\sqrt{2}/2 & \sqrt{2}/2 \\ 0 & \sqrt{2}/2 & \sqrt{2}/2 \\ 0 & 0 & 2 \end{bmatrix}$$

Verify that $A = QR$.

In general, if A is an $m \times n$ matrix of rank n with columns $\mathbf{v}_1, \mathbf{v}_2, \ldots, \mathbf{v}_n$, then the QR-factorization of A is

$$A = QR$$

$$\begin{bmatrix} \mathbf{v}_1 & \mathbf{v}_2 & \cdots & \mathbf{v}_n \end{bmatrix} = \begin{bmatrix} \mathbf{q}_1 & \mathbf{q}_2 & \cdots & \mathbf{q}_n \end{bmatrix} \begin{bmatrix} \mathbf{v}_1 \cdot \mathbf{q}_1 & \mathbf{v}_2 \cdot \mathbf{q}_1 & \cdots & \mathbf{v}_n \cdot \mathbf{q}_1 \\ 0 & \mathbf{v}_2 \cdot \mathbf{q}_2 & \cdots & \mathbf{v}_n \cdot \mathbf{q}_2 \\ \vdots & \vdots & & \vdots \\ 0 & 0 & \cdots & \mathbf{v}_n \cdot \mathbf{q}_n \end{bmatrix}$$

where the columns $\mathbf{q}_1, \mathbf{q}_2, \ldots, \mathbf{q}_n$, of the $m \times n$ matrix Q are the orthonormal vectors that result from the Gram-Schmidt orthonormalization process.

1. Find the QR-factorization of each matrix.

(a) $A = \begin{bmatrix} 1 & 0 \\ 1 & 1 \\ 1 & 1 \end{bmatrix}$ (b) $A = \begin{bmatrix} 1 & 2 \\ 1 & 0 \\ 0 & 0 \\ 0 & 1 \end{bmatrix}$ (c) $A = \begin{bmatrix} 2 & 0 & -2 \\ 2 & 1 & 0 \\ 2 & 1 & 0 \\ 2 & 0 & 0 \end{bmatrix}$

2. Let $A = QR$ be the QR-factorization of the $m \times n$ matrix A of rank n. Show how the least squares problem can be solved using the QR-factorization.

3. Use the result of part 2 to solve the least squares problem $A\mathbf{x} = \mathbf{b}$ when A is the matrix from part 1(a) and $\mathbf{b} = \begin{bmatrix} 1 & -1 & 1 \end{bmatrix}^T$.

REMARK

The *QR*-factorization of a matrix forms the basis for many algorithms of linear algebra. Algorithms for the computation of eigenvalues (see Chapter 7) are based on this factorization, as are algorithms for computing the least squares regression line for a set of data points. It should also be mentioned that, in practice, techniques other than the Gram-Schmidt orthonormalization process are used to compute the *QR*-factorization of a matrix.

2 Orthogonal Matrices and Change of Basis

Let $B = \{\mathbf{v}_1, \mathbf{v}_2, \ldots, \mathbf{v}_n\}$ be an ordered basis for the vector space V. Recall that the coordinate matrix of a vector $\mathbf{x} = c_1\mathbf{v}_1 + c_2\mathbf{v}_2 + \cdots + c_n\mathbf{v}_n$ in V is the column vector

$$[\mathbf{x}]_B = \begin{bmatrix} c_1 \\ c_2 \\ \vdots \\ c_n \end{bmatrix}.$$

If B' is another basis for V, then the transition matrix P from B' to B changes a coordinate matrix relative to B' into a coordinate matrix relative to B,

$$P[\mathbf{x}]_{B'} = [\mathbf{x}]_B.$$

The question you will explore now is whether there are transition matrices P that preserve the length of the coordinate matrix—that is, given $P[\mathbf{x}]_{B'} = [\mathbf{x}]_B$, does $\|[\mathbf{x}]_{B'}\| = \|[\mathbf{x}]_B\|$?

For example, consider the transition matrix

$$P = \begin{bmatrix} 4 & -2 \\ 5 & -2 \end{bmatrix}$$

from B' to B for the bases for R^2,

$$B' = \{(1, -2), (-2, 2)\} \quad \text{and} \quad B = \{(4, -3), (-3, 2)\}.$$

If $\mathbf{x} = (2, -6)$, then $[\mathbf{x}]_{B'} = [4 \quad 1]^T$ and $[\mathbf{x}]_B = P[\mathbf{x}]_{B'} = [14 \quad 18]^T$. (Verify this.) So, using the Euclidean norm for R^2,

$$\|[\mathbf{x}]_{B'}\| = \sqrt{17} \neq 2\sqrt{130} = \|[\mathbf{x}]_B\|.$$

You will see in this project that if the transition matrix P is **orthogonal,** then the norm of the coordinate vector will remain unchanged. You may recall working with orthogonal matrices in Section 3.3 (Exercises 73–82) and Section 5.3 (Exercise 65).

Definition of Orthogonal Matrix

The square matrix P is **orthogonal** when it is invertible and $P^{-1} = P^T$.

---

1. Show that the matrix P defined previously is *not* orthogonal.

2. Show that for any real number θ, the matrix

$$\begin{bmatrix} \cos\theta & \sin\theta \\ -\sin\theta & \cos\theta \end{bmatrix}$$

is orthogonal.

3. Show that a matrix is orthogonal if and only if its columns are pairwise orthogonal.

4. Prove that the inverse of an orthogonal matrix is orthogonal.

5. Is the sum of orthogonal matrices orthogonal? Is the product of orthogonal matrices orthogonal? Illustrate your answers with appropriate examples.

6. Prove that if P is an $n \times n$ orthogonal matrix, then $\|P\mathbf{x}\| = \|\mathbf{x}\|$ for all vectors $\mathbf{x}$ in R^n.

7. Verify the result of part 6 using the bases $B = \{(1, 0), (0, 1)\}$ and

$$B' = \left\{ \left(\frac{2}{\sqrt{13}}, \frac{3}{\sqrt{13}} \right), \left(-\frac{3}{\sqrt{13}}, \frac{2}{\sqrt{13}} \right) \right\}.$$

4 and 5 Cumulative Test

See CalcChat.com for worked-out solutions to odd-numbered exercises.

Take this test to review the material in Chapters 4 and 5. After you are finished, check your work against the answers in the back of the book.

1. Consider the vectors $\mathbf{v} = (1, -2)$ and $\mathbf{w} = (2, -5)$. Find and sketch each vector.
   (a) $\mathbf{v} + \mathbf{w}$ (b) $3\mathbf{v}$ (c) $2\mathbf{v} - 4\mathbf{w}$

2. Write $\mathbf{w} = (7, 2, 4)$ as a linear combination of the vectors $\mathbf{v}_1$, $\mathbf{v}_2$ and $\mathbf{v}_3$ (if possible).
   $$\mathbf{v}_1 = (2, 1, 0), \quad \mathbf{v}_2 = (1, -1, 0), \quad \mathbf{v}_3 = (0, 0, 6)$$

3. Write the third column of the matrix as a linear combination of the first two columns (if possible).
$$\begin{bmatrix} 1 & 0 & -2 \\ 4 & 2 & -2 \\ 7 & 5 & 1 \end{bmatrix}$$

4. Use a software program or a graphing utility to write $\mathbf{v}$ as a linear combination of $\mathbf{u}_1, \mathbf{u}_2, \mathbf{u}_3, \mathbf{u}_4, \mathbf{u}_5,$ and $\mathbf{u}_6$. Then verify your solution.
   $$\mathbf{v} = (10, 30, -13, 14, -7, 27)$$
   $$\mathbf{u}_1 = (1, 2, -3, 4, -1, 2)$$
   $$\mathbf{u}_2 = (1, -2, 1, -1, 2, 1)$$
   $$\mathbf{u}_3 = (0, 2, -1, 2, -1, -1)$$
   $$\mathbf{u}_4 = (1, 0, 3, -4, 1, 2)$$
   $$\mathbf{u}_5 = (1, -2, 1, -1, 2, -3)$$
   $$\mathbf{u}_6 = (3, 2, 1, -2, 3, 0)$$

5. Prove that the set of all singular 3×3 matrices is not a vector space.

6. Determine whether the set is a subspace of R^4.
   $$\{(x, x + y, y, y): x, y \in R\}$$

7. Determine whether the set is a subspace of R^3.
   $$\{(x, xy, y): x, y \in R\}$$

8. Determine whether the columns of matrix A span R^4.
$$A = \begin{bmatrix} 1 & 2 & -1 & 0 \\ 1 & 3 & 0 & 2 \\ 0 & 0 & 1 & -1 \\ 1 & 0 & 0 & 1 \end{bmatrix}$$

9. (a) Explain what it means to say that a set of vectors is *linearly independent*.
   (b) Determine whether the set S is linearly dependent or independent.
   $$S = \{(1, 0, 1, 0), (0, 3, 0, 1), (1, 1, 2, 2), (3, 4, 1, -2)\}$$

10. (a) Define a *basis* for a vector space.
    (b) Determine whether the set $\{\mathbf{v}_1, \mathbf{v}_2\}$ shown in the figure at the left is a basis for R^2.
    (c) Determine whether the set below is a basis for R^3.
    $$\{(1, 2, 1), (0, 1, 2), (2, 1, -3)\}$$

11. Find a basis for the solution space of $A\mathbf{x} = \mathbf{0}$ when
$$A = \begin{bmatrix} 1 & 1 & 0 & 0 \\ -2 & -2 & 0 & 0 \\ 0 & 0 & 1 & 1 \\ 1 & 1 & 0 & 0 \end{bmatrix}.$$

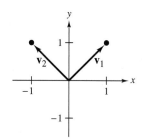

Figure for 10(b)

12. Find the coordinates $[\mathbf{v}]_B$ of the vector $\mathbf{v} = (1, 2, -3)$ relative to the basis $B = \{(0, 1, 1), (1, 1, 1), (1, 0, 1)\}$.

13. Find the transition matrix from the basis $B = \{(2, 1, 0), (1, 0, 0), (0, 1, 1)\}$ to the basis $B' = \{(1, 1, 2), (1, 1, 1), (0, 1, 2)\}$.

14. Let $\mathbf{u} = (1, 2, 0)$ and $\mathbf{v} = (1, -3, 2)$.

(a) Find $\|\mathbf{u}\|$.

(b) Find the distance between $\mathbf{u}$ and $\mathbf{v}$.

(c) Find $\mathbf{u} \cdot \mathbf{v}$.

(d) Find the angle θ between $\mathbf{u}$ and $\mathbf{v}$.

15. Find the inner product of $f(x) = x^2$ and $g(x) = x + 2$ from $C[0, 1]$ using

$$\langle f, g \rangle = \int_0^1 f(x)g(x)\, dx.$$

16. Apply the Gram-Schmidt orthonormalization process to transform the set of vectors into an orthonormal basis for R^3.

$$\{(2, 0, 0), (1, 1, 1), (0, 1, 2)\}$$

17. Let $\mathbf{u} = (1, 2)$ and $\mathbf{v} = (-3, 2)$. Find $\text{proj}_\mathbf{v}\mathbf{u}$, and graph $\mathbf{u}$, $\mathbf{v}$, and $\text{proj}_\mathbf{v}\mathbf{u}$ on the same set of coordinate axes.

18. Find the four fundamental subspaces of the matrix

$$A = \begin{bmatrix} 0 & 1 & 1 & 0 \\ -1 & 0 & 0 & 1 \\ 1 & 1 & 1 & 1 \end{bmatrix}.$$

19. Find the orthogonal complement $S^\perp$ of the set

$$S = \text{span}\left\{ \begin{bmatrix} 1 \\ 0 \\ 1 \end{bmatrix}, \begin{bmatrix} -1 \\ 1 \\ 0 \end{bmatrix} \right\}.$$

20. Consider a set of n linearly independent vectors $S = \{\mathbf{x}_1, \mathbf{x}_2, \ldots, \mathbf{x}_n\}$. Prove that if a vector $\mathbf{y}$ is not in span(S), then the set $S_1 = \{\mathbf{x}_1, \mathbf{x}_2, \ldots, \mathbf{x}_n, \mathbf{y}\}$ is linearly independent.

21. Find the least squares regression line for the points $\{(1, 1), (2, 0), (5, -5)\}$. Graph the points and the line.

22. The two matrices A and B are row-equivalent.

$$A = \begin{bmatrix} 2 & -4 & 0 & 1 & 7 & 11 \\ 1 & -2 & -1 & 1 & 9 & 12 \\ -1 & 2 & 1 & 3 & -5 & 16 \\ 4 & -8 & 1 & -1 & 6 & -2 \end{bmatrix} \quad B = \begin{bmatrix} 1 & -2 & 0 & 0 & 3 & 2 \\ 0 & 0 & 1 & 0 & -5 & -3 \\ 0 & 0 & 0 & 1 & 1 & 7 \\ 0 & 0 & 0 & 0 & 0 & 0 \end{bmatrix}$$

(a) Find the rank of A.

(b) Find a basis for the row space of A.

(c) Find a basis for the column space of A.

(d) Find a basis for the nullspace of A.

(e) Is the last column of A in the span of the first three columns?

(f) Are the first three columns of A linearly independent?

(g) Is the last column of A in the span of columns 1, 3, and 4?

(h) Are columns 1, 3, and 4 linearly dependent?

23. Let S_1 and S_2 be two-dimensional subspaces of R^3. Is it possible that $S_1 \cap S_2 = \{(0, 0, 0)\}$? Explain.

24. Let V be a vector space of dimension n. Prove that any set of less than n vectors cannot span V.

6 Linear Transformations

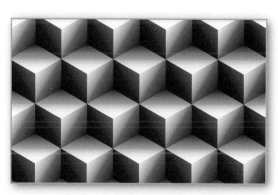

Computer Graphics (p. 338)

Population Age and Growth Distribution (p. 331)

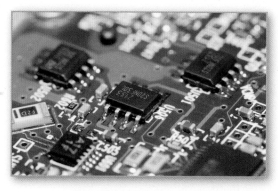

Circuit Design (p. 322)

Control Systems (p. 314)

Multivariate Statistics (p. 304)

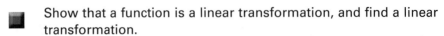

6.1   Introduction to Linear Transformations

- Find the image and preimage of a function.
- Show that a function is a linear transformation, and find a linear transformation.

IMAGES AND PREIMAGES OF FUNCTIONS

In this chapter, you will learn about functions that **map** a vector space V into a vector space W. This type of function is denoted by

$$T: V \rightarrow W.$$

The standard function terminology is used for such functions. For instance, V is the **domain** of T, and W is the **codomain** of T. If $\mathbf{v}$ is in V and $\mathbf{w}$ is in W such that $T(\mathbf{v}) = \mathbf{w}$, then $\mathbf{w}$ is the **image** of $\mathbf{v}$ under T. The set of all images of vectors in V is the **range** of T, and the set of all $\mathbf{v}$ in V such that $T(\mathbf{v}) = \mathbf{w}$ is the **preimage** of $\mathbf{w}$. (See below.)

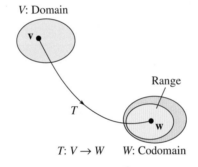

$T: V \rightarrow W$ W: Codomain

REMARK

For a vector

$$\mathbf{v} = (v_1, v_2, \dots, v_n)$$

in R^n, it would be more correct to use double parentheses to denote $T(\mathbf{v})$ as $T(\mathbf{v}) = T((v_1, v_2, \dots, v_n))$. For convenience, however, drop one set of parentheses to produce

$$T(\mathbf{v}) = T(v_1, v_2, \dots, v_n).$$

EXAMPLE 1   A Function from R^2 into R^2

See LarsonLinearAlgebra.com for an interactive version of this type of example.

For any vector $\mathbf{v} = (v_1, v_2)$ in R^2, define $T: R^2 \rightarrow R^2$ by

$$T(v_1, v_2) = (v_1 - v_2, v_1 + 2v_2).$$

a. Find the image of $\mathbf{v} = (-1, 2)$.

b. Find the image of $\mathbf{v} = (0, 0)$.

c. Find the preimage of $\mathbf{w} = (-1, 11)$.

SOLUTION

a. For $\mathbf{v} = (-1, 2)$, you have

$$T(-1, 2) = (-1 - 2, -1 + 2(2)) = (-3, 3).$$

b. If $\mathbf{v} = (0, 0)$, then

$$T(0, 0) = (0 - 0, 0 + 2(0)) = (0, 0).$$

c. If $T(\mathbf{v}) = (v_1 - v_2, v_1 + 2v_2) = (-1, 11)$, then

$$v_1 - v_2 = -1$$
$$v_1 + 2v_2 = 11.$$

This system of equations has the unique solution $v_1 = 3$ and $v_2 = 4$. So, the preimage of $(-1, 11)$ is the set in R^2 consisting of the single vector $(3, 4)$.

LINEAR TRANSFORMATIONS

This chapter centers on functions that map one vector space into another and preserve the operations of vector addition and scalar multiplication. Such functions are called **linear transformations.**

> ### Definition of a Linear Transformation
>
> Let V and W be vector spaces. The function
>
> $$T: V \to W$$
>
> is a **linear transformation** of V into W when the two properties below are true for all $\mathbf{u}$ and $\mathbf{v}$ in V and for any scalar c.
> 1. $T(\mathbf{u} + \mathbf{v}) = T(\mathbf{u}) + T(\mathbf{v})$
> 2. $T(c\mathbf{u}) = cT(\mathbf{u})$

A linear transformation is *operation preserving* because the same result occurs whether you perform the operations of addition and scalar multiplication before or after applying the linear transformation. Although the same symbols denote the vector operations in both V and W, you should note that the operations may be different, as shown in the diagram below.

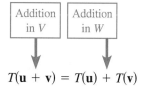

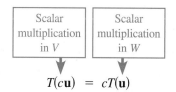

$$T(\mathbf{u} + \mathbf{v}) = T(\mathbf{u}) + T(\mathbf{v}) \qquad T(c\mathbf{u}) = cT(\mathbf{u})$$

EXAMPLE 2 Verifying a Linear Transformation from R^2 into R^2

Show that the function in Example 1 is a linear transformation from R^2 into R^2.

$$T(v_1, v_2) = (v_1 - v_2, v_1 + 2v_2)$$

SOLUTION

To show that the function T is a linear transformation, you must show that it preserves vector addition and scalar multiplication. To do this, let $\mathbf{v} = (v_1, v_2)$ and $\mathbf{u} = (u_1, u_2)$ be vectors in R^2 and let c be any real number. Then, using the properties of vector addition and scalar multiplication, you have the two statements below.

1. $\mathbf{u} + \mathbf{v} = (u_1, u_2) + (v_1, v_2) = (u_1 + v_1, u_2 + v_2)$, so you have

$$
\begin{aligned}
T(\mathbf{u} + \mathbf{v}) &= T(u_1 + v_1, u_2 + v_2) \\
&= ((u_1 + v_1) - (u_2 + v_2), (u_1 + v_1) + 2(u_2 + v_2)) \\
&= ((u_1 - u_2) + (v_1 - v_2), (u_1 + 2u_2) + (v_1 + 2v_2)) \\
&= (u_1 - u_2, u_1 + 2u_2) + (v_1 - v_2, v_1 + 2v_2) \\
&= T(\mathbf{u}) + T(\mathbf{v}).
\end{aligned}
$$

2. $c\mathbf{u} = c(u_1, u_2) = (cu_1, cu_2)$, so you have

$$
\begin{aligned}
T(c\mathbf{u}) &= T(cu_1, cu_2) \\
&= (cu_1 - cu_2, cu_1 + 2cu_2) \\
&= c(u_1 - u_2, u_1 + 2u_2) \\
&= cT(\mathbf{u}).
\end{aligned}
$$

So, T is a linear transformation.

REMARK

A linear transformation $T: V \to V$ from a vector space into itself (as in Example 2) is called a **linear operator.**

Many common functions are not linear transformations, as demonstrated in Example 3.

EXAMPLE 3 **Some Functions That Are Not Linear Transformations**

a. $f(x) = \sin x$ is not a linear transformation from R into R because, in general, $\sin(x_1 + x_2) \neq \sin x_1 + \sin x_2$. For example,

$$\sin[(\pi/2) + (\pi/3)] \neq \sin(\pi/2) + \sin(\pi/3).$$

b. $f(x) = x^2$ is not a linear transformation from R into R because, in general, $(x_1 + x_2)^2 \neq x_1^2 + x_2^2$. For example, $(1 + 2)^2 \neq 1^2 + 2^2$.

c. $f(x) = x + 1$ is not a linear transformation from R into R because

$$f(x_1 + x_2) = x_1 + x_2 + 1$$

whereas

$$f(x_1) + f(x_2) = (x_1 + 1) + (x_2 + 1) = x_1 + x_2 + 2.$$

So $f(x_1 + x_2) \neq f(x_1) + f(x_2)$.

REMARK

The function in Example 3(c) suggests two uses of the term *linear*. The function $f(x) = x + 1$ is a linear function because its graph is a line. It is not a linear transformation from the vector space R into R, however, because it does not preserve vector addition or scalar multiplication.

Two simple linear transformations are the **zero transformation** and the **identity transformation,** which are defined below.

1. $T(\mathbf{v}) = \mathbf{0}$, for all $\mathbf{v}$ Zero transformation ($T: V \to W$)

2. $T(\mathbf{v}) = \mathbf{v}$, for all $\mathbf{v}$ Identity transformation ($T: V \to V$)

You are asked to prove that these are linear transformations in Exercise 77.

Note that the linear transformation in Example 1 has the property that the zero vector maps to itself. That is, $T(\mathbf{0}) = \mathbf{0}$, as shown in Example 1(b). This property is true for all linear transformations, as stated in the first property of the theorem below.

REMARK

One advantage of Theorem 6.1 is that it provides a quick way to identify functions that are not linear transformations. That is, all four conditions of the theorem must be true of a linear transformation, so it follows that if any one of the properties is not satisfied for a function T, then the function is not a linear transformation. For example, the function

$$T(x_1, x_2) = (x_1 + 1, x_2)$$

is not a linear transformation from R^2 into R^2 because $T(0, 0) \neq (0, 0)$.

THEOREM 6.1 Properties of Linear Transformations

Let T be a linear transformation from V into W, where $\mathbf{u}$ and $\mathbf{v}$ are in V. Then the properties listed below are true.

1. $T(\mathbf{0}) = \mathbf{0}$
2. $T(-\mathbf{v}) = -T(\mathbf{v})$
3. $T(\mathbf{u} - \mathbf{v}) = T(\mathbf{u}) - T(\mathbf{v})$
4. If $\mathbf{v} = c_1\mathbf{v}_1 + c_2\mathbf{v}_2 + \cdots + c_n\mathbf{v}_n$, then
$$T(\mathbf{v}) = T(c_1\mathbf{v}_1 + c_2\mathbf{v}_2 + \cdots + c_n\mathbf{v}_n) = c_1T(\mathbf{v}_1) + c_2T(\mathbf{v}_2) + \cdots + c_nT(\mathbf{v}_n).$$

PROOF

To prove the first property, note that $0\mathbf{v} = \mathbf{0}$. Then it follows that

$$T(\mathbf{0}) = T(0\mathbf{v}) = 0T(\mathbf{v}) = \mathbf{0}.$$

The second property follows from $-\mathbf{v} = (-1)\mathbf{v}$, which implies that

$$T(-\mathbf{v}) = T[(-1)\mathbf{v}] = (-1)T(\mathbf{v}) = -T(\mathbf{v}).$$

The third property follows from $\mathbf{u} - \mathbf{v} = \mathbf{u} + (-\mathbf{v})$, which implies that

$$T(\mathbf{u} - \mathbf{v}) = T[\mathbf{u} + (-1)\mathbf{v}] = T(\mathbf{u}) + (-1)T(\mathbf{v}) = T(\mathbf{u}) - T(\mathbf{v}).$$

The proof of the fourth property is left to you.

Property 4 of Theorem 6.1 suggests that a linear transformation $T: V \to W$ is determined completely by its action on a basis for V. In other words, if $\{\mathbf{v}_1, \mathbf{v}_2, \ldots, \mathbf{v}_n\}$ is a basis for the vector space V and if $T(\mathbf{v}_1), T(\mathbf{v}_2), \ldots, T(\mathbf{v}_n)$ are given, then $T(\mathbf{v})$ can be determined for *any* $\mathbf{v}$ in V. Example 4 demonstrates the use of this property.

EXAMPLE 4   **Linear Transformations and Bases**

Let $T: R^3 \to R^3$ be a linear transformation such that

$T(1, 0, 0) = (2, -1, 4)$
$T(0, 1, 0) = (1, 5, -2)$
$T(0, 0, 1) = (0, 3, 1).$

Find $T(2, 3, -2)$.

SOLUTION

$(2, 3, -2) = 2(1, 0, 0) + 3(0, 1, 0) - 2(0, 0, 1)$, so use Property 4 of Theorem 6.1 to write

$$\begin{aligned} T(2, 3, -2) &= 2T(1, 0, 0) + 3T(0, 1, 0) - 2T(0, 0, 1) \\ &= 2(2, -1, 4) + 3(1, 5, -2) - 2(0, 3, 1) \\ &= (7, 7, 0). \end{aligned}$$

In the next example, a matrix defines a linear transformation from R^2 into R^3. The vector $\mathbf{v} = (v_1, v_2)$ is in the matrix form

$$\mathbf{v} = \begin{bmatrix} v_1 \\ v_2 \end{bmatrix}$$

so it can be multiplied *on the left* by a matrix of size 3×2.

EXAMPLE 5   **A Linear Transformation Defined by a Matrix**

Define the function $T: R^2 \to R^3$ as

$$T(\mathbf{v}) = A\mathbf{v} = \begin{bmatrix} 3 & 0 \\ 2 & 1 \\ -1 & -2 \end{bmatrix} \begin{bmatrix} v_1 \\ v_2 \end{bmatrix}.$$

a. Find $T(\mathbf{v})$ when $\mathbf{v} = (2, -1)$.

b. Show that T is a linear transformation from R^2 into R^3.

SOLUTION

a. $\mathbf{v} = (2, -1)$, so you have

$$T(\mathbf{v}) = A\mathbf{v} = \begin{bmatrix} 3 & 0 \\ 2 & 1 \\ -1 & -2 \end{bmatrix} \begin{bmatrix} 2 \\ -1 \end{bmatrix} = \begin{bmatrix} 6 \\ 3 \\ 0 \end{bmatrix}$$

Vector in R^2 Vector in R^3

which means that $T(2, -1) = (6, 3, 0)$.

b. Begin by observing that T maps a vector in R^2 to a vector in R^3. To show that T is a linear transformation, use properties given in Theorem 2.3. For any vectors $\mathbf{u}$ and $\mathbf{v}$ in R^2, the distributive property of matrix multiplication over addition produces

$$T(\mathbf{u} + \mathbf{v}) = A(\mathbf{u} + \mathbf{v}) = A\mathbf{u} + A\mathbf{v} = T(\mathbf{u}) + T(\mathbf{v}).$$

Similarly, for any vector $\mathbf{u}$ in R^2 and any scalar c, the commutative property of scalar multiplication with matrix multiplication produces

$$T(c\mathbf{u}) = A(c\mathbf{u}) = c(A\mathbf{u}) = cT(\mathbf{u}).$$

Example 5 illustrates an important result regarding the representation of linear transformations from R^n into R^m. This result is presented in two stages. Theorem 6.2 below states that every $m \times n$ matrix represents a linear transformation from R^n into R^m. Then, in Section 6.3, you will see the converse—that every linear transformation from R^n into R^m can be represented by an $m \times n$ matrix.

Note that the solution of Example 5(b) makes no reference specifically to the matrix A that defines T. So, this solution serves as a general proof that the function defined by any $m \times n$ matrix is a linear transformation from R^n into R^m.

REMARK

The $m \times n$ zero matrix corresponds to the zero transformation from R^n into R^m, and the $n \times n$ identity matrix I_n corresponds to the identity transformation from R^n into R^n.

THEOREM 6.2   Linear Transformation Given by a Matrix

Let A be an $m \times n$ matrix. The function T defined by

$$T(\mathbf{v}) = A\mathbf{v}$$

is a linear transformation from R^n into R^m. In order to conform to matrix multiplication with an $m \times n$ matrix, $n \times 1$ matrices represent the vectors in R^n and $m \times 1$ matrices represent the vectors in R^m.

Be sure you see that an $m \times n$ matrix A defines a linear transformation from R^n into R^m:

$$A\mathbf{v} = \begin{bmatrix} a_{11} & a_{12} & \cdots & a_{1n} \\ a_{21} & a_{22} & \cdots & a_{2n} \\ \vdots & \vdots & & \vdots \\ a_{m1} & a_{m2} & \cdots & a_{mn} \end{bmatrix} \begin{bmatrix} v_1 \\ v_2 \\ \vdots \\ v_n \end{bmatrix} = \begin{bmatrix} a_{11}v_1 + a_{12}v_2 + \cdots + a_{1n}v_n \\ a_{21}v_1 + a_{22}v_2 + \cdots + a_{2n}v_n \\ \vdots & \vdots & \vdots \\ a_{m1}v_1 + a_{m2}v_2 + \cdots + a_{mn}v_n \end{bmatrix}.$$

Vector in R^n — Vector in R^m

EXAMPLE 6   Linear Transformations Given by Matrices

Consider the linear transformation $T: R^n \to R^m$ defined by $T(\mathbf{v}) = A\mathbf{v}$. Find the dimensions of R^n and R^m for the linear transformation represented by each matrix.

a. $A = \begin{bmatrix} 0 & 1 & -1 \\ 2 & 3 & 0 \\ 4 & 2 & 1 \end{bmatrix}$ b. $A = \begin{bmatrix} 2 & -3 \\ -5 & 0 \\ 0 & -2 \end{bmatrix}$

c. $A = \begin{bmatrix} 1 & 0 & -1 & 2 \\ 3 & 1 & 0 & 0 \end{bmatrix}$

SOLUTION

a. The size of this matrix is 3×3, so it defines a linear transformation from R^3 into R^3.

$$A\mathbf{v} = \begin{bmatrix} 0 & 1 & -1 \\ 2 & 3 & 0 \\ 4 & 2 & 1 \end{bmatrix} \begin{bmatrix} v_1 \\ v_2 \\ v_3 \end{bmatrix} = \begin{bmatrix} u_1 \\ u_2 \\ u_3 \end{bmatrix}$$

Vector in R^3 — Vector in R^3

b. The size of this matrix is 3×2, so it defines a linear transformation from R^2 into R^3.

c. The size of this matrix is 2×4, so it defines a linear transformation from R^4 into R^2.

The next example discusses a common type of linear transformation from R^2 into R^2.

EXAMPLE 7   Rotation in R^2

Show that the linear transformation $T: R^2 \to R^2$ represented by the matrix

$$A = \begin{bmatrix} \cos\theta & -\sin\theta \\ \sin\theta & \cos\theta \end{bmatrix}$$

has the property that it rotates every vector in R^2 counterclockwise about the origin through the angle θ.

SOLUTION

From Theorem 6.2, you know that T is a linear transformation. To show that it rotates every vector in R^2 counterclockwise through the angle θ, let $\mathbf{v} = (x, y)$ be a vector in R^2. Using polar coordinates, you can write $\mathbf{v}$ as

$$\mathbf{v} = (x, y)$$
$$= (r\cos\alpha, r\sin\alpha)$$

where r is the length of $\mathbf{v}$ and α is the angle from the positive x-axis counterclockwise to the vector $\mathbf{v}$. Now, applying the linear transformation T to $\mathbf{v}$ produces

$$T(\mathbf{v}) = A\mathbf{v}$$
$$= \begin{bmatrix} \cos\theta & -\sin\theta \\ \sin\theta & \cos\theta \end{bmatrix}\begin{bmatrix} x \\ y \end{bmatrix}$$
$$= \begin{bmatrix} \cos\theta & -\sin\theta \\ \sin\theta & \cos\theta \end{bmatrix}\begin{bmatrix} r\cos\alpha \\ r\sin\alpha \end{bmatrix}$$
$$= \begin{bmatrix} r\cos\theta\cos\alpha - r\sin\theta\sin\alpha \\ r\sin\theta\cos\alpha + r\cos\theta\sin\alpha \end{bmatrix}$$
$$= \begin{bmatrix} r\cos(\theta + \alpha) \\ r\sin(\theta + \alpha) \end{bmatrix}.$$

Verify that the vector $T(\mathbf{v})$ has the same length as $\mathbf{v}$. Furthermore, the angle from the positive x-axis to $T(\mathbf{v})$ is

$$\theta + \alpha$$

so $T(\mathbf{v})$ is the vector that results from rotating the vector $\mathbf{v}$ counterclockwise through the angle θ, as shown below.

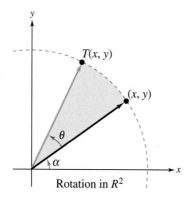

Rotation in R^2

The linear transformation in Example 7 is a **rotation** in R^2. Rotations in R^2 preserve both vector length and the angle between two vectors. That is, $\|T(\mathbf{u})\| = \|\mathbf{u}\|$, $\|T(\mathbf{v})\| = \|\mathbf{v}\|$, and the angle between $T(\mathbf{u})$ and $T(\mathbf{v})$ is equal to the angle between $\mathbf{u}$ and $\mathbf{v}$.

EXAMPLE 8   **A Projection in R^3**

The linear transformation $T: R^3 \rightarrow R^3$ represented by

$$A = \begin{bmatrix} 1 & 0 & 0 \\ 0 & 1 & 0 \\ 0 & 0 & 0 \end{bmatrix}$$

is a **projection** in R^3. If $\mathbf{v} = (x, y, z)$ is a vector in R^3, then $T(\mathbf{v}) = (x, y, 0)$. In other words, T maps every vector in R^3 to its orthogonal projection in the xy-plane, as shown below.

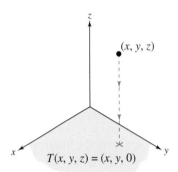

$$T(x, y, z) = (x, y, 0)$$

Projection onto xy-plane

So far, only linear transformations from R^n into R^m or from R^n into R^n have been discussed. The remainder of this section considers some linear transformations involving vector spaces other than R^n.

EXAMPLE 9   **A Linear Transformation from $M_{m,n}$ into $M_{n,m}$**

Let $T: M_{m,n} \rightarrow M_{n,m}$ be the function that maps an $m \times n$ matrix A to its transpose. That is,

$$T(A) = A^T.$$

Show that T is a linear transformation.

SOLUTION

Let A and B be $m \times n$ matrices and let c be a scalar. From Theorem 2.6 you have

$$T(A + B) = (A + B)^T = A^T + B^T = T(A) + T(B)$$

and

$$T(cA) = (cA)^T = c(A^T) = cT(A).$$

So, T is a linear transformation from $M_{m,n}$ into $M_{n,m}$.

LINEAR ALGEBRA APPLIED

Many multivariate statistical methods can use linear transformations. For instance, in a *multiple regression analysis,* there are two or more independent variables and a single dependent variable. A linear transformation is useful for finding weights to be assigned to the independent variables to predict the value of the dependent variable. Also, in a *canonical correlation analysis,* there are two or more independent variables and two or more dependent variables. Linear transformations can help find a linear combination of the independent variables to predict the value of a linear combination of the dependent variables.

EXAMPLE 10 **The Differential Operator (Calculus)**

Let $C'[a, b]$ be the set of all functions whose derivatives are continuous on $[a, b]$. Show that the differential operator D_x defines a linear transformation from $C'[a, b]$ into $C[a, b]$.

SOLUTION

Using operator notation, you can write

$$D_x(f) = \frac{d}{dx}[f]$$

where f is in $C'[a, b]$. To show that D_x is a linear transformation, you must use calculus. Specifically, the derivative of the sum of two differentiable functions is equal to the sum of their derivatives, so you have

$$D_x(f + g) = \frac{d}{dx}[f + g] = \frac{d}{dx}[f] + \frac{d}{dx}[g] = D_x(f) + D_x(g)$$

where g is also in $C'[a, b]$. Similarly, the derivative of a scalar multiple cf of a differentiable function is equal to the scalar multiple of the derivative, so you have

$$D_x(cf) = \frac{d}{dx}[cf] = c\left(\frac{d}{dx}[f]\right) = cD_x(f).$$

The sum of two continuous functions is continuous, and the scalar multiple of a continuous function is continuous, so D_x is a linear transformation from $C'[a, b]$ into $C[a, b]$. ∎

The linear transformation D_x in Example 10 is called the **differential operator.** For polynomials, the differential operator is a linear transformation from P_n into P_{n-1} because the derivative of a polynomial function of degree $n \geq 1$ is a polynomial function of degree $n - 1$. That is,

$$D_x(a_0 + a_1x + \cdots + a_nx^n) = a_1 + \cdots + na_nx^{n-1}.$$

The next example describes a linear transformation from the vector space of polynomial functions P into the vector space of real numbers R.

EXAMPLE 11 **The Definite Integral as a Linear Transformation (Calculus)**

Consider $T: P \rightarrow R$ defined by

$$T(p) = \int_a^b p(x)\, dx$$

where p is a polynomial function. Show that T is a linear transformation from P, the vector space of polynomial functions, into R, the vector space of real numbers.

SOLUTION

Using properties of definite integrals, you can write

$$T(p + q) = \int_a^b [p(x) + q(x)]\, dx = \int_a^b p(x)\, dx + \int_a^b q(x)\, dx = T(p) + T(q)$$

where q is a polynomial function, and

$$T(cp) = \int_a^b [cp(x)]\, dx = c\int_a^b p(x)\, dx = cT(p)$$

where c is a scalar. So, T is a linear transformation. ∎

6.1   Exercises

See CalcChat.com for worked-out solutions to odd-numbered exercises.

Finding an Image and a Preimage   **In Exercises 1–8, use the function to find (a) the image of v and (b) the preimage of w.**

1. $T(v_1, v_2) = (v_1 + v_2, v_1 - v_2)$,
$\mathbf{v} = (3, -4)$, $\mathbf{w} = (3, 19)$

2. $T(v_1, v_2) = (v_1, 2v_2 - v_1, v_2)$,
$\mathbf{v} = (0, 4)$, $\mathbf{w} = (2, 4, 3)$

3. $T(v_1, v_2, v_3) = (2v_1 + v_2, 2v_2 - 3v_1, v_1 - v_3)$,
$\mathbf{v} = (-4, 5, 1)$, $\mathbf{w} = (4, 1, -1)$

4. $T(v_1, v_2, v_3) = (v_2 - v_1, v_1 + v_2, 2v_1)$,
$\mathbf{v} = (2, 3, 0)$, $\mathbf{w} = (-11, -1, 10)$

5. $T(v_1, v_2, v_3) = (4v_2 - v_1, 4v_1 + 5v_2)$,
$\mathbf{v} = (2, -3, -1)$, $\mathbf{w} = (3, 9)$

6. $T(v_1, v_2, v_3) = (2v_1 + v_2, v_1 - v_2)$,
$\mathbf{v} = (2, 1, 4)$, $\mathbf{w} = (-1, 2)$

7. $T(v_1, v_2) = \left(\dfrac{\sqrt{2}}{2}v_1 - \dfrac{\sqrt{2}}{2}v_2, v_1 + v_2, 2v_1 - v_2\right)$,
$\mathbf{v} = (1, 1)$, $\mathbf{w} = (-5\sqrt{2}, -2, -16)$

8. $T(v_1, v_2) = \left(\dfrac{\sqrt{3}}{2}v_1 - \dfrac{1}{2}v_2, v_1 - v_2, v_2\right)$,
$\mathbf{v} = (2, 4)$, $\mathbf{w} = \left(\sqrt{3}, 2, 0\right)$

Linear Transformations   **In Exercises 9–22, determine whether the function is a linear transformation.**

9. $T: R^2 \to R^2$, $T(x, y) = (x, 1)$

10. $T: R^2 \to R^2$, $T(x, y) = (x, y^2)$

11. $T: R^3 \to R^3$, $T(x, y, z) = (x + y, x - y, z)$

12. $T: R^3 \to R^3$, $T(x, y, z) = (x + 1, y + 1, z + 1)$

13. $T: R^2 \to R^3$, $T(x, y) = \left(\sqrt{x}, xy, \sqrt{y}\right)$

14. $T: R^2 \to R^3$, $T(x, y) = (x^2, xy, y^2)$

15. $T: M_{2,2} \to R$, $T(A) = |A|$

16. $T: M_{2,2} \to R$, $T(A) = a + b + c + d$, where
$A = \begin{bmatrix} a & b \\ c & d \end{bmatrix}$.

17. $T: M_{2,2} \to R$, $T(A) = a - b - c - d$, where
$A = \begin{bmatrix} a & b \\ c & d \end{bmatrix}$.

18. $T: M_{2,2} \to R$, $T(A) = b^2$, where $A = \begin{bmatrix} a & b \\ c & d \end{bmatrix}$.

19. $T: M_{3,3} \to M_{3,3}$, $T(A) = \begin{bmatrix} 0 & 0 & 1 \\ 0 & 1 & 0 \\ 1 & 0 & 0 \end{bmatrix} A$

20. $T: M_{3,3} \to M_{3,3}$, $T(A) = \begin{bmatrix} 3 & 0 & 0 \\ 0 & 2 & 0 \\ 0 & 0 & -10 \end{bmatrix} A$

21. $T: P_2 \to P_2$, $T(a_0 + a_1 x + a_2 x^2) = (a_0 + a_1 + a_2) + (a_1 + a_2)x + a_2 x^2$

22. $T: P_2 \to P_2$, $T(a_0 + a_1 x + a_2 x^2) = a_1 + 2a_2 x$

23. Let T be a linear transformation from R^2 into R^2 such that $T(1, 0) = (1, 1)$ and $T(0, 1) = (-1, 1)$. Find $T(1, 4)$ and $T(-2, 1)$.

24. Let T be a linear transformation from R^2 into R^2 such that $T(1, 2) = (1, 0)$ and $T(-1, 1) = (0, 1)$. Find $T(2, 0)$ and $T(0, 3)$.

Linear Transformation and Bases   **In Exercises 25–28, let $T: R^3 \to R^3$ be a linear transformation such that $T(1, 0, 0) = (2, 4, -1)$, $T(0, 1, 0) = (1, 3, -2)$, and $T(0, 0, 1) = (0, -2, 2)$. Find the specified image.**

25. $T(1, -3, 0)$ **26.** $T(2, -1, 0)$

27. $T(2, -4, 1)$ **28.** $T(-2, 4, -1)$

Linear Transformation and Bases   **In Exercises 29–32, let $T: R^3 \to R^3$ be a linear transformation such that $T(1, 1, 1) = (2, 0, -1)$, $T(0, -1, 2) = (-3, 2, -1)$, and $T(1, 0, 1) = (1, 1, 0)$. Find the specified image.**

29. $T(4, 2, 0)$ **30.** $T(0, 2, -1)$

31. $T(2, -1, 1)$ **32.** $T(-2, 1, 0)$

Linear Transformation Given by a Matrix   **In Exercises 33–38, define the linear transformation $T: R^n \to R^m$ by $T(\mathbf{v}) = A\mathbf{v}$. Find the dimensions of R^n and R^m.**

33. $A = \begin{bmatrix} 0 & -1 \\ -1 & 0 \end{bmatrix}$ **34.** $A = \begin{bmatrix} 1 & 2 \\ -2 & 4 \\ -2 & 2 \end{bmatrix}$

35. $A = \begin{bmatrix} 1 & 0 & 0 & 0 \\ 0 & -1 & 0 & 0 \\ 0 & 0 & 1 & 0 \\ 0 & 0 & 0 & 2 \end{bmatrix}$

36. $A = \begin{bmatrix} -1 & 2 & 1 & 3 & 4 \\ 0 & 0 & 2 & -1 & 0 \end{bmatrix}$

37. $A = \begin{bmatrix} 0 & 1 & -2 & 1 \\ -1 & 4 & 5 & 0 \\ 0 & 1 & 3 & 1 \end{bmatrix}$

38. $A = \begin{bmatrix} 0 & 2 & 0 & 2 & 0 \\ 1 & 0 & 1 & 0 & 1 \\ 1 & 2 & 2 & 2 & 1 \end{bmatrix}$

39. For the linear transformation from Exercise 33, find (a) $T(1, 1)$, (b) the preimage of $(1, 1)$, and (c) the preimage of $(0, 0)$.

40. Writing For the linear transformation from Exercise 34, find (a) $T(2, 4)$ and (b) the preimage of $(-1, 2, 2)$. (c) Then explain why the vector $(1, 1, 1)$ has no preimage under this transformation.

41. For the linear transformation from Exercise 35, find (a) $T(2, 1, 2, 1)$ and (b) the preimage of $(-1, -1, -1, -1)$.

42. For the linear transformation from Exercise 36, find (a) $T(1, 0, -1, 3, 0)$ and (b) the preimage of $(-1, 8)$.

43. For the linear transformation from Exercise 37, find (a) $T(1, 0, 2, 3)$ and (b) the preimage of $(0, 0, 0)$.

44. For the linear transformation from Exercise 38, find (a) $T(0, 1, 0, 1, 0)$ (b) the preimage of $(0, 0, 0)$, and (c) the preimage of $(1, -1, 2)$.

45. Let T be a linear transformation from R^2 into R^2 such that $T(x, y) = (x \cos \theta - y \sin \theta, x \sin \theta + y \cos \theta)$. Find (a) $T(4, 4)$ for $\theta = 45°$, (b) $T(4, 4)$ for $\theta = 30°$, and (c) $T(5, 0)$ for $\theta = 120°$.

46. For the linear transformation from Exercise 45, let $\theta = 45°$ and find the preimage of $\mathbf{v} = (1, 1)$.

47. Find the inverse of the matrix A in Example 7. What linear transformation from R^2 into R^2 does A^{-1} represent?

48. For the linear transformation $T: R^2 \to R^2$ given by
$$A = \begin{bmatrix} a & -b \\ b & a \end{bmatrix}$$
find a and b such that $T(12, 5) = (13, 0)$.

Projection in R^3 In Exercises 49 and 50, let the matrix A represent the linear transformation $T: R^3 \to R^3$. Describe the orthogonal projection to which T maps every vector in R^3.

49. $A = \begin{bmatrix} 1 & 0 & 0 \\ 0 & 0 & 0 \\ 0 & 0 & 1 \end{bmatrix}$ **50.** $A = \begin{bmatrix} 0 & 0 & 0 \\ 0 & 1 & 0 \\ 0 & 0 & 1 \end{bmatrix}$

Linear Transformation Given by a Matrix In Exercises 51–54, determine whether the function involving the $n \times n$ matrix A is a linear transformation.

51. $T: M_{n,n} \to M_{n,n}, T(A) = A^{-1}$

52. $T: M_{n,n} \to M_{n,n}, T(A) = AX - XA$, where X is a fixed $n \times m$ matrix

53. $T: M_{n,n} \to M_{n,m}, T(A) = AB$, where B is a fixed $n \times m$ matrix

54. $T: M_{n,n} \to R, T(A) = a_{11} \cdot a_{22} \cdot \cdots \cdot a_{nn}$, where $A = [a_{ij}]$

55. Let T be a linear transformation from P_2 into P_2 such that $T(1) = x$, $T(x) = 1 + x$, and $T(x^2) = 1 + x + x^2$. Find $T(2 - 6x + x^2)$.

56. Let T be a linear transformation from $M_{2,2}$ into $M_{2,2}$ such that
$$T\left(\begin{bmatrix} 1 & 0 \\ 0 & 0 \end{bmatrix}\right) = \begin{bmatrix} 1 & -1 \\ 0 & 2 \end{bmatrix}, \quad T\left(\begin{bmatrix} 0 & 1 \\ 0 & 0 \end{bmatrix}\right) = \begin{bmatrix} 0 & 2 \\ 1 & 1 \end{bmatrix},$$
$$T\left(\begin{bmatrix} 0 & 0 \\ 1 & 0 \end{bmatrix}\right) = \begin{bmatrix} 1 & 2 \\ 0 & 1 \end{bmatrix}, \quad T\left(\begin{bmatrix} 0 & 0 \\ 0 & 1 \end{bmatrix}\right) = \begin{bmatrix} 3 & -1 \\ 1 & 0 \end{bmatrix}.$$
Find $T\left(\begin{bmatrix} 1 & 3 \\ -1 & 4 \end{bmatrix}\right)$.

Calculus In Exercises 57–60, let D_x be the linear transformation from $C'[a, b]$ into $C[a, b]$ from Example 10. Determine whether each statement is true or false. Explain.

57. $D_x(e^{x^2} + 2x) = D_x(e^{x^2}) + 2D_x(x)$

58. $D_x(x^2 - \ln x) = D_x(x^2) - D_x(\ln x)$

59. $D_x(\sin 3x) = 3D_x(\sin x)$

60. $D_x\left(\cos \dfrac{x}{2}\right) = \dfrac{1}{2}D_x(\cos x)$

Calculus In Exercises 61–64, for the linear transformation from Example 10, find the preimage of each function.

61. $D_x(f) = 4x + 3$ **62.** $D_x(f) = e^x$

63. $D_x(f) = \sin x$ **64.** $D_x(f) = \dfrac{1}{x}$

65. Calculus Let T be a linear transformation from P into R such that
$$T(p) = \int_0^1 p(x)\, dx.$$
Find (a) $T(-2 + 3x^2)$, (b) $T(x^3 - x^5)$, and (c) $T(-6 + 4x)$.

66. Calculus Let T be the linear transformation from P_2 into R using the integral in Exercise 65. Find the preimage of 1. That is, find the polynomial function(s) of degree 2 or less such that $T(p) = 1$.

True or False? In Exercises 67 and 68, determine whether each statement is true or false. If a statement is true, give a reason or cite an appropriate statement from the text. If a statement is false, provide an example that shows the statement is not true in all cases or cite an appropriate statement from the text.

67. (a) The function $f(x) = \cos x$ is a linear transformation from R into R.

(b) For polynomials, the differential operator D_x is a linear transformation from P_n into P_{n-1}.

68. (a) The function $g(x) = x^3$ is a linear transformation from R into R.

(b) Any linear function of the form $f(x) = ax + b$ is a linear transformation from R into R.

69. Writing Let $T: R^2 \to R^2$ such that $T(1, 0) = (1, 0)$ and $T(0, 1) = (0, 0)$.

(a) Determine $T(x, y)$ for (x, y) in R^2.

(b) Give a geometric description of T.

70. Writing Let $T: R^2 \to R^2$ such that $T(1, 0) = (0, 1)$ and $T(0, 1) = (1, 0)$.

(a) Determine $T(x, y)$ for (x, y) in R^2.

(b) Give a geometric description of T.

71. Proof Let T be the function that maps R^2 into R^2 such that $T(\mathbf{u}) = \text{proj}_{\mathbf{v}}\mathbf{u}$, where $\mathbf{v} = (1, 1)$.

(a) Find $T(x, y)$.   (b) Find $T(5, 0)$.

(c) Prove that T is a linear transformation from R^2 into R^2.

72. Writing Find $T(3, 4)$ and $T(T(3, 4))$ from Exercise 71 and give geometric descriptions of the results.

73. Show that T from Exercise 71 is represented by the matrix

$$A = \begin{bmatrix} \frac{1}{2} & \frac{1}{2} \\ \frac{1}{2} & \frac{1}{2} \end{bmatrix}.$$

74. CAPSTONE Explain how to determine whether a function $T: V \to W$ is a linear transformation.

75. Proof Use the concept of a fixed point of a linear transformation $T: V \to V$. A vector $\mathbf{u}$ is a **fixed point** when $T(\mathbf{u}) = \mathbf{u}$.

(a) Prove that $\mathbf{0}$ is a fixed point of any linear transformation $T: V \to V$.

(b) Prove that the set of fixed points of a linear transformation $T: V \to V$ is a subspace of V.

(c) Determine all fixed points of the linear transformation $T: R^2 \to R^2$ represented by $T(x, y) = (x, 2y)$.

(d) Determine all fixed points of the linear transformation $T: R^2 \to R^2$ represented by $T(x, y) = (y, x)$.

76. A **translation** in R^2 is a function of the form $T(x, y) = (x - h, y - k)$, where at least one of the constants h and k is nonzero.

(a) Show that a translation in R^2 is not a linear transformation.

(b) For the translation $T(x, y) = (x - 2, y + 1)$, determine the images of $(0, 0)$, $(2, -1)$, and $(5, 4)$.

(c) Show that a translation in R^2 has no fixed points.

77. Proof Prove that (a) the zero transformation and (b) the identity transformation are linear transformations.

78. Let $S = \{\mathbf{v}_1, \mathbf{v}_2, \mathbf{v}_3\}$ be a set of linearly independent vectors in R^3. Find a linear transformation T from R^3 into R^3 such that the set $\{T(\mathbf{v}_1), T(\mathbf{v}_2), T(\mathbf{v}_3)\}$ is linearly dependent.

79. Proof Let $S = \{\mathbf{v}_1, \mathbf{v}_2, \ldots, \mathbf{v}_n\}$ be a set of linearly dependent vectors in V, and let T be a linear transformation from V into V. Prove that the set

$$\{T(\mathbf{v}_1), T(\mathbf{v}_2), \ldots, T(\mathbf{v}_n)\}$$

is linearly dependent.

80. Proof Let V be an inner product space. For a fixed vector $\mathbf{v}_0$ in V, define $T: V \to R$ by $T(\mathbf{v}) = \langle \mathbf{v}, \mathbf{v}_0 \rangle$. Prove that T is a linear transformation.

81. Proof Define $T: M_{n,n} \to R$ by

$$T(A) = a_{11} + a_{22} + \cdots + a_{nn}$$

(the trace of A). Prove that T is a linear transformation.

82. Let V be an inner product space with a subspace W having $B = \{\mathbf{w}_1, \mathbf{w}_2, \ldots, \mathbf{w}_n\}$ as an orthonormal basis. Show that the function $T: V \to W$ represented by

$$T(\mathbf{v}) = \langle \mathbf{v}, \mathbf{w}_1 \rangle \mathbf{w}_1 + \langle \mathbf{v}, \mathbf{w}_2 \rangle \mathbf{w}_2 + \cdots + \langle \mathbf{v}, \mathbf{w}_n \rangle \mathbf{w}_n$$

is a linear transformation. T is called the **orthogonal projection of V onto W.**

83. Guided Proof Let $\{\mathbf{v}_1, \mathbf{v}_2, \ldots, \mathbf{v}_n\}$ be a basis for a vector space V. Prove that if a linear transformation $T: V \to V$ satisfies $T(\mathbf{v}_i) = \mathbf{0}$ for $i = 1, 2, \ldots, n$, then T is the zero transformation.

Getting Started: To prove that T is the zero transformation, you need to show that $T(\mathbf{v}) = \mathbf{0}$ for every vector $\mathbf{v}$ in V.

(i) Let $\mathbf{v}$ be an arbitrary vector in V such that

$$\mathbf{v} = c_1\mathbf{v}_1 + c_2\mathbf{v}_2 + \cdots + c_n\mathbf{v}_n.$$

(ii) Use the definition and properties of linear transformations to rewrite $T(\mathbf{v})$ as a linear combination of $T(\mathbf{v}_i)$.

(iii) Use the fact that $T(\mathbf{v}_i) = \mathbf{0}$ to conclude that $T(\mathbf{v}) = \mathbf{0}$, making T the zero transformation.

84. Guided Proof Prove that $T: V \to W$ is a linear transformation if and only if

$$T(a\mathbf{u} + b\mathbf{v}) = aT(\mathbf{u}) + bT(\mathbf{v})$$

for all vectors $\mathbf{u}$ and $\mathbf{v}$ and all scalars a and b.

Getting Started: This is an "if and only if" statement, so you need to prove the statement in both directions. To prove that T is a linear transformation, you need to show that the function satisfies the definition of a linear transformation. In the other direction, let T be a linear transformation. Use the definition and properties of a linear transformation to prove that $T(a\mathbf{u} + b\mathbf{v}) = aT(\mathbf{u}) + bT(\mathbf{v})$.

(i) Let $T(a\mathbf{u} + b\mathbf{v}) = aT(\mathbf{u}) + bT(\mathbf{v})$. Show that T preserves the properties of vector addition and scalar multiplication by choosing appropriate values of a and b.

(ii) To prove the statement in the other direction, assume that T is a linear transformation. Use the properties and definition of a linear transformation to show that $T(a\mathbf{u} + b\mathbf{v}) = aT(\mathbf{u}) + bT(\mathbf{v})$.

6.2 The Kernel and Range of a Linear Transformation

- Find the kernel of a linear transformation.
- Find a basis for the range, the rank, and the nullity of a linear transformation.
- Determine whether a linear transformation is one-to-one or onto.
- Determine whether vector spaces are isomorphic.

THE KERNEL OF A LINEAR TRANSFORMATION

You know from Theorem 6.1 that for any linear transformation $T: V \rightarrow W$, the zero vector in V maps to the zero vector in W. That is, $T(\mathbf{0}) = \mathbf{0}$. The first question you will consider in this section is whether there are *other* vectors $\mathbf{v}$ such that $T(\mathbf{v}) = \mathbf{0}$. The collection of all such elements is the **kernel** of T. Note that the symbol $\mathbf{0}$ represents the zero vector in both V and W, although these two zero vectors are often different.

> **Definition of Kernel of a Linear Transformation**
>
> Let $T: V \rightarrow W$ be a linear transformation. Then the set of all vectors $\mathbf{v}$ in V that satisfy $T(\mathbf{v}) = \mathbf{0}$ is the **kernel** of T and is denoted by $\ker(T)$.

Sometimes the kernel of a transformation can be found by inspection, as demonstrated in Examples 1, 2, and 3.

EXAMPLE 1   Finding the Kernel of a Linear Transformation

Let $T: M_{3,2} \rightarrow M_{2,3}$ be the linear transformation that maps a 3×2 matrix A to its transpose. That is, $T(A) = A^T$. Find the kernel of T.

SOLUTION

For this linear transformation, the 3×2 zero matrix is clearly the only matrix in $M_{3,2}$ whose transpose is the zero matrix in $M_{2,3}$. So, the kernel of T consists of a single element: the zero matrix in $M_{3,2}$.

EXAMPLE 2   The Kernels of the Zero and Identity Transformations

a. The kernel of the zero transformation $T: V \rightarrow W$ consists of all of V because $T(\mathbf{v}) = \mathbf{0}$ for every $\mathbf{v}$ in V. That is, $\ker(T) = V$.

b. The kernel of the identity transformation $T: V \rightarrow V$ consists of the single element $\mathbf{0}$. That is, $\ker(T) = \{\mathbf{0}\}$.

EXAMPLE 3   Finding the Kernel of a Linear Transformation

Find the kernel of the projection $T: R^3 \rightarrow R^3$ represented by $T(x, y, z) = (x, y, 0)$.

SOLUTION

This linear transformation projects the vector (x, y, z) in R^3 to the vector $(x, y, 0)$ in the xy-plane. The kernel consists of all vectors lying on the z-axis. That is,

$$\ker(T) = \{(0, 0, z): z \text{ is a real number}\}. \text{ (See Figure 6.1.)}$$

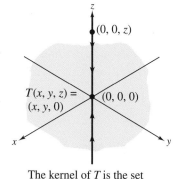

The kernel of T is the set of all vectors on the z-axis.

Figure 6.1

Finding the kernels of the linear transformations in Examples 1, 2, and 3 is relatively easy. Sometimes, the kernel of a linear transformation is not so obvious, as illustrated in the next two examples.

EXAMPLE 4   Finding the Kernel of a Linear Transformation

Find the kernel of the linear transformation $T: R^2 \rightarrow R^3$ represented by

$$T(x_1, x_2) = (x_1 - 2x_2, 0, -x_1).$$

SOLUTION

To find ker(T), you need to find all $\mathbf{x} = (x_1, x_2)$ in R^2 such that

$$T(x_1, x_2) = (x_1 - 2x_2, 0, -x_1) = (0, 0, 0).$$

This leads to the homogeneous system

$$
\begin{aligned}
x_1 - 2x_2 &= 0 \\
0 &= 0 \\
-x_1 \quad\;\; &= 0
\end{aligned}
$$

which has only the trivial solution $(x_1, x_2) = (0, 0)$. So, you have

$$\ker(T) = \{(0, 0)\} = \{\mathbf{0}\}.$$

EXAMPLE 5   Finding the Kernel of a Linear Transformation

Find the kernel of the linear transformation $T: R^3 \rightarrow R^2$ defined by $T(\mathbf{x}) = A\mathbf{x}$, where

$$A = \begin{bmatrix} 1 & -1 & -2 \\ -1 & 2 & 3 \end{bmatrix}.$$

SOLUTION

The kernel of T is the set of all $\mathbf{x} = (x_1, x_2, x_3)$ in R^3 such that $T(x_1, x_2, x_3) = (0, 0)$. From this equation, you can write the homogeneous system

$$\begin{bmatrix} 1 & -1 & -2 \\ -1 & 2 & 3 \end{bmatrix}\begin{bmatrix} x_1 \\ x_2 \\ x_3 \end{bmatrix} = \begin{bmatrix} 0 \\ 0 \end{bmatrix} \quad\longrightarrow\quad \begin{aligned} x_1 - x_2 - 2x_3 &= 0 \\ -x_1 + 2x_2 + 3x_3 &= 0. \end{aligned}$$

Writing the augmented matrix of this system in reduced row-echelon form produces

$$\begin{bmatrix} 1 & 0 & -1 & 0 \\ 0 & 1 & 1 & 0 \end{bmatrix} \quad\longrightarrow\quad \begin{aligned} x_1 &= x_3 \\ x_2 &= -x_3. \end{aligned}$$

Using the parameter $t = x_3$ produces the family of solutions

$$\begin{bmatrix} x_1 \\ x_2 \\ x_3 \end{bmatrix} = \begin{bmatrix} t \\ -t \\ t \end{bmatrix} = t\begin{bmatrix} 1 \\ -1 \\ 1 \end{bmatrix}.$$

So, the kernel of T is

$$\ker(T) = \{t(1, -1, 1): t \text{ is a real number}\} = \operatorname{span}\{(1, -1, 1)\}. \text{ (See Figure 6.2.)}$$

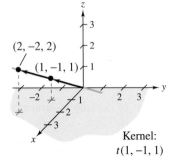

Kernel:
$t(1, -1, 1)$

Figure 6.2

Note in Example 5 that the kernel of T contains infinitely many vectors. Of course, the zero vector is in ker(T), but the kernel also contains such nonzero vectors as $(1, -1, 1)$ and $(2, -2, 2)$, as shown in Figure 6.2. The figure also shows that the kernel is a line passing through the origin, which implies that it is a subspace of R^3. Theorem 6.3 on the next page states that the kernel of every linear transformation $T: V \rightarrow W$ is a subspace of V.

THEOREM 6.3 The Kernel Is a Subspace of V

The kernel of a linear transformation $T: V \to W$ is a subspace of the domain V.

PROOF

From Theorem 6.1, you know that ker(T) is a nonempty subset of V. So, by Theorem 4.5, you can show that ker(T) is a subspace of V by showing that it is closed under vector addition and scalar multiplication. To do so, let $\mathbf{u}$ and $\mathbf{v}$ be vectors in the kernel of T. Then $T(\mathbf{u} + \mathbf{v}) = T(\mathbf{u}) + T(\mathbf{v}) = \mathbf{0} + \mathbf{0} = \mathbf{0}$, which implies that $\mathbf{u} + \mathbf{v}$ is in the kernel. Moreover, if c is any scalar, then $T(c\mathbf{u}) = cT(\mathbf{u}) = c\mathbf{0} = \mathbf{0}$, which implies that $c\mathbf{u}$ is in the kernel.

REMARK

The kernel of T is sometimes called the **nullspace** of T.

The next example shows how to find a basis for the kernel of a transformation defined by a matrix.

EXAMPLE 6 **Finding a Basis for the Kernel**

Define $T: R^5 \to R^4$ by $T(\mathbf{x}) = A\mathbf{x}$, where $\mathbf{x}$ is in R^5 and

$$A = \begin{bmatrix} 1 & 2 & 0 & 1 & -1 \\ 2 & 1 & 3 & 1 & 0 \\ -1 & 0 & -2 & 0 & 1 \\ 0 & 0 & 0 & 2 & 8 \end{bmatrix}.$$

Find a basis for ker(T) as a subspace of R^5.

SOLUTION

Using the procedure shown in Example 5, write the augmented matrix $[A \quad \mathbf{0}]$ in reduced row-echelon form as shown below.

$$\begin{bmatrix} 1 & 0 & 2 & 0 & -1 & 0 \\ 0 & 1 & -1 & 0 & -2 & 0 \\ 0 & 0 & 0 & 1 & 4 & 0 \\ 0 & 0 & 0 & 0 & 0 & 0 \end{bmatrix} \rightarrow \begin{matrix} x_1 = -2x_3 + x_5 \\ x_2 = x_3 + 2x_5 \\ x_4 = -4x_5 \end{matrix}$$

Letting $x_3 = s$ and $x_5 = t$, you have

$$\mathbf{x} = \begin{bmatrix} x_1 \\ x_2 \\ x_3 \\ x_4 \\ x_5 \end{bmatrix} = \begin{bmatrix} -2s + t \\ s + 2t \\ s + 0t \\ 0s - 4t \\ 0s + t \end{bmatrix} = s \begin{bmatrix} -2 \\ 1 \\ 1 \\ 0 \\ 0 \end{bmatrix} + t \begin{bmatrix} 1 \\ 2 \\ 0 \\ -4 \\ 1 \end{bmatrix}.$$

So a basis for the kernel of T is $B = \{(-2, 1, 1, 0, 0), (1, 2, 0, -4, 1)\}$.

DISCOVERY

1. What is the rank of the matrix A in Example 6?

2. Formulate a conjecture relating the dimension of the kernel, the rank, and the number of columns of A.

3. Verify your conjecture for the matrix in Example 5.

In the solution of Example 6, a basis for the kernel of T was found by solving the homogeneous system represented by $A\mathbf{x} = \mathbf{0}$. This procedure is a familiar one—it is the same procedure used to find the *nullspace* of A. In other words, the kernel of T is the solution space of $A\mathbf{x} = \mathbf{0}$, as stated in the corollary to Theorem 6.3 below.

THEOREM 6.3 Corollary

Let $T: R^n \to R^m$ be the linear transformation $T(\mathbf{x}) = A\mathbf{x}$. Then the kernel of T is equal to the solution space of $A\mathbf{x} = \mathbf{0}$.

THE RANGE OF A LINEAR TRANSFORMATION

The kernel is one of two critical subspaces associated with a linear transformation. The other is the range of T, denoted by range(T). Recall from Section 6.1 that the range of $T: V \rightarrow W$ is the set of all vectors **w** in W that are images of vectors in V. That is,

range(T) = {$T(\mathbf{v})$: **v** is in V}.

THEOREM 6.4   The Range of T Is a Subspace of W

The range of a linear transformation $T: V \rightarrow W$ is a subspace of W.

PROOF

The range of T is nonempty because $T(\mathbf{0}) = \mathbf{0}$ implies that the range contains the zero vector. To show that it is closed under vector addition, let $T(\mathbf{u})$ and $T(\mathbf{v})$ be vectors in the range of T. The vectors **u** and **v** are in V, so it follows that $\mathbf{u} + \mathbf{v}$ is also in V, and the sum

$$T(\mathbf{u}) + T(\mathbf{v}) = T(\mathbf{u} + \mathbf{v})$$

is in the range of T.

To show closure under scalar multiplication, let $T(\mathbf{u})$ be a vector in the range of T and let c be a scalar. **u** is in V, so it follows that $c\mathbf{u}$ is also in V, and the scalar multiple $cT(\mathbf{u}) = T(c\mathbf{u})$ is in the range of T. ∎

Note that the kernel and range of a linear transformation $T: V \rightarrow W$ are subspaces of V and W, respectively, as illustrated in Figure 6.3.

To find a basis for the range of a linear transformation $T(\mathbf{x}) = A\mathbf{x}$, observe that the range consists of all vectors **b** such that the system $A\mathbf{x} = \mathbf{b}$ is consistent. Writing the system

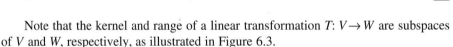

in the form

$$A\mathbf{x} = x_1 \begin{bmatrix} a_{11} \\ a_{21} \\ \vdots \\ a_{m1} \end{bmatrix} + x_2 \begin{bmatrix} a_{12} \\ a_{22} \\ \vdots \\ a_{m2} \end{bmatrix} + \cdots + x_n \begin{bmatrix} a_{1n} \\ a_{2n} \\ \vdots \\ a_{mn} \end{bmatrix} = \begin{bmatrix} b_1 \\ b_2 \\ \vdots \\ b_m \end{bmatrix} = \mathbf{b}$$

shows that **b** is in the range of T if and only if **b** is a linear combination of the column vectors of A. So *the column space of the matrix A is the same as the range of T.*

THEOREM 6.4   Corollary

Let $T: R^n \rightarrow R^m$ be the linear transformation $T(\mathbf{x}) = A\mathbf{x}$. Then the column space of A is equal to the range of T.

In Examples 4 and 5 in Section 4.6, you saw two procedures for finding a basis for the column space of a matrix. The next example uses the procedure from Example 5 in Section 4.6 to find a basis for the range of a linear transformation defined by a matrix.

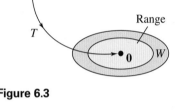

Domain   Kernel

V

T

Range

W

0

Figure 6.3

EXAMPLE 7

Finding a Basis for the Range of a Linear Transformation

See LarsonLinearAlgebra.com for an interactive version of this type of example.

For the linear transformation $R^5 \to R^4$ from Example 6, find a basis for the range of T.

SOLUTION

Use the reduced row-echelon form of A from Example 6.

$$A = \begin{bmatrix} 1 & 2 & 0 & 1 & -1 \\ 2 & 1 & 3 & 1 & 0 \\ -1 & 0 & -2 & 0 & 1 \\ 0 & 0 & 0 & 2 & 8 \end{bmatrix} \Rightarrow \begin{bmatrix} 1 & 0 & 2 & 0 & -1 \\ 0 & 1 & -1 & 0 & -2 \\ 0 & 0 & 0 & 1 & 4 \\ 0 & 0 & 0 & 0 & 0 \end{bmatrix}$$

The leading 1's appear in columns 1, 2, and 4 of the reduced matrix on the right, so the corresponding column vectors of A form a basis for the column space of A. A basis for the range of T is $B = \{(1, 2, -1, 0), (2, 1, 0, 0), (1, 1, 0, 2)\}$.

The next definition gives the dimensions of the kernel and range of a linear transformation.

REMARK

If T is given by a matrix A, then the rank of T is equal to the rank of A, and the nullity of T is equal to the nullity of A, as defined in Section 4.6.

Definition of Rank and Nullity of a Linear Transformation

Let $T: V \to W$ be a linear transformation. The dimension of the kernel of T is called the **nullity** of T and is denoted by **nullity**(T). The dimension of the range of T is called the **rank** of T and is denoted by **rank**(T).

In Examples 6 and 7, the rank and nullity of T are related to the dimension of the domain as shown below.

$$\text{rank}(T) + \text{nullity}(T) = 3 + 2 = 5 = \text{dimension of domain}$$

This relationship is true for any linear transformation from a finite-dimensional vector space, as stated in the next theorem.

THEOREM 6.5 Sum of Rank and Nullity

Let $T: V \to W$ be a linear transformation from an n-dimensional vector space V into a vector space W. Then the sum of the dimensions of the range and kernel is equal to the dimension of the domain. That is,

$$\text{rank}(T) + \text{nullity}(T) = n \quad \text{or} \quad \dim(\text{range}) + \dim(\text{kernel}) = \dim(\text{domain}).$$

PROOF

The proof provided here covers the case in which T is represented by an $m \times n$ matrix A. The general case will follow in the next section, where you will see that any linear transformation from an n-dimensional space into an m-dimensional space can be represented by a matrix. To prove this theorem, assume that the matrix A has a rank of r. Then you have

$$\text{rank}(T) = \dim(\text{range of } T) = \dim(\text{column space}) = \text{rank}(A) = r.$$

From Theorem 4.17, however, you know that

$$\text{nullity}(T) = \dim(\text{kernel of } T) = \dim(\text{solution space of } A\mathbf{x} = \mathbf{0}) = n - r.$$

So, it follows that $\text{rank}(T) + \text{nullity}(T) = r + (n - r) = n.$

EXAMPLE 8 **Finding the Rank and Nullity of a Linear Transformation**

Find the rank and nullity of the linear transformation $T: R^3 \to R^3$ defined by the matrix

$$A = \begin{bmatrix} 1 & 0 & -2 \\ 0 & 1 & 1 \\ 0 & 0 & 0 \end{bmatrix}.$$

SOLUTION

A is in reduced row-echelon form and has two nonzero rows, so it has a rank of 2. This means that the rank of T is also 2, and the nullity is dim(domain) − rank = 3 − 2 = 1.

One way to visualize the relationship between the rank and the nullity of a linear transformation provided by a matrix in row-echelon form is to observe that the number of leading 1's determines the rank, and the number of free variables (columns without leading 1's) determines the nullity. Their sum must be the total number of columns in the matrix, which is the dimension of the domain. In Example 8, the first two columns have leading 1's, indicating that the rank is 2. The third column corresponds to a free variable, indicating that the nullity is 1.

EXAMPLE 9 **Finding the Rank and Nullity of a Linear Transformation**

Let $T: R^5 \to R^7$ be a linear transformation.

a. Find the dimension of the kernel of T when the dimension of the range is 2.

b. Find the rank of T when the nullity of T is 4.

c. Find the rank of T when ker$(T) = \{\mathbf{0}\}$.

SOLUTION

a. By Theorem 6.5, with $n = 5$, you have

$$\dim(\text{kernel}) = n - \dim(\text{range}) = 5 - 2 = 3.$$

b. Again by Theorem 6.5, you have

$$\text{rank}(T) = n - \text{nullity}(T) = 5 - 4 = 1.$$

c. In this case, the nullity of T is 0. So

$$\text{rank}(T) = n - \text{nullity}(T) = 5 - 0 = 5.$$

LINEAR ALGEBRA APPLIED

A control system, such as the one shown for a dairy factory, processes an input signal $\mathbf{x}_k$ and produces an output signal $\mathbf{x}_{k+1}$. Without external feedback, the **difference equation** $\mathbf{x}_{k+1} = A\mathbf{x}_k$, a linear transformation where $\mathbf{x}_i$ is an $n \times 1$ vector and A is an $n \times n$ matrix, can model the relationship between the input and output signals. Typically, however, a control system has external feedback, so the relationship becomes $\mathbf{x}_{k+1} = A\mathbf{x}_k + B\mathbf{u}_k$, where B is an $n \times m$ matrix and $\mathbf{u}_k$ is an $m \times 1$ input, or control, vector. A system is *controllable* when it can reach any desired final state from its initial state in n or fewer steps. If A and B are matrices in a model of a controllable system, then the rank of the *controllability matrix*

$$[B \quad AB \quad A^2B \quad \dots \quad A^{n-1}B]$$

is equal to n.

ONE-TO-ONE AND ONTO LINEAR TRANSFORMATIONS

This section began with a question: Which vectors in the domain of a linear transformation are mapped to the zero vector? Theorem 6.6 (below) states that if the zero vector is the only vector $\mathbf{v}$ such that $T(\mathbf{v}) = \mathbf{0}$, then T is *one-to-one*. A function $T: V \to W$ is **one-to-one** when the preimage of every $\mathbf{w}$ in the range consists of a single vector, as shown below. This is equivalent to saying that T is one-to-one if and only if, for all $\mathbf{u}$ and $\mathbf{v}$ in V, $T(\mathbf{u}) = T(\mathbf{v})$ implies $\mathbf{u} = \mathbf{v}$.

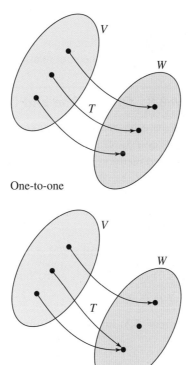

One-to-one

Not one-to-one

THEOREM 6.6   One-to-One Linear Transformations

Let $T: V \to W$ be a linear transformation. Then T is one-to-one if and only if $\ker(T) = \{\mathbf{0}\}$.

PROOF

First assume that T is one-to-one. Then $T(\mathbf{v}) = \mathbf{0}$ can have only one solution: $\mathbf{v} = \mathbf{0}$. In that case, $\ker(T) = \{\mathbf{0}\}$. Conversely, assume that $\ker(T) = \{\mathbf{0}\}$ and $T(\mathbf{u}) = T(\mathbf{v})$. You know that T is a linear transformation, so it follows that

$$T(\mathbf{u} - \mathbf{v}) = T(\mathbf{u}) - T(\mathbf{v}) = \mathbf{0}.$$

This implies that the vector $\mathbf{u} - \mathbf{v}$ lies in the kernel of T and must equal $\mathbf{0}$. So, $\mathbf{u} = \mathbf{v}$, which means that T is one-to-one.

EXAMPLE 10   One-to-One and Not One-to-One Linear Transformations

a. The linear transformation $T: M_{m,n} \to M_{n,m}$ represented by $T(A) = A^T$ is one-to-one because its kernel consists of only the $m \times n$ zero matrix.

b. The zero transformation $T: R^3 \to R^3$ is not one-to-one because its kernel is all of R^3.

A function $T: V \to W$ is **onto** when every element in W has a preimage in V. In other words, T is onto when W is equal to the range of T. The proof of the related theorem below is left as an exercise. (See Exercise 69.)

THEOREM 6.7   Onto Linear Transformations

Let $T: V \to W$ be a linear transformation, where W is finite dimensional. Then T is onto if and only if the rank of T is equal to the dimension of W.

For vector spaces of equal dimensions, you can combine the results of Theorems 6.5, 6.6, and 6.7 to obtain the next theorem relating the concepts of one-to-one and onto.

THEOREM 6.8   One-to-One and Onto Linear Transformations

Let $T: V \to W$ be a linear transformation with vector spaces V and W, *both* of dimension n. Then T is one-to-one if and only if it is onto.

PROOF

If T is one-to-one, then by Theorem 6.6 $\ker(T) = \{\mathbf{0}\}$, and $\dim(\ker(T)) = 0$. In that case, Theorem 6.5 produces

$$\dim(\text{range of } T) = n - \dim(\ker(T)) = n = \dim(W).$$

Consequently, by Theorem 6.7, T is onto. Similarly, if T is onto, then

$$\dim(\text{range of } T) = \dim(W) = n$$

which by Theorem 6.5 implies that $\dim(\ker(T)) = 0$. By Theorem 6.6, T is one-to-one. ∎

The next example brings together several concepts related to the kernel and range of a linear transformation.

EXAMPLE 11   Summarizing Several Results

Consider the linear transformation $T: R^n \to R^m$ represented by $T(\mathbf{x}) = A\mathbf{x}$. Find the nullity and rank of T, and determine whether T is one-to-one, onto, or neither.

a. $A = \begin{bmatrix} 1 & 2 & 0 \\ 0 & 1 & 1 \\ 0 & 0 & 1 \end{bmatrix}$ **b.** $A = \begin{bmatrix} 1 & 2 \\ 0 & 1 \\ 0 & 0 \end{bmatrix}$

c. $A = \begin{bmatrix} 1 & 2 & 0 \\ 0 & 1 & -1 \end{bmatrix}$ **d.** $A = \begin{bmatrix} 1 & 2 & 0 \\ 0 & 1 & 1 \\ 0 & 0 & 0 \end{bmatrix}$

SOLUTION

Note that each matrix is already in row-echelon form, so its rank can be determined by inspection.

$T: R^n \to R^m$	Dim(domain)	Dim(range) Rank(T)	Dim(kernel) Nullity(T)	One-to-One	Onto
a. $T: R^3 \to R^3$	3	3	0	Yes	Yes
b. $T: R^2 \to R^3$	2	2	0	Yes	No
c. $T: R^3 \to R^2$	3	2	1	No	Yes
d. $T: R^3 \to R^3$	3	2	1	No	No

ISOMORPHISMS OF VECTOR SPACES

Distinct vector spaces such as R^3 and $M_{3,1}$ can be thought of as being "essentially the same"—at least with respect to the operations of vector addition and scalar multiplication. Such spaces are **isomorphic** to each other. (The Greek word *isos* means "equal.")

Definition of Isomorphism

A linear transformation $T: V \rightarrow W$ that is one-to-one and onto is called an **isomorphism.** Moreover, if V and W are vector spaces such that there exists an isomorphism from V to W, then V and W are **isomorphic** to each other.

Isomorphic vector spaces are of the same finite dimension, and vector spaces of the same finite dimension are isomorphic, as stated in the next theorem.

THEOREM 6.9 Isomorphic Spaces and Dimension

Two finite-dimensional vector spaces V and W are isomorphic if and only if they are of the same dimension.

PROOF

Assume V is isomorphic to W, where V has dimension n. By the definition of isomorphic spaces, you know there exists a linear transformation $T: V \rightarrow W$ that is one-to-one and onto. T is one-to-one, so it follows that dim(kernel) = 0, which also implies that

$$\dim(\text{range}) = \dim(\text{domain}) = n.$$

In addition, T is onto, so you can conclude that $\dim(\text{range}) = \dim(W) = n$.

To prove the theorem in the other direction, assume V and W both have dimension n. Let $B = \{\mathbf{v}_1, \mathbf{v}_2, \ldots, \mathbf{v}_n\}$ be a basis for V, and let $B' = \{\mathbf{w}_1, \mathbf{w}_2, \ldots, \mathbf{w}_n\}$ be a basis for W. Then an arbitrary vector in V can be represented as

$$\mathbf{v} = c_1\mathbf{v}_1 + c_2\mathbf{v}_2 + \cdots + c_n\mathbf{v}_n$$

and you can define a linear transformation $T: V \rightarrow W$ as shown below.

$$T(\mathbf{v}) = c_1\mathbf{w}_1 + c_2\mathbf{w}_2 + \cdots + c_n\mathbf{w}_n$$

Verify that this linear transformation is both one-to-one and onto. So, V and W are isomorphic. ∎

Example 12 lists some vector spaces that are isomorphic to R^4.

REMARK

Your study of vector spaces has included much greater coverage to R^n than to other vector spaces. This preference for R^n stems from its notational convenience and from the geometric models available for R^2 and R^3.

EXAMPLE 12   Isomorphic Vector Spaces

The vector spaces below are isomorphic to each other.
a. $R^4 = $ 4-space
b. $M_{4,1} = $ space of all 4×1 matrices
c. $M_{2,2} = $ space of all 2×2 matrices
d. $P_3 = $ space of all polynomials of degree 3 or less
e. $V = \{(x_1, x_2, x_3, x_4, 0): x_i$ is a real number$\}$ (subspace of R^5)

Example 12 tells you that the elements in these spaces behave in the same way as an arbitrary vector $\mathbf{v} = (v_1, v_2, v_3, v_4)$.

Finding the Kernel of a Linear Transformation
In Exercises 1–10, find the kernel of the linear transformation.

1. $T: R^3 \to R^3$, $T(x, y, z) = (0, 0, 0)$

2. $T: R^3 \to R^3$, $T(x, y, z) = (x, 0, z)$

3. $T: R^4 \to R^4$, $T(x, y, z, w) = (y, x, w, z)$

4. $T: R^3 \to R^3$, $T(x, y, z) = (-z, -y, -x)$

5. $T: P_3 \to R$,
   $T(a_0 + a_1 x + a_2 x^2 + a_3 x^3) = a_1 + a_2$

6. $T: P_2 \to R$, $T(a_0 + a_1 x + a_2 x^2) = a_0$

7. $T: P_2 \to P_1$, $T(a_0 + a_1 x + a_2 x^2) = a_1 + 2a_2 x$

8. $T: P_3 \to P_2$,
   $T(a_0 + a_1 x + a_2 x^2 + a_3 x^3) = a_1 + 2a_2 x + 3a_3 x^2$

9. $T: R^2 \to R^2$, $T(x, y) = (x + 2y, y - x)$

10. $T: R^2 \to R^2$, $T(x, y) = (x - y, y - x)$

Finding the Kernel and Range In Exercises 11–18, define the linear transformation T by $T(\mathbf{x}) = A\mathbf{x}$. Find (a) the kernel of T and (b) the range of T.

11. $A = \begin{bmatrix} 1 & 2 \\ 3 & 4 \end{bmatrix}$ 12. $A = \begin{bmatrix} 1 & 2 \\ -3 & -6 \end{bmatrix}$

13. $A = \begin{bmatrix} 1 & -1 & 2 \\ 0 & 1 & 2 \end{bmatrix}$ 14. $A = \begin{bmatrix} 1 & -2 & 1 \\ 0 & 2 & 1 \end{bmatrix}$

15. $A = \begin{bmatrix} 1 & 3 \\ -1 & -3 \\ 2 & 2 \end{bmatrix}$ 16. $A = \begin{bmatrix} 1 & 1 \\ -1 & 2 \\ 0 & 1 \end{bmatrix}$

17. $A = \begin{bmatrix} 1 & 2 & -1 & 4 \\ 3 & 1 & 2 & -1 \\ -4 & -3 & -1 & -3 \\ -1 & -2 & 1 & 1 \end{bmatrix}$

18. $A = \begin{bmatrix} -1 & 3 & 2 & 1 & 4 \\ 2 & 3 & 5 & 0 & 0 \\ 2 & 1 & 2 & 1 & 0 \end{bmatrix}$

Finding the Kernel, Nullity, Range, and Rank In Exercises 19–32, define the linear transformation T by $T(\mathbf{x}) = A\mathbf{x}$. Find (a) $\ker(T)$, (b) $\text{nullity}(T)$, (c) $\text{range}(T)$, and (d) $\text{rank}(T)$.

19. $A = \begin{bmatrix} -1 & 1 \\ 1 & 1 \end{bmatrix}$ 20. $A = \begin{bmatrix} 3 & 2 \\ -9 & -6 \end{bmatrix}$

21. $A = \begin{bmatrix} 5 & -3 \\ 1 & 1 \\ 1 & -1 \end{bmatrix}$ 22. $A = \begin{bmatrix} 4 & 1 \\ 0 & 0 \\ 2 & -3 \end{bmatrix}$

23. $A = \begin{bmatrix} \frac{9}{10} & \frac{3}{10} \\ \frac{3}{10} & \frac{1}{10} \end{bmatrix}$ 24. $A = \begin{bmatrix} \frac{1}{26} & -\frac{5}{26} \\ -\frac{5}{26} & \frac{25}{26} \end{bmatrix}$

25. $A = \begin{bmatrix} 1 & 0 & 1 \\ 0 & 1 & 0 \\ 1 & 0 & 1 \end{bmatrix}$ 26. $A = \begin{bmatrix} 1 & 0 & 0 \\ 0 & 0 & 0 \\ 0 & 0 & 1 \end{bmatrix}$

27. $A = \begin{bmatrix} \frac{4}{9} & -\frac{4}{9} & \frac{2}{9} \\ -\frac{4}{9} & \frac{4}{9} & -\frac{2}{9} \\ \frac{2}{9} & -\frac{2}{9} & \frac{1}{9} \end{bmatrix}$ 28. $A = \begin{bmatrix} -\frac{1}{3} & \frac{2}{3} & -\frac{1}{3} \\ \frac{2}{3} & \frac{1}{3} & \frac{2}{3} \\ -\frac{1}{3} & \frac{2}{3} & -\frac{1}{3} \end{bmatrix}$

29. $A = \begin{bmatrix} 0 & -2 & 3 \\ 4 & 0 & 11 \end{bmatrix}$

30. $A = \begin{bmatrix} 1 & 1 & 0 & 0 \\ 0 & 0 & 1 & 1 \end{bmatrix}$

31. $A = \begin{bmatrix} 2 & 2 & -3 & 1 & 13 \\ 1 & 1 & 1 & 1 & -1 \\ 3 & 3 & -5 & 0 & 14 \\ 6 & 6 & -2 & 4 & 16 \end{bmatrix}$

32. $A = \begin{bmatrix} 3 & -2 & 6 & -1 & 15 \\ 4 & 3 & 8 & 10 & -14 \\ 2 & -3 & 4 & -4 & 20 \end{bmatrix}$

Finding the Nullity and Describing the Kernel and Range In Exercises 33–40, let $T: R^3 \to R^3$ be a linear transformation. Find the nullity of T and give a geometric description of the kernel and range of T.

33. $\text{rank}(T) = 2$ 34. $\text{rank}(T) = 1$

35. $\text{rank}(T) = 0$ 36. $\text{rank}(T) = 3$

37. T is the counterclockwise rotation of $45°$ about the z-axis:
$$T(x, y, z) = \left(\frac{\sqrt{2}}{2}x - \frac{\sqrt{2}}{2}y, \frac{\sqrt{2}}{2}x + \frac{\sqrt{2}}{2}y, z \right)$$

38. T is the reflection through the yz-coordinate plane:
$T(x, y, z) = (-x, y, z)$

39. T is the projection onto the vector $\mathbf{v} = (1, 2, 2)$:
$$T(x, y, z) = \frac{x + 2y + 2z}{9}(1, 2, 2)$$

40. T is the projection onto the xy-coordinate plane:
$T(x, y, z) = (x, y, 0)$

Finding the Nullity of a Linear Transformation In Exercises 41–46, find the nullity of T.

41. $T: R^4 \to R^2$, $\text{rank}(T) = 2$

42. $T: R^4 \to R^4$, $\text{rank}(T) = 0$

43. $T: P_5 \to P_2$, $\text{rank}(T) = 3$

44. $T: P_3 \to P_1$, $\text{rank}(T) = 2$

45. $T: M_{2,4} \to M_{4,2}$, $\text{rank}(T) = 4$

46. $T: M_{3,3} \to M_{2,3}$, $\text{rank}(T) = 6$

Verifying That _T_ Is One-to-One and Onto In Exercises 47–50, verify that the matrix defines a linear function _T_ that is one-to-one and onto.

47. $A = \begin{bmatrix} -2 & 0 \\ 0 & 2 \end{bmatrix}$ 48. $A = \begin{bmatrix} 1 & 0 \\ 0 & -1 \end{bmatrix}$

49. $A = \begin{bmatrix} 1 & 0 & 0 \\ 0 & 0 & 1 \\ 0 & 1 & 0 \end{bmatrix}$ 50. $A = \begin{bmatrix} 1 & 2 & 3 \\ -1 & 2 & 4 \\ 0 & 4 & 1 \end{bmatrix}$

Determining Whether _T_ Is One-to-One, Onto, or Neither In Exercises 51–54, determine whether the linear transformation is one-to-one, onto, or neither.

51. _T_ in Exercise 3   52. _T_ in Exercise 10

53. $T: R^2 \to R^3$, $T(\mathbf{x}) = A\mathbf{x}$, where _A_ is given in Exercise 21

54. $T: R^5 \to R^3$, $T(\mathbf{x}) = A\mathbf{x}$, where _A_ is given in Exercise 18

55. Identify the zero element and standard basis for each of the isomorphic vector spaces in Example 12.

56. Which vector spaces are isomorphic to R^6?
    (a) $M_{2,3}$ (b) P_6 (c) $C[0,6]$
    (d) $M_{6,1}$ (e) P_5 (f) $C'[-3,3]$
    (g) $\{(x_1, x_2, x_3, 0, x_5, x_6, x_7): x_i \text{ is a real number}\}$

57. **Calculus** Define $T: P_4 \to P_3$ by $T(p) = p'$. What is the kernel of _T_?

58. **Calculus** Define $T: P_2 \to R$ by
$$T(p) = \int_0^1 p(x)\, dx.$$
What is the kernel of _T_?

59. Let $T: R^3 \to R^3$ be the linear transformation that projects **u** onto $\mathbf{v} = (2, -1, 1)$.
    (a) Find the rank and nullity of _T_.
    (b) Find a basis for the kernel of _T_.

60. **CAPSTONE** Let $T: R^4 \to R^3$ be the linear transformation represented by $T(\mathbf{x}) = A\mathbf{x}$, where
$$A = \begin{bmatrix} 1 & -2 & 1 & 0 \\ 0 & 1 & 2 & 3 \\ 0 & 0 & 0 & 1 \end{bmatrix}.$$
    (a) Find the dimension of the domain.
    (b) Find the dimension of the range.
    (c) Find the dimension of the kernel.
    (d) Is _T_ one-to-one? Explain.
    (e) Is _T_ onto? Explain.
    (f) Is _T_ an isomorphism? Explain.

61. For the transformation $T: R^n \to R^n$ represented by $T(\mathbf{x}) = A\mathbf{x}$, what can be said about the rank of _T_ when (a) $\det(A) \neq 0$ and (b) $\det(A) = 0$?

62. **Writing** Let $T: R^m \to R^n$ be a linear transformation. Explain the differences between the concepts of one-to-one and onto. What can you say about _m_ and _n_ when _T_ is onto? What can you say about _m_ and _n_ when _T_ is one-to-one?

63. Define $T: M_{n,n} \to M_{n,n}$ by $T(A) = A - A^T$. Show that the kernel of _T_ is the set of $n \times n$ symmetric matrices.

64. Determine a relationship among _m_, _n_, _j_, and _k_ such that $M_{m,n}$ is isomorphic to $M_{j,k}$.

True or False? In Exercises 65 and 66, determine whether each statement is true or false. If a statement is true, give a reason or cite an appropriate statement from the text. If a statement is false, provide an example that shows the statement is not true in all cases or cite an appropriate statement from the text.

65. (a) The set of all vectors mapped from a vector space _V_ into another vector space _W_ by a linear transformation _T_ is the kernel of _T_.
    (b) The range of a linear transformation from a vector space _V_ into a vector space _W_ is a subspace of _V_.
    (c) The vector spaces R^3 and $M_{3,1}$ are isomorphic to each other.

66. (a) The dimension of a linear transformation _T_ from a vector space _V_ into a vector space _W_ is the rank of _T_.
    (b) A linear transformation _T_ from _V_ into _W_ is one-to-one when the preimage of every **w** in the range consists of a single vector **v**.
    (c) The vector spaces R^2 and P_1 are isomorphic to each other.

67. **Guided Proof** Let _B_ be an invertible $n \times n$ matrix. Prove that the linear transformation $T: M_{n,n} \to M_{n,n}$ represented by $T(A) = AB$ is an isomorphism.

    Getting Started: To show that the linear transformation is an isomorphism, you need to show that _T_ is both onto and one-to-one.
    (i) _T_ is a linear transformation with vector spaces of equal dimension, so by Theorem 6.8, you only need to show that _T_ is one-to-one.
    (ii) To show that _T_ is one-to-one, you need to determine the kernel of _T_ and show that it is $\{0\}$ (Theorem 6.6). Use the fact that _B_ is an invertible $n \times n$ matrix and that $T(A) = AB$.
    (iii) Conclude that _T_ is an isomorphism.

68. **Proof** Let $T: V \to W$ be a linear transformation. Prove that _T_ is one-to-one if and only if the rank of _T_ equals the dimension of _V_.

69. **Proof** Prove Theorem 6.7.

70. **Proof** Let $T: V \to W$ be a linear transformation, and let _U_ be a subspace of _W_. Prove that the set $T^{-1}(U) = \{\mathbf{v} \in V: T(\mathbf{v}) \in U\}$ is a subspace of _V_. What is $T^{-1}(U)$ when $U = \{0\}$?

6.3   Matrices for Linear Transformations

■ Find the standard matrix for a linear transformation.

■ Find the standard matrix for the composition of linear transformations and find the inverse of an invertible linear transformation.

■ Find the matrix for a linear transformation relative to a nonstandard basis.

THE STANDARD MATRIX FOR A LINEAR TRANSFORMATION

Which representation of $T: R^3 \to R^3$ is better:

$$T(x_1, x_2, x_3) = (2x_1 + x_2 - x_3, \ -x_1 + 3x_2 - 2x_3, \ 3x_2 + 4x_3)$$

or

$$T(\mathbf{x}) = A\mathbf{x} = \begin{bmatrix} 2 & 1 & -1 \\ -1 & 3 & -2 \\ 0 & 3 & 4 \end{bmatrix} \begin{bmatrix} x_1 \\ x_2 \\ x_3 \end{bmatrix}?$$

The second representation is better than the first for at least three reasons: it is simpler to write, simpler to read, and easier to enter into a calculator or math software. Later, you will see that matrix representation of linear transformations also has some theoretical advantages. In this section, you will see that for linear transformations involving finite-dimensional vector spaces, matrix representation is always possible.

The key to representing a linear transformation $T: V \to W$ by a matrix is to determine how it acts on a basis for V. Once you know the image of every vector in the basis, you can use the properties of linear transformations to determine $T(\mathbf{v})$ for any $\mathbf{v}$ in V.

Recall that the standard basis for R^n, written in column vector notation, is

$$B = \{\mathbf{e}_1, \mathbf{e}_2, \ldots, \mathbf{e}_n\}$$

$$= \left\{ \begin{bmatrix} 1 \\ 0 \\ \vdots \\ 0 \end{bmatrix}, \begin{bmatrix} 0 \\ 1 \\ \vdots \\ 0 \end{bmatrix}, \ldots, \begin{bmatrix} 0 \\ 0 \\ \vdots \\ 1 \end{bmatrix} \right\}.$$

THEOREM 6.10   Standard Matrix for a Linear Transformation

Let $T: R^n \to R^m$ be a linear transformation such that, for the standard basis vectors $\mathbf{e}_i$ of R^n,

$$T(\mathbf{e}_1) = \begin{bmatrix} a_{11} \\ a_{21} \\ \vdots \\ a_{m1} \end{bmatrix}, \ T(\mathbf{e}_2) = \begin{bmatrix} a_{12} \\ a_{22} \\ \vdots \\ a_{m2} \end{bmatrix}, \ldots, T(\mathbf{e}_n) = \begin{bmatrix} a_{1n} \\ a_{2n} \\ \vdots \\ a_{mn} \end{bmatrix}.$$

Then the $m \times n$ matrix whose n columns correspond to $T(\mathbf{e}_i)$

$$A = \begin{bmatrix} a_{11} & a_{12} & \cdots & a_{1n} \\ a_{21} & a_{22} & \cdots & a_{2n} \\ \vdots & \vdots & & \vdots \\ a_{m1} & a_{m2} & \cdots & a_{mn} \end{bmatrix}$$

is such that $T(\mathbf{v}) = A\mathbf{v}$ for every $\mathbf{v}$ in R^n. A is called the **standard matrix** for T.

PROOF

To show that $T(\mathbf{v}) = A\mathbf{v}$ for any $\mathbf{v}$ in R^n, you can write

$$\mathbf{v} = [v_1 \ \ v_2 \ \ldots \ v_n]^T = v_1\mathbf{e}_1 + v_2\mathbf{e}_2 + \cdots + v_n\mathbf{e}_n.$$

T is a linear transformation, so you have

$$\begin{aligned}
T(\mathbf{v}) &= T(v_1\mathbf{e}_1 + v_2\mathbf{e}_2 + \cdots + v_n\mathbf{e}_n)\\
&= T(v_1\mathbf{e}_1) + T(v_2\mathbf{e}_2) + \cdots + T(v_n\mathbf{e}_n)\\
&= v_1T(\mathbf{e}_1) + v_2T(\mathbf{e}_2) + \cdots + v_nT(\mathbf{e}_n).
\end{aligned}$$

On the other hand, the matrix product $A\mathbf{v}$ is

$$\begin{aligned}
A\mathbf{v} &= \begin{bmatrix} a_{11} & a_{12} & \cdots & a_{1n}\\ a_{21} & a_{22} & \cdots & a_{2n}\\ \vdots & \vdots & & \vdots\\ a_{m1} & a_{m2} & \cdots & a_{mn} \end{bmatrix} \begin{bmatrix} v_1\\ v_2\\ \vdots\\ v_n \end{bmatrix}\\[2mm]
&= \begin{bmatrix} a_{11}v_1 + a_{12}v_2 + \cdots + a_{1n}v_n\\ a_{21}v_1 + a_{22}v_2 + \cdots + a_{2n}v_n\\ \vdots & & \vdots\\ a_{m1}v_1 + a_{m2}v_2 + \cdots + a_{mn}v_n \end{bmatrix}\\[2mm]
&= v_1\begin{bmatrix} a_{11}\\ a_{21}\\ \vdots\\ a_{m1} \end{bmatrix} + v_2\begin{bmatrix} a_{12}\\ a_{22}\\ \vdots\\ a_{m2} \end{bmatrix} + \cdots + v_n\begin{bmatrix} a_{1n}\\ a_{2n}\\ \vdots\\ a_{mn} \end{bmatrix}\\[2mm]
&= v_1T(\mathbf{e}_1) + v_2T(\mathbf{e}_2) + \cdots + v_nT(\mathbf{e}_n).
\end{aligned}$$

So, $T(\mathbf{v}) = A\mathbf{v}$ for each $\mathbf{v}$ in R^n.

EXAMPLE 1   **Finding the Standard Matrix for a Linear Transformation**

See LarsonLinearAlgebra.com for an interactive version of this type of example.

Find the standard matrix for the linear transformation $T: R^3 \to R^2$ defined by

$$T(x, y, z) = (x - 2y, 2x + y).$$

SOLUTION

Begin by finding the images of $\mathbf{e}_1$, $\mathbf{e}_2$, and $\mathbf{e}_3$.

Vector Notation

$$T(\mathbf{e}_1) = T(1, 0, 0) = (1, 2)$$

$$T(\mathbf{e}_2) = T(0, 1, 0) = (-2, 1)$$

$$T(\mathbf{e}_3) = T(0, 0, 1) = (0, 0)$$

Matrix Notation

$$T(\mathbf{e}_1) = T\left(\begin{bmatrix} 1\\ 0\\ 0 \end{bmatrix}\right) = \begin{bmatrix} 1\\ 2 \end{bmatrix}$$

$$T(\mathbf{e}_2) = T\left(\begin{bmatrix} 0\\ 1\\ 0 \end{bmatrix}\right) = \begin{bmatrix} -2\\ 1 \end{bmatrix}$$

$$T(\mathbf{e}_3) = T\left(\begin{bmatrix} 0\\ 0\\ 1 \end{bmatrix}\right) = \begin{bmatrix} 0\\ 0 \end{bmatrix}$$

REMARK

As a check, note that

$$A\begin{bmatrix} x\\ y\\ z \end{bmatrix} = \begin{bmatrix} 1 & -2 & 0\\ 2 & 1 & 0 \end{bmatrix}\begin{bmatrix} x\\ y\\ z \end{bmatrix}$$
$$= \begin{bmatrix} x - 2y\\ 2x + y \end{bmatrix}$$

which is equivalent to

$$T(x, y, z) = (x - 2y, 2x + y).$$

By Theorem 6.10, the columns of A consist of $T(\mathbf{e}_1)$, $T(\mathbf{e}_2)$, and $T(\mathbf{e}_3)$, and you have

$$A = [T(\mathbf{e}_1) \ \ T(\mathbf{e}_2) \ \ T(\mathbf{e}_3)] = \begin{bmatrix} 1 & -2 & 0\\ 2 & 1 & 0 \end{bmatrix}.$$

A little practice will enable you to determine the standard matrix for a linear transformation, such as the one in Example 1, by inspection. For example, to find the standard matrix for the linear transformation

$$T(x_1, x_2, x_3) = (x_1 - 2x_2 + 5x_3, 2x_1 + 3x_3, 4x_1 + x_2 - 2x_3)$$

use the coefficients of x_1, x_2, and x_3 to form the rows of A, as shown below.

$$A = \begin{bmatrix} 1 & -2 & 5 \\ 2 & 0 & 3 \\ 4 & 1 & -2 \end{bmatrix} \quad \begin{matrix} \leftarrow \\ \leftarrow \\ \leftarrow \end{matrix} \quad \begin{matrix} 1x_1 - 2x_2 + 5x_3 \\ 2x_1 + 0x_2 + 3x_3 \\ 4x_1 + 1x_2 - 2x_3 \end{matrix}$$

EXAMPLE 2   **Finding the Standard Matrix for a Linear Transformation**

The linear transformation $T: R^2 \to R^2$ projects each point in R^2 onto the x-axis, as shown at the right. Find the standard matrix for T.

SOLUTION

This linear transformation is represented by

$$T(x, y) = (x, 0).$$

So, the standard matrix for T is

$$A = [T(1, 0) \quad T(0, 1)]$$

$$= \begin{bmatrix} 1 & 0 \\ 0 & 0 \end{bmatrix}.$$

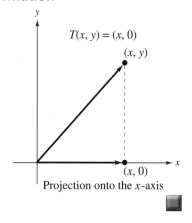

The standard matrix for the zero transformation from R^n into R^m is the $m \times n$ zero matrix, and the standard matrix for the identity transformation from R^n into R^n is I_n.

LINEAR ALGEBRA APPLIED

Ladder networks are useful tools for electrical engineers involved in circuit design. In a ladder network, the output voltage V and current I of one circuit are the input voltage and current of the circuit next to it. In the ladder network shown below, linear transformations can relate the input and output of an individual circuit (enclosed in a dashed box). Using Kirchhoff's Voltage and Current Laws and Ohm's Law,

$$\begin{bmatrix} V_2 \\ I_2 \end{bmatrix} = \begin{bmatrix} 1 & 0 \\ -1/R_1 & 1 \end{bmatrix} \begin{bmatrix} V_1 \\ I_1 \end{bmatrix}$$

and

$$\begin{bmatrix} V_3 \\ I_3 \end{bmatrix} = \begin{bmatrix} 1 & -R_2 \\ 0 & 1 \end{bmatrix} \begin{bmatrix} V_2 \\ I_2 \end{bmatrix}.$$

A *composition* can relate the input and output of the entire ladder network, that is, V_1 and I_1 to V_3 and I_3. Discussion on the composition of linear transformations begins on the next page.

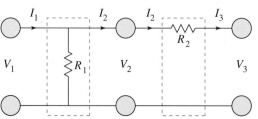

COMPOSITION OF LINEAR TRANSFORMATIONS

The **composition,** T, of $T_1 \colon R^n \to R^m$ with $T_2 \colon R^m \to R^p$ is

$$T(\mathbf{v}) = T_2(T_1(\mathbf{v}))$$

where $\mathbf{v}$ is a vector in R^n. This composition is denoted by

$$T = T_2 \circ T_1.$$

The domain of T is the domain of T_1. Moreover, the composition is not defined unless the range of T_1 lies within the domain of T_2, as shown below.

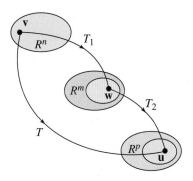

The next theorem emphasizes the usefulness of matrices for representing linear transformations. This theorem not only states that the composition of two linear transformations is a linear transformation, but also says that the standard matrix for the composition is the product of the standard matrices for the two original linear transformations.

THEOREM 6.11   Composition of Linear Transformations

Let $T_1 \colon R^n \to R^m$ and $T_2 \colon R^m \to R^p$ be linear transformations with standard matrices A_1 and A_2, respectively. The **composition** $T \colon R^n \to R^p$, defined by $T(\mathbf{v}) = T_2(T_1(\mathbf{v}))$, is a linear transformation. Moreover, the standard matrix A for T is the matrix product

$$A = A_2 A_1.$$

PROOF

To show that T is a linear transformation, let $\mathbf{u}$ and $\mathbf{v}$ be vectors in R^n and let c be any scalar. T_1 and T_2 are linear transformations, so you can write

$$\begin{aligned}
T(\mathbf{u} + \mathbf{v}) &= T_2(T_1(\mathbf{u} + \mathbf{v})) \\
&= T_2(T_1(\mathbf{u}) + T_1(\mathbf{v})) \\
&= T_2(T_1(\mathbf{u})) + T_2(T_1(\mathbf{v})) \\
&= T(\mathbf{u}) + T(\mathbf{v}) \\
T(c\mathbf{v}) &= T_2(T_1(c\mathbf{v})) \\
&= T_2(cT_1(\mathbf{v})) \\
&= cT_2(T_1(\mathbf{v})) \\
&= cT(\mathbf{v}).
\end{aligned}$$

Now, to show that $A_2 A_1$ is the standard matrix for T, use the associative property of matrix multiplication to write

$$T(\mathbf{v}) = T_2(T_1(\mathbf{v})) = T_2(A_1 \mathbf{v}) = A_2(A_1 \mathbf{v}) = (A_2 A_1)\mathbf{v}.$$

Theorem 6.11 can be generalized to cover the composition of n linear transformations. That is, if the standard matrices of $T_1, T_2, \ldots, T_n$ are $A_1, A_2, \ldots, A_n$, respectively, then the standard matrix for the composition $T(\mathbf{v}) = T_n(T_{n-1} \cdots (T_2(T_1(\mathbf{v}))) \cdots)$ is represented by $A = A_n A_{n-1} \cdots A_2 A_1$.

Matrix multiplication is not commutative, so order is important when forming the compositions of linear transformations. In general, the composition $T_2 \circ T_1$ is not the same as $T_1 \circ T_2$, as demonstrated in the next example.

EXAMPLE 3   The Standard Matrix for a Composition

Let T_1 and T_2 be linear transformations from R^3 into R^3 such that

$$T_1(x, y, z) = (2x + y, 0, x + z) \quad \text{and} \quad T_2(x, y, z) = (x - y, z, y).$$

Find the standard matrices for the compositions $T = T_2 \circ T_1$ and $T' = T_1 \circ T_2$.

SOLUTION

The standard matrices for T_1 and T_2 are

$$A_1 = \begin{bmatrix} 2 & 1 & 0 \\ 0 & 0 & 0 \\ 1 & 0 & 1 \end{bmatrix} \quad \text{and} \quad A_2 = \begin{bmatrix} 1 & -1 & 0 \\ 0 & 0 & 1 \\ 0 & 1 & 0 \end{bmatrix}.$$

By Theorem 6.11, the standard matrix for T is

$$\begin{aligned} A &= A_2 A_1 \\ &= \begin{bmatrix} 1 & -1 & 0 \\ 0 & 0 & 1 \\ 0 & 1 & 0 \end{bmatrix} \begin{bmatrix} 2 & 1 & 0 \\ 0 & 0 & 0 \\ 1 & 0 & 1 \end{bmatrix} \\ &= \begin{bmatrix} 2 & 1 & 0 \\ 1 & 0 & 1 \\ 0 & 0 & 0 \end{bmatrix} \end{aligned}$$

and the standard matrix for T' is

$$\begin{aligned} A' &= A_1 A_2 \\ &= \begin{bmatrix} 2 & 1 & 0 \\ 0 & 0 & 0 \\ 1 & 0 & 1 \end{bmatrix} \begin{bmatrix} 1 & -1 & 0 \\ 0 & 0 & 1 \\ 0 & 1 & 0 \end{bmatrix} \\ &= \begin{bmatrix} 2 & -2 & 1 \\ 0 & 0 & 0 \\ 1 & 0 & 0 \end{bmatrix}. \end{aligned}$$

Another benefit of matrix representation is that it can represent the **inverse** of a linear transformation. Before seeing how this works, consider the next definition.

Definition of Inverse Linear Transformation

If $T_1: R^n \to R^n$ and $T_2: R^n \to R^n$ are linear transformations such that for every $\mathbf{v}$ in R^n,

$$T_2(T_1(\mathbf{v})) = \mathbf{v} \quad \text{and} \quad T_1(T_2(\mathbf{v})) = \mathbf{v}$$

then T_2 is the **inverse** of T_1, and T_1 is said to be **invertible.**

Not every linear transformation has an inverse. If the transformation T_1 is invertible, however, then the inverse is unique and is denoted by T_1^{-1}.

Just as the inverse of a function of a real variable can be thought of as undoing what the function did, the inverse of a linear transformation T can be thought of as undoing the mapping done by T. For example, if T is a linear transformation from R^3 into R^3 such that

$$T(1, 4, -5) = (2, 3, 1)$$

and if T^{-1} exists, then T^{-1} maps $(2, 3, 1)$ back to its preimage under T. That is,

$$T^{-1}(2, 3, 1) = (1, 4, -5).$$

The next theorem states that a linear transformation is invertible if and only if it is an isomorphism (one-to-one and onto). You are asked to prove this theorem in Exercise 56.

REMARK

Several other conditions are equivalent to the three listed in Theorem 6.12; see the summary of equivalent conditions for square matrices in Section 4.6.

THEOREM 6.12 Existence of an Inverse Transformation

Let $T: R^n \to R^n$ be a linear transformation with standard matrix A. Then the conditions listed below are equivalent.

1. T is invertible.
2. T is an isomorphism.
3. A is invertible.

If T is invertible with standard matrix A, then the standard matrix for T^{-1} is A^{-1}.

EXAMPLE 4   **Finding the Inverse of a Linear Transformation**

Consider the linear transformation $T: R^3 \to R^3$ defined by

$$T(x_1, x_2, x_3) = (2x_1 + 3x_2 + x_3, 3x_1 + 3x_2 + x_3, 2x_1 + 4x_2 + x_3).$$

Show that T is invertible, and find its inverse.

SOLUTION

The standard matrix for T is

$$A = \begin{bmatrix} 2 & 3 & 1 \\ 3 & 3 & 1 \\ 2 & 4 & 1 \end{bmatrix}.$$

Using the method presented in Section 2.3 or a graphing calculator, you can find that A is invertible, and its inverse is

$$A^{-1} = \begin{bmatrix} -1 & 1 & 0 \\ -1 & 0 & 1 \\ 6 & -2 & -3 \end{bmatrix}.$$

So, T is invertible and the standard matrix for T^{-1} is A^{-1}.

Using the standard matrix for the inverse, you can find the rule for T^{-1} by computing the image of an arbitrary vector $\mathbf{x} = (x_1, x_2, x_3)$.

$$A^{-1}\mathbf{x} = \begin{bmatrix} -1 & 1 & 0 \\ -1 & 0 & 1 \\ 6 & -2 & -3 \end{bmatrix} \begin{bmatrix} x_1 \\ x_2 \\ x_3 \end{bmatrix}$$

$$= \begin{bmatrix} -x_1 + x_2 \\ -x_1 \quad\quad + x_3 \\ 6x_1 - 2x_2 - 3x_3 \end{bmatrix}$$

Or,

$$T^{-1}(x_1, x_2, x_3) = (-x_1 + x_2, -x_1 + x_3, 6x_1 - 2x_2 - 3x_3).$$

NONSTANDARD BASES AND GENERAL VECTOR SPACES

You will now consider the more general problem of finding a matrix for a linear transformation $T: V \to W$, where B and B' are ordered bases for V and W, respectively. Recall that the coordinate matrix of $\mathbf{v}$ relative to B is denoted by $[\mathbf{v}]_B$. To represent the linear transformation T, multiply A by a *coordinate matrix relative to B* to obtain a *coordinate matrix relative to B'*. That is, $[T(\mathbf{v})]_{B'} = A[\mathbf{v}]_B$. The matrix A is called the **matrix of T relative to the bases B and B'.**

To find the matrix A, you will use a procedure similar to the one used to find the standard matrix for T. That is, the images of the vectors in B are written as coordinate matrices relative to the basis B'. These coordinate matrices form the columns of A.

Transformation Matrix for Nonstandard Bases

Let V and W be finite-dimensional vector spaces with bases B and B', respectively, where

$$B = \{\mathbf{v}_1, \mathbf{v}_2, \ldots, \mathbf{v}_n\}.$$

If $T: V \to W$ is a linear transformation such that

$$[T(\mathbf{v}_1)]_{B'} = \begin{bmatrix} a_{11} \\ a_{21} \\ \vdots \\ a_{m1} \end{bmatrix}, [T(\mathbf{v}_2)]_{B'} = \begin{bmatrix} a_{12} \\ a_{22} \\ \vdots \\ a_{m2} \end{bmatrix}, \ldots, [T(\mathbf{v}_n)]_{B'} = \begin{bmatrix} a_{1n} \\ a_{2n} \\ \vdots \\ a_{mn} \end{bmatrix}$$

then the $m \times n$ matrix whose n columns correspond to $[T(\mathbf{v}_i)]_{B'}$

$$A = \begin{bmatrix} a_{11} & a_{12} & \cdots & a_{1n} \\ a_{21} & a_{22} & \cdots & a_{2n} \\ \vdots & \vdots & & \vdots \\ a_{m1} & a_{m2} & \cdots & a_{mn} \end{bmatrix}$$

is such that $[T(\mathbf{v})]_{B'} = A[\mathbf{v}]_B$ for every $\mathbf{v}$ in V.

EXAMPLE 5   **Finding a Matrix Relative to Nonstandard Bases**

Let $T: R^2 \to R^2$ be a linear transformation defined by $T(x_1, x_2) = (x_1 + x_2, 2x_1 - x_2)$. Find the matrix for T relative to the bases

$$B = \{\underset{\mathbf{v}_1}{(1, 2)}, \underset{\mathbf{v}_2}{(-1, 1)}\} \quad \text{and} \quad B' = \{\underset{\mathbf{w}_1}{(1, 0)}, \underset{\mathbf{w}_2}{(0, 1)}\}.$$

SOLUTION

By the definition of T, you have

$$T(\mathbf{v}_1) = T(1, 2) = (3, 0) = 3\mathbf{w}_1 + 0\mathbf{w}_2$$
$$T(\mathbf{v}_2) = T(-1, 1) = (0, -3) = 0\mathbf{w}_1 - 3\mathbf{w}_2.$$

The coordinate matrices for $T(\mathbf{v}_1)$ and $T(\mathbf{v}_2)$ relative to B' are

$$[T(\mathbf{v}_1)]_{B'} = \begin{bmatrix} 3 \\ 0 \end{bmatrix} \quad \text{and} \quad [T(\mathbf{v}_2)]_{B'} = \begin{bmatrix} 0 \\ -3 \end{bmatrix}.$$

Form the matrix for T relative to B and B' by using these coordinate matrices as columns to produce

$$A = \begin{bmatrix} 3 & 0 \\ 0 & -3 \end{bmatrix}.$$

EXAMPLE 6 — Using a Matrix to Represent a Linear Transformation

For the linear transformation $T: R^2 \to R^2$ in Example 5, use the matrix A to find $T(\mathbf{v})$, where $\mathbf{v} = (2, 1)$.

SOLUTION

Using the basis $B = \{(1, 2), (-1, 1)\}$, you find that $\mathbf{v} = (2, 1) = 1(1, 2) - 1(-1, 1)$, which implies

$$[\mathbf{v}]_B = [1 \quad -1]^T.$$

So, $[T(\mathbf{v})]_{B'}$ is

$$A[\mathbf{v}]_B = \begin{bmatrix} 3 & 0 \\ 0 & -3 \end{bmatrix} \begin{bmatrix} 1 \\ -1 \end{bmatrix} = \begin{bmatrix} 3 \\ 3 \end{bmatrix}.$$

Finally, $B' = \{(1, 0), (0, 1)\}$, so it follows that

$$T(\mathbf{v}) = 3(1, 0) + 3(0, 1) = (3, 3).$$

Check this result by directly calculating $T(\mathbf{v})$ using the definition of T in Example 5: $T(2, 1) = (2 + 1, 2(2) - 1) = (3, 3).$

For the special case where $V = W$ and $B = B'$, the matrix A is called the **matrix of T relative to the basis B.** In this case, the matrix of the identity transformation is simply I_n. To see this, let $B = \{\mathbf{v}_1, \mathbf{v}_2, \ldots, \mathbf{v}_n\}$. The identity transformation maps each $\mathbf{v}_i$ to itself, so you have $[T(\mathbf{v}_1)]_B = [1 \ 0 \ldots 0]^T, [T(\mathbf{v}_2)]_B = [0 \ 1 \ldots 0]^T, \ldots, [T(\mathbf{v}_n)]_B = [0 \ 0 \ldots 1]^T$, and it follows that $A = I_n$.

In the next example, you will construct a matrix representing the differential operator discussed in Example 10 in Section 6.1.

EXAMPLE 7 — A Matrix for the Differential Operator (Calculus)

Let $D_x: P_2 \to P_1$ be the differential operator that maps a polynomial p of degree 2 or less onto its derivative p'. Find the matrix for D_x using the bases

$$B = \{1, x, x^2\} \quad \text{and} \quad B' = \{1, x\}.$$

SOLUTION

The derivatives of the basis vectors are

$$D_x(1) = 0 = 0(1) + 0(x)$$
$$D_x(x) = 1 = 1(1) + 0(x)$$
$$D_x(x^2) = 2x = 0(1) + 2(x).$$

So, the coordinate matrices relative to B' are

$$[D_x(1)]_{B'} = \begin{bmatrix} 0 \\ 0 \end{bmatrix}, \quad [D_x(x)]_{B'} = \begin{bmatrix} 1 \\ 0 \end{bmatrix}, \quad [D_x(x^2)]_{B'} = \begin{bmatrix} 0 \\ 2 \end{bmatrix}$$

and the matrix for D_x is

$$A = \begin{bmatrix} 0 & 1 & 0 \\ 0 & 0 & 2 \end{bmatrix}.$$

Note that this matrix *does* produce the derivative of a quadratic polynomial $p(x) = a + bx + cx^2$.

$$Ap = \begin{bmatrix} 0 & 1 & 0 \\ 0 & 0 & 2 \end{bmatrix} \begin{bmatrix} a \\ b \\ c \end{bmatrix} = \begin{bmatrix} b \\ 2c \end{bmatrix} \implies b + 2cx = D_x[a + bx + cx^2]$$

6.3   Exercises   See CalcChat.com for worked-out solutions to odd-numbered exercises.

The Standard Matrix for a Linear Transformation **In Exercises 1–6, find the standard matrix for the linear transformation T.**

1. $T(x, y) = (x + 2y, x - 2y)$

2. $T(x, y) = (2x - 3y, x - y, y - 4x)$

3. $T(x, y, z) = (x + y, x - y, z - x)$

4. $T(x, y) = (5x + y, 0, 4x - 5y)$

5. $T(x, y, z) = (3x - 2z, 2y - z)$

6. $T(x_1, x_2, x_3, x_4) = (0, 0, 0, 0)$

Finding the Image of a Vector   In Exercises 7–10, use the standard matrix for the linear transformation T to find the image of the vector v.

7. $T(x, y, z) = (2x + y, 3y - z), \quad \mathbf{v} = (0, 1, -1)$

8. $T(x, y) = (x + y, x - y, 2x, 2y), \quad \mathbf{v} = (3, -3)$

9. $T(x, y) = (x - 3y, 2x + y, y), \quad \mathbf{v} = (-2, 4)$

10. $T(x_1, x_2, x_3, x_4) = (x_1 - x_3, x_2 - x_4, x_3 - x_1, x_2 + x_4)$,
    $\mathbf{v} = (1, 2, 3, -2)$

Finding the Standard Matrix and the Image   In Exercises 11–22, (a) find the standard matrix A for the linear transformation T, (b) use A to find the image of the vector v, and (c) sketch the graph of v and its image.

11. T is the reflection in the origin in R^2: $T(x, y) = (-x, -y)$,
    $\mathbf{v} = (3, 4)$.

12. T is the reflection in the line $y = x$ in R^2: $T(x, y) = (y, x)$,
    $\mathbf{v} = (3, 4)$.

13. T is the reflection in the y-axis in R^2: $T(x, y) = (-x, y)$,
    $\mathbf{v} = (2, -3)$.

14. T is the reflection in the x-axis in R^2: $T(x, y) = (x, -y)$,
    $\mathbf{v} = (4, -1)$.

15. T is the counterclockwise rotation of $45°$ in R^2,
    $\mathbf{v} = (2, 2)$.

16. T is the counterclockwise rotation of $120°$ in R^2,
    $\mathbf{v} = (2, 2)$.

17. T is the clockwise rotation (θ is negative) of $60°$ in R^2,
    $\mathbf{v} = (1, 2)$.

18. T is the clockwise rotation (θ is negative) of $30°$ in R^2,
    $\mathbf{v} = (2, 1)$.

19. T is the reflection in the xy-coordinate plane in R^3: $T(x, y, z) = (x, y, -z)$, $\mathbf{v} = (3, 2, 2)$.

20. T is the reflection in the yz-coordinate plane in R^3: $T(x, y, z) = (-x, y, z)$, $\mathbf{v} = (2, 3, 4)$.

21. T is the projection onto the vector $\mathbf{w} = (3, 1)$ in R^2: $T(\mathbf{v}) = \text{proj}_\mathbf{w}\mathbf{v}$, $\mathbf{v} = (1, 4)$.

22. T is the reflection in the vector $\mathbf{w} = (3, 1)$ in R^2: $T(\mathbf{v}) = 2\,\text{proj}_\mathbf{w}\mathbf{v} - \mathbf{v}$, $\mathbf{v} = (1, 4)$.

Finding the Standard Matrix and the Image   In Exercises 23–26, (a) find the standard matrix A for the linear transformation T and (b) use A to find the image of the vector v. Use a software program or a graphing utility to verify your result.

23. $T(x, y, z) = (2x + 3y - z, 3x - 2z, 2x - y + z)$,
    $\mathbf{v} = (1, 2, -1)$

24. $T(x, y, z) = (x + 2y - 3z, 3x - 5y, y - 3z)$,
    $\mathbf{v} = (3, 13, 4)$

25. $T(x_1, x_2, x_3, x_4) = (x_1 - x_2, x_3, x_1 + 2x_2 - x_4, x_4)$,
    $\mathbf{v} = (1, 0, 1, -1)$

26. $T(x_1, x_2, x_3, x_4) = (x_1 + 2x_2, x_2 - x_1, 2x_3 - x_4, x_1)$,
    $\mathbf{v} = (0, 1, -1, 1)$

Finding Standard Matrices for Compositions   In Exercises 27–30, find the standard matrices A and A' for $T = T_2 \circ T_1$ and $T' = T_1 \circ T_2$.

27. $T_1: R^2 \to R^2, T_1(x, y) = (x - 2y, 2x + 3y)$
    $T_2: R^2 \to R^2, T_2(x, y) = (y, 0)$

28. $T_1: R^3 \to R^3, T_1(x, y, z) = (x, y, z)$
    $T_2: R^3 \to R^3, T_2(x, y, z) = (0, x, 0)$

29. $T_1: R^2 \to R^3, T_1(x, y) = (-2x + 3y, x + y, x - 2y)$
    $T_2: R^3 \to R^2, T_2(x, y, z) = (x - 2y, z + 2x)$

30. $T_1: R^2 \to R^3, T_1(x, y) = (x, y, y)$
    $T_2: R^3 \to R^2, T_2(x, y, z) = (y, z)$

Finding the Inverse of a Linear Transformation In Exercises 31–36, determine whether the linear transformation is invertible. If it is, find its inverse.

31. $T(x, y) = (-4x, 4y)$ 32. $T(x, y) = (2x, 0)$

33. $T(x, y) = (x + y, 3x + 3y)$

34. $T(x, y) = (x + y, x - y)$

35. $T(x_1, x_2, x_3) = (x_1, x_1 + x_2, x_1 + x_2 + x_3)$

36. $T(x_1, x_2, x_3, x_4) = (x_1 - 2x_2, x_2, x_3 + x_4, x_3)$

Finding the Image Two Ways   In Exercises 37–42, find $T(\mathbf{v})$ by using (a) the standard matrix and (b) the matrix relative to B and B'.

37. $T: R^2 \to R^3, T(x, y) = (x + y, x, y)$, $\mathbf{v} = (5, 4)$,
    $B = \{(1, -1), (0, 1)\}$,
    $B' = \{(1, 1, 0), (0, 1, 1), (1, 0, 1)\}$

38. $T: R^3 \to R^2, T(x, y, z) = (x - y, y - z)$, $\mathbf{v} = (2, 4, 6)$,
    $B = \{(1, 1, 1), (1, 1, 0), (0, 1, 1)\}$, $B' = \{(1, 1), (2, 1)\}$

39. $T: R^3 \to R^4, T(x, y, z) = (2x, x + y, y + z, x + z)$,
    $\mathbf{v} = (1, -5, 2)$, $B = \{(2, 0, 1), (0, 2, 1), (1, 2, 1)\}$,
    $B' = \{(1, 0, 0, 1), (0, 1, 0, 1), (1, 0, 1, 0), (1, 1, 0, 0)\}$

40. $T: R^4 \to R^2$,

$T(x_1, x_2, x_3, x_4) = (x_1 + x_2 + x_3 + x_4, x_4 - x_1)$,

$\mathbf{v} = (4, -3, 1, 1)$,

$B = \{(1, 0, 0, 1), (0, 1, 0, 1), (1, 0, 1, 0), (1, 1, 0, 0)\}$,

$B' = \{(1, 1), (2, 0)\}$

41. $T: R^3 \to R^3$, $T(x, y, z) = (x + y + z, 2z - x, 2y - z)$,

$\mathbf{v} = (4, -5, 10)$, $B = \{(2, 0, 1), (0, 2, 1), (1, 2, 1)\}$,

$B' = \{(1, 1, 1), (1, 1, 0), (0, 1, 1)\}$

42. $T: R^2 \to R^2$, $T(x, y) = (3x - 13y, x - 4y)$, $\mathbf{v} = (4, 8)$,

$B = B' = \{(2, 1), (5, 1)\}$

43. Let $T: P_2 \to P_3$ be the linear transformation $T(p) = xp$. Find the matrix for T relative to the bases $B = \{1, x, x^2\}$ and $B' = \{1, x, x^2, x^3\}$.

44. Let $T: P_2 \to P_4$ be the linear transformation $T(p) = x^2p$. Find the matrix for T relative to the bases $B = \{1, x, x^2\}$ and $B' = \{1, x, x^2, x^3, x^4\}$.

45. Calculus Let $B = \{1, x, e^x, xe^x\}$ be a basis for a subspace W of the space of continuous functions, and let D_x be the differential operator on W. Find the matrix for D_x relative to the basis B.

46. Calculus Repeat Exercise 45 for $B = \{e^{2x}, xe^{2x}, x^2e^{2x}\}$.

47. Calculus Use the matrix from Exercise 45 to evaluate $D_x[4x - 3xe^x]$.

48. Calculus Use the matrix from Exercise 46 to evaluate $D_x[5e^{2x} - 3xe^{2x} + x^2e^{2x}]$.

49. Calculus Let $B = \{1, x, x^2, x^3\}$ be a basis for P_3, and $T: = P_3 \to P_4$ be the linear transformation represented by

$$T(x^k) = \int_0^x t^k \, dt.$$

(a) Find the matrix A for T with respect to B and the standard basis for P_4.

(b) Use A to integrate $p(x) = 8 - 4x + 3x^3$.

50. CAPSTONE

(a) Explain how to find the standard matrix for a linear transformation.

(b) Explain how to find a composition of linear transformations.

(c) Explain how to find the inverse of a linear transformation.

(d) Explain how to find the transformation matrix relative to nonstandard bases.

51. Define $T: M_{2,3} \to M_{3,2}$ by $T(A) = A^T$.

(a) Find the matrix for T relative to the standard bases for $M_{2,3}$ and $M_{3,2}$.

(b) Show that T is an isomorphism.

(c) Find the matrix for the inverse of T.

52. Let T be a linear transformation such that $T(\mathbf{v}) = k\mathbf{v}$ for $\mathbf{v}$ in R^n. Find the standard matrix for T.

True or False? In Exercises 53 and 54, determine whether each statement is true or false. If a statement is true, give a reason or cite an appropriate statement from the text. If a statement is false, provide an example that shows the statement is not true in all cases or cite an appropriate statement from the text.

53. (a) If $T: R^n \to R^m$ is a linear transformation such that

$T(\mathbf{e}_1) = [a_{11} \ a_{21} \ \ldots \ a_{m1}]^T$

$T(\mathbf{e}_2) = [a_{12} \ a_{22} \ \ldots \ a_{m2}]^T$

$\vdots$

$T(\mathbf{e}_n) = [a_{1n} \ a_{2n} \ \ldots \ a_{mn}]^T$

then the $m \times n$ matrix $A = [a_{ij}]$ whose columns correspond to $T(\mathbf{e}_i)$ and is such that $T(\mathbf{v}) = A\mathbf{v}$ for every $\mathbf{v}$ in R^n is called the standard matrix for T.

(b) All linear transformations T have a unique inverse T^{-1}.

54. (a) The composition T of linear transformations T_1 and T_2, represented by $T(\mathbf{v}) = T_2(T_1(\mathbf{v}))$, is defined when the range of T_1 lies within the domain of T_2.

(b) In general, the compositions $T_2 \circ T_1$ and $T_1 \circ T_2$ have the same standard matrix A.

55. Guided Proof Let $T_1: V \to V$ and $T_2: V \to V$ be one-to-one linear transformations. Prove that the composition $T = T_2 \circ T_1$ is one-to-one and that T^{-1} exists and is equal to $T_1^{-1} \circ T_2^{-1}$.

Getting Started: To show that T is one-to-one, use the definition of a one-to-one transformation and show that $T(\mathbf{u}) = T(\mathbf{v})$ implies $\mathbf{u} = \mathbf{v}$. For the second statement, you first need to use Theorems 6.8 and 6.12 to show that T is invertible, and then show that $T \circ (T_1^{-1} \circ T_2^{-1})$ and $(T_1^{-1} \circ T_2^{-1}) \circ T$ are identity transformations.

(i) Let $T(\mathbf{u}) = T(\mathbf{v})$. Recall that $(T_2 \circ T_1)(\mathbf{v}) = T_2(T_1(\mathbf{v}))$ for all vectors $\mathbf{v}$. Now use the fact that T_2 and T_1 are one-to-one to conclude that $\mathbf{u} = \mathbf{v}$.

(ii) Use Theorems 6.8 and 6.12 to show that T_1, T_2, and T are all invertible transformations. So, T_1^{-1} and T_2^{-1} exist.

(iii) Form the composition $T' = T_1^{-1} \circ T_2^{-1}$. It is a linear transformation from V into V. To show that it is the inverse of T, you need to determine whether the composition of T with T' on both sides gives an identity transformation.

56. Proof Prove Theorem 6.12.

57. Writing Is it always preferable to use the standard basis for R^n? Discuss the advantages and disadvantages of using different bases.

58. Writing Look back at Theorem 4.19 and rephrase it in terms of what you have learned in this chapter.

6.4   Transition Matrices and Similarity

■ Find and use a matrix for a linear transformation.

■ Show that two matrices are similar and use the properties of similar matrices.

THE MATRIX FOR A LINEAR TRANSFORMATION

In Section 6.3, you saw that the matrix for a linear transformation $T: V \rightarrow V$ depends on the basis for V. In other words, the matrix for T relative to a basis B is different from the matrix for T relative to another basis B'.

A classical problem in linear algebra is determining whether it is possible to find a basis B such that the matrix for T relative to B is diagonal. The solution of this problem is discussed in Chapter 7. This section lays a foundation for solving the problem. You will see how the matrices for a linear transformation relative to two different bases are related. In this section, A, A', P, and P^{-1} represent the four square matrices listed below.

1. Matrix for T relative to B: A

2. Matrix for T relative to B': A'

3. Transition matrix from B' to B: P

4. Transition matrix from B to B': P^{-1}

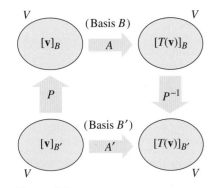

Figure 6.4

Note that in Figure 6.4, there are two ways to get from the coordinate matrix $[\mathbf{v}]_{B'}$ to the coordinate matrix $[T(\mathbf{v})]_{B'}$. One way is direct, using the matrix A' to obtain

$$A'[\mathbf{v}]_{B'} = [T(\mathbf{v})]_{B'}.$$

The other way is indirect, using the matrices P, A, and P^{-1} to obtain

$$P^{-1}AP[\mathbf{v}]_{B'} = [T(\mathbf{v})]_{B'}.$$

This implies that $A' = P^{-1}AP$. Example 1 demonstrates this relationship.

EXAMPLE 1   Finding a Matrix for a Linear Transformation

Find the matrix A' for $T: R^2 \rightarrow R^2$, where $T(x_1, x_2) = (2x_1 - 2x_2, -x_1 + 3x_2)$, relative to the basis $B' = \{(1, 0), (1, 1)\}$.

SOLUTION

The standard matrix for T is $A = \begin{bmatrix} 2 & -2 \\ -1 & 3 \end{bmatrix}$.

Furthermore, using the techniques of Section 4.7, the transition matrix from B' to the standard basis $B = \{(1, 0), (0, 1)\}$ is

$$P = \begin{bmatrix} 1 & 1 \\ 0 & 1 \end{bmatrix}.$$

The inverse of this matrix is the transition matrix from B to B',

$$P^{-1} = \begin{bmatrix} 1 & -1 \\ 0 & 1 \end{bmatrix}.$$

So, the matrix for T relative to B' is

$$A' = P^{-1}AP = \begin{bmatrix} 1 & -1 \\ 0 & 1 \end{bmatrix} \begin{bmatrix} 2 & -2 \\ -1 & 3 \end{bmatrix} \begin{bmatrix} 1 & 1 \\ 0 & 1 \end{bmatrix} = \begin{bmatrix} 3 & -2 \\ -1 & 2 \end{bmatrix}.$$

In Example 1, the basis B is the standard basis for R^2. In the next example, both B and B' are nonstandard bases.

EXAMPLE 2 Finding a Matrix for a Linear Transformation

Let $B = \{(-3, 2), (4, -2)\}$ and $B' = \{(-1, 2), (2, -2)\}$ be bases for R^2, and let

$$A = \begin{bmatrix} -2 & 7 \\ -3 & 7 \end{bmatrix}$$

be the matrix for $T: R^2 \to R^2$ relative to B. Find A', the matrix of T relative to B'.

SOLUTION

In Example 5 in Section 4.7, you found that $P = \begin{bmatrix} 3 & -2 \\ 2 & -1 \end{bmatrix}$ and $P^{-1} = \begin{bmatrix} -1 & 2 \\ -2 & 3 \end{bmatrix}$.
So, the matrix of T relative to B' is

$$A' = P^{-1}AP = \begin{bmatrix} -1 & 2 \\ -2 & 3 \end{bmatrix}\begin{bmatrix} -2 & 7 \\ -3 & 7 \end{bmatrix}\begin{bmatrix} 3 & -2 \\ 2 & -1 \end{bmatrix} = \begin{bmatrix} 2 & 1 \\ -1 & 3 \end{bmatrix}.$$

Figure 6.4 should help you to remember the roles of the matrices A, A', P, and P^{-1}.

EXAMPLE 3 Using a Matrix for a Linear Transformation

For the linear transformation $T: R^2 \to R^2$ from Example 2, find $[\mathbf{v}]_B$, $[T(\mathbf{v})]_B$, and $[T(\mathbf{v})]_{B'}$ for the vector $\mathbf{v}$ whose coordinate matrix is $[\mathbf{v}]_{B'} = [-3 \quad -1]^T$.

SOLUTION

To find $[\mathbf{v}]_B$, use the transition matrix P from B' to B.

$$[\mathbf{v}]_B = P[\mathbf{v}]_{B'} = \begin{bmatrix} 3 & -2 \\ 2 & -1 \end{bmatrix}\begin{bmatrix} -3 \\ -1 \end{bmatrix} = \begin{bmatrix} -7 \\ -5 \end{bmatrix}$$

To find $[T(\mathbf{v})]_B$, multiply $[\mathbf{v}]_B$ on the left by the matrix A to obtain

$$[T(\mathbf{v})]_B = A[\mathbf{v}]_B = \begin{bmatrix} -2 & 7 \\ -3 & 7 \end{bmatrix}\begin{bmatrix} -7 \\ -5 \end{bmatrix} = \begin{bmatrix} -21 \\ -14 \end{bmatrix}.$$

To find $[T(\mathbf{v})]_{B'}$, multiply $[T(\mathbf{v})]_B$ on the left by P^{-1} to obtain

$$[T(\mathbf{v})]_{B'} = P^{-1}[T(\mathbf{v})]_B = \begin{bmatrix} -1 & 2 \\ -2 & 3 \end{bmatrix}\begin{bmatrix} -21 \\ -14 \end{bmatrix} = \begin{bmatrix} -7 \\ 0 \end{bmatrix}$$

or multiply $[\mathbf{v}]_{B'}$ on the left by A' to obtain

$$[T(\mathbf{v})]_{B'} = A'[\mathbf{v}]_{B'} = \begin{bmatrix} 2 & 1 \\ -1 & 3 \end{bmatrix}\begin{bmatrix} -3 \\ -1 \end{bmatrix} = \begin{bmatrix} -7 \\ 0 \end{bmatrix}.$$

REMARK

It is instructive to note that the rule $T(x, y) = \left(x - \frac{3}{2}y, 2x + 4y\right)$ represents the transformation T in Examples 2 and 3. Verify the results of Example 3 by showing that $\mathbf{v} = (1, -4)$ and $T(\mathbf{v}) = (7, -14)$.

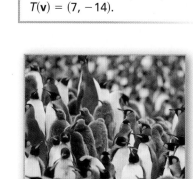

LINEAR ALGEBRA APPLIED

A **Leslie matrix,** named after British mathematician Patrick H. Leslie (1900–1974), can be used to find the age and growth distributions of a population over time. The entries in the first row of an $n \times n$ Leslie matrix L are the average numbers of offspring per member for each of n age classes. The entries in subsequent rows are p_i in row $i + 1$, column i and 0 elsewhere, where p_i is the probability that an ith age class member will survive to become an $(i + 1)$th age class member. If $\mathbf{x}_j$ is the age distribution vector for the jth time period, then the age distribution vector for the $(j + 1)$th time period can be found using the linear transformation $\mathbf{x}_{j+1} = L\mathbf{x}_j$. You will study population growth models using Leslie matrices in more detail in Section 7.4.

Rich Lindie/Shutterstock.com

SIMILAR MATRICES

Two square matrices A and A' that are related by an equation $A' = P^{-1}AP$ are called **similar** matrices, as stated in the next definition.

Definition of Similar Matrices

For square matrices A and A' of order n, A' is **similar** to A when there exists an invertible matrix P such that $A' = P^{-1}AP$.

If A' is similar to A, then it is also true that A is similar to A', as stated in the next theorem. So, it makes sense to simply say that **A and A' are similar.**

THEOREM 6.13   Properties of Similar Matrices

Let A, B, and C be square matrices of order n. Then the properties below are true.
1. A is similar to A.
2. If A is similar to B, then B is similar to A.
3. If A is similar to B and B is similar to C, then A is similar to C.

PROOF

The first property follows from the fact that $A = I_nAI_n$. To prove the second property, write

$$A = P^{-1}BP$$
$$PAP^{-1} = P(P^{-1}BP)P^{-1}$$
$$PAP^{-1} = B$$
$$Q^{-1}AQ = B, \text{ where } Q = P^{-1}.$$

The proof of the third property is left to you. (See Exercise 33.)

From the definition of similarity, it follows that any two matrices that represent the same linear transformation $T: V \to V$ with respect to different bases must be similar.

EXAMPLE 4   **Similar Matrices**

See LarsonLinearAlgebra.com for an interactive version of this type of example.

a. From Example 1, the matrices

$$A = \begin{bmatrix} 2 & -2 \\ -1 & 3 \end{bmatrix} \quad \text{and} \quad A' = \begin{bmatrix} 3 & -2 \\ -1 & 2 \end{bmatrix}$$

are similar because $A' = P^{-1}AP$, where $P = \begin{bmatrix} 1 & 1 \\ 0 & 1 \end{bmatrix}$.

b. From Example 2, the matrices

$$A = \begin{bmatrix} -2 & 7 \\ -3 & 7 \end{bmatrix} \quad \text{and} \quad A' = \begin{bmatrix} 2 & 1 \\ -1 & 3 \end{bmatrix}$$

are similar because $A' = P^{-1}AP$, where $P = \begin{bmatrix} 3 & -2 \\ 2 & -1 \end{bmatrix}$.

You have seen that the matrix for a linear transformation $T: V \to V$ depends on the basis used for V. This observation leads naturally to the question: What choice of basis will make the matrix for T as simple as possible? Is it always the *standard* basis? Not necessarily, as the next example demonstrates.

| EXAMPLE 5 | A Comparison of Two Matrices for a Linear Transformation |

Let

$$A = \begin{bmatrix} 1 & 3 & 0 \\ 3 & 1 & 0 \\ 0 & 0 & -2 \end{bmatrix}$$

be the matrix for $T: R^3 \to R^3$ relative to the standard basis. Find the matrix for T relative to the basis $B' = \{(1, 1, 0), (1, -1, 0), (0, 0, 1)\}$.

SOLUTION

The transition matrix from B' to the standard basis has columns consisting of the vectors in B',

$$P = \begin{bmatrix} 1 & 1 & 0 \\ 1 & -1 & 0 \\ 0 & 0 & 1 \end{bmatrix}$$

and it follows that

$$P^{-1} = \begin{bmatrix} \frac{1}{2} & \frac{1}{2} & 0 \\ \frac{1}{2} & -\frac{1}{2} & 0 \\ 0 & 0 & 1 \end{bmatrix}.$$

So, the matrix for T relative to B' is

$$\begin{aligned} A' &= P^{-1}AP \\ &= \begin{bmatrix} \frac{1}{2} & \frac{1}{2} & 0 \\ \frac{1}{2} & -\frac{1}{2} & 0 \\ 0 & 0 & 1 \end{bmatrix} \begin{bmatrix} 1 & 3 & 0 \\ 3 & 1 & 0 \\ 0 & 0 & -2 \end{bmatrix} \begin{bmatrix} 1 & 1 & 0 \\ 1 & -1 & 0 \\ 0 & 0 & 1 \end{bmatrix} \\ &= \begin{bmatrix} 4 & 0 & 0 \\ 0 & -2 & 0 \\ 0 & 0 & -2 \end{bmatrix}. \end{aligned}$$

Note that matrix A' is diagonal.

Diagonal matrices have many computational advantages over nondiagonal matrices. For example, for the diagonal matrix

$$D = \begin{bmatrix} d_1 & 0 & \ldots & 0 \\ 0 & d_2 & \ldots & 0 \\ \vdots & \vdots & & \vdots \\ 0 & 0 & \ldots & d_n \end{bmatrix}$$

the kth power of D is

$$D^k = \begin{bmatrix} d_1^k & 0 & \ldots & 0 \\ 0 & d_2^k & \ldots & 0 \\ \vdots & \vdots & & \vdots \\ 0 & 0 & \ldots & d_n^k \end{bmatrix}.$$

Also, a diagonal matrix is its own transpose. Moreover, if all of the main diagonal entries of a diagonal matrix are nonzero, then the inverse of the matrix is also a diagonal matrix, whose main diagonal entries are the reciprocals of corresponding entries in the original matrix. With such computational advantages, it is important to find ways (if possible) to choose a basis for V such that the transformation matrix is diagonal, as it is in Example 5. You will pursue this problem in the next chapter.

Finding a Matrix for a Linear Transformation In Exercises 1–12, find the matrix A' for T relative to the basis B'.

1. $T: R^2 \to R^2$, $T(x, y) = (2x - y, y - x)$,
   $B' = \{(1, -2), (0, 3)\}$

2. $T: R^2 \to R^2$, $T(x, y) = (2x + y, x - 2y)$,
   $B' = \{(1, 2), (0, 4)\}$

3. $T: R^2 \to R^2$, $T(x, y) = (x + y, 4y)$,
   $B' = \{(-4, 1), (1, -1)\}$

4. $T: R^2 \to R^2$, $T(x, y) = (x - 2y, 4x)$,
   $B' = \{(-2, 1), (-1, 1)\}$

5. $T: R^2 \to R^2$, $T(x, y) = (-3x + y, 3x - y)$,
   $B' = \{(1, -1), (-1, 5)\}$

6. $T: R^2 \to R^2$, $T(x, y) = (5x + 4y, 4x + 5y)$,
   $B' = \{(12, -13), (13, -12)\}$

7. $T: R^3 \to R^3$, $T(x, y, z) = (x, y, z)$,
   $B' = \{(1, 1, 0), (1, 0, 1), (0, 1, 1)\}$

8. $T: R^3 \to R^3$, $T(x, y, z) = (0, 0, 0)$,
   $B' = \{(1, 1, 0), (1, 0, 1), (0, 1, 1)\}$

9. $T: R^3 \to R^3$, $T(x, y, z) = (y + z, x + z, x + y)$,
   $B' = \{(5, 0, -1), (-3, 2, -1), (4, -6, 5)\}$

10. $T: R^3 \to R^3$, $T(x, y, z) = (-x, x - y, y - z)$,
    $B' = \{(0, -1, 2), (-2, 0, 3), (1, 3, 0)\}$

11. $T: R^3 \to R^3$,
    $T(x, y, z) = (x - y + 2z, 2x + y - z, x + 2y + z)$,
    $B' = \{(1, 0, 1), (0, 2, 2), (1, 2, 0)\}$

12. $T: R^3 \to R^3$,
    $T(x, y, z) = (x, x + 2y, x + y + 3z)$,
    $B' = \{(1, -1, 0), (0, 0, 1), (0, 1, -1)\}$

13. Let $B = \{(1, 3), (-2, -2)\}$ and $B' = \{(-12, 0), (-4, 4)\}$ be bases for R^2, and let

$$A = \begin{bmatrix} 3 & 2 \\ 0 & 4 \end{bmatrix}$$

be the matrix for $T: R^2 \to R^2$ relative to B.

(a) Find the transition matrix P from B' to B.

(b) Use the matrices P and A to find $[\mathbf{v}]_B$ and $[T(\mathbf{v})]_B$, where $[\mathbf{v}]_{B'} = [-1 \quad 2]^T$.

(c) Find P^{-1} and A' (the matrix for T relative to B').

(d) Find $[T(\mathbf{v})]_{B'}$ two ways.

14. Repeat Exercise 13 for $B = \{(1, 1), (-2, 3)\}$, $B' = \{(1, -1), (0, 1)\}$, and $[\mathbf{v}]_{B'} = [1 \quad -3]^T$.
    (Use matrix A in Exercise 13.)

15. Let $B = \{(1, 2), (-1, -1)\}$ and $B' = \{(-4, 1), (0, 2)\}$ be bases for R^2, and let

$$A = \begin{bmatrix} 2 & 1 \\ 0 & -1 \end{bmatrix}$$

be the matrix for $T: R^2 \to R^2$ relative to B.

(a) Find the transition matrix P from B' to B.

(b) Use the matrices P and A to find $[\mathbf{v}]_B$ and $[T(\mathbf{v})]_B$, where $[\mathbf{v}]_{B'} = [-1 \quad 4]^T$.

(c) Find P^{-1} and A' (the matrix for T relative to B').

(d) Find $[T(\mathbf{v})]_{B'}$ two ways.

16. Repeat Exercise 15 for $B = \{(1, -1), (-2, 1)\}$, $B' = \{(-1, 1), (1, 2)\}$, and $[\mathbf{v}]_{B'} = [1 \quad -4]^T$.
    (Use matrix A in Exercise 15.)

17. Let $B = \{(1, 1, 0), (1, 0, 1), (0, 1, 1)\}$ and $B' = \{(1, 0, 0), (0, 1, 0), (0, 0, 1)\}$ be bases for R^3, and let

$$A = \begin{bmatrix} \frac{3}{2} & -1 & -\frac{1}{2} \\ -\frac{1}{2} & 2 & \frac{1}{2} \\ \frac{1}{2} & 1 & \frac{5}{2} \end{bmatrix}$$

be the matrix for $T: R^3 \to R^3$ relative to B.

(a) Find the transition matrix P from B' to B.

(b) Use the matrices P and A to find $[\mathbf{v}]_B$ and $[T(\mathbf{v})]_B$, where $[\mathbf{v}]_{B'} = [1 \quad 0 \quad -1]^T$.

(c) Find P^{-1} and A' (the matrix for T relative to B').

(d) Find $[T(\mathbf{v})]_{B'}$ two ways.

18. Repeat Exercise 17 for
    $B = \{(1, 1, -1), (1, -1, 1), (-1, 1, 1)\}$,
    $B' = \{(1, 0, 0), (0, 1, 0), (0, 0, 1)\}$, and
    $[\mathbf{v}]_{B'} = [2 \quad 1 \quad 1]^T$.
    (Use matrix A in Exercise 17.)

Similar Matrices In Exercises 19–22, use the matrix P to show that the matrices A and A' are similar.

19. $P = \begin{bmatrix} -1 & -1 \\ 1 & 2 \end{bmatrix}$, $A = \begin{bmatrix} 12 & 7 \\ -20 & -11 \end{bmatrix}$, $A' = \begin{bmatrix} 1 & -2 \\ 4 & 0 \end{bmatrix}$

20. $P = A = A' = \begin{bmatrix} 1 & -12 \\ 0 & 1 \end{bmatrix}$

21. $P = \begin{bmatrix} 5 & 0 & 0 \\ 0 & 4 & 0 \\ 0 & 0 & 3 \end{bmatrix}$, $A = \begin{bmatrix} 5 & 10 & 0 \\ 8 & 4 & 0 \\ 0 & 9 & 6 \end{bmatrix}$, $A' = \begin{bmatrix} 5 & 8 & 0 \\ 10 & 4 & 0 \\ 0 & 12 & 6 \end{bmatrix}$

22. $P = \begin{bmatrix} 1 & 1 & 1 \\ 0 & 1 & 1 \\ 0 & 0 & 1 \end{bmatrix}$, $A = \begin{bmatrix} 5 & 0 & 0 \\ 0 & 3 & 0 \\ 0 & 0 & 1 \end{bmatrix}$, $A' = \begin{bmatrix} 5 & 2 & 2 \\ 0 & 3 & 2 \\ 0 & 0 & 1 \end{bmatrix}$

Diagonal Matrix for a Linear Transformation In Exercises 23 and 24, let A be the matrix for $T: R^3 \to R^3$ relative to the standard basis. Find the diagonal matrix A' for T relative to the basis B'.

23. $A = \begin{bmatrix} 0 & 2 & 0 \\ 1 & -1 & 0 \\ 0 & 0 & 1 \end{bmatrix}$,

$B' = \{(-1, 1, 0), (2, 1, 0), (0, 0, 1)\}$

24. $A = \begin{bmatrix} \frac{3}{2} & -1 & -\frac{1}{2} \\ -\frac{1}{2} & 2 & \frac{1}{2} \\ \frac{1}{2} & 1 & \frac{5}{2} \end{bmatrix}$,

$B' = \{(1, 1, -1), (1, -1, 1), (-1, 1, 1)\}$

25. **Proof** Prove that if A and B are similar matrices, then $|A| = |B|$.

Is the converse true?

26. Illustrate the result of Exercise 25 using the matrices

$A = \begin{bmatrix} 1 & 0 & 0 \\ 0 & -2 & 0 \\ 0 & 0 & 3 \end{bmatrix}$, $B = \begin{bmatrix} 11 & 7 & 10 \\ 10 & 8 & 10 \\ -18 & -12 & -17 \end{bmatrix}$,

$P = \begin{bmatrix} -1 & 1 & 0 \\ 2 & 1 & 2 \\ 1 & 1 & 1 \end{bmatrix}$, $P^{-1} = \begin{bmatrix} -1 & -1 & 2 \\ 0 & -1 & 2 \\ 1 & 2 & -3 \end{bmatrix}$,

where $B = P^{-1}AP$.

27. **Proof** Prove that if A and B are similar matrices, then there exists a matrix P such that $B^k = P^{-1}A^kP$.

28. Use the result of Exercise 27 to find B^4, where

$B = P^{-1}AP$

for the matrices

$A = \begin{bmatrix} 1 & 0 \\ 0 & 2 \end{bmatrix}$, $B = \begin{bmatrix} -4 & -15 \\ 2 & 7 \end{bmatrix}$,

$P = \begin{bmatrix} 2 & 5 \\ 1 & 3 \end{bmatrix}$, $P^{-1} = \begin{bmatrix} 3 & -5 \\ -1 & 2 \end{bmatrix}$.

29. Determine all $n \times n$ matrices that are similar to I_n.

30. **Proof** Prove that if A is an idempotent matrix and B is similar to A, then B is idempotent. (Recall that an $n \times n$ matrix A is idempotent when $A = A^2$.)

31. **Proof** Let A be an $n \times n$ matrix such that $A^2 = O$. Prove that if B is similar to A, then $B^2 = O$.

32. **Proof** Consider the matrix equation $B = P^{-1}AP$. Prove that if $A\mathbf{x} = \mathbf{x}$, then $PBP^{-1}\mathbf{x} = \mathbf{x}$.

33. **Proof** Prove Property 3 of Theorem 6.13: For square matrices A, B, and C of order n, if A is similar to B and B is similar to C, then A is similar to C.

34. **Writing** Explain why two similar matrices have the same rank.

35. **Proof** Prove that if A and B are similar matrices, then A^T and B^T are similar matrices.

36. **Proof** Prove that if A and B are similar matrices and A is nonsingular, then B is also nonsingular and A^{-1} and B^{-1} are similar matrices.

37. **Proof** Let $A = CD$, where C and D are $n \times n$ matrices and C is invertible. Prove that the matrix product DC is similar to A.

38. **Proof** Let $B = P^{-1}AP$, where $A = [a_{ij}]$, $P = [p_{ij}]$, and B is a diagonal matrix with main diagonal entries $b_{11}, b_{22}, \ldots, b_{nn}$. Prove that

$\begin{bmatrix} a_{11} & a_{12} & \cdots & a_{1n} \\ a_{21} & a_{22} & \cdots & a_{2n} \\ \vdots & \vdots & & \vdots \\ a_{n1} & a_{n2} & \cdots & a_{nn} \end{bmatrix} \begin{bmatrix} p_{1i} \\ p_{2i} \\ \vdots \\ p_{ni} \end{bmatrix} = b_{ii} \begin{bmatrix} p_{1i} \\ p_{2i} \\ \vdots \\ p_{ni} \end{bmatrix}$

for $i = 1, 2, \ldots, n$.

39. **Writing** Let $B = \{\mathbf{v}_1, \mathbf{v}_2, \ldots, \mathbf{v}_n\}$ be a basis for the vector space V, let B' be the standard basis, and consider the identity transformation $I: V \to V$. What can you say about the matrix for I relative to B? relative to B'? when the domain has the basis B and the range has the basis B'?

40. **CAPSTONE**

(a) Consider two bases B and B' for a vector space V and the matrix A for the linear transformation $T: V \to V$ relative to B. Explain how to obtain the coordinate matrix $[T(\mathbf{v})]_{B'}$ from the coordinate matrix $[\mathbf{v}]_{B'}$, where $\mathbf{v}$ is a vector in V.

(b) Explain how to determine whether two square matrices A and A' of order n are similar.

True or False? In Exercises 41 and 42, determine whether each statement is true or false. If a statement is true, give a reason or cite an appropriate statement from the text. If a statement is false, provide an example that shows the statement is not true in all cases or cite an appropriate statement from the text.

41. (a) The matrix for a linear transformation A' relative to the basis B' is equal to the product $P^{-1}AP$, where P^{-1} is the transition matrix from B to B', A is the matrix for the linear transformation relative to basis B, and P is the transition matrix from B' to B.

(b) Two matrices that represent the same linear transformation $T: V \to V$ with respect to different bases are not necessarily similar.

42. (a) The matrix for a linear transformation A relative to the basis B is equal to the product $PA'P^{-1}$, where P is the transition matrix from B' to B, A' is the matrix for the linear transformation relative to basis B', and P^{-1} is the transition matrix from B to B'.

(b) The standard basis for R^n will always make the coordinate matrix for the linear transformation T the simplest matrix possible.

6.5   Applications of Linear Transformations

◻ Identify linear transformations defined by reflections, expansions, contractions, or shears in R^2.

◻ Use a linear transformation to rotate a figure in R^3.

THE GEOMETRY OF LINEAR TRANSFORMATIONS IN R^2

The first part of this section gives geometric interpretations of linear transformations represented by 2×2 elementary matrices. After a summary of the various types of 2×2 elementary matrices are examples that examine each type of matrix in more detail.

Elementary Matrices for Linear Transformations in R^2

Reflection in y-Axis
$$A = \begin{bmatrix} -1 & 0 \\ 0 & 1 \end{bmatrix}$$

Reflection in x-Axis
$$A = \begin{bmatrix} 1 & 0 \\ 0 & -1 \end{bmatrix}$$

Reflection in Line $y = x$
$$A = \begin{bmatrix} 0 & 1 \\ 1 & 0 \end{bmatrix}$$

Horizontal Expansion ($k > 1$) or Contraction ($0 < k < 1$)
$$A = \begin{bmatrix} k & 0 \\ 0 & 1 \end{bmatrix}$$

Vertical Expansion ($k > 1$) or Contraction ($0 < k < 1$)
$$A = \begin{bmatrix} 1 & 0 \\ 0 & k \end{bmatrix}$$

Horizontal Shear
$$A = \begin{bmatrix} 1 & k \\ 0 & 1 \end{bmatrix}$$

Vertical Shear
$$A = \begin{bmatrix} 1 & 0 \\ k & 1 \end{bmatrix}$$

EXAMPLE 1   Reflections in R^2

The transformations below are **reflections.** These have the effect of mapping a point in the xy-plane to its "mirror image" with respect to one of the coordinate axes or the line $y = x$, as shown in Figure 6.5.

a. Reflection in the y-axis:
$$T(x, y) = (-x, y)$$
$$\begin{bmatrix} -1 & 0 \\ 0 & 1 \end{bmatrix}\begin{bmatrix} x \\ y \end{bmatrix} = \begin{bmatrix} -x \\ y \end{bmatrix}$$

b. Reflection in the x-axis:
$$T(x, y) = (x, -y)$$
$$\begin{bmatrix} 1 & 0 \\ 0 & -1 \end{bmatrix}\begin{bmatrix} x \\ y \end{bmatrix} = \begin{bmatrix} x \\ -y \end{bmatrix}$$

c. Reflection in the line $y = x$:
$$T(x, y) = (y, x)$$
$$\begin{bmatrix} 0 & 1 \\ 1 & 0 \end{bmatrix}\begin{bmatrix} x \\ y \end{bmatrix} = \begin{bmatrix} y \\ x \end{bmatrix}$$

a.
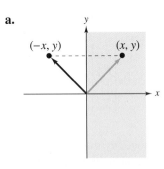

b.
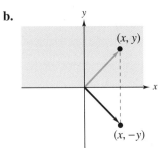

c.
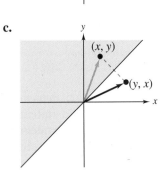

Reflections in R^2
Figure 6.5

EXAMPLE 2   **Expansions and Contractions in R^2**

The transformations below are **expansions** or **contractions,** depending on the value of the positive scalar k.

a. Horizontal expansions and contractions:

$$T(x, y) = (kx, y)$$

$$\begin{bmatrix} k & 0 \\ 0 & 1 \end{bmatrix} \begin{bmatrix} x \\ y \end{bmatrix} = \begin{bmatrix} kx \\ y \end{bmatrix}$$

b. Vertical expansions and contractions:

$$T(x, y) = (x, ky)$$

$$\begin{bmatrix} 1 & 0 \\ 0 & k \end{bmatrix} \begin{bmatrix} x \\ y \end{bmatrix} = \begin{bmatrix} x \\ ky \end{bmatrix}$$

Note in the figures below that the distance the point (x, y) moves by a contraction or an expansion is proportional to its x- or y-coordinate. For example, under the transformation represented by

$$T(x, y) = (2x, y)$$

the point $(1, 3)$ would move one unit to the right, but the point $(4, 3)$ would move four units to the right. Under the transformation represented by

$$T(x, y) = \left(x, \tfrac{1}{2}y\right)$$

the point $(1, 4)$ would move two units down, but the point $(1, 2)$ would move one unit down.

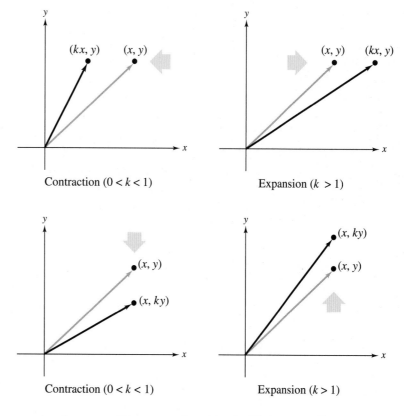

Contraction $(0 < k < 1)$ Expansion $(k > 1)$

Contraction $(0 < k < 1)$ Expansion $(k > 1)$

Another type of linear transformation in R^2 corresponding to an elementary matrix is a *shear*, as described in Example 3.

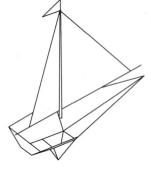

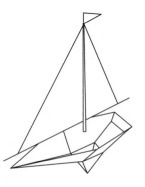

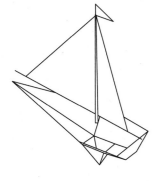

EXAMPLE 3   **Shears in R^2**

The transformations below are **shears**.

$$T(x, y) = (x + ky, y)$$
$$\begin{bmatrix} 1 & k \\ 0 & 1 \end{bmatrix}\begin{bmatrix} x \\ y \end{bmatrix} = \begin{bmatrix} x + ky \\ y \end{bmatrix}$$

$$T(x, y) = (x, y + kx)$$
$$\begin{bmatrix} 1 & 0 \\ k & 1 \end{bmatrix}\begin{bmatrix} x \\ y \end{bmatrix} = \begin{bmatrix} x \\ kx + y \end{bmatrix}$$

a. A horizontal shear represented by

$$T(x, y) = (x + 2y, y)$$

is shown at the right. Under this transformation, points in the upper half-plane "shear" to the right by amounts proportional to their y-coordinates. Points in the lower half-plane "shear" to the left by amounts proportional to the absolute values of their y-coordinates. Points on the x-axis do not move by this transformation.

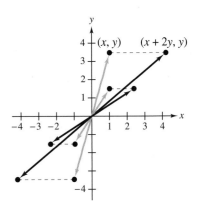

b. A vertical shear represented by

$$T(x, y) = (x, y + 2x)$$

is shown below. Here, points in the right half-plane "shear" upward by amounts proportional to their x-coordinates. Points in the left half-plane "shear" downward by amounts proportional to the absolute values of their x-coordinates. Points on the y-axis do not move.

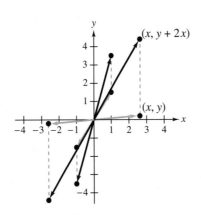

LINEAR ALGEBRA APPLIED

The use of computer graphics is common in many fields. By using graphics software, a designer can "see" an object before it is physically created. Linear transformations can be useful in computer graphics. To illustrate, consider a simplified example. Only 23 points in R^3 were used to generate images of the toy boat shown at the left. Most graphics software can use such minimal information to generate views of an image from any perspective, as well as color, shade, and render as appropriate. Linear transformations, specifically those that produce rotations in R^3, can represent the different views. The remainder of this section discusses rotation in R^3.

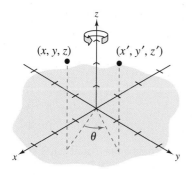

Figure 6.6

ROTATION IN R^3

In Example 7 in Section 6.1, you saw how a linear transformation can be used to rotate figures in R^2. Here you will see how linear transformations can be used to rotate figures in R^3.

Say you want to rotate the point (x, y, z) counterclockwise about the z-axis through an angle θ, as shown in Figure 6.6. Letting the coordinates of the rotated point be (x', y', z'), you have

$$\begin{bmatrix} x' \\ y' \\ z' \end{bmatrix} = \begin{bmatrix} \cos\theta & -\sin\theta & 0 \\ \sin\theta & \cos\theta & 0 \\ 0 & 0 & 1 \end{bmatrix} \begin{bmatrix} x \\ y \\ z \end{bmatrix} = \begin{bmatrix} x\cos\theta - y\sin\theta \\ x\sin\theta + y\cos\theta \\ z \end{bmatrix}.$$

Example 4 uses this matrix to rotate a figure in three-dimensional space.

EXAMPLE 4   Rotation About the z-Axis

The eight vertices of the rectangular prism shown at the right are

$$V_1(0, 0, 0) \qquad V_2(1, 0, 0)$$
$$V_3(1, 2, 0) \qquad V_4(0, 2, 0)$$
$$V_5(0, 0, 3) \qquad V_6(1, 0, 3)$$
$$V_7(1, 2, 3) \qquad V_8(0, 2, 3).$$

Find the coordinates of the vertices after the prism is rotated counterclockwise about the z-axis through (a) $\theta = 60°$, (b) $\theta = 90°$, and (c) $\theta = 120°$.

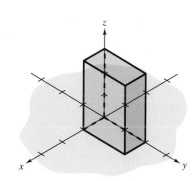

SOLUTION

a. The matrix that yields a rotation of 60° is

$$A = \begin{bmatrix} \cos 60° & -\sin 60° & 0 \\ \sin 60° & \cos 60° & 0 \\ 0 & 0 & 1 \end{bmatrix} = \begin{bmatrix} 1/2 & -\sqrt{3}/2 & 0 \\ \sqrt{3}/2 & 1/2 & 0 \\ 0 & 0 & 1 \end{bmatrix}.$$

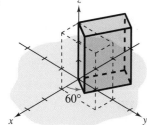

Multiplying this matrix by the column vectors corresponding to each vertex produces the rotated vertices listed below.

$$V_1'(0, 0, 0) \qquad V_2'(0.5, 0.87, 0) \qquad V_3'(-1.23, 1.87, 0) \qquad V_4'(-1.73, 1, 0)$$
$$V_5'(0, 0, 3) \qquad V_6'(0.5, 0.87, 3) \qquad V_7'(-1.23, 1.87, 3) \qquad V_8'(-1.73, 1, 3)$$

Figure 6.7(a) shows a graph of the rotated prism.

b. The matrix that yields a rotation of 90° is

$$A = \begin{bmatrix} \cos 90° & -\sin 90° & 0 \\ \sin 90° & \cos 90° & 0 \\ 0 & 0 & 1 \end{bmatrix} = \begin{bmatrix} 0 & -1 & 0 \\ 1 & 0 & 0 \\ 0 & 0 & 1 \end{bmatrix}$$

and Figure 6.7(b) shows a graph of the rotated prism.

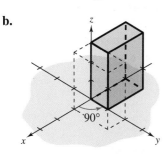

c. The matrix that yields a rotation of 120° is

$$A = \begin{bmatrix} \cos 120° & -\sin 120° & 0 \\ \sin 120° & \cos 120° & 0 \\ 0 & 0 & 1 \end{bmatrix} = \begin{bmatrix} -1/2 & -\sqrt{3}/2 & 0 \\ \sqrt{3}/2 & -1/2 & 0 \\ 0 & 0 & 1 \end{bmatrix}$$

and Figure 6.7(c) shows a graph of the rotated prism.

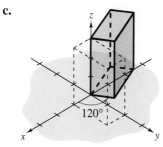

Figure 6.7

REMARK

To illustrate the right-hand rule, imagine the thumb of your right hand pointing in the positive direction of an axis. The cupped fingers will point in the direction of counterclockwise rotation. The figure below shows counterclockwise rotation about the *z*-axis.

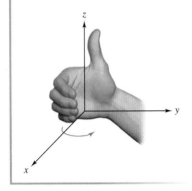

Example 4 uses matrices to perform rotations about the *z*-axis. Similarly, you can use matrices to rotate figures about the *x*- or *y*-axis. A summary of all three types of rotations is below.

Rotation About the *x*-Axis   Rotation About the *y*-Axis   Rotation About the *z*-Axis

$$\begin{bmatrix} 1 & 0 & 0 \\ 0 & \cos\theta & -\sin\theta \\ 0 & \sin\theta & \cos\theta \end{bmatrix} \qquad \begin{bmatrix} \cos\theta & 0 & \sin\theta \\ 0 & 1 & 0 \\ -\sin\theta & 0 & \cos\theta \end{bmatrix} \qquad \begin{bmatrix} \cos\theta & -\sin\theta & 0 \\ \sin\theta & \cos\theta & 0 \\ 0 & 0 & 1 \end{bmatrix}$$

In each case, the rotation is oriented counterclockwise (using the "right-hand rule") relative to the specified axis, as shown below.

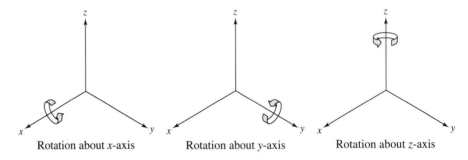

Rotation about *x*-axis   Rotation about *y*-axis   Rotation about *z*-axis

EXAMPLE 5   **Rotation About the *x*-Axis and *y*-Axis**

See LarsonLinearAlgebra.com for an interactive version of this type of example.

a. The matrix that yields a rotation of 90° about the *x*-axis is

$$A = \begin{bmatrix} 1 & 0 & 0 \\ 0 & \cos 90° & -\sin 90° \\ 0 & \sin 90° & \cos 90° \end{bmatrix} = \begin{bmatrix} 1 & 0 & 0 \\ 0 & 0 & -1 \\ 0 & 1 & 0 \end{bmatrix}.$$

Figure 6.8(a) shows the prism from Example 4 rotated 90° about the *x*-axis.

b. The matrix that yields a rotation of 90° about the *y*-axis is

$$A = \begin{bmatrix} \cos 90° & 0 & \sin 90° \\ 0 & 1 & 0 \\ -\sin 90° & 0 & \cos 90° \end{bmatrix} = \begin{bmatrix} 0 & 0 & 1 \\ 0 & 1 & 0 \\ -1 & 0 & 0 \end{bmatrix}.$$

Figure 6.8(b) shows the prism from Example 4 rotated 90° about the *y*-axis.

a.

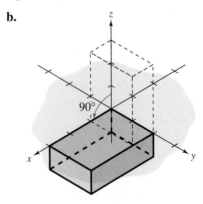

Figure 6.8

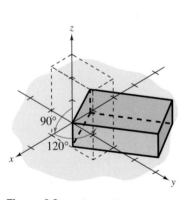

Figure 6.9

Rotations about the coordinate axes can be combined to produce any desired view of a figure. For example, Figure 6.9 shows the prism from Example 4 rotated 90° about the *y*-axis and then 120° about the *z*-axis.

6.5 Exercises

See CalcChat.com for worked-out solutions to odd-numbered exercises.

1. Let $T: R^2 \rightarrow R^2$ be a reflection in the x-axis. Find the image of each vector.
   (a) $(3, 5)$ (b) $(2, -1)$ (c) $(a, 0)$
   (d) $(0, b)$ (e) $(-c, d)$ (f) $(f, -g)$

2. Let $T: R^2 \rightarrow R^2$ be a reflection in the y-axis. Find the image of each vector.
   (a) $(5, 2)$ (b) $(-1, -6)$ (c) $(a, 0)$
   (d) $(0, b)$ (e) $(c, -d)$ (f) (f, g)

3. Let $T: R^2 \rightarrow R^2$ be a reflection in the line $y = x$. Find the image of each vector.
   (a) $(0, 1)$ (b) $(-1, 3)$ (c) $(a, 0)$
   (d) $(0, b)$ (e) $(-c, d)$ (f) $(f, -g)$

4. Let $T: R^2 \rightarrow R^2$ be a reflection in the line $y = -x$. Find the image of each vector.
   (a) $(-1, 2)$ (b) $(2, 3)$ (c) $(a, 0)$
   (d) $(0, b)$ (e) $(e, -d)$ (f) $(-f, g)$

5. Let $T(1, 0) = (2, 0)$ and $T(0, 1) = (0, 1)$.
   (a) Determine $T(x, y)$ for any (x, y).
   (b) Give a geometric description of T.

6. Let $T(1, 0) = (1, 1)$ and $T(0, 1) = (0, 1)$.
   (a) Determine $T(x, y)$ for any (x, y).
   (b) Give a geometric description of T.

Identifying and Representing a Transformation In Exercises 7–14, (a) identify the transformation, and (b) graphically represent the transformation for an arbitrary vector in R^2.

7. $T(x, y) = (x, y/2)$ 8. $T(x, y) = (x/4, y)$

9. $T(x, y) = (12x, y)$ 10. $T(x, y) = (x, 3y)$

11. $T(x, y) = (x + 3y, y)$ 12. $T(x, y) = (x + 4y, y)$

13. $T(x, y) = (x, 5x + y)$

14. $T(x, y) = (x, 9x + y)$

Finding Fixed Points of a Linear Transformation In Exercises 15–22, find all fixed points of the linear transformation. Recall that the vector **v** is a fixed point of T when $T(\mathbf{v}) = \mathbf{v}$.

15. A reflection in the y-axis

16. A reflection in the x-axis

17. A reflection in the line $y = x$

18. A reflection in the line $y = -x$

19. A vertical contraction

20. A horizontal expansion

21. A horizontal shear

22. A vertical shear

Sketching an Image of the Unit Square In Exercises 23–30, sketch the image of the unit square [a square with vertices at $(0, 0)$, $(1, 0)$, $(1, 1)$, and $(0, 1)$] under the specified transformation.

23. T is a reflection in the x-axis.

24. T is a reflection in the line $y = x$.

25. T is the contraction represented by $T(x, y) = (x/2, y)$.

26. T is the contraction represented by $T(x, y) = (x, y/4)$.

27. T is the expansion represented by $T(x, y) = (x, 3y)$.

28. T is the expansion represented by $T(x, y) = (5x, y)$.

29. T is the shear represented by $T(x, y) = (x + 2y, y)$.

30. T is the shear represented by $T(x, y) = (x, y + 3x)$.

Sketching an Image of a Rectangle In Exercises 31–38, sketch the image of the rectangle with vertices at $(0, 0)$, $(1, 0)$, $(1, 2)$, and $(0, 2)$ under the specified transformation.

31. T is a reflection in the y-axis.

32. T is a reflection in the line $y = x$.

33. T is the contraction represented by $T(x, y) = (x/3, y)$.

34. T is the contraction represented by $T(x, y) = (x, y/2)$.

35. T is the expansion represented by $T(x, y) = (x, 6y)$.

36. T is the expansion represented by $T(x, y) = (2x, y)$.

37. T is the shear represented by $T(x, y) = (x + y, y)$.

38. T is the shear represented by $T(x, y) = (x, y + 2x)$.

Sketching an Image of a Figure In Exercises 39–44, sketch each of the images under the specified transformation.

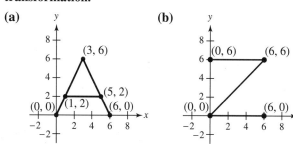

39. T is a reflection in the x-axis.

40. T is a reflection in the line $y = x$.

41. T is the shear represented by $T(x, y) = (x + y, y)$.

42. T is the shear represented by $T(x, y) = (x, x + y)$.

43. T is the expansion and contraction represented by $T(x, y) = \left(2x, \frac{1}{2}y\right)$.

44. T is the expansion and contraction represented by $T(x, y) = \left(\frac{1}{2}x, 2y\right)$.

Giving a Geometric Description In Exercises 45–50, give a geometric description of the linear transformation defined by the elementary matrix.

45. $A = \begin{bmatrix} 2 & 0 \\ 0 & 1 \end{bmatrix}$ **46.** $A = \begin{bmatrix} 1 & 0 \\ 2 & 1 \end{bmatrix}$

47. $A = \begin{bmatrix} 1 & 5 \\ 0 & 1 \end{bmatrix}$ **48.** $A = \begin{bmatrix} 1 & 0 \\ 0 & \frac{1}{2} \end{bmatrix}$

49. $A = \begin{bmatrix} 1 & 0 \\ 0 & -2 \end{bmatrix}$ **50.** $A = \begin{bmatrix} -\frac{1}{4} & 0 \\ 0 & 1 \end{bmatrix}$

Giving a Geometric Description In Exercises 51 and 52, give a geometric description of the linear transformation defined by the matrix product.

51. $A = \begin{bmatrix} 2 & 0 \\ 2 & 1 \end{bmatrix} = \begin{bmatrix} 2 & 0 \\ 0 & 1 \end{bmatrix}\begin{bmatrix} 1 & 0 \\ 2 & 1 \end{bmatrix}$

52. $A = \begin{bmatrix} 0 & 3 \\ 1 & 0 \end{bmatrix} = \begin{bmatrix} 0 & 1 \\ 1 & 0 \end{bmatrix}\begin{bmatrix} 1 & 0 \\ 0 & 3 \end{bmatrix}$

53. The linear transformation defined by a diagonal matrix with positive main diagonal elements is called a **magnification.** Find the images of $(1, 0)$, $(0, 1)$, and $(2, 2)$ under the linear transformation defined by A and graphically interpret your result.

$$A = \begin{bmatrix} 2 & 0 \\ 0 & 3 \end{bmatrix}$$

---

54. CAPSTONE Describe the transformation defined by each matrix. Assume k and θ are positive scalars.

(a) $\begin{bmatrix} -1 & 0 \\ 0 & 1 \end{bmatrix}$ (b) $\begin{bmatrix} 1 & 0 \\ 0 & -1 \end{bmatrix}$

(c) $\begin{bmatrix} 0 & 1 \\ 1 & 0 \end{bmatrix}$ (d) $\begin{bmatrix} k & 0 \\ 0 & 1 \end{bmatrix}, k > 1$

(e) $\begin{bmatrix} k & 0 \\ 0 & 1 \end{bmatrix}, 0 < k < 1$ (f) $\begin{bmatrix} 1 & 0 \\ 0 & k \end{bmatrix}, k > 1$

(g) $\begin{bmatrix} 1 & 0 \\ 0 & k \end{bmatrix}, 0 < k < 1$ (h) $\begin{bmatrix} 1 & k \\ 0 & 1 \end{bmatrix}$

(i) $\begin{bmatrix} 1 & 0 \\ k & 1 \end{bmatrix}$ (j) $\begin{bmatrix} 1 & 0 & 0 \\ 0 & \cos\theta & -\sin\theta \\ 0 & \sin\theta & \cos\theta \end{bmatrix}$

(k) $\begin{bmatrix} \cos\theta & 0 & \sin\theta \\ 0 & 1 & 0 \\ -\sin\theta & 0 & \cos\theta \end{bmatrix}$ (l) $\begin{bmatrix} \cos\theta & -\sin\theta & 0 \\ \sin\theta & \cos\theta & 0 \\ 0 & 0 & 1 \end{bmatrix}$

Finding a Matrix to Produce a Rotation In Exercises 55–58, find the matrix that produces the rotation.

55. 30° about the z-axis   **56.** 60° about the x-axis

57. 120° about the x-axis   **58.** 60° about the y-axis

Finding the Image of a Vector In Exercises 59–62, find the image of the vector $(1, 1, 1)$ for the rotation.

59. 30° about the z-axis   **60.** 60° about the x-axis

61. 120° about the x-axis   **62.** 60° about the y-axis

Determining a Rotation In Exercises 63–68, determine which single counterclockwise rotation about the x-, y-, or z-axis produces the rotated tetrahedron. The figure at the right shows the tetrahedron before rotation.

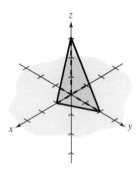

63. 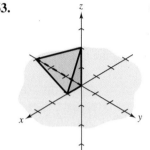   **64.**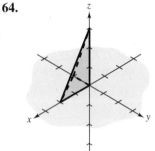

65. 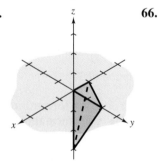   **66.**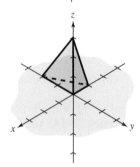

67. 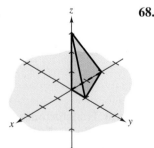   **68.**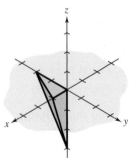

Determining a Matrix to Produce a Pair of Rotations In Exercises 69–72, determine the matrix that produces the pair of rotations. Then find the image of the vector $(1, 1, 1)$ under these rotations.

69. 90° about the x-axis and then 90° about the y-axis

70. 30° about the z-axis and then 60° about the y-axis

71. 45° about the z-axis and then 135° about the x-axis

72. 120° about the x-axis and then 135° about the z-axis

6 Review Exercises

See CalcChat.com for worked-out solutions to odd-numbered exercises.

Finding an Image and a Preimage In Exercises 1–6, find (a) the image of v and (b) the preimage of w for the linear transformation.

1. $T: R^2 \to R^2$, $T(v_1, v_2) = (v_1, v_1 + 2v_2)$, $\mathbf{v} = (2, -3)$, $\mathbf{w} = (4, 12)$

2. $T: R^2 \to R^2$, $T(v_1, v_2) = (v_1 + v_2, 2v_2)$, $\mathbf{v} = (4, -1)$, $\mathbf{w} = (8, 4)$

3. $T: R^3 \to R^3$, $T(v_1, v_2, v_3) = (0, v_1 + v_2, v_2 + v_3)$, $\mathbf{v} = (-3, 2, 5)$, $\mathbf{w} = (0, 2, 5)$

4. $T: R^3 \to R^3$, $T(v_1, v_2, v_3) = (v_1 + v_2, v_2 + v_3, v_3)$, $\mathbf{v} = (-2, 1, 2)$, $\mathbf{w} = (0, 1, 2)$

5. $T: R^2 \to R^3$, $T(v_1, v_2) = (v_1 + v_2, v_1 - v_2, 2v_1 + 3v_2)$, $\mathbf{v} = (2, -3)$, $\mathbf{w} = (1, -3, 4)$

6. $T: R^2 \to R$, $T(v_1, v_2) = (2v_1 - v_2)$, $\mathbf{v} = (2, -3)$, $\mathbf{w} = 4$

Linear Transformations and Standard Matrices In Exercises 7–18, determine whether the function is a linear transformation. If it is, find its standard matrix A.

7. $T: R \to R^2$, $T(x) = (x, x + 2)$

8. $T: R^2 \to R$, $T(x_1, x_2) = (x_1 + x_2)$

9. $T: R^2 \to R^2$, $T(x_1, x_2) = (x_1 + 2x_2, -x_1 - x_2)$

10. $T: R^2 \to R^2$, $T(x_1, x_2) = (x_1 + 3, x_2)$

11. $T: R^2 \to R^2$, $T(x, y) = (x - 2y, 2y - x)$

12. $T: R^2 \to R^2$, $T(x, y) = (x + y, y)$

13. $T: R^2 \to R^2$, $T(x, y) = (x + h, y + k)$, $h \neq 0$ or $k \neq 0$ (translation in R^2)

14. $T: R^2 \to R^2$, $T(x, y) = (|x|, |y|)$

15. $T: R^3 \to R^3$, $T(x_1, x_2, x_3) = (x_1 + x_2, 2, x_3 - x_1)$

16. $T: R^3 \to R^3$, $T(x_1, x_2, x_3) = (x_1 - x_2, x_2 - x_3, x_3 - x_1)$

17. $T: R^3 \to R^3$, $T(x, y, z) = (z, y, x)$

18. $T: R^3 \to R^3$, $T(x, y, z) = (x, 0, -y)$

19. Let T be a linear transformation from R^2 into R^2 such that $T(2, 0) = (1, 1)$ and $T(0, 3) = (3, 3)$. Find $T(1, 1)$ and $T(0, 1)$.

20. Let T be a linear transformation from R^3 into R such that $T(1, 1, 1) = 1$, $T(1, 1, 0) = 2$, and $T(1, 0, 0) = 3$. Find $T(0, 1, 1)$.

21. Let T be a linear transformation from R^2 into R^2 such that $T(4, -2) = (2, -2)$ and $T(3, 3) = (-3, 3)$. Find $T(-7, 2)$.

22. Let T be a linear transformation from R^2 into R^2 such that $T(1, -1) = (2, -3)$ and $T(0, 2) = (0, 8)$. Find $T(2, 4)$.

Linear Transformation Given by a Matrix In Exercises 23–28, define the linear transformation $T: R^n \to R^m$ by $T(\mathbf{v}) = A\mathbf{v}$. Use the matrix A to (a) determine the dimensions of R^n and R^m, (b) find the image of v, and (c) find the preimage of w.

23. $A = \begin{bmatrix} 0 & 1 & 2 \\ -2 & 0 & 0 \end{bmatrix}$, $\mathbf{v} = (6, 1, 1)$, $\mathbf{w} = (3, 5)$

24. $A = \begin{bmatrix} 1 & 2 & -1 \\ 1 & 0 & 1 \end{bmatrix}$, $\mathbf{v} = (5, 2, 2)$, $\mathbf{w} = (4, 2)$

25. $A = \begin{bmatrix} 1 & 1 & 1 \\ 0 & 1 & 1 \\ 0 & 0 & 1 \end{bmatrix}$, $\mathbf{v} = (2, 1, -5)$, $\mathbf{w} = (6, 4, 2)$

26. $A = \begin{bmatrix} 2 & 1 \\ 0 & 1 \end{bmatrix}$, $\mathbf{v} = (8, 4)$, $\mathbf{w} = (5, 2)$

27. $A = \begin{bmatrix} 4 & 0 \\ 0 & 5 \\ 1 & 1 \end{bmatrix}$, $\mathbf{v} = (2, 2)$, $\mathbf{w} = (4, -5, 0)$

28. $A = \begin{bmatrix} -1 & 0 \\ 0 & 1 \\ -1 & -3 \end{bmatrix}$, $\mathbf{v} = (3, 5)$, $\mathbf{w} = (5, 2, -1)$

29. Use the standard matrix for counterclockwise rotation in R^2 to rotate the triangle with vertices $(3, 5)$, $(5, 3)$, and $(3, 0)$ counterclockwise $90°$ about the origin. Graph the triangles.

30. Rotate the triangle in Exercise 29 counterclockwise $90°$ about the point $(5, 3)$. Graph the triangles.

Finding the Kernel and Range In Exercises 31–34, find (a) ker(T) and (b) range(T).

31. $T: R^4 \to R^3$,
$$T(w, x, y, z) = (2w + 4x + 6y + 5z, \\ -w - 2x + 2y, 8y + 4z)$$

32. $T: R^3 \to R^3$, $T(x, y, z) = (x + 2y, y + 2z, z + 2x)$

33. $T: R^3 \to R^3$, $T(x, y, z) = (x, y, z + 3y)$

34. $T: R^3 \to R^3$, $T(x, y, z) = (x + y, y + z, x - z)$

Finding the Kernel, Nullity, Range, and Rank In Exercises 35–38, define the linear transformation T by $T(\mathbf{v}) = A\mathbf{v}$. Find (a) ker($T$), (b) nullity($T$), (c) range($T$), and (d) rank($T$).

35. $A = \begin{bmatrix} 1 & 2 \\ -1 & 0 \\ 1 & 1 \end{bmatrix}$

36. $A = \begin{bmatrix} -1 & 2 \\ 0 & -1 \\ -2 & 2 \end{bmatrix}$

37. $A = \begin{bmatrix} 2 & 1 & 3 \\ 1 & 1 & 0 \\ 0 & 1 & -3 \end{bmatrix}$

38. $A = \begin{bmatrix} 1 & 1 & -1 \\ 1 & 2 & 1 \\ 0 & 1 & 0 \end{bmatrix}$

39. For $T: R^5 \to R^3$ and nullity$(T) = 2$, find rank(T).

40. For $T: P_5 \to P_3$ and nullity$(T) = 4$, find rank(T).

41. For $T: P_4 \to R^5$ and rank$(T) = 3$, find nullity(T).

42. For $T: M_{3,3} \to M_{3,3}$ and rank$(T) = 5$, find nullity(T).

Finding a Power of a Standard Matrix   In Exercises 43–46, find the specified power of A, the standard matrix for T.

43. $T: R^3 \to R^3$, reflection in the xy-plane. Find A^2.

44. $T: R^3 \to R^3$, projection onto the xy-plane. Find A^2.

45. $T: R^2 \to R^2$, counterclockwise rotation through the angle θ. Find A^3.

46. Calculus $T: P_3 \to P_3$, differential operator D_x. Find A^2.

Finding Standard Matrices for Compositions   In Exercises 47 and 48, find the standard matrices for $T = T_2 \circ T_1$ and $T' = T_1 \circ T_2$.

47. $T_1: R^2 \to R^3$, $T_1(x, y) = (x, x + y, y)$

$T_2: R^3 \to R^2$, $T_2(x, y, z) = (0, y)$

48. $T_1: R \to R^2$, $T_1(x) = (x, 4x)$

$T_2: R^2 \to R$, $T_2(x, y) = (y + 3x)$

Finding the Inverse of a Linear Transformation In Exercises 49–52, determine whether the linear transformation is invertible. If it is, find its inverse.

49. $T: R^2 \to R^2$, $T(x, y) = (0, y)$

50. $T: R^2 \to R^2$,

$T(x, y) = (x \cos \theta - y \sin \theta, x \sin \theta + y \cos \theta)$

51. $T: R^2 \to R^2$, $T(x, y) = (x, -y)$

52. $T: R^3 \to R^2$, $T(x, y, z) = (x + y, y - z)$

One-to-One, Onto, and Invertible Transformations In Exercises 53–56, determine whether the linear transformation represented by the matrix A is (a) one-to-one, (b) onto, and (c) invertible.

53. $A = \begin{bmatrix} 6 & 0 \\ 0 & -1 \end{bmatrix}$

54. $A = \begin{bmatrix} 1 & \frac{1}{4} \\ 0 & 1 \end{bmatrix}$

55. $A = \begin{bmatrix} 1 & 1 & 1 \\ 0 & 1 & 1 \end{bmatrix}$

56. $A = \begin{bmatrix} 4 & 0 & 7 \\ 5 & 5 & 1 \\ 0 & 0 & 2 \end{bmatrix}$

Finding the Image Two Ways   In Exercises 57 and 58, find $T(\mathbf{v})$ by using (a) the standard matrix and (b) the matrix relative to B and B'.

57. $T: R^2 \to R^3$,

$T(x, y) = (-x, y, x + y)$, $\mathbf{v} = (0, 1)$,

$B = \{(1, 1), (1, -1)\}$, $B' = \{(0, 1, 0), (0, 0, 1), (1, 0, 0)\}$

58. $T: R^2 \to R^2$,

$T(x, y) = (2y, 0)$, $\mathbf{v} = (-1, 3)$,

$B = \{(2, 1), (-1, 0)\}$, $B' = \{(-1, 0), (2, 2)\}$

Finding a Matrix for a Linear Transformation   In Exercises 59 and 60, find the matrix A' for T relative to the basis B'.

59. $T: R^2 \to R^2$, $T(x, y) = (x - 3y, y - x)$,

$B' = \{(1, -1), (1, 1)\}$

60. $T: R^3 \to R^3$, $T(x, y, z) = (x + 3y, 3x + y, -2z)$,

$B' = \{(1, 1, 0), (1, -1, 0), (0, 0, 1)\}$

Similar Matrices   In Exercises 61 and 62, use the matrix P to show that the matrices A and A' are similar.

61. $P = \begin{bmatrix} 2 & -1 \\ 3 & 5 \end{bmatrix}$, $A = \begin{bmatrix} 6 & -3 \\ 2 & -2 \end{bmatrix}$, $A' = \begin{bmatrix} 1 & -9 \\ -1 & 3 \end{bmatrix}$

62. $P = \begin{bmatrix} 1 & 2 & 0 \\ 0 & 1 & -1 \\ 1 & 0 & 0 \end{bmatrix}$, $A = \begin{bmatrix} 1 & 0 & 1 \\ -1 & 3 & 1 \\ 0 & 0 & 2 \end{bmatrix}$, $A' = \begin{bmatrix} 2 & 0 & 0 \\ 0 & 1 & 0 \\ 0 & 0 & 3 \end{bmatrix}$

63. Define $T: R^3 \to R^3$ by $T(\mathbf{v}) = \text{proj}_{\mathbf{u}}\mathbf{v}$, where $\mathbf{u} = (0, 1, 2)$.

(a) Find A, the standard matrix for T.

(b) Let S be the linear transformation represented by $I - A$. Show that S is of the form

$S(\mathbf{v}) = \text{proj}_{\mathbf{w}_1}\mathbf{v} + \text{proj}_{\mathbf{w}_2}\mathbf{v}$

where $\mathbf{w}_1$ and $\mathbf{w}_2$ are fixed vectors in R^3.

(c) Show that the kernel of T is equal to the range of S.

64. Define $T: R^2 \to R^2$ by $T(\mathbf{v}) = \text{proj}_{\mathbf{u}}\mathbf{v}$, where $\mathbf{u} = (4, 3)$.

(a) Find A, the standard matrix for T, and show that $A^2 = A$.

(b) Show that $(I - A)^2 = I - A$.

(c) Find $A\mathbf{v}$ and $(I - A)\mathbf{v}$ for $\mathbf{v} = (5, 0)$.

(d) Sketch the graphs of $\mathbf{u}$, $\mathbf{v}$, $A\mathbf{v}$, and $(I - A)\mathbf{v}$.

65. Let S and T be linear transformations from V into W. Show that $S + T$ and kT are both linear transformations, where $(S + T)(\mathbf{v}) = S(\mathbf{v}) + T(\mathbf{v})$ and $(kT)(\mathbf{v}) = kT(\mathbf{v})$.

66. Proof   Let $T: R^2 \to R^2$ such that $T(\mathbf{v}) = A\mathbf{v} + \mathbf{b}$, where A is a 2×2 matrix. (Such a transformation is called an **affine transformation**.) Prove that T is a linear transformation if and only if $\mathbf{b} = \mathbf{0}$.

Sum of Two Linear Transformations   In Exercises 67 and 68, consider the sum $S + T$ of two linear transformations $S: V \to W$ and $T: V \to W$, defined as $(S + T)(\mathbf{v}) = S(\mathbf{v}) + T(\mathbf{v})$.

67. Proof   Prove that rank$(S + T) \le$ rank$(S) +$ rank(T).

68. Give an example for each.

(a) Rank$(S + T) =$ rank$(S) +$ rank(T)

(b) Rank$(S + T) <$ rank$(S) +$ rank(T)

69. Proof   Let $T: P_3 \to R$ such that

$T(a_0 + a_1 x + a_2 x^2 + a_3 x^3) = a_0 + a_1 + a_2 + a_3$.

(a) Prove that T is a linear transformation.

(b) Find the rank and nullity of T.

(c) Find a basis for the kernel of T.

70. Proof Let

$$T: V \to U \quad \text{and} \quad S: U \to W$$

be linear transformations.

(a) Prove that $\ker(T)$ is contained in $\ker(S \circ T)$.

(b) Prove that if $S \circ T$ is onto, then so is S.

71. Let V be an inner product space. For a fixed nonzero vector $\mathbf{v}_0$ in V, let $T: V \to R$ be the linear transformation $T(\mathbf{v}) = \langle \mathbf{v}, \mathbf{v}_0 \rangle$. Find the kernel, range, rank, and nullity of T.

72. Calculus Let $B = \{1, x, \sin x, \cos x\}$ be a basis for a subspace W of the space of continuous functions, and let D_x be the differential operator on W. Find the matrix for D_x relative to the basis B. Find the range and kernel of D_x.

73. Writing Are the vector spaces R^4, $M_{2,2}$, and $M_{1,4}$ exactly the same? Describe their similarities and differences.

74. Calculus Define $T: P_3 \to P_3$ by

$$T(p) = p(x) + p'(x).$$

Find the rank and nullity of T.

Identifying and Representing a Transformation In Exercises 75–80, (a) identify the transformation, and (b) graphically represent the transformation for an arbitrary vector in R^2.

75. $T(x, y) = (x, 2y)$

76. $T(x, y) = (x + y, y)$

77. $T(x, y) = (x, y + 3x)$

78. $T(x, y) = (5x, y)$

79. $T(x, y) = (x + 5y, y)$

80. $T(x, y) = \left(x, y + \frac{3}{2}x\right)$

Sketching an Image of a Triangle In Exercises 81–84, sketch the image of the triangle with vertices $(0, 0)$, $(1, 0)$, and $(0, 1)$ under the specified transformation.

81. T is a reflection in the x-axis.

82. T is the expansion represented by $T(x, y) = (2x, y)$.

83. T is the shear represented by $T(x, y) = (x + 3y, y)$.

84. T is the shear represented by $T(x, y) = (x, y + 2x)$.

Giving a Geometric Description In Exercises 85 and 86, give a geometric description of the linear transformation defined by the matrix product.

85. $\begin{bmatrix} 0 & 12 \\ 1 & 0 \end{bmatrix} = \begin{bmatrix} 12 & 0 \\ 0 & 1 \end{bmatrix} \begin{bmatrix} 0 & 1 \\ 1 & 0 \end{bmatrix}$

86. $\begin{bmatrix} 1 & 0 \\ 6 & 2 \end{bmatrix} = \begin{bmatrix} 1 & 0 \\ 0 & 2 \end{bmatrix} \begin{bmatrix} 1 & 0 \\ 3 & 1 \end{bmatrix}$

Finding a Matrix to Produce a Rotation In Exercises 87–90, find the matrix that produces the rotation. Then find the image of the vector $(1, -1, 1)$.

87. $45°$ about the z-axis

88. $90°$ about the x-axis

89. $60°$ about the x-axis

90. $30°$ about the y-axis

Determining a Matrix to Produce a Pair of Rotations In Exercises 91–94, determine the matrix that produces the pair of rotations.

91. $60°$ about the x-axis and then $30°$ about the z-axis

92. $120°$ about the y-axis and then $45°$ about the z-axis

93. $30°$ about the y-axis and then $45°$ about the z-axis

94. $60°$ about the x-axis and then $60°$ about the z-axis

Finding an Image of a Unit Cube In Exercises 95–98, find the image of the unit cube with vertices $(0, 0, 0)$, $(1, 0, 0)$, $(1, 1, 0)$, $(0, 1, 0)$, $(0, 0, 1)$, $(1, 0, 1)$, $(1, 1, 1)$, and $(0, 1, 1)$ when it rotates through the given angle.

95. $45°$ about the z-axis

96. $90°$ about the x-axis

97. $30°$ about the x-axis

98. $120°$ about the z-axis

True or False? In Exercises 99–102, determine whether each statement is true or false. If a statement is true, give a reason or cite an appropriate statement from the text. If a statement is false, provide an example that shows the statement is not true in all cases or cite an appropriate statement from the text.

99. (a) Reflections that map a point in the xy-plane to its mirror image across the line $y = x$ are linear transformations that are defined by the matrix $\begin{bmatrix} 1 & 0 \\ 0 & 1 \end{bmatrix}$.

(b) Horizontal expansions or contractions are linear transformations that are defined by the matrix $\begin{bmatrix} k & 0 \\ 0 & 1 \end{bmatrix}$.

100. (a) Reflections that map a point in the xy-plane to its mirror image across the x-axis are linear transformations that are defined by the matrix $\begin{bmatrix} 1 & 0 \\ 0 & -1 \end{bmatrix}$.

(b) Vertical expansions or contractions are linear transformations that are defined by the matrix $\begin{bmatrix} 1 & 0 \\ 0 & k \end{bmatrix}$.

101. (a) In calculus, any linear function is also a linear transformation from R^2 to R^2.

(b) A linear transformation is onto if and only if, for all $\mathbf{u}$ and $\mathbf{v}$ in V, $T(\mathbf{u}) = T(\mathbf{v})$ implies $\mathbf{u} = \mathbf{v}$.

(c) For ease of computation, it is best to choose a basis for V such that the transformation matrix is diagonal.

102. (a) For polynomials, the differential operator D_x is a linear transformation from P_n into P_{n-1}.

(b) The set of all vectors $\mathbf{v}$ in V that satisfy $T(\mathbf{v}) = \mathbf{v}$ is the kernel of T.

(c) The standard matrix A of the composition of two linear transformations $T(\mathbf{v}) = T_2(T_1(\mathbf{v}))$ is the product of the standard matrix for T_2 and the standard matrix for T_1.

6   Projects

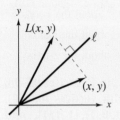

Let ℓ be the line $ax + by = 0$ in R^2. The linear transformation $L: R^2 \to R^2$ that maps a point (x, y) to its mirror image with respect to ℓ is called the **reflection** in ℓ. (See Figure 6.10.) The goal of these two projects is to find the matrix for this reflection relative to the standard basis.

1 Reflections in R^2 (I)

In this project, you will use transition matrices to determine the standard matrix for the reflection L in the line $ax + by = 0$.

1. Find the standard matrix for L for the line $x = 0$.

2. Find the standard matrix for L for the line $y = 0$.

3. Find the standard matrix for L for the line $x - y = 0$.

4. Consider the line ℓ represented by $x - 3y = 0$. Find a vector $\mathbf{v}$ parallel to ℓ and another vector $\mathbf{w}$ orthogonal to ℓ. Determine the matrix A for the reflection in ℓ relative to the ordered basis $\{\mathbf{v}, \mathbf{w}\}$. Finally, use the appropriate transition matrix to find the matrix for the reflection relative to the standard basis. Use this matrix to find the images of the points $(3, 1)$, $(2, -1)$, and $(0, 5)$.

5. Consider the general line ℓ represented by $ax + by = 0$. Find a vector $\mathbf{v}$ parallel to ℓ and another vector $\mathbf{w}$ orthogonal to ℓ. Determine the matrix A for the reflection in ℓ relative to the ordered basis $\{\mathbf{v}, \mathbf{w}\}$. Finally, use the appropriate transition matrix to find the matrix for the reflection relative to the standard basis.

6. Find the standard matrix for the reflection in the line $2x + 3y = 0$. Use this matrix to find the images of the points $(2, 3)$, $(-3, 2)$, and $(13, 0)$.

2 Reflections in R^2 (II)

In this project, you will use projections to determine the standard matrix for the reflection L in the line $ax + by = 0$. Recall that the projection of the vector $\mathbf{u}$ onto the vector $\mathbf{v}$ (shown at the right) is

$$\text{proj}_{\mathbf{v}}\mathbf{u} = \frac{\mathbf{u} \cdot \mathbf{v}}{\mathbf{v} \cdot \mathbf{v}}\mathbf{v}.$$

1. Find the standard matrix for the projection onto the y-axis. That is, find the standard matrix for $\text{proj}_{\mathbf{v}}\mathbf{u}$ when $\mathbf{v} = (0, 1)$.

2. Find the standard matrix for the projection onto the x-axis.

3. Consider the line ℓ represented by $x - 3y = 0$. Find a vector $\mathbf{v}$ parallel to ℓ and another vector $\mathbf{w}$ orthogonal to ℓ. Determine the matrix A for the projection onto ℓ relative to the ordered basis $\{\mathbf{v}, \mathbf{w}\}$. Finally, use the appropriate transition matrix to find the matrix for the projection relative to the standard basis. Use this matrix to find $\text{proj}_{\mathbf{v}}\mathbf{u}$ for $\mathbf{u} = (3, 1)$, $\mathbf{u} = (2, -1)$, and $\mathbf{u} = (0, 5)$.

4. Consider the general line ℓ represented by $ax + by = 0$. Find a vector $\mathbf{v}$ parallel to ℓ and another vector $\mathbf{w}$ orthogonal to ℓ. Determine the matrix A for the projection onto ℓ relative to the ordered basis $\{\mathbf{v}, \mathbf{w}\}$. Finally, use the appropriate transition matrix to find the matrix for the projection relative to the standard basis.

5. Use Figure 6.11 to show that $\text{proj}_{\mathbf{v}}\mathbf{u} = \frac{1}{2}(\mathbf{u} + L(\mathbf{u}))$, where L is the reflection in the line ℓ. Solve this equation for L and compare your answer with the formula from the first project.

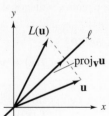

Figure 6.11

7 Eigenvalues and Eigenvectors

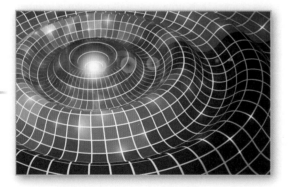

Population of Rabbits (p. 379)

Architecture (p. 388)

Relative Maxima and Minima (p. 375)

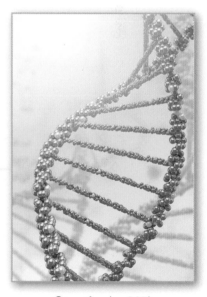

Genetics (p. 365)

Diffusion (p. 354)

7.1   Eigenvalues and Eigenvectors

- Verify eigenvalues and corresponding eigenvectors.
- Find eigenvalues and corresponding eigenspaces.
- Use the characteristic equation to find eigenvalues and eigenvectors, and find the eigenvalues and eigenvectors of a triangular matrix.
- Find the eigenvalues and eigenvectors of a linear transformation.

THE EIGENVALUE PROBLEM

This section presents one of the most important problems in linear algebra, the **eigenvalue problem.** Its central question is "when A is an $n \times n$ matrix, do nonzero vectors $\mathbf{x}$ in R^n exist such that $A\mathbf{x}$ is a scalar multiple of $\mathbf{x}$?" The scalar, denoted by the Greek letter lambda (λ), is called an **eigenvalue** of the matrix A, and the nonzero vector $\mathbf{x}$ is called an **eigenvector** of A corresponding to λ. The origins of the terms *eigenvalue* and *eigenvector* are from the German word *Eigenwert,* meaning "proper value." So, you have

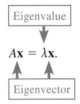

Eigenvalues and eigenvectors have many important applications, many of which are discussed throughout this chapter. For now, you will consider a geometric interpretation of the problem in R^2. If λ is an eigenvalue of a matrix A and $\mathbf{x}$ is an eigenvector of A corresponding to λ, then multiplication of $\mathbf{x}$ by the matrix A produces a vector $\lambda\mathbf{x}$ that is parallel to $\mathbf{x}$, as shown below.

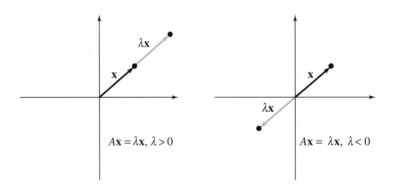

REMARK

Only eigenvectors of real eigenvalues are presented in this chapter.

Definitions of Eigenvalue and Eigenvector

Let A be an $n \times n$ matrix. The scalar λ is an **eigenvalue** of A when there is a *nonzero* vector $\mathbf{x}$ such that $A\mathbf{x} = \lambda\mathbf{x}$. The vector $\mathbf{x}$ is an **eigenvector** of A corresponding to λ.

Note that an eigen*vector* cannot be zero. Allowing $\mathbf{x}$ to be the zero vector would render the definition meaningless, because $A\mathbf{0} = \lambda\mathbf{0}$ is true for all real values of λ. An eigen*value* of $\lambda = 0$, however, is possible. (See Example 2.)

A matrix can have more than one eigenvalue, as demonstrated in Examples 1 and 2.

EXAMPLE 1   Verifying Eigenvectors and Eigenvalues

For the matrix

$$A = \begin{bmatrix} 2 & 0 \\ 0 & -1 \end{bmatrix}$$

verify that $\mathbf{x}_1 = (1, 0)$ is an eigenvector of A corresponding to the eigenvalue $\lambda_1 = 2$, and that $\mathbf{x}_2 = (0, 1)$ is an eigenvector of A corresponding to the eigenvalue $\lambda_2 = -1$.

SOLUTION

Multiplying $\mathbf{x}_1$ on the left by A produces

$$A\mathbf{x}_1 = \begin{bmatrix} 2 & 0 \\ 0 & -1 \end{bmatrix}\begin{bmatrix} 1 \\ 0 \end{bmatrix} = \begin{bmatrix} 2 \\ 0 \end{bmatrix} = 2\begin{bmatrix} 1 \\ 0 \end{bmatrix}.$$

Eigenvalue Eigenvector

So, $\mathbf{x}_1 = (1, 0)$ is an eigenvector of A corresponding to the eigenvalue $\lambda_1 = 2$. Similarly, multiplying $\mathbf{x}_2$ on the left by A produces

$$A\mathbf{x}_2 = \begin{bmatrix} 2 & 0 \\ 0 & -1 \end{bmatrix}\begin{bmatrix} 0 \\ 1 \end{bmatrix} = \begin{bmatrix} 0 \\ -1 \end{bmatrix} = -1\begin{bmatrix} 0 \\ 1 \end{bmatrix}.$$

So, $\mathbf{x}_2 = (0, 1)$ is an eigenvector of A corresponding to the eigenvalue $\lambda_2 = -1$.

EXAMPLE 2   Verifying Eigenvectors and Eigenvalues

For the matrix

$$A = \begin{bmatrix} 1 & -2 & 1 \\ 0 & 0 & 0 \\ 0 & 1 & 1 \end{bmatrix}$$

verify that

$$\mathbf{x}_1 = (-3, -1, 1) \quad \text{and} \quad \mathbf{x}_2 = (1, 0, 0)$$

are eigenvectors of A and find their corresponding eigenvalues.

SOLUTION

Multiplying $\mathbf{x}_1$ on the left by A produces

$$A\mathbf{x}_1 = \begin{bmatrix} 1 & -2 & 1 \\ 0 & 0 & 0 \\ 0 & 1 & 1 \end{bmatrix}\begin{bmatrix} -3 \\ -1 \\ 1 \end{bmatrix} = \begin{bmatrix} 0 \\ 0 \\ 0 \end{bmatrix} = 0\begin{bmatrix} -3 \\ -1 \\ 1 \end{bmatrix}.$$

So, $\mathbf{x}_1 = (-3, -1, 1)$ is an eigenvector of A corresponding to the eigenvalue $\lambda_1 = 0$. Similarly, multiplying $\mathbf{x}_2$ on the left by A produces

$$A\mathbf{x}_2 = \begin{bmatrix} 1 & -2 & 1 \\ 0 & 0 & 0 \\ 0 & 1 & 1 \end{bmatrix}\begin{bmatrix} 1 \\ 0 \\ 0 \end{bmatrix} = \begin{bmatrix} 1 \\ 0 \\ 0 \end{bmatrix} = 1\begin{bmatrix} 1 \\ 0 \\ 0 \end{bmatrix}.$$

So, $\mathbf{x}_2 = (1, 0, 0)$ is an eigenvector of A corresponding to the eigenvalue $\lambda_2 = 1$.

DISCOVERY

1. In Example 2, $\lambda_2 = 1$ is an eigenvalue of the matrix A. Calculate the determinant of the matrix $\lambda_2 I - A$, where I is the 3×3 identity matrix.

2. Repeat for the other eigenvalue, $\lambda_1 = 0$.

3. In general, when λ is an eigenvalue of the matrix A, what is the value of $|\lambda I - A|$?

EIGENSPACES

Although Examples 1 and 2 list only one eigenvector for each eigenvalue, each of the four eigenvalues in Examples 1 and 2 has infinitely many eigenvectors. For instance, in Example 1, the vectors $(2, 0)$ and $(-3, 0)$ are eigenvectors of A corresponding to the eigenvalue 2. In fact, if A is an $n \times n$ matrix with an eigenvalue λ and a corresponding eigenvector $\mathbf{x}$, then every nonzero scalar multiple of $\mathbf{x}$ is also an eigenvector of A. To see this, let c be a nonzero scalar, which then produces

$$A(c\mathbf{x}) = c(A\mathbf{x}) = c(\lambda\mathbf{x}) = \lambda(c\mathbf{x}).$$

It is also true that if $\mathbf{x}_1$ and $\mathbf{x}_2$ are eigenvectors corresponding to the *same* eigenvalue λ, then their sum is also an eigenvector corresponding to λ, because

$$A(\mathbf{x}_1 + \mathbf{x}_2) = A\mathbf{x}_1 + A\mathbf{x}_2 = \lambda\mathbf{x}_1 + \lambda\mathbf{x}_2 = \lambda(\mathbf{x}_1 + \mathbf{x}_2).$$

In other words, the set of all eigenvectors of an eigenvalue λ, together with the zero vector, is a subspace of R^n. This special subspace of R^n is called the **eigenspace** of λ.

THEOREM 7.1   Eigenvectors of λ Form a Subspace

If A is an $n \times n$ matrix with an eigenvalue λ, then the set of all eigenvectors of λ, together with the zero vector

$$\{\mathbf{x}: \mathbf{x} \text{ is an eigenvector of } \lambda\} \cup \{\mathbf{0}\}$$

is a subspace of R^n. This subspace is the **eigenspace** of λ.

Determining the eigenvalues and corresponding eigenspaces of a matrix can involve algebraic manipulation. Occasionally, however, it is possible to find eigenvalues and eigenspaces by inspection, as demonstrated in Example 3.

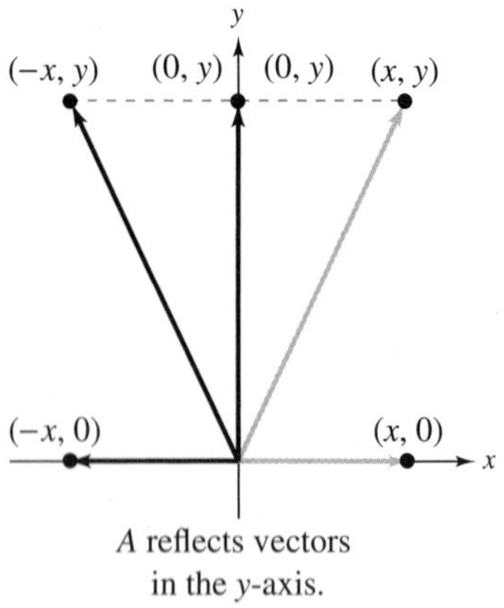

EXAMPLE 3   Finding Eigenspaces in R^2 Geometrically

Find the eigenvalues and corresponding eigenspaces of $A = \begin{bmatrix} -1 & 0 \\ 0 & 1 \end{bmatrix}$.

SOLUTION

Geometrically, multiplying a vector (x, y) in R^2 by the matrix A corresponds to a reflection in the y-axis. That is, if $\mathbf{v} = (x, y)$, then

$$A\mathbf{v} = \begin{bmatrix} -1 & 0 \\ 0 & 1 \end{bmatrix}\begin{bmatrix} x \\ y \end{bmatrix} = \begin{bmatrix} -x \\ y \end{bmatrix}.$$

Figure 7.1 illustrates that the only vectors reflected onto scalar multiples of themselves are those lying on either the x-axis or the y-axis.

For a vector on the x-axis
$$\begin{bmatrix} -1 & 0 \\ 0 & 1 \end{bmatrix}\begin{bmatrix} x \\ 0 \end{bmatrix} = \begin{bmatrix} -x \\ 0 \end{bmatrix} = -1\begin{bmatrix} x \\ 0 \end{bmatrix}$$
Eigenvalue is $\lambda_1 = -1$.

For a vector on the y-axis
$$\begin{bmatrix} -1 & 0 \\ 0 & 1 \end{bmatrix}\begin{bmatrix} 0 \\ y \end{bmatrix} = \begin{bmatrix} 0 \\ y \end{bmatrix} = 1\begin{bmatrix} 0 \\ y \end{bmatrix}$$
Eigenvalue is $\lambda_2 = 1$.

So, the eigenvectors corresponding to $\lambda_1 = -1$ are the nonzero vectors on the x-axis, and the eigenvectors corresponding to $\lambda_2 = 1$ are the nonzero vectors on the y-axis. This implies that the eigenspace corresponding to $\lambda_1 = -1$ is the x-axis, and that the eigenspace corresponding to $\lambda_2 = 1$ is the y-axis.

A reflects vectors in the y-axis.

Figure 7.1

REMARK

The geometric solution in Example 3 is not typical of the general eigenvalue problem. A more general approach follows.

FINDING EIGENVALUES AND EIGENVECTORS

To find the eigenvalues and eigenvectors of an $n \times n$ matrix A, let I be the $n \times n$ identity matrix. Rewriting $A\mathbf{x} = \lambda\mathbf{x}$ as $\lambda I\mathbf{x} = A\mathbf{x}$ and rearranging gives $(\lambda I - A)\mathbf{x} = \mathbf{0}$. This homogeneous system of equations has nonzero solutions if and only if the coefficient matrix $(\lambda I - A)$ is *not* invertible—that is, if and only if its determinant is zero. The next theorem formally states this.

> **THEOREM 7.2 Eigenvalues and Eigenvectors of a Matrix**
>
> Let A be an $n \times n$ matrix.
>
> 1. An eigenvalue of A is a scalar λ such that $\det(\lambda I - A) = 0$.
> 2. The eigenvectors of A corresponding to λ are the nonzero solutions of $(\lambda I - A)\mathbf{x} = \mathbf{0}$.

The equation $\det(\lambda I - A) = 0$ is the **characteristic equation** of A. Moreover, when expanded to polynomial form, the polynomial

$$|\lambda I - A| = \lambda^n + c_{n-1}\lambda^{n-1} + \cdots + c_2\lambda^2 + c_1\lambda + c_0$$

is the **characteristic polynomial** of A. So, the eigenvalues of an $n \times n$ matrix A correspond to the roots of the characteristic polynomial of A.

EXAMPLE 4   **Finding Eigenvalues and Eigenvectors**

See LarsonLinearAlgebra.com for an interactive version of this type of example.

Find the eigenvalues and corresponding eigenvectors of $A = \begin{bmatrix} 2 & -12 \\ 1 & -5 \end{bmatrix}$.

SOLUTION

The characteristic polynomial of A is

$$|\lambda I - A| = \begin{vmatrix} \lambda - 2 & 12 \\ -1 & \lambda + 5 \end{vmatrix} = \lambda^2 + 3\lambda - 10 + 12 = (\lambda + 1)(\lambda + 2).$$

So, the characteristic equation is $(\lambda + 1)(\lambda + 2) = 0$, which gives $\lambda_1 = -1$ and $\lambda_2 = -2$ as the eigenvalues of A. To find the corresponding eigenvectors, solve the homogeneous linear system represented by $(\lambda I - A)\mathbf{x} = \mathbf{0}$ twice: first for $\lambda = \lambda_1 = -1$, and then for $\lambda = \lambda_2 = -2$. For $\lambda_1 = -1$, the coefficient matrix is

$$(-1)I - A = \begin{bmatrix} -1-2 & 12 \\ -1 & -1+5 \end{bmatrix} = \begin{bmatrix} -3 & 12 \\ -1 & 4 \end{bmatrix}$$

which row reduces to $\begin{bmatrix} 1 & -4 \\ 0 & 0 \end{bmatrix}$, showing that $x_1 - 4x_2 = 0$. Letting $x_2 = t$, you can conclude that every eigenvector of λ_1 is of the form

$$\mathbf{x} = \begin{bmatrix} x_1 \\ x_2 \end{bmatrix} = \begin{bmatrix} 4t \\ t \end{bmatrix} = t\begin{bmatrix} 4 \\ 1 \end{bmatrix}, \quad t \neq 0.$$

For $\lambda_2 = -2$, you have

$$(-2)I - A = \begin{bmatrix} -2-2 & 12 \\ -1 & -2+5 \end{bmatrix} = \begin{bmatrix} -4 & 12 \\ -1 & 3 \end{bmatrix} \longrightarrow \begin{bmatrix} 1 & -3 \\ 0 & 0 \end{bmatrix}.$$

Letting $x_2 = t$, you can conclude that every eigenvector of λ_2 is of the form

$$\mathbf{x} = \begin{bmatrix} x_1 \\ x_2 \end{bmatrix} = \begin{bmatrix} 3t \\ t \end{bmatrix} = t\begin{bmatrix} 3 \\ 1 \end{bmatrix}, \quad t \neq 0.$$

The homogeneous systems that arise when you are finding eigenvectors will always row reduce to a matrix having at least one row of zeros, because the systems must have nontrivial solutions. A summary of the steps used to find the eigenvalues and corresponding eigenvectors of a matrix is below.

Finding Eigenvalues and Eigenvectors

Let A be an $n \times n$ matrix.

1. Form the characteristic equation $|\lambda I - A| = 0$. It will be a polynomial equation of degree n in the variable λ.
2. Find the real roots of the characteristic equation. These are the eigenvalues of A.
3. For each eigenvalue λ_i, find the eigenvectors corresponding to λ_i by solving the homogeneous system $(\lambda_i I - A)\mathbf{x} = \mathbf{0}$. This can require row reducing an $n \times n$ matrix. The reduced row-echelon form must have at least one row of zeros.

Finding the eigenvalues of an $n \times n$ matrix can involve the factorization of an nth-degree polynomial. Once you have found an eigenvalue, you can find the corresponding eigenvectors by any appropriate method, such as Gauss-Jordan elimination.

EXAMPLE 5   Finding Eigenvalues and Eigenvectors

Find the eigenvalues and corresponding eigenvectors of

$$A = \begin{bmatrix} 2 & 1 & 0 \\ 0 & 2 & 0 \\ 0 & 0 & 2 \end{bmatrix}.$$

What is the dimension of the eigenspace of each eigenvalue?

SOLUTION

The characteristic polynomial of A is

$$|\lambda I - A| = \begin{vmatrix} \lambda - 2 & -1 & 0 \\ 0 & \lambda - 2 & 0 \\ 0 & 0 & \lambda - 2 \end{vmatrix}$$

$$= (\lambda - 2)^3.$$

So, the characteristic equation is $(\lambda - 2)^3 = 0$, and the only eigenvalue is $\lambda = 2$. To find the eigenvectors of $\lambda = 2$, solve the homogeneous linear system represented by $(2I - A)\mathbf{x} = \mathbf{0}$.

$$2I - A = \begin{bmatrix} 0 & -1 & 0 \\ 0 & 0 & 0 \\ 0 & 0 & 0 \end{bmatrix}$$

This implies that $x_2 = 0$. Using the parameters $s = x_1$ and $t = x_3$, you can conclude that the eigenvectors of $\lambda = 2$ are of the form

$$\mathbf{x} = \begin{bmatrix} x_1 \\ x_2 \\ x_3 \end{bmatrix} = \begin{bmatrix} s \\ 0 \\ t \end{bmatrix} = s \begin{bmatrix} 1 \\ 0 \\ 0 \end{bmatrix} + t \begin{bmatrix} 0 \\ 0 \\ 1 \end{bmatrix}, \; s \text{ and } t \text{ not both zero.}$$

$\lambda = 2$ has two linearly independent eigenvectors, so the dimension of its eigenspace is 2.

If an eigenvalue λ_i occurs as a *multiple root* (k times) of the characteristic polynomial, then λ_i has **multiplicity** k. This implies that $(\lambda - \lambda_i)^k$ is a factor of the characteristic polynomial and $(\lambda - \lambda_i)^{k+1}$ is not a factor of the characteristic polynomial. For instance, in Example 5, the eigenvalue $\lambda = 2$ has a multiplicity of 3.

Also note that in Example 5, the dimension of the eigenspace of $\lambda = 2$ is 2. In general, the multiplicity of an eigenvalue is greater than or equal to the dimension of its eigenspace. (In Exercise 63, you are asked to prove this.)

EXAMPLE 6   Finding Eigenvalues and Eigenvectors

Find the eigenvalues of

$$A = \begin{bmatrix} 1 & 0 & 0 & 0 \\ 0 & 1 & 5 & -10 \\ 1 & 0 & 2 & 0 \\ 1 & 0 & 0 & 3 \end{bmatrix}$$

and find a basis for each of the corresponding eigenspaces.

SOLUTION

The characteristic polynomial of A is

$$|\lambda I - A| = \begin{vmatrix} \lambda - 1 & 0 & 0 & 0 \\ 0 & \lambda - 1 & -5 & 10 \\ -1 & 0 & \lambda - 2 & 0 \\ -1 & 0 & 0 & \lambda - 3 \end{vmatrix}$$

$$= (\lambda - 1)^2(\lambda - 2)(\lambda - 3).$$

So, the characteristic equation is $(\lambda - 1)^2(\lambda - 2)(\lambda - 3) = 0$ and the eigenvalues are $\lambda_1 = 1$, $\lambda_2 = 2$, and $\lambda_3 = 3$. (Note that $\lambda_1 = 1$ has a multiplicity of 2.)

You can find a basis for the eigenspace of $\lambda_1 = 1$ as shown below.

$$(1)I - A = \begin{bmatrix} 0 & 0 & 0 & 0 \\ 0 & 0 & -5 & 10 \\ -1 & 0 & -1 & 0 \\ -1 & 0 & 0 & -2 \end{bmatrix} \longrightarrow \begin{bmatrix} 1 & 0 & 0 & 2 \\ 0 & 0 & 1 & -2 \\ 0 & 0 & 0 & 0 \\ 0 & 0 & 0 & 0 \end{bmatrix}$$

TECHNOLOGY

Use a graphing utility or a software program to find the eigenvalues and eigenvectors in Example 6. When finding the eigenvectors, the technology you use may produce a matrix in which the columns are scalar multiples of the eigenvectors you would obtain by hand calculations. The **Technology Guide** at *CengageBrain.com* can help you use technology to find eigenvalues and eigenvectors.

Letting $s = x_2$ and $t = x_4$ produces

$$\mathbf{x} = \begin{bmatrix} x_1 \\ x_2 \\ x_3 \\ x_4 \end{bmatrix} = \begin{bmatrix} 0s - 2t \\ s + 0t \\ 0s + 2t \\ 0s + t \end{bmatrix} = s\begin{bmatrix} 0 \\ 1 \\ 0 \\ 0 \end{bmatrix} + t\begin{bmatrix} -2 \\ 0 \\ 2 \\ 1 \end{bmatrix}.$$

So, a basis for the eigenspace corresponding to $\lambda_1 = 1$ is

$$B_1 = \{(0, 1, 0, 0), (-2, 0, 2, 1)\}. \qquad \text{Basis for } \lambda_1 = 1$$

For $\lambda_2 = 2$ and $\lambda_3 = 3$, use the same procedure to obtain the eigenspace bases

$$B_2 = \{(0, 5, 1, 0)\} \qquad \text{Basis for } \lambda_2 = 2$$
$$B_3 = \{(0, -5, 0, 1)\}. \qquad \text{Basis for } \lambda_3 = 3$$

Finding eigenvalues and eigenvectors of matrices of order $n \geq 4$ can be tedious. Moreover, using the procedure shown in Example 6 on a computer can introduce roundoff errors. Consequently, it can be more efficient to use numerical methods of approximating eigenvalues. One of these numerical methods appears in Section 10.3. Other methods appear in texts on advanced linear algebra and numerical analysis.

The next theorem states that the eigenvalues of an $n \times n$ triangular matrix are simply the entries on the main diagonal. The proof of this theorem follows from the fact that the determinant of a triangular matrix is the product of its main diagonal entries.

THEOREM 7.3   Eigenvalues of Triangular Matrices

If A is an $n \times n$ triangular matrix, then its eigenvalues are the entries on its main diagonal.

EXAMPLE 7   Finding Eigenvalues of Triangular and Diagonal Matrices

Find the eigenvalues of each matrix.

a. $A = \begin{bmatrix} 2 & 0 & 0 \\ -1 & 1 & 0 \\ 5 & 3 & -3 \end{bmatrix}$

b. $A = \begin{bmatrix} -1 & 0 & 0 & 0 & 0 \\ 0 & 2 & 0 & 0 & 0 \\ 0 & 0 & 0 & 0 & 0 \\ 0 & 0 & 0 & -4 & 0 \\ 0 & 0 & 0 & 0 & 3 \end{bmatrix}$

SOLUTION

a. Without using Theorem 7.3,

$$|\lambda I - A| = \begin{vmatrix} \lambda - 2 & 0 & 0 \\ 1 & \lambda - 1 & 0 \\ -5 & -3 & \lambda + 3 \end{vmatrix} = (\lambda - 2)(\lambda - 1)(\lambda + 3).$$

So, the eigenvalues are $\lambda_1 = 2$, $\lambda_2 = 1$, and $\lambda_3 = -3$, which are the main diagonal entries of A.

b. In this case, use Theorem 7.3 to conclude that the eigenvalues are the main diagonal entries $\lambda_1 = -1$, $\lambda_2 = 2$, $\lambda_3 = 0$, $\lambda_4 = -4$, and $\lambda_5 = 3$.

LINEAR ALGEBRA APPLIED

Eigenvalues and eigenvectors are useful for modeling real-life phenomena. For example, consider an experiment to determine the diffusion of a fluid from one flask to another through a permeable membrane and then out of the second flask. If researchers determine that the flow rate between flasks is twice the volume of fluid in the first flask and the flow rate out of the second flask is three times the volume of fluid in the second flask, then the system of linear differential equations below, where y_i represents the volume of fluid in flask i, models this situation.

$$y_1' = -2y_1$$
$$y_2' = 2y_1 - 3y_2$$

In Section 7.4, you will use eigenvalues and eigenvectors to solve such systems of linear differential equations. For now, verify that the solution of this system is

$$y_1 = C_1 e^{-2t}$$
$$y_2 = 2C_1 e^{-2t} + C_2 e^{-3t}.$$

EIGENVALUES AND EIGENVECTORS
OF LINEAR TRANSFORMATIONS

This section began with definitions of eigenvalues and eigenvectors in terms of matrices. Eigenvalues and eigenvectors can also be defined in terms of linear transformations. A number λ is an **eigenvalue** of a linear transformation $T: V \to V$ when there is a nonzero vector $\mathbf{x}$ such that $T(\mathbf{x}) = \lambda \mathbf{x}$. The vector $\mathbf{x}$ is an **eigenvector** of T corresponding to λ, and the set of all eigenvectors of λ (with the zero vector) is the **eigenspace** of λ.

Consider $T: R^3 \to R^3$, whose matrix relative to the standard basis is

$$A = \begin{bmatrix} 1 & 3 & 0 \\ 3 & 1 & 0 \\ 0 & 0 & -2 \end{bmatrix}.$$

Standard basis:
$B = \{(1, 0, 0), (0, 1, 0), (0, 0, 1)\}$

In Example 5 in Section 6.4, you found that the matrix of T relative to the basis $B' = \{(1, 1, 0), (1, -1, 0), (0, 0, 1)\}$ is the diagonal matrix

$$A' = \begin{bmatrix} 4 & 0 & 0 \\ 0 & -2 & 0 \\ 0 & 0 & -2 \end{bmatrix}.$$

Nonstandard basis:
$B' = \{(1, 1, 0), (1, -1, 0), (0, 0, 1)\}$

For a linear transformation T, can you find a basis B' whose corresponding matrix is diagonal? The next example illustrates the answer.

EXAMPLE 8   Finding Eigenvalues and Eigenspaces

Find the eigenvalues and a basis for each corresponding eigenspace of

$$A = \begin{bmatrix} 1 & 3 & 0 \\ 3 & 1 & 0 \\ 0 & 0 & -2 \end{bmatrix}.$$

SOLUTION

$$\begin{aligned} |\lambda I - A| &= \begin{vmatrix} \lambda - 1 & -3 & 0 \\ -3 & \lambda - 1 & 0 \\ 0 & 0 & \lambda + 2 \end{vmatrix} \\ &= [(\lambda - 1)^2 - 9](\lambda + 2) \\ &= (\lambda - 4)(\lambda + 2)^2 \end{aligned}$$

so the eigenvalues of A are $\lambda_1 = 4$ and $\lambda_2 = -2$. Bases for the eigenspaces are $B_1 = \{(1, 1, 0)\}$ and $B_2 = \{(1, -1, 0), (0, 0, 1)\}$, respectively (verify these).

Example 8 illustrates two results. If $T: R^3 \to R^3$ is the linear transformation whose standard matrix is A, and B' is a basis for R^3 made up of the three linearly independent eigenvectors corresponding to the eigenvalues of A, then the matrix A' for T relative to the basis B' is diagonal. Also, the main diagonal entries of the matrix A' are the eigenvalues of A.

$$A' = \begin{bmatrix} 4 & 0 & 0 \\ 0 & -2 & 0 \\ 0 & 0 & -2 \end{bmatrix}$$

Nonstandard basis:
$B' = \{(1, 1, 0), (1, -1, 0), (0, 0, 1)\}$

Eigenvalues of A ← → Eigenvectors of A

The next section discusses these results in more detail.

7.1   Exercises   See CalcChat.com for worked-out solutions to odd-numbered exercises.

Verifying Eigenvalues and Eigenvectors   **In Exercises 1–6, verify that** λ_i **is an eigenvalue of** A **and that** x_i **is a corresponding eigenvector.**

1. $A = \begin{bmatrix} 2 & 0 \\ 0 & -2 \end{bmatrix}$, $\begin{aligned} \lambda_1 &= 2, x_1 = (1, 0) \\ \lambda_2 &= -2, x_2 = (0, 1) \end{aligned}$

2. $A = \begin{bmatrix} 4 & -5 \\ 2 & -3 \end{bmatrix}$, $\begin{aligned} \lambda_1 &= -1, x_1 = (1, 1) \\ \lambda_2 &= 2, x_2 = (5, 2) \end{aligned}$

3. $A = \begin{bmatrix} 2 & 3 & 1 \\ 0 & -1 & 2 \\ 0 & 0 & 3 \end{bmatrix}$, $\begin{aligned} \lambda_1 &= 2, x_1 = (1, 0, 0) \\ \lambda_2 &= -1, x_2 = (1, -1, 0) \\ \lambda_3 &= 3, x_3 = (5, 1, 2) \end{aligned}$

4. $A = \begin{bmatrix} -2 & 2 & -3 \\ 2 & 1 & -6 \\ -1 & -2 & 0 \end{bmatrix}$, $\begin{aligned} \lambda_1 &= 5, x_1 = (1, 2, -1) \\ \lambda_2 &= -3, x_2 = (-2, 1, 0) \\ \lambda_3 &= -3, x_3 = (3, 0, 1) \end{aligned}$

5. $A = \begin{bmatrix} 0 & 1 & 0 \\ 0 & 0 & 1 \\ 1 & 0 & 0 \end{bmatrix}$, $\lambda_1 = 1, x_1 = (1, 1, 1)$

6. $A = \begin{bmatrix} 4 & -1 & 3 \\ 0 & 2 & 1 \\ 0 & 0 & 3 \end{bmatrix}$, $\begin{aligned} \lambda_1 &= 4, x_1 = (1, 0, 0) \\ \lambda_2 &= 2, x_2 = (1, 2, 0) \\ \lambda_3 &= 3, x_3 = (-2, 1, 1) \end{aligned}$

7. Use A, λ_i, and x_i from Exercise 1 to show that
   (a) $A(cx_1) = 2(cx_1)$ for any real number c.
   (b) $A(cx_2) = -2(cx_2)$ for any real number c.

8. Use A, λ_i, and x_i from Exercise 4 to show that
   (a) $A(cx_1) = 5(cx_1)$ for any real number c.
   (b) $A(cx_2) = -3(cx_2)$ for any real number c.
   (c) $A(cx_3) = -3(cx_3)$ for any real number c.

Determining Eigenvectors   **In Exercises 9–12, determine whether x is an eigenvector of** A.

9. $A = \begin{bmatrix} 7 & 2 \\ 2 & 4 \end{bmatrix}$
   (a) $x = (1, 2)$
   (b) $x = (2, 1)$
   (c) $x = (1, -2)$
   (d) $x = (-1, 0)$

10. $A = \begin{bmatrix} -3 & 10 \\ 5 & 2 \end{bmatrix}$
    (a) $x = (4, 4)$
    (b) $x = (-8, 4)$
    (c) $x = (-4, 8)$
    (d) $x = (5, -3)$

11. $A = \begin{bmatrix} -1 & -1 & 1 \\ -2 & 0 & -2 \\ 3 & -3 & 1 \end{bmatrix}$
    (a) $x = (2, -4, 6)$
    (b) $x = (2, 0, 6)$
    (c) $x = (2, 2, 0)$
    (d) $x = (-1, 0, 1)$

12. $A = \begin{bmatrix} 1 & 0 & 5 \\ 0 & -2 & 4 \\ 1 & -2 & 9 \end{bmatrix}$
    (a) $x = (1, 1, 0)$
    (b) $x = (-5, 2, 1)$
    (c) $x = (0, 0, 0)$
    (d) $x = (2\sqrt{6} - 3, -2\sqrt{6} + 6, 3)$

Finding Eigenspaces in R^2 **Geometrically**   **In Exercises 13 and 14, use the method shown in Example 3 to find the eigenvalue(s) and corresponding eigenspace(s) of** A.

13. $A = \begin{bmatrix} 1 & 0 \\ 0 & -1 \end{bmatrix}$

14. $A = \begin{bmatrix} 1 & k \\ 0 & 1 \end{bmatrix}$

Characteristic Equation, Eigenvalues, and Eigenvectors   **In Exercises 15–28, find (a) the characteristic equation and (b) the eigenvalues (and corresponding eigenvectors) of the matrix.**

15. $\begin{bmatrix} 6 & -3 \\ -2 & 1 \end{bmatrix}$

16. $\begin{bmatrix} 1 & -4 \\ -2 & 8 \end{bmatrix}$

17. $\begin{bmatrix} 1 & 2 \\ 2 & 1 \end{bmatrix}$

18. $\begin{bmatrix} -2 & 4 \\ 1 & 1 \end{bmatrix}$

19. $\begin{bmatrix} 1 & -\frac{3}{2} \\ \frac{1}{2} & -1 \end{bmatrix}$

20. $\begin{bmatrix} \frac{1}{4} & \frac{1}{4} \\ \frac{1}{2} & 0 \end{bmatrix}$

21. $\begin{bmatrix} 2 & -2 & 3 \\ 0 & 3 & -2 \\ 0 & -1 & 2 \end{bmatrix}$

22. $\begin{bmatrix} 3 & 2 & 1 \\ 0 & 0 & 2 \\ 0 & 2 & 0 \end{bmatrix}$

23. $\begin{bmatrix} 1 & 2 & -2 \\ -2 & 5 & -2 \\ -6 & 6 & -3 \end{bmatrix}$

24. $\begin{bmatrix} 3 & 2 & -3 \\ -3 & -4 & 9 \\ -1 & -2 & 5 \end{bmatrix}$

25. $\begin{bmatrix} 0 & -3 & 5 \\ -4 & 4 & -10 \\ 0 & 0 & 4 \end{bmatrix}$

26. $\begin{bmatrix} 1 & -\frac{3}{2} & \frac{5}{2} \\ -2 & \frac{13}{2} & -10 \\ \frac{3}{2} & -\frac{9}{2} & 8 \end{bmatrix}$

27. $\begin{bmatrix} 2 & 0 & 0 & 0 \\ 0 & 2 & 0 & 0 \\ 0 & 0 & 3 & 1 \\ 0 & 0 & 4 & 0 \end{bmatrix}$

28. $\begin{bmatrix} 5 & 0 & 0 & 0 \\ 1 & 4 & 0 & 0 \\ 0 & 0 & 1 & 3 \\ 0 & 0 & 0 & 4 \end{bmatrix}$

Finding Eigenvalues   **In Exercises 29–40, use a software program or a graphing utility to find the eigenvalues of the matrix.**

29. $\begin{bmatrix} -4 & 5 \\ -2 & 3 \end{bmatrix}$

30. $\begin{bmatrix} 2 & 3 \\ 3 & -6 \end{bmatrix}$

31. $\begin{bmatrix} \frac{1}{2} & \frac{1}{3} \\ -\frac{1}{3} & -\frac{1}{3} \end{bmatrix}$

32. $\begin{bmatrix} \frac{1}{2} & -\frac{1}{2} \\ -\frac{1}{2} & -\frac{1}{2} \end{bmatrix}$

33. $\begin{bmatrix} 2 & 4 & 2 \\ 1 & 0 & 1 \\ 1 & -4 & 5 \end{bmatrix}$

34. $\begin{bmatrix} 1 & 2 & -1 \\ 1 & 0 & 1 \\ 1 & -1 & 2 \end{bmatrix}$

35. $\begin{bmatrix} 3 & -\frac{1}{2} & 5 \\ -\frac{1}{3} & -\frac{1}{6} & -\frac{1}{4} \\ 0 & 0 & 4 \end{bmatrix}$

36. $\begin{bmatrix} \frac{1}{2} & 0 & 5 \\ -2 & \frac{1}{5} & \frac{1}{4} \\ 1 & 0 & 3 \end{bmatrix}$

37. $\begin{bmatrix} 1 & 1 & 2 & 3 \\ 2 & 2 & 4 & 6 \\ 3 & 3 & 6 & 9 \\ 4 & 4 & 8 & 12 \end{bmatrix}$ **38.** $\begin{bmatrix} 1 & 1 & 0 & 0 \\ 4 & 4 & 0 & 0 \\ 0 & 0 & 1 & 1 \\ 0 & 0 & 2 & 2 \end{bmatrix}$

39. $\begin{bmatrix} 1 & 0 & -1 & 1 \\ 0 & 1 & 0 & 1 \\ -2 & 0 & 2 & -2 \\ 0 & 2 & 0 & 2 \end{bmatrix}$

40. $\begin{bmatrix} 1 & -3 & 3 & 3 \\ -1 & 4 & -3 & -3 \\ -2 & 0 & 1 & 1 \\ 1 & 0 & 0 & 0 \end{bmatrix}$

Eigenvalues of Triangular and Diagonal Matrices In Exercises 41–44, find the eigenvalues of the triangular or diagonal matrix.

41. $\begin{bmatrix} 2 & 0 & 1 \\ 0 & 3 & 4 \\ 0 & 0 & 1 \end{bmatrix}$ **42.** $\begin{bmatrix} -5 & 0 & 0 \\ 3 & 7 & 0 \\ 4 & -2 & 3 \end{bmatrix}$

43. $\begin{bmatrix} -6 & 0 & 0 & 0 \\ 0 & 5 & 0 & 0 \\ 0 & 0 & -4 & 0 \\ 0 & 0 & 0 & -4 \end{bmatrix}$ **44.** $\begin{bmatrix} \frac{1}{2} & 0 & 0 & 0 \\ 0 & \frac{5}{4} & 0 & 0 \\ 0 & 0 & 0 & 0 \\ 0 & 0 & 0 & \frac{3}{4} \end{bmatrix}$

Eigenvalues and Eigenvectors of Linear Transformations In Exercises 45–48, consider the linear transformation $T: R^n \to R^n$ whose matrix A relative to the standard basis is given. Find (a) the eigenvalues of A, (b) a basis for each of the corresponding eigenspaces, and (c) the matrix A' for T relative to the basis B', where B' is made up of the basis vectors found in part (b).

45. $\begin{bmatrix} 2 & -2 \\ 1 & 5 \end{bmatrix}$ **46.** $\begin{bmatrix} -8 & 16 \\ 1 & -2 \end{bmatrix}$

47. $\begin{bmatrix} 0 & 2 & -1 \\ -1 & 3 & 1 \\ 0 & 0 & -1 \end{bmatrix}$ **48.** $\begin{bmatrix} 3 & 1 & 4 \\ 2 & 4 & 0 \\ 5 & 5 & 6 \end{bmatrix}$

Cayley-Hamilton Theorem In Exercises 49–52, demonstrate the Cayley-Hamilton Theorem for the matrix A. The Cayley-Hamilton Theorem states that a matrix satisfies its characteristic equation. For example, the characteristic equation of
$$A = \begin{bmatrix} 1 & -3 \\ 2 & 5 \end{bmatrix}$$
is $\lambda^2 - 6\lambda + 11 = 0$, and by the theorem you have $A^2 - 6A + 11I_2 = O$.

49. $A = \begin{bmatrix} 5 & 0 \\ -7 & 3 \end{bmatrix}$ **50.** $A = \begin{bmatrix} 6 & -1 \\ 1 & 5 \end{bmatrix}$

51. $A = \begin{bmatrix} 1 & 0 & -4 \\ 0 & 3 & 1 \\ 2 & 0 & 1 \end{bmatrix}$ **52.** $A = \begin{bmatrix} -3 & 1 & 0 \\ -1 & 3 & 2 \\ 0 & 4 & 3 \end{bmatrix}$

53. Perform each computational check on the eigenvalues found in Exercises 15–27 odd.
(a) The sum of the n eigenvalues equals the trace of the matrix. (Recall that the **trace** of a matrix is the sum of the main diagonal entries of the matrix.)
(b) The product of the n eigenvalues equals $|A|$.
(When λ is an eigenvalue of multiplicity k, remember to use it k times in the sum or product of these checks.)

54. Perform each computational check on the eigenvalues found in Exercises 16–28 even.
(a) The sum of the n eigenvalues equals the trace of the matrix. (Recall that the **trace** of a matrix is the sum of the main diagonal entries of the matrix.)
(b) The product of the n eigenvalues equals $|A|$.
(When λ is an eigenvalue of multiplicity k, remember to use it k times in the sum or product of these checks.)

55. Show that if A is an $n \times n$ matrix whose ith row is identical to the ith row of I, then 1 is an eigenvalue of A.

56. Proof Prove that $\lambda = 0$ is an eigenvalue of A if and only if A is singular.

57. Proof For an invertible matrix A, prove that A and A^{-1} have the same eigenvectors. How are the eigenvalues of A related to the eigenvalues of A^{-1}?

58. Proof Prove that A and A^T have the same eigenvalues. Are the eigenspaces the same?

59. Proof Prove that the constant term of the characteristic polynomial is $\pm |A|$.

60. Define $T: R^2 \to R^2$ by
$$T(\mathbf{v}) = \text{proj}_{\mathbf{u}}\mathbf{v}$$
where $\mathbf{u}$ is a fixed vector in R^2. Show that the eigenvalues of A (the standard matrix of T) are 0 and 1.

61. Guided Proof Prove that a triangular matrix is nonsingular if and only if its eigenvalues are real and nonzero.

Getting Started: This is an "if and only if" statement, so you must prove that the statement is true in both directions. Review Theorems 3.2 and 3.7.
(i) To prove the statement in one direction, assume that the triangular matrix A is nonsingular. Use your knowledge of nonsingular and triangular matrices and determinants to conclude that the entries on the main diagonal of A are nonzero.
(ii) A is triangular, so use Theorem 7.3 and part (i) to conclude that the eigenvalues are real and nonzero.
(iii) To prove the statement in the other direction, assume that the eigenvalues of the triangular matrix A are real and nonzero. Repeat parts (i) and (ii) in reverse order to prove that A is nonsingular.

62. Guided Proof Prove that if $A^2 = O$, then 0 is the only eigenvalue of A.

Getting Started: You need to show that if there exists a nonzero vector $\mathbf{x}$ and a real number λ such that $A\mathbf{x} = \lambda\mathbf{x}$, then if $A^2 = O$, λ must be zero.

(i) $A^2 = A \cdot A$, so you can write $A^2\mathbf{x}$ as $A(A\mathbf{x})$.

(ii) Use the fact that $A\mathbf{x} = \lambda\mathbf{x}$ and the properties of matrix multiplication to show that $A^2\mathbf{x} = \lambda^2\mathbf{x}$.

(iii) A^2 is a zero matrix, so you can conclude that λ must be zero.

63. Proof Prove that the multiplicity of an eigenvalue is greater than or equal to the dimension of its eigenspace.

64. CAPSTONE An $n \times n$ matrix A has the characteristic equation

$$|\lambda I - A| = (\lambda + 2)(\lambda - 1)(\lambda - 3)^2 = 0.$$

(a) What are the eigenvalues of A?

(b) What is the order of A? Explain.

(c) Is $\lambda I - A$ singular? Explain.

(d) Is A singular? Explain. (*Hint:* Use the result of Exercise 56.)

65. When the eigenvalues of

$$A = \begin{bmatrix} a & b \\ 0 & d \end{bmatrix}$$

are $\lambda_1 = 0$ and $\lambda_2 = 1$, what are the possible values of a and d?

66. Show that

$$A = \begin{bmatrix} 0 & 1 \\ -1 & 0 \end{bmatrix}$$

has no real eigenvalues.

True or False? In Exercises 67 and 68, determine whether each statement is true or false. If a statement is true, give a reason or cite an appropriate statement from the text. If a statement is false, provide an example that shows the statement is not true in all cases or cite an appropriate statement from the text.

67. (a) The scalar λ is an eigenvalue of an $n \times n$ matrix A when there exists a vector $\mathbf{x}$ such that $A\mathbf{x} = \lambda\mathbf{x}$.

(b) To find the eigenvalue(s) of an $n \times n$ matrix A, you can solve the characteristic equation $\det(\lambda I - A) = 0$.

68. (a) Geometrically, if λ is an eigenvalue of a matrix A and $\mathbf{x}$ is an eigenvector of A corresponding to λ, then multiplying $\mathbf{x}$ by A produces a vector $\lambda\mathbf{x}$ parallel to $\mathbf{x}$.

(b) If A is an $n \times n$ matrix with an eigenvalue λ, then the set of all eigenvectors of λ is a subspace of R^n.

Finding the Dimension of an Eigenspace In Exercises 69–72, find the dimension of the eigenspace corresponding to the eigenvalue $\lambda = 3$.

69. $A = \begin{bmatrix} 3 & 0 & 0 \\ 0 & 3 & 0 \\ 0 & 0 & 3 \end{bmatrix}$ **70.** $A = \begin{bmatrix} 3 & 1 & 0 \\ 0 & 3 & 0 \\ 0 & 0 & 3 \end{bmatrix}$

71. $A = \begin{bmatrix} 3 & 1 & 0 \\ 0 & 3 & 1 \\ 0 & 0 & 3 \end{bmatrix}$ **72.** $A = \begin{bmatrix} 3 & 1 & 1 \\ 0 & 3 & 1 \\ 0 & 0 & 3 \end{bmatrix}$

73. Calculus Let $T: C'[0, 1] \to C[0, 1]$ be the linear transformation $T(f) = f'$. Show that $\lambda = 1$ is an eigenvalue of T with corresponding eigenvector $f(x) = e^x$.

74. Calculus For the linear transformation in Exercise 73, find the eigenvalue corresponding to the eigenvector $f(x) = e^{-2x}$.

75. Define $T: P_2 \to P_2$ by

$$T(a_0 + a_1x + a_2x^2) = (-3a_1 + 5a_2) + (-4a_0 + 4a_1 - 10a_2)x + 4a_2x^2.$$

Find the eigenvalues and the eigenvectors of T relative to the standard basis $\{1, x, x^2\}$.

76. Define $T: P_2 \to P_2$ by

$$T(a_0 + a_1x + a_2x^2) = (2a_0 + a_1 - a_2) + (-a_1 + 2a_2)x - a_2x^2.$$

Find the eigenvalues and eigenvectors of T relative to the standard basis $\{1, x, x^2\}$.

77. Define $T: M_{2,2} \to M_{2,2}$ by

$$T\left(\begin{bmatrix} a & b \\ c & d \end{bmatrix}\right) = \begin{bmatrix} a - c + d & b + d \\ -2a + 2c - 2d & 2b + 2d \end{bmatrix}.$$

Find the eigenvalues and eigenvectors of T relative to the standard basis

$$B = \left\{ \begin{bmatrix} 1 & 0 \\ 0 & 0 \end{bmatrix}, \begin{bmatrix} 0 & 1 \\ 0 & 0 \end{bmatrix}, \begin{bmatrix} 0 & 0 \\ 1 & 0 \end{bmatrix}, \begin{bmatrix} 0 & 0 \\ 0 & 1 \end{bmatrix} \right\}.$$

78. Find all values of the angle θ for which the matrix

$$A = \begin{bmatrix} \cos\theta & -\sin\theta \\ \sin\theta & \cos\theta \end{bmatrix}$$

has real eigenvalues. Interpret your answer geometrically.

79. What are the possible eigenvalues of an idempotent matrix? (Recall that a square matrix A is **idempotent** when $A^2 = A$.)

80. What are the possible eigenvalues of a nilpotent matrix? (Recall that a square matrix A is **nilpotent** when there exists a positive integer k such that $A^k = 0$.)

81. Proof Let A be an $n \times n$ matrix such that the sum of the entries in each row is a fixed constant r. Prove that r is an eigenvalue of A. Illustrate this result with an example.

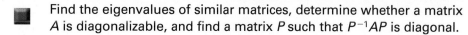

7.2 Diagonalization

■ Find the eigenvalues of similar matrices, determine whether a matrix A is diagonalizable, and find a matrix P such that $P^{-1}AP$ is diagonal.

■ Find, for a linear transformation $T: V \to V$, a basis B for V such that the matrix for T relative to B is diagonal.

THE DIAGONALIZATION PROBLEM

The preceding section discussed the eigenvalue problem. In this section, you will look at another classic problem in linear algebra called the **diagonalization problem.** Expressed in terms of matrices*, the problem is "for a square matrix A, does there exist an invertible matrix P such that $P^{-1}AP$ is diagonal?"

Recall from Section 6.4 that two square matrices A and B are similar when there exists an invertible matrix P such that $B = P^{-1}AP$.

Matrices that are similar to diagonal matrices are called **diagonalizable.**

Definition of a Diagonalizable Matrix

An $n \times n$ matrix A is **diagonalizable** when A is similar to a diagonal matrix. That is, A is diagonalizable when there exists an invertible matrix P such that $P^{-1}AP$ is a diagonal matrix.

With this definition, the diagonalization problem can be stated as "which square matrices are diagonalizable?" Clearly, every diagonal matrix D is diagonalizable, because $D = I^{-1}DI$, where I is the identity matrix. Example 1 shows another example of a diagonalizable matrix.

EXAMPLE 1   A Diagonalizable Matrix

The matrix from Example 5 in Section 6.4

$$A = \begin{bmatrix} 1 & 3 & 0 \\ 3 & 1 & 0 \\ 0 & 0 & -2 \end{bmatrix}$$

is diagonalizable because

$$P = \begin{bmatrix} 1 & 1 & 0 \\ 1 & -1 & 0 \\ 0 & 0 & 1 \end{bmatrix}$$

has the property that

$$P^{-1}AP = \begin{bmatrix} 4 & 0 & 0 \\ 0 & -2 & 0 \\ 0 & 0 & -2 \end{bmatrix}.$$

As suggested in Example 8 in the preceding section, the eigenvalue problem is related closely to the diagonalization problem. The next two theorems shed more light on this relationship. The first theorem tells you that similar matrices have the same eigenvalues.

---

*At the end of this section, the diagonalization problem will be expressed in terms of linear transformations.

> ### THEOREM 7.4   Similar Matrices Have the Same Eigenvalues
>
> If A and B are similar $n \times n$ matrices, then they have the same eigenvalues.

PROOF

A and B are similar, so there exists an invertible matrix P such that $B = P^{-1}AP$. By the properties of determinants, it follows that

$$
\begin{aligned}
|\lambda I - B| &= |\lambda I - P^{-1}AP| \\
&= |P^{-1}\lambda IP - P^{-1}AP| \\
&= |P^{-1}(\lambda I - A)P| \\
&= |P^{-1}||\lambda I - A||P| \\
&= |P^{-1}P||\lambda I - A| \\
&= |\lambda I - A|.
\end{aligned}
$$

This means that A and B have the same characteristic polynomial. So, they must have the same eigenvalues.

REMARK

Check that A and D are similar by showing that they satisfy the matrix equation $D = P^{-1}AP$, where

$$
P = \begin{bmatrix} 1 & 0 & 0 \\ 1 & 1 & 1 \\ 1 & 1 & 2 \end{bmatrix}.
$$

In fact, the columns of P are eigenvectors of A corresponding to the eigenvalues 1, 2, and 3. (Verify this.)

EXAMPLE 2   **Finding Eigenvalues of Similar Matrices**

The matrices A and D are similar.

$$
A = \begin{bmatrix} 1 & 0 & 0 \\ -1 & 1 & 1 \\ -1 & -2 & 4 \end{bmatrix} \quad \text{and} \quad D = \begin{bmatrix} 1 & 0 & 0 \\ 0 & 2 & 0 \\ 0 & 0 & 3 \end{bmatrix}
$$

Use Theorem 7.4 to find the eigenvalues of A.

SOLUTION

D is a diagonal matrix, so its eigenvalues are the entries on its main diagonal—that is, $\lambda_1 = 1$, $\lambda_2 = 2$, and $\lambda_3 = 3$. Matrices A and D are similar, so you know from Theorem 7.4 that A has the same eigenvalues. Check this by showing that the characteristic polynomial of A is $|\lambda I - A| = (\lambda - 1)(\lambda - 2)(\lambda - 3)$.

The two diagonalizable matrices in Examples 1 and 2 provide a clue to the diagonalization problem. Each of these matrices has a set of three linearly independent eigenvectors. (See Example 3.) This is characteristic of diagonalizable matrices, as stated in Theorem 7.5.

> ### THEOREM 7.5   Condition for Diagonalization
>
> An $n \times n$ matrix A is diagonalizable if and only if it has n linearly independent eigenvectors.

PROOF

First, assume A is diagonalizable. Then there exists an invertible matrix P such that $P^{-1}AP = D$ is diagonal. Letting the column vectors of P be $\mathbf{p}_1, \mathbf{p}_2, \ldots, \mathbf{p}_n$, and the main diagonal entries of D be $\lambda_1, \lambda_2, \ldots, \lambda_n$, produces

$$
PD = \begin{bmatrix} \mathbf{p}_1 & \mathbf{p}_2 & \cdots & \mathbf{p}_n \end{bmatrix} \begin{bmatrix} \lambda_1 & 0 & \cdots & 0 \\ 0 & \lambda_2 & \cdots & 0 \\ \vdots & \vdots & & \vdots \\ 0 & 0 & \cdots & \lambda_n \end{bmatrix} = \begin{bmatrix} \lambda_1\mathbf{p}_1 & \lambda_2\mathbf{p}_2 & \cdots & \lambda_n\mathbf{p}_n \end{bmatrix}.
$$

$P^{-1}AP = D$, so $AP = PD$, which implies

$$[A\mathbf{p}_1 \quad A\mathbf{p}_2 \quad \ldots \quad A\mathbf{p}_n] = [\lambda_1\mathbf{p}_1 \quad \lambda_2\mathbf{p}_2 \quad \ldots \quad \lambda_n\mathbf{p}_n].$$

In other words, $A\mathbf{p}_i = \lambda_i\mathbf{p}_i$ for each column vector $\mathbf{p}_i$. This means that the column vectors $\mathbf{p}_i$ of P are eigenvectors of A. Moreover, P is invertible, so its column vectors are linearly independent. So, A has n linearly independent eigenvectors.

Conversely, assume A has n linearly independent eigenvectors $\mathbf{p}_1, \mathbf{p}_2, \ldots, \mathbf{p}_n$ with corresponding eigenvalues $\lambda_1, \lambda_2, \ldots, \lambda_n$. Let P be the matrix whose columns are these n eigenvectors. That is, $P = [\mathbf{p}_1 \quad \mathbf{p}_2 \quad \ldots \quad \mathbf{p}_n]$. Each $\mathbf{p}_i$ is an eigenvector of A, so you have $A\mathbf{p}_i = \lambda_i\mathbf{p}_i$ and

$$AP = A[\mathbf{p}_1 \quad \mathbf{p}_2 \quad \ldots \quad \mathbf{p}_n] = [\lambda_1\mathbf{p}_1 \quad \lambda_2\mathbf{p}_2 \quad \ldots \quad \lambda_n\mathbf{p}_n].$$

The right-hand matrix in this equation can be written as the matrix product below.

$$[\lambda_1\mathbf{p}_1 \quad \lambda_2\mathbf{p}_2 \quad \ldots \quad \lambda_n\mathbf{p}_n] = [\mathbf{p}_1 \quad \mathbf{p}_2 \quad \ldots \quad \mathbf{p}_n] \begin{bmatrix} \lambda_1 & 0 & \ldots & 0 \\ 0 & \lambda_2 & \ldots & 0 \\ \vdots & \vdots & & \vdots \\ 0 & 0 & \ldots & \lambda_n \end{bmatrix} = PD$$

Finally, the vectors $\mathbf{p}_1, \mathbf{p}_2, \ldots, \mathbf{p}_n$ are linearly independent, so P is invertible and you can write the equation $AP = PD$ as $P^{-1}AP = D$, which means that A is diagonalizable. ∎

A key result of this proof is the fact that for diagonalizable matrices, *the columns of P consist of n linearly independent eigenvectors*. Example 3 verifies this important property for the matrices in Examples 1 and 2.

EXAMPLE 3   Diagonalizable Matrices

a. The matrix A in Example 1 has the eigenvalues and corresponding eigenvectors below.

$$\lambda_1 = 4, \mathbf{p}_1 = \begin{bmatrix} 1 \\ 1 \\ 0 \end{bmatrix}; \quad \lambda_2 = -2, \mathbf{p}_2 = \begin{bmatrix} 1 \\ -1 \\ 0 \end{bmatrix}; \quad \lambda_3 = -2, \mathbf{p}_3 = \begin{bmatrix} 0 \\ 0 \\ 1 \end{bmatrix}$$

The matrix P whose columns correspond to these eigenvectors is

$$P = \begin{bmatrix} 1 & 1 & 0 \\ 1 & -1 & 0 \\ 0 & 0 & 1 \end{bmatrix}.$$

Moreover, P is row-equivalent to the identity matrix, so the eigenvectors $\mathbf{p}_1$, $\mathbf{p}_2$, and $\mathbf{p}_3$ are linearly independent.

b. The matrix A in Example 2 has the eigenvalues and corresponding eigenvectors below.

$$\lambda_1 = 1, \mathbf{p}_1 = \begin{bmatrix} 1 \\ 1 \\ 1 \end{bmatrix}; \quad \lambda_2 = 2, \mathbf{p}_2 = \begin{bmatrix} 0 \\ 1 \\ 1 \end{bmatrix}; \quad \lambda_3 = 3, \mathbf{p}_3 = \begin{bmatrix} 0 \\ 1 \\ 2 \end{bmatrix}$$

The matrix P whose columns correspond to these eigenvectors is

$$P = \begin{bmatrix} 1 & 0 & 0 \\ 1 & 1 & 1 \\ 1 & 1 & 2 \end{bmatrix}.$$

Again, P is row-equivalent to the identity matrix, so the eigenvectors $\mathbf{p}_1$, $\mathbf{p}_2$, and $\mathbf{p}_3$ are linearly independent.

The second part of the proof of Theorem 7.5 and Example 3 suggest the steps for diagonalizing a matrix listed below.

Steps for Diagonalizing a Square Matrix

Let A be an $n \times n$ matrix.

1. Find n linearly independent eigenvectors $\mathbf{p}_1, \mathbf{p}_2, \ldots, \mathbf{p}_n$ for A (if possible) with corresponding eigenvalues $\lambda_1, \lambda_2, \ldots, \lambda_n$. If n linearly independent eigenvectors do not exist, then A is not diagonalizable.
2. Let P be the $n \times n$ matrix whose columns consist of these eigenvectors. That is, $P = [\mathbf{p}_1 \quad \mathbf{p}_2 \cdots \mathbf{p}_n]$.
3. The diagonal matrix $D = P^{-1}AP$ will have the eigenvalues $\lambda_1, \lambda_2, \ldots, \lambda_n$ on its main diagonal. Note that the order of the eigenvectors used to form P will determine the order in which the eigenvalues appear on the main diagonal of D.

EXAMPLE 4   **Diagonalizing a Matrix**

Show that the matrix A is diagonalizable.

$$A = \begin{bmatrix} 1 & -1 & -1 \\ 1 & 3 & 1 \\ -3 & 1 & -1 \end{bmatrix}$$

Then find a matrix P such that $P^{-1}AP$ is diagonal.

SOLUTION

The characteristic polynomial of A is $|\lambda I - A| = (\lambda - 2)(\lambda + 2)(\lambda - 3)$. (Verify this.) So, the eigenvalues of A are $\lambda_1 = 2$, $\lambda_2 = -2$, and $\lambda_3 = 3$. From these eigenvalues, you obtain the reduced row-echelon forms and corresponding eigenvectors below.

Eigenvector

$$2I - A = \begin{bmatrix} 1 & 1 & 1 \\ -1 & -1 & -1 \\ 3 & -1 & 3 \end{bmatrix} \rightarrow \begin{bmatrix} 1 & 0 & 1 \\ 0 & 1 & 0 \\ 0 & 0 & 0 \end{bmatrix} \quad \begin{bmatrix} -1 \\ 0 \\ 1 \end{bmatrix}$$

$$-2I - A = \begin{bmatrix} -3 & 1 & 1 \\ -1 & -5 & -1 \\ 3 & -1 & -1 \end{bmatrix} \rightarrow \begin{bmatrix} 1 & 0 & -\frac{1}{4} \\ 0 & 1 & \frac{1}{4} \\ 0 & 0 & 0 \end{bmatrix} \quad \begin{bmatrix} 1 \\ -1 \\ 4 \end{bmatrix}$$

$$3I - A = \begin{bmatrix} 2 & 1 & 1 \\ -1 & 0 & -1 \\ 3 & -1 & 4 \end{bmatrix} \rightarrow \begin{bmatrix} 1 & 0 & 1 \\ 0 & 1 & -1 \\ 0 & 0 & 0 \end{bmatrix} \quad \begin{bmatrix} -1 \\ 1 \\ 1 \end{bmatrix}$$

Form the matrix P whose columns are the eigenvectors just obtained.

$$P = \begin{bmatrix} -1 & 1 & -1 \\ 0 & -1 & 1 \\ 1 & 4 & 1 \end{bmatrix}$$

This matrix is nonsingular (check this), which implies that the eigenvectors are linearly independent and A is diagonalizable. So, it follows that

$$P^{-1}AP = \begin{bmatrix} 2 & 0 & 0 \\ 0 & -2 & 0 \\ 0 & 0 & 3 \end{bmatrix}.$$

EXAMPLE 5 **Diagonalizing a Matrix**

Show that the matrix A is diagonalizable.

$$A = \begin{bmatrix} 1 & 0 & 0 & 0 \\ 0 & 1 & 5 & -10 \\ 1 & 0 & 2 & 0 \\ 1 & 0 & 0 & 3 \end{bmatrix}$$

Then find a matrix P such that $P^{-1}AP$ is diagonal.

SOLUTION

In Example 6 in Section 7.1, you found that the three eigenvalues of A are $\lambda_1 = 1$, $\lambda_2 = 2$, and $\lambda_3 = 3$, and that they have the eigenvectors listed below.

$$\lambda_1: \begin{bmatrix} 0 \\ 1 \\ 0 \\ 0 \end{bmatrix}, \begin{bmatrix} -2 \\ 0 \\ 2 \\ 1 \end{bmatrix} \quad \lambda_2: \begin{bmatrix} 0 \\ 5 \\ 1 \\ 0 \end{bmatrix} \quad \lambda_3: \begin{bmatrix} 0 \\ -5 \\ 0 \\ 1 \end{bmatrix}$$

The matrix whose columns consist of these eigenvectors is

$$P = \begin{bmatrix} 0 & -2 & 0 & 0 \\ 1 & 0 & 5 & -5 \\ 0 & 2 & 1 & 0 \\ 0 & 1 & 0 & 1 \end{bmatrix}.$$

P is invertible (check this), so its column vectors form a linearly independent set. This means that A is diagonalizable, and

$$P^{-1}AP = \begin{bmatrix} 1 & 0 & 0 & 0 \\ 0 & 1 & 0 & 0 \\ 0 & 0 & 2 & 0 \\ 0 & 0 & 0 & 3 \end{bmatrix}.$$

EXAMPLE 6 **A Matrix That Is Not Diagonalizable**

Show that the matrix A is not diagonalizable.

$$A = \begin{bmatrix} 1 & 2 \\ 0 & 1 \end{bmatrix}$$

SOLUTION

A is triangular, so the eigenvalues are the entries on the main diagonal. The only eigenvalue is $\lambda = 1$. The matrix $(I - A)$ has the reduced row-echelon form below.

$$I - A = \begin{bmatrix} 0 & -2 \\ 0 & 0 \end{bmatrix} \longrightarrow \begin{bmatrix} 0 & 1 \\ 0 & 0 \end{bmatrix}$$

This implies that $x_2 = 0$, and letting $x_1 = t$, you can write every eigenvector of A in the form

$$\mathbf{x} = \begin{bmatrix} x_1 \\ x_2 \end{bmatrix} = \begin{bmatrix} t \\ 0 \end{bmatrix} = t \begin{bmatrix} 1 \\ 0 \end{bmatrix}.$$

So, A does not have two linearly independent eigenvectors, and you can conclude that A is not diagonalizable.

REMARK

The condition in Theorem 7.6 is sufficient but not necessary for diagonalization, as demonstrated in Example 5. In other words, a diagonalizable matrix need not have distinct eigenvalues.

For a square matrix A of order n to be diagonalizable, the sum of the dimensions of the eigenspaces must be equal to n. This can happen when A has n distinct eigenvalues. So, you have the next theorem.

THEOREM 7.6   Sufficient Condition for Diagonalization

If an $n \times n$ matrix A has n *distinct* eigenvalues, then the corresponding eigenvectors are linearly independent and A is diagonalizable.

PROOF

Let $\lambda_1, \lambda_2, \ldots, \lambda_n$ be n distinct eigenvalues of A with corresponding eigenvectors $\mathbf{x}_1$, $\mathbf{x}_2, \ldots, \mathbf{x}_n$. To begin, assume the set of eigenvectors is linearly dependent. Moreover, consider the eigenvectors to be ordered so that the first m eigenvectors are linearly independent, but the first $m + 1$ are linearly dependent, where $m < n$. Then $\mathbf{x}_{m+1}$ can be written as a linear combination of the first m eigenvectors:

$$\mathbf{x}_{m+1} = c_1\mathbf{x}_1 + c_2\mathbf{x}_2 + \cdots + c_m\mathbf{x}_m \qquad \text{Equation 1}$$

where the c_i's are not all zero. Multiplication of both sides of Equation 1 by A yields

$$A\mathbf{x}_{m+1} = Ac_1\mathbf{x}_1 + Ac_2\mathbf{x}_2 + \cdots + Ac_m\mathbf{x}_m.$$

Now $A\mathbf{x}_i = \lambda_i\mathbf{x}_i$, $i = 1, 2, \ldots, m + 1$, so you have

$$\lambda_{m+1}\mathbf{x}_{m+1} = c_1\lambda_1\mathbf{x}_1 + c_2\lambda_2\mathbf{x}_2 + \cdots + c_m\lambda_m\mathbf{x}_m. \qquad \text{Equation 2}$$

Multiplication of Equation 1 by λ_{m+1} yields

$$\lambda_{m+1}\mathbf{x}_{m+1} = c_1\lambda_{m+1}\mathbf{x}_1 + c_2\lambda_{m+1}\mathbf{x}_2 + \cdots + c_m\lambda_{m+1}\mathbf{x}_m. \qquad \text{Equation 3}$$

Subtracting Equation 2 from Equation 3 produces

$$c_1(\lambda_{m+1} - \lambda_1)\mathbf{x}_1 + c_2(\lambda_{m+1} - \lambda_2)\mathbf{x}_2 + \cdots + c_m(\lambda_{m+1} - \lambda_m)\mathbf{x}_m = \mathbf{0}$$

and, using the fact that the first m eigenvectors are linearly independent, all coefficients of this equation must be zero. That is,

$$c_1(\lambda_{m+1} - \lambda_1) = c_2(\lambda_{m+1} - \lambda_2) = \cdots = c_m(\lambda_{m+1} - \lambda_m) = 0.$$

All the eigenvalues are distinct, so it follows that $c_i = 0$, $i = 1, 2, \ldots, m$. But this result contradicts our assumption that $\mathbf{x}_{m+1}$ can be written as a linear combination of the first m eigenvectors. So, the set of eigenvectors is linearly independent, and from Theorem 7.5, you can conclude that A is diagonalizable. ∎

EXAMPLE 7   Determining Whether a Matrix Is Diagonalizable

See LarsonLinearAlgebra.com for an interactive version of this type of example.

Determine whether the matrix A is diagonalizable.

$$A = \begin{bmatrix} 1 & -2 & 1 \\ 0 & 0 & 1 \\ 0 & 0 & -3 \end{bmatrix}$$

SOLUTION

A is a triangular matrix, so its eigenvalues are the main diagonal entries $\lambda_1 = 1, \lambda_2 = 0$, and $\lambda_3 = -3$. Moreover, these three values are distinct, so you can conclude from Theorem 7.6 that A is diagonalizable.

DIAGONALIZATION AND LINEAR TRANSFORMATIONS

So far in this section, the diagonalization problem has been in terms of matrices. In terms of linear transformations, the diagonalization problem can be stated as: For a linear transformation

$$T: V \rightarrow V$$

does there exist a basis B for V such that the matrix for T relative to B is diagonal? The answer is "yes" when the standard matrix for T is diagonalizable.

EXAMPLE 8   Finding a Basis

Let $T: R^3 \rightarrow R^3$ be the linear transformation represented by

$$T(x_1, x_2, x_3) = (x_1 - x_2 - x_3, x_1 + 3x_2 + x_3, -3x_1 + x_2 - x_3).$$

If possible, find a basis B for R^3 such that the matrix for T relative to B is diagonal.

SOLUTION

The standard matrix for T is

$$A = \begin{bmatrix} 1 & -1 & -1 \\ 1 & 3 & 1 \\ -3 & 1 & -1 \end{bmatrix}.$$

From Example 4, you know that A is diagonalizable. So, the three linearly independent eigenvectors found in Example 4 can be used to form the basis B. That is,

$$B = \{(-1, 0, 1), (1, -1, 4), (-1, 1, 1)\}.$$

The matrix for T relative to this basis is

$$D = \begin{bmatrix} 2 & 0 & 0 \\ 0 & -2 & 0 \\ 0 & 0 & 3 \end{bmatrix}.$$

LINEAR ALGEBRA APPLIED

Genetics is the science of heredity. A mixture of chemistry and biology, genetics attempts to explain hereditary evolution and gene movement between generations based on the deoxyribonucleic acid (DNA) of a species. Research in the area of genetics called *population genetics*, which focuses on genetic structures of specific populations, is especially popular today. Such research has led to a better understanding of the types of genetic inheritance. For example, in humans, one type of genetic inheritance is called *X-linked inheritance* (or *sex-linked inheritance*), which refers to recessive genes on the *X* chromosome. Males have one *X* and one *Y* chromosome, and females have two *X* chromosomes. If a male has a defective gene on the *X* chromosome, then its corresponding trait will be expressed because there is not a normal gene on the *Y* chromosome to suppress its activity. With females, the trait will not be expressed unless it is present on both *X* chromosomes, which is rare. This is why inherited diseases or conditions are usually found in males, hence the term *sex-linked inheritance*. Some of these include hemophilia A, Duchenne muscular dystrophy, red-green color blindness, and male pattern baldness. Matrix eigenvalues and diagonalization can be useful for coming up with mathematical models to describe *X*-linked inheritance in a population.

7.2   Exercises

See CalcChat.com for worked-out solutions to odd-numbered exercises.

Diagonalizable Matrices and Eigenvalues In Exercises 1–6, (a) verify that A is diagonalizable by finding $P^{-1}AP$, and (b) use the result of part (a) and Theorem 7.4 to find the eigenvalues of A.

1. $A = \begin{bmatrix} -11 & 36 \\ -3 & 10 \end{bmatrix}$, $P = \begin{bmatrix} -3 & -4 \\ -1 & -1 \end{bmatrix}$

2. $A = \begin{bmatrix} 1 & 3 \\ -1 & 5 \end{bmatrix}$, $P = \begin{bmatrix} 3 & 1 \\ 1 & 1 \end{bmatrix}$

3. $A = \begin{bmatrix} 3 & -2 \\ 2 & -2 \end{bmatrix}$, $P = \begin{bmatrix} 1 & 2 \\ 2 & 1 \end{bmatrix}$

4. $A = \begin{bmatrix} 4 & -5 \\ 2 & -3 \end{bmatrix}$, $P = \begin{bmatrix} 1 & 5 \\ 1 & 2 \end{bmatrix}$

5. $A = \begin{bmatrix} -1 & 1 & 0 \\ 0 & 3 & 0 \\ 4 & -2 & 5 \end{bmatrix}$, $P = \begin{bmatrix} 0 & 1 & -3 \\ 0 & 4 & 0 \\ 1 & 2 & 2 \end{bmatrix}$

6. $A = \begin{bmatrix} 0.80 & 0.10 & 0.05 & 0.05 \\ 0.10 & 0.80 & 0.05 & 0.05 \\ 0.05 & 0.05 & 0.80 & 0.10 \\ 0.05 & 0.05 & 0.10 & 0.80 \end{bmatrix}$,

$P = \begin{bmatrix} 1 & -1 & 0 & 1 \\ 1 & -1 & 0 & -1 \\ 1 & 1 & 1 & 0 \\ 1 & 1 & -1 & 0 \end{bmatrix}$

Diagonalizing a Matrix In Exercises 7–14, find (if possible) a nonsingular matrix P such that $P^{-1}AP$ is diagonal. Verify that $P^{-1}AP$ is a diagonal matrix with the eigenvalues on the main diagonal.

7. $A = \begin{bmatrix} 6 & -3 \\ -2 & 1 \end{bmatrix}$
(See Exercise 15, Section 7.1.)

8. $A = \begin{bmatrix} \frac{1}{4} & \frac{1}{4} \\ \frac{1}{2} & 0 \end{bmatrix}$
(See Exercise 20, Section 7.1.)

9. $A = \begin{bmatrix} 2 & -2 & 3 \\ 0 & 3 & -2 \\ 0 & -1 & 2 \end{bmatrix}$
(See Exercise 21, Section 7.1.)

10. $A = \begin{bmatrix} 3 & 2 & 1 \\ 0 & 0 & 2 \\ 0 & 2 & 0 \end{bmatrix}$
(See Exercise 22, Section 7.1.)

11. $A = \begin{bmatrix} 1 & 2 & -2 \\ -2 & 5 & -2 \\ -6 & 6 & -3 \end{bmatrix}$
(See Exercise 23, Section 7.1.)

12. $A = \begin{bmatrix} 3 & 2 & -3 \\ -3 & -4 & 9 \\ -1 & -2 & 5 \end{bmatrix}$
(See Exercise 24, Section 7.1.)

13. $A = \begin{bmatrix} 1 & 0 & 0 \\ 1 & 2 & 1 \\ 1 & 0 & 2 \end{bmatrix}$

14. $A = \begin{bmatrix} 2 & 0 & 0 \\ 4 & 4 & 0 \\ 0 & 4 & 4 \end{bmatrix}$

Showing That a Matrix Is Not Diagonalizable In Exercises 15–22, show that the matrix is not diagonalizable.

15. $\begin{bmatrix} 0 & 0 \\ 5 & 0 \end{bmatrix}$

16. $\begin{bmatrix} 1 & \frac{1}{2} \\ -2 & -1 \end{bmatrix}$

17. $\begin{bmatrix} 7 & 7 \\ 0 & 7 \end{bmatrix}$

18. $\begin{bmatrix} 1 & 0 \\ -2 & 1 \end{bmatrix}$

19. $\begin{bmatrix} 1 & -2 & 1 \\ 0 & 1 & 4 \\ 0 & 0 & 2 \end{bmatrix}$

20. $\begin{bmatrix} 3 & 2 & -2 \\ 0 & -2 & 3 \\ 0 & 0 & -2 \end{bmatrix}$

21. $\begin{bmatrix} 1 & 0 & -1 & 1 \\ 0 & 1 & 0 & 1 \\ -2 & 0 & 2 & -2 \\ 0 & 2 & 0 & 2 \end{bmatrix}$
(See Exercise 39, Section 7.1.)

22. $\begin{bmatrix} 1 & -3 & 3 & 3 \\ -1 & 4 & -3 & -3 \\ -2 & 0 & 1 & 1 \\ 1 & 0 & 0 & 0 \end{bmatrix}$
(See Exercise 40, Section 7.1.)

Determining a Sufficient Condition for Diagonalization In Exercises 23–26, find the eigenvalues of the matrix and determine whether there is a sufficient number of eigenvalues to guarantee that the matrix is diagonalizable by Theorem 7.6.

23. $\begin{bmatrix} 1 & 1 \\ 1 & 1 \end{bmatrix}$

24. $\begin{bmatrix} 2 & 0 \\ 5 & 2 \end{bmatrix}$

25. $\begin{bmatrix} -3 & -2 & 3 \\ 3 & 4 & -9 \\ 1 & 2 & -5 \end{bmatrix}$

26. $\begin{bmatrix} 4 & 3 & -2 \\ 0 & 1 & 1 \\ 0 & 0 & -2 \end{bmatrix}$

Finding a Basis In Exercises 27–30, find a basis B for the domain of T such that the matrix for T relative to B is diagonal.

27. $T: R^2 \to R^2: T(x, y) = (x + y, x + y)$

28. $T: R^3 \to R^3:$
$T(x, y, z) = (-2x + 2y - 3z, 2x + y - 6z, -x - 2y)$

29. $T: P_1 \to P_1: T(a + bx) = a + (a + 2b)x$

30. $T: P_2 \to P_2:$
$T(c + bx + ax^2) = (3c + a) + (2b + 3a)x + ax^2$

31. **Proof** Let A be a diagonalizable $n \times n$ matrix and let P be an invertible $n \times n$ matrix such that $B = P^{-1}AP$ is the diagonal form of A. Prove that $A^k = PB^kP^{-1}$, where k is a positive integer.

32. Let $\lambda_1, \lambda_2, \ldots, \lambda_n$ be n distinct eigenvalues of an $n \times n$ matrix A. Use the result of Exercise 31 to find the eigenvalues of A^k.

Finding a Power of a Matrix In Exercises 33–36, use the result of Exercise 31 to find the power of A shown.

33. $A = \begin{bmatrix} 10 & 18 \\ -6 & -11 \end{bmatrix}, A^6$ **34.** $A = \begin{bmatrix} 1 & 3 \\ 2 & 0 \end{bmatrix}, A^7$

35. $A = \begin{bmatrix} 2 & 0 & -2 \\ 0 & 2 & -2 \\ 3 & 0 & -3 \end{bmatrix}, A^5$

36. $A = \begin{bmatrix} 2 & 3 & -2 \\ -2 & -5 & 0 \\ -2 & -1 & 4 \end{bmatrix}, A^8$

True or False? In Exercises 37 and 38, determine whether each statement is true or false. If a statement is true, give a reason or cite an appropriate statement from the text. If a statement is false, provide an example that shows the statement is not true in all cases or cite an appropriate statement from the text.

37. (a) If A and B are similar $n \times n$ matrices, then they always have the same characteristic polynomial equation.

(b) The fact that an $n \times n$ matrix A has n distinct eigenvalues does not guarantee that A is diagonalizable.

38. (a) If A is a diagonalizable matrix, then it has n linearly independent eigenvectors.

(b) If an $n \times n$ matrix A is diagonalizable, then it must have n distinct eigenvalues.

39. Are the two matrices similar? If so, find a matrix P such that $B = P^{-1}AP$.

$$A = \begin{bmatrix} 1 & 0 & 0 \\ 0 & 2 & 0 \\ 0 & 0 & 3 \end{bmatrix} \quad B = \begin{bmatrix} 3 & 0 & 0 \\ 0 & 2 & 0 \\ 0 & 0 & 1 \end{bmatrix}$$

40. Calculus For a real number x, you can define e^x by the series

$$e^x = 1 + x + \frac{x^2}{2!} + \frac{x^3}{3!} + \frac{x^4}{4!} + \cdots.$$

In a similar way, for a square matrix X, you can define e^X by the series

$$e^X = I + X + \frac{1}{2!}X^2 + \frac{1}{3!}X^3 + \frac{1}{4!}X^4 + \cdots.$$

Evaluate e^X, where X is the square matrix shown.

(a) $X = \begin{bmatrix} 1 & 0 \\ 0 & 1 \end{bmatrix}$ (b) $X = \begin{bmatrix} 1 & 0 \\ 1 & 0 \end{bmatrix}$

(c) $X = \begin{bmatrix} 0 & 1 \\ 1 & 0 \end{bmatrix}$ (d) $X = \begin{bmatrix} 2 & 0 \\ 0 & -2 \end{bmatrix}$

41. Writing Can a matrix be similar to two different diagonal matrices? Explain.

42. Proof Prove that if matrix A is diagonalizable, then A^T is diagonalizable.

43. Proof Prove that if matrix A is diagonalizable with n real eigenvalues $\lambda_1, \lambda_2, \ldots, \lambda_n$, then $|A| = \lambda_1\lambda_2 \cdots \lambda_n$.

44. Proof Prove that the matrix

$$A = \begin{bmatrix} a & b \\ c & d \end{bmatrix}$$

is diagonalizable when $-4bc < (a - d)^2$ and is not diagonalizable when $-4bc > (a - d)^2$.

45. Guided Proof Prove that if the eigenvalues of a diagonalizable matrix A are all ± 1, then the matrix is equal to its inverse.

Getting Started: To show that the matrix is equal to its inverse, use the fact that there exists an invertible matrix P such that $D = P^{-1}AP$, where D is a diagonal matrix with ± 1 along its main diagonal.

(i) Let $D = P^{-1}AP$, where D is a diagonal matrix with ± 1 along its main diagonal.

(ii) Find A in terms of P, P^{-1}, and D.

(iii) Use the properties of the inverse of a product of matrices and the fact that D is diagonal to expand to find A^{-1}.

(iv) Conclude that $A^{-1} = A$.

46. Guided Proof Prove that nonzero nilpotent matrices are not diagonalizable.

Getting Started: From Exercise 80 in Section 7.1, you know that 0 is the only eigenvalue of the nilpotent matrix A. Show that it is impossible for A to be diagonalizable.

(i) Assume A is diagonalizable, so there exists an invertible matrix P such that $P^{-1}AP = D$, where D is the zero matrix.

(ii) Find A in terms of P, P^{-1}, and D.

(iii) Find a contradiction and conclude that nonzero nilpotent matrices are not diagonalizable.

47. Proof Prove that if A is a nonsingular diagonalizable matrix, then A^{-1} is also diagonalizable.

48. CAPSTONE Explain how to determine whether an $n \times n$ matrix A is diagonalizable using (a) similar matrices, (b) eigenvectors, and (c) distinct eigenvalues.

Showing That a Matrix Is Not Diagonalizable In Exercises 49 and 50, show that the matrix is not diagonalizable.

49. $\begin{bmatrix} 4 & k \\ 0 & 4 \end{bmatrix}, k \neq 0$ **50.** $\begin{bmatrix} 0 & 0 \\ k & 0 \end{bmatrix}, k \neq 0$

7.3   Symmetric Matrices and Orthogonal Diagonalization

   ■ Recognize, and apply properties of, symmetric matrices.

   ■ Recognize, and apply properties of, orthogonal matrices.

   ■ Find an orthogonal matrix *P* that orthogonally diagonalizes a symmetric matrix *A*.

SYMMETRIC MATRICES

For most matrices, you must go through much of the diagonalization process before determining whether diagonalization is possible. One exception is with a triangular matrix that has distinct entries on the main diagonal. Such a matrix can be recognized as diagonalizable by inspection. In this section, you will study another type of matrix that is guaranteed to be diagonalizable: a **symmetric** matrix. Recall the definition below.

Definition of a Symmetric Matrix

A square matrix *A* is **symmetric** when it is equal to its transpose: $A = A^T$.

DISCOVERY

1. Pick an arbitrary nonsymmetric square matrix and calculate its eigenvalues.

2. Can you find a nonsymmetric square matrix for which the eigenvalues are not real?

3. Now pick an arbitrary symmetric matrix and calculate its eigenvalues.

4. Can you find a symmetric matrix for which the eigenvalues are not real?

5. What can you conclude about the eigenvalues of a symmetric matrix?

See LarsonLinearAlgebra.com for an interactive version of this type of exercise.

EXAMPLE 1   Symmetric Matrices and Nonsymmetric Matrices

The matrices *A* and *B* are symmetric, but the matrix *C* is not.

$$A = \begin{bmatrix} 0 & 1 & -2 \\ 1 & 3 & 0 \\ -2 & 0 & 5 \end{bmatrix} \qquad B = \begin{bmatrix} 4 & 3 \\ 3 & 1 \end{bmatrix} \qquad C = \begin{bmatrix} 3 & 2 & 1 \\ 1 & -4 & 0 \\ 1 & 0 & 5 \end{bmatrix}$$

  Nonsymmetric matrices have properties that are not exhibited by symmetric matrices, as listed below.

1. A nonsymmetric matrix may not be diagonalizable.

2. A nonsymmetric matrix can have eigenvalues that are not real. For example, the matrix

$$A = \begin{bmatrix} 0 & -1 \\ 1 & 0 \end{bmatrix}$$

has a characteristic equation of $\lambda^2 + 1 = 0$. So, its eigenvalues are the imaginary numbers $\lambda_1 = i$ and $\lambda_2 = -i$.

3. For a nonsymmetric matrix, the number of linearly independent eigenvectors corresponding to an eigenvalue can be less than the multiplicity of the eigenvalue. (See Example 6, Section 7.2.)

Theorem 7.7 lists properties of *symmetric* matrices.

REMARK

Theorem 7.7 is called the **Real Spectral Theorem,** and the set of eigenvalues of *A* is called the **spectrum** of *A*.

THEOREM 7.7   Properties of Symmetric Matrices

If *A* is an $n \times n$ symmetric matrix, then the properties listed below are true.

1. *A* is diagonalizable.

2. All eigenvalues of *A* are real.

3. If λ is an eigenvalue of *A* with multiplicity *k*, then λ has *k* linearly independent eigenvectors. That is, the eigenspace of λ has dimension *k*.

A proof of Theorem 7.7 is beyond the scope of this text. The next example proves that every 2×2 symmetric matrix is diagonalizable.

EXAMPLE 2 **Every 2 × 2 Symmetric Matrix Is Diagonalizable**

Prove that a symmetric matrix

$$A = \begin{bmatrix} a & c \\ c & b \end{bmatrix}$$

is diagonalizable.

SOLUTION

The characteristic polynomial of A is

$$|\lambda I - A| = \begin{vmatrix} \lambda - a & -c \\ -c & \lambda - b \end{vmatrix}$$
$$= \lambda^2 - (a + b)\lambda + ab - c^2.$$

As a quadratic in λ, this polynomial has a discriminant of

$$(a + b)^2 - 4(ab - c^2) = a^2 + 2ab + b^2 - 4ab + 4c^2$$
$$= a^2 - 2ab + b^2 + 4c^2$$
$$= (a - b)^2 + 4c^2.$$

This discriminant is the sum of two squares, so it must be either zero or positive. If $(a - b)^2 + 4c^2 = 0$, then $a = b$ and $c = 0$, so A is already diagonal. That is,

$$A = \begin{bmatrix} a & 0 \\ 0 & a \end{bmatrix}.$$

On the other hand, if $(a - b)^2 + 4c^2 > 0$, then by the Quadratic Formula the characteristic polynomial of A has two distinct real roots, which means that A has two distinct real eigenvalues. So, A is diagonalizable in this case as well. ■

EXAMPLE 3 **Dimensions of the Eigenspaces of a Symmetric Matrix**

Find the eigenvalues of the symmetric matrix

$$A = \begin{bmatrix} 1 & -2 & 0 & 0 \\ -2 & 1 & 0 & 0 \\ 0 & 0 & 1 & -2 \\ 0 & 0 & -2 & 1 \end{bmatrix}$$

and determine the dimensions of the corresponding eigenspaces.

SOLUTION

The characteristic polynomial of A is

$$|\lambda I - A| = \begin{vmatrix} \lambda - 1 & 2 & 0 & 0 \\ 2 & \lambda - 1 & 0 & 0 \\ 0 & 0 & \lambda - 1 & 2 \\ 0 & 0 & 2 & \lambda - 1 \end{vmatrix} = (\lambda + 1)^2(\lambda - 3)^2.$$

So, the eigenvalues of A are $\lambda_1 = -1$ and $\lambda_2 = 3$. Each of these eigenvalues has a multiplicity of 2, so you know from Theorem 7.7 that the corresponding eigenspaces also have dimension 2. Specifically, the eigenspace of $\lambda_1 = -1$ has a basis of $B_1 = \{(1, 1, 0, 0), (0, 0, 1, 1)\}$ and the eigenspace of $\lambda_2 = 3$ has a basis of $B_2 = \{(1, -1, 0, 0), (0, 0, 1, -1)\}$. (Verify these.) ■

ORTHOGONAL MATRICES

To diagonalize a square matrix A, you need to find an *invertible* matrix P such that $P^{-1}AP$ is diagonal. For symmetric matrices, the matrix P can be chosen to have the special property that $P^{-1} = P^T$. This unusual matrix property is defined below.

Definition of an Orthogonal Matrix

A square matrix P is **orthogonal** when it is invertible and $P^{-1} = P^T$.

EXAMPLE 4   Orthogonal Matrices

a. The matrix $P = \begin{bmatrix} 0 & 1 \\ -1 & 0 \end{bmatrix}$ is orthogonal because $P^{-1} = P^T = \begin{bmatrix} 0 & -1 \\ 1 & 0 \end{bmatrix}$.

b. The matrix

$$P = \begin{bmatrix} \frac{3}{5} & 0 & -\frac{4}{5} \\ 0 & 1 & 0 \\ \frac{4}{5} & 0 & \frac{3}{5} \end{bmatrix}$$

is orthogonal because

$$P^{-1} = P^T = \begin{bmatrix} \frac{3}{5} & 0 & \frac{4}{5} \\ 0 & 1 & 0 \\ -\frac{4}{5} & 0 & \frac{3}{5} \end{bmatrix}.$$

In Example 4, the columns of the matrices P form orthonormal sets in R^2 and R^3, respectively (verify this), which suggests the next theorem.

THEOREM 7.8   Property of Orthogonal Matrices

An $n \times n$ matrix P is orthogonal if and only if its column vectors form an orthonormal set.

PROOF

To prove the theorem in one direction, assume that the column vectors of P form an orthonormal set:

$$P = \begin{bmatrix} \mathbf{p}_1 & \mathbf{p}_2 & \cdots & \mathbf{p}_n \end{bmatrix}$$

$$= \begin{bmatrix} p_{11} & p_{12} & \cdots & p_{1n} \\ p_{21} & p_{22} & \cdots & p_{2n} \\ \vdots & \vdots & & \vdots \\ p_{n1} & p_{n2} & \cdots & p_{nn} \end{bmatrix}.$$

Then the product P^TP has the form

$$P^TP = \begin{bmatrix} \mathbf{p}_1 \cdot \mathbf{p}_1 & \mathbf{p}_1 \cdot \mathbf{p}_2 & \cdots & \mathbf{p}_1 \cdot \mathbf{p}_n \\ \mathbf{p}_2 \cdot \mathbf{p}_1 & \mathbf{p}_2 \cdot \mathbf{p}_2 & \cdots & \mathbf{p}_2 \cdot \mathbf{p}_n \\ \vdots & \vdots & & \vdots \\ \mathbf{p}_n \cdot \mathbf{p}_1 & \mathbf{p}_n \cdot \mathbf{p}_2 & \cdots & \mathbf{p}_n \cdot \mathbf{p}_n \end{bmatrix}.$$

The set

$$\{\mathbf{p}_1, \mathbf{p}_2, \ldots, \mathbf{p}_n\}$$

is orthonormal, so you have

$$\mathbf{p}_i \cdot \mathbf{p}_j = 0, i \neq j \quad \text{and} \quad \mathbf{p}_i \cdot \mathbf{p}_i = \|\mathbf{p}_i\|^2 = 1.$$

So, the matrix composed of dot products has the form

$$P^T P = \begin{bmatrix} 1 & 0 & \cdots & 0 \\ 0 & 1 & \cdots & 0 \\ \vdots & \vdots & & \vdots \\ 0 & 0 & \cdots & 1 \end{bmatrix} = I_n.$$

This implies that $P^T = P^{-1}$, so P is orthogonal.

Conversely, if P is orthogonal, then reverse the steps above to verify that the column vectors of P form an orthonormal set.

EXAMPLE 5   **An Orthogonal Matrix**

Show that

$$P = \begin{bmatrix} \dfrac{1}{3} & \dfrac{2}{3} & \dfrac{2}{3} \\[2mm] -\dfrac{2}{\sqrt{5}} & \dfrac{1}{\sqrt{5}} & 0 \\[2mm] -\dfrac{2}{3\sqrt{5}} & -\dfrac{4}{3\sqrt{5}} & \dfrac{5}{3\sqrt{5}} \end{bmatrix}$$

is orthogonal by showing that $P^T = P^{-1}$. Then show that the column vectors of P form an orthonormal set.

SOLUTION

$$PP^T = \begin{bmatrix} \dfrac{1}{3} & \dfrac{2}{3} & \dfrac{2}{3} \\[2mm] -\dfrac{2}{\sqrt{5}} & \dfrac{1}{\sqrt{5}} & 0 \\[2mm] -\dfrac{2}{3\sqrt{5}} & -\dfrac{4}{3\sqrt{5}} & \dfrac{5}{3\sqrt{5}} \end{bmatrix} \begin{bmatrix} \dfrac{1}{3} & -\dfrac{2}{\sqrt{5}} & -\dfrac{2}{3\sqrt{5}} \\[2mm] \dfrac{2}{3} & \dfrac{1}{\sqrt{5}} & -\dfrac{4}{3\sqrt{5}} \\[2mm] \dfrac{2}{3} & 0 & \dfrac{5}{3\sqrt{5}} \end{bmatrix} = I_3$$

so it follows that $P^T = P^{-1}$. Moreover, letting

$$\mathbf{p}_1 = \begin{bmatrix} \dfrac{1}{3} \\[2mm] -\dfrac{2}{\sqrt{5}} \\[2mm] -\dfrac{2}{3\sqrt{5}} \end{bmatrix}, \; \mathbf{p}_2 = \begin{bmatrix} \dfrac{2}{3} \\[2mm] \dfrac{1}{\sqrt{5}} \\[2mm] -\dfrac{4}{3\sqrt{5}} \end{bmatrix}, \; \text{and} \; \mathbf{p}_3 = \begin{bmatrix} \dfrac{2}{3} \\[2mm] 0 \\[2mm] \dfrac{5}{3\sqrt{5}} \end{bmatrix}$$

produces

$$\mathbf{p}_1 \cdot \mathbf{p}_2 = \mathbf{p}_1 \cdot \mathbf{p}_3 = \mathbf{p}_2 \cdot \mathbf{p}_3 = 0$$

and

$$\|\mathbf{p}_1\| = \|\mathbf{p}_2\| = \|\mathbf{p}_3\| = 1.$$

So, $\{\mathbf{p}_1, \mathbf{p}_2, \mathbf{p}_3\}$ is an orthonormal set, as guaranteed by Theorem 7.8.

It can be shown that for a symmetric matrix, the eigenvectors corresponding to distinct eigenvalues are orthogonal. The next theorem states this property.

THEOREM 7.9   Property of Symmetric Matrices

Let A be an $n \times n$ symmetric matrix. If λ_1 and λ_2 are distinct eigenvalues of A, then their corresponding eigenvectors $\mathbf{x}_1$ and $\mathbf{x}_2$ are orthogonal.

PROOF

Let λ_1 and λ_2 be distinct eigenvalues of A with corresponding eigenvectors $\mathbf{x}_1$ and $\mathbf{x}_2$. So, $A\mathbf{x}_1 = \lambda_1\mathbf{x}_1$ and $A\mathbf{x}_2 = \lambda_2\mathbf{x}_2$. To prove the theorem, use the matrix form of the dot product, $\mathbf{x}_1 \cdot \mathbf{x}_2 = \mathbf{x}_1^T\mathbf{x}_2$. (See Section 5.1.) Now you can write

$$
\begin{aligned}
\lambda_1(\mathbf{x}_1 \cdot \mathbf{x}_2) &= (\lambda_1\mathbf{x}_1) \cdot \mathbf{x}_2 \\
&= (A\mathbf{x}_1) \cdot \mathbf{x}_2 \\
&= (A\mathbf{x}_1)^T\mathbf{x}_2 \\
&= (\mathbf{x}_1^T A^T)\mathbf{x}_2 \\
&= (\mathbf{x}_1^T A)\mathbf{x}_2 \qquad A \text{ is symmetric, so } A = A^T. \\
&= \mathbf{x}_1^T(A\mathbf{x}_2) \\
&= \mathbf{x}_1^T(\lambda_2\mathbf{x}_2) \\
&= \mathbf{x}_1 \cdot (\lambda_2\mathbf{x}_2) \\
&= \lambda_2(\mathbf{x}_1 \cdot \mathbf{x}_2).
\end{aligned}
$$

This implies that $(\lambda_1 - \lambda_2)(\mathbf{x}_1 \cdot \mathbf{x}_2) = 0$. Also, $\lambda_1 \neq \lambda_2$, so it follows that $\mathbf{x}_1 \cdot \mathbf{x}_2 = 0$, which means that $\mathbf{x}_1$ and $\mathbf{x}_2$ are orthogonal.

EXAMPLE 6   **Eigenvectors of a Symmetric Matrix**

Show that any two eigenvectors of

$$
A = \begin{bmatrix} 3 & 1 \\ 1 & 3 \end{bmatrix}
$$

corresponding to distinct eigenvalues are orthogonal.

SOLUTION

The characteristic polynomial of A is

$$
|\lambda I - A| = \begin{vmatrix} \lambda - 3 & -1 \\ -1 & \lambda - 3 \end{vmatrix} = (\lambda - 2)(\lambda - 4)
$$

which implies that the eigenvalues of A are $\lambda_1 = 2$ and $\lambda_2 = 4$. Verify that every eigenvector corresponding to $\lambda_1 = 2$ is of the form

$$
\mathbf{x}_1 = \begin{bmatrix} s \\ -s \end{bmatrix}, \quad s \neq 0
$$

and every eigenvector corresponding to $\lambda_2 = 4$ is of the form

$$
\mathbf{x}_2 = \begin{bmatrix} t \\ t \end{bmatrix}, \quad t \neq 0.
$$

So,

$$
\mathbf{x}_1 \cdot \mathbf{x}_2 = st - st = 0
$$

which means that $\mathbf{x}_1$ and $\mathbf{x}_2$ are orthogonal.

ORTHOGONAL DIAGONALIZATION

A matrix A is **orthogonally diagonalizable** when there exists an orthogonal matrix P such that $P^{-1}AP = D$ is diagonal. The important theorem below states that the set of orthogonally diagonalizable matrices is precisely the set of symmetric matrices.

> **THEOREM 7.10 Fundamental Theorem of Symmetric Matrices**
>
> Let A be an $n \times n$ matrix. Then A is orthogonally diagonalizable (and has real eigenvalues) if and only if A is symmetric.

PROOF

The proof of the theorem in one direction is fairly straightforward. That is, if you assume A is orthogonally diagonalizable, then there exists an orthogonal matrix P such that $D = P^{-1}AP$ is diagonal. Moreover, $P^{-1} = P^T$, so you have

$$A = PDP^{-1}$$
$$= PDP^T$$

which implies that

$$A^T = (PDP^T)^T$$
$$= (P^T)^T D^T P^T$$
$$= PDP^T$$
$$= A.$$

So, A is symmetric.

The proof of the theorem in the other direction is more involved, but it is important because it is constructive. Assume A is symmetric. If A has an eigenvalue λ of multiplicity k, then by Theorem 7.7, λ has k linearly independent eigenvectors. Through the Gram-Schmidt orthonormalization process, use this set of k vectors to form an orthonormal basis of eigenvectors for the eigenspace corresponding to λ. Repeat this procedure for each eigenvalue of A. The collection of all resulting eigenvectors is orthogonal by Theorem 7.9, and you know from the orthonormalization process that the collection is also orthonormal. Now let P be the matrix whose columns consist of these n orthonormal eigenvectors. By Theorem 7.8, P is an orthogonal matrix. Finally, by Theorem 7.5, $P^{-1}AP$ is diagonal. So, A is orthogonally diagonalizable.

EXAMPLE 7 **Determining Whether a Matrix Is Orthogonally Diagonalizable**

Which matrices are orthogonally diagonalizable?

$$A_1 = \begin{bmatrix} 1 & 1 & 1 \\ 1 & 0 & 1 \\ 1 & 1 & 1 \end{bmatrix} \qquad A_2 = \begin{bmatrix} 5 & 2 & 1 \\ 2 & 1 & 8 \\ -1 & 8 & 0 \end{bmatrix}$$

$$A_3 = \begin{bmatrix} 3 & 2 & 0 \\ 2 & 0 & 1 \end{bmatrix} \qquad A_4 = \begin{bmatrix} 0 & 0 \\ 0 & -2 \end{bmatrix}$$

SOLUTION

By Theorem 7.10, the orthogonally diagonalizable matrices are the symmetric ones: A_1 and A_4.

As mentioned above, the second part of the proof of Theorem 7.10 is *constructive*. That is, it gives you steps to follow to diagonalize a symmetric matrix orthogonally. A summary of these steps is on the next page.

Orthogonal Diagonalization of a Symmetric Matrix

Let A be an $n \times n$ symmetric matrix.

1. Find all eigenvalues of A and determine the multiplicity of each.
2. For *each* eigenvalue of multiplicity 1, find a unit eigenvector. (Find any eigenvector and then normalize it.)
3. For each eigenvalue of multiplicity $k \geq 2$, find a set of k linearly independent eigenvectors. (You know from Theorem 7.7 that this is possible.) If this set is not orthonormal, then apply the Gram-Schmidt orthonormalization process.
4. The results of Steps 2 and 3 produce an orthonormal set of n eigenvectors. Use these eigenvectors to form the columns of P. The matrix $P^{-1}AP = P^TAP = D$ will be diagonal. (The main diagonal entries of D are the eigenvalues of A.)

EXAMPLE 8   **Orthogonal Diagonalization**

Find a matrix P that orthogonally diagonalizes $A = \begin{bmatrix} -2 & 2 \\ 2 & 1 \end{bmatrix}$.

SOLUTION

1. The characteristic polynomial of A is

$$|\lambda I - A| = \begin{vmatrix} \lambda + 2 & -2 \\ -2 & \lambda - 1 \end{vmatrix} = (\lambda + 3)(\lambda - 2).$$

So the eigenvalues are $\lambda_1 = -3$ and $\lambda_2 = 2$.

2. For each eigenvalue, find an eigenvector by converting the matrix $\lambda I - A$ to reduced row-echelon form.

Eigenvector

$$-3I - A = \begin{bmatrix} -1 & -2 \\ -2 & -4 \end{bmatrix} \rightarrow \begin{bmatrix} 1 & 2 \\ 0 & 0 \end{bmatrix} \rightarrow \begin{bmatrix} -2 \\ 1 \end{bmatrix}$$

$$2I - A = \begin{bmatrix} 4 & -2 \\ -2 & 1 \end{bmatrix} \rightarrow \begin{bmatrix} 1 & -\frac{1}{2} \\ 0 & 0 \end{bmatrix} \rightarrow \begin{bmatrix} 1 \\ 2 \end{bmatrix}$$

The eigenvectors $(-2, 1)$ and $(1, 2)$ form an *orthogonal* basis for R^2. Normalize these eigenvectors to produce an *orthonormal* basis.

$$\mathbf{p}_1 = \frac{(-2, 1)}{\|(-2, 1)\|} = \left(-\frac{2}{\sqrt{5}}, \frac{1}{\sqrt{5}} \right) \qquad \mathbf{p}_2 = \frac{(1, 2)}{\|(1, 2)\|} = \left(\frac{1}{\sqrt{5}}, \frac{2}{\sqrt{5}} \right)$$

3. Each eigenvalue has a multiplicity of 1, so go directly to step 4.

4. Using $\mathbf{p}_1$ and $\mathbf{p}_2$ as column vectors, construct the matrix P.

$$P = \begin{bmatrix} -\dfrac{2}{\sqrt{5}} & \dfrac{1}{\sqrt{5}} \\ \dfrac{1}{\sqrt{5}} & \dfrac{2}{\sqrt{5}} \end{bmatrix}$$

Verify that P orthogonally diagonalizes A by finding $P^{-1}AP = P^TAP$.

$$P^TAP = \begin{bmatrix} -\dfrac{2}{\sqrt{5}} & \dfrac{1}{\sqrt{5}} \\ \dfrac{1}{\sqrt{5}} & \dfrac{2}{\sqrt{5}} \end{bmatrix} \begin{bmatrix} -2 & 2 \\ 2 & 1 \end{bmatrix} \begin{bmatrix} -\dfrac{2}{\sqrt{5}} & \dfrac{1}{\sqrt{5}} \\ \dfrac{1}{\sqrt{5}} & \dfrac{2}{\sqrt{5}} \end{bmatrix} = \begin{bmatrix} -3 & 0 \\ 0 & 2 \end{bmatrix}$$

EXAMPLE 9   Orthogonal Diagonalization

See LarsonLinearAlgebra.com for an interactive version of this type of example.

Find a matrix P that orthogonally diagonalizes $A = \begin{bmatrix} 2 & 2 & -2 \\ 2 & -1 & 4 \\ -2 & 4 & -1 \end{bmatrix}$.

SOLUTION

1. The characteristic polynomial of A, $|\lambda I - A| = (\lambda + 6)(\lambda - 3)^2$, yields the eigenvalues $\lambda_1 = -6$ and $\lambda_2 = 3$. The eigenvalue λ_1 has a multiplicity of 1 and the eigenvalue λ_2 has a multiplicity of 2.

2. An eigenvector for λ_1 is $\mathbf{v}_1 = (1, -2, 2)$, which normalizes to

$$\mathbf{u}_1 = \frac{\mathbf{v}_1}{\|\mathbf{v}_1\|} = \left(\frac{1}{3}, -\frac{2}{3}, \frac{2}{3}\right).$$

3. Two eigenvectors for λ_2 are $\mathbf{v}_2 = (2, 1, 0)$ and $\mathbf{v}_3 = (-2, 0, 1)$. Note that $\mathbf{v}_1$ is orthogonal to $\mathbf{v}_2$ and $\mathbf{v}_3$ by Theorem 7.9. The eigenvectors $\mathbf{v}_2$ and $\mathbf{v}_3$, however, are not orthogonal to each other. To find two orthonormal eigenvectors for λ_2, use the Gram-Schmidt process as shown below.

$$\mathbf{w}_2 = \mathbf{v}_2 = (2, 1, 0)$$

$$\mathbf{w}_3 = \mathbf{v}_3 - \left(\frac{\mathbf{v}_3 \cdot \mathbf{w}_2}{\mathbf{w}_2 \cdot \mathbf{w}_2}\right)\mathbf{w}_2 = \left(-\frac{2}{5}, \frac{4}{5}, 1\right)$$

These vectors normalize to

$$\mathbf{u}_2 = \frac{\mathbf{w}_2}{\|\mathbf{w}_2\|} = \left(\frac{2}{\sqrt{5}}, \frac{1}{\sqrt{5}}, 0\right)$$

$$\mathbf{u}_3 = \frac{\mathbf{w}_3}{\|\mathbf{w}_3\|} = \left(-\frac{2}{3\sqrt{5}}, \frac{4}{3\sqrt{5}}, \frac{5}{3\sqrt{5}}\right).$$

4. The matrix P has $\mathbf{u}_1$, $\mathbf{u}_2$, and $\mathbf{u}_3$ as its column vectors.

$$P = \begin{bmatrix} \dfrac{1}{3} & \dfrac{2}{\sqrt{5}} & -\dfrac{2}{3\sqrt{5}} \\[2mm] -\dfrac{2}{3} & \dfrac{1}{\sqrt{5}} & \dfrac{4}{3\sqrt{5}} \\[2mm] \dfrac{2}{3} & 0 & \dfrac{5}{3\sqrt{5}} \end{bmatrix}$$

A check shows that $P^{-1}AP = P^TAP = \begin{bmatrix} -6 & 0 & 0 \\ 0 & 3 & 0 \\ 0 & 0 & 3 \end{bmatrix}$.

LINEAR ALGEBRA APPLIED

The *Hessian matrix* is a symmetric matrix that can be helpful in finding relative maxima and minima of functions of several variables. For a function f of two variables x and y—that is, a surface in R^3—the Hessian matrix has the form

$$\begin{bmatrix} f_{xx} & f_{xy} \\ f_{yx} & f_{yy} \end{bmatrix}.$$

The determinant of this matrix, evaluated at a point for which f_x and f_y are zero, is the expression used in the Second Partials Test for relative extrema.

7.3 Exercises

See CalcChat.com for worked-out solutions to odd-numbered exercises.

Determining Whether a Matrix Is Symmetric In Exercises 1 and 2, determine whether the matrix is symmetric.

1. $\begin{bmatrix} 4 & -2 & 1 \\ 3 & 1 & 2 \\ 1 & 2 & 1 \end{bmatrix}$
2. $\begin{bmatrix} 2 & 0 & 3 & 5 \\ 0 & 11 & 0 & -2 \\ 3 & 0 & 5 & 0 \\ 5 & -2 & 0 & 1 \end{bmatrix}$

Proof In Exercises 3–6, prove that the symmetric matrix is diagonalizable.

3. $A = \begin{bmatrix} 0 & 0 & a \\ 0 & a & 0 \\ a & 0 & 0 \end{bmatrix}$
4. $A = \begin{bmatrix} 0 & a & 0 \\ a & 0 & a \\ 0 & a & 0 \end{bmatrix}$
5. $A = \begin{bmatrix} a & 0 & a \\ 0 & a & 0 \\ a & 0 & a \end{bmatrix}$
6. $A = \begin{bmatrix} a & a & a \\ a & a & a \\ a & a & a \end{bmatrix}$

Finding Eigenvalues and Dimensions of Eigenspaces In Exercises 7–18, find the eigenvalues of the symmetric matrix. For each eigenvalue, find the dimension of the corresponding eigenspace.

7. $\begin{bmatrix} 2 & 1 \\ 1 & 2 \end{bmatrix}$
8. $\begin{bmatrix} 3 & 0 \\ 0 & 3 \end{bmatrix}$
9. $\begin{bmatrix} 3 & 0 & 0 \\ 0 & 2 & 0 \\ 0 & 0 & 2 \end{bmatrix}$
10. $\begin{bmatrix} 2 & 1 & 1 \\ 1 & 2 & 1 \\ 1 & 1 & 2 \end{bmatrix}$
11. $\begin{bmatrix} 0 & 2 & 2 \\ 2 & 0 & 2 \\ 2 & 2 & 0 \end{bmatrix}$
12. $\begin{bmatrix} 0 & 4 & 4 \\ 4 & 2 & 0 \\ 4 & 0 & -2 \end{bmatrix}$
13. $\begin{bmatrix} 0 & 1 & 1 \\ 1 & 0 & 1 \\ 1 & 1 & 1 \end{bmatrix}$
14. $\begin{bmatrix} 2 & -1 & -1 \\ -1 & 2 & -1 \\ -1 & -1 & 2 \end{bmatrix}$
15. $\begin{bmatrix} 3 & 0 & 0 & 0 \\ 0 & 3 & 0 & 0 \\ 0 & 0 & 3 & 5 \\ 0 & 0 & 5 & 3 \end{bmatrix}$
16. $\begin{bmatrix} -1 & 2 & 0 & 0 \\ 2 & -1 & 0 & 0 \\ 0 & 0 & -1 & 2 \\ 0 & 0 & 2 & -1 \end{bmatrix}$
17. $\begin{bmatrix} 2 & -1 & 0 & 0 & 0 \\ -1 & 2 & 0 & 0 & 0 \\ 0 & 0 & 2 & 0 & 0 \\ 0 & 0 & 0 & 2 & 0 \\ 0 & 0 & 0 & 0 & 2 \end{bmatrix}$
18. $\begin{bmatrix} 1 & -1 & 0 & 0 & 0 \\ -1 & 1 & 0 & 0 & 0 \\ 0 & 0 & 1 & 0 & 0 \\ 0 & 0 & 0 & 1 & -1 \\ 0 & 0 & 0 & -1 & 1 \end{bmatrix}$

Determining Whether a Matrix Is Orthogonal In Exercises 19–32, determine whether the matrix is orthogonal. If the matrix is orthogonal, then show that the column vectors of the matrix form an orthonormal set.

19. $\begin{bmatrix} \frac{\sqrt{2}}{2} & \frac{\sqrt{2}}{2} \\ -\frac{\sqrt{2}}{2} & \frac{\sqrt{2}}{2} \end{bmatrix}$
20. $\begin{bmatrix} \frac{4}{9} & -\frac{4}{9} \\ \frac{4}{9} & \frac{3}{9} \end{bmatrix}$
21. $\begin{bmatrix} -0.936 & -0.352 \\ 0.352 & -0.936 \end{bmatrix}$
22. $\begin{bmatrix} \frac{1}{2} & \frac{\sqrt{3}}{2} \\ -\frac{\sqrt{3}}{2} & \frac{1}{2} \end{bmatrix}$
23. $\begin{bmatrix} 0 & 0 & 0 \\ 0 & 1 & 0 \\ 1 & 0 & 1 \end{bmatrix}$
24. $\begin{bmatrix} 1 & 0 & 0 \\ 0 & 1 & 0 \\ 0 & 0 & 1 \end{bmatrix}$
25. $\begin{bmatrix} \frac{2}{3} & -\frac{2}{3} & \frac{1}{3} \\ \frac{2}{3} & \frac{1}{3} & -\frac{2}{3} \\ \frac{1}{3} & \frac{2}{3} & \frac{2}{3} \end{bmatrix}$
26. $\begin{bmatrix} -\frac{4}{5} & 0 & \frac{3}{5} \\ 0 & 1 & 0 \\ \frac{3}{5} & 0 & \frac{4}{5} \end{bmatrix}$
27. $\begin{bmatrix} -4 & 0 & 3 \\ 0 & 1 & 0 \\ 3 & 0 & 4 \end{bmatrix}$
28. $\begin{bmatrix} 4 & -1 & -4 \\ -1 & 0 & -17 \\ 1 & 4 & -1 \end{bmatrix}$
29. $\begin{bmatrix} \frac{\sqrt{2}}{2} & -\frac{\sqrt{6}}{6} & \frac{\sqrt{3}}{3} \\ 0 & \frac{\sqrt{6}}{3} & \frac{\sqrt{3}}{3} \\ \frac{\sqrt{2}}{2} & \frac{\sqrt{6}}{6} & -\frac{\sqrt{3}}{3} \end{bmatrix}$
30. $\begin{bmatrix} \frac{\sqrt{2}}{3} & 0 & \frac{\sqrt{5}}{2} \\ 0 & \frac{2\sqrt{5}}{5} & 0 \\ -\frac{\sqrt{2}}{6} & -\frac{\sqrt{5}}{5} & \frac{1}{2} \end{bmatrix}$
31. $\begin{bmatrix} \frac{1}{8} & 0 & 0 & \frac{3}{8}\sqrt{7} \\ 0 & 1 & 0 & 0 \\ 0 & 0 & 1 & 0 \\ \frac{3}{8}\sqrt{7} & 0 & 0 & \frac{1}{8} \end{bmatrix}$
32. $\begin{bmatrix} \frac{1}{10}\sqrt{10} & 0 & 0 & -\frac{3}{10}\sqrt{10} \\ 0 & 0 & 1 & 0 \\ 0 & 1 & 0 & 0 \\ \frac{3}{10}\sqrt{10} & 0 & 0 & \frac{1}{10}\sqrt{10} \end{bmatrix}$

Eigenvectors of a Symmetric Matrix In Exercises 33–38, show that any two eigenvectors of the symmetric matrix corresponding to distinct eigenvalues are orthogonal.

33. $\begin{bmatrix} 3 & 3 \\ 3 & 3 \end{bmatrix}$

34. $\begin{bmatrix} -1 & -2 \\ -2 & 2 \end{bmatrix}$

35. $\begin{bmatrix} 1 & 0 & 0 \\ 0 & 1 & 0 \\ 0 & 0 & 2 \end{bmatrix}$

36. $\begin{bmatrix} 3 & 0 & 0 \\ 0 & -3 & 0 \\ 0 & 0 & 2 \end{bmatrix}$

37. $\begin{bmatrix} 0 & \sqrt{3} & 0 \\ \sqrt{3} & 0 & -1 \\ 0 & -1 & 0 \end{bmatrix}$

38. $\begin{bmatrix} 1 & 0 & 1 \\ 0 & 1 & 0 \\ 1 & 0 & -1 \end{bmatrix}$

Orthogonally Diagonalizable Matrices In Exercises 39–42, determine whether the matrix is orthogonally diagonalizable.

39. $\begin{bmatrix} 4 & 5 \\ 0 & 1 \end{bmatrix}$

40. $\begin{bmatrix} 3 & 2 & -3 \\ -2 & -1 & 2 \\ -3 & 2 & 3 \end{bmatrix}$

41. $\begin{bmatrix} 5 & -3 & 8 \\ -3 & -3 & -3 \\ 8 & -3 & 8 \end{bmatrix}$

42. $\begin{bmatrix} 0 & 1 & 0 & -1 \\ 1 & 0 & -1 & 0 \\ 0 & -1 & 0 & -1 \\ -1 & 0 & -1 & 0 \end{bmatrix}$

Orthogonal Diagonalization In Exercises 43–52, find a matrix P such that $P^T A P$ orthogonally diagonalizes A. Verify that $P^T A P$ gives the correct diagonal form.

43. $A = \begin{bmatrix} 1 & 1 \\ 1 & 1 \end{bmatrix}$

44. $A = \begin{bmatrix} 4 & 2 \\ 2 & 4 \end{bmatrix}$

45. $A = \begin{bmatrix} 2 & \sqrt{2} \\ \sqrt{2} & 1 \end{bmatrix}$

46. $A = \begin{bmatrix} 0 & 1 & 1 \\ 1 & 0 & 1 \\ 1 & 1 & 0 \end{bmatrix}$

47. $A = \begin{bmatrix} 0 & 10 & 10 \\ 10 & 5 & 0 \\ 10 & 0 & -5 \end{bmatrix}$

48. $A = \begin{bmatrix} 0 & 3 & 0 \\ 3 & 0 & 4 \\ 0 & 4 & 0 \end{bmatrix}$

49. $A = \begin{bmatrix} 1 & -1 & 2 \\ -1 & 1 & 2 \\ 2 & 2 & 2 \end{bmatrix}$

50. $A = \begin{bmatrix} -2 & 2 & 4 \\ 2 & -2 & 4 \\ 4 & 4 & 4 \end{bmatrix}$

51. $A = \begin{bmatrix} 4 & 2 & 0 & 0 \\ 2 & 4 & 0 & 0 \\ 0 & 0 & 4 & 2 \\ 0 & 0 & 2 & 4 \end{bmatrix}$

52. $A = \begin{bmatrix} 1 & 1 & 0 & 0 \\ 1 & 1 & 0 & 0 \\ 0 & 0 & 1 & 1 \\ 0 & 0 & 1 & 1 \end{bmatrix}$

True or False? In Exercises 53 and 54, determine whether each statement is true or false. If a statement is true, give a reason or cite an appropriate statement from the text. If a statement is false, provide an example that shows the statement is not true in all cases or cite an appropriate statement from the text.

53. (a) Let A be an $n \times n$ matrix. Then A is symmetric if and only if A is orthogonally diagonalizable.

(b) The eigenvectors corresponding to distinct eigenvalues are orthogonal for symmetric matrices.

54. (a) A square matrix P is orthogonal when it is invertible—that is, when $P^{-1} = P^T$.

(b) If A is an $n \times n$ symmetric matrix, then A has real eigenvalues.

55. **Proof** Prove that if A and B are $n \times n$ orthogonal matrices, then AB and BA are orthogonal.

56. **Proof** Prove that if a symmetric matrix A has only one eigenvalue λ, then $A = \lambda I$.

57. **Proof** Prove that if A is an orthogonal matrix, then so are A^T and A^{-1}.

58. **CAPSTONE** Consider the matrix below.
$$A = \begin{bmatrix} -1 & 0 & -1 & 0 & 1 \\ 0 & 1 & 0 & -1 & 0 \\ -1 & 0 & 1 & 0 & -1 \\ 0 & -1 & 0 & -1 & 0 \\ 1 & 0 & -1 & 0 & -1 \end{bmatrix}$$
(a) Is A symmetric? Explain.
(b) Is A diagonalizable? Explain.
(c) Are the eigenvalues of A real? Explain.
(d) The eigenvalues of A are distinct. What are the dimensions of the corresponding eigenspaces? Explain.
(e) Is A orthogonal? Explain.
(f) For the eigenvalues of A, are the corresponding eigenvectors orthogonal? Explain.
(g) Is A orthogonally diagonalizable? Explain.

59. **Proof** Prove that the matrix below is orthogonal for any value of θ.
$$\begin{bmatrix} \cos\theta & -\sin\theta & 0 \\ \sin\theta & \cos\theta & 0 \\ 0 & 0 & 1 \end{bmatrix}$$

60. Find $A^T A$ and $A A^T$ for the matrix below. What do you observe?
$$A = \begin{bmatrix} 1 & -3 & 2 \\ 4 & -6 & 1 \end{bmatrix}$$

7.4   Applications of Eigenvalues and Eigenvectors

- Model population growth using an age transition matrix and an age distribution vector, and find a stable age distribution vector.
- Use a matrix equation to solve a system of first-order linear differential equations.
- Find the matrix of a quadratic form and use the Principal Axes Theorem to perform a rotation of axes for a conic and a quadric surface.
- Solve a constrained optimization problem.

POPULATION GROWTH

Matrices can be used to form models for population growth. The first step in this process is to group the population into age classes of equal duration. For example, if the maximum life span of a member is M years, then the n intervals below represent the age classes.

$$\left[0, \frac{M}{n}\right)$$ First age class

$$\left[\frac{M}{n}, \frac{2M}{n}\right)$$ Second age class

$$\vdots \qquad\qquad \vdots$$

$$\left[\frac{(n-1)M}{n}, M\right]$$ nth age class

The **age distribution vector x** represents the number of population members in each age class, where

$$\mathbf{x} = \begin{bmatrix} x_1 \\ x_2 \\ \vdots \\ x_n \end{bmatrix}.$$

Number in first age class
Number in second age class
$\vdots$
Number in nth age class

Over a period of M/n years, the *probability* that a member of the ith age class will survive to become a member of the $(i+1)$th age class is p_i, where

$$0 \le p_i \le 1, i = 1, 2, \ldots, n-1.$$

The *average number* of offspring produced by a member of the ith age class is b_i, where $0 \le b_i, i = 1, 2, \ldots, n$. These numbers can be written in matrix form, as shown below.

REMARK

Recall from Section 6.4 that the age transition matrix L is also called a **Leslie matrix** after mathematician Patrick H. Leslie.

$$L = \begin{bmatrix} b_1 & b_2 & \cdots & b_{n-1} & b_n \\ p_1 & 0 & \cdots & 0 & 0 \\ 0 & p_2 & \cdots & 0 & 0 \\ \vdots & \vdots & & \vdots & \vdots \\ 0 & 0 & \cdots & p_{n-1} & 0 \end{bmatrix}$$

Multiplying this **age transition matrix** by the age distribution vector for a specific time period produces the age distribution vector for the next time period. That is,

$$L\mathbf{x}_j = \mathbf{x}_{j+1}.$$

Example 1 illustrates this procedure.

EXAMPLE 1   **A Population Growth Model**

A population of rabbits has the characteristics below.
a. Half of the rabbits survive their first year. Of those, half survive their second year.
   The maximum life span is 3 years.
b. During the first year, the rabbits produce no offspring. The average number of
   offspring is 6 during the second year and 8 during the third year.

The population now consists of 24 rabbits in the first age class, 24 in the second,
and 20 in the third. How many rabbits will there be in each age class in 1 year?

SOLUTION

The current age distribution vector is

$$\mathbf{x}_1 = \begin{bmatrix} 24 \\ 24 \\ 20 \end{bmatrix} \quad \begin{matrix} 0 \le \text{age} < 1 \\ 1 \le \text{age} < 2 \\ 2 \le \text{age} \le 3 \end{matrix}$$

and the age transition matrix is

$$L = \begin{bmatrix} 0 & 6 & 8 \\ 0.5 & 0 & 0 \\ 0 & 0.5 & 0 \end{bmatrix}.$$

After 1 year, the age distribution vector will be

$$\mathbf{x}_2 = L\mathbf{x}_1 = \begin{bmatrix} 0 & 6 & 8 \\ 0.5 & 0 & 0 \\ 0 & 0.5 & 0 \end{bmatrix}\begin{bmatrix} 24 \\ 24 \\ 20 \end{bmatrix} = \begin{bmatrix} 304 \\ 12 \\ 12 \end{bmatrix}. \quad \begin{matrix} 0 \le \text{age} < 1 \\ 1 \le \text{age} < 2 \\ 2 \le \text{age} \le 3 \end{matrix}$$

EXAMPLE 2   **Finding a Stable Age Distribution Vector**

Find a stable age distribution vector for the population in Example 1.

SOLUTION

To solve this problem, find an eigenvalue λ and a corresponding eigenvector $\mathbf{x}$ such that
$L\mathbf{x} = \lambda\mathbf{x}$. The characteristic polynomial of L is

$$|\lambda I - L| = (\lambda + 1)^2(\lambda - 2)$$

(check this), which implies that the eigenvalues are -1 and 2. Choosing the positive
value, let $\lambda = 2$. Verify that the corresponding eigenvectors are of the form

$$\mathbf{x} = \begin{bmatrix} x_1 \\ x_2 \\ x_3 \end{bmatrix} = \begin{bmatrix} 16t \\ 4t \\ t \end{bmatrix} = t\begin{bmatrix} 16 \\ 4 \\ 1 \end{bmatrix}.$$

For example, if $t = 2$, then the initial age distribution vector is

$$\mathbf{x}_1 = \begin{bmatrix} 32 \\ 8 \\ 2 \end{bmatrix} \quad \begin{matrix} 0 \le \text{age} < 1 \\ 1 \le \text{age} < 2 \\ 2 \le \text{age} \le 3 \end{matrix}$$

and the age distribution vector for the next year is

$$\mathbf{x}_2 = L\mathbf{x}_1 = \begin{bmatrix} 0 & 6 & 8 \\ 0.5 & 0 & 0 \\ 0 & 0.5 & 0 \end{bmatrix}\begin{bmatrix} 32 \\ 8 \\ 2 \end{bmatrix} = \begin{bmatrix} 64 \\ 16 \\ 4 \end{bmatrix}. \quad \begin{matrix} 0 \le \text{age} < 1 \\ 1 \le \text{age} < 2 \\ 2 \le \text{age} \le 3 \end{matrix}$$

Notice that the ratio of the three age classes is still $16 : 4 : 1$, and so the percent of
the population in each age class remains the same.

SYSTEMS OF LINEAR DIFFERENTIAL EQUATIONS (CALCULUS)

A **system of first-order linear differential equations** has the form

$$y_1' = a_{11}y_1 + a_{12}y_2 + \cdots + a_{1n}y_n$$
$$y_2' = a_{21}y_1 + a_{22}y_2 + \cdots + a_{2n}y_n$$
$$\vdots$$
$$y_n' = a_{n1}y_1 + a_{n2}y_2 + \cdots + a_{nn}y_n$$

where each y_i is a function of t and $y_i' = \dfrac{dy_i}{dt}$. If you let

$$\mathbf{y}' = \begin{bmatrix} y_1' \\ y_2' \\ \vdots \\ y_n' \end{bmatrix}, \quad \mathbf{y} = \begin{bmatrix} y_1 \\ y_2 \\ \vdots \\ y_n \end{bmatrix}, \quad \text{and} \quad A = \begin{bmatrix} a_{11} & a_{12} & \cdots & a_{1n} \\ a_{21} & a_{22} & \cdots & a_{2n} \\ \vdots & \vdots & & \vdots \\ a_{n1} & a_{n2} & \cdots & a_{nn} \end{bmatrix}$$

then the system can be written in matrix form as

$$\mathbf{y}' = A\mathbf{y}.$$

EXAMPLE 3   **Solving a System of Linear Differential Equations**

Solve the system of linear differential equations.

$$y_1' = 4y_1$$
$$y_2' = -y_2$$
$$y_3' = 2y_3$$

SOLUTION

From calculus, you know that the solution of the differential equation $y' = ky$ is

$$y = Ce^{kt}.$$

So, the solution of the system is

$$y_1 = C_1 e^{4t}$$
$$y_2 = C_2 e^{-t}$$
$$y_3 = C_3 e^{2t}.$$

The matrix form of the system of linear differential equations in Example 3 is $\mathbf{y}' = A\mathbf{y}$, or

$$\begin{bmatrix} y_1' \\ y_2' \\ y_3' \end{bmatrix} = \begin{bmatrix} 4 & 0 & 0 \\ 0 & -1 & 0 \\ 0 & 0 & 2 \end{bmatrix} \begin{bmatrix} y_1 \\ y_2 \\ y_3 \end{bmatrix}.$$

So, the coefficients of t in the solutions $y_i = C_i e^{\lambda_i t}$ are the *eigenvalues* of the matrix A.

If A is a *diagonal* matrix, then the solution of

$$\mathbf{y}' = A\mathbf{y}$$

can be obtained immediately, as in Example 3. If A is *not* diagonal, then the solution requires more work. First, find a matrix P that diagonalizes A. Then, the change of variables $\mathbf{y} = P\mathbf{w}$ and $\mathbf{y}' = P\mathbf{w}'$ produces

$$P\mathbf{w}' = \mathbf{y}' = A\mathbf{y} = AP\mathbf{w} \quad \Longrightarrow \quad \mathbf{w}' = P^{-1}AP\mathbf{w}$$

where $P^{-1}AP$ is a diagonal matrix. Example 4 demonstrates this procedure.

| EXAMPLE 4 | Solving a System of Linear Differential Equations |

See LarsonLinearAlgebra.com for an interactive version of this type of example.

To solve the system of linear differential equations

$$y_1' = 3y_1 + 2y_2$$
$$y_2' = 6y_1 - y_2$$

first find a matrix P that diagonalizes $A = \begin{bmatrix} 3 & 2 \\ 6 & -1 \end{bmatrix}$. Verify that the eigenvalues of A are $\lambda_1 = -3$ and $\lambda_2 = 5$, and that the corresponding eigenvectors are $\mathbf{p}_1 = \begin{bmatrix} 1 & -3 \end{bmatrix}^T$ and $\mathbf{p}_2 = \begin{bmatrix} 1 & 1 \end{bmatrix}^T$. Diagonalize A using the matrix P whose columns consist of $\mathbf{p}_1$ and $\mathbf{p}_2$ to obtain

$$P = \begin{bmatrix} 1 & 1 \\ -3 & 1 \end{bmatrix}, \quad P^{-1} = \begin{bmatrix} \frac{1}{4} & -\frac{1}{4} \\ \frac{3}{4} & \frac{1}{4} \end{bmatrix}, \quad \text{and} \quad P^{-1}AP = \begin{bmatrix} -3 & 0 \\ 0 & 5 \end{bmatrix}.$$

The system $\mathbf{w}' = P^{-1}AP\mathbf{w}$ has the form below.

$$\begin{bmatrix} w_1' \\ w_2' \end{bmatrix} = \begin{bmatrix} -3 & 0 \\ 0 & 5 \end{bmatrix}\begin{bmatrix} w_1 \\ w_2 \end{bmatrix} \quad \longrightarrow \quad \begin{matrix} w_1' = -3w_1 \\ w_2' = 5w_2 \end{matrix}$$

The solution of this system of equations is

$$w_1 = C_1 e^{-3t}$$
$$w_2 = C_2 e^{5t}.$$

To return to the original variables y_1 and y_2, use the substitution $\mathbf{y} = P\mathbf{w}$ and write

$$\begin{bmatrix} y_1 \\ y_2 \end{bmatrix} = \begin{bmatrix} 1 & 1 \\ -3 & 1 \end{bmatrix}\begin{bmatrix} w_1 \\ w_2 \end{bmatrix}$$

which implies that the solution is

$$y_1 = w_1 + w_2 = C_1 e^{-3t} + C_2 e^{5t}$$
$$y_2 = -3w_1 + w_2 = -3C_1 e^{-3t} + C_2 e^{5t}.$$

If A has eigenvalues with multiplicity greater than 1 or if A has complex eigenvalues, then the technique for solving the system must be modified.

1. *Eigenvalues with multiplicity greater than 1:* The coefficient matrix of the system

$$\begin{matrix} y_1' = & y_2 \\ y_2' = -4y_1 + 4y_2 \end{matrix} \quad \text{is} \quad A = \begin{bmatrix} 0 & 1 \\ -4 & 4 \end{bmatrix}.$$

The only eigenvalue of A is $\lambda = 2$, and the solution of the system is

$$y_1 = C_1 e^{2t} + C_2 t e^{2t}$$
$$y_2 = (2C_1 + C_2)e^{2t} + 2C_2 t e^{2t}.$$

2. *Complex eigenvalues:* The coefficient matrix of the system

$$\begin{matrix} y_1' = -y_2 \\ y_2' = y_1 \end{matrix} \quad \text{is} \quad A = \begin{bmatrix} 0 & -1 \\ 1 & 0 \end{bmatrix}.$$

The eigenvalues of A are $\lambda_1 = i$ and $\lambda_2 = -i$, and the solution of the system is

$$y_1 = C_1 \cos t + C_2 \sin t$$
$$y_2 = -C_2 \cos t + C_1 \sin t.$$

Check these solutions by differentiating and substituting into the original systems of equations.

QUADRATIC FORMS

Eigenvalues and eigenvectors can be used to solve the rotation of axes problem introduced in Section 4.8. Recall that classifying the graph of the quadratic equation

$$ax^2 + bxy + cy^2 + dx + ey + f = 0 \qquad \text{Quadratic equation}$$

is fairly straightforward as long as the equation has no xy-term (that is, $b = 0$). If the equation has an xy-term, however, then the classification is accomplished most easily by first performing a rotation of axes that eliminates the xy-term. The resulting equation (relative to the new $x'y'$-axes) will then be of the form

$$a'(x')^2 + c'(y')^2 + d'x' + e'y' + f' = 0.$$

You will see that the coefficients a' and c' are eigenvalues of the matrix

$$A = \begin{bmatrix} a & b/2 \\ b/2 & c \end{bmatrix}.$$

The expression

$$ax^2 + bxy + cy^2 \qquad \text{Quadratic form}$$

is the **quadratic form** associated with the quadratic equation

$$ax^2 + bxy + cy^2 + dx + ey + f = 0$$

and the matrix A is the **matrix of the quadratic form.** Note that the matrix A *is symmetric.* Moreover, the matrix A will be diagonal if and only if its corresponding quadratic form has no xy-term, as illustrated in Example 5.

EXAMPLE 5   Finding the Matrix of the Quadratic Form

Find the matrix of the quadratic form associated with each quadratic equation.
a. $4x^2 + 9y^2 - 36 = 0$ **b.** $13x^2 - 10xy + 13y^2 - 72 = 0$

SOLUTION

a. $a = 4, b = 0$, and $c = 9$, so the matrix is

$$A = \begin{bmatrix} 4 & 0 \\ 0 & 9 \end{bmatrix}. \qquad \text{Diagonal matrix (no } xy\text{-term)}$$

b. $a = 13, b = -10$, and $c = 13$, so the matrix is

$$A = \begin{bmatrix} 13 & -5 \\ -5 & 13 \end{bmatrix}. \qquad \text{Nondiagonal matrix (}xy\text{-term)}$$

In standard form, the equation $4x^2 + 9y^2 - 36 = 0$ is

$$\frac{x^2}{3^2} + \frac{y^2}{2^2} = 1$$

which is the equation of the ellipse shown in Figure 7.2. Although it is not apparent by inspection, the graph of the equation $13x^2 - 10xy + 13y^2 - 72 = 0$ is similar. In fact, when you rotate the x- and y-axes counterclockwise $45°$ to form a new $x'y'$-coordinate system, this equation takes the form

$$\frac{(x')^2}{3^2} + \frac{(y')^2}{2^2} = 1$$

(verify this) which is the equation of the ellipse shown in Figure 7.3.

To see how to use the matrix of a quadratic form to perform a rotation of axes, let

$$X = [x \quad y]^T.$$

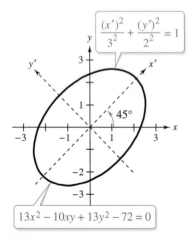

Figure 7.2

$$\frac{x^2}{3^2} + \frac{y^2}{2^2} = 1$$

Figure 7.3

$$\frac{(x')^2}{3^2} + \frac{(y')^2}{2^2} = 1$$

$45°$

$$13x^2 - 10xy + 13y^2 - 72 = 0$$

Then the quadratic expression $ax^2 + bxy + cy^2 + dx + ey + f$ can be written in matrix form as shown below.

$$X^T A X + [d \quad e]X + f = [x \quad y]\begin{bmatrix} a & b/2 \\ b/2 & c \end{bmatrix}\begin{bmatrix} x \\ y \end{bmatrix} + [d \quad e]\begin{bmatrix} x \\ y \end{bmatrix} + f$$

$$= ax^2 + bxy + cy^2 + dx + ey + f$$

If $b = 0$, then no rotation is necessary. But if $b \neq 0$, then use the fact that A is symmetric and apply Theorem 7.10 to conclude that there exists an orthogonal matrix P such that $P^T A P = D$ is diagonal. So, if you let

$$P^T X = X' = \begin{bmatrix} x' \\ y' \end{bmatrix}$$

then it follows that $X = PX'$, and $X^T A X = (PX')^T A(PX') = (X')^T P^T A P X' = (X')^T D X'$.

The choice of the matrix P must be made with care. P is orthogonal, so its determinant will be ± 1. It can be shown (see Exercise 67) that if P is chosen so that $|P| = 1$, then P will be of the form

$$P = \begin{bmatrix} \cos\theta & -\sin\theta \\ \sin\theta & \cos\theta \end{bmatrix}$$

where θ is the angle of rotation of the conic measured from the positive x-axis to the positive x'-axis. This leads to the **Principal Axes Theorem.**

Principal Axes Theorem

For a conic whose equation is $ax^2 + bxy + cy^2 + dx + ey + f = 0$, the rotation $X = PX'$ eliminates the xy-term when P is an orthogonal matrix, with $|P| = 1$, that diagonalizes the matrix of the quadratic form A. That is,

$$P^T A P = \begin{bmatrix} \lambda_1 & 0 \\ 0 & \lambda_2 \end{bmatrix}$$

where λ_1 and λ_2 are eigenvalues of A. The equation of the rotated conic is

$$\lambda_1(x')^2 + \lambda_2(y')^2 + [d \quad e]PX' + f = 0.$$

EXAMPLE 6   **Rotation of a Conic**

Perform a rotation of axes to eliminate the xy-term in the quadratic equation

$$13x^2 - 10xy + 13y^2 - 72 = 0.$$

SOLUTION

The matrix of the quadratic form associated with this equation is

$$A = \begin{bmatrix} 13 & -5 \\ -5 & 13 \end{bmatrix}.$$

The characteristic polynomial of A is $(\lambda - 8)(\lambda - 18)$ (check this), so it follows that the eigenvalues of A are $\lambda_1 = 8$ and $\lambda_2 = 18$. Then, the equation of the rotated conic is

$$8(x')^2 + 18(y')^2 - 72 = 0$$

which, when written in the standard form

$$\frac{(x')^2}{3^2} + \frac{(y')^2}{2^2} = 1$$

is the equation of an ellipse. (See Figure 7.3.)

In Example 6, the eigenvectors of the matrix A are

$$\mathbf{x}_1 = \begin{bmatrix} 1 \\ 1 \end{bmatrix} \quad \text{and} \quad \mathbf{x}_2 = \begin{bmatrix} -1 \\ 1 \end{bmatrix}$$

which you can normalize to form the columns of P, as shown below.

$$P = \begin{bmatrix} \dfrac{1}{\sqrt{2}} & -\dfrac{1}{\sqrt{2}} \\ \dfrac{1}{\sqrt{2}} & \dfrac{1}{\sqrt{2}} \end{bmatrix} = \begin{bmatrix} \cos\theta & -\sin\theta \\ \sin\theta & \cos\theta \end{bmatrix}$$

Note first that $|P| = 1$, which implies that P is a rotation. Moreover, $\cos 45° = 1/\sqrt{2} = \sin 45°$, so the angle of rotation is $45°$ as shown in Figure 7.3.

The orthogonal matrix P specified in the Principal Axes Theorem is not unique. Its entries depend on the ordering of the eigenvalues λ_1 and λ_2 *and* on the subsequent choice of eigenvectors $\mathbf{x}_1$ and $\mathbf{x}_2$. For instance, in the solution of Example 6, any of the choices of P shown below would have worked.

$$\begin{array}{ccc} \mathbf{x}_1 \quad \mathbf{x}_2 & \mathbf{x}_1 \quad \mathbf{x}_2 & \mathbf{x}_1 \quad \mathbf{x}_2 \end{array}$$

$$\begin{bmatrix} -\dfrac{1}{\sqrt{2}} & \dfrac{1}{\sqrt{2}} \\ -\dfrac{1}{\sqrt{2}} & -\dfrac{1}{\sqrt{2}} \end{bmatrix} \quad \begin{bmatrix} -\dfrac{1}{\sqrt{2}} & -\dfrac{1}{\sqrt{2}} \\ \dfrac{1}{\sqrt{2}} & -\dfrac{1}{\sqrt{2}} \end{bmatrix} \quad \begin{bmatrix} \dfrac{1}{\sqrt{2}} & \dfrac{1}{\sqrt{2}} \\ -\dfrac{1}{\sqrt{2}} & \dfrac{1}{\sqrt{2}} \end{bmatrix}$$

$$\begin{array}{ccc} \lambda_1 = 8,\ \lambda_2 = 18 & \lambda_1 = 18,\ \lambda_2 = 8 & \lambda_1 = 18,\ \lambda_2 = 8 \\ \theta = 225° & \theta = 135° & \theta = 315° \end{array}$$

For any of these choices of P, the graph of the rotated conic will, of course, be the same. (See below.)

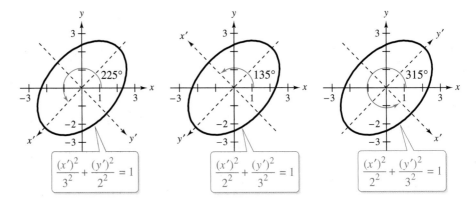

The list below summarizes the steps used to apply the Principal Axes Theorem.

1. Form the matrix A and find its eigenvalues λ_1 and λ_2.

2. Find eigenvectors corresponding to λ_1 and λ_2. Normalize these eigenvectors to form the columns of P.

3. If $|P| = -1$, then multiply one of the columns of P by -1 to obtain a matrix of the form

$$P = \begin{bmatrix} \cos\theta & -\sin\theta \\ \sin\theta & \cos\theta \end{bmatrix}$$

4. The angle θ represents the angle of rotation of the conic.

5. The equation of the rotated conic is $\lambda_1(x')^2 + \lambda_2(y')^2 + [d \quad e]PX' + f = 0$.

Example 7 shows how to apply the Principal Axes Theorem to rotate a conic whose center is translated away from the origin.

EXAMPLE 7   Rotation of a Conic

Perform a rotation of axes to eliminate the xy-term in the quadratic equation

$$3x^2 - 10xy + 3y^2 + 16\sqrt{2}x - 32 = 0.$$

SOLUTION

The matrix of the quadratic form associated with this equation is

$$A = \begin{bmatrix} 3 & -5 \\ -5 & 3 \end{bmatrix}.$$

The eigenvalues of A are

$$\lambda_1 = 8 \quad \text{and} \quad \lambda_2 = -2$$

with corresponding eigenvectors of

$$\mathbf{x}_1 = (-1, 1) \quad \text{and} \quad \mathbf{x}_2 = (-1, -1).$$

This implies that the matrix P is

$$P = \begin{bmatrix} -\dfrac{1}{\sqrt{2}} & -\dfrac{1}{\sqrt{2}} \\ \dfrac{1}{\sqrt{2}} & -\dfrac{1}{\sqrt{2}} \end{bmatrix}$$

$$= \begin{bmatrix} \cos\theta & -\sin\theta \\ \sin\theta & \cos\theta \end{bmatrix}, \text{ where } |P| = 1.$$

$\cos 135° = -1/\sqrt{2}$ and $\sin 135° = 1/\sqrt{2}$, so the angle of rotation is $135°$. Finally, from the matrix product

$$[d \quad e]PX' = [16\sqrt{2} \quad 0] \begin{bmatrix} -\dfrac{1}{\sqrt{2}} & -\dfrac{1}{\sqrt{2}} \\ \dfrac{1}{\sqrt{2}} & -\dfrac{1}{\sqrt{2}} \end{bmatrix} \begin{bmatrix} x' \\ y' \end{bmatrix}$$

$$= -16x' - 16y'$$

the equation of the rotated conic is

$$8(x')^2 - 2(y')^2 - 16x' - 16y' - 32 = 0.$$

In standard form, the equation is

$$\frac{(x' - 1)^2}{1^2} - \frac{(y' + 4)^2}{2^2} = 1$$

which is the equation of a hyperbola. Its graph is shown in Figure 7.4.

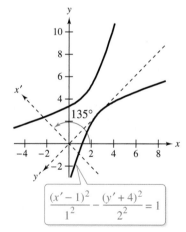

$$\frac{(x' - 1)^2}{1^2} - \frac{(y' + 4)^2}{2^2} = 1$$

Figure 7.4

Quadratic forms can also be used to analyze equations of quadric surfaces in R^3, which are the three-dimensional analogs of conic sections. The equation of a quadric surface in R^3 is a second-degree polynomial of the form

$$ax^2 + by^2 + cz^2 + dxy + exz + fyz + gx + hy + iz + j = 0.$$

There are six basic types of quadric surfaces: ellipsoids, hyperboloids of one sheet, hyperboloids of two sheets, elliptic cones, elliptic paraboloids, and hyperbolic paraboloids. The intersection of a surface with a plane, called the **trace** of the surface in the plane, is useful to help visualize the graph of the surface in R^3. The six basic types of quadric surfaces, together with their traces, are shown on the next two pages.

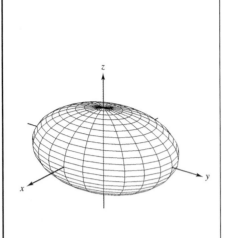

Ellipsoid

$$\frac{x^2}{a^2} + \frac{y^2}{b^2} + \frac{z^2}{c^2} = 1$$

Trace	Plane
Ellipse	Parallel to xy-plane
Ellipse	Parallel to xz-plane
Ellipse	Parallel to yz-plane

The surface is a sphere when $a = b = c \neq 0$.

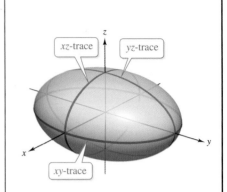

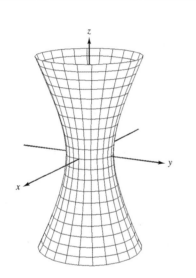

Hyperboloid of One Sheet

$$\frac{x^2}{a^2} + \frac{y^2}{b^2} - \frac{z^2}{c^2} = 1$$

Trace	Plane
Ellipse	Parallel to xy-plane
Hyperbola	Parallel to xz-plane
Hyperbola	Parallel to yz-plane

The axis of the hyperboloid corresponds to the variable whose coefficient is negative.

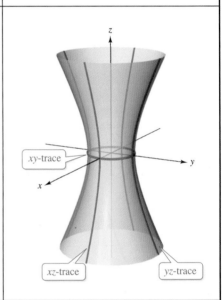

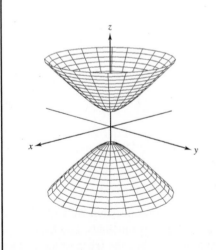

Hyperboloid of Two Sheets

$$\frac{z^2}{c^2} - \frac{x^2}{a^2} - \frac{y^2}{b^2} = 1$$

Trace	Plane
Ellipse	Parallel to xy-plane
Hyperbola	Parallel to xz-plane
Hyperbola	Parallel to yz-plane

The axis of the hyperboloid corresponds to the variable whose coefficient is positive. There is no trace in the coordinate plane perpendicular to this axis.

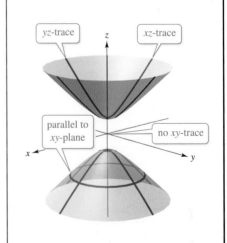

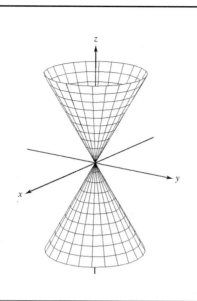

Elliptic Cone

$$\frac{x^2}{a^2} + \frac{y^2}{b^2} - \frac{z^2}{c^2} = 0$$

Trace	Plane
Ellipse	Parallel to xy-plane
Hyperbola	Parallel to xz-plane
Hyperbola	Parallel to yz-plane

The axis of the cone corresponds to the variable whose coefficient is negative. The traces in the coordinate planes parallel to this axis are intersecting lines.

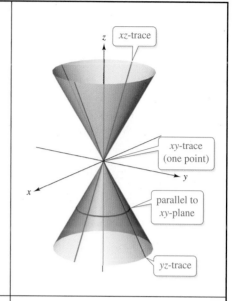

Elliptic Paraboloid

$$z = \frac{x^2}{a^2} + \frac{y^2}{b^2}$$

Trace	Plane
Ellipse	Parallel to xy-plane
Parabola	Parallel to xz-plane
Parabola	Parallel to yz-plane

The axis of the paraboloid corresponds to the variable raised to the first power.

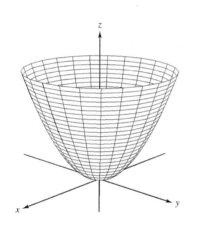

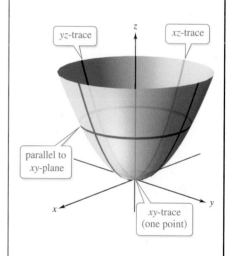

Hyperbolic Paraboloid

$$z = \frac{y^2}{b^2} - \frac{x^2}{a^2}$$

Trace	Plane
Hyperbola	Parallel to xy-plane
Parabola	Parallel to xz-plane
Parabola	Parallel to yz-plane

The axis of the paraboloid corresponds to the variable raised to the first power.

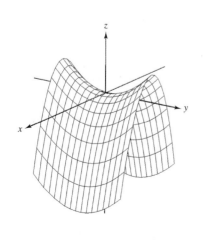

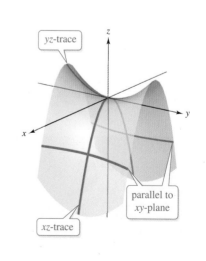

The quadratic form of the equation

$$ax^2 + by^2 + cz^2 + dxy + exz + fyz + gx + hy + iz + j = 0 \qquad \text{Quadric surface}$$

is

$$ax^2 + by^2 + cz^2 + dxy + exz + fyz. \qquad \text{Quadratic form}$$

The corresponding matrix is

REMARK

In general, the matrix A of the quadratic form will always be symmetric.

$$A = \begin{bmatrix} a & \dfrac{d}{2} & \dfrac{e}{2} \\ \dfrac{d}{2} & b & \dfrac{f}{2} \\ \dfrac{e}{2} & \dfrac{f}{2} & c \end{bmatrix}.$$

In its three-dimensional version, the Principal Axes Theorem relates the eigenvalues and eigenvectors of A to the equation of the rotated surface, as shown in Example 8.

EXAMPLE 8   Rotation of a Quadric Surface

Perform a rotation of axes to eliminate the xz-term in the quadratic equation

$$5x^2 + 4y^2 + 5z^2 + 8xz - 36 = 0.$$

SOLUTION

The matrix A associated with this quadratic equation is

$$A = \begin{bmatrix} 5 & 0 & 4 \\ 0 & 4 & 0 \\ 4 & 0 & 5 \end{bmatrix}$$

which has eigenvalues of $\lambda_1 = 1$, $\lambda_2 = 4$, and $\lambda_3 = 9$ (verify this). So, in the rotated $x'y'z'$-system, the quadratic equation is $(x')^2 + 4(y')^2 + 9(z')^2 - 36 = 0$, which in standard form is

$$\frac{(x')^2}{6^2} + \frac{(y')^2}{3^2} + \frac{(z')^2}{2^2} = 1.$$

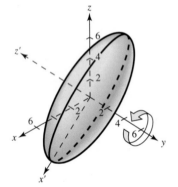

Figure 7.5

The graph of this equation is an ellipsoid. As shown in Figure 7.5, the $x'y'z'$-axes represent a counterclockwise rotation of $45°$ about the y-axis. Verify that the columns of

$$P = \begin{bmatrix} \dfrac{1}{\sqrt{2}} & 0 & \dfrac{1}{\sqrt{2}} \\ 0 & 1 & 0 \\ -\dfrac{1}{\sqrt{2}} & 0 & \dfrac{1}{\sqrt{2}} \end{bmatrix}$$

are the normalized eigenvectors of A, that P is orthogonal, and that $P^T A P$ is diagonal.

LINEAR ALGEBRA APPLIED

Some of the world's most unusual architecture makes use of quadric surfaces. For example, *Catedral Metropolitana Nossa Senhora Aparecida*, a cathedral located in Brasilia, Brazil, is in the shape of a hyperboloid of one sheet. It was designed by Pritzker Prize winning architect Oscar Niemeyer, and dedicated in 1970. The sixteen identical curved steel columns are intended to represent two hands reaching up to the sky. In the triangular gaps formed by the columns, semitransparent stained glass allows light inside for nearly the entire height of the columns.

CONSTRAINED OPTIMIZATION

Many real-life applications require you to determine the maximum or minimum value of a quantity subject to a *constraint*. For instance, consider a simplified example in which you need to find the maximum and minimum values of the quadric surface $f(x, y) = 9x^2 + 5y^2$ along the curve formed by intersection of the surface with the unit cylinder $x^2 + y^2 = 1$, as shown in Figure 7.6. The constraint is the unit cylinder $x^2 + y^2 = 1$. By inspection, the maximum value of f is 9 when $x = \pm 1$ and $y = 0$, and the minimum value of f is 5 when $x = 0$ and $y = \pm 1$.

The theorem below allows you to use the eigenvalues and eigenvectors of a symmetric matrix to solve a constrained optimization problem.

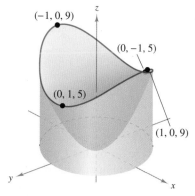

Figure 7.6

> ### Constrained Optimization Theorem
>
> For a quadratic form f in n variables with matrix of the quadratic form A subject to the constraint $\|\mathbf{x}\|^2 = 1$, the maximum value of f is the greatest eigenvalue of A and the minimum value of f is the least eigenvalue of A.

PROOF

The quadratic form f can be written as

$$f(x_1, x_2, \ldots, x_n) = \mathbf{x}^T A \mathbf{x}.$$

The matrix of the quadratic form A is symmetric, so A has n real eigenvalues (counting multiplicities). Call these $\lambda_1, \lambda_2, \ldots \lambda_n$, and assume that $\lambda_1 \geq \lambda_2 \geq \cdots \geq \lambda_n$. Now consider a change of variables $\mathbf{x} = P\mathbf{x}'$, where $\mathbf{x}' = [x_1' \quad x_2' \quad \ldots \quad x_n']^T$ and P is an orthogonal matrix that diagonalizes A. Then

$$
\begin{aligned}
f(x_1, x_2, \ldots, x_n) &= \mathbf{x}^T A \mathbf{x} \\
&= (P\mathbf{x}')^T A (P\mathbf{x}') \\
&= (\mathbf{x}')^T P^T A P \mathbf{x}' \\
&= \lambda_1 (x_1')^2 + \lambda_2 (x_2')^2 + \cdots + \lambda_n (x_n')^2
\end{aligned}
$$

and

$$
\begin{aligned}
\|\mathbf{x}\|^2 &= \|P\mathbf{x}'\|^2 \\
&= (P\mathbf{x}')^T (P\mathbf{x}') \\
&= (\mathbf{x}')^T P^T P \mathbf{x}' \\
&= (\mathbf{x}')^T \mathbf{x}' \\
&= (x_1')^2 + (x_2')^2 + \cdots + (x_n')^2 \\
&= \|\mathbf{x}'\|^2.
\end{aligned}
$$

$\|\mathbf{x}\|^2 = 1$, so $\|\mathbf{x}'\|^2 = 1$, and

$$
\begin{aligned}
\lambda_1 &= \lambda_1 [(x_1')^2 + (x_2')^2 + \cdots + (x_n')^2] \\
&\geq \lambda_1 (x_1')^2 + \lambda_2 (x_2')^2 + \cdots + \lambda_n (x_n')^2 \\
&\geq \lambda_n [(x_1')^2 + (x_2')^2 + \cdots + (x_n')^2] \\
&= \lambda_n.
\end{aligned}
$$

This shows that $\lambda_1 \geq \mathbf{x}^T A \mathbf{x} \geq \lambda_n$. So, all values of $f(x_1, x_2, \ldots, x_n) = \mathbf{x}^T A \mathbf{x}$ for which $\|\mathbf{x}\|^2 = 1$ lie between λ_1 and λ_n. If $\mathbf{x}$ is a normalized eigenvector that corresponds to λ_1, then

$$f(x_1, x_2, \ldots, x_n) = \mathbf{x}^T A \mathbf{x} = \mathbf{x}^T (\lambda_1 \mathbf{x}) = \lambda_1 \|\mathbf{x}\|^2 = \lambda_1.$$

If $\mathbf{x}$ is a normalized eigenvector that corresponds to λ_n, then

$$f(x_1, x_2, \ldots, x_n) = \mathbf{x}^T A \mathbf{x} = \mathbf{x}^T (\lambda_n \mathbf{x}) = \lambda_n \|\mathbf{x}\|^2 = \lambda_n.$$

So, f has a constrained maximum of λ_1 and a constrained minimum of λ_n.

| EXAMPLE 9 | Finding Maximum and Minimum Values |

Find the maximum and minimum values of $f(x_1, x_2) = 9x_1^2 + 5x_2^2$ subject to the constrait $\|\mathbf{x}\|^2 = 1$.

SOLUTION

The matrix of the quadratic form is the diagonal matrix

$$A = \begin{bmatrix} 9 & 0 \\ 0 & 5 \end{bmatrix}.$$

REMARK

With the substitutions $x = x_1$ and $y = x_2$, this is the same problem as that given at the top of the preceding page.

By inspection, the eigenvalues of A are $\lambda_1 = 9$ and $\lambda_2 = 5$. So by the Constrained Optimization Theorem, the maximum value of z is 9 and the minimum value of z is 5.

| EXAMPLE 10 | Finding Maximum and Minimum Values |

Find the maximum and minimum values, and the corresponding normalized eigenvectors, of $z = 7x_1^2 + 6x_1x_2 + 7x_2^2$ subject to the costraint $\|\mathbf{x}\|^2 = 1$.

SOLUTION

The quadratic form f can be written using matrix notation as

$$f(x_1, x_2) = \mathbf{x}^T A \mathbf{x} = \begin{bmatrix} x_1 & x_2 \end{bmatrix} \begin{bmatrix} 7 & 3 \\ 3 & 7 \end{bmatrix} \begin{bmatrix} x_1 \\ x_2 \end{bmatrix}.$$

Verify that the eigenvalues of $A = \begin{bmatrix} 7 & 3 \\ 3 & 7 \end{bmatrix}$ are $\lambda_1 = 10$ and $\lambda_2 = 4$, with corresponding eigenvectors

$$\begin{bmatrix} 1 \\ 1 \end{bmatrix} \quad \text{and} \quad \begin{bmatrix} -1 \\ 1 \end{bmatrix}.$$

So, the constrained maximum of 10 occurs when $(x_1, x_2) = \dfrac{1}{\sqrt{2}}(1, 1) = \left(\dfrac{1}{\sqrt{2}}, \dfrac{1}{\sqrt{2}} \right)$

and the constrained minimum of 4 occurs when $(x_1, x_2) = \dfrac{1}{\sqrt{2}}(-1, 1) = \left(-\dfrac{1}{\sqrt{2}}, \dfrac{1}{\sqrt{2}} \right)$.

| EXAMPLE 11 | Using a Change of Variables |

To find the maximum and minimum values of

$$z = 4xy$$

subject to the constraint $9x^2 + 4y^2 = 36$, you cannot use the Constrained Optimization Theorem directly because the constraint is not $\|\mathbf{x}\|^2 = 1$. However, with the change of variables

$$x = 2x' \quad \text{and} \quad y = 3y'$$

the problem becomes finding the maximum and minimum values of

$$z = 24x'y'$$

subject to the constraint $(x')^2 + (y')^2 = 1$. Verify that the maximum value of 12 occurs when $(x', y') = \left(1/\sqrt{2}, 1/\sqrt{2} \right)$, or $(x, y) = \left(\sqrt{2}, 3/\sqrt{2} \right)$, and the minimum value of -12 occurs when $(x', y') = \left(1/\sqrt{2}, -1/\sqrt{2} \right)$, or $(x, y) = \left(\sqrt{2}, -3/\sqrt{2} \right)$.

7.4 Exercises

See CalcChat.com for worked-out solutions to odd-numbered exercises.

Finding Age Distribution Vectors In Exercises 1–6, use the age transition matrix L and age distribution vector x_1 to find the age distribution vectors x_2 and x_3. Then find a stable age distribution vector.

1. $L = \begin{bmatrix} 0 & 2 \\ \frac{1}{2} & 0 \end{bmatrix}$, $x_1 = \begin{bmatrix} 10 \\ 10 \end{bmatrix}$

2. $L = \begin{bmatrix} 0 & 4 \\ \frac{1}{16} & 0 \end{bmatrix}$, $x_1 = \begin{bmatrix} 160 \\ 160 \end{bmatrix}$

3. $L = \begin{bmatrix} 0 & 3 & 4 \\ 1 & 0 & 0 \\ 0 & \frac{1}{2} & 0 \end{bmatrix}$, $x_1 = \begin{bmatrix} 12 \\ 12 \\ 12 \end{bmatrix}$

4. $L = \begin{bmatrix} 0 & 2 & 0 \\ \frac{1}{2} & 0 & 0 \\ 0 & \frac{1}{2} & 0 \end{bmatrix}$, $x_1 = \begin{bmatrix} 8 \\ 8 \\ 8 \end{bmatrix}$

5. $L = \begin{bmatrix} 0 & 2 & 2 & 0 \\ \frac{1}{4} & 0 & 0 & 0 \\ 0 & 1 & 0 & 0 \\ 0 & 0 & \frac{1}{2} & 0 \end{bmatrix}$, $x_1 = \begin{bmatrix} 100 \\ 100 \\ 100 \\ 100 \end{bmatrix}$

6. $L = \begin{bmatrix} 0 & 6 & 4 & 0 & 0 \\ \frac{1}{2} & 0 & 0 & 0 & 0 \\ 0 & 1 & 0 & 0 & 0 \\ 0 & 0 & \frac{1}{2} & 0 & 0 \\ 0 & 0 & 0 & \frac{1}{2} & 0 \end{bmatrix}$, $x_1 = \begin{bmatrix} 24 \\ 24 \\ 24 \\ 24 \\ 24 \end{bmatrix}$

7. **Population Growth Model** A population has the characteristics below.

   (a) A total of 75% of the population survives the first year. Of that 75%, 25% survives the second year. The maximum life span is 3 years.

   (b) The average number of offspring for each member of the population is 2 the first year, 4 the second year, and 2 the third year.

   The population now consists of 160 members in each of the three age classes. How many members will there be in each age class in 1 year? in 2 years?

8. **Population Growth Model** A population has the characteristics below.

   (a) A total of 80% of the population survives the first year. Of that 80%, 25% survives the second year. The maximum life span is 3 years.

   (b) The average number of offspring for each member of the population is 3 the first year, 6 the second year, and 3 the third year.

   The population now consists of 120 members in each of the three age classes. How many members will there be in each age class in 1 year? in 2 years?

9. **Population Growth Model** A population has the characteristics below.

   (a) A total of 60% of the population survives the first year. Of that 60%, 50% survives the second year. The maximum life span is 3 years.

   (b) The average number of offspring for each member of the population is 2 the first year, 5 the second year, and 2 the third year.

   The population now consists of 100 members in each of the three age classes. How many members will there be in each age class in 1 year? in 2 years?

10. Find the limit (if it exists) of $A^n x_1$ as n approaches infinity, where
$$A = \begin{bmatrix} 0 & 2 \\ \frac{1}{2} & 0 \end{bmatrix} \quad \text{and} \quad x_1 = \begin{bmatrix} a \\ a \end{bmatrix}.$$

Solving a System of Linear Differential Equations In Exercises 11–20, solve the system of first-order linear differential equations.

11. $y_1' = 2y_1$
$\; y_2' = y_2$

12. $y_1' = -5y_1$
$\; y_2' = 4y_2$

13. $y_1' = -4y_1$
$\; y_2' = -\frac{1}{2}y_2$

14. $y_1' = \frac{1}{2}y_1$
$\; y_2' = \frac{1}{8}y_2$

15. $y_1' = -y_1$
$\; y_2' = 6y_2$
$\; y_3' = y_3$

16. $y_1' = 5y_1$
$\; y_2' = -2y_2$
$\; y_3' = -3y_3$

17. $y_1' = -0.3y_1$
$\; y_2' = 0.4y_2$
$\; y_3' = -0.6y_3$

18. $y_1' = -\frac{2}{3}y_1$
$\; y_2' = -\frac{3}{5}y_2$
$\; y_3' = -8y_3$

19. $y_1' = 7y_1$
$\; y_2' = 9y_2$
$\; y_3' = -7y_3$
$\; y_4' = -9y_4$

20. $y_1' = -0.1y_1$
$\; y_2' = -\frac{7}{4}y_2$
$\; y_3' = -2\pi y_3$
$\; y_4' = \sqrt{5}y_4$

Solving a System of Linear Differential Equations In Exercises 21–28, solve the system of first-order linear differential equations.

21. $y_1' = y_1 - 4y_2$
$\; y_2' = 2y_2$

22. $y_1' = y_1 - 4y_2$
$\; y_2' = -2y_1 + 8y_2$

23. $y_1' = y_1 + 2y_2$
$\; y_2' = 2y_1 + y_2$

24. $y_1' = y_1 - y_2$
$\; y_2' = 2y_1 + 4y_2$

25. $y_1' = y_1 - 2y_2 + y_3$
$\; y_2' = 2y_2 + 4y_3$
$\; y_3' = 3y_3$

26. $y_1' = 2y_1 + y_2 + y_3$
$\; y_2' = y_1 + y_2$
$\; y_3' = y_1 + y_3$

27. $y_1' = 3y_2 - 5y_3$
$\; y_2' = 4y_1 - 4y_2 + 10y_3$
$\; y_3' = - 4y_3$

28. $y_1' = -2y_1 + y_3$
$\; y_2' = 3y_2 + 4y_3$
$\; y_3' = y_3$

Writing a System and Verifying the General Solution
In Exercises 29–32, write the system of first-order linear differential equations represented by the matrix equation $y' = Ay$. Then verify the general solution.

29. $A = \begin{bmatrix} 1 & 1 \\ 0 & 1 \end{bmatrix}$, $\begin{aligned} y_1 &= C_1 e^t + C_2 t e^t \\ y_2 &= C_2 e^t \end{aligned}$

30. $A = \begin{bmatrix} 1 & -1 \\ 1 & 1 \end{bmatrix}$, $\begin{aligned} y_1 &= C_1 e^t \cos t + C_2 e^t \sin t \\ y_2 &= -C_2 e^t \cos t + C_1 e^t \sin t \end{aligned}$

31. $A = \begin{bmatrix} 0 & 1 & 0 \\ 0 & 0 & 1 \\ 0 & -4 & 0 \end{bmatrix}$,

$\begin{aligned} y_1 &= C_1 + C_2 \cos 2t + C_3 \sin 2t \\ y_2 &= 2C_3 \cos 2t - 2C_2 \sin 2t \\ y_3 &= -4C_2 \cos 2t - 4C_3 \sin 2t \end{aligned}$

32. $A = \begin{bmatrix} 0 & 1 & 0 \\ 0 & 0 & 1 \\ 1 & -3 & 3 \end{bmatrix}$,

$\begin{aligned} y_1 &= C_1 e^t + C_2 t e^t + C_3 t^2 e^t \\ y_2 &= (C_1 + C_2)e^t + (C_2 + 2C_3)t e^t + C_3 t^2 e^t \\ y_3 &= (C_1 + 2C_2 + 2C_3)e^t + (C_2 + 4C_3)t e^t + C_3 t^2 e^t \end{aligned}$

Finding the Matrix of a Quadratic Form In Exercises 33–38, find the matrix A of the quadratic form associated with the equation.

33. $x^2 + y^2 - 4 = 0$ **34.** $x^2 - 4xy + y^2 - 4 = 0$

35. $9x^2 + 10xy - 4y^2 - 36 = 0$

36. $12x^2 - 5xy - x + 2y - 20 = 0$

37. $10xy - 10y^2 + 4x - 48 = 0$

38. $16x^2 - 4xy + 20y^2 - 72 = 0$

Finding the Matrix of a Quadratic Form In Exercises 39–44, find the matrix A of the quadratic form associated with the equation. Then find the eigenvalues of A and an orthogonal matrix P such that $P^T A P$ is diagonal.

39. $2x^2 - 3xy - 2y^2 + 10 = 0$

40. $5x^2 - 2xy + 5y^2 + 10x - 17 = 0$

41. $13x^2 + 6\sqrt{3}xy + 7y^2 - 16 = 0$

42. $3x^2 - 2\sqrt{3}xy + y^2 + 2x + 2\sqrt{3}y = 0$

43. $16x^2 - 24xy + 9y^2 - 60x - 80y + 100 = 0$

44. $17x^2 + 32xy - 7y^2 - 75 = 0$

Rotation of a Conic In Exercises 45–52, use the Principal Axes Theorem to perform a rotation of axes to eliminate the xy-term in the quadratic equation. Identify the resulting rotated conic and give its equation in the new coordinate system.

45. $13x^2 - 8xy + 7y^2 - 45 = 0$

46. $x^2 + 4xy + y^2 - 9 = 0$

47. $2x^2 - 4xy + 5y^2 - 36 = 0$

48. $7x^2 + 32xy - 17y^2 - 50 = 0$

49. $2x^2 + 4xy + 2y^2 + 6\sqrt{2}x + 2\sqrt{2}y + 4 = 0$

50. $8x^2 + 8xy + 8y^2 + 10\sqrt{2}x + 26\sqrt{2}y + 31 = 0$

51. $xy + x - 2y + 3 = 0$

52. $5x^2 - 2xy + 5y^2 + 10\sqrt{2}x = 0$

Rotation of a Quadric Surface In Exercises 53–56, find the matrix A of the quadratic form associated with the equation. Then find the equation of the quadric surface in the rotated $x'y'z'$-system.

53. $3x^2 - 2xy + 3y^2 + 8z^2 - 16 = 0$

54. $2x^2 + 2y^2 + 2z^2 + 2xy + 2xz + 2yz - 1 = 0$

55. $x^2 + 2y^2 + 2z^2 + 2yz - 1 = 0$

56. $x^2 + y^2 + z^2 + 2xy - 8 = 0$

Constrained Optimization In Exercises 57–66, find the maximum and minimum values, and a vector where each occurs, of the quadratic form subject to the constraint.

57. $z = 3x_1^2 + 2x_2^2; \|\mathbf{x}\|^2 = 1$

58. $z = 11x_1^2 + 4x_2^2; \|\mathbf{x}\|^2 = 1$

59. $z = x_1^2 + 12x_2^2; 4x_1^2 + 25x_2^2 = 100$

60. $z = -5x^2 + 9y^2; x^2 + 9y^2 = 9$

61. $z = 5x^2 + 12xy + 5y^2; x^2 + y^2 = 1$

62. $z = 5x_1^2 + 12x_1x_2; \|\mathbf{x}\|^2 = 1$

63. $z = 6x_1x_2; \|\mathbf{x}\|^2 = 1$

64. $z = 9xy; 9x^2 + 16y^2 = 144$

65. $w = x^2 + 3y^2 + z^2 + 2xy + 2xz + 2yz; x^2 + y^2 + z^2 = 1$

66. $w = 2x^2 - y^2 - z^2 + 4xy - 4xz + 8yz; x^2 + y^2 + z^2 = 1$

67. Let P be a 2×2 orthogonal matrix such that $|P| = 1$. Show that there exists a number $\theta, 0 \le \theta < 2\pi$, such that

$$P = \begin{bmatrix} \cos \theta & -\sin \theta \\ \sin \theta & \cos \theta \end{bmatrix}.$$

68. CAPSTONE

(a) Explain how to model population growth using an age transition matrix and an age distribution vector, and how to find a stable age distribution vector.

(b) Explain how to use a matrix equation to solve a system of first-order linear differential equations.

(c) Explain how to use the Principal Axes Theorem to perform a rotation of axes for a conic and a quadric surface.

(d) Explain how to solve a constrained optimization problem.

69. Use your school's library, the Internet, or some other reference source to find real-life applications of constrained optimization.

7 Review Exercises
See CalcChat.com for worked-out solutions to odd-numbered exercises.

Characteristic Equation, Eigenvalues, and Basis In Exercises 1–6, find (a) the characteristic equation of A, (b) the eigenvalues of A, and (c) a basis for the eigenspace corresponding to each eigenvalue.

1. $A = \begin{bmatrix} 2 & 1 \\ 5 & -2 \end{bmatrix}$ **2.** $A = \begin{bmatrix} 2 & 1 \\ -4 & -2 \end{bmatrix}$

3. $A = \begin{bmatrix} 9 & 4 & -3 \\ -2 & 0 & 6 \\ -1 & -4 & 11 \end{bmatrix}$ **4.** $A = \begin{bmatrix} -4 & 1 & 2 \\ 0 & 1 & 1 \\ 0 & 0 & 3 \end{bmatrix}$

5. $A = \begin{bmatrix} 2 & 0 & 1 \\ 0 & 3 & 4 \\ 0 & 0 & 1 \end{bmatrix}$ **6.** $A = \begin{bmatrix} 1 & 0 & 4 \\ 0 & 1 & -2 \\ 1 & 0 & -2 \end{bmatrix}$

Characteristic Equation, Eigenvalues, and Basis In Exercises 7 and 8, use a software program or a graphing utility to find (a) the characteristic equation of A, (b) the eigenvalues of A, and (c) a basis for the eigenspace corresponding to each eigenvalue.

7. $A = \begin{bmatrix} 2 & 1 & 0 & 0 \\ 1 & 2 & 0 & 0 \\ 0 & 0 & 2 & 1 \\ 0 & 0 & 1 & 2 \end{bmatrix}$ **8.** $A = \begin{bmatrix} 3 & 0 & 2 & 0 \\ 1 & 3 & 1 & 0 \\ 0 & 1 & 1 & 0 \\ 0 & 0 & 0 & 4 \end{bmatrix}$

Determining Whether a Matrix Is Diagonalizable In Exercises 9–14, determine whether A is diagonalizable. If it is, find a nonsingular matrix P such that $P^{-1}AP$ is diagonal.

9. $A = \begin{bmatrix} 1 & -4 \\ -2 & 8 \end{bmatrix}$ **10.** $A = \begin{bmatrix} \frac{1}{6} & \frac{1}{4} \\ \frac{2}{3} & 0 \end{bmatrix}$

11. $A = \begin{bmatrix} -2 & -1 & 3 \\ 0 & 1 & 2 \\ 0 & 0 & 1 \end{bmatrix}$ **12.** $A = \begin{bmatrix} 3 & -2 & 2 \\ -2 & 0 & -1 \\ 2 & -1 & 0 \end{bmatrix}$

13. $A = \begin{bmatrix} 1 & 0 & 2 \\ 0 & 1 & 0 \\ 2 & 0 & 1 \end{bmatrix}$ **14.** $A = \begin{bmatrix} 2 & -1 & 1 \\ -2 & 3 & -2 \\ -1 & 1 & 0 \end{bmatrix}$

15. For what value(s) of a does the matrix

$$A = \begin{bmatrix} 0 & 1 \\ a & 1 \end{bmatrix}$$

have the characteristics below?

(a) A has an eigenvalue of multiplicity 2.

(b) A has -1 and 2 as eigenvalues.

(c) A has real eigenvalues.

16. Show that if $0 < \theta < \pi$, then the transformation for a counterclockwise rotation through an angle θ has no real eigenvalues.

Writing In Exercises 17–20, explain why the matrix is not diagonalizable.

17. $A = \begin{bmatrix} 0 & 9 \\ 0 & 0 \end{bmatrix}$ **18.** $A = \begin{bmatrix} -1 & 2 \\ 0 & -1 \end{bmatrix}$

19. $A = \begin{bmatrix} 3 & 0 & 0 \\ 1 & 3 & 0 \\ 0 & 0 & 3 \end{bmatrix}$ **20.** $A = \begin{bmatrix} -2 & 3 & 1 \\ 0 & 4 & 3 \\ 0 & 0 & -2 \end{bmatrix}$

Determining Whether Two Matrices Are Similar In Exercises 21–24, determine whether the matrices are similar. If they are, find a matrix P such that $A = P^{-1}BP$.

21. $A = \begin{bmatrix} 1 & 0 \\ 0 & 2 \end{bmatrix}$, $B = \begin{bmatrix} 2 & 0 \\ 0 & 1 \end{bmatrix}$

22. $A = \begin{bmatrix} 5 & 0 \\ 0 & 3 \end{bmatrix}$, $B = \begin{bmatrix} 7 & 2 \\ -4 & 1 \end{bmatrix}$

23. $A = \begin{bmatrix} 1 & 1 & 0 \\ 0 & 1 & 1 \\ 0 & 0 & 1 \end{bmatrix}$, $B = \begin{bmatrix} 1 & 1 & 0 \\ 0 & 1 & 0 \\ 0 & 0 & 1 \end{bmatrix}$

24. $A = \begin{bmatrix} 1 & 0 & 0 \\ 0 & -2 & 0 \\ 0 & 0 & -2 \end{bmatrix}$, $B = \begin{bmatrix} 1 & -3 & -3 \\ 3 & -5 & -3 \\ -3 & 3 & 1 \end{bmatrix}$

Determining Symmetric and Orthogonal Matrices In Exercises 25–32, determine whether the matrix is symmetric, orthogonal, both, or neither.

25. $A = \begin{bmatrix} -\dfrac{\sqrt{2}}{2} & \dfrac{\sqrt{2}}{2} \\ \dfrac{\sqrt{2}}{2} & \dfrac{\sqrt{2}}{2} \end{bmatrix}$ **26.** $A = \begin{bmatrix} \dfrac{2\sqrt{5}}{5} & \dfrac{\sqrt{5}}{5} \\ \dfrac{\sqrt{5}}{5} & -\dfrac{2\sqrt{5}}{5} \end{bmatrix}$

27. $A = \begin{bmatrix} 0 & 0 & 1 \\ 0 & 1 & 0 \\ 1 & 0 & 0 \end{bmatrix}$ **28.** $A = \begin{bmatrix} 0 & 0 & 1 \\ 0 & 1 & 0 \\ 1 & 0 & 1 \end{bmatrix}$

29. $A = \begin{bmatrix} \frac{1}{3} & \frac{1}{2} & \frac{1}{3} \\ \frac{1}{3} & 0 & \frac{1}{3} \\ \frac{1}{3} & \frac{1}{2} & \frac{1}{3} \end{bmatrix}$ **30.** $A = \begin{bmatrix} \frac{4}{5} & 0 & \frac{3}{5} \\ 0 & 1 & 0 \\ -\frac{3}{5} & 0 & \frac{4}{5} \end{bmatrix}$

31. $A = \begin{bmatrix} -\frac{2}{3} & \frac{1}{3} & -\frac{2}{3} \\ \frac{2}{3} & \frac{2}{3} & -\frac{1}{3} \\ \frac{1}{3} & -\frac{2}{3} & \frac{2}{3} \end{bmatrix}$

32. $A = \begin{bmatrix} \dfrac{\sqrt{3}}{3} & \dfrac{\sqrt{3}}{3} & \dfrac{\sqrt{3}}{3} \\ \dfrac{\sqrt{3}}{3} & \dfrac{2\sqrt{3}}{3} & 0 \\ \dfrac{\sqrt{3}}{3} & 0 & \dfrac{\sqrt{3}}{3} \end{bmatrix}$

Eigenvectors of a Symmetric Matrix In Exercises 33–36, show that any two eigenvectors of the symmetric matrix corresponding to distinct eigenvalues are orthogonal.

33. $\begin{bmatrix} 2 & 0 \\ 0 & -3 \end{bmatrix}$

34. $\begin{bmatrix} 4 & -2 \\ -2 & 1 \end{bmatrix}$

35. $\begin{bmatrix} -1 & 0 & -1 \\ 0 & -1 & 0 \\ -1 & 0 & 1 \end{bmatrix}$

36. $\begin{bmatrix} 2 & 0 & 0 \\ 0 & 2 & 0 \\ 0 & 0 & 5 \end{bmatrix}$

Orthogonally Diagonalizable Matrices In Exercises 37–40, determine whether the matrix is orthogonally diagonalizable.

37. $\begin{bmatrix} -3 & -1 \\ -1 & -2 \end{bmatrix}$

38. $\begin{bmatrix} -4 & 1 \\ -1 & 3 \end{bmatrix}$

39. $\begin{bmatrix} 4 & 1 & 2 \\ 0 & -1 & 0 \\ 2 & 1 & -5 \end{bmatrix}$

40. $\begin{bmatrix} 5 & 4 & -1 \\ 4 & 1 & 3 \\ -1 & 3 & -2 \end{bmatrix}$

Orthogonal Diagonalization In Exercises 41–46, find a matrix P that orthogonally diagonalizes A. Verify that $P^T A P$ gives the correct diagonal form.

41. $A = \begin{bmatrix} 3 & 4 \\ 4 & -3 \end{bmatrix}$

42. $A = \begin{bmatrix} 8 & 15 \\ 15 & -8 \end{bmatrix}$

43. $A = \begin{bmatrix} 1 & 1 & 0 \\ 1 & 1 & 0 \\ 0 & 0 & 0 \end{bmatrix}$

44. $A = \begin{bmatrix} 3 & 0 & -3 \\ 0 & -3 & 0 \\ -3 & 0 & 3 \end{bmatrix}$

45. $A = \begin{bmatrix} 2 & 0 & -1 \\ 0 & 1 & 0 \\ -1 & 0 & 2 \end{bmatrix}$

46. $A = \begin{bmatrix} 1 & 2 & 0 \\ 2 & 1 & 0 \\ 0 & 0 & 5 \end{bmatrix}$

Steady State Probability Vector In Exercises 47–54, find the steady state probability vector for the matrix. An eigenvector $\mathbf{v}$ of an $n \times n$ matrix A is a steady state probability vector when $A\mathbf{v} = \mathbf{v}$ and the components of $\mathbf{v}$ sum to 1.

47. $A = \begin{bmatrix} \frac{2}{3} & \frac{1}{2} \\ \frac{1}{3} & \frac{1}{2} \end{bmatrix}$

48. $A = \begin{bmatrix} \frac{1}{2} & 1 \\ \frac{1}{2} & 0 \end{bmatrix}$

49. $A = \begin{bmatrix} 0.8 & 0.3 \\ 0.2 & 0.7 \end{bmatrix}$

50. $A = \begin{bmatrix} 0.4 & 0.2 \\ 0.6 & 0.8 \end{bmatrix}$

51. $A = \begin{bmatrix} \frac{1}{2} & \frac{1}{4} & 0 \\ \frac{1}{2} & \frac{1}{2} & \frac{1}{2} \\ 0 & \frac{1}{4} & \frac{1}{2} \end{bmatrix}$

52. $A = \begin{bmatrix} \frac{1}{3} & \frac{2}{3} & \frac{1}{3} \\ \frac{1}{3} & \frac{1}{3} & 0 \\ \frac{1}{3} & 0 & \frac{2}{3} \end{bmatrix}$

53. $A = \begin{bmatrix} 0.7 & 0.1 & 0.1 \\ 0.2 & 0.7 & 0.1 \\ 0.1 & 0.2 & 0.8 \end{bmatrix}$

54. $A = \begin{bmatrix} 0.3 & 0.1 & 0.4 \\ 0.2 & 0.4 & 0.0 \\ 0.5 & 0.5 & 0.6 \end{bmatrix}$

55. **Proof** Prove that if A is an $n \times n$ symmetric matrix, then $P^T A P$ is symmetric for any $n \times n$ matrix P.

56. Show that the characteristic polynomial of

$$A = \begin{bmatrix} 0 & 1 & 0 & 0 & \cdots & 0 \\ 0 & 0 & 1 & 0 & \cdots & 0 \\ \vdots & \vdots & \vdots & \vdots & & \vdots \\ 0 & 0 & 0 & 0 & \cdots & 1 \\ -a_0 & -a_1 & -a_2 & -a_3 & \cdots & -a_{n-1} \end{bmatrix}$$

is $p(\lambda) = \lambda^n + a_{n-1}\lambda^{n-1} + \cdots + a_2\lambda^2 + a_1\lambda + a_0$. A is called the **companion matrix** of the polynomial p.

Finding the Companion Matrix and Eigenvalues In Exercises 57 and 58, use the result of Exercise 56 to find the companion matrix A of the polynomial and find the eigenvalues of A.

57. $p(\lambda) = 4\lambda^2 - 9\lambda$

58. $p(\lambda) = 2\lambda^3 - 7\lambda^2 - 120\lambda + 189$

59. The characteristic equation of

$$A = \begin{bmatrix} 8 & -4 \\ 2 & 2 \end{bmatrix}$$

is $\lambda^2 - 10\lambda + 24 = 0$. Using $A^2 - 10A + 24I_2 = O$, you can find powers of A by the process below.

$A^2 = 10A - 24I_2$, $A^3 = 10A^2 - 24A$,
$A^4 = 10A^3 - 24A^2, \ldots$

Use this process to find the matrices A^2, A^3, and A^4.

60. Repeat Exercise 59 for the matrix

$$A = \begin{bmatrix} 9 & 4 & -3 \\ -2 & 0 & 6 \\ -1 & -4 & 11 \end{bmatrix}.$$

61. **Proof** Let A be an $n \times n$ matrix.

(a) Prove or disprove that an eigenvector of A is also an eigenvector of A^2.

(b) Prove or disprove that an eigenvector of A^2 is also an eigenvector of A.

62. **Proof** Let A be an $n \times n$ matrix. Prove that if $A\mathbf{x} = \lambda\mathbf{x}$, then $\mathbf{x}$ is an eigenvector of $(A + cI)$, where λ and c are scalars. What is the corresponding eigenvalue?

63. **Proof** Let A and B be $n \times n$ matrices. Prove that if A is nonsingular, then AB is similar to BA.

64. (a) Find a symmetric matrix B such that $B^2 = A$ for

$$A = \begin{bmatrix} 2 & 1 \\ 1 & 2 \end{bmatrix}.$$

(b) Generalize the result of part (a) by proving that if A is an $n \times n$ symmetric matrix with positive eigenvalues, then there exists a symmetric matrix B such that $B^2 = A$.

65. Determine all $n \times n$ symmetric matrices that have 0 as their only eigenvalue.

66. Find an orthogonal matrix P such that $P^{-1}AP$ is diagonal for the matrix
$$A = \begin{bmatrix} a & b \\ b & a \end{bmatrix}.$$

67. Writing Let A be an $n \times n$ idempotent matrix (that is, $A^2 = A$). Describe the eigenvalues of A.

68. Writing The matrix below has an eigenvalue $\lambda = 2$ of multiplicity 4.
$$A = \begin{bmatrix} 2 & a & 0 & 0 \\ 0 & 2 & b & 0 \\ 0 & 0 & 2 & c \\ 0 & 0 & 0 & 2 \end{bmatrix}$$

(a) Under what conditions is A diagonalizable?

(b) Under what conditions does the eigenspace of $\lambda = 2$ have dimension 1? 2? 3?

True or False? In Exercises 69 and 70, determine whether each statement is true or false. If a statement is true, give a reason or cite an appropriate statement from the text. If a statement is false, provide an example that shows the statement is not true in all cases or cite an appropriate statement from the text.

69. (a) An eigenvector of an $n \times n$ matrix A is a nonzero vector $\mathbf{x}$ in R^n such that $A\mathbf{x}$ is a scalar multiple of $\mathbf{x}$.

(b) Similar matrices may or may not have the same eigenvalues.

(c) To diagonalize a square matrix A, you need to find an invertible matrix P such that $P^{-1}AP$ is diagonal.

70. (a) An eigenvalue of a matrix A is a scalar λ such that $\det(\lambda I - A) = 0$.

(b) An eigenvector may be the zero vector $\mathbf{0}$.

(c) A matrix A is orthogonally diagonalizable when there exists an orthogonal matrix P such that $P^{-1}AP = D$ is diagonal.

Finding Age Distribution Vectors In Exercises 71–74, use the age transition matrix L and the age distribution vector $\mathbf{x}_1$ to find the age distribution vectors $\mathbf{x}_2$ and $\mathbf{x}_3$. Then find a stable age distribution vector.

71. $L = \begin{bmatrix} 0 & 1 \\ \frac{1}{4} & 0 \end{bmatrix}$, $\mathbf{x}_1 = \begin{bmatrix} 100 \\ 100 \end{bmatrix}$

72. $L = \begin{bmatrix} 0 & 1 \\ \frac{3}{4} & 0 \end{bmatrix}$, $\mathbf{x}_1 = \begin{bmatrix} 32 \\ 32 \end{bmatrix}$

73. $L = \begin{bmatrix} 0 & 3 & 12 \\ 1 & 0 & 0 \\ 0 & \frac{1}{6} & 0 \end{bmatrix}$, $\mathbf{x}_1 = \begin{bmatrix} 300 \\ 300 \\ 300 \end{bmatrix}$

74. $L = \begin{bmatrix} 0 & 2 & 2 \\ \frac{1}{2} & 0 & 0 \\ 0 & 0 & 0 \end{bmatrix}$, $\mathbf{x}_1 = \begin{bmatrix} 240 \\ 240 \\ 240 \end{bmatrix}$

75. Population Growth Model A population has the characteristics below.

(a) A total of 90% of the population survives the first year. Of that 90%, 75% survives the second year. The maximum life span is 3 years.

(b) The average number of offspring for each member of the population is 4 the first year, 6 the second year, and 2 the third year.

The population now consists of 120 members in each of the three age classes. How many members will there be in each age class in 1 year? in 2 years?

76. Population Growth Model A population has the characteristics below.

(a) A total of 75% of the population survives the first year. Of that 75%, 60% survives the second year. The maximum life span is 3 years.

(b) The average number of offspring for each member of the population is 4 the first year, 8 the second year, and 2 the third year.

The population now consists of 120 members in each of the three age classes. How many members will there be in each age class in 1 year? in 2 years?

Solving a System of Linear Differential Equations In Exercises 77–80, solve the system of first-order linear differential equations.

77. $y_1' = 3y_1$
$y_2' = y_1 - y_2$

78. $y_1' = y_2$
$y_2' = y_1$

79. $y_1' = 3y_1$
$y_2' = 8y_2$
$y_3' = -8y_3$

80. $y_1' = 6y_1 - y_2 + 2y_3$
$y_2' = 3y_2 - y_3$
$y_3' = y_3$

Rotation of a Conic In Exercises 81–84, (a) find the matrix A of the quadratic form associated with the equation, (b) find an orthogonal matrix P such that P^TAP is diagonal, (c) use the Principal Axes Theorem to perform a rotation of axes to eliminate the xy-term in the quadratic equation, and (d) sketch the graph of each equation.

81. $x^2 + 3xy + y^2 - 3 = 0$

82. $x^2 - \sqrt{3}xy + 2y^2 - 10 = 0$

83. $xy - 2 = 0$

84. $9x^2 - 24xy + 16y^2 - 400x - 300y = 0$

Constrained Optimization In Exercises 85–88, find the maximum and minimum values, and a vector where each occurs, of the quadratic form subject to the constraint.

85. $z = x^2 - y^2$; $x^2 + y^2 = 1$

86. $z = x_1x_2$; $25x_1^2 + 4x_2^2 = 100$

87. $z = 15x_1^2 - 4x_1x_2 + 15x_2^2$; $\|\mathbf{x}\|^2 = 1$

88. $z = -11x^2 + 10xy - 11y^2$; $x^2 + y^2 = 1$

7   Projects

1 Population Growth and Dynamical Systems (I)

Systems of differential equations often arise in biological applications of population growth of various species of animals. These equations are called **dynamical systems** because they describe the changes in a system as functions of time. Assume that a biologist studies the populations of predator sharks $y_1(t)$ and their small fish prey $y_2(t)$ over time t. One model for the relative growths of these populations is

$$y_1'(t) = ay_1(t) + by_2(t) \qquad \text{Predator}$$
$$y_2'(t) = cy_1(t) + dy_2(t) \qquad \text{Prey}$$

where $a, b, c,$ and d are constants. The constants a and d are positive, reflecting the growth rates of the species. In a predator-prey relationship, $b > 0$ and $c < 0$ because an increase in prey fish y_2 would cause an increase in predator sharks y_1, whereas an increase in y_1 would cause a decrease in y_2.

The system of linear differential equations below models the populations of sharks $y_1(t)$ and prey fish $y_2(t)$ with the populations at time $t = 0$.

$$y_1'(t) = 0.3y_1(t) + 0.5y_2(t) \qquad y_1(0) = 40$$
$$y_2'(t) = -0.6y_1(t) + 2.0y_2(t) \qquad y_2(0) = 107$$

1. Use the diagonalization techniques of this chapter to find the populations $y_1(t)$ and $y_2(t)$ at any time $t > 0$.

2. Interpret the solutions in terms of the long-term population trends for the two species. Does one species ultimately disappear? Why or why not?

3. Graph the solutions $y_1(t)$ and $y_2(t)$ over the domain $0 \le t \le 5$.

4. Explain why the quotient $y_2(t)/y_1(t)$ approaches a limit as t increases.

2 The Fibonacci Sequence

The **Fibonacci sequence** is named after the Italian mathematician Leonard Fibonacci of Pisa (1170–1250). To form this sequence, define the first two terms as $x_1 = 1$ and $x_2 = 1$, and then define the nth term as the sum of its two immediate predecessors. That is, $x_n = x_{n-1} + x_{n-2}$. So, the third term is $2 = 1 + 1$. the fourth term is $3 = 2 + 1$, and so on. The formula $x_n = x_{n-1} + x_{n-2}$ is called *recursive* because the first $n - 1$ terms must be calculated before the nth term can be calculated. In this project, you will use eigenvalues and diagonalization to derive an explicit formula for the nth term of the Fibonacci sequence.

1. Calculate the first 15 terms of the Fibonacci sequence.

2. Explain how the matrix identity $\begin{bmatrix} 1 & 1 \\ 1 & 0 \end{bmatrix} \begin{bmatrix} x_{n-1} \\ x_{n-2} \end{bmatrix} = \begin{bmatrix} x_{n-1} + x_{n-2} \\ x_{n-1} \end{bmatrix}$ can be used to generate the Fibonacci sequence recursively.

3. Starting with $\begin{bmatrix} x_1 \\ x_2 \end{bmatrix} = \begin{bmatrix} 1 \\ 1 \end{bmatrix}$, show that $A^{n-2} \begin{bmatrix} 1 \\ 1 \end{bmatrix} = \begin{bmatrix} x_n \\ x_{n-1} \end{bmatrix}$, where $A = \begin{bmatrix} 1 & 1 \\ 1 & 0 \end{bmatrix}$.

4. Find a matrix P that diagonalizes A.

5. Derive an explicit formula for the nth term of the Fibonacci sequence. Use this formula to calculate $x_1, x_2,$ and x_3.

6. Determine the limit of x_n/x_{n-1} as n approaches infinity. Do you recognize this number?

REMARK

You can learn more about dynamical systems and population modeling in most books on differential equations. You can learn more about Fibonacci numbers in most books on number theory. You might find it interesting to look at the *Fibonacci Quarterly,* the official journal of the Fibonacci Association.

6 and 7   Cumulative Test

See CalcChat.com for worked-out solutions
to odd-numbered exercises.

Take this test to review the material in Chapters 6 and 7. After you are finished,
check your work against the answers in the back of the book.

In Exercises 1 and 2, determine whether the function is a linear transformation.

1. $T: R^3 \to R^2$, $T(x, y, z) = (2x, x + y)$ 2. $T: M_{2,2} \to R$, $T(A) = |A + A^T|$

3. Let $T: R^n \to R^m$ be the linear transformation defined by $T(\mathbf{v}) = A\mathbf{v}$, where

$$A = \begin{bmatrix} 3 & 0 & 1 & 0 \\ 0 & 3 & 0 & 2 \end{bmatrix}.$$

   Find the dimensions of R^n and R^m.

4. Let $T: R^2 \to R^3$ be the linear transformation defined by $T(\mathbf{v}) = A\mathbf{v}$, where

$$A = \begin{bmatrix} -2 & 0 \\ 1 & 0 \\ 0 & 0 \end{bmatrix}.$$

   Find (a) $T(2, -1)$ and (b) the preimage of $(-6, 3, 0)$.

5. Find the kernel of the linear transformation

   $T: R^4 \to R^4$, $T(x_1, x_2, x_3, x_4) = (x_1 - x_2, x_2 - x_1, 0, x_3 + x_4)$.

6. Let $T: R^4 \to R^2$ be the linear transformation defined by $T(\mathbf{v}) = A\mathbf{v}$, where

$$A = \begin{bmatrix} 1 & 0 & 1 & 0 \\ 0 & -1 & 0 & -1 \end{bmatrix}.$$

   Find a basis for (a) the kernel of T and (b) the range of T. (c) Determine the rank
   and nullity of T.

In Exercises 7–10, find the standard matrix for the linear transformation T.

7. $T(x, y) = (3x + 2y, 2y - x)$ 8. $T(x, y, z) = (x + y, y + z, x - z)$

9. $T(x, y, z) = (3z - 2y, 4x + 11z)$ 10. $T(x_1, x_2, x_3) = (0, 0, 0)$

11. Find the standard matrix A for the linear transformation $\text{proj}_{\mathbf{v}}\mathbf{u}: R^2 \to R^2$ that
    projects an arbitrary vector $\mathbf{u}$ onto the vector $\mathbf{v} = \begin{bmatrix} 1 & -1 \end{bmatrix}^T$, as shown in the figure.
    Use this matrix to find the images of the vectors $(1, 1)$ and $(-2, 2)$.

12. Let $T: R^2 \to R^2$ be the linear transformation defined by a counterclockwise rotation
    of $30°$ in R^2.

    (a) Find the standard matrix A for the linear transformation.

    (b) Use A to find the image of the vector $\mathbf{v} = (1, 2)$.

    (c) Sketch the graph of $\mathbf{v}$ and its image.

In Exercises 13 and 14, find the standard matrices for $T = T_2 \circ T_1$ and $T' = T_1 \circ T_2$.

13. $T_1: R^2 \to R^2$, $T_1(x, y) = (x - 2y, 2x + 3y)$
    $T_2: R^2 \to R^2$, $T_2(x, y) = (2x, x - y)$

14. $T_1: R^3 \to R^3$, $T_1(x, y, z) = (x + 2y, y - z, -2x + y + 2z)$
    $T_2: R^3 \to R^3$, $T_2(x, y, z) = (y + z, x + z, 2y - 2z)$

15. Find the inverse of the linear transformation $T: R^2 \to R^2$ defined by
    $T(x, y) = (x - y, 2x + y)$. Verify that $(T^{-1} \circ T)(3, -2) = (3, -2)$.

16. Determine whether the linear transformation $T: R^3 \to R^3$ defined by
    $T(x_1, x_2, x_3) = (x_1 + x_2, x_2 + x_3, x_1 + x_3)$ is invertible. If it is, find its inverse.

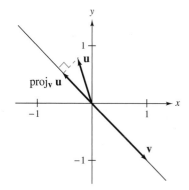

Figure for 11

17. Find the matrix of the linear transformation $T(x, y) = (y, 2x, x + y)$ relative to the bases $B = \{(1, 1), (1, 0)\}$ for R^2 and $B' = \{(1, 0, 0), (1, 1, 0), (1, 1, 1)\}$ for R^3. Use this matrix to find the image of the vector $(0, 1)$.

18. Let $B = \{(1, 0), (0, 1)\}$ and $B' = \{(1, 1), (1, 2)\}$ be bases for R^2.

    (a) Find the matrix A of $T: R^2 \to R^2$, $T(x, y) = (x - 2y, x + 4y)$, relative to the basis B.

    (b) Find the transition matrix P from B' to B.

    (c) Find the matrix A' of T relative to the basis B'.

    (d) Find $\left[T(\mathbf{v})\right]_{B'}$ when $\left[\mathbf{v}\right]_{B'} = \begin{bmatrix} 3 \\ -2 \end{bmatrix}$.

    (e) Verify your answer in part (d) by finding $\left[\mathbf{v}\right]_B$ and $\left[T(\mathbf{v})\right]_B$.

In Exercises 19–22, find the eigenvalues and the corresponding eigenvectors of the matrix.

19. $\begin{bmatrix} 7 & 2 \\ -2 & 3 \end{bmatrix}$
20. $\begin{bmatrix} -15 & -5 \\ 0 & 5 \end{bmatrix}$

21. $\begin{bmatrix} 1 & 2 & 1 \\ 0 & 3 & 1 \\ 0 & -3 & -1 \end{bmatrix}$
22. $\begin{bmatrix} 1 & -1 & 1 \\ 0 & 1 & 2 \\ 0 & 0 & 1 \end{bmatrix}$

In Exercises 23 and 24, find a nonsingular matrix P such that $P^{-1}AP$ is diagonal.

23. $A = \begin{bmatrix} 2 & 3 & 1 \\ 0 & -1 & 2 \\ 0 & 0 & 3 \end{bmatrix}$
24. $A = \begin{bmatrix} 0 & -3 & 5 \\ -4 & 4 & -10 \\ 0 & 0 & 4 \end{bmatrix}$

25. Find a basis B for R^3 such that the matrix for the linear transformation $T: R^3 \to R^3$, $T(x, y, z) = (2x - 2z, 2y - 2z, 3x - 3z)$, relative to B is diagonal.

26. Find an orthogonal matrix P such that P^TAP diagonalizes the symmetric matrix $A = \begin{bmatrix} 1 & 3 \\ 3 & 1 \end{bmatrix}$.

27. Use the Gram-Schmidt orthonormalization process to find an orthogonal matrix P such that P^TAP diagonalizes the symmetric matrix $A = \begin{bmatrix} 0 & 2 & 2 \\ 2 & 0 & 2 \\ 2 & 2 & 0 \end{bmatrix}$.

28. Solve the system of differential equations.

    $y_1' = y_1$
    $y_2' = 9y_2$

29. Find the matrix of the quadratic form associated with the quadratic equation

    $3x^2 - 16xy + 3y^2 - 13 = 0$.

30. A population has the following characteristics.

    (a) A total of 80% of the population survives the first year. Of that 80%, 40% survives the second year. The maximum life span is 3 years.

    (b) The average number of offspring for each member of the population is 3 the first year, 6 the second year, and 3 the third year.

    The population now consists of 150 members in each of the three age classes. How many members will there be in each age class in 1 year? in 2 years?

31. Define an *orthogonal matrix*.

32. Prove that if A is similar to B and A is diagonalizable, then B is diagonalizable.

Appendix   Mathematical Induction and Other Forms of Proofs

■ Use the Principle of Mathematical Induction to prove statements involving a positive integer n.

■ Prove by contradiction that a mathematical statement is true.

■ Use a counterexample to show that a mathematical statement is false.

MATHEMATICAL INDUCTION

In this appendix, you will study some basic strategies for writing mathematical proofs—mathematical induction, proof by contradiction, and the use of counterexamples. Example 1 illustrates the logical need for using mathematical induction.

EXAMPLE 1   Sum of Odd Integers

Use the pattern to propose a formula for the sum of the first n odd integers.

$$1 = 1$$
$$1 + 3 = 4$$
$$1 + 3 + 5 = 9$$
$$1 + 3 + 5 + 7 = 16$$
$$1 + 3 + 5 + 7 + 9 = 25$$

SOLUTION

Notice that the sums on the right are equal to the squares 1^2, 2^2, 3^2, 4^2, and 5^2. From this pattern, it appears that the sum S_n of the first n odd integers is

$$S_n = 1 + 3 + 5 + 7 + \cdots + (2n - 1) = n^2.$$

Although this particular formula *is* valid, it is important for you to see that recognizing a pattern and then simply *jumping to the conclusion* that the pattern must be true for all values of n is not a logically valid method of proof. There are many examples in which a pattern appears to be developing for small values of n and then at some point the pattern fails. One of the most famous cases of this was the conjecture by the French mathematician Pierre de Fermat (1601–1665), who speculated that all numbers of the form

$$F_n = 2^{2^n} + 1, \quad n = 0, 1, 2, \ldots$$

are prime. For $n = 0$, 1, 2, 3, and 4, the conjecture is true.

$$F_0 = 3 \qquad F_1 = 5 \qquad F_2 = 17 \qquad F_3 = 257 \qquad F_4 = 65,537$$

The size of the next Fermat number ($F_5 = 4{,}294{,}967{,}297$) is so great that it was difficult for Fermat to determine whether it was prime or not. However, another well-known mathematician, Leonhard Euler (1707–1783), later found the factorization

$$F_5 = 4{,}294{,}967{,}297 = (641)(6{,}700{,}417)$$

which proved that F_5 is not prime and Fermat's conjecture was false.

Just because a rule, pattern, or formula seems to work for several values of n, you cannot simply decide that it is valid for all values of n without going through a *legitimate proof.* One legitimate method of proof for such conjectures is the **Principle of Mathematical Induction.**

The Principle of Mathematical Induction

Let P_n be a statement involving the positive integer n. If
1. P_1 is true, and
2. for every positive integer k, the truth of P_k implies the truth of P_{k+1}

then the statement P_n must be true for all positive integers n.

The next example uses the Principle of Mathematical Induction to prove the conjecture from Example 1.

EXAMPLE 2   Using Mathematical Induction

Use mathematical induction to prove the formula below.

$$S_n = 1 + 3 + 5 + 7 + \cdots + (2n - 1) = n^2$$

SOLUTION

Mathematical induction consists of two distinct parts. First, you must show that the formula is true when $n = 1$.

1. When $n = 1$, the formula is valid because $S_1 = 1 = 1^2$.

   The second part of mathematical induction has two steps. The first step is to *assume* that the formula is valid for some integer k (the **induction hypothesis**). The second step is to use this assumption to prove that the formula is valid for the *next* integer, $k + 1$.

2. Assuming that the formula

   $$S_k = 1 + 3 + 5 + 7 + \cdots + (2k - 1) = k^2$$

   is true, you must show that the formula $S_{k+1} = (k + 1)^2$ is true.

   $$\begin{aligned}
   S_{k+1} &= 1 + 3 + 5 + 7 + \cdots + (2k - 1) + [2(k + 1) - 1] \\
   &= [1 + 3 + 5 + 7 + \cdots + (2k - 1)] + (2k + 2 - 1) \\
   &= S_k + (2k + 1) \qquad \text{Group terms to form } S_k. \\
   &= k^2 + 2k + 1 \qquad \text{Substitute } k^2 \text{ for } S_k. \\
   &= (k + 1)^2
   \end{aligned}$$

Combining the results of parts (1) and (2), you can conclude by mathematical induction that the formula is valid for *all* positive integers n.

Figure A.1

A well-known illustration used to explain why the Principle of Mathematical Induction works is the unending line of dominoes shown in Figure A.1. If the line contains infinitely many dominoes, then it is clear that you could not knock down the entire line by knocking down only *one domino* at a time. However, if it were true that each domino would knock down the next one as it fell, then you could knock them all down simply by pushing the first one and starting a chain reaction.

Mathematical induction works in the same way. If the truth of P_k implies the truth of P_{k+1} and if P_1 is true, then the chain reaction proceeds as shown below:

P_1 implies P_2

P_2 implies P_3

P_3 implies P_4, and so on.

In the next example, you will see the proof of a formula that is often used in calculus.

EXAMPLE 3   Using Mathematical Induction

Use mathematical induction to prove the formula for the sum of the first n squares.

$$S_n = 1^2 + 2^2 + 3^2 + 4^2 + \cdots + n^2 = \frac{n(n + 1)(2n + 1)}{6}$$

SOLUTION

1. When $n = 1$, the formula is valid, because

$$S_1 = 1^2 = \frac{1(1 + 1)[2(1) + 1]}{6} = \frac{1(2)(3)}{6} = 1.$$

2. Assuming the formula is true for k,

$$S_k = 1^2 + 2^2 + 3^2 + 4^2 + \cdots + k^2 = \frac{k(k + 1)(2k + 1)}{6}$$

you must show that it is true for $k + 1$,

$$S_{k+1} = \frac{(k + 1)[(k + 1) + 1][2(k + 1) + 1]}{6} = \frac{(k + 1)(k + 2)(2k + 3)}{6}.$$

To do this, write S_{k+1} as the sum of S_k and the $(k + 1)$st term, $(k + 1)^2$.

$$S_{k+1} = (1^2 + 2^2 + 3^2 + 4^2 + \cdots + k^2) + (k + 1)^2$$

$$= \frac{k(k + 1)(2k + 1)}{6} + (k + 1)^2 \qquad \text{Induction hypothesis}$$

$$= \frac{(k + 1)(2k^2 + 7k + 6)}{6} \qquad \text{Combine fractions and simplify.}$$

$$= \frac{(k + 1)(k + 2)(2k + 3)}{6} \qquad S_k \text{ implies } S_{k+1}.$$

Combining the results of parts (1) and (2), you can conclude by mathematical induction that the formula is valid for *all* positive integers n.

Many of the proofs in linear algebra use mathematical induction. Here is an example from Chapter 2.

EXAMPLE 4   Using Mathematical Induction in Linear Algebra

If $A_1, A_2, \ldots, A_n$ are invertible matrices, then prove the generalization of Theorem 2.9.

$$(A_1 A_2 A_3 \cdots A_n)^{-1} = A_n^{-1} \cdots A_3^{-1} A_2^{-1} A_1^{-1}$$

SOLUTION

1. The formula is valid trivially when $n = 1$ because $A_1^{-1} = A_1^{-1}$.

2. Assuming the formula is valid for k, $(A_1 A_2 A_3 \cdots A_k)^{-1} = A_k^{-1} \cdots A_3^{-1} A_2^{-1} A_1^{-1}$, you must show that it is valid for $k + 1$. To do this, use Theorem 2.9, which states that the inverse of a product of two invertible matrices is the product of their inverses in reverse order.

$$(A_1 A_2 A_3 \cdots A_k A_{k+1})^{-1} = [(A_1 A_2 A_3 \cdots A_k)A_{k+1}]^{-1}$$

$$= A_{k+1}^{-1}(A_1 A_2 A_3 \cdots A_k)^{-1} \qquad \text{Theorem 2.9}$$

$$= A_{k+1}^{-1}(A_k^{-1} \cdots A_3^{-1} A_2^{-1} A_1^{-1}) \qquad \text{Induction hypothesis}$$

$$= A_{k+1}^{-1} A_k^{-1} \cdots A_3^{-1} A_2^{-1} A_1^{-1} \qquad S_k \text{ implies } S_{k+1}.$$

Combining the results of parts (1) and (2), you can conclude by mathematical induction that the formula is valid for *all* positive integers n.

PROOF BY CONTRADICTION

Another basic strategy for writing a proof is *proof by contradiction*. In mathematical logic, you describe proof by contradiction by the equivalence below.

     p implies q if and only if   not q implies not p.

One way to prove that q is a true statement is to assume that q is not true. If this leads you to a statement that you know is false, then you have proved that q must be true.

Example 5 shows how to use proof by contradiction to prove that $\sqrt{2}$ is irrational.

EXAMPLE 5   Using Proof by Contradiction

Prove that $\sqrt{2}$ is an irrational number.

SOLUTION

Begin by assuming that $\sqrt{2}$ is *not* an irrational number. Then $\sqrt{2}$ is rational and can be written as the quotient of two integers a and b ($b \neq 0$) that have no common factors.

     $\sqrt{2} = \dfrac{a}{b}$ Assume that $\sqrt{2}$ is a rational number.

     $2b^2 = a^2$ Square each side and multiply by b^2.

This implies that 2 is a factor of a^2. So, 2 is also a factor of a. Let $a = 2c$.

     $2b^2 = (2c)^2$ Substitute $2c$ for a.

     $b^2 = 2c^2$ Simplify and divide each side by 2.

This implies that 2 is a factor of b^2, and it is also a factor of b. So, 2 is a factor of both a and b. But this is impossible because a and b have no common factors. It must be impossible that $\sqrt{2}$ is a rational number. You can conclude that $\sqrt{2}$ is an irrational number.

EXAMPLE 6   Using Proof by Contradiction

An integer greater than 1 is *prime* when its only positive factors are 1 and itself and *composite* when it has at least one other factor that is prime. Prove that there are infinitely many prime numbers.

SOLUTION

Assume there are only finitely many prime numbers, $p_1, p_2, \ldots, p_n$. Consider the number $N = p_1 p_2 \cdots p_n + 1$. This number is either prime or composite. N is not prime because $N \neq p_i$. But, N is not composite because none of the primes $(p_1, p_2, \ldots, p_n)$ divide evenly into N. This is a contradiction, so the assumption is false.

It follows that there are infinitely many prime numbers.

You can use proof by contradiction to prove many theorems in linear algebra.

EXAMPLE 7   Using Proof by Contradiction in Linear Algebra

Let A and B be $n \times n$ matrices such that AB is singular. Prove that either A or B is singular.

SOLUTION

Assume that neither A nor B is singular. You know that a matrix is singular if and only if its determinant is zero, so $\det(A)$ and $\det(B)$ are both nonzero real numbers. By Theorem 3.5, $\det(AB) = \det(A)\det(B)$. So, $\det(AB)$ is not zero because it is a product of two nonzero real numbers. But this contradicts that AB is a singular matrix. So, you can conclude that the assumption was wrong and that either A or B is singular.

USING COUNTEREXAMPLES

Often you can disprove a statement using a *counterexample*. For instance, when Euler disproved Fermat's conjecture about prime numbers of the form $F_n = 2^{2^n} + 1$, $n = 0, 1, 2, \ldots$, he used the counterexample $F_5 = 4{,}294{,}967{,}297$, which is not prime.

EXAMPLE 8   **Using a Counterexample**

Use a counterexample to show that the statement is false.

Every odd number is prime.

SOLUTION

Certainly, you can list many odd numbers that are prime (3, 5, 7, 11), but the statement above is not true, because 9 is odd but it is not a prime number. The number 9 is a counterexample.

Counterexamples can be used to disprove statements in linear algebra, as shown in the next two examples.

EXAMPLE 9   **Using a Counterexample in Linear Algebra**

Use a counterexample to show that the statement is false.

If A and B are square singular matrices of order n, then $A + B$ is a singular matrix of order n.

SOLUTION

Let $A = \begin{bmatrix} 1 & 0 \\ 0 & 0 \end{bmatrix}$ and $B = \begin{bmatrix} 0 & 0 \\ 0 & 1 \end{bmatrix}$. Both A and B are singular of order 2, but

$$A + B = \begin{bmatrix} 1 & 0 \\ 0 & 1 \end{bmatrix}$$

is the identity matrix of order 2, which is not singular.

EXAMPLE 10   **Using a Counterexample in Linear Algebra**

Use a counterexample to show that the statement is false.

The set of all 2×2 matrices of the form

$$\begin{bmatrix} 1 & b \\ c & d \end{bmatrix}$$

with the standard operations is a vector space.

SOLUTION

To show that this set of matrices is not a vector space, let

$$A = \begin{bmatrix} 1 & 2 \\ 3 & 4 \end{bmatrix} \quad \text{and} \quad B = \begin{bmatrix} 1 & 5 \\ 6 & 7 \end{bmatrix}.$$

Both A and B are of the stated form, but the sum of these matrices,

$$A + B = \begin{bmatrix} 2 & 7 \\ 9 & 11 \end{bmatrix}$$

is not. This means that the set does not have closure under addition, so it does not satisfy the first axiom in the definition.

REMARK

Recall that in order for a set to be a vector space, it must satisfy *each* of the ten axioms in the definition of a vector space. (See Section 4.2.)

Exercises

Using Mathematical Induction   In Exercises 1–4, use mathematical induction to prove the formula for every positive integer n.

1. $1 + 2 + 3 + \cdots + n = \dfrac{n(n + 1)}{2}$

2. $1^3 + 2^3 + 3^3 + \cdots + n^3 = \dfrac{n^2(n + 1)^2}{4}$

3. $3 + 7 + 11 + \cdots + (4n - 1) = n(2n + 1)$

4. $\left(1 + \dfrac{1}{1}\right)\left(1 + \dfrac{1}{2}\right)\left(1 + \dfrac{1}{3}\right) \cdots \left(1 + \dfrac{1}{n}\right) = n + 1$

Proposing a Formula and Using Mathematical Induction   In Exercises 5 and 6, propose a formula for the sum of the first n terms of the sequence. Then use mathematical induction to prove the formula.

5. $2^1, 2^2, 2^3, \ldots$

6. $\dfrac{1}{1 \cdot 2}, \dfrac{1}{2 \cdot 3}, \dfrac{1}{3 \cdot 4}, \ldots$

Using Mathematical Induction   In Exercises 7–14, use mathematical induction to prove the statement.

7. $n! > 2^n, \quad n \geq 4$

8. $\dfrac{1}{\sqrt{1}} + \dfrac{1}{\sqrt{2}} + \dfrac{1}{\sqrt{3}} + \cdots + \dfrac{1}{\sqrt{n}} > \sqrt{n}, \quad n \geq 2$

9. For all integers $n > 0$,

$$a^0 + a^1 + a^2 + \cdots + a^n = \dfrac{1 - a^{n+1}}{1 - a}, \quad a \neq 1.$$

10. If $x_1 \neq 0, x_2 \neq 0, \ldots, x_n \neq 0$, then $(x_1 x_2 x_3 \cdots x_n)^{-1} = x_1^{-1} x_2^{-1} x_3^{-1} \cdots x_n^{-1}$.

11. (From Chapter 2) If A is an invertible matrix and k is a positive integer, then

$$(A^k)^{-1} = \underbrace{A^{-1} A^{-1} \cdots A^{-1}}_{k \text{ factors}} = (A^{-1})^k$$

12. (From Chapter 2) $(A_1 A_2 A_3 \cdots A_n)^T = A_n^T \cdots A_3^T A_2^T A_1^T$, assuming that $A_1, A_2, A_3, \ldots, A_n$ are matrices with sizes such that the multiplications are defined.

13. (From Chapter 3)

$$|A_1 A_2 A_3 \cdots A_n| = |A_1||A_2||A_3| \cdots |A_n|$$

where $A_1, A_2, A_3, \ldots, A_n$ are square matrices of the same order.

14. (From Chapter 6) If the standard matrices of the linear transformations $T_1, T_2, T_3, \ldots, T_n$ are $A_1, A_2, A_3, \ldots, A_n$ respectively, then the standard matrix for the composition

$$T(\mathbf{v}) = T_n(T_{n-1} \cdots (T_3(T_2(T_1(\mathbf{v})))) \cdots)$$

is $A = A_n A_{n-1} \cdots A_3 A_2 A_1$.

Using Proof by Contradiction   In Exercises 15–26, use proof by contradiction to prove the statement.

15. If p is an integer and p^2 is odd, then p is odd. (*Hint:* An odd number can be written as $2n + 1$, where n is an integer.)

16. If p is a positive integer and p^2 is divisible by 2, then p is divisible by 2.

17. If a and b are real numbers and $a \leq b$, then $a + c \leq b + c$.

18. If a, b, and c are real numbers such that $ac \geq bc$ and $c > 0$, then $a \geq b$.

19. If a and b are real numbers and $1 < a < b$, then $a^{-1} > b^{-1}$.

20. If a and b are real numbers and $(a + b)^2 = a^2 + b^2$, then $a = 0$ or $b = 0$ or $a = b = 0$.

21. If a is a real number and $0 < a < 1$, then $a^2 < a$.

22. The sum of a rational number and an irrational number is irrational.

23. (From Chapter 3) If A and B are square matrices of order n such that $\det(AB) = 1$, then both A and B are nonsingular.

24. (From Chapter 4) In a vector space, the zero vector is unique.

25. (From Chapter 4) Let $S = \{\mathbf{u}, \mathbf{v}\}$ be a linearly independent set. Prove that the set $\{\mathbf{u} - \mathbf{v}, \mathbf{u} + \mathbf{v}\}$ is linearly independent.

26. (From Chapter 5) Let $S = \{\mathbf{x}_1, \mathbf{x}_2, \ldots, \mathbf{x}_n\}$ be a linearly independent set. Prove that if a vector $\mathbf{y}$ is not in $\text{span}(S)$, then the set $S_1 = \{\mathbf{x}_1, \mathbf{x}_2, \ldots, \mathbf{x}_n, \mathbf{y}\}$ is linearly independent.

Using a Counterexample   In Exercises 27–33, use a counterexample to show that the statement is false.

27. If a and b are real numbers and $a < b$, then $a^2 < b^2$.

28. The product of two irrational numbers is irrational.

29. If f is a polynomial function and $f(a) = f(b)$, then $a = b$.

30. If f and g are differentiable functions and $y = f(x)g(x)$, then $y' = f'(x)g'(x)$.

31. The set of all 2×2 matrices of the form

$$\begin{bmatrix} 0 & a \\ b & 2 \end{bmatrix}$$

with the standard operations is a vector space.

32. $T: R^2 \rightarrow R^2, \quad T(x_1, x_2) = (x_1 + 4, x_2)$ is a linear transformation.

33. (From Chapter 2) If A, B, and C are matrices and $AC = BC$, then $A = B$.

Answers to Odd-Numbered Exercises and Tests

Chapter 1

Section 1.1 *(page 10)*

1. Linear   **3.** Not linear   **5.** Not linear

7. $x = 2t$ **9.** $x = 1 - s - t$
$\quad\; y = t$ $\qquad y = s$
$\qquad\qquad\qquad z = t$

11.

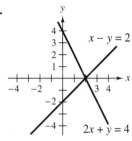
$x = 2$
$y = 0$

13.

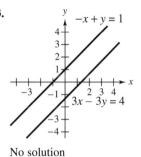
No solution

15.

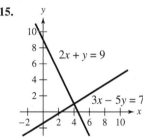
$x = 4$
$y = 1$

17.

$x = 2$
$y = -1$

19.

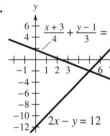
$x = 5$
$y = -2$

21.

$x = 2$
$y = 1$
$x = \frac{18}{5}$
$y = \frac{3}{5}$

23.

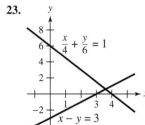

25. $x_1 = 5$ **27.** $x = \frac{3}{2}$ **29.** $x_1 = -t$
$\quad\;\; x_2 = 3$ $\quad y = \frac{3}{2}$ $\quad\;\; x_2 = 2t$
$\qquad\qquad\qquad z = 0$ $\qquad x_3 = t$

31. (a)   (b) Inconsistent

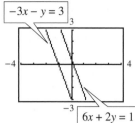

33. (a)   (b) Consistent

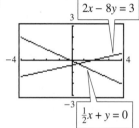
(c) $x = \frac{1}{2}$
$\quad\; y = -\frac{1}{4}$
(d) $x = \frac{1}{2}$
$\quad\; y = -\frac{1}{4}$
(e) The solutions are
the same.

35. (a)   (b) Consistent

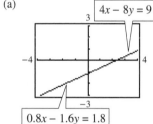
(c) There are infinitely
many solutions.
(d) $x = \frac{9}{4} + 2t$
$\quad\; y = t$
(e) The solutions are
consistent.

37. $x_1 = -1$ **39.** $u = 60$ **41.** $x = -\frac{1}{3}$
$\quad\;\; x_2 = -1$ $\quad v = 60$ $\quad y = -\frac{2}{3}$

43. $x = 14$ **45.** $x_1 = 8$ **47.** $x = 3$
$\quad\; y = -2$ $\quad\;\; x_2 = 7$ $\quad y = 2$
$\qquad\qquad\qquad\qquad\qquad z = 1$

49. No solution   **51.** $x_1 = \frac{5}{2} - \frac{1}{2}t$ **53.** No solution
$\qquad\qquad\qquad\;\; x_2 = 4t - 1$
$\qquad\qquad\qquad\;\; x_3 = t$

55. $x = 1$ **57.** $x = -1.2$ **59.** $x_1 = -15$
$\quad\; y = 0$ $\quad\; y = -0.6$ $\quad\;\; x_2 = 40$
$\quad\; z = 3$ $\quad\; z = 2.4$ $\quad\;\; x_3 = 45$
$\quad\; w = 2$ $\quad\;\; x_4 = -75$

61. $x_1 = \frac{1}{5}$
$\quad\;\; x_2 = -\frac{4}{5}$
$\quad\;\; x_3 = \frac{1}{2}$

63. This system must have at least one solution because
$x = y = z = 0$ is an obvious solution.
Solution: $x = 0$
$\qquad\qquad\; y = 0$
$\qquad\qquad\; z = 0$
This system has exactly one solution.

65. This system must have at least one solution because
$x = y = z = 0$ is an obvious solution.
Solution: $x = -\frac{3}{5}t$
$\qquad\qquad\; y = \frac{4}{5}t$
$\qquad\qquad\; z = t$
This system has an infinite number of solutions.

67. Apple juice: 103 mg
Orange juice: 124 mg

69. (a) True. You can describe the entire solution set using parametric representation.
$ax + by = c$
Choosing $y = t$ as the free variable, the solution is
$x = \dfrac{c}{a} - \dfrac{b}{a}t,\ y = t$, where t is any real number.

(b) False. For example, consider the system
$$x_1 + x_2 + x_3 = 1$$
$$x_1 + x_2 + x_3 = 2$$
which is an inconsistent system.

(c) False. A consistent system may have only one solution.

71. $3x_1 - x_2 = 4$
$-3x_1 + x_2 = -4$
(The answer is not unique.)

73. $x = 3$ **75.** $x = \dfrac{2}{5 - t}$
$y = -4$
$y = \dfrac{1}{4t - 1}$
$z = \dfrac{1}{t}$, where $t \neq 5, \dfrac{1}{4}, 0$.

77. $x = \cos\theta$ **79.** $k = \pm 1$
$y = \sin\theta$

81. All $k \neq 0$ **83.** $k = -2$ **85.** $k = 1, -2$

87. (a) Three lines intersecting at one point
(b) Three coincident lines
(c) Three lines having no common point

89. Answers will vary. (*Hint:* Choose three different values of x and solve the resulting system of linear equations in the variables a, b, and c.)

91. $x - 4y = -3$ $x - 4y = -3$
$5x - 6y = 13$ $14y = 28$

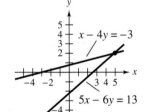

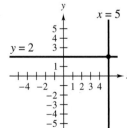

$x - 4y = -3$ $x = 5$
$y = 2$ $y = 2$

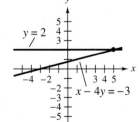

The intersection points are all the same.

93. $x = 39,600$
$y = 398$
The graphs are misleading because, while they appear parallel, when the equations are solved for y, they have slightly different slopes.

Section 1.2   *(page 22)*

1. 3×3 **3.** 2×4 **5.** 4×5

7. Add 5 times the second row to the first row.

9. Interchange the first and second rows, add 3 times the new first row to the third row.

11. $x_1 = 0$ **13.** $x_1 = 2$
$x_2 = 2$ $x_2 = -1$
$x_3 = -1$

15. $x_1 = 1$ **17.** $x_1 = -26$
$x_2 = 1$ $x_2 = 13$
$x_3 = 0$ $x_3 = -7$
$x_4 = 4$

19. Reduced row-echelon form

21. Not in row-echelon form

23. Not in row-echelon form

25. $x = 2$ **27.** No solution
$y = 3$

29. $x = 4$ **31.** $x_1 = 4$
$y = -2$ $x_2 = -3$
$x_3 = 2$

33. No solution   **35.** $x = 100 + 96t - 3s$
$y = s$
$z = 54 + 52t$
$w = t$

37. $x = 0$ **39.** $x_1 = 23.5361 + 0.5278t$
$y = 2 - 4t$ $x_2 = 18.5444 + 4.1111t$
$z = t$ $x_3 = 7.4306 + 2.1389t$
$x_4 = t$

41. $x_1 = 2$ **43.** $x_1 = 0$
$x_2 = -2$ $x_2 = -t$
$x_3 = 3$ $x_3 = t$
$x_4 = -5$
$x_5 = 1$

45. $x_1 = -t$ **47.** \$100,000 at 3%
$x_2 = s$ \$250,000 at 4%
$x_3 = 0$ \$150,000 at 5%
$x_4 = t$

49. Augmented
(a) Two equations in two variables
(b) All real $k \neq -\dfrac{4}{3}$
Coefficient
(a) Two equations in three variables
(b) All real k

51. (a) $a + b + c = 0$
(b) $a + b + c \neq 0$
(c) Not possible

53. (a) $x = \dfrac{8}{3} - \dfrac{5}{6}t$ (b) $x = \dfrac{18}{7} - \dfrac{11}{14}t$
$y = -\dfrac{8}{3} + \dfrac{5}{6}t$ $y = -\dfrac{20}{7} + \dfrac{13}{14}t$
$z = t$ $z = t$
(c) $x = 3 - t$ (d) Each system has an
$y = -3 + t$ infinite number of
$z = t$ solutions.

55. $\begin{bmatrix} 1 & 0 \\ 0 & 1 \end{bmatrix}$

57. $\begin{bmatrix} 1 & 0 \\ 0 & 1 \end{bmatrix}, \begin{bmatrix} 1 & k \\ 0 & 0 \end{bmatrix}, \begin{bmatrix} 0 & 1 \\ 0 & 0 \end{bmatrix}, \begin{bmatrix} 0 & 0 \\ 0 & 0 \end{bmatrix}$

59. (a) True. In the notation $m \times n$, m is the number of rows of the matrix. So, a 6×3 matrix has six rows.

(b) True. On page 16, the sentence reads, "Every matrix is row-equivalent to a matrix in row-echelon form."

(c) False. Consider the row-echelon form

$$\begin{bmatrix} 1 & 0 & 0 & 0 & 0 \\ 0 & 1 & 0 & 0 & 1 \\ 0 & 0 & 1 & 0 & 2 \\ 0 & 0 & 0 & 1 & 3 \end{bmatrix}$$

which gives the solution $x_1 = 0$, $x_2 = 1$, $x_3 = 2$, and $x_4 = 3$.

(d) True. Theorem 1.1 states that if a homogeneous system has fewer equations than variables, then it must have an infinite number of solutions.

61. Yes, it is possible:
$$x_1 + x_2 + x_3 = 0$$
$$x_1 + x_2 + x_3 = 1$$

63. $ad - bc \neq 0$ **65.** $\lambda = 1, 3$

67. Sample answer: $x + 3z = -2$
$$y + 4z = 1$$
$$2y + 8z = 2$$

69. The rows have been interchanged. The first elementary row operation is redundant, so you can just use the second and third elementary row operations.

Section 1.3 *(page 32)*

1. (a) $p(x) = 29 - 18x + 3x^2$

(b)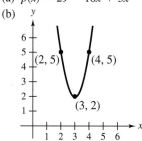

3. (a) $p(x) = 2x$

(b)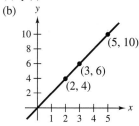

5. (a) $p(x) = -\frac{3}{2}x + 2x^2 + \frac{1}{2}x^3$

(b)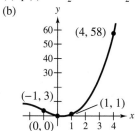

7. (a) $p(x) = -6 - 3x + x^2 - x^3 + x^4$

(b)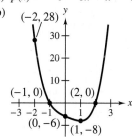

9. (a) Let $z = x - 2014$.
$$p(z) = 7 + \tfrac{7}{2}z + \tfrac{3}{2}z^2$$
$$p(x) = 7 + \tfrac{7}{2}(x - 2014) + \tfrac{3}{2}(x - 2014)^2$$

(b)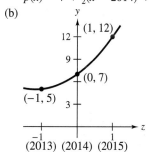

11. (a) $p(x) = 0.254 - 1.579x + 12.022x^2$

(b)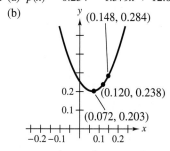

13. $p(x) = -\dfrac{4}{\pi^2}x^2 + \dfrac{4}{\pi}x$

$\sin \dfrac{\pi}{3} \approx \dfrac{8}{9} \approx 0.889$

(Actual value is $\sqrt{3}/2 \approx 0.866$.)

15. $(x - 5) + (y - 10)^2 = 65$

17. $p(x) = 282 + 3(x - 2000) - 0.03(x - 2000)^2$
2020: 330 million; 2030: 345 million

19. (a) Using $z = x - 2000$
$$a_0 + 7a_1 + 49a_2 + 343a_3 = 14{,}065$$
$$a_0 + 8a_1 + 64a_2 + 512a_3 = 17{,}681$$
$$a_0 + 9a_1 + 81a_2 + 729a_3 = 14{,}569$$
$$a_0 + 10a_1 + 100a_2 + 1000a_3 = 18{,}760$$

(b) $p(x) = -1{,}378{,}235 + 500{,}729.5(x - 2000)$
$$- 59{,}488(x - 2000)^2 + 2338.5(x - 2000)^3$$

No. Answers will vary. Sample answer: The model does not produce reasonable outcomes after 2010.

21. (a) $x_1 = 700 - s - t$ (b) $x_1 = 600$ (c) $x_1 = 500$
$x_2 = 300 - s - t$ $x_2 = 200$ $x_2 = 100$
$x_3 = s$ $x_3 = 0$ $x_3 = 100$
$x_4 = 100 - t$ $x_4 = 0$ $x_4 = 0$
$x_5 = t$ $x_5 = 100$ $x_5 = 100$

23. (a) $x_1 = 100 + t$ (b) $x_1 =\quad 100$ (c) $x_1 = 200$

$\quad\quad x_2 = -100 + t$ $\quad\quad x_2 = -100$ $\quad\quad x_2 =\quad\ 0$

$\quad\quad x_3 = 200 + t$ $\quad\quad x_3 =\quad 200$ $\quad\quad x_3 = 300$

$\quad\quad x_4 = t$ $\quad\quad\quad x_4 =\quad\quad 0$ $\quad\quad x_4 = 100$

$\quad$ (d) $x_1 = 400$

$\quad\quad x_2 = 200$

$\quad\quad x_3 = 500$

$\quad\quad x_4 = 300$

25. $I_1 = 0$

$\quad I_2 = 1$

$\quad I_3 = 1$

27. (a) $I_1 = 1$ (b) $I_1 = 0$

$\quad\quad I_2 = 2$ $\quad\quad I_2 = 1$

$\quad\quad I_3 = 1$ $\quad\quad I_3 = 1$

29. $T_1 = 37.5°, T_2 = 45°, T_3 = 25°, T_4 = 32.5°$

31. $A = 1, B = 3, C = -2$

33. $A = 1, B = 2, C = 1$

35. $x = 2$

$\quad y = 2$

$\quad \lambda = -4$

37. $p(x) = 1 - 2x + 2x^2$

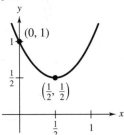

39. Solve the system:

$\quad p(-1) = a_0 - a_1 + a_2 = 0$

$\quad p(0) = a_0 \quad\quad\quad\quad = 0$

$\quad p(1) = a_0 + a_1 + a_2 = 0$

$\quad a_0 = a_1 = a_2 = 0$

41. (a) $p(x) = 1 - \frac{7}{15}x + \frac{1}{15}x^2$

$\quad$ (b) $p(x) = 1 = x$

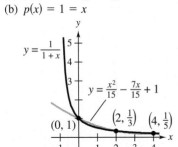

Review Exercises *(page 35)*

1. Not linear   **3.** Linear   **5.** Not linear

7. $x = -\frac{1}{3} + \frac{4}{3}s - \frac{2}{3}t$

$\quad y = s$

$\quad z = t$

9. $x = \frac{1}{2}$ **11.** $x = -12$ **13.** $x = 0$

$\quad y = \frac{3}{2}$ $\quad\quad y = -8$ $\quad\quad y = 0$

15. No solution

17. $x = 0$ **19.** $x_1 = -\frac{1}{2}$ **21.** 2×3

$\quad y = 0$ $\quad\quad x_2 = \frac{4}{5}$

23. $x_1 = 5$ **25.** $x_1 = -2t$

$\quad x_2 = -5$ $\quad\quad x_2 = t$

$\quad\quad\quad\quad\quad\quad x_3 = 0$

27. Reduced row-echelon form

29. Not in row-echelon form

31. $x =\quad 2$ **33.** $x =\quad \frac{1}{2}$ **35.** $x = 4 + 3t$

$\quad y = -3$ $\quad\quad y = -\frac{1}{3}$ $\quad\quad y = 5 + 2t$

$\quad z =\quad 3$ $\quad\quad z =\quad 1$ $\quad\quad z = t$

37. No solution   **39.** $x_1 =\quad 1$ **41.** $x_1 =\quad 21.6$

$\quad\quad\quad\quad\quad\quad\quad\quad x_2 =\quad 4$ $\quad\quad x_2 = -6.1$

$\quad\quad\quad\quad\quad\quad\quad\quad x_3 = -3$ $\quad\quad x_3 = -0.1$

$\quad\quad\quad\quad\quad\quad\quad\quad x_4 = -2$

43. $x = 0$ **45.** No solution   **47.** $x_1 =\quad 2t$

$\quad y = 2 - 4t$ $\quad\quad\quad\quad\quad\quad\quad x_2 = -3t$

$\quad z = t$ $\quad\quad\quad\quad\quad\quad\quad\quad x_3 =\quad\ t$

49. $x_1 = 0$ **51.** $k = \pm 1$

$\quad x_2 = 0$

$\quad x_3 = 0$

53. (a) $b = 2a$ and $a \neq -3$

$\quad$ (b) $b \neq 2a$

$\quad$ (c) $a = -3$ and $b = -6$

55. Use an elimination method to get both matrices in reduced row-echelon form. The two matrices are row-equivalent because each is row-equivalent to

$$\begin{bmatrix} 1 & 0 & 0 \\ 0 & 1 & 0 \\ 0 & 0 & 1 \end{bmatrix}.$$

57. $\begin{bmatrix} 1 & 0 & -1 & -2 & \ldots & 2-n \\ 0 & 1 & 2 & 3 & \ldots & n-1 \\ 0 & 0 & 0 & 0 & \ldots & 0 \\ \vdots & \vdots & \vdots & \vdots & & \vdots \\ 0 & 0 & 0 & 0 & \ldots & 0 \end{bmatrix}$

59. (a) False. See page 3, following Example 2.

$\quad$ (b) True. See page 5, Example 4(b).

61. 6 touchdowns, 6 extra-point kicks, 1 field goal

63. $A = 2, B = 6, C = 4$

65. (a) $p(x) = 90 - \frac{135}{2}x + \frac{25}{2}x^2$

$\quad$ (b)

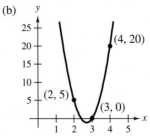

67. $p(x) = 50 + \frac{15}{2}x + \frac{5}{2}x^2$

$\quad$ (First year is represented by $x = 0$.)

$\quad$ Fourth-year sales: $p(3) = 95$

69. (a) $a_0 \qquad\qquad\qquad = 80$
$\qquad a_0 + 4a_1 + \quad 16a_2 = 68$
$\qquad a_0 + 80a_1 + 6400a_2 = 30$

(b) and (c) $a_0 = \quad 80$
$\qquad\qquad a_1 = -\frac{25}{8}$
$\qquad\qquad a_2 = \quad \frac{1}{32}$
$\qquad$ So, $y = \frac{1}{32}x^2 - \frac{25}{8}x + 80$.

(d) The results of parts (b) and (c) are the same.

(e) There is precisely one polynomial function of degree $n-1$ (or less) that fits n distinct points.

71. (a) $x_1 = 100 - r + t$ \qquad (b) $x_1 = \quad 50$
$\qquad x_2 = 300 - r + s$ \qquad\qquad $x_2 = 250$
$\qquad x_3 = r$ \qquad\qquad\qquad\quad $x_3 = 100$
$\qquad x_4 = -s + t$ \qquad\qquad\quad $x_4 = \quad 0$
$\qquad x_5 = s$ \qquad\qquad\qquad\quad $x_5 = \quad 50$
$\qquad x_6 = t$ \qquad\qquad\qquad\quad $x_6 = \quad 50$

Chapter 2

Section 2.1 *(page 48)*

1. $x = -4, y = 22$

3. $x = 2, y = 3$

5. (a) $\begin{bmatrix} -2 & 0 \\ 6 & 3 \end{bmatrix}$ (b) $\begin{bmatrix} 4 & 4 \\ -2 & -1 \end{bmatrix}$ (c) $\begin{bmatrix} 2 & 4 \\ 4 & 2 \end{bmatrix}$

(d) $\begin{bmatrix} 5 & 6 \\ 0 & 0 \end{bmatrix}$ (e) $\begin{bmatrix} -\frac{5}{2} & -1 \\ 5 & \frac{5}{2} \end{bmatrix}$

7. (a) $\begin{bmatrix} 4 & -2 & 5 \\ -4 & 0 & 2 \end{bmatrix}$ (b) $\begin{bmatrix} 0 & 4 & -3 \\ 2 & -2 & 6 \end{bmatrix}$

(c) $\begin{bmatrix} 4 & 2 & 2 \\ -2 & -2 & 8 \end{bmatrix}$ (d) $\begin{bmatrix} 2 & 5 & -2 \\ 1 & -3 & 10 \end{bmatrix}$

(e) $\begin{bmatrix} 3 & -\frac{5}{2} & \frac{9}{2} \\ -\frac{7}{2} & \frac{1}{2} & 0 \end{bmatrix}$

9. (a), (b), (d), and (e) Not possible

(c) $\begin{bmatrix} 12 & 0 & 6 \\ -2 & -8 & 0 \end{bmatrix}$

11. (a) $c_{21} = -6$ (b) $c_{13} = 29$

13. $x = 3, y = 2, z = 1$

15. (a) $\begin{bmatrix} 0 & 15 \\ 6 & 12 \end{bmatrix}$ (b) $\begin{bmatrix} -2 & 2 \\ 31 & 14 \end{bmatrix}$

17. (a) $\begin{bmatrix} -8 & -2 & -5 \\ 4 & 8 & 17 \\ -20 & 1 & 4 \end{bmatrix}$ (b) $\begin{bmatrix} 9 & 5 & 4 \\ 3 & 11 & -5 \\ -17 & -1 & -16 \end{bmatrix}$

19. (a) Not possible (b) $\begin{bmatrix} 3 & -4 \\ 10 & 16 \\ 26 & 46 \end{bmatrix}$

21. (a) $[12]$ (b) $\begin{bmatrix} 6 & 4 & 2 \\ 9 & 6 & 3 \\ 0 & 0 & 0 \end{bmatrix}$

23. (a) $\begin{bmatrix} -1 & 19 \\ 4 & -27 \\ 0 & 14 \end{bmatrix}$ (b) Not possible

25. (a) $\begin{bmatrix} 3 \\ 10 \\ 26 \end{bmatrix}$ (b) Not possible

27. (a) $\begin{bmatrix} 60 & 72 \\ -20 & -24 \\ 10 & 12 \\ 60 & 72 \end{bmatrix}$ (b) Not possible

29. 3×4 \quad **31.** 4×2 \quad **33.** 3×2

35. Not possible, sizes do not match.

37. $x_1 = t, x_2 = \frac{5}{4}t, x_3 = \frac{3}{4}t$

39. $\begin{bmatrix} -1 & 1 \\ -2 & 1 \end{bmatrix}\begin{bmatrix} x_1 \\ x_2 \end{bmatrix} = \begin{bmatrix} 4 \\ 0 \end{bmatrix}$ \quad **41.** $\begin{bmatrix} -2 & -3 \\ 6 & 1 \end{bmatrix}\begin{bmatrix} x_1 \\ x_2 \end{bmatrix} = \begin{bmatrix} -4 \\ -36 \end{bmatrix}$
$\qquad\quad \begin{bmatrix} x_1 \\ x_2 \end{bmatrix} = \begin{bmatrix} 4 \\ 8 \end{bmatrix}$ \qquad\qquad\qquad $\begin{bmatrix} x_1 \\ x_2 \end{bmatrix} = \begin{bmatrix} -7 \\ 6 \end{bmatrix}$

43. $\begin{bmatrix} 1 & -2 & 3 \\ -1 & 3 & -1 \\ 2 & -5 & 5 \end{bmatrix}\begin{bmatrix} x_1 \\ x_2 \\ x_3 \end{bmatrix} = \begin{bmatrix} 9 \\ -6 \\ 17 \end{bmatrix}$
$\qquad\qquad\qquad \begin{bmatrix} x_1 \\ x_2 \\ x_3 \end{bmatrix} = \begin{bmatrix} 1 \\ -1 \\ 2 \end{bmatrix}$

45. $\begin{bmatrix} 1 & -5 & 2 \\ -3 & 1 & -1 \\ 0 & -2 & 5 \end{bmatrix}\begin{bmatrix} x_1 \\ x_2 \\ x_3 \end{bmatrix} = \begin{bmatrix} -20 \\ 8 \\ -16 \end{bmatrix}$
$\qquad\qquad\qquad \begin{bmatrix} x_1 \\ x_2 \\ x_3 \end{bmatrix} = \begin{bmatrix} -1 \\ 3 \\ -2 \end{bmatrix}$

47. $\begin{bmatrix} 2 & -1 & 0 & 1 \\ 0 & 3 & -1 & -1 \\ 1 & 0 & 1 & -3 \\ 1 & 1 & 2 & 0 \end{bmatrix}\begin{bmatrix} x_1 \\ x_2 \\ x_3 \\ x_4 \end{bmatrix} = \begin{bmatrix} 3 \\ -3 \\ -4 \\ 0 \end{bmatrix}$
$\qquad\qquad\qquad\quad \begin{bmatrix} x_1 \\ x_2 \\ x_3 \\ x_4 \end{bmatrix} = \begin{bmatrix} \frac{1}{2} \\ -\frac{1}{2} \\ 0 \\ \frac{3}{2} \end{bmatrix}$

49. $\mathbf{b} = 3\begin{bmatrix} 1 \\ 3 \end{bmatrix} + 0\begin{bmatrix} -1 \\ -3 \end{bmatrix} - 2\begin{bmatrix} 2 \\ 1 \end{bmatrix} = \begin{bmatrix} -1 \\ 7 \end{bmatrix}$
(The answer is not unique.)

51. $\mathbf{b} = 1\begin{bmatrix} 1 \\ 1 \\ 2 \end{bmatrix} + 2\begin{bmatrix} 1 \\ 0 \\ -1 \end{bmatrix} + 0\begin{bmatrix} -5 \\ -1 \\ -1 \end{bmatrix} = \begin{bmatrix} 3 \\ 1 \\ 0 \end{bmatrix}$

53. $\begin{bmatrix} -5 & 2 \\ 3 & -1 \end{bmatrix}$ \quad **55.** $a = 7, b = -4, c = -\frac{1}{2}, d = \frac{7}{2}$

57. $\begin{bmatrix} 1 & 0 & 0 \\ 0 & 4 & 0 \\ 0 & 0 & 9 \end{bmatrix}$

59. $AB = \begin{bmatrix} -10 & 0 \\ 0 & -12 \end{bmatrix}$
$\quad BA = \begin{bmatrix} -10 & 0 \\ 0 & -12 \end{bmatrix}$

61. Proof \quad **63.** 2 \quad **65.** 4

67. Proof \quad **69.** $w = z, x = -y$

71. Let $A = \begin{bmatrix} a_{11} & a_{12} \\ a_{21} & a_{22} \end{bmatrix}$.

Then the given matrix equation expands to

$$\begin{bmatrix} a_{11} + a_{21} & a_{12} + a_{22} \\ a_{11} + a_{21} & a_{12} + a_{22} \end{bmatrix} = \begin{bmatrix} 1 & 0 \\ 0 & 1 \end{bmatrix}.$$

Because $a_{11} + a_{21} = 1$ and $a_{11} + a_{21} = 0$ cannot both be true, you can conclude that there is no solution.

73. (a) $A^2 = \begin{bmatrix} i^2 & 0 \\ 0 & i^2 \end{bmatrix} = \begin{bmatrix} -1 & 0 \\ 0 & -1 \end{bmatrix}$

$A^3 = \begin{bmatrix} i^3 & 0 \\ 0 & i^3 \end{bmatrix} = \begin{bmatrix} -i & 0 \\ 0 & -i \end{bmatrix}$

$A^4 = \begin{bmatrix} i^4 & 0 \\ 0 & i^4 \end{bmatrix} = \begin{bmatrix} 1 & 0 \\ 0 & 1 \end{bmatrix}$

(b) $B^2 = \begin{bmatrix} -i^2 & 0 \\ 0 & -i^2 \end{bmatrix} = \begin{bmatrix} 1 & 0 \\ 0 & 1 \end{bmatrix} = A^4$

75. Proof   **77.** Proof

79. [\$1037.50   \$1400.00   \$1012.50]

Each entry represents the total profit at each outlet.

81. $\begin{bmatrix} 0.40 & 0.15 & 0.15 \\ 0.28 & 0.53 & 0.17 \\ 0.32 & 0.32 & 0.68 \end{bmatrix}$

P^2 gives the proportions of the voting population that changed parties or remained loyal to their parties from the first election to the third.

83. $\left[\begin{array}{cc|c} -1 & 4 & 0 \\ -1 & 1 & 0 \\ 0 & 0 & 5 \end{array}\right]$

85. (a) True. On page 43, "... for the product of two matrices to be defined, the number of columns of the first matrix must equal the number of rows of the second matrix."

(b) True. On page 46, "... the system $A\mathbf{x} = \mathbf{b}$ is consistent if and only if $\mathbf{b}$ can be expressed as ... a linear combination, where the coefficients of the linear combination are a solution of the system."

87. (a) $AT = \begin{bmatrix} -1 & -4 & -2 \\ 1 & 2 & 3 \end{bmatrix}$

$AAT = \begin{bmatrix} -1 & -2 & -3 \\ -1 & -4 & -2 \end{bmatrix}$

Triangle associated with T Triangle associated with AT

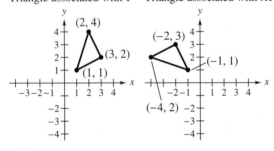

Triangle associated with AAT

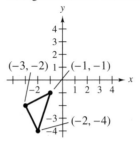

The transformation matrix A rotates the triangle $90°$ counterclockwise about the origin.

(b) Given the triangle associated with AAT, the transformation that would produce the triangle associated with AT would be a $90°$ clockwise rotation about the origin. Another such rotation would produce the triangle associated with T.

Section 2.2 *(page 59)*

1. $\begin{bmatrix} -8 & -7 \\ 15 & -1 \end{bmatrix}$ **3.** $\begin{bmatrix} -24 & -4 & 12 \\ -12 & 32 & 12 \end{bmatrix}$ **5.** $\begin{bmatrix} 10 & 8 \\ -59 & 9 \end{bmatrix}$

7. $\begin{bmatrix} 3 & 2 \\ 13 & 4 \end{bmatrix}$ **9.** $\begin{bmatrix} 0 & -12 \\ 12 & -24 \end{bmatrix}$ **11.** $\begin{bmatrix} 7 & 7 \\ 28 & 14 \end{bmatrix}$

13. (a) $\begin{bmatrix} 3 & \frac{2}{3} \\ -\frac{4}{3} & \frac{11}{3} \\ \frac{10}{3} & 0 \end{bmatrix}$ (b) $\begin{bmatrix} -\frac{13}{3} & -\frac{10}{3} \\ 4 & -5 \\ -\frac{26}{3} & -\frac{16}{3} \end{bmatrix}$

(c) $\begin{bmatrix} -14 & -4 \\ 7 & -17 \\ -17 & -2 \end{bmatrix}$ (d) $\begin{bmatrix} -\frac{13}{6} & 1 \\ -\frac{1}{3} & -\frac{17}{6} \\ 0 & \frac{10}{3} \end{bmatrix}$

15. $\begin{bmatrix} -2 & -10 & 0 \\ 2 & 0 & 10 \end{bmatrix}$ **17.** $\begin{bmatrix} -3 & -5 & -10 \\ -2 & -5 & -5 \end{bmatrix}$

19. $\begin{bmatrix} 1 & 6 & -1 \\ -2 & -2 & -8 \end{bmatrix}$ **21.** $\begin{bmatrix} 12 & -4 \\ 8 & 4 \end{bmatrix}$

23. (a) $\begin{bmatrix} 12 & 7 \\ 24 & 15 \end{bmatrix}$ (b) $\begin{bmatrix} 12 & 7 \\ 24 & 15 \end{bmatrix}$

25. $AB = \begin{bmatrix} -9 & 2 \\ -3 & 6 \end{bmatrix}$, $BA = \begin{bmatrix} -8 & 4 \\ 2 & 5 \end{bmatrix}$

27. $AC = BC = \begin{bmatrix} 2 & 3 \\ 2 & 3 \end{bmatrix}$ **29.** Proof

31. $\begin{bmatrix} 1 & 2 \\ 0 & -1 \end{bmatrix}$ **33.** $\begin{bmatrix} 2 & 2 \\ 0 & 0 \end{bmatrix}$ **35.** $\begin{bmatrix} 1 & 0 \\ 0 & 1 \end{bmatrix}$

37. $(A + B)(A - B) = A^2 + BA - AB - B^2$, which is not necessarily equal to $A^2 - B^2$ because AB is not necessarily equal to BA.

39. $\begin{bmatrix} 1 & -3 & 5 \\ -2 & 4 & -1 \end{bmatrix}$ **41.** $(AB)^T = B^TA^T = \begin{bmatrix} 2 & -5 \\ 4 & -1 \end{bmatrix}$

43. $(AB)^T = B^TA^T = \begin{bmatrix} 4 & 0 & -4 \\ 10 & 4 & -2 \\ 1 & -1 & -3 \end{bmatrix}$

45. (a) $\begin{bmatrix} 16 & 8 & 4 \\ 8 & 8 & 0 \\ 4 & 0 & 2 \end{bmatrix}$ (b) $\begin{bmatrix} 21 & 3 \\ 3 & 5 \end{bmatrix}$

47. (a) $\begin{bmatrix} 68 & 26 & -10 & 6 \\ 26 & 41 & 3 & -1 \\ -10 & 3 & 43 & 5 \\ 6 & -1 & 5 & 10 \end{bmatrix}$ (b) $\begin{bmatrix} 29 & -14 & 5 & -5 \\ -14 & 81 & -3 & 2 \\ 5 & -3 & 39 & -13 \\ -5 & 2 & -13 & 13 \end{bmatrix}$

49. $\begin{bmatrix} 1 & 0 & 0 & 0 & 0 \\ 0 & 1 & 0 & 0 & 0 \\ 0 & 0 & 1 & 0 & 0 \\ 0 & 0 & 0 & 1 & 0 \\ 0 & 0 & 0 & 0 & 1 \end{bmatrix}$

51. $\begin{bmatrix} 1 & 0 & 0 & 0 & 0 \\ 0 & -1 & 0 & 0 & 0 \\ 0 & 0 & 1 & 0 & 0 \\ 0 & 0 & 0 & -1 & 0 \\ 0 & 0 & 0 & 0 & 1 \end{bmatrix}$ **53.** $\begin{bmatrix} \pm 3 & 0 \\ 0 & \pm 2 \end{bmatrix}$

55. (a) True. See Theorem 2.1, part 1.
(b) False. See Theorem 2.6, part 4, or Example 9.
(c) True. See Example 10.

57. (a) $a = 3$ and $b = -1$
(b) $a + b = 1$
$\quad\quad b = 1$
$\quad a\quad = 1$
No solution
(c) $a + b + c = 0$
$\quad\quad b + c = 0$
$\quad a\quad + c = 0$
$a = -c \rightarrow b = 0 \rightarrow c = 0 \rightarrow a = 0$
(d) $a = -3t$
$b = t$
$c = t$
Let $t = 1$: $a = -3, b = 1, c = 1$

59. $\begin{bmatrix} -4 & 0 \\ 8 & 2 \end{bmatrix}$ **61–69.** Proofs

71. Skew-symmetric   **73.** Symmetric   **75.** Proof

77. (a) $\frac{1}{2}(A + A^T)$

$= \frac{1}{2}\left(\begin{bmatrix} a_{11} & a_{12} & \cdots & a_{1n} \\ a_{21} & a_{22} & \cdots & a_{2n} \\ \vdots & \vdots & & \vdots \\ a_{n1} & a_{n2} & \cdots & a_{nn} \end{bmatrix} + \begin{bmatrix} a_{11} & a_{21} & \cdots & a_{n1} \\ a_{12} & a_{22} & \cdots & a_{n2} \\ \vdots & \vdots & & \vdots \\ a_{1n} & a_{2n} & \cdots & a_{nn} \end{bmatrix}\right)$

$= \frac{1}{2}\begin{bmatrix} 2a_{11} & a_{12} + a_{21} & \cdots & a_{1n} + a_{n1} \\ a_{21} + a_{12} & 2a_{22} & \cdots & a_{2n} + a_{n2} \\ \vdots & \vdots & & \vdots \\ a_{n1} + a_{1n} & a_{n2} + a_{2n} & \cdots & 2a_{nn} \end{bmatrix}$

(b) $\frac{1}{2}(A - A^T)$

$= \frac{1}{2}\left(\begin{bmatrix} a_{11} & a_{12} & \cdots & a_{1n} \\ a_{21} & a_{22} & \cdots & a_{2n} \\ \vdots & \vdots & & \vdots \\ a_{n1} & a_{n2} & \cdots & a_{nn} \end{bmatrix} - \begin{bmatrix} a_{11} & a_{21} & \cdots & a_{n1} \\ a_{12} & a_{22} & \cdots & a_{n2} \\ \vdots & \vdots & & \vdots \\ a_{1n} & a_{2n} & \cdots & a_{nn} \end{bmatrix}\right)$

$= \frac{1}{2}\begin{bmatrix} 0 & a_{12} - a_{21} & \cdots & a_{1n} - a_{n1} \\ a_{21} - a_{12} & 0 & \cdots & a_{2n} - a_{n2} \\ \vdots & \vdots & & \vdots \\ a_{n1} - a_{1n} & a_{n2} - a_{2n} & \cdots & 0 \end{bmatrix}$

(c) Proof
(d) $A = \frac{1}{2}(A - A^T) + \frac{1}{2}(A + A^T)$

$= \begin{bmatrix} 0 & 4 & -\frac{1}{2} \\ -4 & 0 & -\frac{1}{2} \\ \frac{1}{2} & \frac{1}{2} & 0 \end{bmatrix} + \begin{bmatrix} 2 & 1 & \frac{7}{2} \\ 1 & 6 & \frac{1}{2} \\ \frac{7}{2} & \frac{1}{2} & 1 \end{bmatrix}$

$\quad\quad$ Skew-symmetric $\quad\quad$ Symmetric

79. Sample answers:
(a) An example of a 2×2 matrix of the given form is

$A_2 = \begin{bmatrix} 0 & 1 \\ 0 & 0 \end{bmatrix}.$

An example of a 4×4 matrix of the given form is

$A_3 = \begin{bmatrix} 0 & 1 & 2 \\ 0 & 0 & 3 \\ 0 & 0 & 0 \end{bmatrix}.$

(b) $A_2^2 = \begin{bmatrix} 0 & 0 \\ 0 & 0 \end{bmatrix}$

$A_3^2 = \begin{bmatrix} 0 & 0 & 3 \\ 0 & 0 & 0 \\ 0 & 0 & 0 \end{bmatrix}$ and $A_3^3 = \begin{bmatrix} 0 & 0 & 0 \\ 0 & 0 & 0 \\ 0 & 0 & 0 \end{bmatrix}$

(c) The conjecture is that if A is a 4×4 matrix of the given form, then A^4 is the 4×4 zero matrix. A graphing utility shows this to be true.
(d) If A is an $n \times n$ matrix of the given form, then A^n is the $n \times n$ zero matrix.

Section 2.3 *(page 71)*

1. $AB = \begin{bmatrix} 1 & 0 \\ 0 & 1 \end{bmatrix} = BA$ **3.** $AB = \begin{bmatrix} 1 & 0 \\ 0 & 1 \end{bmatrix} = BA$

5. $AB = \begin{bmatrix} 1 & 0 & 0 \\ 0 & 1 & 0 \\ 0 & 0 & 1 \end{bmatrix} = BA$ **7.** $\begin{bmatrix} \frac{1}{2} & 0 \\ 0 & \frac{1}{3} \end{bmatrix}$

9. $\begin{bmatrix} 7 & -2 \\ -3 & 1 \end{bmatrix}$ **11.** $\begin{bmatrix} -19 & -33 \\ -4 & -7 \end{bmatrix}$ **13.** $\begin{bmatrix} 1 & 1 & -1 \\ -3 & 2 & -1 \\ 3 & -3 & 2 \end{bmatrix}$

15. Singular   **17.** $\begin{bmatrix} -\frac{3}{2} & \frac{3}{2} & 1 \\ \frac{9}{2} & -\frac{7}{2} & -3 \\ -1 & 1 & 1 \end{bmatrix}$ **19.** $\begin{bmatrix} \frac{1}{2} & 0 & 0 \\ 0 & \frac{1}{3} & 0 \\ 0 & 0 & \frac{1}{5} \end{bmatrix}$

21. $\begin{bmatrix} 3.75 & 0 & -1.25 \\ 3.45\overline{83} & -1 & -1.375 \\ 4.1\overline{6} & 0 & -2.5 \end{bmatrix}$ **23.** $\begin{bmatrix} 1 & 0 & 0 \\ -\frac{3}{4} & \frac{1}{4} & 0 \\ \frac{7}{20} & -\frac{1}{4} & \frac{1}{5} \end{bmatrix}$

25. Singular   **27.** $\begin{bmatrix} -24 & 7 & 1 & -2 \\ -10 & 3 & 0 & -1 \\ -29 & 7 & 3 & -2 \\ 12 & -3 & -1 & 1 \end{bmatrix}$ **29.** Singular

31. $\begin{bmatrix} \frac{5}{13} & -\frac{3}{13} \\ \frac{1}{13} & \frac{2}{13} \end{bmatrix}$ **33.** Does not exist   **35.** $\begin{bmatrix} \frac{16}{59} & \frac{15}{59} \\ -\frac{4}{59} & \frac{70}{59} \end{bmatrix}$

37. $\begin{bmatrix} \frac{11}{4} & \frac{3}{2} \\ \frac{3}{4} & \frac{1}{2} \end{bmatrix}$ **39.** $\begin{bmatrix} \frac{1}{4} & 0 & 0 \\ 0 & 1 & 0 \\ 0 & 0 & \frac{1}{9} \end{bmatrix}$

41. (a) $\begin{bmatrix} 35 & 17 \\ 4 & 10 \end{bmatrix}$ (b) $\begin{bmatrix} 2 & -7 \\ 5 & 6 \end{bmatrix}$ (c) $\begin{bmatrix} 1 & \frac{5}{2} \\ -\frac{7}{2} & 3 \end{bmatrix}$

43. (a) $\frac{1}{16}\begin{bmatrix} 138 & 56 & -84 \\ 37 & 26 & -71 \\ 24 & 34 & 3 \end{bmatrix}$ (b) $\frac{1}{4}\begin{bmatrix} 4 & 6 & 1 \\ -2 & 2 & 4 \\ 3 & -8 & 2 \end{bmatrix}$

(c) $\frac{1}{8}\begin{bmatrix} 4 & -2 & 3 \\ 6 & 2 & -8 \\ 1 & 4 & 2 \end{bmatrix}$

45. (a) $x = 1$ (b) $x = 2$
$\quad\ \ y = -1$ $\qquad\ \ y = 4$

47. (a) $x_1 = 1$ (b) $x_1 = 0$
$\qquad x_2 = 1$ $\qquad\ x_2 = 1$
$\qquad x_3 = -1$ $\qquad x_3 = -1$

49. $x_1 = 0$ **51.** $x_1 = 1$
$\quad\ \ x_2 = 1$ $\qquad\quad\ x_2 = -2$
$\quad\ \ x_3 = 2$ $\qquad\quad\ x_3 = 3$
$\quad\ \ x_4 = -1$ $\qquad\quad\ x_4 = 0$
$\quad\ \ x_5 = 0$ $\qquad\quad\ x_5 = 1$
$\qquad\qquad\qquad\quad x_6 = -2$

53. $x = 4$ **55.** $x = 6$

57. $\begin{bmatrix} -1 & \frac{1}{2} \\ \frac{3}{4} & -\frac{1}{4} \end{bmatrix}$ **59.** Proof; $A^{-1} = \begin{bmatrix} \sin\theta & -\cos\theta \\ \cos\theta & \sin\theta \end{bmatrix}$

61. $F^{-1} \approx \begin{bmatrix} 32.8725 & -20.4602 & -2.0925 \\ -20.4602 & 56.3552 & -20.4602 \\ -2.0925 & -20.4602 & 32.8725 \end{bmatrix}$;

$\mathbf{w} \approx \begin{bmatrix} 111.4719 \\ 178.1528 \\ 333.4999 \end{bmatrix}$

63–69. Proofs
71. (a) True. See Theorem 2.10, part 1.
(b) False. See Theorem 2.9.
(c) True. See "Finding the Inverse of a Matrix by Gauss-Jordan Elimination," part 2, page 64.
73. The sum of two invertible matrices is not necessarily invertible. For example, let
$A = \begin{bmatrix} 1 & 0 \\ 0 & 1 \end{bmatrix}$ and $B = \begin{bmatrix} -1 & 0 \\ 0 & -1 \end{bmatrix}$.

75. (a) $\begin{bmatrix} -1 & 0 & 0 \\ 0 & \frac{1}{3} & 0 \\ 0 & 0 & \frac{1}{2} \end{bmatrix}$ (b) $\begin{bmatrix} 2 & 0 & 0 \\ 0 & 3 & 0 \\ 0 & 0 & 4 \end{bmatrix}$

77. (a) Proof   (b) $H = \begin{bmatrix} 0 & -1 & 0 \\ -1 & 0 & 0 \\ 0 & 0 & 1 \end{bmatrix}$

79. $A = PDP^{-1}$
No, A is not necessarily equal to D.
81. Answers will vary. Sample answer: For an $n \times n$ matrix A, set up the matrix $[A \quad I]$ and row reduce it until you have $[I \quad A^{-1}]$. If this is not possible or if A is not square, then A has no inverse. If it is possible, then the inverse is A^{-1}.

83. Answers will vary. Sample answer: For the system of equations
$a_{11}x_1 + a_{12}x_2 + a_{13}x_3 = b_1$
$a_{21}x_1 + a_{22}x_2 + a_{23}x_3 = b_2$
$a_{31}x_1 + a_{32}x_2 + a_{33}x_3 = b_3$
write as the matrix equation
$$A\mathbf{x} = \mathbf{b}$$
$$\begin{bmatrix} a_{11} & a_{12} & a_{13} \\ a_{21} & a_{22} & a_{23} \\ a_{31} & a_{32} & a_{33} \end{bmatrix}\begin{bmatrix} x_1 \\ x_2 \\ x_3 \end{bmatrix} = \begin{bmatrix} b_1 \\ b_2 \\ b_3 \end{bmatrix}.$$
If A is invertible, then the solution is $\mathbf{x} = A^{-1}\mathbf{b}$.

Section 2.4 (page 82)

1. Elementary, multiply Row 2 by 2.
3. Elementary, add 2 times Row 1 to Row 2.
5. Not elementary
7. Elementary, add -5 times Row 2 to Row 3.
9. $\begin{bmatrix} 0 & 0 & 1 \\ 0 & 1 & 0 \\ 1 & 0 & 0 \end{bmatrix}$ **11.** $\begin{bmatrix} 0 & 0 & 1 \\ 0 & 1 & 0 \\ 1 & 0 & 0 \end{bmatrix}$
13. Sample answer:
$\begin{bmatrix} \frac{1}{5} & 0 \\ 0 & 1 \end{bmatrix}\begin{bmatrix} 0 & 1 \\ 1 & 0 \end{bmatrix}\begin{bmatrix} 0 & 1 & 7 \\ 5 & 10 & -5 \end{bmatrix} = \begin{bmatrix} 1 & 2 & -1 \\ 0 & 1 & 7 \end{bmatrix}$
15. Sample answer:
$\begin{bmatrix} 1 & 0 & 0 \\ 0 & 1 & 0 \\ 0 & 0 & \frac{1}{2} \end{bmatrix}\begin{bmatrix} 1 & 0 & 0 \\ 0 & \frac{1}{4} & 0 \\ 0 & 0 & 1 \end{bmatrix}\begin{bmatrix} 1 & 0 & 0 \\ 0 & 1 & 0 \\ 6 & 0 & 1 \end{bmatrix}$
$\cdot \begin{bmatrix} 1 & -2 & -1 & 0 \\ 0 & 4 & 8 & -4 \\ -6 & 12 & 8 & 1 \end{bmatrix} = \begin{bmatrix} 1 & -2 & -1 & 0 \\ 0 & 1 & 2 & -1 \\ 0 & 0 & 1 & \frac{1}{2} \end{bmatrix}$
17. Sample answer:
$\begin{bmatrix} 1 & 0 & 0 & 0 \\ 0 & 1 & 0 & 0 \\ 0 & 0 & -\frac{1}{5} & 0 \\ 0 & 0 & 0 & 1 \end{bmatrix}\begin{bmatrix} 1 & 0 & 0 & 0 \\ 0 & 1 & 0 & 0 \\ 0 & 3 & 1 & 0 \\ 0 & 0 & 0 & 1 \end{bmatrix}\begin{bmatrix} 1 & 0 & 0 & 0 \\ 0 & 1 & 0 & 0 \\ 0 & 0 & 1 & 0 \\ 1 & 0 & 0 & 1 \end{bmatrix}$
$\cdot \begin{bmatrix} 1 & 0 & 0 & 0 \\ 0 & 1 & 0 & 0 \\ 2 & 0 & 1 & 0 \\ 0 & 0 & 0 & 1 \end{bmatrix}\begin{bmatrix} 1 & 0 & 0 & 0 \\ 0 & \frac{1}{2} & 0 & 0 \\ 0 & 0 & 1 & 0 \\ 0 & 0 & 0 & 1 \end{bmatrix}\begin{bmatrix} 1 & 0 & 0 & 0 \\ -3 & 1 & 0 & 0 \\ 0 & 0 & 1 & 0 \\ 0 & 0 & 0 & 1 \end{bmatrix}$
$\cdot \begin{bmatrix} 0 & 0 & 1 & 0 \\ 0 & 1 & 0 & 0 \\ 1 & 0 & 0 & 0 \\ 0 & 0 & 0 & 1 \end{bmatrix}\begin{bmatrix} -2 & 1 & 0 \\ 3 & -4 & 0 \\ 1 & -2 & 2 \\ -1 & 2 & -2 \end{bmatrix} = \begin{bmatrix} 1 & -2 & 2 \\ 0 & 1 & -3 \\ 0 & 0 & 1 \\ 0 & 0 & 0 \end{bmatrix}$
19. $\begin{bmatrix} 0 & 1 \\ 1 & 0 \end{bmatrix}$ **21.** $\begin{bmatrix} 0 & 0 & 1 \\ 0 & 1 & 0 \\ 1 & 0 & 0 \end{bmatrix}$
23. $\begin{bmatrix} \frac{1}{k} & 0 & 0 \\ 0 & 1 & 0 \\ 0 & 0 & 1 \end{bmatrix}, k \neq 0$

25. $\begin{bmatrix} 0 & 1 \\ -\frac{1}{2} & \frac{3}{2} \end{bmatrix}$ **27.** $\begin{bmatrix} 1 & 0 & \frac{1}{4} \\ 0 & \frac{1}{6} & \frac{1}{24} \\ 0 & 0 & \frac{1}{4} \end{bmatrix}$

29. $\begin{bmatrix} 1 & 0 \\ 1 & 1 \end{bmatrix}\begin{bmatrix} 1 & -1 \\ 0 & 1 \end{bmatrix}\begin{bmatrix} 1 & 0 \\ 0 & -2 \end{bmatrix}$

(The answer is not unique.)

31. $\begin{bmatrix} 1 & 1 \\ 0 & 1 \end{bmatrix}\begin{bmatrix} 1 & 0 \\ 3 & 1 \end{bmatrix}\begin{bmatrix} 1 & 0 \\ 0 & -1 \end{bmatrix}$

(The answer is not unique.)

33. $\begin{bmatrix} 1 & 0 & 0 \\ -1 & 1 & 0 \\ 0 & 0 & 1 \end{bmatrix}\begin{bmatrix} 1 & -2 & 0 \\ 0 & 1 & 0 \\ 0 & 0 & 1 \end{bmatrix}$

(The answer is not unique.)

35. $\begin{bmatrix} 1 & 0 & 0 & 0 \\ 0 & -1 & 0 & 0 \\ 0 & 0 & 1 & 0 \\ 0 & 0 & 0 & 1 \end{bmatrix}\begin{bmatrix} 1 & 0 & 0 & 0 \\ 0 & 1 & 0 & 0 \\ 0 & 0 & 2 & 0 \\ 0 & 0 & 0 & 1 \end{bmatrix}$

$\begin{bmatrix} 1 & 0 & 0 & 0 \\ 0 & 1 & 0 & 0 \\ 0 & 0 & 1 & 0 \\ 0 & 0 & 0 & -1 \end{bmatrix}\begin{bmatrix} 1 & 0 & 0 & 0 \\ 0 & 1 & 0 & 0 \\ 0 & 0 & 1 & 0 \\ 0 & 0 & -1 & 1 \end{bmatrix}$

$\begin{bmatrix} 1 & 0 & 0 & 1 \\ 0 & 1 & 0 & 0 \\ 0 & 0 & 1 & 0 \\ 0 & 0 & 0 & 1 \end{bmatrix}\begin{bmatrix} 1 & 0 & 0 & 0 \\ 0 & 1 & -3 & 0 \\ 0 & 0 & 1 & 0 \\ 0 & 0 & 0 & 1 \end{bmatrix}$

(The answer is not unique.)

37. No. For example, $\begin{bmatrix} 1 & 0 \\ 2 & 1 \end{bmatrix}\begin{bmatrix} 1 & 1 \\ 0 & 1 \end{bmatrix} = \begin{bmatrix} 1 & 1 \\ 2 & 3 \end{bmatrix}$.

39. $\begin{bmatrix} 1 & 0 & 0 \\ 0 & 1 & 0 \\ -\frac{a}{c} & -\frac{b}{c} & \frac{1}{c} \end{bmatrix}$

41. (a) True. See "Remark" next to "Definition of an Elementary Matrix," page 74.

(b) False. Multiplication of a matrix by a scalar is not a single elementary row operation, so it cannot be represented by a corresponding elementary matrix.

(c) True. See Theorem 2.13.

43. $\begin{bmatrix} 1 & 0 \\ -2 & 1 \end{bmatrix}\begin{bmatrix} 1 & 0 \\ 0 & 1 \end{bmatrix}$

(The answer is not unique.)

45. $\begin{bmatrix} 1 & 0 & 0 \\ 2 & 1 & 0 \\ -1 & 1 & 1 \end{bmatrix}\begin{bmatrix} 3 & 0 & 1 \\ 0 & 1 & -1 \\ 0 & 0 & 2 \end{bmatrix}$

(The answer is not unique.)

47. $x = \frac{1}{3}$
$y = \frac{1}{3}$
$z = -\frac{5}{3}$

49. Idempotent　　**51.** Not idempotent

53. *Case 1:* $b = 1, a = 0$
Case 2: $b = 0, a =$ any real number

55–59. Proofs　　**61.** Answers will vary.

Section 2.5 *(page 91)*

1. Not stochastic　　**3.** Stochastic　　**5.** Stochastic
7. Los Angeles: 25 planes, St. Louis: 13 planes, Dallas: 12 planes

9. $X_1 = \begin{bmatrix} 0.15 \\ 0.17 \\ 0.68 \end{bmatrix}$, $X_2 = \begin{bmatrix} 0.175 \\ 0.217 \\ 0.608 \end{bmatrix}$, $X_3 = \begin{bmatrix} 0.1875 \\ 0.2477 \\ 0.5648 \end{bmatrix}$

11. (a) 350　　(b) 475
13. (a) 25　　(b) 44　　(c) 40
15. (a) Nonsmokers: 5025; smokers of 1 pack/day or less: 2500; smokers of more than 1 pack/day: 2475
(b) Nonsmokers: 5047; smokers of 1 pack/day or less: about 2499; smokers of 1 pack/day or more: about 2454
(c) Nonsmokers: about 5159; smokers of 1 pack/day or less: about 2478; smokers of 1 pack/day or more: about 2363

17. Regular; $\begin{bmatrix} \frac{1}{6} \\ \frac{5}{6} \end{bmatrix}$　　**19.** Not regular; $\begin{bmatrix} 1 \\ 0 \end{bmatrix}$

21. Regular; $\begin{bmatrix} \frac{2}{5} \\ \frac{3}{5} \end{bmatrix}$　　**23.** Regular; $\begin{bmatrix} \frac{43}{101} \\ \frac{16}{101} \\ \frac{42}{101} \end{bmatrix}$

25. Not regular; $\begin{bmatrix} 1-t \\ t \\ 0 \end{bmatrix}$, $0 \le t \le 1$

27. Regular; $\begin{bmatrix} \frac{145}{499} \\ \frac{260}{499} \\ \frac{94}{499} \end{bmatrix}$　　**29.** Regular; $\begin{bmatrix} 0.4 \\ 0.3 \\ 0.2 \\ 0.1 \end{bmatrix}$

31. (a) $\begin{bmatrix} 0.2 \\ 0.3 \\ 0.5 \end{bmatrix}$　　(b) $\begin{bmatrix} \frac{1}{7} \\ \frac{2}{7} \\ \frac{4}{7} \end{bmatrix}$

33. $\begin{bmatrix} 0.2 \\ 0.8 \end{bmatrix}$

Eventually, 20% of the members of the community will make contributions and 80% will not.

35. $\begin{bmatrix} \frac{4}{17} \\ \frac{11}{17} \\ \frac{2}{17} \end{bmatrix}$

Eventually, 200 stockholders will be invested in Stock A, 550 will be invested in Stock B, and 100 will be invested in Stock C.

37. Absorbing; S_3 is absorbing and it is possible to move from S_1 to S_3 in two transitions and from S_2 to S_3 in one transition.
39. Absorbing; S_3 is absorbing and it is possible to move from S_1 or S_2 to S_3 in one transition and from S_4 to S_3 in two transitions.

41. $\begin{bmatrix} 0 \\ 1 \\ 0 \end{bmatrix}$　　**43.** $\begin{bmatrix} 1 \\ 0 \\ 0 \\ 0 \end{bmatrix}$　　**45.** 16,875 people

47. Sample answer: The entries corresponding to nonabsorbing states are 0.

49. (a) $\bar{X} \approx \begin{bmatrix} 0 \\ 0.5536 \\ 0 \\ 0.4464 \end{bmatrix}$ (b) $\bar{X} \approx \begin{bmatrix} 0 \\ 0.6554 \\ 0 \\ 0.3446 \end{bmatrix}$

51. Yes; $\begin{bmatrix} 0 \\ 1 - \frac{11}{6}t \\ \frac{5}{6}t \\ t \end{bmatrix}$, $0 \le t \le \frac{6}{11}$

53. Proof   **55.** Answers will vary.

Section 2.6 (page 102)

1. Uncoded: $[19 \quad 5 \quad 12], [12 \quad 0 \quad 3], [15 \quad 14 \quad 19],$
$[15 \quad 12 \quad 9], [4 \quad 1 \quad 20], [5 \quad 4 \quad 0]$

Encoded: $-48, 5, 31, -6, -6, 9, -85, 23, 43,$
$-27, 3, 15, -115, 36, 59, 9, -5, -4$

3. HAPPY_NEW_YEAR   **5.** ICEBERG_DEAD_AHEAD

7. MEET_ME_TONIGHT_RON

9. _SEPTEMBER_THE_ELEVENTH_WE_WILL_ALWAYS_
REMEMBER

11. $D = \begin{bmatrix} 0.1 & 0.2 \\ 0.8 & 0.1 \end{bmatrix} \begin{matrix} \text{Coal} \\ \text{Steel} \end{matrix}$ $X = \begin{bmatrix} 20{,}000 \\ 40{,}000 \end{bmatrix} \begin{matrix} \text{Coal} \\ \text{Steel} \end{matrix}$

with column headers Coal Steel

13. $X = \begin{bmatrix} 8622.0 \\ 4685.0 \\ 3661.4 \end{bmatrix} \begin{matrix} \text{Farmer} \\ \text{Baker} \\ \text{Grocer} \end{matrix}$

15. (a)   **17.** (a)

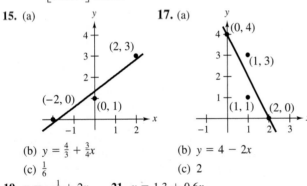

(b) $y = \frac{4}{3} + \frac{3}{4}x$ (b) $y = 4 - 2x$
(c) $\frac{1}{6}$ (c) 2

19. $y = -\frac{1}{3} + 2x$ **21.** $y = 1.3 + 0.6x$
23. $y = 0.412x + 3$ **25.** $y = -0.5x + 7.5$
27. (a) $y = -1.78x + 127.6$
(b) about 69
29. (a) $y = -0.5x + 126$
(b)

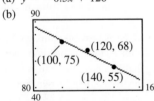

(c)

Number	100	120	140
Actual	75	68	55
Estimated	76	66	56

The estimated values are close to the actual values.
(d) 41%   (e) 172
31. Answers will vary.

Review Exercises (page 104)

1. $\begin{bmatrix} -13 & -8 & 18 \\ 0 & 11 & -19 \end{bmatrix}$ **3.** $\begin{bmatrix} 14 & -2 & 8 \\ 14 & -10 & 40 \\ 36 & -12 & 48 \end{bmatrix}$

5. $\begin{bmatrix} 4 & 6 & 3 \\ 0 & 6 & -10 \\ 0 & 0 & 6 \end{bmatrix}$ **7.** $\begin{bmatrix} 2 & 1 \\ 1 & 4 \end{bmatrix} \begin{bmatrix} x_1 \\ x_2 \end{bmatrix} = \begin{bmatrix} -8 \\ -4 \end{bmatrix}$, $x = \begin{bmatrix} -4 \\ 0 \end{bmatrix}$

9. $\begin{bmatrix} -3 & -1 & 1 \\ 2 & 4 & -5 \\ 1 & -2 & 3 \end{bmatrix} \begin{bmatrix} x_1 \\ x_2 \\ x_3 \end{bmatrix} = \begin{bmatrix} 0 \\ -3 \\ 1 \end{bmatrix}$, $x = \begin{bmatrix} \frac{2}{3} \\ -\frac{17}{3} \\ -\frac{11}{3} \end{bmatrix}$

11. $A^T = \begin{bmatrix} 1 & 0 \\ 2 & 1 \\ -3 & 2 \end{bmatrix}$, $A^T A = \begin{bmatrix} 1 & 2 & -3 \\ 2 & 5 & -4 \\ -3 & -4 & 13 \end{bmatrix}$,

$AA^T = \begin{bmatrix} 14 & -4 \\ -4 & 5 \end{bmatrix}$

13. $A^T = \begin{bmatrix} 1 & 3 & -1 \end{bmatrix}$, $A^T A = [11]$

$AA^T = \begin{bmatrix} 1 & 3 & -1 \\ 3 & 9 & -3 \\ -1 & -3 & 1 \end{bmatrix}$

15. $\begin{bmatrix} 1 & -1 \\ 2 & -3 \end{bmatrix}$ **17.** $\begin{bmatrix} \frac{3}{20} & \frac{3}{20} & \frac{1}{10} \\ \frac{3}{10} & -\frac{1}{30} & -\frac{2}{15} \\ -\frac{1}{5} & -\frac{1}{5} & \frac{1}{5} \end{bmatrix}$ **19.** $\begin{bmatrix} x_1 \\ x_2 \end{bmatrix} = \begin{bmatrix} 10 \\ -12 \end{bmatrix}$

21. $\begin{bmatrix} x_1 \\ x_2 \\ x_3 \end{bmatrix} = \begin{bmatrix} 2 \\ -3 \\ 3 \end{bmatrix}$ **23.** $\begin{bmatrix} 1 \\ -5 \end{bmatrix}$ **25.** $\begin{bmatrix} 0 \\ -\frac{1}{7} \\ \frac{3}{7} \end{bmatrix}$

27. $\begin{bmatrix} \frac{1}{14} & \frac{1}{42} \\ -\frac{1}{21} & \frac{2}{21} \end{bmatrix}$ **29.** $x \ne -3$ **31.** $\begin{bmatrix} 1 & 0 & -4 \\ 0 & 1 & 0 \\ 0 & 0 & 1 \end{bmatrix}$

33. $\begin{bmatrix} 1 & 3 \\ 0 & 1 \end{bmatrix} \begin{bmatrix} 2 & 0 \\ 0 & 1 \end{bmatrix}$
(The answer is not unique.)

35. $\begin{bmatrix} 1 & 0 & 0 \\ 0 & 1 & 0 \\ 0 & 0 & 4 \end{bmatrix} \begin{bmatrix} 1 & 0 & 0 \\ 0 & 1 & -2 \\ 0 & 0 & 1 \end{bmatrix} \begin{bmatrix} 1 & 0 & 1 \\ 0 & 1 & 0 \\ 0 & 0 & 1 \end{bmatrix}$
(The answer is not unique.)

37. $\begin{bmatrix} -1 & 0 \\ 0 & -1 \end{bmatrix}$ and $\begin{bmatrix} 1 & 0 \\ 0 & 1 \end{bmatrix}$
(The answer is not unique.)

39. $\begin{bmatrix} 0 & 0 \\ 0 & 0 \end{bmatrix}, \begin{bmatrix} 1 & 0 \\ 0 & 1 \end{bmatrix}$, and $\begin{bmatrix} 1 & 0 \\ 0 & 0 \end{bmatrix}$
(The answer is not unique.)

41. (a) $a = -1$ (b) and (c) Proofs
$b = -1$
$c = 1$

43. $\begin{bmatrix} 1 & 0 \\ 3 & 1 \end{bmatrix} \begin{bmatrix} 2 & 5 \\ 0 & -1 \end{bmatrix}$
(The answer is not unique.)

45. $\begin{bmatrix} 1 & 0 & 0 \\ 0 & 1 & 0 \\ -4 & 5 & 1 \end{bmatrix} \begin{bmatrix} 4 & 1 & 0 \\ 0 & 3 & -7 \\ 0 & 0 & 36 \end{bmatrix}$
(The answer is not unique.)

47. $x = 4, y = 1, z = -1$

49. (a) $\begin{bmatrix} 418 & 454 \\ 90 & 100 \end{bmatrix}$ (b) $\begin{bmatrix} 209 & 227 \\ 45 & 50 \end{bmatrix}$

51. (a) $\begin{bmatrix} 2200b_{11} + 3180b_{21} + 1210b_{31} & 121.90 \\ 2120b_{11} + 1590b_{21} + 610b_{31} & 71.30 \\ 3250b_{11} + 3860b_{21} + 2040b_{31} & 170.90 \end{bmatrix}$

The first column gives the total sales for gas on each day and the second column gives the total profit for each day.

(b) $364.10

53. $\begin{bmatrix} 0 & 0 \\ 0 & 0 \end{bmatrix}$ **55.** Stochastic   **57.** Not stochastic

59. $X_1 = \begin{bmatrix} \frac{5}{12} \\ \frac{7}{12} \end{bmatrix}$, $X_2 = \begin{bmatrix} \frac{17}{48} \\ \frac{31}{48} \end{bmatrix}$, $X_3 = \begin{bmatrix} \frac{65}{192} \\ \frac{127}{192} \end{bmatrix}$

61. $X_1 = \begin{bmatrix} 0.375 \\ 0.475 \\ 0.150 \end{bmatrix}$, $X_2 = \begin{bmatrix} 0.3063 \\ 0.4488 \\ 0.2450 \end{bmatrix}$, $X_3 \approx \begin{bmatrix} 0.2653 \\ 0.4274 \\ 0.3073 \end{bmatrix}$

63. (a) 120   (b) 144

65. Regular; $\begin{bmatrix} \frac{5}{7} \\ \frac{2}{7} \end{bmatrix}$ **67.** Not regular; $\begin{bmatrix} 0 \\ 0 \\ 1 \end{bmatrix}$

69. $\begin{bmatrix} \frac{3}{7} \\ \frac{4}{7} \end{bmatrix}$

Eventually, $\frac{3}{7}$ of the customers will turn in their tickets and $\frac{4}{7}$ will not.

71. Not absorbing; no state is absorbing.

73. (a) False. See Theorem 2.1, part 1, page 52.

(b) True. See Theorem 2.6, part 2, page 57.

75. (a) False. The entries must be between 0 and 1 inclusive.

(b) True. See page 90, Example 7(a).

77. Uncoded: $[15 \quad 14][5 \quad 0][9 \quad 6][0 \quad 2][25 \quad 0]$
$[12 \quad 1][14 \quad 4]$

Encoded: 103 44 25 10 57 24 4 2 125 50 62 25 78 32

79. $A^{-1} = \begin{bmatrix} 3 & 2 \\ 4 & 3 \end{bmatrix}$; ALL_SYSTEMS_GO

81. _CAN_YOU_HEAR_ME_NOW

83. $D = \begin{bmatrix} 0.20 & 0.50 \\ 0.30 & 0.10 \end{bmatrix}$, $X \approx \begin{bmatrix} 133{,}333 \\ 133{,}333 \end{bmatrix}$

85. $y = \frac{20}{3} - \frac{3}{2}x$ **87.** $y = 2.5x$

89. (a) $y = 13.4x + 164$

(b) $y = 13.4x + 164$; They are the same.

(c)

Year	2008	2009	2010	2011	2012	2013
Actual	270	286	296	316	326	336
Estimated	271	285	298	311	325	338

The estimated values are close to the actual values.

Chapter 3

Section 3.1 (page 116)

1. 1   **3.** 5   **5.** 27   **7.** -24 **9.** 0

11. $\lambda^2 - 4\lambda - 5$

13. (a) $M_{11} = 4$ (b) $C_{11} = 4$
$M_{12} = 3$ $C_{12} = -3$
$M_{21} = 2$ $C_{21} = -2$
$M_{22} = 1$ $C_{22} = 1$

15. (a) $M_{11} = 23$ $M_{12} = -8$ $M_{13} = -22$
$M_{21} = 5$ $M_{22} = -5$ $M_{23} = 5$
$M_{31} = 7$ $M_{32} = -22$ $M_{33} = -23$

(b) $C_{11} = 23$ $C_{12} = 8$ $C_{13} = -22$
$C_{21} = -5$ $C_{22} = -5$ $C_{23} = -5$
$C_{31} = 7$ $C_{32} = 22$ $C_{33} = -23$

17. (a) $4(-5) + 5(-5) + 6(-5) = -75$

(b) $2(8) + 5(-5) - 3(22) = -75$

19. -58 **21.** -30 **23.** 0.002   **25.** $2x - 3y - 1$

27. 0   **29.** $65{,}644w + 62{,}256x + 12{,}294y - 24{,}672z$

31. -100 **33.** 29   **35.** 0.281   **37.** 19

39. -24 **41.** 0

43. (a) False. See "Definition of the Determinant of a 2×2 Matrix," page 110.

(b) True. See "Remark," page 112.

(c) False. See "Minors and Cofactors of a Square Matrix," page 111.

45. $x = -1, -4$ **47.** $x = -1, 4$ **49.** $\lambda = -1 \pm \sqrt{3}$

51. $\lambda = -2, 0,$ or 1 **53.** Proof   **55.** $18uv - 1$

57. e^{5x} **59.** $1 - \ln x$ **61.** r

63. $wz - xy$ **65.** $wz - xy$

67. $xy^2 - xz^2 + yz^2 - x^2y + x^2z - y^2z$

69. (a) Proof

(b) $\begin{vmatrix} x & 0 & 0 & d \\ -1 & x & 0 & c \\ 0 & -1 & x & b \\ 0 & 0 & -1 & a \end{vmatrix}$

Section 3.2 (page 124)

1. The first row is 2 times the second row. If one row of a matrix is a multiple of another row, then the determinant of the matrix is zero.

3. The second row consists entirely of zeros. If one row of a matrix consists entirely of zeros, then the determinant of the matrix is zero.

5. The second and third columns are interchanged. If two columns of a matrix are interchanged, then the determinant of the matrix changes sign.

7. The first row of the matrix is multiplied by 5. If a row in a matrix is multiplied by a scalar, then the determinant of the matrix is multiplied by that scalar.

9. A 4 is factored out of the second column and a 3 is factored out of the third column. If a column of a matrix is multiplied by a scalar, then the determinant of the matrix is multiplied by that scalar.

11. A 5 is factored out of each column. If a column matrix is multiplied by a scalar, then the determinant of the matrix is multiplied by that scalar.

13. -4 times the first row is added to the second row. If a scalar multiple of one row of a matrix is added to another row, then the determinant of the matrix is unchanged.

15. A multiple of the first row is added to the second row. If a scalar multiple of one row of a matrix is added to another row, then the determinants are equal.

17. The second row of the matrix is multiplied by -1. If a row of a matrix is multiplied by a scalar, then the determinant is multiplied by that scalar.

19. The fifth column is 2 times the first column. If one column of a matrix is a multiple of another column, then the determinant of the matrix is zero.

21. -1 **23.** 8   **25.** 28   **27.** 0   **29.** -59

31. -1344 **33.** 136   **35.** -1100

37. (a) True. See Theorem 3.3, part 1, page 119.
  (b) True. See Theorem 3.3, part 3, page 119.
  (c) True. See Theorem 3.4, part 2, page 121.

39. k **41.** 1   **43.** Proof

45. (a) $\cos^2 \theta + \sin^2 \theta = 1$ (b) $\sin^2 \theta - 1 = -\cos^2 \theta$

47. Proof

Section 3.3   *(page 131)*

1. (a) 0   (b) -1 (c) $\begin{bmatrix} -2 & -3 \\ 4 & 6 \end{bmatrix}$ (d) 0

3. (a) 2   (b) -6 (c) $\begin{bmatrix} 1 & 4 & 3 \\ -1 & 0 & 3 \\ 0 & 2 & 0 \end{bmatrix}$ (d) -12

5. (a) 3   (b) 6   (c) $\begin{bmatrix} 6 & 3 & -2 & 2 \\ 2 & 1 & 0 & -1 \\ 9 & 4 & -3 & 8 \\ 8 & 5 & -4 & 5 \end{bmatrix}$ (d) 18

7. -250 **9.** 54   **11.** 0   **13.** -3125

15. (a) -2 (b) -2 (c) $\begin{bmatrix} 0 & 0 \\ 0 & 0 \end{bmatrix}$ (d) 0

17. (a) 1   (b) -1 (c) $\begin{bmatrix} 0 & 1 & 3 \\ -1 & 2 & 3 \\ 1 & 2 & 1 \end{bmatrix}$ (d) -8

19. Singular   **21.** Nonsingular

23. Singular   **25.** $\frac{1}{5}$ **27.** $-\frac{1}{3}$ **29.** $\frac{1}{24}$

31. The solution is unique because the determinant of the coefficient matrix is nonzero.

33. The solution is not unique because the determinant of the coefficient matrix is zero.

35. The solution is unique because the determinant of the coefficient matrix is nonzero.

37. $k = -1, 4$ **39.** $k = 24$ **41.** $k = \pm\dfrac{\sqrt{2}}{2}$

43. (a) 14   (b) 196   (c) 196   (d) 56   (e) $\frac{1}{14}$

45. (a) -30 (b) 900   (c) 900   (d) -240 (e) $-\frac{1}{30}$

47. (a) 29   (b) 841   (c) 841   (d) 232   (e) $\frac{1}{29}$

49. (a) -30 (b) 900   (c) 900   (d) -480 (e) $-\frac{1}{30}$

51. (a) 22   (b) 22   (c) 484   (d) 88   (e) $\frac{1}{22}$

53. (a) -26 (b) -26 (c) 676   (d) -208 (e) $-\frac{1}{26}$

55. (a) -115 (b) -115 (c) 13,225   (d) -1840 (e) $-\frac{1}{115}$

57. (a) 25   (b) 9   (c) -125 (d) 81

59. Proof

61. $\begin{bmatrix} 0 & 1 \\ 0 & 0 \end{bmatrix}$ and $\begin{bmatrix} 1 & 0 \\ 0 & 0 \end{bmatrix}$

  (The answer is not unique.)

63. 0   **65.** Proof

67. No; in general, $P^{-1}AP \neq A$. For example, let

$$P = \begin{bmatrix} 1 & 2 \\ 3 & 5 \end{bmatrix}, \quad P^{-1} = \begin{bmatrix} -5 & 2 \\ 3 & -1 \end{bmatrix}, \text{ and } A = \begin{bmatrix} 2 & 1 \\ -1 & 0 \end{bmatrix}.$$

Then you have

$$P^{-1}AP = \begin{bmatrix} -27 & -49 \\ 16 & 29 \end{bmatrix} \neq A.$$

The equation $|P^{-1}AP| = |A|$ is true in general because

$$|P^{-1}AP| = |P^{-1}||A||P| = |P^{-1}||P||A| = \frac{1}{|P|}|P||A| = |A|.$$

69. Proof

71. (a) False. See Theorem 3.6, page 127.
  (b) True. See Theorem 3.8, page 128.
  (c) True. See "Equivalent Conditions for a Nonsingular Matrix," parts 1 and 2, page 129.

73. Orthogonal   **75.** Not orthogonal   **77.** Orthogonal

79. Proof   **81.** Orthogonal   **83.** Proof

Section 3.4   *(page 142)*

1. $\text{adj}(A) = \begin{bmatrix} 4 & -2 \\ -3 & 1 \end{bmatrix}$, $A^{-1} = \begin{bmatrix} -2 & 1 \\ \frac{3}{2} & -\frac{1}{2} \end{bmatrix}$

3. $\text{adj}(A) = \begin{bmatrix} 0 & 0 & 0 \\ 0 & -12 & -6 \\ 0 & 4 & 2 \end{bmatrix}$, A^{-1} does not exist.

5. $\text{adj}(A) = \begin{bmatrix} -7 & -12 & 13 \\ 2 & 3 & -5 \\ 2 & 3 & -2 \end{bmatrix}$, $A^{-1} = \begin{bmatrix} \frac{7}{3} & 4 & -\frac{13}{3} \\ -\frac{2}{3} & -1 & \frac{5}{3} \\ -\frac{2}{3} & -1 & \frac{2}{3} \end{bmatrix}$

7. $\text{adj}(A) = \begin{bmatrix} 7 & 1 & 9 & -13 \\ 7 & 1 & 0 & -4 \\ -4 & 2 & -9 & 10 \\ 2 & -1 & 9 & -5 \end{bmatrix}$,

$A^{-1} = \begin{bmatrix} \frac{7}{9} & \frac{1}{9} & 1 & -\frac{13}{9} \\ \frac{7}{9} & \frac{1}{9} & 0 & -\frac{4}{9} \\ -\frac{4}{9} & \frac{2}{9} & -1 & \frac{10}{9} \\ \frac{2}{9} & -\frac{1}{9} & 1 & -\frac{5}{9} \end{bmatrix}$

9. $x_1 = 1$ **11.** $x = 2$ **13.** $x = \frac{3}{4}$
$x_2 = 2$ $y = -2$ $y = -\frac{1}{2}$

15. Cramer's Rule does not apply because the coefficient matrix has a determinant of zero.

17. $x = 1$ **19.** $x = 1$
$y = 1$ $y = \frac{1}{2}$
$z = 2$ $z = \frac{3}{2}$

21. $x_1 = -1, x_2 = 3, x_3 = 2$ **23.** $x_1 = -12, x_2 = 10$

25. $x_1 = 5, x_2 = -3, x_3 = 2, x_4 = -1$

27. $x = \dfrac{4k - 3}{2k - 1}, y = \dfrac{4k - 1}{2k - 1}$

  The system will be inconsistent if $k = \frac{1}{2}$.

29. 3   **31.** 3   **33.** Collinear   **35.** Not collinear

37. $3y - 4x = 0$ **39.** $x = -2$ **41.** $\frac{1}{3}$ **43.** 2   **45.** 10

47. Not coplanar   **49.** Coplanar   **51.** Not coplanar

53. $4x - 10y + 3z = 27$ **55.** $x + y + z = 0$ **57.** $z = -4$

59. Incorrect. The numerator and denominator should be interchanged.

61. (a) $a + b + c = 156.8$
$$4a + 2b + c = 161.7$$
$$9a + 3b + c = 177.2$$
   (b) $a = 5.3$, $b = -11$, $c = 162.5$
   (c)

   (d) The polynomial fits the data exactly.

63. Proof   **65.** Proof

67. Sample answer: $\left| \mathrm{adj}(A) \right| = \begin{vmatrix} -2 & 0 \\ -1 & 1 \end{vmatrix} = -2,$

$$|A|^{2-1} = \begin{vmatrix} 1 & 0 \\ 1 & -2 \end{vmatrix}^{2-1} = -2$$

69. Proof

Review Exercises   (page 144)

1. 10   **3.** 0   **5.** 14   **7.** -6 **9.** 1620

11. 82   **13.** -64 **15.** -1 **17.** -1

19. Because the second row is a multiple of the first row, the determinant is zero.

21. A -4 has been factored out of the second column and a 3 has been factored out of the third column. If a column of a matrix is multiplied by a scalar, then the determinant of the matrix is also multiplied by that scalar.

23. (a) -1 (b) -5 (c) $\begin{bmatrix} 1 & -2 \\ 2 & 1 \end{bmatrix}$ (d) 5

25. (a) -35 (b) $-42{,}875$ (c) 1225   (d) -875

27. (a) -20 (b) $-\frac{1}{20}$ **29.** $-\frac{1}{6}$ **31.** $-\frac{1}{10}$

33. $x_1 = 0$ **35.** $x_1 = -3$
     $x_2 = -\frac{1}{2}$ $x_2 = -1$
     $x_3 = \frac{1}{2}$ $x_3 = 2$

37. Unique solution   **39.** Unique solution

41. Not a unique solution

43. (a) 8   (b) 4   (c) 64   (d) 8   (e) $\frac{1}{2}$

45. Proof   **47.** 0   **49.** $-\frac{1}{2}$ **51.** u **53.** $-uv$

55. Row reduction is generally preferred for matrices with few zeros. For a matrix with many zeros, it is often easier to expand along a row or column having many zeros.

57. $x = \pi/4 + n\pi/2$, where n is an integer.   **59.** $\begin{bmatrix} 1 & -1 \\ 2 & 0 \end{bmatrix}$

61. Unique solution: $x = 0.6$
                 $y = 0.5$

63. Unique solution: $x_1 = \frac{1}{2}$
                    $x_2 = -\frac{1}{3}$
                    $x_3 = 1$

65. $x_1 = 6, x_2 = -2$ **67.** 16   **69.** $x - 2y = -4$

71. $9x + 4y - 3z = 0$

73. Incorrect. In the numerator, the column of constants,

$$\begin{bmatrix} -1 \\ 6 \\ 1 \end{bmatrix}$$

should replace the third column of the coefficient matrix, not the first column.

75. (a) False. See "Minors and Cofactors of a Square Matrix," page 111.
   (b) False. See Theorem 3.3, part 1, page 119.
   (c) True. See Theorem 3.4, part 3, page 121.
   (d) False. See Theorem 3.9, page 130.

77. (a) False. See Theorem 3.11, page 137.
   (b) False. See "Test for Collinear Points in the xy-Plane," page 139.

Cumulative Test Chapters 1–3   (page 149)

1. Not linear   **2.** Linear   **3.** $x = 1, y = -2$

4. $x_1 = 2, x_2 = -3, x_3 = -2$

5. $x = -10, y = 20, z = -40, w = 12$

6. $x_1 = s - 2t, x_2 = 2 + t, x_3 = t, x_4 = s$

7. $x_1 = -2s, x_2 = s, x_3 = 2t, x_4 = t$

8. $k = 12$ **9.** $x = -3, y = 4$

10. $A^T A = \begin{bmatrix} 29 & 23 & 17 \\ 23 & 25 & 27 \\ 17 & 27 & 37 \end{bmatrix}$ **11.** $\begin{bmatrix} -\frac{1}{4} & \frac{1}{8} \\ \frac{1}{6} & \frac{1}{12} \end{bmatrix}$

12. $\begin{bmatrix} -\frac{2}{7} & \frac{1}{7} \\ \frac{1}{7} & \frac{2}{21} \end{bmatrix}$ **13.** $\begin{bmatrix} -1 & 0 & 0 \\ 0 & 2 & 0 \\ 0 & 0 & \frac{1}{3} \end{bmatrix}$ **14.** $\begin{bmatrix} 1 & 0 & -1 \\ 0 & 0 & 1 \\ \frac{3}{5} & \frac{1}{5} & -\frac{9}{5} \end{bmatrix}$

15. $x = \frac{4}{3}, y = -\frac{2}{3}$ **16.** $x = 4, y = 2$

17. $\begin{bmatrix} 0 & 1 \\ 1 & 0 \end{bmatrix} \begin{bmatrix} 1 & 0 \\ 2 & 1 \end{bmatrix} \begin{bmatrix} 1 & 0 \\ 0 & -4 \end{bmatrix}$ **18.** -6

   (The answer is not unique.)

19. (a) 14   (b) -10 (c) $\begin{bmatrix} -2 & -14 \\ -8 & 14 \end{bmatrix}$ (d) -140

20. (a) 84   (b) $\frac{1}{84}$

21. (a) 567   (b) 7   (c) $\frac{1}{7}$ (d) 343

22. $\begin{bmatrix} \frac{4}{11} & -\frac{10}{11} & \frac{7}{11} \\ -\frac{1}{11} & -\frac{3}{11} & \frac{1}{11} \\ -\frac{2}{11} & \frac{5}{11} & \frac{2}{11} \end{bmatrix}$ **23.** $a = 1, b = 0, c = 2$

24. $y = \frac{7}{6}x^2 + \frac{1}{6}x + 1$ **25.** $3x + 2y = 11$ **26.** 35

27. $I_1 = 3, I_2 = 4, I_3 = 1$

28. $BA = \begin{bmatrix} 13{,}275.00 & 15{,}500.00 \end{bmatrix}$
   The entries represent the total values (in dollars) of the products sent to the two warehouses.

29. No; Sample answer:

$$A = \begin{bmatrix} 2 & 3 \\ 1 & 4 \end{bmatrix}, B = \begin{bmatrix} 6 & -1 \\ 5 & 0 \end{bmatrix}, C = \begin{bmatrix} 1 & 1 \\ 1 & 1 \end{bmatrix}$$

Chapter 4

Section 4.1   *(page 159)*

1. $\mathbf{v} = (4, 5)$

3.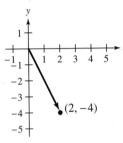

5.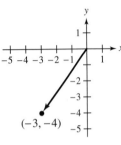

7. $\mathbf{u} + \mathbf{v} = (3, 1)$

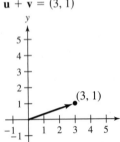

9. $\mathbf{u} + \mathbf{v} = (-1, -4)$

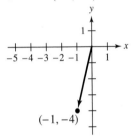

11. $\mathbf{v} = \left(-3, \frac{9}{2}\right)$

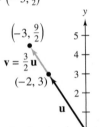

13. $\mathbf{v} = (-8, -1)$

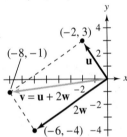

15. $\mathbf{v} = \left(-\frac{9}{2}, \frac{7}{2}\right)$

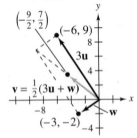

17. (a)

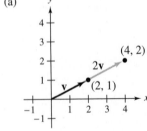

(b)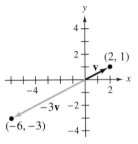

(c)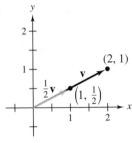

19. $\mathbf{u} - \mathbf{v} = (-1, 0, 4)$
   $\mathbf{v} - \mathbf{u} = (1, 0, -4)$

21. $(6, 12, 6)$ **23.** $\left(-\frac{1}{4}, \frac{3}{2}, \frac{13}{4}\right)$

25. (a)

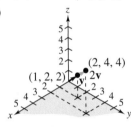

(b)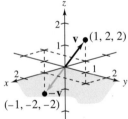

(c)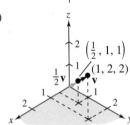

27. (a) Scalar multiple   (b) Not a scalar multiple

29. (a) $(4, -2, -8, 1)$ (b) $(8, 12, 24, 34)$ (c) $(-4, 4, 13, 3)$

31. (a) $(-9, 3, 2, -3, 6)$ (b) $(-2, -18, -12, 18, 36)$
   (c) $(11, -6, -4, 6, -3)$

33. (a) $(1, 6, -5, -3)$ (b) $(-1, -8, 10, 0)$
   (c) $\left(-\frac{3}{2}, 11, -\frac{13}{2}, -\frac{21}{2}\right)$

35. $\left(\frac{1}{3}, -\frac{5}{3}, -2, 1\right)$ **37.** $(4, 8, 18, -2)$ **39.** $\left(-1, \frac{5}{3}, 6, \frac{2}{3}\right)$

41. $\mathbf{v} = \mathbf{u} + \mathbf{w}$ **43.** $\mathbf{v} = 2\mathbf{u} + \mathbf{w}$ **45.** $\mathbf{v} = -\mathbf{u}$

47. $\mathbf{v} = \mathbf{u}_1 + 2\mathbf{u}_2 - 3\mathbf{u}_3$

49. It is not possible to write $\mathbf{v}$ as a linear combination of $\mathbf{u}_1, \mathbf{u}_2$, and $\mathbf{u}_3$.

51. $\begin{bmatrix} 3 \\ 9 \\ 6 \end{bmatrix} = (-1)\begin{bmatrix} 1 \\ 7 \\ 4 \end{bmatrix} + 2\begin{bmatrix} 2 \\ 8 \\ 5 \end{bmatrix}$

53. $\mathbf{v} = 2\mathbf{u}_1 + \mathbf{u}_2 - 2\mathbf{u}_3 + \mathbf{u}_4 - \mathbf{u}_5$ **55.** No

57. (a) True. Two vectors in R^n are equal if and only if their corresponding components are equal, that is, $\mathbf{u} = \mathbf{v}$ if and only if $u_1 = v_1, u_2 = v_2, \ldots, u_n = v_n$.

(b) False. The vector $-\mathbf{v}$ is called the additive inverse of the vector $\mathbf{v}$.

59. If $\mathbf{b} = x_1\mathbf{a}_1 + \cdots + x_n\mathbf{a}_n$ is a linear combination of the columns of A, then a solution to $A\mathbf{x} = \mathbf{b}$ is

$$\mathbf{x} = \begin{bmatrix} x_1 \\ \vdots \\ x_n \end{bmatrix}.$$

The system $A\mathbf{x} = \mathbf{b}$ is inconsistent if $\mathbf{b}$ is not a linear combination of the columns of A.

61. Answers will vary.   **63.** Proof

65. (a) Additive identity
(b) Distributive property
(c) Add $-0\mathbf{v}$ to both sides.
(d) Additive inverse and associative property
(e) Additive inverse
(f) Additive identity

67. (a) Multiply both sides by c^{-1}.
(b) Associative property and Theorem 4.3, property 4
(c) Multiplicative inverse
(d) Multiplicative identity

Section 4.2   *(page 166)*

1. $(0, 0, 0, 0)$

3. $\begin{bmatrix} 0 & 0 & 0 \\ 0 & 0 & 0 \\ 0 & 0 & 0 \\ 0 & 0 & 0 \end{bmatrix}$ **5.** $0 + 0x + 0x^2 + 0x^3$

7. $-(v_1, v_2, v_3) = (-v_1, -v_2, -v_3)$

9. $-\begin{bmatrix} a_{11} & a_{12} & a_{13} \\ a_{21} & a_{22} & a_{23} \end{bmatrix} = \begin{bmatrix} -a_{11} & -a_{12} & -a_{13} \\ -a_{21} & -a_{22} & -a_{23} \end{bmatrix}$

11. $-(a_0 + a_1x + a_2x^2 + a_3x^3 + a_4x^4)$
$= -a_0 - a_1x - a_2x^2 - a_3x^3 - a_4x^4$

13. Vector space
15. Not a vector space; Sample answer: Axiom 1 fails.
17. Not a vector space; Axiom 4 fails.
19. Vector space
21. Not a vector space; Sample answer: Axiom 6 fails.
23. Vector space   **25.** Vector space   **27.** Vector space
29. Not a vector space; Sample answer: Axiom 1 fails.
31. Not a vector space; Axiom 1 fails.
33. Vector space   **35.** Vector space
37. Proof   **39.** Proof
41. (a) The set is not a vector space. Axiom 8 fails because
$(1 + 2)(1, 1) = 3(1, 1) = (3, 1)$
$1(1, 1) + 2(1, 1) = (1, 1) + (2, 1) = (3, 2)$.

(b) The set is not a vector space. Axiom 2 fails because
$(1, 2) + (2, 1) = (1, 0)$
$(2, 1) + (1, 2) = (2, 0)$.
(Axioms 4, 5, and 8 also fail.)

(c) The set is not a vector space. Axiom 6 fails because
$(-1)(1, 1) = \left(\sqrt{-1}, \sqrt{-1}\right)$, which is not in R^2. (Axioms 8 and 9 also fail.)

43. Proof   **45.** Answers will vary.

47. (a) Add $-\mathbf{w}$ to both sides.
(b) Associative property
(c) Additive inverse
(d) Additive identity

49. (a) True. See page 161.
(b) False. See example 6, page 165.
(c) False. With standard operations on R^3, the additive inverse axiom is not satisfied.

51. Proof

Section 4.3   *(page 173)*

1. W is nonempty and $W \subset R^4$, so you need only check that W is closed under addition and scalar multiplication. Given
$(x_1, x_2, x_3, 0) \in W$ and $(y_1, y_2, y_3, 0) \in W$
it follows that
$(x_1, x_2, x_3, 0) + (y_1, y_2, y_3, 0)$
$= (x_1 + y_1, x_2 + y_2, x_3 + y_3, 0) \in W$.
Also, for any real number c and $(x_1, x_2, x_3, 0) \in W$, it follows that
$c(x_1, x_2, x_3, 0) = (cx_1, cx_2, cx_3, 0) \in W$.

3. W is nonempty and $W \subset M_{2,2}$, so you need only check that W is closed under addition and scalar multiplication. Given
$\begin{bmatrix} 0 & a_1 \\ b_1 & 0 \end{bmatrix} \in W$ and $\begin{bmatrix} 0 & a_2 \\ b_2 & 0 \end{bmatrix} \in W$
it follows that
$\begin{bmatrix} 0 & a_1 \\ b_1 & 0 \end{bmatrix} + \begin{bmatrix} 0 & a_2 \\ b_2 & 0 \end{bmatrix} = \begin{bmatrix} 0 & a_1 + a_2 \\ b_1 + b_2 & 0 \end{bmatrix} \in W$.
Also, for any real number c and
$\begin{bmatrix} 0 & a \\ b & 0 \end{bmatrix} \in W$, it follows that
$c\begin{bmatrix} 0 & a \\ b & 0 \end{bmatrix} = \begin{bmatrix} 0 & ca \\ cb & 0 \end{bmatrix} \in W$.

5. Recall from calculus that continuity implies integrability; $W \subset V$. So, because W is nonempty, you need only check that W is closed under addition and scalar multiplication. Given continuous functions $f, g \in W$, it follows that $f + g$ is continuous and $f + g \in W$. Also, for any real number c and for a continuous function $f \in W$, cf is continuous. So, $cf \in W$.

7. Not closed under addition:
$(0, 0, -1) + (0, 0, -1) = (0, 0, -2)$

9. Not closed under scalar multiplication:
$\sqrt{2}(1, 1) = \left(\sqrt{2}, \sqrt{2}\right)$

11. Not closed under scalar multiplication:
$(-1)e^x = -e^x$

13. Not closed under scalar multiplication:
$(-2)(1, 1, 1) = (-2, -2, -2)$

15. Not closed under scalar multiplication:

$$2\begin{bmatrix} 1 & 0 & 0 \\ 0 & 1 & 0 \\ 0 & 0 & 0 \end{bmatrix} = \begin{bmatrix} 2 & 0 & 0 \\ 0 & 2 & 0 \\ 0 & 0 & 0 \end{bmatrix}$$

17. Not closed under addition:

$$\begin{bmatrix} 1 & 0 & 0 \\ 0 & 1 & 0 \\ 0 & 0 & 1 \end{bmatrix} + \begin{bmatrix} 1 & 0 & 1 \\ 0 & 1 & 0 \\ 0 & 0 & 1 \end{bmatrix} = \begin{bmatrix} 2 & 0 & 1 \\ 0 & 2 & 0 \\ 0 & 0 & 2 \end{bmatrix}$$

19. Not closed under addition:
$(2, 8) + (3, 27) = (5, 35)$

21. Not a subspace; not closed under scalar multiplication

23. Subspace; nonempty and closed under addition and scalar multiplication

25. Subspace; nonempty and closed under addition and scalar multiplication

27. Subspace; nonempty and closed under addition and scalar multiplication

29. Subspace; nonempty and closed under addition and scalar multiplication

31. Not a subspace; not closed under scalar multiplication

33. Not a subspace; not closed under addition

35. Subspace; nonempty and closed under addition and scalar multiplication

37. Subspace; nonempty and closed under addition and scalar multiplication

39. Subspace; nonempty and closed under addition and scalar multiplication

41. Not a subspace; not closed under addition

43. (a) True. See Theorem 4.5, part 2, page 168.
(b) True. See Theorem 4.6, page 170.
(c) False. There may be elements of W that are not elements of U, or vice-versa.

45–59. Proofs

Section 4.4   (page 184)

1. (a) $\mathbf{z} = 2(2, -1, 3) - (5, 0, 4)$
(b) $\mathbf{v} = \frac{1}{4}(2, -1, 3) + \frac{3}{2}(5, 0, 4)$
(c) $\mathbf{w} = 8(2, -1, 3) - 3(5, 0, 4)$
(d) $\mathbf{u}$ cannot be written as a linear combination of the given vectors.

3. (a) $\mathbf{u} = -\frac{7}{4}(2, 0, 7) + \frac{5}{4}(2, 4, 5) + 0(2, -12, 13)$
(b) $\mathbf{v}$ cannot be written as a linear combination of the given vectors.
(c) $\mathbf{w} = -\frac{1}{6}(2, 0, 7) + \frac{1}{3}(2, 4, 5) + 0(2, -12, 13)$
(d) $\mathbf{z} = -4(2, 0, 7) + 5(2, 4, 5) + 0(2, -12, 13)$

5. $\begin{bmatrix} 6 & -19 \\ 10 & 7 \end{bmatrix} = 3A - 2B$

7. $\begin{bmatrix} -2 & 23 \\ 0 & -9 \end{bmatrix} = -A + 4B$

9. S spans R^2.   **11.** S spans R^2.
13. S does not span R^2; line   **15.** S does not span R^2; line
17. S does not span R^2; line   **19.** S spans R^3.
21. S does not span R^3; plane   **23.** S does not span R^3; plane
25. S does not span P_2.   **27.** Linearly independent
29. Linearly dependent   **31.** Linearly independent

33. Linearly dependent   **35.** Linearly independent
37. Linearly dependent   **39.** Linearly independent
41. Linearly dependent   **43.** Linearly independent
45. Linearly dependent   **47.** Linearly independent
49. Linearly dependent   **51.** Linearly independent

53. $(3, 4) - 4(-1, 1) - \frac{7}{2}(2, 0) = (0, 0)$,
$(3, 4) = 4(-1, 1) + \frac{7}{2}(2, 0)$
(The answer is not unique.)

55. $(1, 1, 1) - (1, 1, 0) - (0, 0, 1) - 0(0, 1, 1) = (0, 0, 0)$
$(1, 1, 1) = (1, 1, 0) + (0, 0, 1) - 0(0, 1, 1)$
(The answer is not unique.)

57. (a) All $t \neq 1, -2$ (b) All $t \neq \frac{1}{2}$

59. Proof

61. Because the matrix

$$\begin{bmatrix} 1 & 2 & -1 \\ 0 & 1 & 1 \\ 2 & 5 & -1 \end{bmatrix} \text{ row reduces to } \begin{bmatrix} 1 & 0 & -3 \\ 0 & 1 & 1 \\ 0 & 0 & 0 \end{bmatrix} \text{ and}$$

$$\begin{bmatrix} -2 & -6 & 0 \\ 1 & 1 & -2 \end{bmatrix} \text{ row reduces to } \begin{bmatrix} 1 & 0 & -3 \\ 0 & 1 & 1 \end{bmatrix},$$

S_1 and S_2 span the same subspace.

63. (a) False. See "Definition of Linear Dependence and Linear Independence," page 179.
(b) True. Any vector $\mathbf{u} = (u_1, u_2, u_3, u_4)$ in R^4 can be written as
$\mathbf{u} = u_1(1, 0, 0, 0) - u_2(0, -1, 0, 0) + u_3(0, 0, 1, 0)$
$\qquad + u_4(0, 0, 0, 1)$.

65–77. Proofs

Section 4.5   (page 193)

1. $\{(1, 0, 0, 0, 0, 0), (0, 1, 0, 0, 0, 0), (0, 0, 1, 0, 0, 0),$
$(0, 0, 0, 1, 0, 0), (0, 0, 0, 0, 1, 0), (0, 0, 0, 0, 0, 1)\}$

3. $\left\{ \begin{bmatrix} 1 & 0 & 0 \\ 0 & 0 & 0 \\ 0 & 0 & 0 \end{bmatrix}, \begin{bmatrix} 0 & 1 & 0 \\ 0 & 0 & 0 \\ 0 & 0 & 0 \end{bmatrix}, \begin{bmatrix} 0 & 0 & 1 \\ 0 & 0 & 0 \\ 0 & 0 & 0 \end{bmatrix}, \right.$

$\begin{bmatrix} 0 & 0 & 0 \\ 1 & 0 & 0 \\ 0 & 0 & 0 \end{bmatrix}, \begin{bmatrix} 0 & 0 & 0 \\ 0 & 1 & 0 \\ 0 & 0 & 0 \end{bmatrix}, \begin{bmatrix} 0 & 0 & 0 \\ 0 & 0 & 1 \\ 0 & 0 & 0 \end{bmatrix},$

$\left. \begin{bmatrix} 0 & 0 & 0 \\ 0 & 0 & 0 \\ 1 & 0 & 0 \end{bmatrix}, \begin{bmatrix} 0 & 0 & 0 \\ 0 & 0 & 0 \\ 0 & 1 & 0 \end{bmatrix}, \begin{bmatrix} 0 & 0 & 0 \\ 0 & 0 & 0 \\ 0 & 0 & 1 \end{bmatrix} \right\}$

5. $\{1, x, x^2, x^3, x^4\}$

7. S is linearly dependent and does not span R^2.
9. S does not span R^2.
11. S is linearly dependent.
13. S is linearly dependent and does not span R^2.
15. S is linearly dependent and does not span R^3.
17. S does not span R^3.
19. S is linearly dependent and does not span R^3.
21. S is linearly dependent.
23. S is linearly dependent.
25. S does not span P_2.
27. S does not span P_2.
29. S is linearly dependent and does not span P_2.
31. S does not span $M_{2,2}$.
33. S is linearly dependent and does not span $M_{2,2}$.

35. Basis   **37.** Not a basis   **39.** Basis
41. Basis   **43.** Not a basis   **45.** Basis
47. Basis   **49.** Not a basis   **51.** Basis
53. Basis; $(8, 3, 8) = 2(4, 3, 2) - (0, 3, 2) + 3(0, 0, 2)$
55. Not a basis   **57.** 6   **59.** 8
61. 6   **63.** $3m$
65. $\begin{bmatrix} 1 & 0 & 0 \\ 0 & 0 & 0 \\ 0 & 0 & 0 \end{bmatrix}, \begin{bmatrix} 0 & 0 & 0 \\ 0 & 1 & 0 \\ 0 & 0 & 0 \end{bmatrix}, \begin{bmatrix} 0 & 0 & 0 \\ 0 & 0 & 0 \\ 0 & 0 & 1 \end{bmatrix}$; 3
67. $\{(1, 0), (0, 1)\}, \{(1, 0), (1, 1)\}, \{(0, 1), (1, 1)\}$
69. $\{(2, 2,), (1, 0)\}$
71. (a) Line   (b) $\{(2, 1)\}$ (c) 1
73. (a) Line   (b) $\{(2, 1, -1)\}$ (c) 1
75. (a) $\{(2, 1, 0, 1), (-1, 0, 1, 0)\}$ (b) 2
77. (a) $\{(0, 6, 1, -1)\}$ (b) 1
79. (a) False. If the dimension of V is n, then every spanning set of V must have at least n vectors.
　　 (b) True. Find a set of n basis vectors in V that will span V and add any other vector.
81–85. Proofs

Section 4.6 *(page 205)*

1. (a) $(0, -2), (1, -3)$ (b) $\begin{bmatrix} 0 \\ 1 \end{bmatrix}, \begin{bmatrix} -2 \\ -3 \end{bmatrix}$
3. (a) $(4, 3, 1), (1, -4, 0)$ (b) $\begin{bmatrix} 4 \\ 1 \end{bmatrix}, \begin{bmatrix} 3 \\ -4 \end{bmatrix}, \begin{bmatrix} 1 \\ 0 \end{bmatrix}$
5. (a) $\{(1, 0), (0, 1)\}$ (b) 2
7. (a) $\{(1, 0, \frac{1}{2}), (0, 1, -\frac{1}{2})\}$ (b) 2
9. (a) $\{(1, 0, 0), (0, 1, 0), (0, 0, 1)\}$ (b) 3
11. (a) $\{(1, 2, -2, 0), (0, 0, 0, 1)\}$ (b) 2
13. $\{(1, 0, 0), (0, 1, 0), (0, 0, 1)\}$ **15.** $\{(1, 1, 0), (0, 0, 1)\}$
17. $\{(1, 0, -1, 0), (0, 1, 0, 0), (0, 0, 0, 1)\}$
19. $\{(1, 0, 0, 0), (0, 1, 0, 0), (0, 0, 1, 0), (0, 0, 0, 1)\}$
21. (a) $\left\{\begin{bmatrix} 1 \\ 0 \end{bmatrix}, \begin{bmatrix} 0 \\ 1 \end{bmatrix}\right\}$ (b) 2   **23.** (a) $\left\{\begin{bmatrix} 1 \\ 0 \end{bmatrix}, \begin{bmatrix} 0 \\ 1 \end{bmatrix}\right\}$ (b) 2
25. (a) $\left\{\begin{bmatrix} 1 \\ 0 \\ \frac{5}{9} \\ \frac{2}{9} \end{bmatrix}, \begin{bmatrix} 0 \\ 1 \\ -\frac{4}{9} \\ \frac{2}{9} \end{bmatrix}\right\}$ (b) 2   **27.** $\left\{t\begin{bmatrix} 1 \\ 2 \end{bmatrix}\right\}$
29. $\left\{t\begin{bmatrix} -2 \\ 1 \\ 0 \end{bmatrix} + s\begin{bmatrix} -3 \\ 0 \\ 1 \end{bmatrix}\right\}$ **31.** $\left\{t\begin{bmatrix} -3 \\ 0 \\ 1 \end{bmatrix}\right\}$
33. $\left\{t\begin{bmatrix} -1 \\ 2 \\ 1 \end{bmatrix}\right\}$ **35.** $\left\{\begin{bmatrix} 0 \\ 0 \end{bmatrix}\right\}$
37. $\left\{t\begin{bmatrix} 2 \\ -2 \\ 0 \\ 1 \end{bmatrix} + s\begin{bmatrix} -1 \\ 1 \\ 1 \\ 0 \end{bmatrix}\right\}$ **39.** $\left\{\begin{bmatrix} 0 \\ 0 \\ 0 \\ 0 \end{bmatrix}\right\}$

41. (a) $\text{rank}(A) = 3$
　　 $\text{nullity}(A) = 2$
　　 (b) $\left\{\begin{bmatrix} -3 \\ 1 \\ 1 \\ 0 \\ 0 \end{bmatrix}, \begin{bmatrix} 4 \\ -2 \\ 0 \\ 2 \\ 1 \end{bmatrix}\right\}$
　　 (c) $\{(1, 0, 3, 0, -4), (0, 1, -1, 0, 2), (0, 0, 0, 1, -2)\}$
　　 (d) $\left\{\begin{bmatrix} 1 \\ 2 \\ 3 \\ 4 \end{bmatrix}, \begin{bmatrix} 2 \\ 5 \\ 7 \\ 9 \end{bmatrix}, \begin{bmatrix} 0 \\ 1 \\ 2 \\ -1 \end{bmatrix}\right\}$
　　 (e) Linearly dependent   (f) (i) Yes (ii) No (iii) Yes
43. (a) $\{(-1, -3, 2)\}$ (b) 1
45. (a) $\left\{(-4, -1, 1, 0), \left(-3, -\frac{2}{3}, 0, 1\right)\right\}$ (b) 2
47. (a) $\{(8, -9, -6, 6)\}$ (b) 1
49. Consistent; $\begin{bmatrix} 17 \\ 0 \end{bmatrix} + t\begin{bmatrix} 4 \\ 1 \end{bmatrix}$
51. Consistent; $\begin{bmatrix} 3 \\ 5 \\ 0 \end{bmatrix} + t\begin{bmatrix} 2 \\ -4 \\ 1 \end{bmatrix}$ **53.** Inconsistent
55. Consistent; $\begin{bmatrix} 1 \\ 0 \\ 2 \\ -3 \\ 0 \end{bmatrix} + t\begin{bmatrix} 5 \\ 0 \\ -6 \\ -4 \\ 1 \end{bmatrix} + s\begin{bmatrix} -2 \\ 1 \\ 0 \\ 0 \\ 0 \end{bmatrix}$
57. $\begin{bmatrix} -1 \\ 4 \end{bmatrix} + 2\begin{bmatrix} 2 \\ 0 \end{bmatrix} = \begin{bmatrix} 3 \\ 4 \end{bmatrix}$
59. Not in the column space
61. $3\begin{bmatrix} -1 \\ 0 \\ -2 \end{bmatrix} + 3\begin{bmatrix} 1 \\ 1 \\ 1 \end{bmatrix} = \begin{bmatrix} 0 \\ 3 \\ -3 \end{bmatrix}$ **63.** Proof
65. (a) $\begin{bmatrix} 1 & 0 \\ 0 & 1 \end{bmatrix}, \begin{bmatrix} 0 & 1 \\ 1 & 0 \end{bmatrix}$ (b) $\begin{bmatrix} 1 & 0 \\ 0 & 0 \end{bmatrix}, \begin{bmatrix} 0 & 1 \\ 0 & 0 \end{bmatrix}$
　　 (c) $\begin{bmatrix} 1 & 0 \\ 0 & 0 \end{bmatrix}, \begin{bmatrix} 0 & 0 \\ 0 & 1 \end{bmatrix}$
67. (a) m (b) r (c) r (d) R^n (e) R^m
69. Answers will vary.
71. (a) Proof   (b) Proof   (c) Proof
73. (a) True. The nullspace of A is the solution space of the homogeneous system $A\mathbf{x} = \mathbf{0}$.
　　 (b) True. See Theorem 4.16, page 200.
75. (a) False. See "Remark," page 196.
　　 (b) False. See Theorem 4.19, page 204.
　　 (c) True. The columns of A become the rows of A^T, so the columns of A span the same space as the rows of A^T.
77. (a) $0, n$ (b) Proof
79. (a) Proof   (b) Proof
81. The rank of the matrix is at most 3, so the four row vectors form a linearly dependent set.

Section 4.7 *(page 216)*

1. $\begin{bmatrix} 5 \\ -2 \end{bmatrix}$ **3.** $\begin{bmatrix} 7 \\ -4 \\ -1 \\ 2 \end{bmatrix}$ **5.** $\begin{bmatrix} 8 \\ -3 \end{bmatrix}$ **7.** $\begin{bmatrix} 5 \\ 4 \\ 3 \end{bmatrix}$ **9.** $\begin{bmatrix} -1 \\ 2 \\ 0 \\ 1 \end{bmatrix}$

11. $\begin{bmatrix} 3 \\ 2 \end{bmatrix}$ **13.** $\begin{bmatrix} 1 \\ -1 \\ 2 \end{bmatrix}$ **15.** $\begin{bmatrix} 0 \\ -1 \\ 2 \end{bmatrix}$ **17.** $\begin{bmatrix} \frac{3}{2} & -\frac{1}{2} \\ -2 & 1 \end{bmatrix}$

19. $\begin{bmatrix} 2 & -1 \\ 4 & 3 \end{bmatrix}$ **21.** $\begin{bmatrix} \frac{1}{5} & \frac{1}{20} & -\frac{1}{2} \\ 0 & \frac{1}{4} & 0 \\ -\frac{1}{5} & -\frac{1}{20} & 0 \end{bmatrix}$ **23.** $\begin{bmatrix} 3 & -2 & 1 \\ 4 & -1 & 0 \\ 0 & 1 & -3 \end{bmatrix}$

25. $\begin{bmatrix} \frac{9}{5} & \frac{4}{5} \\ \frac{8}{5} & \frac{3}{5} \end{bmatrix}$ **27.** $\begin{bmatrix} -\frac{2}{41} & \frac{11}{82} \\ -\frac{19}{41} & \frac{43}{82} \end{bmatrix}$ **29.** $\begin{bmatrix} 1 & 1 & -1 \\ -3 & 2 & -1 \\ 3 & -3 & 2 \end{bmatrix}$

31. $\begin{bmatrix} -7 & 3 & 10 \\ 5 & -1 & -6 \\ 11 & -3 & -10 \end{bmatrix}$ **33.** $\begin{bmatrix} -24 & 7 & 1 & -2 \\ -10 & 3 & 0 & -1 \\ -29 & 7 & 3 & -2 \\ 12 & -3 & -1 & 1 \end{bmatrix}$

35. $\begin{bmatrix} 1 & -\frac{3}{11} & \frac{5}{11} & 0 & -\frac{7}{11} \\ 0 & -\frac{2}{11} & \frac{3}{22} & 0 & -\frac{1}{11} \\ -\frac{5}{4} & \frac{9}{22} & -\frac{19}{44} & -\frac{1}{4} & \frac{21}{22} \\ -\frac{3}{4} & \frac{1}{2} & -\frac{1}{4} & \frac{1}{4} & \frac{1}{2} \\ 0 & -\frac{1}{11} & -\frac{2}{11} & 0 & \frac{5}{11} \end{bmatrix}$

37. (a) $\begin{bmatrix} -\frac{1}{3} & \frac{1}{3} \\ \frac{3}{4} & -\frac{1}{2} \end{bmatrix}$ (b) $\begin{bmatrix} 6 & 4 \\ 9 & 4 \end{bmatrix}$ (c) Verify. (d) $\begin{bmatrix} 6 \\ 3 \end{bmatrix}$

39. (a) $\begin{bmatrix} 4 & 5 & 1 \\ -7 & -10 & -1 \\ -2 & -2 & 0 \end{bmatrix}$ (b) $\begin{bmatrix} \frac{1}{2} & \frac{1}{2} & -\frac{5}{4} \\ -\frac{1}{2} & -\frac{1}{2} & \frac{3}{4} \\ \frac{3}{2} & \frac{1}{2} & \frac{5}{4} \end{bmatrix}$

(c) Verify. (d) $\begin{bmatrix} \frac{11}{4} \\ -\frac{9}{4} \\ \frac{5}{4} \end{bmatrix}$

41. (a) $\begin{bmatrix} -\frac{48}{5} & -24 & \frac{4}{5} \\ 4 & 10 & \frac{1}{2} \\ -\frac{6}{5} & -5 & -\frac{2}{5} \end{bmatrix}$ (b) $\begin{bmatrix} \frac{3}{32} & \frac{17}{20} & \frac{5}{4} \\ -\frac{1}{16} & -\frac{3}{10} & -\frac{1}{2} \\ \frac{1}{2} & \frac{6}{5} & 0 \end{bmatrix}$

(c) Verify. (d) $\begin{bmatrix} \frac{279}{160} \\ -\frac{61}{80} \\ -\frac{7}{10} \end{bmatrix}$

43. (a) $\begin{bmatrix} \frac{19}{39} & -\frac{9}{13} & \frac{44}{39} \\ -\frac{3}{13} & -\frac{6}{13} & -\frac{9}{13} \\ -\frac{23}{39} & \frac{2}{13} & -\frac{4}{39} \end{bmatrix}$ (b) $\begin{bmatrix} -\frac{2}{7} & -\frac{4}{21} & -\frac{13}{7} \\ -\frac{5}{7} & -\frac{8}{7} & -\frac{1}{7} \\ \frac{4}{7} & -\frac{13}{21} & \frac{5}{7} \end{bmatrix}$

(c) Verify. (d) $\begin{bmatrix} \frac{22}{7} \\ \frac{6}{7} \\ \frac{19}{7} \end{bmatrix}$

45. $\begin{bmatrix} 1 \\ 5 \\ -2 \\ 1 \end{bmatrix}$ **47.** $\begin{bmatrix} 13 \\ 114 \\ 3 \\ 0 \end{bmatrix}$ **49.** $\begin{bmatrix} 0 \\ 3 \\ 2 \end{bmatrix}$ **51.** $\begin{bmatrix} 1 \\ 2 \\ -1 \end{bmatrix}$

53. Yes; When $B = B'$, $P^{-1} = I_n$.
55. (a) False. See Theorem 4.20, page 210.
  (b) True. See the discussion before Example 5, page 214.
  (c) True. See paragraph before Example 1, page 208.
57. QP

Section 4.8 *(page 225)*

1. b, c, d **3.** c **5.** a, b, d **7.** b **9.** c
11. b, c **13.** $-x\cos x + \sin x$ **15.** -2 **17.** $-x$
19. 0 **21.** $2e^{3x}$ **23.** 12 **25.** $e^{-x}(\cos x - \sin x)$
27. $W = (b-a)e^{(a+b)x} \neq 0$ **29.** $W = a \neq 0$
31. (a) Verify. (b) Linearly independent
  (c) $y = C_1 \sin 4x + C_2 \cos 4x$
33. (a) Verify. (b) Linearly dependent (c) Not applicable
35. (a) Verify. (b) Linearly independent
  (c) $y = C_1 + C_2 \sin 2x + C_3 \cos 2x$
37. (a) Verify. (b) Linearly dependent
  (c) Not applicable
39. (a) Verify.

  (b) $\theta(t) = C_1 \sin\sqrt{\dfrac{g}{L}}t + C_2 \cos\sqrt{\dfrac{g}{L}}t$; proof

41. No. For instance, consider $y'' = 1$. Two solutions are
  $y = \dfrac{x^2}{2}$ and $y = \dfrac{x^2}{2} + 1$. Their sum is not a solution.

43. Parabola

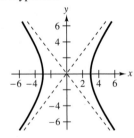

45. Ellipse

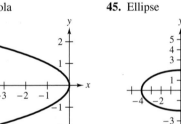

47. Hyperbola

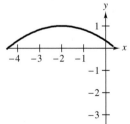

49. Parabola

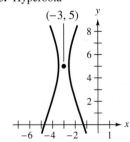

51. Point

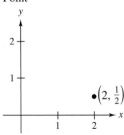

53. Hyperbola

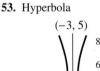

55. Ellipse

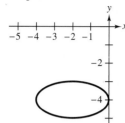

57. Hyperbola

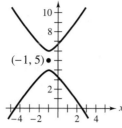

59. c   **60.** b   **61.** a   **62.** d

63. $\dfrac{(y')^2}{2} - \dfrac{(x')^2}{2} = 1$

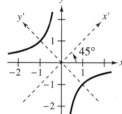

65. $\dfrac{(x')^2}{3} + \dfrac{(y')^2}{5} = 1$

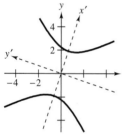

67. $\dfrac{(x')^2}{4} - \dfrac{(y')^2}{4} = 1$

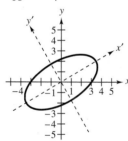

69. $(x' - 1)^2 = 6\left(y' + \frac{1}{6}\right)$

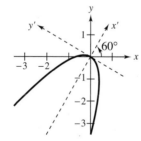

71. $\dfrac{(x')^2}{16} + \dfrac{(y')^2}{4} = 1$

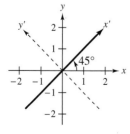

73. $x' = -(y')^2$

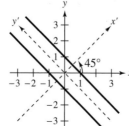

75. $y' = 0$

77. $x' = \pm\dfrac{\sqrt{2}}{2}$

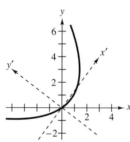

79. Proof   **81.** (a) Proof   (b) Proof
83. Answers will vary.

Review Exercises *(page 227)*

1. (a) $(4, -1, 3)$ **3.** (a) $(3, 1, 4, 4)$
  (b) $(6, 2, 0)$ (b) $(0, 4, 4, 2)$
  (c) $(-2, -3, -3)$ (c) $(3, -3, 0, 2)$
  (d) $(-3, -8, -9)$ (d) $(9, -7, 2, 7)$

5. $\left(\frac{1}{2}, -4, -4\right)$ **7.** $\left(\frac{5}{2}, -6, 0\right)$

9. $\mathbf{v} = 2\mathbf{u}_1 - \mathbf{u}_2 + 3\mathbf{u}_3$ **11.** $\mathbf{v} = \frac{9}{8}\mathbf{u}_1 + \frac{1}{8}\mathbf{u}_2 + 0\mathbf{u}_3$

13. $O_{4,2} = \begin{bmatrix} 0 & 0 \\ 0 & 0 \\ 0 & 0 \\ 0 & 0 \end{bmatrix}, \quad -A = \begin{bmatrix} -a_{11} & -a_{12} \\ -a_{21} & -a_{22} \\ -a_{31} & -a_{32} \\ -a_{41} & -a_{42} \end{bmatrix}$

15. $\mathbf{0} = (0, 0, 0, 0, 0)$
  $-\mathbf{v} = (-v_1, -v_2, -v_3, -v_4, -v_5)$

17. Subspace   **19.** Not a subspace
21. Subspace   **23.** Not a subspace
25. (a) Subspace   (b) Not a subspace
27. (a) Yes   (b) Yes   (c) Yes
29. (a) No   (b) No   (c) No
31. (a) Yes   (b) No   (c) No
33. Basis   **35.** Not a basis

37. (a) $\left\{ t\begin{bmatrix} 3 \\ 4 \end{bmatrix} \right\}$ (b) 1   (c) 1

39. (a) $\left\{ t\begin{bmatrix} 3 \\ 0 \\ 1 \\ 0 \end{bmatrix} + s\begin{bmatrix} -1 \\ -2 \\ 0 \\ 1 \end{bmatrix} \right\}$ (b) 2   (c) 2

41. (a) $\left\{ t\begin{bmatrix} 4 \\ -2 \\ 1 \end{bmatrix} \right\}$ (b) 1   (c) 2

43. (a) $\{(1, 0), (0, 1)\}$ (b) 2
45. (a) $\{(1, 0, 0), (0, 1, 0), (0, 0, 1)\}$ (b) 3

47. (a) $\left\{ \begin{bmatrix} -3 \\ 0 \\ 4 \\ 1 \end{bmatrix}, \begin{bmatrix} -2 \\ 1 \\ 0 \\ 0 \end{bmatrix} \right\}$ (b) 2

49. (a) $\left\{ \begin{bmatrix} 2 \\ 3 \\ 7 \\ 0 \end{bmatrix}, \begin{bmatrix} -1 \\ 0 \\ 0 \\ 1 \end{bmatrix} \right\}$ (b) 2

51. $\begin{bmatrix} -2 \\ 8 \end{bmatrix}$ **53.** $\begin{bmatrix} \frac{3}{4} \\ \frac{1}{4} \end{bmatrix}$ **55.** $\begin{bmatrix} 2 \\ -1 \\ -1 \end{bmatrix}$ **57.** $\begin{bmatrix} \frac{2}{5} \\ -\frac{1}{4} \end{bmatrix}$

59. $\begin{bmatrix} -1 \\ 4 \\ \frac{3}{2} \end{bmatrix}$ **61.** $\begin{bmatrix} 3 \\ 1 \\ 0 \\ 1 \end{bmatrix}$ **63.** $\begin{bmatrix} 1 & 3 \\ -1 & 1 \end{bmatrix}$

65. $\begin{bmatrix} 0 & 0 & 1 \\ 0 & 1 & 0 \\ 1 & 0 & 0 \end{bmatrix}$ **67.** $\begin{bmatrix} 10 & 23 & 21 \\ 11 & 26 & 24 \\ 12 & 27 & 24 \end{bmatrix}$

69. (a) $\begin{bmatrix} \frac{3}{2} & -1 \\ -2 & 1 \end{bmatrix}$ (b) $\begin{bmatrix} -2 & -2 \\ -4 & -3 \end{bmatrix}$

  (c) Verify.   (d) $\begin{bmatrix} 15 \\ -18 \end{bmatrix}$

71. (a) $\begin{bmatrix} 0 & -1 & 0 \\ -1 & 0 & 0 \\ 1 & 1 & 1 \end{bmatrix}$ (b) $\begin{bmatrix} 0 & -1 & 0 \\ -1 & 0 & 0 \\ 1 & 1 & 1 \end{bmatrix}$

(c) Verify. (d) $\begin{bmatrix} -2 \\ 1 \\ -2 \end{bmatrix}$

73. Basis for W: $\{x, x^2, x^3\}$
Basis for U: $\{(x-1), x(x-1), x^2(x-1)\}$
Basis for $W \cap U$: $\{x(x-1), x^2(x-1)\}$
75. No. For example, the set
$\{x^2 + x, x^2 - x, 1\}$
is a basis for P_2.
77. Yes; proof **79.** Proof **81.** Answers will vary.
83. (a) True. See discussion above "Definitions of Vector
Addition and Scalar Multiplication in R^n," page 155.
(b) False. See Theorem 4.3, part 2, page 157.
(c) True. See "Definition of a Vector Space" and the
discussion following, page 161.
85. (a) True. See discussion under "Vectors in R^n," page 155.
(b) False. See "Definition of a Vector Space," part 4, page 161.
(c) True. See discussion following "Summary of Important
Vector Spaces," page 163.
87. a, d **89.** a **91.** e^x **93.** -8
95. (a) Verify. (b) Linearly independent
(c) $y(t) = C_1 e^{-3x} + C_2 x e^{-3x}$
97. (a) Verify. (b) Linearly dependent (c) Not applicable
99. Circle **101.** Hyperbola

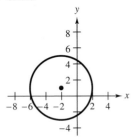 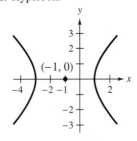

103. Parabola **105.** Ellipse

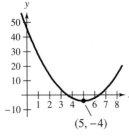 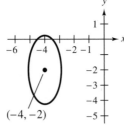

107. $\dfrac{(x')^2}{6} - \dfrac{(y')^2}{6} = 1$ **109.** $(x')^2 = 4(y'-1)$

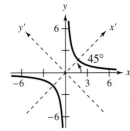 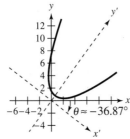 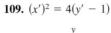

Chapter 5
Section 5.1 *(page 241)*

1. 5 **3.** $5\sqrt{2}$ **5.** (a) $\dfrac{\sqrt{17}}{4}$ (b) $\dfrac{5\sqrt{41}}{8}$ (c) $\dfrac{\sqrt{577}}{8}$
7. (a) $\sqrt{19}$ (b) $\sqrt{2}$ (c) 5
9. (a) $\left(-\frac{5}{13}, \frac{12}{13}\right)$ (b) $\left(\frac{5}{13}, -\frac{12}{13}\right)$
11. (a) $\left(\dfrac{3}{\sqrt{38}}, \dfrac{2}{\sqrt{38}}, -\dfrac{5}{\sqrt{38}}\right)$ (b) $\left(-\dfrac{3}{\sqrt{38}}, -\dfrac{2}{\sqrt{38}}, \dfrac{5}{\sqrt{38}}\right)$
13. $(2\sqrt{2}, 2\sqrt{2})$ **15.** $\left(\dfrac{5}{\sqrt{6}}, \dfrac{5\sqrt{5}}{\sqrt{6}}, 0\right)$
17. (a) $\left(-\frac{1}{2}, \frac{3}{2}, 0, 2\right)$ (b) $(2, -6, 0, -8)$ **19.** $2\sqrt{2}$
21. 3
23. (a) -6 (b) 13 (c) 25 (d) $(-12, 18)$ (e) -30
25. (a) 0 (b) 41 (c) 9 (d) **0** (e) 0 **27.** -7
29. (a) $\|\mathbf{u}\| \approx 1.0843, \|\mathbf{v}\| \approx 0.3202$ (b) $(0, 0.7809, 0.6247)$
(c) $(-0.9223, -0.1153, -0.3689)$ (d) 0.1113
(e) 1.1756 (f) 0.1025
31. (a) $\|\mathbf{u}\| \approx 1.7321, \|\mathbf{v}\| = 2$ (b) $(-0.5, 0.7071, -0.5)$
(c) $(0, -0.5774, -0.8165)$ (d) 0 (e) 3 (f) 4
33. (a) $\|\mathbf{u}\| \approx 3.4641, \|\mathbf{v}\| \approx 3.3166$
(b) $(-0.6030, 0.4264, -0.5222, -0.4264)$
(c) $(-0.5774, -0.5, -0.4082, -0.5)$
(d) -6.4495 (e) 12 (f) 11
35. $|(6, 8) \cdot (3, -2)| \le \|(6, 8)\| \, \|(3, -2)\|$
$2 \le 10\sqrt{13}$
37. $|(1, 1, -2) \cdot (1, -3, -2)| \le \|(1, 1, -2)\| \, \|(1, -3, -2)\|$
$2 \le 2\sqrt{21}$
39. 1.713 rad (98.13°) **41.** $\dfrac{7\pi}{12}$ rad (105°)
43. 1.080 rad (61.87°) **45.** $\dfrac{\pi}{4}$ rad (45°) **47.** Orthogonal
49. Parallel **51.** Neither **53.** Neither **55.** $\mathbf{v} = (t, 0)$
57. $\mathbf{v} = (t, s, -2t + s)$
59. $\|(5, 1)\| \le \|(4, 0)\| + \|(1, 1)\|$
$\sqrt{26} \le 4 + \sqrt{2}$
61. $\|(1, 2, -1)\| \le \|(1, 1, 1)\| + \|(0, 1, -2)\|$
$\sqrt{6} \le \sqrt{3} + \sqrt{5}$
63. $\|(2, 0)\|^2 = \|(1, -1)\|^2 + \|(1, 1)\|^2$
$4 = \left(\sqrt{2}\right)^2 + \left(\sqrt{2}\right)^2$
65. $\|(7, 1, -2)\|^2 = \|(3, 4, -2)\|^2 + \|(4, -3, 0)\|^2$
$54 = \left(\sqrt{29}\right)^2 + 5^2$
67. (a) -6 (b) 13 (c) 25 (d) $\begin{bmatrix} -12 \\ 18 \end{bmatrix}$ (e) -30
69. (a) 0 (b) 14 (c) 6 (d) $\begin{bmatrix} 0 \\ 0 \\ 0 \end{bmatrix}$ (e) 0
71. Orthogonal; $\mathbf{u} \cdot \mathbf{v} = 0$
73. (a) False. See "Definition of Length of a Vector in R^n,"
page 232.
(b) False. See "Definition of Dot Product in R^n," page 235.
75. (a) $(\mathbf{u} \cdot \mathbf{v}) - \mathbf{v}$ is meaningless because $\mathbf{u} \cdot \mathbf{v}$ is a scalar and $\mathbf{v}$
is a vector.
(b) $\mathbf{u} + (\mathbf{u} \cdot \mathbf{v})$ is meaningless because $\mathbf{u}$ is a vector and $\mathbf{u} \cdot \mathbf{v}$
is a scalar.

77. $\left(-\frac{5}{13}, \frac{12}{13}\right), \left(\frac{5}{13}, -\frac{12}{13}\right)$

79. \$11,877.50

This value gives the total revenue earned from selling the hamburgers and hot dogs.

81. $54.7°$ **83–87.** Proofs

89. $A\mathbf{x} = \mathbf{0}$ means that the dot product of each row of A with the column vector $\mathbf{x}$ is zero. So, $\mathbf{x}$ is orthogonal to the row vectors of A.

Section 5.2   *(page 251)*

1–7. Proofs

9. Axiom 4 fails. $\langle (0, 1), (0, 1) \rangle = 0$, but $(0, 1) \neq \mathbf{0}$.

11. Axiom 4 fails. $\langle (1, 1), (1, 1) \rangle = 0$, but $(1, 1) \neq \mathbf{0}$.

13. Axiom 1 fails. If $\mathbf{u} = (1, 1, 1)$ and $\mathbf{v} = (1, 0, 0)$ $\langle \mathbf{u}, \mathbf{v} \rangle = 1$ and $\langle \mathbf{v}, \mathbf{u} \rangle = 0$.

15. Axiom 3 fails. If $\mathbf{u} = (1, 1, 1)$, $\mathbf{v} = (1, 0, 0)$, and $c = 2$, $c\langle \mathbf{u}, \mathbf{v} \rangle = 2$ and $\langle c\mathbf{u}, \mathbf{v} \rangle = 4$.

17. (a) -33 (b) 5 (c) 13 (d) $2\sqrt{65}$

19. (a) 15 (b) $\sqrt{57}$ (c) 5 (d) $2\sqrt{13}$

21. (a) -25 (b) $\sqrt{53}$ (c) $\sqrt{94}$ (d) $\sqrt{197}$

23. (a) 0 (b) $8\sqrt{3}$ (c) $\sqrt{411}$ (d) $3\sqrt{67}$

25. (a) 4 (b) $\sqrt{6}$ (c) 3 (d) $\sqrt{7}$ **27.** Proof

29. (a) -15 (b) $\sqrt{35}$ (c) $\sqrt{10}$ (d) $5\sqrt{3}$

31. (a) -5 (b) $\sqrt{39}$ (c) $\sqrt{5}$ (d) $3\sqrt{6}$ **33.** Proof

35. (a) -4 (b) $\sqrt{11}$ (c) $\sqrt{2}$ (d) $\sqrt{21}$

37. (a) 0 (b) $\sqrt{2}$ (c) $\sqrt{2}$ (d) 2

39. (a) $\frac{2}{3}$ (b) $\sqrt{2}$ (c) $\frac{\sqrt{46}}{\sqrt{15}}$ (d) $\frac{2\sqrt{14}}{\sqrt{15}}$

41. (a) $\frac{2}{e} \approx 0.736$ (b) $\frac{\sqrt{6}}{3} \approx 0.816$

(c) $\sqrt{\dfrac{e^2}{2} - \dfrac{1}{2e^2}} \approx 1.904$

(d) $\sqrt{\dfrac{e^2}{2} + \dfrac{2}{3} - \dfrac{1}{2e^2} - \dfrac{4}{e}} \approx 1.680$

43. 2.103 rad $(120.5°)$ **45.** 1.16 rad $(66.59°)$

47. $\dfrac{\pi}{2}$ rad $(90°)$ **49.** 1.23 rad $(70.53°)$ **51.** $\dfrac{\pi}{2}$ rad $(90°)$

53. (a) $|\langle (5, 12), (3, 4) \rangle| \leq \|(5, 12)\| \, \|(3, 4)\|$
$$63 \leq (13)(5)$$

(b) $\|(5, 12) + (3, 4)\| \leq \|(5, 12)\| + \|(3, 4)\|$
$$8\sqrt{5} \leq 13 + 5$$

55. (a) $|(0, 1, 5) \cdot (-4, 3, 3)| \leq \|(0, 1, 5)\| \, \|(-4, 3, 3)\|$
$$18 \leq 2\sqrt{221}$$

(b) $\|(0, 1, 5) + (-4, 3, 3)\| \leq \|(0, 1, 5)\| + \|(-4, 3, 3)\|$
$$4\sqrt{6} \leq \sqrt{26} + \sqrt{34}$$

57. (a) $|\langle 2x, 1 + 3x^2 \rangle| \leq \|2x\| \, \|1 + 3x^2\|$
$$0 \leq (2)(\sqrt{10})$$

(b) $\|2x + 1 + 3x^2\| \leq \|2x\| + \|1 + 3x^2\|$
$$\sqrt{14} \leq 2 + \sqrt{10}$$

59. (a) $|0(-3) + 3(1) + 2(4) + 1(3)| \leq \sqrt{14}\sqrt{35}$
$$14 \leq \sqrt{14}\sqrt{35}$$

(b) $\left\| \begin{bmatrix} -3 & 4 \\ 6 & 4 \end{bmatrix} \right\| \leq \sqrt{14} + \sqrt{35}$
$$\sqrt{77} \leq \sqrt{14} + \sqrt{35}$$

61. (a) $|\langle \sin x, \cos x \rangle| \leq \|\sin x\| \, \|\cos x\|$
$$\frac{1}{4} \leq \left(\sqrt{\frac{\pi}{8} - \frac{1}{4}} \right) \left(\sqrt{\frac{\pi}{8} + \frac{1}{4}} \right)$$

(b) $\|\sin x + \cos x\| \leq \|\sin x\| + \|\cos x\|$
$$\sqrt{\frac{\pi}{4} + \frac{1}{2}} \leq \sqrt{\frac{\pi}{8} - \frac{1}{4}} + \sqrt{\frac{\pi}{8} + \frac{1}{4}}$$

63. (a) $|\langle x, e^x \rangle| \leq \|x\| \, \|e^x\|$
$$1 \leq \sqrt{\tfrac{1}{3}} \cdot \sqrt{\tfrac{1}{2}e^2 - \tfrac{1}{2}}$$

(b) $\|x + e^x\| \leq \|x\| + \|e^x\|$
$$\sqrt{\tfrac{11}{6} + \tfrac{1}{2}e^2} \leq \sqrt{\tfrac{1}{3}} + \sqrt{\tfrac{1}{2}e^2 - \tfrac{1}{2}}$$

65. Because
$$\langle f, g \rangle = \int_{-\pi/2}^{\pi/2} \cos x \sin x \, dx$$
$$= \frac{1}{2} \sin^2 x \, \Big]_{-\pi/2}^{\pi/2} = 0$$

f and g are orthogonal.

67. The functions $f(x) = x$ and $g(x) = \frac{1}{2}(5x^3 - 3x)$ are orthogonal because
$$\langle f, g \rangle = \int_{-1}^{1} x \frac{1}{2}(5x^3 - 3x) \, dx$$
$$= \frac{1}{2} \int_{-1}^{1} (5x^4 - 3x^2) \, dx = \frac{1}{2}(x^5 - x^3) \, \Big]_{-1}^{1} = 0.$$

69. (a) $\left(\frac{8}{5}, \frac{4}{5}\right)$ (b) $\left(\frac{4}{5}, \frac{8}{5}\right)$

(c)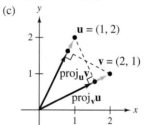

71. (a) $(1, 1)$ (b) $\left(-\frac{4}{5}, \frac{12}{5}\right)$

(c)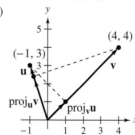

73. (a) $(4, -4, 0)$ (b) $\left(\frac{8}{7}, -\frac{24}{35}, \frac{8}{35}\right)$

75. (a) $\left(\frac{1}{2}, -\frac{1}{2}, -1, -1\right)$ (b) $\left(0, -\frac{5}{46}, -\frac{15}{46}, \frac{15}{23}\right)$

77. $\text{proj}_g f = 0$ **79.** $\text{proj}_g f = \dfrac{2e^x}{e^2 - 1}$

81. $\text{proj}_g f = 0$ **83.** $\text{proj}_g f = -\sin 2x$

85. (a) False. See the introduction to this section, page 243.

(b) False. $\|\mathbf{v}\| = 0$ if and only if $\mathbf{v} = \mathbf{0}$.

87. (a) $\langle \mathbf{u}, \mathbf{v} \rangle = 4(2) + 2(2)(-2) = 0 \Rightarrow \mathbf{u}$ and $\mathbf{v}$ are orthogonal.

(b)

Not orthogonal in the Euclidean sense

89–95. Proofs   **97.** $c_1 = \frac{1}{4},\ c_2 = 1$

99. $c_1 = \frac{1}{4},\ c_2 = \frac{1}{16}$ **101.** Proof

Section 5.3 *(page 263)*

1. (a) Yes (b) No (c) Yes
3. (a) Yes (b) Yes (c) Yes
5. (a) Yes (b) No (c) Yes
7. (a) No (b) No (c) Yes
9. (a) Yes (b) No (c) No
11. (a) Yes (b) Yes (c) No

13. (a) Proof (b) $\left(-\dfrac{1}{\sqrt{10}}, \dfrac{3}{\sqrt{10}}\right), \left(\dfrac{3}{\sqrt{10}}, \dfrac{1}{\sqrt{10}}\right)$

15. (a) Proof (b) $\left(\dfrac{\sqrt{3}}{3}, \dfrac{\sqrt{3}}{3}, \dfrac{\sqrt{3}}{3}\right), \left(-\dfrac{\sqrt{2}}{2}, 0, \dfrac{\sqrt{2}}{2}\right)$

17. The set $\{1, x, x^2, x^3\}$ is orthogonal because
$\langle 1, x \rangle = 0, \langle 1, x^2 \rangle = 0, \langle 1, x^3 \rangle = 0, \langle x, x^2 \rangle = 0,$
$\langle x, x^3 \rangle = 0, \langle x^2, x^3 \rangle = 0.$
Furthermore, the set is orthonormal because
$\|1\| = 1, \|x\| = 1, \|x^2\| = 1,$ and $\|x^3\| = 1.$
So, $\{1, x, x^2, x^3\}$ is an orthonormal basis for P_3.

19. $\begin{bmatrix} \dfrac{4\sqrt{13}}{13} \\ \dfrac{7\sqrt{13}}{13} \end{bmatrix}$ **21.** $\begin{bmatrix} \dfrac{\sqrt{10}}{2} \\ -2 \\ -\dfrac{\sqrt{10}}{2} \end{bmatrix}$ **23.** $\begin{bmatrix} 11 \\ 2 \\ 15 \end{bmatrix}$

25. $\left\{\left(\frac{3}{5}, \frac{4}{5}\right), \left(\frac{4}{5}, -\frac{3}{5}\right)\right\}$ **27.** $\{(0, 1), (1, 0)\}$

29. $\left\{\left(\frac{2}{3}, \frac{1}{3}, -\frac{2}{3}\right), \left(\frac{1}{3}, \frac{2}{3}, \frac{2}{3}\right), \left(\frac{2}{3}, -\frac{2}{3}, \frac{1}{3}\right)\right\}$

31. $\left\{\left(\frac{4}{5}, -\frac{3}{5}, 0\right), \left(\frac{3}{5}, \frac{4}{5}, 0\right), (0, 0, 1)\right\}$

33. $\left\{\left(0, \dfrac{\sqrt{2}}{2}, \dfrac{\sqrt{2}}{2}\right), \left(\dfrac{\sqrt{6}}{3}, \dfrac{\sqrt{6}}{6}, -\dfrac{\sqrt{6}}{6}\right), \left(\dfrac{\sqrt{3}}{3}, -\dfrac{\sqrt{3}}{3}, \dfrac{\sqrt{3}}{3}\right)\right\}$

35. $\left\{\left(-\dfrac{4\sqrt{2}}{7}, \dfrac{3\sqrt{2}}{14}, \dfrac{5\sqrt{2}}{14}\right)\right\}$

37. $\left\{\left(\frac{3}{5}, \frac{4}{5}, 0\right), \left(\frac{4}{5}, -\frac{3}{5}, 0\right)\right\}$

39. $\left\{\left(\dfrac{\sqrt{6}}{6}, \dfrac{\sqrt{6}}{3}, -\dfrac{\sqrt{6}}{6}, 0\right), \left(\dfrac{\sqrt{3}}{3}, 0, \dfrac{\sqrt{3}}{3}, \dfrac{\sqrt{3}}{3}\right),\right.$
$\left.\left(\dfrac{\sqrt{3}}{3}, -\dfrac{\sqrt{3}}{3}, -\dfrac{\sqrt{3}}{3}, 0\right)\right\}$

41. $\left\{\left(\frac{2}{3}, -\frac{1}{3}\right), \left(\dfrac{\sqrt{2}}{6}, \dfrac{2\sqrt{2}}{3}\right)\right\}$

43. $\langle x, 1 \rangle = \displaystyle\int_{-1}^{1} x\, dx = \left.\dfrac{x^2}{2}\right]_{-1}^{1} = 0$

45. $\langle x^2, 1 \rangle = \displaystyle\int_{-1}^{1} x^2\, dx = \left.\dfrac{x^3}{3}\right]_{-1}^{1} = \dfrac{2}{3}$

47. $\langle x, x \rangle = \displaystyle\int_{-1}^{1} x^2\, dx = \left.\dfrac{x^3}{3}\right]_{-1}^{1} = \dfrac{2}{3}$

49. $\left\{\left(\dfrac{2\sqrt{5}}{5}, \dfrac{\sqrt{5}}{5}, 0\right), \left(-\dfrac{\sqrt{30}}{30}, \dfrac{\sqrt{30}}{15}, \dfrac{\sqrt{30}}{6}\right)\right\}$

51. $\left\{\left(-\dfrac{\sqrt{2}}{2}, 0, \dfrac{\sqrt{2}}{2}, 0\right), \left(-\dfrac{\sqrt{6}}{6}, 0, \dfrac{\sqrt{6}}{6}, \dfrac{\sqrt{6}}{3}\right)\right\}$

53. $\left\{\left(\dfrac{3\sqrt{10}}{10}, 0, \dfrac{\sqrt{10}}{10}, 0\right), \left(0, -\dfrac{2\sqrt{5}}{5}, 0, \dfrac{\sqrt{5}}{5}\right)\right\}$

55. (a) True. See "Definitions of Orthogonal and Orthonormal Sets," page 254.
(b) False. See "Remark," page 260.

57. Orthonormal

59. $\left\{\dfrac{\sqrt{2}}{2}(-1 + x^2), -\dfrac{\sqrt{6}}{6}(1 - 2x + x^2)\right\}$

61. Orthonormal   **63.** Proof   **65.** Proof

67. $N(A)$ basis: $\{(3, -1, 2)\}$
$N(A^T)$ basis: $\{(-1, -1, 1)\}$
$R(A)$ basis: $\{(1, 0, 1), (1, 2, 3)\}$
$R(A^T)$ basis: $\{(1, 1, -1), (0, 2, 1)\}$

69. Proof

71. $\left\{\left(\dfrac{1}{\sqrt{2}}, 0, \dfrac{1}{\sqrt{2}}, 0\right), \left(0, -\dfrac{1}{\sqrt{2}}, 0, \dfrac{1}{\sqrt{2}}\right), \left(\dfrac{1}{\sqrt{2}}, 0, -\dfrac{1}{\sqrt{2}}, 0\right),\right.$
$\left.\left(0, \dfrac{1}{\sqrt{2}}, 0, \dfrac{1}{\sqrt{2}}\right)\right\}$

Section 5.4 *(page 275)*

1. $y = 1 + 2x$ **3.** Not collinear   **5.** Not orthogonal

7. Orthogonal   **9.** (a) span $\left\{\begin{bmatrix} 1 \\ 0 \\ -2 \end{bmatrix}\right\}$ (b) R^3

11. (a) span $\left\{\begin{bmatrix} 1 \\ 0 \\ 0 \\ 0 \end{bmatrix}, \begin{bmatrix} 0 \\ 1 \\ 1 \\ 0 \end{bmatrix}, \begin{bmatrix} 0 \\ 1 \\ 0 \\ -1 \end{bmatrix}\right\}$ (b) R^4

13. (a) span $\left\{\begin{bmatrix} 0 \\ 1 \\ 0 \end{bmatrix}\right\}$ (b) R^3

15. span $\left\{\begin{bmatrix} 0 \\ 1 \\ -1 \\ 1 \end{bmatrix}\right\}$ **17.** $\begin{bmatrix} 0 \\ \frac{2}{3} \\ \frac{2}{3} \\ \frac{2}{3} \end{bmatrix}$ **19.** $\begin{bmatrix} \frac{5}{3} \\ \frac{8}{3} \\ \frac{13}{3} \end{bmatrix}$

21. $N(A)$ basis: $\left\{\begin{bmatrix} -3 \\ 0 \\ 1 \end{bmatrix}\right\}$

$N(A^T) = \left\{\begin{bmatrix} 0 \\ 0 \end{bmatrix}\right\}$

$R(A)$ basis: $\left\{\begin{bmatrix} 1 \\ 0 \end{bmatrix}, \begin{bmatrix} 2 \\ 1 \end{bmatrix}\right\}$

$R(A^T)$ basis: $\left\{\begin{bmatrix} 1 \\ 2 \\ 3 \end{bmatrix}, \begin{bmatrix} 0 \\ 1 \\ 0 \end{bmatrix}\right\}$

23. $N(A)$ basis: $\left\{ \begin{bmatrix} 0 \\ -1 \\ 1 \end{bmatrix} \right\}$

$N(A^T)$ basis: $\left\{ \begin{bmatrix} -1 \\ -1 \\ 1 \\ 0 \end{bmatrix}, \begin{bmatrix} 0 \\ -1 \\ -1 \\ 1 \end{bmatrix} \right\}$

$R(A)$ basis: $\left\{ \begin{bmatrix} 1 \\ 0 \\ 1 \\ 1 \end{bmatrix}, \begin{bmatrix} 0 \\ 1 \\ 1 \\ 2 \end{bmatrix} \right\}$

$R(A^T)$ basis: $\left\{ \begin{bmatrix} 1 \\ 0 \\ 0 \end{bmatrix}, \begin{bmatrix} 0 \\ 1 \\ 1 \end{bmatrix} \right\}$

25. $\begin{bmatrix} 1 \\ -1 \end{bmatrix}$ **27.** $\begin{bmatrix} 2 \\ -2 \\ 1 \end{bmatrix}$ **29.** $\begin{bmatrix} 1 \\ -1 \\ 2 \end{bmatrix}$

31. $y = -x + \frac{1}{3}$

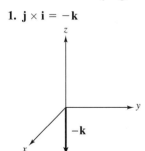

33.

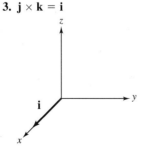

35. $y = x^2 - x$ **37.** $y = \frac{3}{7}x^2 + \frac{6}{5}x + \frac{26}{35}$

39. $y = 380.4 + 31.39t$; 976,800

41. $\ln y = 5.595 - 0.140 \ln x$ or $y = 269.1x^{-0.14}$

43. (a) False. The orthogonal complement of R^n is $\{\mathbf{0}\}$.
 (b) True. See "Definition of Direct Sum," page 267.

45. Proof   **47.** Proof

Section 5.5   *(page 288)*

1. $\mathbf{j} \times \mathbf{i} = -\mathbf{k}$

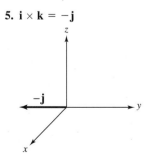

3. $\mathbf{j} \times \mathbf{k} = \mathbf{i}$

5. $\mathbf{i} \times \mathbf{k} = -\mathbf{j}$

7. (a) $-\mathbf{i} - \mathbf{j} + \mathbf{k}$
 (b) $\mathbf{i} + \mathbf{j} - \mathbf{k}$
 (c) $\mathbf{0}$

9. (a) $5\mathbf{i} - 3\mathbf{j} - \mathbf{k}$
 (b) $-5\mathbf{i} + 3\mathbf{j} + \mathbf{k}$
 (c) $\mathbf{0}$

11. (a) $(0, -2, -2)$
 (b) $(0, 2, 2)$
 (c) $(0, 0, 0)$

13. (a) $(-14, 13, 17)$
 (b) $(14, -13, -17)$
 (c) $(0, 0, 0)$

15. $(-2, -2, -1)$ **17.** $(-8, -14, 54)$ **19.** $(-1, -1, -1)$

21. $-\mathbf{i} + 12\mathbf{j} - 2\mathbf{k}$ **23.** $-2\mathbf{i} + 3\mathbf{j} - \mathbf{k}$

25. $-8\mathbf{i} - 2\mathbf{j} + 7\mathbf{k}$ **27.** $(5, -4, -3)$

29. $(2, -1, -1)$ **31.** $\mathbf{i} - \mathbf{j} - 3\mathbf{k}$ **33.** $\mathbf{i} - 5\mathbf{j} - 3\mathbf{k}$

35. $\left(\frac{2}{3}, \frac{2}{3}, -\frac{1}{3} \right)$ **37.** $\frac{1}{\sqrt{19}}\mathbf{i} - \frac{3}{\sqrt{19}}\mathbf{j} + \frac{3}{\sqrt{19}}\mathbf{k}$

39. $-\frac{71}{\sqrt{7602}}\mathbf{i} - \frac{44}{\sqrt{7602}}\mathbf{j} + \frac{25}{\sqrt{7602}}\mathbf{k}$ **41.** $\frac{1}{\sqrt{2}}\mathbf{i} + \frac{1}{\sqrt{2}}\mathbf{k}$

43. 1   **45.** $6\sqrt{5}$ **47.** $2\sqrt{83}$

49. $\frac{5\sqrt{174}}{2}$ **51.** 1   **53.** -3 **55–65.** Proofs

67. (a) $g(x) = -\frac{1}{6} + x$
 (b)

69. (a) $g(x) = \frac{1}{2}(e^2 - 7) + 6x$
 (b)

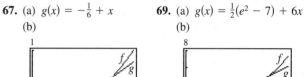

71. (a) $g(x) = \frac{12}{\pi^3}(\pi - 2x)$
 (b)

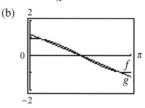

73. (a) $g(x) = 0.05 - 0.6x + 1.5x^2$
 (b)

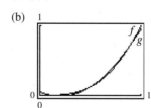

75. (a) $g(x) = \frac{24}{\pi^3}x$
 (b)

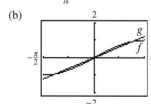

77. $g(x) = 2\sin x + \sin 2x + \frac{2}{3}\sin 3x$

79. $g(x) = \frac{\pi^2}{3} + 4\cos x + \cos 2x + \frac{4}{9}\cos 3x$

81. $g(x) = \dfrac{1}{2\pi}(1 - e^{-2\pi})(1 + \cos x + \sin x)$

83. $g(x) = \dfrac{1 - e^{-4\pi}}{20\pi}(5 + 8\cos x + 4\sin x)$

85. $g(x) = (1 + \pi) - 2\sin x - \sin 2x - \frac{2}{3}\sin 3x$

87. $g(x) = \sin 2x$

89. $g(x) = 2\left(\sin x + \dfrac{\sin 2x}{2} + \dfrac{\sin 3x}{3} + \cdots + \dfrac{\sin nx}{n}\right)$

91. $\dfrac{1 - e^{-2\pi}}{2\pi} + \dfrac{1 - e^{-2\pi}}{\pi}\displaystyle\sum_{j=1}^{n}\left(\dfrac{1}{j^2 + 1}\cos jx + \dfrac{j}{j^2 + 1}\sin jx\right)$

93. Answers will vary.

Review Exercises   *(page 290)*

1. (a) $\sqrt{17}$ (b) $\sqrt{5}$ (c) 6   (d) $\sqrt{10}$

3. (a) $\sqrt{6}$ (b) $\sqrt{14}$ (c) 7   (d) $\sqrt{6}$

5. (a) $\sqrt{6}$ (b) $\sqrt{3}$ (c) -1 (d) $\sqrt{11}$

7. (a) $\sqrt{7}$ (b) $\sqrt{7}$ (c) 6   (d) $\sqrt{2}$

9. $\|\mathbf{v}\| = \sqrt{38}; \ \mathbf{u} = \left(\dfrac{5}{\sqrt{38}}, \dfrac{3}{\sqrt{38}}, -\dfrac{2}{\sqrt{38}}\right)$

11. $\|\mathbf{v}\| = \sqrt{6}; \ \mathbf{u} = \left(-\dfrac{1}{\sqrt{6}}, \dfrac{1}{\sqrt{6}}, \dfrac{2}{\sqrt{6}}\right)$

13. (a) $(4, 4, 3)$ (b) $\left(-2, -2, -\frac{3}{2}\right)$ (c) $(-16, -16, -12)$

15. $\dfrac{\pi}{2}$ rad $(90°)$ **17.** $\dfrac{\pi}{12}$ rad $(15°)$ **19.** π rad $(180°)$

21. $(s, 3t, 4t)$ **23.** $\left(\frac{1}{2}r - \frac{1}{2}s - t, r, s, t\right)$

25. (a) -10 (b) $\dfrac{\sqrt{259}}{2}$

27. Triangle Inequality:
$$\left\|\left(4, -\tfrac{3}{2}, -1\right) + \left(\tfrac{1}{2}, 3, 1\right)\right\| \le \left\|\left(4, -\tfrac{3}{2}, -1\right)\right\| + \left\|\left(\tfrac{1}{2}, 3, 1\right)\right\|$$
$$\dfrac{3\sqrt{11}}{2} \le \dfrac{\sqrt{47}}{\sqrt{2}} + \dfrac{\sqrt{85}}{2}$$

Cauchy-Schwarz Inequality:
$$\left|\left\langle\left(4, -\tfrac{3}{2}, -1\right), \left(\tfrac{1}{2}, 3, 1\right)\right\rangle\right| \le \left\|\left(4, -\tfrac{3}{2}, -1\right)\right\| \left\|\left(\tfrac{1}{2}, 3, 1\right)\right\|$$
$$10 \le \dfrac{\sqrt{47}}{\sqrt{2}}\dfrac{\sqrt{85}}{2} \approx 22.347$$

29. (a) 0   (b) Orthogonal
(c) Because $\langle f, g \rangle = 0$, it follows that $|\langle f, g \rangle| \le \|f\|\|g\|$.

31. $\left(-\frac{9}{13}, \frac{45}{13}\right)$ **33.** $(0, 5)$ **35.** $\left(\frac{18}{29}, \frac{12}{29}, \frac{24}{29}\right)$

37. $\left\{\left(\dfrac{1}{\sqrt{2}}, \dfrac{1}{\sqrt{2}}\right), \left(-\dfrac{1}{\sqrt{2}}, \dfrac{1}{\sqrt{2}}\right)\right\}$

39. $\left\{\left(0, \frac{3}{5}, \frac{4}{5}\right), (1, 0, 0), \left(0, \frac{4}{5}, -\frac{3}{5}\right)\right\}$

41. (a) $(-1, 4, -2) = 2(0, 2, -2) - (1, 0, -2)$

(b) $\left\{\left(0, \dfrac{1}{\sqrt{2}}, -\dfrac{1}{\sqrt{2}}\right), \left(\dfrac{1}{\sqrt{3}}, -\dfrac{1}{\sqrt{3}}, -\dfrac{1}{\sqrt{3}}\right)\right\}$

(c) $(-1, 4, -2) = 3\sqrt{2}\left(0, \dfrac{1}{\sqrt{2}}, -\dfrac{1}{\sqrt{2}}\right)$
$$-\sqrt{3}\left(\dfrac{1}{\sqrt{3}}, -\dfrac{1}{\sqrt{3}}, -\dfrac{1}{\sqrt{3}}\right)$$

43. $\langle f, g \rangle = \displaystyle\int_0^{\pi} \sin x \cos x \, dx$
$$= \dfrac{1}{2}\sin^2 x \Big]_0^{\pi} = 0$$

45. (a) $\dfrac{1}{5}$ (b) $\dfrac{1}{\sqrt{7}}$ (c) $\dfrac{2\sqrt{2}}{\sqrt{105}}$ (d) $\left\{\sqrt{3}x, \dfrac{\sqrt{7}}{2}(-3x + 5x^3)\right\}$

47. $\left\{\left(-\dfrac{1}{\sqrt{2}}, 0, \dfrac{1}{\sqrt{2}}\right), \left(-\dfrac{1}{\sqrt{6}}, \dfrac{2}{\sqrt{6}}, -\dfrac{1}{\sqrt{6}}\right)\right\}$
(The answer is not unique.)

49–57. Proofs   **59.** span $\left\{\begin{bmatrix} 2 \\ -1 \\ 3 \end{bmatrix}\right\}$

61. $N(A)$ basis: $\left\{\begin{bmatrix} 1 \\ 0 \\ -1 \end{bmatrix}\right\}$

$N(A^T)$ basis: $\left\{\begin{bmatrix} 3 \\ 1 \\ 0 \end{bmatrix}\right\}$

$R(A)$ basis: $\left\{\begin{bmatrix} 0 \\ 0 \\ 1 \end{bmatrix}, \begin{bmatrix} 1 \\ -3 \\ 0 \end{bmatrix}\right\}$

$R(A^T)$ basis: $\left\{\begin{bmatrix} 0 \\ 1 \\ 0 \end{bmatrix}, \begin{bmatrix} 1 \\ 0 \\ 1 \end{bmatrix}\right\}$

63. $y = -65.5 + 24.65t - 2.688t^2 + 0.1184t^3$; \$197.8 billion

65. $(0, 0, 3)$ **67.** $13\mathbf{i} + 6\mathbf{j} - \mathbf{k}$ **69.** 1   **71.** 6   **73.** 7

75. (a) $g(x) = \frac{3}{5}x$
(b)

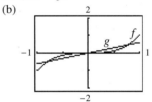

77. (a) $g(x) = \dfrac{2}{\pi}$
(b)

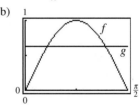

79. (a) $g(x) = \frac{2}{35}(3 + 24x - 10x^2)$
(b)

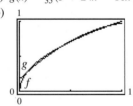

81. $g(x) = \dfrac{\pi^2}{3} - 4\cos x$

83. (a) True. See Theorem 5.18, page 279.
(b) False. See Theorem 5.17, page 278.
(c) True. See discussion before Theorem 5.19, page 283

Cumulative Test for Chapters 4 and 5   *(page 295)*

1. (a) $(3, -7)$ (b) $(3, -6)$

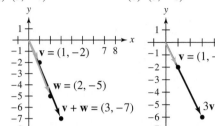

(c) $(-6, 16)$

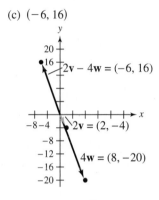

2. $\mathbf{w} = 3\mathbf{v}_1 + \mathbf{v}_2 + \frac{2}{3}\mathbf{v}_3$

3. $\begin{bmatrix} -2 \\ -2 \\ 1 \end{bmatrix} = -2\begin{bmatrix} 1 \\ 4 \\ 7 \end{bmatrix} + 3\begin{bmatrix} 0 \\ 2 \\ 5 \end{bmatrix}$

4. $\mathbf{v} = 5\mathbf{u}_1 - \mathbf{u}_2 + \mathbf{u}_3 + 2\mathbf{u}_4 - 5\mathbf{u}_5 + 3\mathbf{u}_6$ **5.** Proof

6. Yes   **7.** No   **8.** Yes

9. (a) A set of vectors $\{\mathbf{v}_1, \ldots, \mathbf{v}_n\}$ is linearly independent if the vector equation $c_1\mathbf{v}_1 + \cdots + c_n\mathbf{v}_n = \mathbf{0}$ has only the trivial solution.

(b) Linearly dependent

10. (a) A set of vectors $\{\mathbf{v}_1, \ldots, \mathbf{v}_n\}$ in a vector space V is a basis for V if the set is linearly independent and spans V.

(b) Yes   (c) Yes

11. $\left\{ \begin{bmatrix} 1 \\ -1 \\ 0 \\ 0 \end{bmatrix}, \begin{bmatrix} 0 \\ 0 \\ 1 \\ -1 \end{bmatrix} \right\}$ **12.** $\begin{bmatrix} -4 \\ 6 \\ -5 \end{bmatrix}$ **13.** $\begin{bmatrix} 0 & 1 & -1 \\ 2 & 0 & 1 \\ -1 & -1 & 1 \end{bmatrix}$

14. (a) $\sqrt{5}$ (b) $\sqrt{29}$ (c) -5 (d) 2.21 rad $(126.7°)$

15. $\dfrac{11}{12}$ **16.** $\left\{ (1, 0, 0), \left(0, \dfrac{\sqrt{2}}{2}, \dfrac{\sqrt{2}}{2}\right), \left(0, -\dfrac{\sqrt{2}}{2}, \dfrac{\sqrt{2}}{2}\right) \right\}$

17. $\frac{1}{13}(-3, 2)$

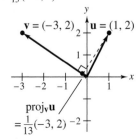

18. $N(A)$ basis: $\left\{ \begin{bmatrix} 0 \\ 1 \\ -1 \\ 0 \end{bmatrix} \right\}$

$N(A^T)$ basis: $\left\{ \begin{bmatrix} 0 \\ 0 \\ 0 \end{bmatrix} \right\}$

$R(A) = R^3$

$R(A^T)$ basis: $\left\{ \begin{bmatrix} 0 \\ 1 \\ 1 \\ 0 \end{bmatrix}, \begin{bmatrix} -1 \\ 0 \\ 0 \\ 1 \end{bmatrix}, \begin{bmatrix} 1 \\ 1 \\ 1 \\ 1 \end{bmatrix} \right\}$

19. span $\left\{ \begin{bmatrix} -1 \\ -1 \\ 1 \end{bmatrix} \right\}$ **20.** Proof

21. $y = \frac{36}{13} - \frac{20}{13}x$

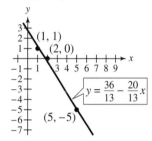

22. (a) 3   (b) One basis consists of the first three rows of A.

(c) One basis consists of columns 1, 3, and 4 of A.

(d) $\left\{ \begin{bmatrix} 2 \\ 1 \\ 0 \\ 0 \\ 0 \\ 0 \end{bmatrix}, \begin{bmatrix} -3 \\ 0 \\ 5 \\ -1 \\ 1 \\ 0 \end{bmatrix}, \begin{bmatrix} -2 \\ 0 \\ 3 \\ -7 \\ 0 \\ 1 \end{bmatrix} \right\}$

(e) No   (f) No   (g) Yes   (h) No

23. No. Two planes can intersect in a line, but not in a single point.

24. Proof

Chapter 6

Section 6.1   *(page 306)*

1. (a) $(-1, 7)$ (b) $(11, -8)$

3. (a) $(-3, 22, -5)$ (b) $(1, 2, 2)$

5. (a) $(-14, -7)$ (b) $(1, 1, t)$

7. (a) $(0, 2, 1)$ (b) $(-6, 4)$

9. Not linear   **11.** Linear   **13.** Not linear

15. Not linear   **17.** Linear   **19.** Linear   **21.** Linear

23. $T(1, 4) = (-3, 5)$
$T(-2, 1) = (-3, -1)$

25. $(-1, -5, 5)$

27. $(0, -6, 8)$ **29.** $(10, 0, 2)$ **31.** $\left(2, \frac{5}{2}, 2\right)$

33. $T: R^2 \to R^2$ **35.** $T: R^4 \to R^4$ **37.** $T: R^4 \to R^3$

39. (a) $(-1, -1)$ (b) $(-1, -1)$ (c) $(0, 0)$

41. (a) $(2, -1, 2, 2)$ (b) $\left(-1, 1, -1, -\frac{1}{2}\right)$

43. (a) $(-1, 9, 9)$ (b) $(-4t, -t, 0, t)$

45. (a) $(0, 4\sqrt{2})$ (b) $(2\sqrt{3} - 2, 2\sqrt{3} + 2)$

(c) $\left(-\dfrac{5}{2}, \dfrac{5\sqrt{3}}{2}\right)$

47. $A^{-1} = \begin{bmatrix} \cos\theta & \sin\theta \\ -\sin\theta & \cos\theta \end{bmatrix}$; clockwise rotation through θ

49. Projection onto the xz-plane

51. Not a linear transformation

53. Linear transformation

55. $x^2 - 3x - 5$

57. True. D_x is a linear transformation and preserves addition and scalar multiplication.

59. False, because $3\cos 3x \ne 3\cos x$.

61. $g(x) = 2x^2 + 3x + C$ **63.** $g(x) = -\cos x + C$

65. (a) -1 (b) $\frac{1}{12}$ (c) -4

67. (a) False, because $\cos(x_1 + x_2) \ne \cos x_1 + \cos x_2$.
(b) True. See discussion following Example 10, page 305.

69. (a) $(x, 0)$ (b) Projection onto the x-axis

71. (a) $\left(\frac{1}{2}(x+y), \frac{1}{2}(x+y)\right)$ (b) $\left(\frac{5}{2}, \frac{5}{2}\right)$ (c) Proof

73. $A\mathbf{u} = \begin{bmatrix} \frac{1}{2} & \frac{1}{2} \\ \frac{1}{2} & \frac{1}{2} \end{bmatrix} \begin{bmatrix} x \\ y \end{bmatrix} = \begin{bmatrix} \frac{1}{2}x + \frac{1}{2}y \\ \frac{1}{2}x + \frac{1}{2}y \end{bmatrix} = T(\mathbf{u})$

75. (a) Proof   (b) Proof   (c) $(t, 0)$ (d) (t, t)

77–83. Proofs

Section 6.2 *(page 318)*

1. R^3 **3.** $\{(0, 0, 0, 0)\}$

5. $\{a_0 - a_2x + a_2x^2 + a_3x^3 : a_0, a_2, a_3 \text{ are real}\}$

7. $\{a_0 : a_0 \text{ is real}\}$ **9.** $\{(0, 0)\}$

11. (a) $\{(0, 0)\}$ (b) R^2

13. (a) $\text{span}\{(-4, -2, 1)\}$ (b) R^2

15. (a) $\{(0, 0)\}$ (b) $\text{span}\{(1, -1, 0), (0, 0, 1)\}$

17. (a) $\text{span}\{(-1, 1, 1, 0)\}$
(b) $\text{span}\{(1, 0, -1, 0), (0, 1, -1, 0), (0, 0, 0, 1)\}$

19. (a) $\{(0, 0)\}$ (b) 0 (c) R^2 (d) 2

21. (a) $\{(0, 0)\}$ (b) 0
(c) $\{(4s, 4t, s - t) : s \text{ and } t \text{ are real}\}$ (d) 2

23. (a) $\{(t, -3t) : t \text{ is real}\}$ (b) 1
(c) $\{(3t, t) : t \text{ is real}\}$ (d) 1

25. (a) $\{(-t, 0, t) : t \text{ is real}\}$ (b) 1
(c) $\{(s, t, s) : s \text{ and } t \text{ are real}\}$ (d) 2

27. (a) $\{(s + t, s, -2t) : s \text{ and } t \text{ are real}\}$ (b) 2
(c) $\{(2t, -2t, t) : t \text{ is real}\}$ (d) 1

29. (a) $\{(-11t, 6t, 4t) : t \text{ is real}\}$ (b) 1 (c) R^2 (d) 2

31. (a) $\{(2s - t, t, 4s, -5s, s) : s \text{ and } t \text{ are real}\}$ (b) 2
(c) $\{(7r, 7s, 7t, 8r + 20s + 2t) : r, s, \text{ and } t \text{ are real}\}$ (d) 3

33. Nullity $= 1$
Kernel: a line
Range: a plane

35. Nullity $= 3$
Kernel: R^3
Range: $\{(0, 0, 0)\}$

37. Nullity $= 0$
Kernel: $\{(0, 0, 0)\}$
Range: R^3

39. Nullity $= 2$
Kernel: $\{(x, y, z) : x + 2y + 2z = 0\}$ (plane)
Range: $\{(t, 2t, 2t), t \text{ is real}\}$ (line)

41. 2 **43.** 3 **45.** 4

47. Because $|A| = -4 \ne 0$, the homogeneous equation $A\mathbf{x} = \mathbf{0}$ has only the trivial solution. So, $\ker(T) = \{(0, 0)\}$ and T is one-to-one (by Theorem 6.6). Furthermore, because $\text{rank}(T) = \dim(R^2) - \text{nullity}(T) = 2 - 0 = 2 = \dim(R^2)$, T is onto (by Theorem 6.7).

49. Because $|A| = -1 \ne 0$, the homogeneous equation $A\mathbf{x} = \mathbf{0}$ has only the trivial solution. So, $\ker(T) = \{(0, 0, 0)\}$ and T is one-to-one (by Theorem 6.6). Furthermore, because $\text{rank}(T) = \dim(R^3) - \text{nullity}(T) = 3 - 0 = 3 = \dim(R^3)$, T is onto (by Theorem 6.7).

51. One-to-one and onto   **53.** One-to-one

	Zero	Standard Basis

55. (a) $(0, 0, 0, 0)$ $\{(1, 0, 0, 0), (0, 1, 0, 0),$
$(0, 0, 1, 0), (0, 0, 0, 1)\}$

(b) $\begin{bmatrix} 0 \\ 0 \\ 0 \\ 0 \end{bmatrix}$ $\left\{ \begin{bmatrix} 1 \\ 0 \\ 0 \\ 0 \end{bmatrix}, \begin{bmatrix} 0 \\ 1 \\ 0 \\ 0 \end{bmatrix}, \begin{bmatrix} 0 \\ 0 \\ 1 \\ 0 \end{bmatrix}, \begin{bmatrix} 0 \\ 0 \\ 0 \\ 1 \end{bmatrix} \right\}$

(c) $\begin{bmatrix} 0 & 0 \\ 0 & 0 \end{bmatrix}$ $\left\{ \begin{bmatrix} 1 & 0 \\ 0 & 0 \end{bmatrix}, \begin{bmatrix} 0 & 1 \\ 0 & 0 \end{bmatrix}, \begin{bmatrix} 0 & 0 \\ 1 & 0 \end{bmatrix}, \begin{bmatrix} 0 & 0 \\ 0 & 1 \end{bmatrix} \right\}$

(d) $p(x) = 0$ $\{1, x, x^2, x^3\}$

(e) $(0, 0, 0, 0, 0)$ $\{(1, 0, 0, 0, 0), (0, 1, 0, 0, 0),$
$(0, 0, 1, 0, 0), (0, 0, 0, 1, 0)\}$

57. The set of constant functions: $p(x) = a_0$

59. (a) Rank $= 1$, nullity $= 2$ (b) $\{(1, 0, -2), (1, 2, 0)\}$

61. (a) Rank $= n$ (b) Rank $< n$

63. $T(A) = \mathbf{0} \implies A - A^T = \mathbf{0} \implies A = A^T$
So, $\ker(T) = \{A : A = A^T\}$.

65. (a) False. See "Definition of Kernel of a Linear Transformation," page 309.
(b) False. See Theorem 6.4, page 312.
(c) True. See discussion before "Definition of Isomorphism," page 317.

67. Proof   **69.** Proof

Section 6.3 *(page 328)*

1. $\begin{bmatrix} 1 & 2 \\ 1 & -2 \end{bmatrix}$ **3.** $\begin{bmatrix} 1 & 1 & 0 \\ 1 & -1 & 0 \\ -1 & 0 & 1 \end{bmatrix}$ **5.** $\begin{bmatrix} 3 & 0 & -2 \\ 0 & 2 & -1 \end{bmatrix}$

7. $(1, 4)$ **9.** $(-14, 0, 4)$

11. (a) $\begin{bmatrix} -1 & 0 \\ 0 & -1 \end{bmatrix}$ (b) $(-3, -4)$

(c)

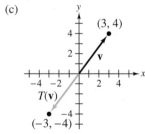

13. (a) $\begin{bmatrix} -1 & 0 \\ 0 & 1 \end{bmatrix}$ (b) $(-2, -3)$

(c)
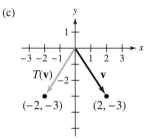

15. (a) $\begin{bmatrix} \dfrac{\sqrt{2}}{2} & -\dfrac{\sqrt{2}}{2} \\ \dfrac{\sqrt{2}}{2} & \dfrac{\sqrt{2}}{2} \end{bmatrix}$ (b) $(0, 2\sqrt{2})$

(c)
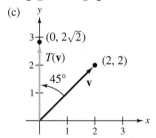

17. (a) $\begin{bmatrix} \dfrac{1}{2} & \dfrac{\sqrt{3}}{2} \\ -\dfrac{\sqrt{3}}{2} & \dfrac{1}{2} \end{bmatrix}$ (b) $\left(\dfrac{1}{2} + \sqrt{3},\ 1 - \dfrac{\sqrt{3}}{2}\right)$

(c)
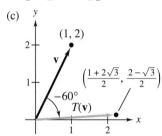

19. (a) $\begin{bmatrix} 1 & 0 & 0 \\ 0 & 1 & 0 \\ 0 & 0 & -1 \end{bmatrix}$ (b) $(3, 2, -2)$

(c)
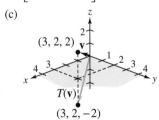

21. (a) $\begin{bmatrix} \frac{9}{10} & \frac{3}{10} \\ \frac{3}{10} & \frac{1}{10} \end{bmatrix}$ (c)
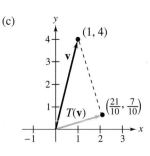

(b) $\left(\frac{21}{10}, \frac{7}{10}\right)$

23. (a) $\begin{bmatrix} 2 & 3 & -1 \\ 3 & 0 & -2 \\ 2 & -1 & 1 \end{bmatrix}$ (b) $\begin{bmatrix} 9 \\ 5 \\ -1 \end{bmatrix}$

25. (a) $\begin{bmatrix} 1 & -1 & 0 & 0 \\ 0 & 0 & 1 & 0 \\ 1 & 2 & 0 & -1 \\ 0 & 0 & 0 & 1 \end{bmatrix}$ (b) $\begin{bmatrix} 1 \\ 1 \\ 2 \\ -1 \end{bmatrix}$

27. $A = \begin{bmatrix} 2 & 3 \\ 0 & 0 \end{bmatrix}, A' = \begin{bmatrix} 0 & 1 \\ 0 & 2 \end{bmatrix}$

29. $A = \begin{bmatrix} -4 & 1 \\ -3 & 4 \end{bmatrix}, A' = \begin{bmatrix} 4 & 4 & 3 \\ 3 & -2 & 1 \\ -3 & -2 & -2 \end{bmatrix}$

31. $T^{-1}(x, y) = \left(-\frac{1}{4}x, \frac{1}{4}y\right)$ **33.** T is not invertible.

35. $T^{-1}(x_1, x_2, x_3) = (x_1, -x_1 + x_2, -x_2 + x_3)$

37. (a) and (b) $(9, 5, 4)$ **39.** (a) and (b) $(2, -4, -3, 3)$

41. (a) and (b) $(9, 16, -20)$

43. $\begin{bmatrix} 0 & 0 & 0 \\ 1 & 0 & 0 \\ 0 & 1 & 0 \\ 0 & 0 & 1 \end{bmatrix}$ **45.** $\begin{bmatrix} 0 & 1 & 0 & 0 \\ 0 & 0 & 0 & 0 \\ 0 & 0 & 1 & 1 \\ 0 & 0 & 0 & 1 \end{bmatrix}$

47. $4 - 3e^x - 3xe^x$

49. (a) $\begin{bmatrix} 0 & 0 & 0 & 0 \\ 1 & 0 & 0 & 0 \\ 0 & \frac{1}{2} & 0 & 0 \\ 0 & 0 & \frac{1}{3} & 0 \\ 0 & 0 & 0 & \frac{1}{4} \end{bmatrix}$ (b) $8x - 2x^2 + \frac{3}{4}x^4$

51. (a) $\begin{bmatrix} 1 & 0 & 0 & 0 & 0 & 0 \\ 0 & 0 & 0 & 1 & 0 & 0 \\ 0 & 1 & 0 & 0 & 0 & 0 \\ 0 & 0 & 0 & 0 & 1 & 0 \\ 0 & 0 & 1 & 0 & 0 & 0 \\ 0 & 0 & 0 & 0 & 0 & 1 \end{bmatrix}$ (b) Proof

(c) $\begin{bmatrix} 1 & 0 & 0 & 0 & 0 & 0 \\ 0 & 0 & 1 & 0 & 0 & 0 \\ 0 & 0 & 0 & 0 & 1 & 0 \\ 0 & 1 & 0 & 0 & 0 & 0 \\ 0 & 0 & 0 & 1 & 0 & 0 \\ 0 & 0 & 0 & 0 & 0 & 1 \end{bmatrix}$

53. (a) True. See Theorem 6.10 on page 320.
(b) False. See sentence after "Definition of Inverse Linear Transformation," page 324.

55. Proof

57. Sometimes it is preferable to use a nonstandard basis. For example, some linear transformations have diagonal matrix representations relative to a nonstandard basis.

Section 6.4 *(page 334)*

1. $A' = \begin{bmatrix} 4 & -3 \\ \frac{5}{3} & -1 \end{bmatrix}$ **3.** $A' = \begin{bmatrix} -\frac{1}{3} & \frac{4}{3} \\ -\frac{13}{3} & \frac{16}{3} \end{bmatrix}$

5. $A' = \begin{bmatrix} -4 & 8 \\ 0 & 0 \end{bmatrix}$ **7.** $A' = \begin{bmatrix} 1 & 0 & 0 \\ 0 & 1 & 0 \\ 0 & 0 & 1 \end{bmatrix}$

9. $A' = \begin{bmatrix} 9 & -5 & \frac{15}{2} \\ 26 & -14 & \frac{39}{2} \\ 8 & -4 & 5 \end{bmatrix}$ **11.** $A' = \begin{bmatrix} \frac{7}{3} & \frac{10}{3} & -\frac{1}{3} \\ -\frac{1}{6} & \frac{4}{3} & \frac{8}{3} \\ \frac{2}{3} & -\frac{4}{3} & -\frac{2}{3} \end{bmatrix}$

13. (a) $\begin{bmatrix} 6 & 4 \\ 9 & 4 \end{bmatrix}$ (b) $[\mathbf{v}]_B = \begin{bmatrix} 2 \\ -1 \end{bmatrix}$, $[T(\mathbf{v})]_B = \begin{bmatrix} 4 \\ -4 \end{bmatrix}$

(c) $A' = \begin{bmatrix} 0 & -\frac{4}{3} \\ 9 & 7 \end{bmatrix}$, $P^{-1} = \begin{bmatrix} -\frac{1}{3} & \frac{1}{3} \\ \frac{3}{4} & -\frac{1}{2} \end{bmatrix}$ (d) $\begin{bmatrix} -\frac{8}{3} \\ 5 \end{bmatrix}$

15. (a) $\begin{bmatrix} 5 & 2 \\ 9 & 2 \end{bmatrix}$ (b) $[\mathbf{v}]_B = \begin{bmatrix} 3 \\ -1 \end{bmatrix}$, $[T(\mathbf{v})]_B = \begin{bmatrix} 5 \\ 1 \end{bmatrix}$

(c) $A' = \begin{bmatrix} -7 & -2 \\ 27 & 8 \end{bmatrix}$, $P^{-1} = \begin{bmatrix} -\frac{1}{4} & \frac{1}{4} \\ \frac{9}{8} & -\frac{5}{8} \end{bmatrix}$ (d) $\begin{bmatrix} -1 \\ 5 \end{bmatrix}$

17. (a) $\begin{bmatrix} \frac{1}{2} & \frac{1}{2} & -\frac{1}{2} \\ \frac{1}{2} & -\frac{1}{2} & \frac{1}{2} \\ -\frac{1}{2} & \frac{1}{2} & \frac{1}{2} \end{bmatrix}$ (b) $[\mathbf{v}]_B = \begin{bmatrix} 1 \\ 0 \\ -1 \end{bmatrix}$, $[T(\mathbf{v})]_B = \begin{bmatrix} 2 \\ -1 \\ -2 \end{bmatrix}$

(c) $A' = \begin{bmatrix} 1 & 0 & 0 \\ 0 & 2 & 0 \\ 0 & 0 & 3 \end{bmatrix}$, $P^{-1} = \begin{bmatrix} 1 & 1 & 0 \\ 1 & 0 & 1 \\ 0 & 1 & 1 \end{bmatrix}$ (d) $\begin{bmatrix} 1 \\ 0 \\ -3 \end{bmatrix}$

19. $\begin{bmatrix} 1 & -2 \\ 4 & 0 \end{bmatrix} = \begin{bmatrix} -2 & -1 \\ 1 & 1 \end{bmatrix} \begin{bmatrix} 12 & 7 \\ -20 & -11 \end{bmatrix} \begin{bmatrix} -1 & -1 \\ 1 & 2 \end{bmatrix}$

21. $\begin{bmatrix} 5 & 8 & 0 \\ 10 & 4 & 0 \\ 0 & 12 & 6 \end{bmatrix} = \begin{bmatrix} \frac{1}{5} & 0 & 0 \\ 0 & \frac{1}{4} & 0 \\ 0 & 0 & \frac{1}{3} \end{bmatrix} \begin{bmatrix} 5 & 10 & 0 \\ 8 & 4 & 0 \\ 0 & 9 & 6 \end{bmatrix} \begin{bmatrix} 5 & 0 & 0 \\ 0 & 4 & 0 \\ 0 & 0 & 3 \end{bmatrix}$

23. $\begin{bmatrix} -2 & 0 & 0 \\ 0 & 1 & 0 \\ 0 & 0 & 1 \end{bmatrix}$

25. Proof   **27.** Proof   **29.** I_n **31–37.** Proofs

39. The matrix for I relative to B, or relative to B', is the identity matrix. The matrix for I relative to B and B' is the square matrix whose columns are the coordinates of $\mathbf{v}_1, \ldots, \mathbf{v}_n$ relative to the standard basis.

41. (a) True. See discussion before Example 1, page 330.
(b) False. See sentence following the proof of Theorem 6.13, page 332.

Section 6.5 *(page 341)*

1. (a) $(3, -5)$ (b) $(2, 1)$ (c) $(a, 0)$
(d) $(0, -b)$ (e) $(-c, -d)$ (f) (f, g)

3. (a) $(1, 0)$ (b) $(3, -1)$ (c) $(0, a)$
(d) $(b, 0)$ (e) $(d, -c)$ (f) $(-g, f)$

5. (a) $(2x, y)$ (b) Horizontal expansion

7. (a) Vertical contraction   **9.** (a) Horizontal expansion
(b)
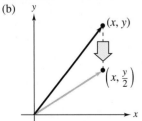
(b)
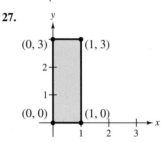

11. (a) Horizontal shear
(b)
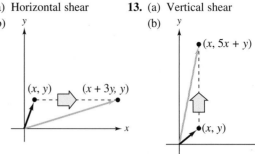

13. (a) Vertical shear
(b)

15. $\{(0, t): t \text{ is real}\}$ **17.** $\{(t, t): t \text{ is real}\}$
19. $\{(t, 0): t \text{ is real}\}$ **21.** $\{(t, 0): t \text{ is real}\}$

23.   **25.**
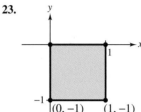

27.   **29.**
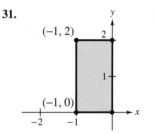

31.   **33.**
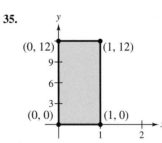

35.   **37.**

39. (a)   (b)
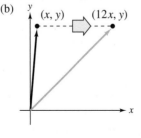
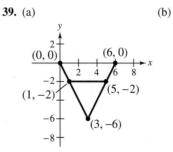
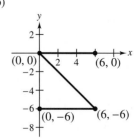

41. (a)   (b)

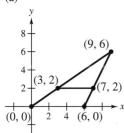

43. (a)   (b)

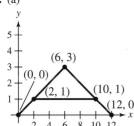

45. Horizontal expansion   **47.** Horizontal shear
49. Reflection in the x-axis and a vertical expansion (in either order)
51. Vertical shear followed by a horizontal expansion
53. $T(1, 0) = (2, 0)$, $T(0, 1) = (0, 3)$, $T(2, 2) = (4, 6)$

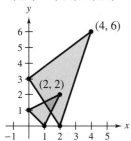

55. $\begin{bmatrix} \dfrac{\sqrt{3}}{2} & -\dfrac{1}{2} & 0 \\ \dfrac{1}{2} & \dfrac{\sqrt{3}}{2} & 0 \\ 0 & 0 & 1 \end{bmatrix}$ **57.** $\begin{bmatrix} 1 & 0 & 0 \\ 0 & -\dfrac{1}{2} & -\dfrac{\sqrt{3}}{2} \\ 0 & \dfrac{\sqrt{3}}{2} & -\dfrac{1}{2} \end{bmatrix}$

59. $\left((\sqrt{3} - 1)/2, (\sqrt{3} + 1)/2, 1\right)$

61. $\left(1, \dfrac{-1 - \sqrt{3}}{2}, \dfrac{-1 + \sqrt{3}}{2}\right)$

63. 90° about the x-axis   **65.** 180° about the y-axis
67. 90° about the z-axis

69. $\begin{bmatrix} 0 & 1 & 0 \\ 0 & 0 & -1 \\ -1 & 0 & 0 \end{bmatrix}$; $(1, -1, -1)$

71. $\begin{bmatrix} \dfrac{\sqrt{2}}{2} & -\dfrac{\sqrt{2}}{2} & 0 \\ -\dfrac{1}{2} & -\dfrac{1}{2} & -\dfrac{\sqrt{2}}{2} \\ \dfrac{1}{2} & \dfrac{1}{2} & -\dfrac{\sqrt{2}}{2} \end{bmatrix}$; $\left(0, \dfrac{-2 - \sqrt{2}}{2}, \dfrac{2 - \sqrt{2}}{2}\right)$

Review Exercises   *(page 343)*

1. (a) $(2, -4)$ (b) $(4, 4)$
3. (a) $(0, -1, 7)$ (b) $\{(t - 3, 5 - t, t): t \text{ is real}\}$
5. (a) $(-1, 5, -5)$ (b) $(-1, 2)$
7. Not linear

9. Linear, $A = \begin{bmatrix} 1 & 2 \\ -1 & -1 \end{bmatrix}$ **11.** Linear, $A = \begin{bmatrix} 1 & -2 \\ -1 & 2 \end{bmatrix}$

13. Not linear
15. Not linear

17. Linear, $A = \begin{bmatrix} 0 & 0 & 1 \\ 0 & 1 & 0 \\ 1 & 0 & 0 \end{bmatrix}$

19. $T(1, 1) = \left(\frac{3}{2}, \frac{3}{2}\right)$, $T(0, 1) = (1, 1)$
21. $T(-7, 2) = (-2, 2)$
23. (a) $T: R^3 \to R^2$ (b) $(3, -12)$
   (c) $\left\{\left(-\frac{5}{2}, 3 - 2t, t\right): t \text{ is real}\right\}$
25. (a) $T: R^3 \to R^3$ (b) $(-2, -4, -5)$ (c) $(2, 2, 2)$
27. (a) $T: R^2 \to R^3$ (b) $(8, 10, 4)$ (c) $(1, -1)$
29.

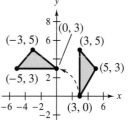

31. (a) $\text{span}\{(-2, 1, 0, 0), (2, 0, 1, -2)\}$
   (b) $\text{span}\{(5, 0, 4), (0, 5, 8)\}$
33. (a) $\{\mathbf{0}\}$ (b) R^3
35. (a) $\{(0, 0)\}$ (b) 0   (c) $\text{span}\left\{\left(1, 0, \frac{1}{2}\right), \left(0, 1, -\frac{1}{2}\right)\right\}$ (d) 2
37. (a) $\{(-3t, 3t, t)\}$ (b) 1
   (c) $\text{span}\{(1, 0, -1), (0, 1, 2)\}$ (d) 2
39. 3   **41.** 2   **43.** $A^2 = I$ **45.** $A^3 = \begin{bmatrix} \cos 3\theta & -\sin 3\theta \\ \sin 3\theta & \cos 3\theta \end{bmatrix}$

47. $A' = \begin{bmatrix} 0 & 0 & 0 \\ 0 & 1 & 0 \\ 0 & 1 & 0 \end{bmatrix}$, $A = \begin{bmatrix} 0 & 0 \\ 1 & 1 \end{bmatrix}$

49. T is not invertible.   **51.** $T^{-1}(x, y) = (x, -y)$
53. (a) One-to-one   (b) Onto   (c) Invertible
55. (a) Not one-to-one   (b) Onto   (c) Not invertible

57. (a) and (b) $(0, 1, 1)$ **59.** $A' = \begin{bmatrix} 3 & -1 \\ 1 & -1 \end{bmatrix}$

61. $\begin{bmatrix} 1 & -9 \\ -1 & 3 \end{bmatrix} = \begin{bmatrix} \frac{5}{13} & \frac{1}{13} \\ -\frac{3}{13} & \frac{2}{13} \end{bmatrix} \begin{bmatrix} 6 & -3 \\ 2 & -2 \end{bmatrix} \begin{bmatrix} 2 & -1 \\ 3 & 5 \end{bmatrix}$

63. (a) $A = \begin{bmatrix} 0 & 0 & 0 \\ 0 & \frac{1}{5} & \frac{2}{5} \\ 0 & \frac{2}{5} & \frac{4}{5} \end{bmatrix}$ (b) Answers will vary.
   (c) Answers will vary.
65. Proof   **67.** Proof
69. (a) Proof   (b) Rank = 1, nullity = 3
   (c) $\{1 - x, 1 - x^2, 1 - x^3\}$

71. $\text{Ker}(T) = \{\mathbf{v}: \langle \mathbf{v}, \mathbf{v}_0 \rangle = 0\}$
Range $= R$
Rank $= 1$
Nullity $= \dim(V) - 1$

73. Although they are not the same, they have the same dimension (4) and are isomorphic.

75. (a) Vertical expansion   **77.** (a) Vertical shear
(b) 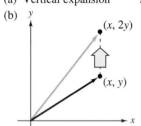   (b)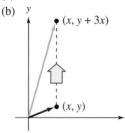

79. (a) Horizontal shear
(b)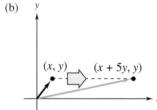

81. 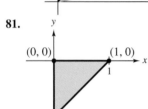   **83.**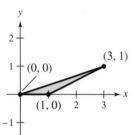

85. Reflection in the line $y = x$ followed by a horizontal expansion

87. $\begin{bmatrix} \dfrac{\sqrt{2}}{2} & -\dfrac{\sqrt{2}}{2} & 0 \\ \dfrac{\sqrt{2}}{2} & \dfrac{\sqrt{2}}{2} & 0 \\ 0 & 0 & 1 \end{bmatrix}, (\sqrt{2}, 0, 1)$

89. $\begin{bmatrix} 1 & 0 & 0 \\ 0 & \dfrac{1}{2} & -\dfrac{\sqrt{3}}{2} \\ 0 & \dfrac{\sqrt{3}}{2} & \dfrac{1}{2} \end{bmatrix}, \left(1, \dfrac{-1-\sqrt{3}}{2}, \dfrac{-\sqrt{3}+1}{2}\right)$

91. $\begin{bmatrix} \dfrac{\sqrt{3}}{2} & -\dfrac{1}{4} & \dfrac{\sqrt{3}}{4} \\ \dfrac{1}{2} & \dfrac{\sqrt{3}}{4} & -\dfrac{3}{4} \\ 0 & \dfrac{\sqrt{3}}{2} & \dfrac{1}{2} \end{bmatrix}$ **93.** $\begin{bmatrix} \dfrac{\sqrt{6}}{4} & -\dfrac{\sqrt{2}}{2} & \dfrac{\sqrt{2}}{4} \\ \dfrac{\sqrt{6}}{4} & \dfrac{\sqrt{2}}{2} & \dfrac{\sqrt{2}}{4} \\ -\dfrac{1}{2} & 0 & \dfrac{\sqrt{3}}{2} \end{bmatrix}$

95. $(0, 0, 0), \left(\dfrac{\sqrt{2}}{2}, \dfrac{\sqrt{2}}{2}, 0\right), (0, \sqrt{2}, 0),$
$\left(-\dfrac{\sqrt{2}}{2}, \dfrac{\sqrt{2}}{2}, 0\right), (0, 0, 1), \left(\dfrac{\sqrt{2}}{2}, \dfrac{\sqrt{2}}{2}, 1\right),$
$(0, \sqrt{2}, 1), \left(-\dfrac{\sqrt{2}}{2}, \dfrac{\sqrt{2}}{2}, 1\right)$

97. $(0, 0, 0), (1, 0, 0), \left(1, \dfrac{\sqrt{3}}{2}, \dfrac{1}{2}\right), \left(0, \dfrac{\sqrt{3}}{2}, \dfrac{1}{2}\right),$
$\left(0, -\dfrac{1}{2}, \dfrac{\sqrt{3}}{2}\right), \left(1, -\dfrac{1}{2}, \dfrac{\sqrt{3}}{2}\right),$
$\left(1, \dfrac{-1+\sqrt{3}}{2}, \dfrac{1+\sqrt{3}}{2}\right), \left(0, \dfrac{-1+\sqrt{3}}{2}, \dfrac{1+\sqrt{3}}{2}\right)$

99. (a) False. See "Elementary Matrices for Linear Transformations in R^2," page 336.
(b) True. See "Elementary Matrices for Linear Transformations in R^2," page 336.

101. (a) False. See "Remark," page 300.
(b) False. See Theorem 6.7, page 316.

Chapter 7

Section 7.1   *(page 356)*

1. $\begin{bmatrix} 2 & 0 \\ 0 & -2 \end{bmatrix}\begin{bmatrix} 1 \\ 0 \end{bmatrix} = 2\begin{bmatrix} 1 \\ 0 \end{bmatrix}, \begin{bmatrix} 2 & 0 \\ 0 & -2 \end{bmatrix}\begin{bmatrix} 0 \\ 1 \end{bmatrix} = -2\begin{bmatrix} 0 \\ 1 \end{bmatrix}$

3. $\begin{bmatrix} 2 & 3 & 1 \\ 0 & -1 & 2 \\ 0 & 0 & 3 \end{bmatrix}\begin{bmatrix} 1 \\ 0 \\ 0 \end{bmatrix} = 2\begin{bmatrix} 1 \\ 0 \\ 0 \end{bmatrix},$
$\begin{bmatrix} 2 & 3 & 1 \\ 0 & -1 & 2 \\ 0 & 0 & 3 \end{bmatrix}\begin{bmatrix} 1 \\ -1 \\ 0 \end{bmatrix} = -1\begin{bmatrix} 1 \\ -1 \\ 0 \end{bmatrix},$
$\begin{bmatrix} 2 & 3 & 1 \\ 0 & -1 & 2 \\ 0 & 0 & 3 \end{bmatrix}\begin{bmatrix} 5 \\ 1 \\ 2 \end{bmatrix} = 3\begin{bmatrix} 5 \\ 1 \\ 2 \end{bmatrix}$

5. $\begin{bmatrix} 0 & 1 & 0 \\ 0 & 0 & 1 \\ 1 & 0 & 0 \end{bmatrix}\begin{bmatrix} 1 \\ 1 \\ 1 \end{bmatrix} = 1\begin{bmatrix} 1 \\ 1 \\ 1 \end{bmatrix}$

7. (a) $\begin{bmatrix} 2 & 0 \\ 0 & -2 \end{bmatrix}\begin{bmatrix} c \\ 0 \end{bmatrix} = 2\begin{bmatrix} c \\ 0 \end{bmatrix}$
(b) $\begin{bmatrix} 2 & 0 \\ 0 & -2 \end{bmatrix}\begin{bmatrix} 0 \\ c \end{bmatrix} = -2\begin{bmatrix} 0 \\ c \end{bmatrix}$

9. (a) No   (b) Yes   (c) Yes   (d) No
11. (a) Yes   (b) No   (c) Yes   (d) Yes
13. $\lambda = 1, (t, 0); \lambda = -1, (0, t)$
15. (a) $\lambda(\lambda - 7) = 0$ (b) $\lambda = 0, (1, 2); \lambda = 7, (3, -1)$
17. (a) $(\lambda + 1)(\lambda - 3) = 0$
(b) $\lambda = -1, (-1, 1); \lambda = 3, (1, 1)$
19. (a) $\lambda^2 - \dfrac{1}{4} = 0$ (b) $\lambda = -\dfrac{1}{2}, (1, 1); \lambda = \dfrac{1}{2}, (3, 1)$
21. (a) $(\lambda - 2)(\lambda - 4)(\lambda - 1) = 0$
(b) $\lambda = 4, (7, -4, 2); \lambda = 2, (1, 0, 0); \lambda = 1, (-1, 1, 1)$
23. (a) $(\lambda + 3)(\lambda - 3)^2 = 0$
(b) $\lambda = -3, (1, 1, 3); \lambda = 3, (1, 0, -1), (1, 1, 0)$
25. (a) $(\lambda - 4)(\lambda - 6)(\lambda + 2) = 0$
(b) $\lambda = -2, (3, 2, 0); \lambda = 4, (5, -10, -2);$
$\lambda = 6, (1, -2, 0)$
27. (a) $(\lambda - 2)^2(\lambda - 4)(\lambda + 1) = 0$
(b) $\lambda = 2, (1, 0, 0, 0), (0, 1, 0, 0); \lambda = 4, (0, 0, 1, 1);$
$\lambda = -1, (0, 0, 1, -4)$
29. $\lambda = -2, 1$ **31.** $\lambda = -\dfrac{1}{6}, \dfrac{1}{3}$ **33.** $\lambda = -1, 4, 4$
35. $\lambda = 4, \dfrac{17 \pm \sqrt{385}}{12}$

37. $\lambda = 0, 0, 0, 21$ **39.** $\lambda = 0, 0, 3, 3$ **41.** $\lambda = 2, 3, 1$

43. $\lambda = -6, 5, -4, -4$

45. (a) $\lambda_1 = 3, \lambda_2 = 4$

(b) $B_1 = \{(2, -1)\}, B_2 = \{(1, -1)\}$

(c) $\begin{bmatrix} 3 & 0 \\ 0 & 4 \end{bmatrix}$

47. (a) $\lambda_1 = -1, \lambda_2 = 1, \lambda_3 = 2$

(b) $B_1 = \{(1, 0, 1)\}, B_2 = \{(2, 1, 0)\}, B_3 = \{(1, 1, 0)\}$

(c) $\begin{bmatrix} -1 & 0 & 0 \\ 0 & 1 & 0 \\ 0 & 0 & 2 \end{bmatrix}$

49. $\lambda^2 - 8\lambda + 15$ **51.** $\lambda^3 - 5\lambda^2 + 15\lambda - 27$

53.

Exercise	(a) Trace of A	(b) Determinant of A
15	7	0
17	2	-3
19	0	$-\frac{1}{4}$
21	7	8
23	3	-27
25	8	-48
27	7	-16

55–63. Proofs   **65.** $a = 0, d = 1$ or $a = 1, d = 0$

67. (a) False. **x** must be nonzero.

(b) True. See Theorem 7.2, page 351.

69. Dim $= 3$ **71.** Dim $= 1$

73. $T(e^x) = \dfrac{d}{dx}[e^x] = e^x = 1(e^x)$

75. $\lambda = -2, 3 + 2x; \lambda = 4, -5 + 10x + 2x^2; \lambda = 6, -1 + 2x$

77. $\lambda = 0, \begin{bmatrix} 1 & 0 \\ 1 & 0 \end{bmatrix}, \begin{bmatrix} 1 & 1 \\ 0 & -1 \end{bmatrix}; \lambda = 3, \begin{bmatrix} 1 & 0 \\ -2 & 0 \end{bmatrix}$

79. $\lambda = 0, 1$ **81.** Proof

Section 7.2 *(page 366)*

1. (a) $P^{-1} = \begin{bmatrix} 1 & -4 \\ -1 & 3 \end{bmatrix}, P^{-1}AP = \begin{bmatrix} 1 & 0 \\ 0 & -2 \end{bmatrix}$

(b) $\lambda = 1, -2$

3. (a) $P^{-1} = \begin{bmatrix} -\frac{1}{3} & \frac{2}{3} \\ \frac{2}{3} & -\frac{1}{3} \end{bmatrix}, P^{-1}AP = \begin{bmatrix} -1 & 0 \\ 0 & 2 \end{bmatrix}$

(b) $\lambda = -1, 2$

5. (a) $P^{-1} = \begin{bmatrix} \frac{2}{3} & -\frac{2}{3} & 1 \\ 0 & \frac{1}{4} & 0 \\ -\frac{1}{3} & \frac{1}{12} & 0 \end{bmatrix}, P^{-1}AP = \begin{bmatrix} 5 & 0 & 0 \\ 0 & 3 & 0 \\ 0 & 0 & -1 \end{bmatrix}$

(b) $\lambda = 5, 3, -1$

7. $P = \begin{bmatrix} 1 & 3 \\ 2 & -1 \end{bmatrix}$ (The answer is not unique.)

9. $P = \begin{bmatrix} 7 & 1 & -1 \\ -4 & 0 & 1 \\ 2 & 0 & 1 \end{bmatrix}$ (The answer is not unique.)

11. $P = \begin{bmatrix} 1 & -1 & 1 \\ 1 & 0 & 1 \\ 3 & 1 & 0 \end{bmatrix}$ (The answer is not unique.)

13. A is not diagonalizable.

15. There is only one eigenvalue, $\lambda = 0$, and the dimension of its eigenspace is 1.

17. There is only one eigenvalue, $\lambda = 7$, and the dimension of its eigenspace is 1.

19. There are two eigenvalues, 1 and 2. The dimension of the eigenspace for the repeated eigenvalue 1 is 1.

21. There are two repeated eigenvalues, 0 and 3. The eigenspace associated with 3 is of dimension 1.

23. $\lambda = 0, 2$; The matrix is diagonalizable.

25. $\lambda = 0, -2$; Insufficient number of eigenvalues to guarantee diagonalization

27. $\{(1, -1), (1, 1)\}$ **29.** $\{(-1 + x), x\}$

31. Proof   **33.** $\begin{bmatrix} -188 & -378 \\ 126 & 253 \end{bmatrix}$

35. $\begin{bmatrix} 2 & 0 & -2 \\ -30 & 32 & -2 \\ 3 & 0 & -3 \end{bmatrix}$

37. (a) True. See the proof of Theorem 7.4, page 360.

(b) False. See Theorem 7.6, page 364.

39. Yes. $P = \begin{bmatrix} 0 & 0 & 1 \\ 0 & 1 & 0 \\ 1 & 0 & 0 \end{bmatrix}$

41. Yes, the order of elements on the main diagonal may change.

43–47. Proofs

49. $\lambda = 4$ is the only eigenvalue, and a basis for the eigenspace is $\{(1, 0)\}$, so the matrix does not have two linearly independent eigenvectors. By Theorem 7.5, the matrix is not diagonalizable.

Section 7.3 *(page 376)*

1. Not symmetric

3. $P = \begin{bmatrix} 1 & 1 & 0 \\ 0 & 0 & 1 \\ -1 & 1 & 0 \end{bmatrix}, P^{-1}AP = \begin{bmatrix} -a & 0 & 0 \\ 0 & a & 0 \\ 0 & 0 & a \end{bmatrix}$

5. $P = \begin{bmatrix} 1 & 0 & 1 \\ 0 & 1 & 0 \\ -1 & 0 & 1 \end{bmatrix}, P^{-1}AP = \begin{bmatrix} 0 & 0 & 0 \\ 0 & a & 0 \\ 0 & 0 & 2a \end{bmatrix}$

7. $\lambda = 1$, dim $= 1$ **9.** $\lambda = 2$, dim $= 2$
$\lambda = 3$, dim $= 1$ $\lambda = 3$, dim $= 1$

11. $\lambda = -2$, dim $= 2$ **13.** $\lambda = -1$, dim $= 1$
$\lambda = 4$, dim $= 1$ $\lambda = 1 + \sqrt{2}$, dim $= 1$
$\lambda = 1 - \sqrt{2}$, dim $= 1$

15. $\lambda = -2$, dim $= 1$ **17.** $\lambda = 1$, dim $= 1$
$\lambda = 3$, dim $= 2$ $\lambda = 2$, dim $= 3$
$\lambda = 8$, dim $= 1$ $\lambda = 3$, dim $= 1$

19. Orthogonal   **21.** Orthogonal   **23.** Not orthogonal

25. Orthogonal   **27.** Not orthogonal   **29.** Orthogonal

31. Not orthogonal   **33–37.** Proofs

39. Not orthogonally diagonalizable

41. Orthogonally diagonalizable

43. $P = \begin{bmatrix} \sqrt{2}/2 & \sqrt{2}/2 \\ -\sqrt{2}/2 & \sqrt{2}/2 \end{bmatrix}$ **45.** $P = \begin{bmatrix} \sqrt{3}/3 & \sqrt{6}/3 \\ -\sqrt{6}/3 & \sqrt{3}/3 \end{bmatrix}$

(The answer is not unique.)   (The answer is not unique.)

47. $P = \begin{bmatrix} -\frac{2}{3} & -\frac{1}{3} & \frac{2}{3} \\ \frac{1}{3} & \frac{2}{3} & \frac{2}{3} \\ \frac{2}{3} & -\frac{2}{3} & \frac{1}{3} \end{bmatrix}$ (The answer is not unique.)

49. $\begin{bmatrix} -\sqrt{3}/3 & -\sqrt{2}/2 & \sqrt{6}/6 \\ -\sqrt{3}/3 & \sqrt{2}/2 & \sqrt{6}/6 \\ \sqrt{3}/3 & 0 & \sqrt{6}/3 \end{bmatrix}$

(The answer is not unique.)

51. $P = \begin{bmatrix} \sqrt{2}/2 & 0 & \sqrt{2}/2 & 0 \\ -\sqrt{2}/2 & 0 & \sqrt{2}/2 & 0 \\ 0 & \sqrt{2}/2 & 0 & \sqrt{2}/2 \\ 0 & -\sqrt{2}/2 & 0 & \sqrt{2}/2 \end{bmatrix}$

(The answer is not unique.)

53. (a) True. See Theorem 7.10, page 373.
(b) True. See Theorem 7.9, page 372.

55–59. Proofs

Section 7.4 (page 391)

1. $\mathbf{x}_2 = \begin{bmatrix} 20 \\ 5 \end{bmatrix}, \mathbf{x}_3 = \begin{bmatrix} 10 \\ 10 \end{bmatrix}; t\begin{bmatrix} 2 \\ 1 \end{bmatrix}$

3. $\mathbf{x}_2 = \begin{bmatrix} 84 \\ 12 \\ 6 \end{bmatrix}, \mathbf{x}_3 = \begin{bmatrix} 60 \\ 84 \\ 6 \end{bmatrix}; t\begin{bmatrix} 8 \\ 4 \\ 1 \end{bmatrix}$

5. $\mathbf{x}_2 = \begin{bmatrix} 400 \\ 25 \\ 100 \\ 50 \end{bmatrix}, \mathbf{x}_3 = \begin{bmatrix} 250 \\ 100 \\ 25 \\ 50 \end{bmatrix}; t\begin{bmatrix} 8 \\ 2 \\ 2 \\ 1 \end{bmatrix}$

7. $\mathbf{x}_2 = \begin{bmatrix} 1280 \\ 120 \\ 40 \end{bmatrix}, \mathbf{x}_3 = \begin{bmatrix} 3120 \\ 960 \\ 30 \end{bmatrix}$

9. $\mathbf{x}_2 = \begin{bmatrix} 900 \\ 60 \\ 50 \end{bmatrix}, \mathbf{x}_3 = \begin{bmatrix} 2200 \\ 540 \\ 30 \end{bmatrix}$

11. $y_1 = C_1 e^{2t}$
$y_2 = C_2 e^{t}$

13. $y_1 = C_1 e^{-4t}$
$y_2 = C_2 e^{t/2}$

15. $y_1 = C_1 e^{-t}$
$y_2 = C_2 e^{6t}$
$y_3 = C_3 e^{t}$

17. $y_1 = C_1 e^{-0.3t}$
$y_2 = C_2 e^{0.4t}$
$y_3 = C_3 e^{-0.6t}$

19. $y_1 = C_1 e^{7t}$
$y_2 = C_2 e^{9t}$
$y_3 = C_3 e^{-7t}$
$y_4 = C_4 e^{-9t}$

21. $y_1 = C_1 e^{t} - 4C_2 e^{2t}$
$y_2 = C_2 e^{2t}$

23. $y_1 = C_1 e^{-t} + C_2 e^{3t}$
$y_2 = -C_1 e^{-t} + C_2 e^{3t}$

25. $y_1 = C_1 e^{t} - 2C_2 e^{2t} - 7C_3 e^{3t}$
$y_2 = \qquad C_2 e^{2t} + 8C_3 e^{3t}$
$y_3 = \qquad\qquad 2C_3 e^{3t}$

27. $y_1 = 3C_1 e^{2t} - 5C_2 e^{-4t} - C_3 e^{-6t}$
$y_2 = 2C_1 e^{2t} + 10C_2 e^{-4t} + 2C_3 e^{-6t}$
$y_3 = \qquad\qquad 2C_2 e^{-4t}$

29. $y_1' = y_1 + y_2$
$y_2' = \qquad y_2$

31. $y_1' = y_2$
$y_2' = y_3$
$y_3' = -4y_2$

33. $\begin{bmatrix} 1 & 0 \\ 0 & 1 \end{bmatrix}$

35. $\begin{bmatrix} 9 & 5 \\ 5 & -4 \end{bmatrix}$

37. $\begin{bmatrix} 0 & 5 \\ 5 & -10 \end{bmatrix}$

39. $A = \begin{bmatrix} 2 & -\frac{3}{2} \\ -\frac{3}{2} & -2 \end{bmatrix}, \lambda_1 = -\frac{5}{2}, \lambda_2 = \frac{5}{2}, P = \begin{bmatrix} \frac{1}{\sqrt{10}} & -\frac{3}{\sqrt{10}} \\ \frac{3}{\sqrt{10}} & \frac{1}{\sqrt{10}} \end{bmatrix}$

41. $A = \begin{bmatrix} 13 & 3\sqrt{3} \\ 3\sqrt{3} & 7 \end{bmatrix}, \lambda_1 = 4, \lambda_2 = 16, P = \begin{bmatrix} \frac{1}{2} & \frac{\sqrt{3}}{2} \\ -\frac{\sqrt{3}}{2} & \frac{1}{2} \end{bmatrix}$

43. $A = \begin{bmatrix} 16 & -12 \\ -12 & 9 \end{bmatrix}, \lambda_1 = 0, \lambda_2 = 25, P = \begin{bmatrix} \frac{3}{5} & -\frac{4}{5} \\ \frac{4}{5} & \frac{3}{5} \end{bmatrix}$

45. Ellipse, $5(x')^2 + 15(y')^2 - 45 = 0$

47. Ellipse, $(x')^2 + 6(y')^2 - 36 = 0$

49. Parabola, $4(y')^2 + 4x' + 8y' + 4 = 0$

51. Hyperbola, $\frac{1}{2}[-(x')^2 + (y')^2 - 3\sqrt{2}x' - \sqrt{2}y' + 6] = 0$

53. $A = \begin{bmatrix} 3 & -1 & 0 \\ -1 & 3 & 0 \\ 0 & 0 & 8 \end{bmatrix}, 2(x')^2 + 4(y')^2 + 8(z')^2 - 16 = 0$

55. $A = \begin{bmatrix} 1 & 0 & 0 \\ 0 & 2 & 1 \\ 0 & 1 & 2 \end{bmatrix}, (x')^2 + (y')^2 + 3(z')^2 - 1 = 0$

57. Maximum: $3; \begin{bmatrix} 1 \\ 0 \end{bmatrix}$
Minimum: $2; \begin{bmatrix} 0 \\ 1 \end{bmatrix}$

59. Maximum: $48; \begin{bmatrix} 0 \\ 2 \end{bmatrix}$
Minimum: $25; \begin{bmatrix} 5 \\ 0 \end{bmatrix}$

61. Maximum: $11; \begin{bmatrix} \frac{1}{\sqrt{2}} \\ \frac{1}{\sqrt{2}} \end{bmatrix}$
Minimum: $-1; \begin{bmatrix} -\frac{1}{\sqrt{2}} \\ \frac{1}{\sqrt{2}} \end{bmatrix}$

63. Maximum: $3; \begin{bmatrix} \frac{1}{\sqrt{2}} \\ \frac{1}{\sqrt{2}} \end{bmatrix}$
Minimum: $-3; \begin{bmatrix} -\frac{1}{\sqrt{2}} \\ \frac{1}{\sqrt{2}} \end{bmatrix}$

65. Maximum: $4; \begin{bmatrix} \frac{1}{\sqrt{6}} \\ \frac{2}{\sqrt{6}} \\ \frac{1}{\sqrt{6}} \end{bmatrix}$; Minimum: $0; \begin{bmatrix} -\frac{1}{\sqrt{2}} \\ 0 \\ \frac{1}{\sqrt{2}} \end{bmatrix}$

67. Let $P = \begin{bmatrix} a & b \\ c & d \end{bmatrix}$ be a 2×2 orthogonal matrix such that
$|P| = 1$. Define $\theta \in (0, 2\pi)$ as follows.
(i) If $a = 1$, then $c = 0, b = 0$, and $d = 1$, so let $\theta = 0$.
(ii) If $a = -1$, then $c = 0, b = 0$ and $d = -1$, so let $\theta = \pi$.
(iii) If $a \geq 0$ and $c > 0$, let $\theta = \arccos(a), 0 < \theta \leq \pi/2$.
(iv) If $a \geq 0$ and $c < 0$, let $\theta = 2\pi - \arccos(a)$,
$3\pi/2 \leq \theta < 2\pi$.
(v) If $a \leq 0$ and $c > 0$, let $\theta = \arccos(a), \pi/2 \leq \theta < \pi$.
(vi) If $a \leq 0$ and $c < 0$, let $\theta = 2\pi - \arccos(a)$,
$\pi < \theta \leq 3\pi/2$.
In each of these cases, confirm that
$P = \begin{bmatrix} a & b \\ c & d \end{bmatrix} = \begin{bmatrix} \cos\theta & -\sin\theta \\ \sin\theta & \cos\theta \end{bmatrix}$.

69. Answers will vary.

Review Exercises *(page 393)*

1. (a) $\lambda^2 - 9 = 0$ (b) $\lambda = -3, \lambda = 3$
(c) A basis for $\lambda = -3$ is $\{(1, -5)\}$ and a basis for $\lambda = 3$
is $\{(1, 1)\}$.

3. (a) $(\lambda - 4)(\lambda - 8)^2 = 0$ (b) $\lambda = 4, \lambda = 8$
(c) A basis for $\lambda = 4$ is $\{(1, -2, -1)\}$ and a basis for $\lambda = 8$
is $\{(4, -1, 0), (3, 0, 1)\}$.

5. (a) $(\lambda - 2)(\lambda - 3)(\lambda - 1) = 0$
(b) $\lambda = 1, \lambda = 2, \lambda = 3$
(c) A basis for $\lambda = 1$ is $\{(1, 2, -1)\}$ a basis for $\lambda = 2$ is
$\{(1, 0, 0)\}$, and a basis for $\lambda = 3$ is $\{(0, 1, 0)\}$.

7. (a) $(\lambda - 1)^2(\lambda - 3)^2 = 0$ (b) $\lambda = 1, \lambda = 3$
(c) A basis for $\lambda = 1$ is $\{(1, -1, 0, 0), (0, 0, 1, -1)\}$ and a
basis for $\lambda = 3$ is $\{(1, 1, 0, 0), (0, 0, 1, 1)\}$.

9. $P = \begin{bmatrix} 4 & -1 \\ 1 & 2 \end{bmatrix}$ (The answer is not unique.)

11. Not diagonalizable

13. $P = \begin{bmatrix} 1 & 0 & 1 \\ 0 & 1 & 0 \\ 1 & 0 & -1 \end{bmatrix}$ (The answer is not unique.)

15. (a) $a = -\frac{1}{4}$ (b) $a = 2$ (c) $a \geq -\frac{1}{4}$

17. A has only one eigenvalue, $\lambda = 0$, and the dimension of its
eigenspace is 1.

19. A has only one eigenvalue, $\lambda = 3$, and the dimension of its
eigenspace is 2.

21. $P = \begin{bmatrix} 0 & 1 \\ 1 & 0 \end{bmatrix}$

23. The eigenspace corresponding to $\lambda = 1$ of a matrix A has
dimension 1, while that of matrix B has dimension 2, so the
matrices are not similar.

25. Both symmetric and orthogonal
27. Both symmetric and orthogonal
29. Neither   **31.** Neither   **33.** Proof
35. Proof   **37.** Orthogonally diagonalizable
39. Not orthogonally diagonalizable

41. $P = \begin{bmatrix} \dfrac{2}{\sqrt{5}} & -\dfrac{1}{\sqrt{5}} \\ \dfrac{1}{\sqrt{5}} & \dfrac{2}{\sqrt{5}} \end{bmatrix}$ (The answer is not unique.)

43. $P = \begin{bmatrix} 0 & \dfrac{1}{\sqrt{2}} & \dfrac{1}{\sqrt{2}} \\ 0 & -\dfrac{1}{\sqrt{2}} & \dfrac{1}{\sqrt{2}} \\ 1 & 0 & 0 \end{bmatrix}$ (The answer is not unique.)

45. $P = \begin{bmatrix} \dfrac{1}{\sqrt{2}} & 0 & \dfrac{1}{\sqrt{2}} \\ 0 & 1 & 0 \\ -\dfrac{1}{\sqrt{2}} & 0 & \dfrac{1}{\sqrt{2}} \end{bmatrix}$ (The answer is not unique.)

47. $\left(\frac{3}{5}, \frac{2}{5}\right)$ **49.** $\left(\frac{3}{5}, \frac{2}{5}\right)$ **51.** $\left(\frac{1}{4}, \frac{1}{2}, \frac{1}{4}\right)$ **53.** $\left(\frac{4}{16}, \frac{5}{16}, \frac{7}{16}\right)$

55. Proof   **57.** $A = \begin{bmatrix} 0 & 1 \\ 0 & \frac{9}{4} \end{bmatrix}$, $\lambda_1 = 0, \lambda_2 = \frac{9}{4}$

59. $A^2 = \begin{bmatrix} 56 & -40 \\ 20 & -4 \end{bmatrix}$, $A^3 = \begin{bmatrix} 368 & -304 \\ 152 & -88 \end{bmatrix}$, $A^4 = \begin{bmatrix} 2336 & -2080 \\ 1040 & -784 \end{bmatrix}$

61. (a) and (b) Proofs   **63.** Proof
65. $A = O$ **67.** $\lambda = 0$ or 1
69. (a) True. See "Definitions of Eigenvalue and Eigenvector,"
page 348.
(b) False. See Theorem 7.4, page 360.
(c) True. See "Definition of a Diagonalizable Matrix,"
page 359.

71. $\mathbf{x}_2 = \begin{bmatrix} 100 \\ 25 \end{bmatrix}$, $\mathbf{x}_3 = \begin{bmatrix} 25 \\ 25 \end{bmatrix}$; $t\begin{bmatrix} 2 \\ 1 \end{bmatrix}$

73. $\mathbf{x}_2 = \begin{bmatrix} 4500 \\ 300 \\ 50 \end{bmatrix}$, $\mathbf{x}_3 = \begin{bmatrix} 1500 \\ 4500 \\ 50 \end{bmatrix}$; $t\begin{bmatrix} 24 \\ 12 \\ 1 \end{bmatrix}$

75. $\mathbf{x}_2 = \begin{bmatrix} 1440 \\ 108 \\ 90 \end{bmatrix}$, $\mathbf{x}_3 = \begin{bmatrix} 6588 \\ 1296 \\ 81 \end{bmatrix}$

77. $y_1 = 4C_1e^{3t}$
$y_2 = C_1e^{3t} + C_2e^{-t}$

79. $y_1 = C_1e^{3t}$
$y_2 = C_2e^{8t}$
$y_3 = C_3e^{-8t}$

81. (a) $A = \begin{bmatrix} 1 & \dfrac{3}{2} \\ \dfrac{3}{2} & 1 \end{bmatrix}$

(b) $P = \begin{bmatrix} \dfrac{1}{\sqrt{2}} & -\dfrac{1}{\sqrt{2}} \\ \dfrac{1}{\sqrt{2}} & \dfrac{1}{\sqrt{2}} \end{bmatrix}$

(c) $5(x')^2 - (y')^2 = 6$

(d)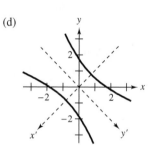

83. (a) $A = \begin{bmatrix} 0 & \dfrac{1}{2} \\ \dfrac{1}{2} & 0 \end{bmatrix}$

(b) $P = \begin{bmatrix} \dfrac{1}{\sqrt{2}} & -\dfrac{1}{\sqrt{2}} \\ \dfrac{1}{\sqrt{2}} & \dfrac{1}{\sqrt{2}} \end{bmatrix}$

(c) $(x')^2 - (y')^2 = 4$

(d)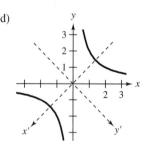

85. Maximum: 1; $\begin{bmatrix} 1 \\ 0 \end{bmatrix}$

Minimum: -1; $\begin{bmatrix} 0 \\ 1 \end{bmatrix}$

87. Maximum: 17; $\begin{bmatrix} -\dfrac{1}{\sqrt{2}} \\ \dfrac{1}{\sqrt{2}} \end{bmatrix}$

Minimum: 13; $\begin{bmatrix} \dfrac{1}{\sqrt{2}} \\ \dfrac{1}{\sqrt{2}} \end{bmatrix}$

Cumulative Test for Chapters 6 and 7
(page 397)

1. Linear transformation   **2.** Not a linear transformation

3. $\dim(R^n) = 4$; $\dim(R^m) = 2$

4. (a) $(-4, 2, 0)$ (b) $(3, t)$

5. $\{(s, s, -t, t): s, t \text{ are real}\}$

6. (a) $\text{span}\{(0, -1, 0, 1), (1, 0, -1, 0)\}$
   (b) $\text{span}\{(1, 0), (0, 1)\}$ (c) Rank = 2, nullity = 2

7. $\begin{bmatrix} 3 & 2 \\ -1 & 2 \end{bmatrix}$ **8.** $\begin{bmatrix} 1 & 1 & 0 \\ 0 & 1 & 1 \\ 1 & 0 & -1 \end{bmatrix}$

9. $\begin{bmatrix} 0 & -2 & 3 \\ 4 & 0 & 11 \end{bmatrix}$ **10.** $\begin{bmatrix} 0 & 0 & 0 \\ 0 & 0 & 0 \\ 0 & 0 & 0 \end{bmatrix}$

11. $\begin{bmatrix} \frac{1}{2} & -\frac{1}{2} \\ -\frac{1}{2} & \frac{1}{2} \end{bmatrix}$, $T(1, 1) = (0, 0)$, $T(-2, 2) = (-2, 2)$

12. (a) $\begin{bmatrix} \frac{\sqrt{3}}{2} & -\frac{1}{2} \\ \frac{1}{2} & \frac{\sqrt{3}}{2} \end{bmatrix}$ (b) $\begin{bmatrix} \frac{\sqrt{3}}{2} - 1 \\ \frac{1}{2} + \sqrt{3} \end{bmatrix}$

(c)

13. $T = \begin{bmatrix} 2 & -4 \\ -1 & -5 \end{bmatrix}$, $T' = \begin{bmatrix} 0 & 2 \\ 7 & -3 \end{bmatrix}$

14. $T = \begin{bmatrix} -2 & 2 & 1 \\ -1 & 3 & 2 \\ 4 & 0 & -6 \end{bmatrix}$, $T' = \begin{bmatrix} 2 & 1 & 3 \\ 1 & -2 & 3 \\ 1 & 2 & -5 \end{bmatrix}$

15. $T^{-1}(x, y) = \left(\frac{1}{3}x + \frac{1}{3}y, -\frac{2}{3}x + \frac{1}{3}y\right)$

16. $T^{-1}(x_1, x_2, x_3) = \left(\frac{x_1 - x_2 + x_3}{2}, \frac{x_1 + x_2 - x_3}{2}, \frac{-x_1 + x_2 + x_3}{2}\right)$

17. $\begin{bmatrix} -1 & -2 \\ 0 & 1 \\ 2 & 1 \end{bmatrix}$, $T(0, 1) = (1, 0, 1)$

18. (a) $A = \begin{bmatrix} 1 & -2 \\ 1 & 4 \end{bmatrix}$ (b) $P = \begin{bmatrix} 1 & 1 \\ 1 & 2 \end{bmatrix}$

(c) $A' = \begin{bmatrix} -7 & -15 \\ 6 & 12 \end{bmatrix}$ (d) $\begin{bmatrix} 9 \\ -6 \end{bmatrix}$

(e) $[\mathbf{v}]_B = \begin{bmatrix} 1 \\ -1 \end{bmatrix}$, $[T(\mathbf{v})]_B = \begin{bmatrix} 3 \\ -3 \end{bmatrix}$

19. $\lambda = 5$ (repeated), $\begin{bmatrix} 1 \\ -1 \end{bmatrix}$

20. $\lambda = 5$, $\begin{bmatrix} -1 \\ 4 \end{bmatrix}$; $\lambda = -15$, $\begin{bmatrix} 1 \\ 0 \end{bmatrix}$

21. $\lambda = 1$, $\begin{bmatrix} 1 \\ 0 \\ 0 \end{bmatrix}$; $\lambda = 0$, $\begin{bmatrix} -1 \\ -1 \\ 3 \end{bmatrix}$; $\lambda = 2$, $\begin{bmatrix} 1 \\ 1 \\ -1 \end{bmatrix}$

22. $\lambda = 1$ (three times), $\begin{bmatrix} 1 \\ 0 \\ 0 \end{bmatrix}$

23. $P = \begin{bmatrix} 1 & 1 & 5 \\ 0 & -1 & 1 \\ 0 & 0 & 2 \end{bmatrix}$ **24.** $P = \begin{bmatrix} 3 & -1 & -5 \\ 2 & 2 & 10 \\ 0 & 0 & 2 \end{bmatrix}$

25. $\{(0, 1, 0), (1, 1, 1), (2, 2, 3)\}$

26. $P = \begin{bmatrix} \frac{1}{\sqrt{2}} & \frac{1}{\sqrt{2}} \\ -\frac{1}{\sqrt{2}} & \frac{1}{\sqrt{2}} \end{bmatrix}$ **27.** $P = \begin{bmatrix} \frac{1}{\sqrt{3}} & \frac{1}{\sqrt{2}} & \frac{1}{\sqrt{6}} \\ \frac{1}{\sqrt{3}} & 0 & -\frac{2}{\sqrt{6}} \\ \frac{1}{\sqrt{3}} & -\frac{1}{\sqrt{2}} & \frac{1}{\sqrt{6}} \end{bmatrix}$

28. $y_1 = C_1 e^t$
   $y_2 = C_2 e^{9t}$

29. $\begin{bmatrix} 3 & -8 \\ -8 & 3 \end{bmatrix}$ **30.** $\mathbf{x}_2 = \begin{bmatrix} 1800 \\ 120 \\ 60 \end{bmatrix}$, $\mathbf{x}_3 = \begin{bmatrix} 6300 \\ 1440 \\ 48 \end{bmatrix}$

31. P is orthogonal when $P^{-1} = P^T$.   **32.** Proof

Index

Properties of Matrix Addition and Scalar Multiplication

If A, B, and C are $m \times n$ matrices, and c and d are scalars, then the properties below are true.

1. $A + B = B + A$ — **Commutative property of addition**
2. $A + (B + C) = (A + B) + C$ — **Associative property of addition**
3. $(cd)A = c(dA)$ — **Associative property of multiplication**
4. $1A = A$ — **Multiplicative identity**
5. $c(A + B) = cA + cB$ — **Distributive property**
6. $(c + d)A = cA + dA$ — **Distributive property**

Properties of Matrix Multiplication

If A, B, and C are matrices (with sizes such that the matrix products are defined), and c is a scalar, then the properties below are true.

1. $A(BC) = (AB)C$ — **Associative property of multiplication**
2. $A(B + C) = AB + AC$ — **Distributive property**
3. $(A + B)C = AC + BC$ — **Distributive property**
4. $c(AB) = (cA)B = A(cB)$

Properties of the Identity Matrix

If A is a matrix of size $m \times n$, then the properties below are true.

1. $AI_n = A$
2. $I_m A = A$

Properties of Vector Addition and Scalar Multiplication in R^n

Let $\mathbf{u}$, $\mathbf{v}$, and $\mathbf{w}$ be vectors in R^n, and let c and d be scalars.

1. $\mathbf{u} + \mathbf{v}$ is a vector in R^n — **Closure under addition**
2. $\mathbf{u} + \mathbf{v} = \mathbf{v} + \mathbf{u}$ — **Commutative property of addition**
3. $(\mathbf{u} + \mathbf{v}) + \mathbf{w} = \mathbf{u} + (\mathbf{v} + \mathbf{w})$ — **Associative property of addition**
4. $\mathbf{u} + \mathbf{0} = \mathbf{u}$ — **Additive identity property**
5. $\mathbf{u} + (-\mathbf{u}) = \mathbf{0}$ — **Additive inverse property**
6. $c\mathbf{u}$ is a vector in R^n. — **Closure under scalar multiplication**
7. $c(\mathbf{u} + \mathbf{v}) = c\mathbf{u} + c\mathbf{v}$ — **Distributive property**
8. $(c + d)\mathbf{u} = c\mathbf{u} + d\mathbf{u}$ — **Distributive property**
9. $c(d\mathbf{u}) = (cd)\mathbf{u}$ — **Associative property of multiplication**
10. $1(\mathbf{u}) = \mathbf{u}$ — **Multiplicative identity property**

Summary of Important Vector Spaces

R = set of all real numbers
R^2 = set of all ordered pairs
R^3 = set of all ordered triples
R^n = set of all n-tuples
$C(-\infty, \infty)$ = set of all continuous functions defined on the real line
$C[a, b]$ = set of all continuous functions defined on a closed interval $[a, b]$, where $a \neq b$
P = set of all polynomials
P_n = set of all polynomials of degree $\leq n$ (together with the zero polynomial)
$M_{m,n}$ = set of all $m \times n$ matrices
$M_{n,n}$ = set of all $n \times n$ square matrices